# Gamete Biology

## Emerging Frontiers on Fertility and Contraceptive Development

Society of Reproduction and Fertility Volume 63

Proceedings of the International Congress
held at New Delhi, India
February 2006

Edited by:
Satish K. Gupta, Koji Koyama and Joanne F. Murray

**NOTTINGHAM**
University Press

First published by Nottingham University Press
This reissued original edition published 2023 by 5m Books Ltd www.5mbooks.com

**British Library Cataloguing in Publication Data**
Gamete Biology: Emerging Frontiers on Fertility and Contraceptive Development
S.K. Gupta, K. Koyama and J.F. Murray (Eds)

ISBN 9781789182835

*Disclaimer*
Every reasonable effort has been made to ensure that the material in this book
is true, correct, complete and appropriate at the time of writing. Nevertheless
the publishers and the author do not accept responsibility for any omission or
error, or for any injury, damage, loss or financial consequences arising from
the use of the book. Views expressed in the articles are those of the author and
not of the Editor or Publisher.

Typeset by Nottingham University Press, Nottingham

EU GPSR Authorised Representative
LOGOS EUROPE, 9 rue Nicolas Poussin, 17000, LA ROCHELLE, France
E-mail: Contact@logoseurope.eu

# Preface

The increase in human population is unabated in spite of the introduction of oral steroidal contraceptives in the late fifties and approximately one billion people are being added every 12 years. It is projected that by the year 2050, the human population may reach a phenomenal 10 billion. The major escalation in population is expected to occur in developing third world countries, further imposing a daunting challenge for these countries to provide for the ever increasing demands of people for fulfillment of their basic necessities of food, education, employment, health care, as well as for the overall economic and social development of the people. On the other hand, globally infertility in humans is becoming a major issue. Epidemiological studies suggest that 1 out of 7 couples are classed as sub-fertile. The success rate of assisted reproduction is still below 30%, although the treatments and techniques advance and improve continuously. Considering the above, it is imperative that we should have a better understanding of the development and differentiation of the male as well as the female gametes and their union (fertilization) leading to procreation.

In recent years, impressive progress has been made in our understanding of the developmental processes involved in the formation of the spermatozoa and egg in various species. Interestingly, using stem cell biology, researchers have created both male and female gametes from totipotent cells. These two highly differentiated haploid cells have been further characterized at biochemical and molecular levels to understand the intricacies involved in their maturation and activation thereby making them competent to fertilize. Further, progress has been made in our understanding of the cellular and molecular basis of the sperm-egg interaction leading to fertilization. These developments have enabled us to devise various technologies to overcome infertility. At the same time, these investigations will lead to the development of novel, safer and effective contraceptives.

In the present Proceedings, a comprehensive overview of the recent research findings and technological developments on these aspects is presented. It has been covered under the 6 major themes namely, *Regulatory events involved in the genesis and development of gamete; Spermatozoa associated proteins and their functions; Molecular characterization and functional significance of oocyte-specific proteins; Signaling pathways associated with attainment of fertilization competence in spermatozoa and oocyte; Molecular basis of sperm-oocyte interaction;* and *Translation of knowledge pertaining to gamete biology to either overcome infertility or development of novel contraceptives.* Prominent and leading scientists and clinicians have provided an insight on these issues and shared the significant advances that have taken place in these important areas of Gamete Biology.

It is our pleasure to thank Drs. T.C. Anand Kumar (Hope Infertility Clinic and Research Foundation, Bangalore, India), M. Rajalakshmi (National Academy of Medical Sciences, New Delhi, India), P.R. Adiga (Indian Institute of Science, Bangalore, India), K. Muralidhar (Delhi University, New Delhi, India), Shinzo Isojima (Hyogo College of Medicine, Hyogo, Japan) and G.P. Talwar (Talwar Research Foundation, New Delhi, India) for chairing the 6 major themes of the Congress. We would also like to express our gratitude to Prof. M.K. Bhan, Secretary, Department of Biotechnology, Government of India and to Prof. N.K. Ganguly, Director General, Indian Council of Medical Research, Government of India for delivering the Inaugural and Valedictory addresses respectively. The organization of the International Congress was made possible by the financial and administrative support of the National Institute of

Immunology, New Delhi, India and Hyogo College of Medicine, Hyogo, Japan. It was organized in an ordered fashion by the sincere and untiring work by members of the Local Organizing Committee (Chair: Mr. B. Bose; Members, partial list: Dr. Suraj K, Ms. Sanchita Chakravarty, Dr. Sangeeta Choudhury, Mr. Pankaj Bansal, Ms. Beena Bhandari, Mr. Manish Jain, Dr. Anurag Mitra, Mr. S. Kannan, Mr. P.L. Dahra, Mr. A.B. Ray, Mr. G.C. Verma, Mr. Jaskaran Singh, Mr. Pradeep Chawla, Mr. Rajkamal Singh and Mr. M.S. Rao). The organization of the Congress was made feasible by the financial support from various agencies such as "Hitech Research Center" Project (2004-2008), Ministry of Education, Culture, Sports, Science and Technology, Japan; Department of Biotechnology, Government of India; Department of Science and Technology, Government of India; Council of Scientific and Industrial Research, Government of India; Indian Council of Medical Research, Government of India and Indian National Science Academy, Government of India, US Government under Indo-US joint collaborative programme on Contraceptive and Reproductive Health Research (CRHR), and sponsorship by several industrial houses.

We would like to acknowledge the Society for Reproduction and Fertility, UK for enabling us in the dissemination of the updated information in the field of Gamete Biology to researchers, clinicians and educationists. The credit for the timely publication of the Proceedings of this nature goes to all the speakers who have contributed their manuscripts promptly. Mrs. Sarah Keeling, Production Manager, Nottingham University Press, UK needs special thanks for her valiant efforts to put various manuscripts in the form of the present Proceedings. SKG would like to convey his appreciation for the help rendered by Mrs. Anasua Ganguly during preparations of the Proceedings. Last but not the least, Mrs. Rita Gupta, better half of SKG and Mrs. Yoko Koyama better half of KK deserve special thanks for their moral support and willingness to extend their help to ensure successful organization of the Congress.


Satish K. Gupta
Koji Koyama
Joanne F. Murray

# Contents

## REGULATORY EVENTS INVOLVED IN THE GENESIS AND DEVELOPMENT OF GAMETE

## SPERMATOZOA ASSOCIATED PROTEINS AND THEIR FUNCTIONS

## MOLECULAR CHARACTERIZATION AND FUNCTIONAL SIGNIFICANCE OF OOCYTE-SPECIFIC PROTEINS

## SIGNALLING PATHWAYS ASSOCIATED WITH ATTAINMENT OF FERTILIZATION COMPETENCE IN SPERMATOZOA AND OOCYTE

## MOLECULAR BASIS OF SPERM-OOCYTE INTERACTION

## TRANSLATION OF KNOWLEDGE PERTAINING TO GAMETE BIOLOGY TO EITHER OVERCOME INFERTILITY OR DEVELOPMENT OF NOVEL CONTRACEPTIVES

# Chromatin remodeling during mammalian spermatogenesis: role of testis specific histone variants and transition proteins

MM Pradeepa[1] and MRS Rao[1, 2]

[1]Jawaharlal Nehru Centre for Advanced Scientific Research, Jakkur, Bangalore-560 064; [2]Department of Biochemistry, Indian Institute of Science, Bangalore-560 012, India

The structure of chromatin undergoes extensive alteration during mammalian spermatogenesis. Several testis specific histone subtypes are synthesised and replace their somatic counterparts during pre-meiotic, meiotic and post-meiotic stages of germ cell differentiation. Early work from our laboratory showed that pachytene spermatocyte nuclei as well as nucleosome core particle are more accessible to DNase1 than the interphase liver nuclei. The higher order structure of chromatin in pachytene spermatocytes is also loosely packed due to the poor DNA and chromatin condensing property of the testis specific linker histone H1t. A careful analysis of the amino acid sequence of histone H1t revealed the absence of the DNA condensing domain containing SPKK/TPKK motifs in the C-terminus of the histone H1t. The spermiogenesis process following the meiotic division is characterised by extensive remodeling of chromatin. Transition proteins, TP1 and TP2, unique to mammalian spermatogenesis play an important role in this spermiogenesis process. We have shown that TP1 is a DNA melting protein while TP2 is a DNA condensing protein. We have delineated the molecular anatomy of TP2 including the presence of two novel zinc finger modules, which are essential for the recognition of CpG islands in the genome. TP2 is also phosphorylated by sperm specific protein kinase A and the phosphorylation/dephosphorylation cycle plays an important role in the chromatin condensation process.

## Introduction

In eukaryotic cells the genome is packaged inside the nucleus due to the existence of a highly ordered nucleoprotein architecture called chromatin. In the first order of organization of chromatin structure, 146 bp of DNA is packaged into a nucleosome core particle around a histone octamer (van Holde, 1988; Wolffe, 1998). The linear array of the 10 nm polynucleosomal filament is folded into 30 nm irregular fibres (either solenoid or zig-zag structure) facilitated and/or stabilised by the linker histone H1.

E-mail: mrsrao@jncasr.ac.in

Spermatogenesis is a process in which spermatogonial stem cells undergo a series of bio-chemical and morphological changes resulting in the production of highly differentiated hap-loid cells called spermatozoa. The entire process of spermatogenesis can be divided into three phases: 1) stem cell renewal and differentiation, 2) meiosis, and 3) spermiogenesis. In mam-mals, spermatogenesis is characterised by a unique chromatin remodeling process, in which somatic histones are sequentially replaced by testis specific variants, followed by the replace-ment of both somatic and testis specific histones with a class of basic proteins, transition pro-teins (TP1, TP2 and TP4). These transition proteins appear in a brief period of 2-3 days during stages 12-15 of spermiogenesis. Finally these transition proteins are replaced by protamines during stages 16-19. Although it was initially believed that spermatozoa contained only prota-mine, recent evidence shows that 10% sperm chromatin still retains nucleosomal histones in both rodents and humans (Pittogi et al., 1999). The packing of DNA in mammalian spermato-zoa approaches the physical limits of molecular compaction making mammalian sperm chro-matin the most condensed eukaryotic nucleus.

Our laboratory has been studying for the past two decades the influence of testis specific histone variants on chromatin structure during meiotic prophase at pachytene interval as well as the biochemical properties of transition proteins, TP1 and TP2, using rat spermatogenesis as the model system. A brief review of contributions from our laboratory is presented in this article.

### Testis specific histone variants and chromatin structure of pachytene spermatocytes

Several testis specific variants of histones are expressed and assembled to nucleosome during meiotic prophase as well as in post-meiotic cells (Fig. 1). The testis specific histone TH2A gene is expressed only in testis. The H2A and TH2A histones differ in eight amino acid residues in the first half of the molecule and three consecutive changes are present in the C-terminal region. A testis specific variant of H2B (TH2B) differs by addition of three potential phosphory-lation sites (Ser12, Thr23 and Thr34) and repositioning of two others (Ser5 and Ser60), resulting in a different phosphorylation map of the N-terminal tail (Kimmins and Sassone-Corsi, 2005). Among the five-histone classes, H1 histones exhibit the most diversity in its amino acid se-quences. In mice and humans, there are eight previously described H1 subtypes, including the five somatic subtypes H1a–H1e, the replacement subtype H1o, the testis-specific linker-his-tone H1t (Lennox and Cohen, 1988) and the oocyte-specific H1 linker histone, H1foo (Tanaka et al., 2001). H1t is first detected in mid-pachytene spermatocytes, where it rapidly integrates into the chromatin, replaces about 40 percent of the other somatic H1 subtypes and persists until the elongating spermatid stage (Meistrich et al., 1985). H1-like protein in spermatids 1 (HILS1), has been found recently in human and mouse (Iguchi et al., 2003; Yan et al., 2003). H1LS is detected later in the nucleai of elongating and condensing spermatids. Recently, Nishimune's group has identified and characterised a novel haploid germ cell-specific nuclear protein (HANP1) in the mouse testis, also designated HANP1/H1T2, which is implicated in the replacement of histones by protamines during spermiogenesis (Tanaka et al., 2005).

### Histone TH2B

The early work from the laboratory concentrated on studying the structural alterations in pachytene chromatin brought about by replacement of somatic histones by the testis specific nucleosomal core histones. Biophysical studies employing circular dichroism spectroscopy and thermal de-naturation techniques revealed that nucleosomal core particle isolated from rat pachytene sper-matocytes were less compact compared to nucleosomal core particle isolated from rat liver.

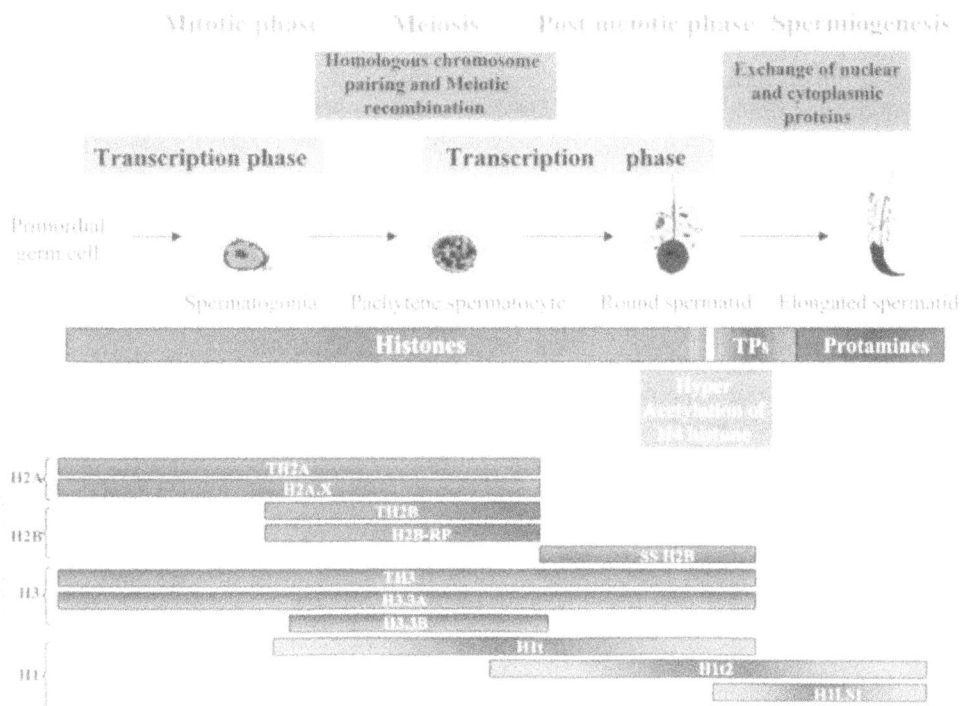

**Fig.1** Chromatin dynamics during spermatogenesis: Spermatogenesis is characterized by sequential replacement of somatic histones by the testis specific histone variants at different stages of germ cell differentiation. The type of histone variants replacing the somatic counterpart are depicted in this figure including the recently discovered histone H1 variants in spermatids.

The nucleosomal core histones of pachytene chromatin had 80% of H2B replaced by the testis specific TH2B while about 10% of H2A was replaced by TH2A. It was concluded that the presence of testis specific core histones contributed to the less compact nature of the pachytene nucleosomal core particle (Rao et al., 1983). This conclusion was further substantiated by DNase1 foot printing of pachytene nucleosome core particle showing increased sensitivity of pachytene nucleosomal core particle (Rao and Rao, 1987). The sensitive sites exactly mapped to H2B/TH2B interacting site in the nucleosomal core DNA. Additional evidence came from the observation that sub-nucleosomal particles were generated, after micrococcal nuclease digestion, at a much faster rate from the pachytene chromatin than from liver chromatin (Rao and Rao, 1987). Based on these observations, it was suggested that the nucleosome core is rendered less compact by the appearance of testis specific histone TH2B to facilitate DNA disentanglement for the DNA recombination between the paired homologous chromosomes. The amino acid sequence of testis specific histone H2B is now available. A comparison of sequences of H2B and TH2B reveals minor changes only in N-terminal eleven amino acids, which differs by addition of three potential phosphorylation sites (Ser12, Thr23 and Thr34) and repositioning of two others (Ser5 and Ser60), resulting in a different phosphorylation map of the N-terminal tail. It remains to be seen how such small changes in the N-terminal region can influence the compaction of the nucleosomal core particle. We also have to keep in mind the recent developments in our knowledge of various covalent modifications that occur on histone tails. It remains to be seen, if such covalent modifications if any, might also influence compaction of pachytene chromatin at the nucleosomal level.

As mentioned in the introduction, the fifth linker histone H1 is necessary for stabilization of higher order structure of chromatin at the 30 nm chromatin fibre level. Histone H1t, the testis specific linker histone H1, is expressed during mid-pachytene spermatocytes and comprises of almost 40% of total histone H1 in pachytene spermatocytes. The histone H1t gene has its own promoter elements to specify a tissue and stage specific expression pattern (Wilkerson et al., 2002). A series of biochemical and biophysical experiments have shown that histone H1t isolated from rat testis is a poor condenser of DNA and chromatin as compared to somatic histone H1bdec from rat liver (Khadake and Rao, 1995). This observation is also corroborated by other studies (DeLucia et al., 1994; Talasz et al., 1998). We have argued that there has to be some information missing in the amino acid sequence of histone H1t, which is present in somatic histone H1s, to render this with different condensing properties (Khadake and Rao, 1995). The structure of histone H1 can be divided into 3-sub domains: a) N-terminal tail; b) globular domain; and c) C-terminal tail. The N-terminal tail is implicated in influencing the higher order structure of chromatin while the globular domain is shown to be responsible for interaction of histone H1 with the nucleosome core particle. Among these structural domains, the globular domain is highly conserved among all histone H1s, while N-terminal tail shows some sequence heterogeneity. The C-terminal tail comprising about 100-120 amino acids is highly basic in nature and varies considerably in its amino acid sequence among all H1 histone subtypes.

The C-terminal domain of histone H1 is absolutely essential for generating 30 nm chromatin fibre and DNA condensation mediated by histone H1 is mainly due to its C-terminal domain (Allan et al., 1986). Based on these facts, we compared the amino acid sequence of the C-terminal tail of rat H1t and H1d and were surprised to see the presence of SPKK motifs only in the Histone H1d but not in H1t. Three SPKK motifs in Histone H1d are present within a stretch of 32 amino acids. SPKK motif has been shown to be a minor DNA binding motif (Suzuki, 1989). It was surmised that SPKK motifs might contribute to different DNA condensing properties of H1t and H1d. This prediction proved to be correct by showing that 16 mer synthetic peptide sequence containing two SPKK units could mimic all the DNA and chromatin condensing property of Histone H1d (Khadake and Rao, 1997). We have further shown that this domain indeed represents chromatin-condensing domain by several deletion mutants of histone H1d (Bharath et al., 2002). We have analysed amino acid sequence of several histone H1 sub-types and found the presence of S/TPKK motif in most of histone H1 except H1t. A representative comparison of the sequence pattern is shown in Fig. 2. We have included in this comparison the recently described oocyte specific H1 histone H1oo, histone H1 variant H1X, as well as testis specific H1 variants, H1LS and H1T2. Interestingly, we find all the testis specific H1 histones lack the 16 mer (S/TPKK) motif known to have DNA condensation properties.

The significance of absence of these motifs in this newly discovered H1 variant needs to be investigated. In summary, the presence of histone H1t in pachytene chromatin seems to loosen the chromatin structure even at higher order structure level. Thus it can be concluded that specific replacement of both core histones and linker histone H1 in pachytene chromatin renders chromatin to possess a much more open structure thereby probably influencing genetic recombination related events.

**A.**

**B.**

**Comparison of sequences of C-terminus of histone H1 in mammals**

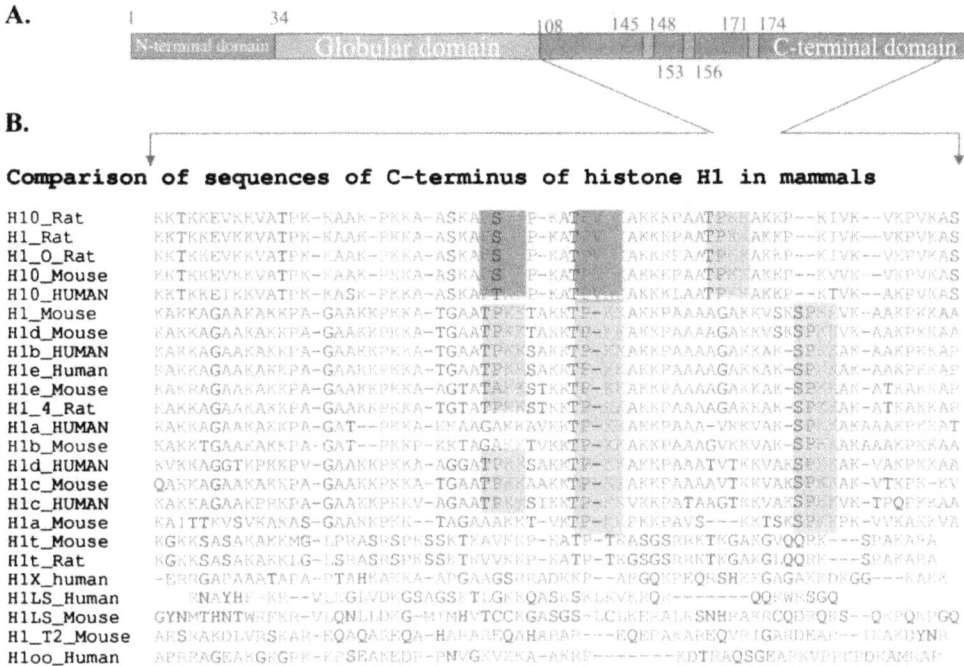

**Fig. 2** A. Structural domains of histone H1. B. Multiple sequence alignment of the C-termnial domain of histone H1 subtypes highlighting the SPKK/TPKK motifs.

## Chromatin dynamics during spermiogenesis

*Role of Transition proteins (TPs) during final stages of spermiogenesis*

The process of spermiogenesis in mammals, wherein the haploid round spermatids mature into highly condensed spermatozoa, can be broadly divided into three phases (Fig. 3). In the first phase, encompassing stages 1–10, the round spermatids are transcriptionally active and contain nucleosomal chromatin. The second phase (stages 12–15) involves the replacement of nucleosomal histones by transition proteins TP1, TP2, and TP4. These transition proteins are exclusively localized to nuclei of elongating and condensing spermatids (Meistrich, 1989) and they constitute about 90% of the chromatin basic proteins, with the level of TP1 being about 2.5 times those of TP2 (Yu *et al.*, 2000). Finally, in the third phase, the transition proteins are replaced by protamines P1 and P2 during stages 16–19 (Meistrich, 1989). The biological significance of the evolution of transition protein genes and their physiological roles are not yet clearly understood. Both TP1$^{-/-}$ and TP2$^{-/-}$ knock out mice have been generated which are less fertile than normal and show abnormal chromatin condensation (Yu *et al.*, 2000; Zhao *et al.*, 2001). TP1 and TP2 double knockout mice are, however, sterile and spermatogenesis is severely impaired suggesting their important role in spermiogenesis (Zhao *et al.*, 2004). Rat TP4 is a minor basic protein of 138 amino acid residues (2%) present in the step 12-15 spermatids. Akama *et al.* (1995) from their studies on the DNA binding properties of boar TP4, suggested that TP4 induces local destabilization of DNA and speculated that TP4 might contribute to chromatin organization during spermiogenesis by helping the relaxation of the negatively super coiled DNA.

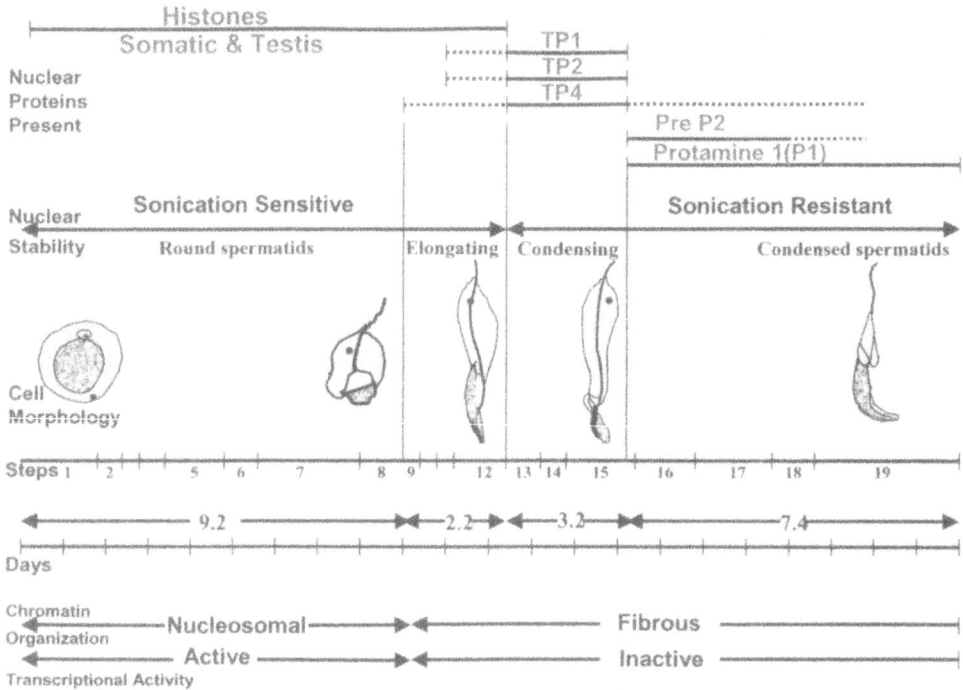

**Fig. 3** Schematic representation of the stages of mammalian spermiogenesis: Morphological changes and the types of histones present at different stages of spermiogenesis are described. Stages 12-15 represent the interval during which transition proteins replace the histones.

## Transition protein 1 (TP1)

In order to gain an insight into the significance of the appearance of TP1, we have analysed the DNA binding properties using biophysical approaches like fluorescence quenching, thermal melting and UV absorption. From these DNA-protein interaction studies, we concluded that TP1 behaves as a DNA melting protein (Singh and Rao, 1987). It was speculated that the 2-tyrosine residues of TP1 might intercalate between the nucleic acid bases resulting in local melting of the DNA duplex. Subsequently, it was shown that addition of TP1, decreases the compactness of DNA around the histone octamer and destabilizes the nucleosome core particles suggesting that TP1 may be involved in displacement of histones from nucleosome type chromatin (Singh and Rao, 1988). Interestingly, TP1 has also been shown to stimulate DNA repair activity both *in vivo* and *in vitro* (Caron et al., 2001), the significance of which needs to be further investigated.

## Transition protein 2 (TP2)

TP2 is a basic protein with a molecular mass of 13 kDa. Results from extensive biochemical and biophysical studies of TP2 from our laboratory have shown that TP2 has both DNA and chromatin condensation properties (Baskaran and Rao, 1990; Brewer et al., 2002). We have also shown that TP2 is a zinc metalloprotein and contains two atoms of zinc per molecule (Baskaran and Rao, 1991). Using circular dichroism spectroscopy technique, TP2 was shown to condense DNA with a preference to GC rich DNA in a zinc dependent manner (Kundu and Rao, 1995). It was therefore interesting to study whether TP2 binds to CpG islands in the genome, which are natural GC rich sequences that are present in the 5' and 3' region of genes. By electrophoretic mobility shift assays, it was confirmed that TP2 preferentially binds to CpG island sequence. This interaction was abolished in the presence of EDTA and also upon methylation of C in the CpG dinucleotide. We have concluded that TP2 recognizes CpG island in a zinc dependent manner and methylation of CpG island abolishes this specific interaction (Kundu and Rao, 1996). More recently by mutational analysis the domain architecture of TP2 was delineated and shown to possess two structural and functional domains. The N-terminal domain having 2 zinc fingers, which is responsible for specific recognition of CpG islands (Meetei et al., 2000), and the C-terminal basic domain is rich in arginine and lysine residues, which is involved in DNA/chromatin condensation (Kundu and Rao, 1999). By employing extensive site directed mutational analysis, we have identified amino acid residues involved in zinc coordination. Based on our results, we have proposed novel zinc finger modules coordinating the two zinc atoms in TP2 (Meetei et al., 2000). We would like to mention here that the identified zinc finger module of TP2 does not correspond to any of the known canonical zinc finger modules reported in the literature.

We were interested in studying the *in vivo* localization of TP2 after ectopic expression. In the absence of suitable *in vitro* cell culture and transfection system available for round spermatids, to study the effect of ectopic expression of TP2, we used COS-7 cells, for transfection studies of different deletion mutants. We have observed that wild type TP2 localizes to nucleolus containing ribosomal DNA, which is GC rich (Meetei et al., 2000). By transfection studies of TP2 deletion mutants, we have delineated the Nuclear Localization Sequence (NLS) of TP2 to 87GKVSKRKAV95, which is rich in basic amino acid residues and closely resembles the consensus monopartite NLS (Meetei et al., 2000). TP2 is known to get phosphorylated immediately after synthesis in the cytosol by protein kinase A. By site-directed mutagenesis the major phosphorylation sites in TP2 were demonstrated to occur at Thr101 and Ser109 residues in the C-terminal third of the protein (Meetei et al., 2002). Further, we have demonstrated that the sperm-specific isoform of the catalytic subunit of protein kinase A (Cs-PKA) is involved in the phosphorylation of TP2 (Meetei et al., 2002). Recently, it was found that the nuclear transport of TP2 is modulated by phosphorylation status of TP2 but is not required for its import into the nucleus (Ullas and Rao, 2003). Since phosphorylation sites are present in the C-terminal domain of protein, which is known for DNA condensation property, we checked for condensing ability of phosphorylated and unphosphorylated TP2. Interestingly, phosphorylated TP2 is a poor condenser of DNA compared to unphosphorylated TP2 (Meetei et al., 2002). Based on the above results, we have proposed a model in which the phosphorylation event temporarily masks the condensation property of the basic C-terminal domain, thus allowing lateral diffusion of TP2 along the chromatin to facilitate its zinc finger modules to search and dock onto the GC-rich CpG island sequences. Subsequent dephosphorylation triggers the initiation of chromatin condensation (Fig. 4).

In conclusion, histone variants and transition proteins play a significant role in mammalian spermatogenesis. In order to have a deeper insight into the chromatin remodeling process, a more comprehensive study is necessary to evaluate the role of several histone modifications like acetylation and methylation and also the chromatin remodelling molecular complexes, which are gaining importance in recent years.

**A.**

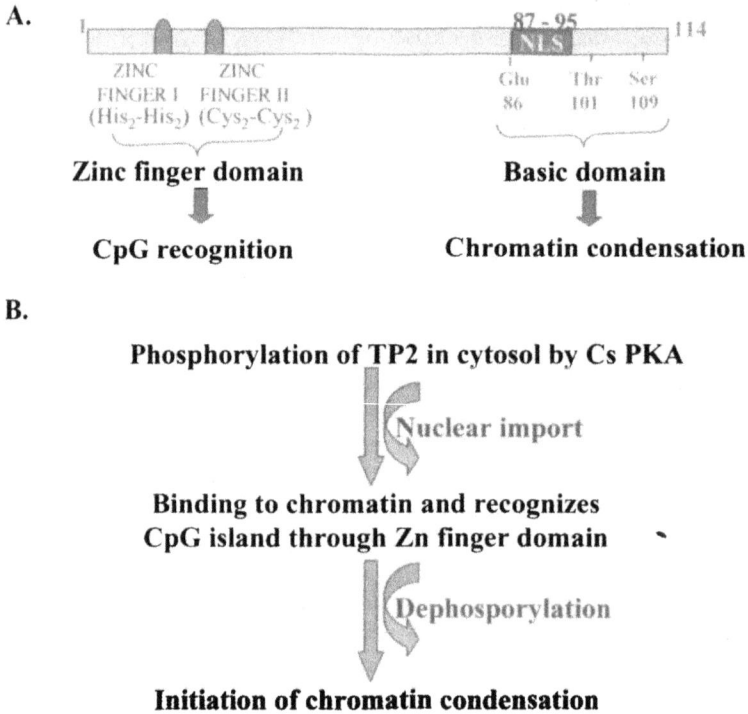

|                              |                              |
| :--------------------------: | :--------------------------: |
| ZINC         ZINC            | Glu    Thr    Ser            |
| FINGER I    FINGER II        | 86     101    109            |
| (His₂-His₂) (Cys₂-Cys₂)      |                              |
| **Zinc finger domain**       | **Basic domain**             |
| ⬇                            | ⬇                            |
| **CpG recognition**          | **Chromatin condensation**   |

**B.**

**Phosphorylation of TP2 in cytosol by Cs PKA**

Nuclear import

**Binding to chromatin and recognizes**
**CpG island through Zn finger domain**

Dephosporylation

**Initiation of chromatin condensation**

**Fig. 4** A. Structural and functional domains of TP2: TP2 has two zinc fingers in the N-terminal region. The first finger involves His2-His2 finger while the second finger involves Cys2-Cys2 finger. These two zinc fingers are implicated to recognize CpG islands in the genome. The C-terminal one third of TP2 is basic in nature having NLS as well as CsPKA phosphorylation sites. B. Sequence of events leading to TP2 mediated chromatin condensation.

## Acknowledgements

We would like to thank the previous graduate students from our laboratory, Drs. Jagmohan Singh, BJ Rao, R Bhaskaran, J Khadake, TK Kundu, S Bharath, AR Meetei and KS Ullas whose work have been reviewed in this article.

## References

Akama K, Ichimura H, Sato H, Kojima S, Miura K, Hayashi H, Komatsu Y and Nakano M (1995) The amino acid sequence and interaction with the nucleosome core DNA of transition protein 4 from boar late spermatid nuclei *European Journal of Biochemistry* **233** 179-185

Allan J, Mitchell T, Harborne N, Bohm L and Crane-Robinson C (1986) Roles of H1 domains in determining higher order chromatin structure and H1 location *Journal of Molecular Biology* **187** 591-601

Baskaran R and Rao MRS (1990) Interaction of spermatid-specific protein TP2 with nucleic acids, *in vitro*. A comparative study with TP1 *Journal of Biological Chemistry* **265** 21039-21047

Baskaran R and Rao MRS (1991) Mammalian spermatid specific protein, TP2, is a zinc metalloprotein with two finger motifs *Biochemical and Biophysical Research Communications* **179** 1491-1499

Bharath MM, Ramesh S, Chandra NR and Rao MRS (2002) Identification of a 34 amino acid stretch in the C-terminal domain of histone H1 as the DNA condensing domain by site directed mutagenesis *Biochemistry* **41** 7617-7627

Brewer L, Corzett M and Balhorn R (2002) Condensation of DNA by spermatid basic nuclear proteins *Journal Biological Chemistry* **277** 38895-38900

Caron N, Veilleux S and Boissonneault G (2001) Stimulation of DNA repair by the spermatidal TP1 protein *Molecular Reproduction and Development* **58** 437-443

**DeLucia F, Farone-Manella MR, D'Erne M, Quesada P, Caifa P and Farina B** (1994) Histone induced condensation of rat testis chromatin: testis specific H1t versus somatic H1 variants *Biochemical and Biophysical Research Communications* **198** 32-39

**Iguchi N, Tanaka H, Yomogida K and Nishimune Y** (2003) Isolation and characterization of a novel cDNA encoding a DNA-binding protein (H1LS1) specifically expressed in testicular haploid germ cells *International Journal of Andrology* **26** 354-365

**Khadake JR and Rao MRS** (1995) DNA- and chromatin-condensing properties of rat testes H1a and H1t compared to those of rat liver H1bdec: H1t Is a poor condenser of chromatin *Biochemistry* **34** 15792-15801

**Khadake JR and Rao MRS** (1997) Condensation of DNA and chromatin by an SPKK containing octapeptide repeat motif present in the C-terminus of histone H1 *Biochemistry* **36** 1041-1051

**Kimmins S and Sassone-Corsi P** (2005) Chromatin remodelling and epigenetic features of germ cells *Nature* **434** 583-589

**Kundu TK and Rao MRS** (1995) DNA condensation by the rat spermatidal protein TP2 shows GC-rich sequence preference and is zinc dependent *Biochemistry* **34** 5143-5150

**Kundu TK and Rao MRS** (1996) Zinc dependent recognition of a human CpG island sequence by the mammalian spermatidal protein TP2 *Biochemistry* **35** 15626-15632

**Kundu TK and Rao MRS** (1999) CpG islands in chromatin organization and gene expression *Journal of Biochemistry (Tokyo)* **125** 217-222

**Lennox RW and Cohen LH** (1988) The production of tissue-specific histone complements during development *Biochemistry and Cell Biology* **66** 636–649

**Meetei AR, Ullas KS and Rao MRS** (2000) Identification of two novel zinc finger modules and nuclear localization signal in rat spermatidal protein TP2 by site-directed mutagenesis *Journal of Biological Chemistry* **275** 38500–38507

**Meetei AR, Ullas KS, Vasupradha V and Rao MRS** (2002) Involvement of protein kinase A in the phosphorylation of spermatidal protein TP2 and its effect on DNA condensation *Biochemistry* **41** 185–195

**Meistrich ML, Bucci LR, Trostle-Weige PK and Brock WA** (1985) Histone variants in rat spermatogonia and primary spermatocytes *Developmental Biology* **112** 230–240

**Meistrich ML** (1989) Histone and basic nuclear protein transitions in mammalian permatogenesis. In *Histones and Other Basic Nuclear Proteins* pp 165–182 Eds LS Hnilica, GS Stein and JL Stein. CRC Press, Orlando

**Pittoggi C, Renzi L, Zaccagnini G, Cimini D, Degrassi F, Giordano R, Magnano AR, Lorenzini R, Lavia P and Spadafora C** (1999) A fraction of mouse sperm chromatin is organized in nucleosomal hypersensitive domains enriched in retroposon DNA *Journal of Cell Science* **112** 3537–3548

**Rao BJ, Brahmachari SK and Rao MRS** (1983) Structural organization of the meiotic prophase chromatin in the rat testis *Journal of Biological Chemistry* **258** 13478-13485

**Rao BJ and Rao MRS** (1987) DNase1 site mapping and micrococcal nuclease digestion of pachytene chromatin reveal novel structural features *Journal of Biological Chemistry* **262** 4472-4476.

**Singh J and Rao MRS** (1987) Interaction of rat testis protein, TP, with nucleic acids *in vitro Journal of Biological Chemistry* **262** 734-740

**Singh J and Rao MRS** (1988) Interaction of rat testis protein, TP, with nucleosome core particle *Biochemistry International* **17** 701-710

**Suzuki M** (1989) SPKK, a new nucleic acid-binding unit of protein found in histone *EMBO Journal* **8** 797-804

**Talasz H, Sapojnikova N, Helliger W, Lindner H and Puschendorf B** (1998) In vitro binding of H1 histone subtypes to nucleosomal organized mouse mammary tumour virus long terminal repeat promoter *Journal of Biological Chemistry* **273** 32236–32243

**Tanaka M, Hennebold JD, Macfarlane J and Adashi EY** (2001) A mammalian oocyte-specific linker histone gene H1oo: homology with the genes for the oocyte-specific cleavage stage histone (cs-H1) of sea urchin and the B4/H1M histone of the frog *Development (Cambridge, U.K.)* **128** 655–664

**Tanaka H, Iguchi N, Isotani A, Kitamura K, Toyama Y, Matsuoka Y, Onishi M, Masai K, Maekawa M, Toshimori K, Okabe M and Nishimune Y** (2005) HANP1/H1T2, a novel histone H1-like protein involved in nuclear formation and sperm fertility *Molecular and Cellular Biology* **25** 7107-7119

**Ullas KS and Rao MRS** (2003) Phosphorylation of rat spermatidal protein TP2 by sperm specific kinase A (Cs-PKA) and modulation of its transport into the haploid nucleus *Journal of Biological Chemistry* **275** 38500–38507

**van Holde KE** (1988) Chromatin. *Springer Series in Molecular Biology* pp 231-241 Ed A Rich. Springer-Verlag, New York

**Wolffe AP** (1998) Chromatin Structure and Function. *Academic Press*, London

**Wilkerson DC, Wolfe SA and Grimes SR** (2002) Sp1 and Sp3 activate the testis-specific histone H1t promoter through the H1t/GC-box *Journal of Cellular Biochemistry* **86** 716-725

**Yan W, Ma L, Burns KH and Matzuk MM** (2003) H1LS1 is a spermatid-specific linker histone H1 like protein implicated in chromatin remodeling during mammalian spermiogenesis *Proceedings of National Academy of Sciences USA* **100** 10546-10551

**Yu YE, Zhang Y, Unni E, Shirley CR, Deng JM, Russell LD, Weil MM, Behringer RR and Meistrich ML** (2000) Abnormal spermatogenesis and reduced fertility in transition nuclear protein 1-deficient mice *Proceedings of National Academy of Sciences USA* **97** 4683–4688

**Zhao M, Shirley CR, Yu YE, Mohapatra B, Zhang Y, Unni E, Deng JM, Arango N A, Terry NH and Weil MM** (2001) Targeted disruption of the transition protein 2 gene affects sperm chromatin structure and reduces fertility in mice *Molecular Cell Biology* **21** 7243–7255

**Zhao M, Shirley CR, Mounsey S and Meistrich ML** (2004) Nucleoprotein transitions during spermiogenesis in mice with transition nuclear protein Tnp1 and Tnp2 mutations *Biology of Reproduction* **71** 1016–1025

# Retinoid signaling during spermatogenesis as revealed by genetic and metabolic manipulations of retinoic acid receptor alpha

D J Wolgemuth[1-5] and S S W Chung[1,2,4]

[1]Department of Genetics and Development, [2]Department of Obstetrics and Gynecology, [3]The Center for Reproductive Sciences, [4]The Institute of Human Nutrition, and [5]The Herbert Irving Comprehensive Cancer Center, Columbia University College of Physicians and Surgeons, New York, NY 10032, USA

The importance of dietary retinol (vitamin A) and retinoid signaling for normal development and differentiation has been recognised for many years. Vitamin A deficiency results in a variety of abnormalities, most of which can be corrected by supplementing the diet with all-*trans*-retinoic acid (ATRA), with the exception of blindness and male sterility. ATRA, an active metabolite of vitamin A, functions primarily by binding to nuclear receptors of the steroid hormone superfamily, the retinoic acid receptors (RARs). Gene targeting studies revealed the importance of ATRA signaling through the RARs for spermatogenesis. Mice that are homozygous for a null mutation in the gene encoding RARα, *Rara*-/-, exhibit defects in spermatogenesis and male sterility. The abnormalities in these RARα-deficient testes have been examined in detail in a series of recent studies from our laboratory and will be summarised in this paper. We also review how dietary, pharmacologic and genetic strategies, alone or in combination, can be used to gain further insight into retinoid function in mammalian spermatogenesis.

## Introduction

*Vitamin A is required for spermatogenesis*

Vitamin A has long been recognised to be essential for spermatogenesis and for maturation of spermatozoa in the epididymis (Wolbach and Howe, 1925; Howell et al., 1963; Chung and Wolgemuth, 2004). Although the many adverse symptoms of vitamin A deficiency (VAD) in animals can be reversed by supplementing the diet with all-*trans*-retinoic acid (ATRA), the animals are still blind and the males remain sterile (Howell et al., 1963). The gene ablation studies discussed below indicate that ATRA, not retinol, is the active retinoid functioning in normal spermatogenesis (Kastner et al., 1995). However, dietary retinol is clearly required, probably because the supplemented ATRA cannot efficiently cross the Sertoli cell barrier, which is formed from specialised junctions in the Sertoli cells and results in sequestering the meiotic and post-meiotic cells from the circulation.

E-mail: djw3@columbia.edu

The term retinoids refers to a group of natural and synthetic compounds that are structurally related to vitamin A but may or may not exhibit the biological activity of vitamin A. Naturally-occurring retinoids can be formed from metabolism of vitamin A from the diet or from stored forms as well. Most of the known actions of retinoids are mediated by two retinoic acid isomers, specifically ATRA and 9-*cis* retinoic acid (9cRA) (Giguere *et al.*, 1987; Petkovich *et al.*, 1987; Heyman *et al.*, 1992; Levin *et al.*, 1992; Mangelsdorf, 1994). However, there are other active retinoids, including 11-*cis* retinaldehyde, which is needed to form rhodopsin (Saari, 1994) and all-*trans*-4-oxo retinoic acid, which is active in modulating positional specification in early *Xenopus* embryos (Pijnappel *et al.*, 1993).

The function of retinoids in regulating gene expression has been the subject of much attention in the last ten years or so, stimulated by the discovery of the retinoid receptors (Mangelsdorf, 1994; Chambon, 1996). ATRA and 9cRA serve as the ligands for two classes of ligand-dependent transcription factors, the retinoic acid receptors (RARs) and the retinoid-X receptors (RXRs). The RARs and RXRs function as heterodimers, binding to specific DNA binding sites termed retinoid response elements (RAREs). Binding of the receptors to the RAREs, which are usually found upstream of the transcription start site, can both enhance or repress transcription (Mangelsdorf, 1994; Chambon, 1996). While the RARs can bind both ATRA and 9cRA as ligands with high affinity, RXRs bind only 9cRA (Heyman *et al.*, 1992; Levin *et al.*, 1992).

There are three distinct classes of RARs and RXRs, RA/XRα, RA/XRβ, and RA/XRγ, each of which is the product of distinct genes (Mangelsdorf, 1994; Chambon, 1996). The ligand binding domains of the receptors are highly conserved (> 75% amino acid identity); however, they are structurally and functionally diverse. This diversity of the function is enhanced by the presence of more than one promoter for each of the genes and differential splicing, which together yield a medley of alternative transcripts. The different transcripts generate receptor isoforms with differing amino terminal regions, which may be functionally distinct. The expression pattern of the receptors is thus complex, with multiple variants produced from each gene and tissue-specific differences in their expression (Leid *et al.*, 1992). Nevertheless, the identity of the RARs as retinoid receptors was confirmed by their ability to bind ATRA with high affinity and their ability to respond in a dose-dependent manner to ATRA in transactivation assays with RARE-containing reporter constructs.


## Approaches to Modulate Vitamin A Signaling in Spermatogenesis

*Dietary regimens*

The defects and abnormalities in the testis during VAD have been most extensively studied in the rat model (Griswold *et al.*, 1989; Eskild and Hansson, 1994). Following the initial loss of body weight (known as the growth retardation phase of VAD), various abnormalities in spermatogenic cells were noted. These included an abrupt decrease in all stages of spermatids by day 2 and their disappearance by day 10; a marked decrease in primary spermatocytes during days 5–12, degeneration of spermatids, delay in spermiation and disruption of Sertoli cell-spermatid association. The rapid appearance of these abnormalities at the onset of the growth retardation phase of VAD suggested that the progression of spermiogenesis, completion of spermiation and differentiation of spermatocytes were particularly sensitive to alteration in vitamin A status and hence, retinoid signaling. There was only a minor drop in the total number of spermatogonia immediately following the onset of growth retardation (Mitranond *et al.*, 1979; Sobhon *et al.*, 1979); however, by day 16–20, only approximately 25% of the spermatogonia remained, with a striking reduction of type A and B spermatogonia. This observation suggested vitamin A is critical in the maintenance of the spermatogonial population as well.

The question of whether the effects of defective retinoid signaling on germ cells in various stages of spermatogenesis are cell-intrinsic or are a result of disrupted Sertoli-germ cell interactions is of great interest. Clearly, the microenvironment in the adluminal compartment of the tubules would be altered as a consequence of breakage of the inter-Sertoli and Sertoli-germ cellular junctions, which in turn, could trigger degeneration of the spermatogenic cells. However, tight junctions have been reported to remain intact during the severe regression, three to four weeks after the onset of growth retardation (Ismail and Morales, 1992). In contrast, a disruption of inter-Sertoli cell tight junctions in rats on a VAD diet was noted as early as at around 10 days after growth retardation phase by Huang and colleagues (Huang et al., 1988). Nonetheless, abnormalities in germ cells were still observed before disruption of the junctions. These observations suggest that germ cell degeneration can occur when inter-Sertoli cell tight junction are still intact and may represent an immediate consequence of the absence of retinoid signaling. Subsequent studies supported the disruptive effect of VAD on Sertoli cell tight junctions (Morales and Cavicchia, 2002). This barrier could be restored by vitamin A replenishment following VAD for up to 7–9 weeks after the onset of growth retardation. Interestingly, when vitamin A was restored and spermatogenesis was resumed, zygotene and pachytene spermatocytes were forming without an intact Sertoli cell barrier, although some of these cells exhibited apoptosis (Morales and Cavicchia, 2002).

Defects at specific stages of spermatogonial proliferation and differentiation and at the entry into meiotic prophase upon VAD have also been extensively analysed (Chung and Wolgemuth, 2004). At the onset of the growth retardation phase of VAD in rats, the production of type $A_2$ spermatogonia was arrested, and there was also a temporary arrest of preleptotene spermatocytes. A massive and synchronous production of $A_1$ spermatogonia was detected on administration of vitamin A or intraperitoneal injection of ATRA. This resumption of spermatogenesis can be seen even after long periods of VAD (de Rooij, 2001). Further, *Rara* mRNA was readily detected by *in situ* hybridization in Sertoli cells and $A_1$ spermatogonia 6 hours after injection of ATRA in VAD mice (de Rooij et al., 1994). The differentiation of $A_{al}$ into $A_1$ spermatogonia appears to be a rather vulnerable stage because it can be blocked in a number of different situations, including VAD, elevated testosterone levels and high testicular temperature (de Rooij and Grootegoed, 1998; de Rooij, 2001). It is not yet known whether the action of ATRA in inducing $A_{al}$ to $A_1$ differentiation is direct, or indirect via Sertoli cells, since both cell types express RARs (Akmal et al., 1997). Collectively, four major defects in spermatogenesis have been identified at the onset of VAD, including failure of the production of $A_2$ spermatogonia from $A_1$ spermatogonia, a delay in the onset of and an abnormality in the progression of meiotic prophase, spermatid degeneration and a breakdown of inter-Sertoli cell tight junctions.

*Pharmacologic strategies—synthetic retinoids as antagonists of RAR function*

To better discriminate among the individual retinoid receptors and ultimately to identify their respective targets, medicinal chemical techniques have been used to synthesise unique ligands specific for the individual receptors (Chen et al., 1995; Chen et al., 1996; Yu et al., 1996; Beard and Chandraratna, 1999; Dawson and Zhang, 2002). Potential ligands were identified through the combined use of receptor binding affinity and transactivation analysis. Several companies have synthesised such compounds; however, as we are collaborating with Dr. Peter Reczek (Roswell Park Institute, USA), our laboratory is focusing on compounds synthesised by Bristol-Myers Squibb (Buffalo, New York, USA); specifically, a series of compounds that includes BMS453, BMS532 and BMS614. Chemically, these compounds are classified as 'arotinoids' and they were in fact the first antagonists of the RARs to be described (Chen *et al.*, 1995). BMS453 was shown to have good oral bioavailability (92-98%) in rats and monkeys (Schulze *et al.*, 2001); the bioavailabilities of both BMS532 and BMS614 are less well-characterised but are predicted to be similar to BMS453.

BMS453 has slight RARß activity and potent antagonist activity for all three RARs. The compound binds well to recombinant receptor protein of all three subtypes with a Kd for binding in the sub-nanomolar range: apparent Kd of 0.4 nM, 0.2 nM and 0.8 nM for RARα, ß and γ, respectively (Chen *et al.*, 1995; P. Reczek, unpublished observations). An unanticipated consequence of this screen was the discovery of novel receptor-selective transcriptional antagonists. That is, while BMS453 functions as a pan-RAR antagonist, BMS532 and BMS614 exhibited specificity for RARα. These compounds, like BMS453, fit the three criteria for antagonism of ATRA: 1) they bound to the RARs with a Kd for binding comparable to that of ATRA; 2) there was no detectable activity of the compounds in a transactivation assay; and 3) the compounds were able to compete with the activity of ATRA in a transactivation assay (antagonist function) (P. Reczek, unpublished observations). The receptor specificity of BMS532 and BMS614 was thought to be due to the amide bonds in place of the double carbon bond found in the pan-antagonist BMS453. This in turn was believed to interact with serine 232 of the RARα protein. Serine is replaced by alanine in RARß and RARγ (Ostrowski *et al.*, 1995; 1998).

The potential therapeutic uses of RAR antagonists in the treatment of retinoid toxicity were obvious (Apfel *et al.*, 1992). However, there was also considerable interest in the possible use of antagonists in treating dermatologic and inflammatory disease because of the pronounced influence of retinoid-receptor stimulation on collagen synthesis and IL-8 production (Wang and Gudas, 1988; Vincenti *et al.*, 1994). Studies were therefore performed to investigate the potential target-organ toxicity of BMS453 in rats following oral administration for various lengths of time (Schulze *et al.*, 2001). One month and then 1-, 3-, or 7-day oral toxicity studies were conducted in rats and a 1-week oral toxicity study was conducted in rabbits.

The outcomes of this study have been published (Schulze *et al.*, 2001) and are summarised below. In the initial studies, Sprague Dawley rats were given daily oral doses of 15, 60 or 240 mg/kg body weight BMS453 for 1 month. Increases in leukocyte counts, alkaline phosphatase and alanine aminotransferase concentrations, and striking testicular degeneration were found at these doses. There were overt signs of toxicity and death at 240 mg/kg, whereas body weight and food consumption decreased at the 60 and 240 mg/kg doses. In studies in which rats were administered doses of BMS453 daily for only one week and at doses ranging from 12.5 to 100 mg/kg, only minimal changes in testicular morphology and no other side effects were observed immediately after the dosing period. However, one month after the dosing ceased, marked testicular degeneration was observed at all doses. The next studies involved dosing at 2, 10 or 50 mg/kg for 1, 3 or 7 consecutive days. Although no changes were observed at the time of cessation of dosing, one month later, testicular atrophy was observed and the degeneration was still obvious if not worse after 4 months. Testicular atrophy was also observed in the rabbits (Schulze *et al.*, 2001). We are currently undertaking a detailed analysis of which stages of spermatogenesis are affected by BMS453, analogous to the studies on VAD, using the mouse model. By using the mouse model, we can then also determine the extent to which the abnormalities mimic those observed in VAD and in loss of receptor-specific function by gene ablation (see below).

*Genetic approaches*

A very powerful approach to understanding the function of a specific gene is to mutate the endogenous gene by homologous recombination in the mouse model. As applied to retinoid signaling, this would include those genes encoding not only the receptors but also enzymes and other proteins involved in the synthesis, transport and metabolism of vitamin A and the retinoids (Chung and Wolgemuth, 2004). Among the first of the gene mutations made was of the gene encoding RARα, *Rara* (Lufkin *et al.*, 1993). Surprisingly, given that *Rara* is expressed at many stages of embryogenesis (Dolle *et al.*, 1990; Ruberte *et al.*, 1991), the animals survived through embryonic development and were born. However, it was reported that there was a neonatal lethality phenotype. Interestingly, in the few mice that survived to adulthood, there was sterility in the males. Although there was no detailed analysis of the etiology of the sterility, it was noted that there was a degeneration of the seminiferous epithelium in the mutant mice and that the kinds of defects observed were similar to that seen in mice that are VAD. We have subsequently undertaken a detailed analysis of the abnormalities in the RARα-deficient testes and have begun to elucidate the cell-type specific functions of RARα in the mouse testis, which we will summarise in the following discussion. These studies became feasible because we observed that the peri-natal lethality previously reported could be allevi-ated by rearing the mutant mice in a pathogen-free environment (Chung *et al.*, 2004; 2005).

Mutations in some of the other receptors have yielded male reproductive defects as well. Mutation of the gene encoding RXRß also results in male sterility, due to oligo-astheno-teratozoospermia (Kastner *et al.*, 1996). Similar to the RARα-deficient mice, there was a failure of spermatid release into the lumen of the seminiferous tubules and there were very few epididymal sperm. Those that were present exhibited abnormalities in the acrosome and tail. The Sertoli cells accumulated lipids that appeared by histochemical staining to be unsaturated triglycerides. Again similar to the RARα-deficient testes, as the RXRß-deficient mice aged, the testes progressively degenerated, eventually containing predominantly acellular, lipid-filled tubules. As *in situ* hybridization analysis showed that *Rxrb* was expressed only in Sertoli cells (Kastner *et al.*, 1996), the authors speculated that the primary site of function for the RXRß protein during spermatogenesis was in the Sertoli cells.

These same investigators recently generated mice carrying a mutation in the *Rxrb* gene that lacked transactivation function of the receptor but was otherwise intact (Mascrez *et al.*, 2004). In striking contrast to null *Rxrb*[-/-] mice, these mutant animals were fertile and did not display the defects in spermiogenesis, at least up to one year of age. They also did not exhibit defects in spermiation. These observations suggested that the role of RXRß in spermatid release was ligand-independent. However, defects in cholesterol homeostasis in the testis were observed, which apparently requires the transactivation domain of RXRß functioning in an RXRß/LXRß heterodimer. Interestingly, studies on the effect of deletion of *Cnot7*, which encoded a CCR4-associated transcriptional co-factor, in mice revealed its role as a specific co-regulator of RXRß in testicular somatic cells (Nakamura *et al.*, 2004). These mice appeared physiologically nor-mal except that they were sterile due to oligo-astheno-teratozoospermia comparable to that seen in the *Rxrb*[-/-] testes. Spermatogenic defects in the *Cnot7*[-/-] male germ cells were restricted to postmeiotic stages. The spermatids were unsynchronised and multiple generations of elon-gated spermatids were seen in the same section of the tubules. Further, lipids were accumu-lated in the cytoplasm of Sertoli cells of the mutant testes as compared to control mice. Cnot7 protein was further shown to bind to the AF-1 domain of RXRß suggesting RXRß malfunctions in the absence of Cnot7.

Finally, mutation of the gene encoding RARγ also resulted in male sterility, but with quite a different phenotype. There was a squamous metaplasia of the glandular epithelia of the prostate and seminal vesicles (Lohnes *et al.*, 1993), which was interestingly also characteristic of VAD.

When animals were made doubly *Rara;Rarg* mutant, the genital ducts were severely abnormal (Kastner *et al.*, 1997). The vas deferens and seminal vesicles were agenic or absent.

## Morphological changes observed in RARα-deficient testes

We have undertaken a series of experiments to characterise in detail the abnormalities observed in the testes from *Rara*<sup>-/-</sup> mice with the goal of understanding the role of retinoid signaling and RARα; in particular, in cellular differentiation. These observations are summarised briefly below.

a.  *Elongated spermatids failed to align at the lumen of stage VIII tubules:* Unlike in control testes, where stage VIII tubules with four layers of cells (mature step 16 spermatids aligned along the tubular lumen of the seminiferous epithelium, step 8 spermatids, pachytene spermatocytes, and pre-leptotene spermatocytes at the basal lamina) were readily observed, such tubules were never seen in *Rara*<sup>-/-</sup> testes (Chung *et al.*, 2004).

b.  *Tubules contained mixed spermatogenic cell types, reflecting an asynchronous cell association:* A prominent observation was the presence of spermatids exemplifying almost the whole series of spermiogenic differentiation, from step 1 to condensed elongated spermatids, within the same plane of a section within a single tubule (Chung *et al.*, 2004). These spermatids were frequently observed along with spermatocytes with diverse maturation states, from early pachytene to diplotene. It was thus difficult to "stage" the *Rara*<sup>-/-</sup> testes according to the criteria developed by Oakberg (Oakberg, 1956) and refined by Russell et al. (Russell *et al.*, 1990). Nonetheless, we attempted to assess the developmental stage of the tubules using the acrosomal system (Russell *et al.*, 1990) and we refer to these approximately-staged tubules with a roman numeral followed by an asterisk (e.g. stage IX*).

c.  *Some tubules were missing a complete layer of cells:* An entire layer of given cell types was missing in some severely defective tubules (Chung *et al.*, 2004). At stage XII-I* for example, MI/MII spermatocytes with chromosomes aligned at the metaphase plate, secondary spermatocytes and step 1 round spermatids were readily seen, however, zygotene-early pachytene spermatocytes and spermatogonia were missing. This may suggest abnormalities in the timing of progression through spermatogenesis or defects in proliferation or both.

d.  *Tubules contained vacuolated spaces:* Approximately 17% of the tubules in *Rara*<sup>-/-</sup> testes contained prominent vacuolar-like spaces, although the tubular lumens were found to be relatively similar in size in *Rara*<sup>-/-</sup> and control testes (Chung *et al.*, 2004).

e.  *Spermatids failed to orient properly toward the basal aspect of Sertoli cell membrane:* Step 8-9 spermatids undergo a characteristic morphological transformation during spermiogenesis in which the spermatid nuclei move toward their inner cell membrane in stage VIII tubules. The acrosomal region of the nucleus abuts the spermatid plasma membrane and rotates within the cell to face the basal aspect of the Sertoli cell, with the spermatid tails projecting into the tubule lumen. The spermatid orientation can be visualised by periodic acid-Schiff (PAS) staining of the acrosomes and determining their position relative to the basal compartment (Chung *et al.*, 2005). In the *Rara*<sup>+/+</sup> testis, step 16 spermatids were released in stage IX tubules (Fig. 1A). Three layers of germ cells were then observed, including leptotene spermatocytes, late pachytene spermatocytes and step 9 elongating spermatids (Fig. 1A). The acrosomes of the very early elongating step 9 spermatids were also exclusively toward the basal aspect of the Sertoli cell (arrows in Fig. 1A). In striking contrast, nuclei in step 8 and 9 spermatids in *Rara*<sup>-/-</sup> testis failed to become orientated toward the basal aspect of Sertoli cells (Fig. 1B, arrows). Rather, approximately 77% of step 8-9 spermatids in stage VIII*-IX* tubules in *Rara*<sup>-/-</sup> testes appeared to be randomly oriented with regard to the basal lamina, as compared to ~ 4% of the controls.

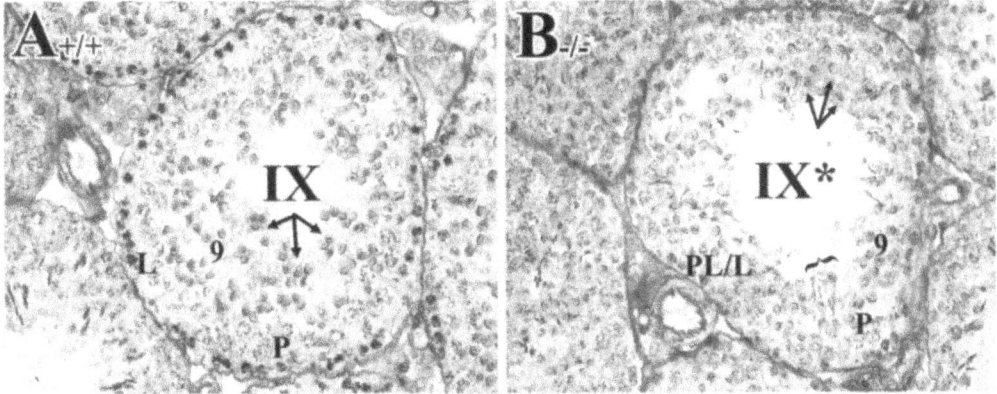

**Fig. 1** Retention of spermatids, failure of spermatid release and aberrant orientation of spermatids of *Rara*-/- testes observed at stage XI within the seminiferous epithelium. The acrosome of spermatids at stage IX in histological sections of testes from *Rara*+/+ (A) and *Rara*-/- males (B) were detected with PAS staining (magenta color). P, pachytene spermatocytes; PL/L, preleptotene or leptotene spermatocytes; Arabic numerals, the step of elongated spermatids. Roman numerals indicate the stage of the seminiferous tubule. Arrows in panels (A) and (B) indicate the orientation of elongated spermatids while the bracket in panel (B) indicates retentions of spermatids. Both panels, 60x.

f.     *Spermatids failed to be released into the lumen:* In stage IX* tubules, four (rather than three) layers of germ cells were observed, with late spermatids retained in the epithelia (Fig. 1B, bracket). As noted above, at stage VIII* of *Rara*-/- testes, mature spermatids failed to align at the surface of tubular lumen. Further, while entrenchment of spermatids within more basal aspects of Sertoli cells was observed in *Rara*+/+ testes at stage IV, it was absent in corresponding *Rara*-/- stage IV*.

### Assessment of kinetics of the progression through spermatogenesis in *Rara*-/- mice

Abnormalities in the timing of spermatogenic cell progression through the mitotic and meiotic cell cycles and various stages of differentiation could, at least in part, account for the striking asynchrony of cell association that was observed. The highly regulated temporal progression of spermatogenic differentiation means that any delays or acceleration in the timing of the interval that occurs between DNA synthesis at pre-leptotene and entry into specific meiotic stages or stages of spermiogenic differentiation might perturb the normal cellular associations. We therefore examined the kinetics of this progression in detail in experiments using BrdU-labeling (Chung et al., 2004). In brief, thirteen days after administration of BrdU to 8-9 day-old control mice, ~ 50% of the step 1 spermatids in *Rara*+/+ testes were labeled. In most of the *Rara*-/- testes examined at this age, the number of labeled step 1 spermatids was much lower. There were relatively higher numbers of tubules with labeled pachytene spermatocytes (~ 42%) in the *Rara*-/- testes, when compared to controls (~ 29%). In post-injection day 18 control testes, step 8-9 round and elongating spermatids along with a few step 13-14 spermatids were labeled. In contrast, only labeled step 8-9 spermatids were detected in *Rara*-/- testes. Thus, although some of the cells in the first spermatogenic wave had proceeded at a comparable pace through early spermiogenesis, most of them arrested after they reached step 8-9, consistent with the morphological observations.

## Apoptosis in elongated spermatids in *Rara*⁻ testes

TUNEL staining was also used in combination with morphological analysis to determine whether cells in the degenerating tubules and the abnormally oriented elongating spermatids, in particular, were undergoing apoptosis (Chung et al., 2005). In normal testis, spermatogonia and early and meiotically dividing spermatocytes are the most frequent TUNEL-positive cells (Henriksen et al., 1995; Rodriguez et al., 1997; Sjoblom et al., 1998). In young *Rara*⁻ adult (8-9 weeks) testes, TUNEL-positive spermatids at the periphery of some of the tubules were noted in *Rara*⁻ testes but never in *Rara*⁺/⁺ testes. These peripherally located spermatids were shown to be at steps 10-11 of spermiogenesis as determined by staining with PAS.

## Activated caspase-3 was involved in apoptosis of spermatogonia and spermatocytes but not elongated spermatids in *Rara*⁻ testes

Caspase-3 is the most significant effector caspase. Its activation hallmarks the point of no return in programmed cell death signaling (Earnshaw et al., 1999). However, the death of spermatozoa, which are cells with a transcriptionally inactive nucleus, was shown to occur via a caspase-independent death program (Weil et al., 1998). Although TUNEL-labelled elongated spermatids at the periphery of the seminiferous tubules were obviously detected in the adult *Rara*⁻ testes, no expression of activated caspase-3 was observed in the corresponding cell types in *Rara*⁻ testes of the next serial section (Chung et al., 2005). This suggested that the apoptosis observed in elongated spermatids of *Rara*⁻ testes may occur via a caspase-3-independent pathway, similar to what was suggested for mature sperm. In contrast, activated caspase-3 was found primarily in early primary spermatocytes and occasionally in spermatogonia, but not in Sertoli cells or Leydig cells, in 2-3-week-old testes from both strains.

## Combined genetic and dietary approaches—fine tuning

The fact that there are high levels of retinyl esters (REs) in the liver and other tissues of mice as compared to other animals has complicated studies on VAD in mice because of the length of time required before the symptoms appear (McCarthy and Cerecedo, 1952). Three mutant strains of mice have now been generated which appear to result in VAD upon dietary regimens much more quickly. For example, retinol binding protein (RBP)-deficient-mice were viable and fertile, with only vision abnormalities that were corrected after the first several months (Quadro et al., 1999). However, they exhibit reduced levels of plasma retinol (12.5% of wild type) and abnormal retinol mobilization from hepatic stores. This suggested the possibility that *RBP*⁻ mice may become VAD more quickly, which would be most useful in studies on embryonic development. Recent studies showed the importance of embryonic RBP in distributing vitamin A to certain developing tissues under restricted diets (Quadro et al., 2005). The fetal offspring from RBP null dams display a phenotype of VAD upon maternal dietary VAD (Quadro et al., 2005). We suggest that such models might also be useful to study the onset of abnormalities of VAD during the first wave of spermatogenesis.

The second mouse model is the null mutation of the cellular retinol-binding protein I (CRBP I) gene, and the mice appear physiologically normal and are fertile. CRBP I-deficient mice exhibited ~ 50% reduction of REs accumulation in hepatic stellate cells (Ghyselinck et al., 1999). Under VAD, these mice exhausted their RE stores six times faster than wild-type mice and developed abnormalities characteristic of hypo-vitaminosis. However, it took 23 weeks for serum retinol concentrations to drop to significantly lower levels, which was much longer than the time required in lecithin retinol acyltransferase (LRAT)-deficient mice upon VAD (discussed below).

LRAT catalyses the esterification of retinol in the liver and in some extrahepatic tissues as well. Recently, the gene encoding LRAT (*Lrat*) was mutated by two groups and the physiological consequences were examined (Batten *et al.*, 2004; Liu and Gudas, 2005; O'Byrne *et al.*, 2005). Similar regions around the ATG site were deleted in both groups' targeting strategies, but the *Lrat*[-/-] mice generated by Gudas and colleagues have the neo[R] cassette removed from the targeted allele to avoid possible deleterious effects resulting from the presence of the *neo* promoter (Liu and Gudas, 2005). *Lrat*[-/-] mice developed normally and the females were fertile while the males exhibited some impaired fertility. Some of the testes were reported to exhibit marked hypoplasia and oligospermia as well as dilated seminiferous tubular lumens and an absence of the seminiferous epithelium at 4 weeks and 3 months of age, but seemed to actually improve with age (Liu and Gudas, 2005). The *Lrat*[-/-] mice from the Palczewski group, although originally reported to have reproductive problems, seemed to do reasonably well when maintained in a pathogen-free environment on breeder chow (O'Byrne *et al.*, 2005; W.S. Blaner, personal communication).

The possible use of this model as a tool to study VAD was carefully examined (Liu and Gudas, 2005). LRAT-deficient mice exhibited a loss of more than 99.5% of hepatic REs even when mice were on a control, vitamin A-sufficient diet. Upon VAD, *Lrat*[-/-] mice became VAD in a period as short as 6 weeks and retinol was not detectable in most of the tissues. In addition, under vitamin A supplementation, the retinol concentrations in serum increased rapidly in the *Lrat*[-/-] mice, indicating that serum retinol concentrations in mutant mice can be conveniently modulated by the quantitative manipulation of dietary retinol. As such, *Lrat*[-/-] mice would be a convenient model for studying the effects of VAD in various tissues and during development. However, as there are some indications of male sub-fertility even in a vitamin A-sufficient diet, the effects of VAD on spermatogenesis may be best studied in the other two models. In any of the models, it is important to note that the dietary regimens are reversible, while the receptor knock-out models represent essentially permanent changes in retinoid-signaling, evidenced from conception.

## Summary and future directions

Our observations along with previous studies on VAD suggest that there are distinct signaling pathways involved in vitamin A function and mediated by RARα that differ between the germ cells and the somatic cells of the testis and further, that distinct cell-cell interactions between Sertoli cells and germ cells are altered in the absence of retinoid signaling. The morphological observations observed in the *Rara*[-/-] testes are summarised in cartoon form in Fig. 2. It is striking that these defects occured primarily in Stage VII-VIII tubules. As noted above, at stage VIII-IX, there was a delay or arrest in pre-leptotene and leptotene spermatocytes in the first, second and third waves of spermatogenesis. Further, there was an abnormal orientation and temporary arrest and accumulation of step 8-9 spermatids in the first wave. Consequently, the spermatozoa showed abnormal alignment to the tubular lumen which led to aberrant spermiation at stage VIII. The high level of asynchrony of spermatogenic progression in the mutant tubules suggested the involvement of ATRA signaling in the transition of randomly cycling populations of undifferentiated spermatogonia to the stages where the differentiating spermatogonia were highly synchronised. Intriguingly, this step occurred when $A_1$ spermatogoina differentiate to form $A_2$ spermatogonia, again from stage VIII to IX. Coincidently, *Rara* mRNA expression was found to be stage-specific, with the highest expression at stage VIII (Kim and Griswold, 1990; Akmal *et al.*, 1997). Although the mechanisms responsible for this stage specificity remain to be determined, it appears that important developmental events that occur at stage VIII are under rigid control, possibly via RARα-mediated signaling.

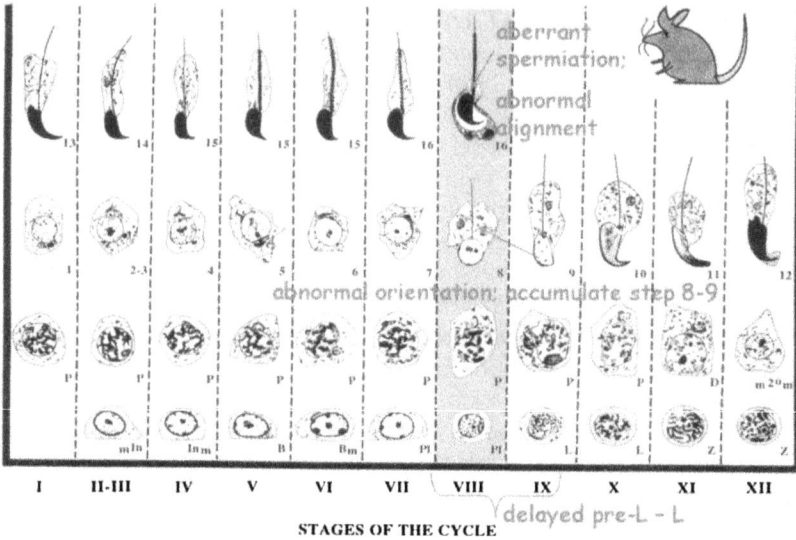

**Fig. 2** Diagrammatic representation of the spermatogenic cycle illustrating the profound abnormalities in RARα-deficient mice clustered in stage VIII-IX tubules. Details of the symbols used in the staging map shown can be found in Russell *et al.* (1990). The green bar line indicates the particular stage, stage VIII, that showed the highest frequency of cellular abnormalities. Red arrows point to the specific cell type at stage VIII, where various abnormalities were found.

The approaches discussed above hold promise not only for examining the precise spatial and temporal requirements for retinoid signaling in spermatogenesis, but also for identifying upstream factors or downstream targets that might be involved in retinoid-dependent signaling cascades. During spermatogenesis, direct modulation of the expression of a male germ cell-associated kinase (*Mak*) mRNA (Wang and Kim, 1993) as well as the levels of mRNA for one of the RARs (*Rara*) (Kim and Griswold, 1990) have been shown in response to the presence or absence of retinol in the diet of rats. Knock-out mutations of the *bone morphogenetic protein 8b* (*Bmp8*) (Zhao *et al.*, 1996) and *desert hedgehog* (Bitgood *et al.*, 1996) genes have demonstrated that each of these secreted factors was required for normal spermatogenesis. There is evidence that other members of these gene families are regulated by ATRA. Expression of *Bmp2* was induced by a RARα agonist in F9 embryonal carcinoma cells, while *Bmp4* was repressed (Rogers, 1996). In the chick limb bud, implantation of an ATRA-soaked bead resulted in ectopic expression of *sonic hedgehog* (Riddle *et al.*, 1993). It is possible that these retinoid-dependent pathways are conserved among various tissues, including the testis.

Finally, the need for a reduction in the toxic side effects associated with retinoid therapies has led to the search for synthetic retinoids with enhanced activity and receptor-specificity (Beard and Chandraratna, 1999; Dawson and Zhang, 2002). Such pharmacologic approaches may also be useful in examining the consequence of inhibiting RARα function at a particular stage of spermatogenesis during post-natal and adult life. For example, RARα-specific antagonists would be able to exert their inhibitory activity quickly, as compared to a dietary regimen of VAD, but would be reversible, unlike the genetic knock-out approach. This could be particularly powerful for looking at the role of retinoid signaling in specific cell types in the first wave of spermatogenesis. In addition, RARα-specific antagonists, such as BMS532 and BMS614, would be predicted to have marked effects on spermatogenesis with a reduction in any possible side effects due to RARβ or RARγ targets.

## Acknowledgements

This work was supported in part by grants from the NIH (P01 DK05077) and the CONRAD Foundation (CIG-05-105 and CIG 05-107).

## References

**Akmal KM, Dufour JM and Kim KH** (1997) Retinoic acid receptor alpha gene expression in the rat testis: potential role during the prophase of meiosis and in the transition from round to elongating spermatids *Biology of Reproduction* 56 549-556

**Apfel C, Bauer F, Crettaz M, Forni L, Kamber M, Kaufmann F, LeMotte P, Pirson W and Klaus M** (1992) A retinoic acid receptor alpha antagonist selectively counteracts retinoic acid effects *Proceedings of the National Academy of Sciences USA* 89 7129-7133

**Batten ML, Imanishi Y, Maeda T, Tu DC, Moise AR, Bronson D, Possin D, Van Gelder RN, Baehr W and Palczewski K** (2004) Lecithin-retinol acyltransferase is essential for accumulation of all-trans-retinyl esters in the eye and in the liver *Journal of Biological Chemistry* 279 10422-10432

**Beard RL and Chandraratna R** (1999) RAR-selective ligands: receptor subtype and function selectivity In *Retinoids: The Biochemical and Molecular Basis of Vitamin A and Retinoid Action* pp 185-208 Eds H Nau and WS Blaner. Springer-Verlag, Berlin Heidelberg New York

**Bitgood MJ, Shen L and McMahon AP** (1996) Sertoli cell signaling by Desert hedgehog regulates the male germline *Current Biology* 6 298-304

**Chambon P** (1996) A decade of molecular biology of retinoic acid receptors *The FASEB journal: official publication of the Federation of American Societies for Experimental Biology* 10 940-954

**Chen JY, Clifford J, Zusi C, Starrett J, Tortolani D, Ostrowski J, Reczek PR, Chambon P and Gronemeyer H** (1996) Two distinct actions of retinoid-receptor ligands *Nature* 382 819-822

**Chen JY, Penco S, Ostrowski J, Balaguer P, Pons M, Starrett JE, Reczek P, Chambon P and Gronemeyer H** (1995) RAR-specific agonist/antagonists which dissociate transactivation and AP1 transrepression inhibit anchorage-independent cell proliferation *EMBO Journal* 14 1187-1197

**Chung SS, Sung W, Wang X and Wolgemuth DJ** (2004) Retinoic acid receptor alpha is required for synchronization of spermatogenic cycles and its absence results in progressive breakdown of the spermatogenic process *Developmental dynamics: an official publication of the American Association of Anatomists* 230 754-766

**Chung SS, Wang X and Wolgemuth DJ** (2005) Male sterility in mice lacking retinoic acid receptor alpha involves specific abnormalities in spermiogenesis *Differentiation* 73 188-198

**Chung SS and Wolgemuth DJ** (2004) Role of retinoid signaling in the regulation of spermatogenesis *Cytogenetic and Genomic Research* 105 189-202

**Dawson MI and Zhang XK** (2002) Discovery and design of retinoic acid receptor and retinoid X receptor class- and subtype-selective synthetic analogs of all-trans-retinoic acid and 9-cis-retinoic acid *Current Medicinal Chemistry* 9 623-637

**de Rooij DG** (2001) Proliferation and differentiation of spermatogonial stem cells *Reproduction* 121 347-354

**de Rooij DG and Grootegoed JA** (1998) Spermatogonial stem cells *Current Opinion in Cell Biology* 10 694-701

**de Rooij DG, van Pelt AMM, Van de Kant HJG, van der Saag PT, Peters AHFM, Heyting C and de Boer P** (1994) Role of retinoids in spermatogonial proliferation and differentiation and the meiotic prophase. In *Function of somatic cells in the testis* pp 345 Ed A. Bartke. Springer, Berlin Heidelberg New York

**Dolle P, Ruberte E, Leroy P, Morriss-Kay G and Chambon P** (1990) Retinoic acid receptors and cellular retinoid binding proteins. I. A systematic study of their differential pattern of transcription during mouse organogenesis *Development* 110 1133-1151

**Earnshaw WC, Martins LM and Kaufmann SH** (1999) Mammalian caspases: structure, activation, substrates, and functions during apoptosis *Annual Review of Biochemistry* 68 383-424

**Eskild W and Hansson V** (1994) Vitamin A functions in the reproductive organs. In *Vitamin A in health and disease* pp 531-559 Eds R Blomhoff. Dekker, New York

**Ghyselinck NB, Bavik C, Sapin V, Mark M, Bonnier D, Hindelang C, Dierich A, Nilsson CB, Hakansson H, Sauvant P, Azais-Braesco V, Frasson M, Picaud S and Chambon P** (1999) Cellular retinol-binding protein I is essential for vitamin A homeostasis *EMBO Journal* 18 4903-4914

Giguere V, Ong ES, Segui P and Evans RM (1987) Identification of a receptor for the morphogen retinoic acid *Nature* **330** 624-629

Griswold MD, Bishop PD, Kim KH, Ping R, Siiteri JE and Morales C (1989) Function of vitamin A in normal and synchronized seminiferous tubules *Annals of the New York Academy of Sciences* **564** 154-172

Henriksen K, Hakovirta H and Parvinen M (1995) In-situ quantification of stage-specific apoptosis in the rat seminiferous epithelium: effects of short-term experimental cryptorchidism *International Journal of Andrology* **18** 256-262

Heyman RA, Mangelsdorf DJ, Dyck JA, Stein RB, Eichele G, Evans RM and Thaller C (1992) 9-cis retinoic acid is a high affinity ligand for the retinoid X receptor *Cell* **68** 397-406

Howell JM, Thompson JN and Pitt GAJ (1963) Histology of the lesions produced in the reproductive tract of animals fed a diet deficient in vitamin A alcohol but containing vitamin A acid, I. The male rat. *Journal of Reproduction and Fertility* **5** 159-167

Huang HF, Yang CS, Meyenhofer M, Gould S and Boccabella AV (1988) Disruption of sustentacular (Sertoli) cell tight junctions and regression of spermatogenesis in vitamin-A-deficient rats *Acta Anatomica (Basel)* **133** 10-15

Ismail N and Morales CR (1992) Effects of vitamin A deficiency on the inter-Sertoli cell tight junctions and on the germ cell population *Microscopy Research and Technique* **20** 43-49

Kastner P, Mark M and Chambon P (1995) Nonsteroid nuclear receptors: what are genetic studies telling us about their role in real life? *Cell* **83** 859-869

Kastner P, Mark M, Ghyselinck N, Krezel W, Dupe V, Grondona JM and Chambon P (1997) Genetic evidence that the retinoid signal is transduced by heterodimeric RXR/RAR functional units during mouse development *Development* **124** 313-326

Kastner P, Mark M, Leid M, Gansmuller A, Chin W, Grondona JM, Decimo D, Krezel W, Dierich A and Chambon P (1996) Abnormal spermatogenesis in RXR beta mutant mice *Genes and Development* **10** 80-92

Kim KH and Griswold MD (1990) The regulation of retinoic acid receptor mRNA levels during spermatogenesis *Molecular Endocrinology* **4** 1679-1688

Leid M, Kastner P and Chambon P (1992) Multiplicity generates diversity in the retinoic acid signalling pathways *Trends in Biochemical Sciences* **17** 427-433

Levin AA, Sturzenbecker LJ, Kazmer S, Bosakowski T, Huselton C, Allenby G, Speck J, Kratzeisen C, Rosenberger M, Lovey A and Grippo JF (1992) 9-cis retinoic acid stereoisomer binds and activates the nuclear receptor RXR alpha *Nature* **355** 359-361

Liu L and Gudas LJ (2005) Disruption of the lecithin:retinol acyltransferase gene makes mice more susceptible to vitamin A deficiency *Journal of Biological Chemistry* **280** 40226-40234

Lohnes D, Kastner P, Dierich A, Mark M, LeMeur M and Chambon P (1993) Function of retinoic acid receptor gamma in the mouse *Cell* **73** 643-658

Lufkin T, Lohnes D, Mark M, Dierich A, Gorry P, Gaub MP, LeMeur M and Chambon P (1993) High postnatal lethality and testis degeneration in retinoic acid receptor alpha mutant mice *Proceedings of the National Academy of Sciences USA* **90** 7225-7229

Mangelsdorf DJ (1994) Vitamin A receptors *Nutrition Reviews* **52** S32-44

Mascrez B, Ghyselinck NB, Watanabe M, Annicotte JS, Chambon P, Auwerx J and Mark M (2004) Ligand-dependent contribution of RXRbeta to cholesterol homeostasis in Sertoli cells *EMBO Reports* **5** 285-290

McCarthy PT and Cerecedo LR (1952) Vitamin A deficiency in the mouse *The Journal of Nutrition* **46** 361-376

Mitranond V, Sobhon P, Tosukhowong P and Chindaduangrat W (1979) Cytological changes in the testes of vitamin-A-deficient rats. I. Quantitation of germinal cells in the seminiferous tubules *Acta Anatomica (Basel)* **103** 159-168

Morales A and Cavicchia JC (2002) Spermatogenesis and blood-testis barrier in rats after long-term Vitamin A deprivation *Tissue & Cell* **34** 349-355

Nakamura T, Yao R, Ogawa T, Suzuki T, Ito C, Tsunekawa N, Inoue K, Ajima R, Miyasaka T, Yoshida Y, Ogura A, Toshimori K, Noce T, Yamamoto T and Noda T (2004) Oligo-astheno-teratozoospermia in mice lacking Cnot7, a regulator of retinoid X receptor beta *Nature Genetics* **36** 528-533

O'Byrne SM, Wongsiriroj N, Libien J, Vogel S, Goldberg IJ, Baehr W, Palczewski K and Blaner WS (2005) Retinoid absorption and storage is impaired in mice lacking lecithin:retinol acyltransferase (LRAT) *Journal of Biological Chemistry* **280** 35647-35657

Oakberg EF (1956) A description of spermiogenesis in the mouse and its use in an analysis of the cycle of the seminiferous epithelium and germ cell renewal *American Journal of Anatomy* **99** 391-414

Ostrowski J, Hammer L, Roalsvig T, Pokornowski K and Reczek PR (1995) The N-terminal portion of domain E of retinoic acid receptors alpha and beta is essential for the recognition of retinoic acid and various analogs *Proceedings of the National Academy of Sciences USA* **92** 1812-1816

Ostrowski J, Roalsvig T, Hammer L, Marinier A, Starrett JE, Jr, Yu KL and Reczek PR (1998) Serine 232 and methionine 272 define the ligand binding pocket in retinoic acid receptor subtypes *Journal of Biological Chemistry* **273** 3490-3495

Petkovich M, Brand NJ, Krust A and Chambon P (1987) A human retinoic acid receptor which belongs to the family of nuclear receptors *Nature* **330** 444-450

Pijnappel WW, Hendriks HF, Folkers GE, van den Brink CE, Dekker EJ, Edelenbosch C, van der Saag PT and Durston AJ (1993) The retinoid ligand 4-oxo-retinoic acid is a highly active modulator of positional specification *Nature* **366** 340-344

Quadro L, Blaner WS, Salchow DJ, Vogel S, Piantedosi R, Gouras P, Freeman S, Cosma MP, Colantuoni V and Gottesman ME (1999) Impaired retinal function and vitamin A availability in mice lacking retinol-binding protein *EMBO Journal* **18** 4633-4644

Quadro L, Hamberger L, Gottesman ME, Wang F, Colantuoni V, Blaner WS and Mendelsohn CL (2005) Pathways of vitamin A delivery to the embryo: insights from a new tunable model of embryonic vitamin A deficiency *Endocrinology* **146** 4479-4490

Riddle RD, Johnson RL, Laufer E and Tabin C (1993) Sonic hedgehog mediates the polarizing activity of the ZPA *Cell* **75** 1401-1416

Rodriguez I, Ody C, Araki K, Garcia I and Vassalli P (1997) An early and massive wave of germinal cell apoptosis is required for the development of functional spermatogenesis *EMBO Journal* **16** 2262-2270

Rogers MB (1996) Receptor-selective retinoids implicate retinoic acid receptor alpha and gamma in the regulation of bmp-2 and bmp-4 in F9 embryonal carcinoma cells *Cell Growth & Differentiation* **7** 115-122

Ruberte E, Dolle P, Chambon P and Morriss-Kay G (1991) Retinoic acid receptors and cellular retinoid binding proteins. II. Their differential pattern of transcription during early morphogenesis in mouse embryos *Development* **111** 45-60

Russell LD, Ettlin RA, SinhaHikim AP and Clegg ED (1990) *Histological and Histopathological Evaluation of the Testis*, Cache River Press, Clearwater, FL

Saari JC (1994) Retinoids in photosensitive systems. In *The Retinoids Biology, Chemistry, and Medicine* pp 351-386 Eds MB Sporn, AB Roberts and DS Goodman. Raven Press, Ltd., New York

Schulze GE, Clay RJ, Mezza LE, Bregman CL, Buroker RA and Frantz JD (2001) BMS-189453, a novel retinoid receptor antagonist, is a potent testicular toxin *Toxicological Sciences* **59** 297-308

Sjoblom T, West A and Lahdetie J (1998) Apoptotic response of spermatogenic cells to the germ cell mutagens etoposide, adriamycin, and diepoxybutane *Environmental and Molecular Mutagenesis* **31** 133-148

Sobhon P, Mitranond V, Tosukhowong P and Chindaduangrat W (1979) Cytological changes in the testes of vitamin-A-deficient rats. II. Ultrastructural study of the seminiferous tubules *Acta Anatomica (Basel)* **103** 169-183

Vincenti MP, Coon CI, Lee O and Brinckerhoff CE (1994) Regulation of collagenase gene expression by IL-1 beta requires transcriptional and post-transcriptional mechanisms *Nucleic Acids Research* **22** 4818-4827

Wang SY and Gudas LJ (1988) Protein synthesis inhibitors prevent the induction of laminin B1, collagen IV (alpha 1), and other differentiation-specific mRNAs by retinoic acid in F9 teratocarcinoma cells *Journal of Cell Physiology* **136** 305-311

Wang Z and Kim KH (1993) Vitamin A-deficient testis germ cells are arrested at the end of S phase of the cell cycle: a molecular study of the origin of synchronous spermatogenesis in regenerated seminiferous tubules *Biology of Reproduction* **48** 1157-1165

Weil M, Jacobson MD and Raff MC (1998) Are caspases involved in the death of cells with a transcriptionally inactive nucleus? Sperm and chicken erythrocytes *Journal of Cell Science* **111** 2707-2715

Wolbach SB and Howe PR (1925) Tissue changes following deprivation of fat-soluble A vitamin. *Journal of Experimental Medicine* **42** 753-777

Yu KL, Spinazze P, Ostrowski J, Currier SJ, Pack EJ, Hammer L, Roalsvig T, Honeyman JA, Tortolani DR, Reczek PR, Mansuri MM and Starrett JE, Jr. (1996) Retinoic acid receptor beta, gamma-selective ligands: synthesis and biological activity of 6-substituted 2-naphthoic acid retinoids *Journal of Medicinal Chemistry* **39** 2411-2421

Zhao GQ, Deng K, Labosky PA, Liaw L and Hogan BL (1996) The gene encoding bone morphogenetic protein 8B is required for the initiation and maintenance of spermatogenesis in the mouse *Genes & Development* **10** 1657-1669

# The role of androgens in spermatogenesis

Shinji Komori[1,2], Hiroyuki Kasumi[1], Kazuko Sakata[1] and Koji Koyama[1,2]

[1]Department of Obstetrics and Gynecology, Hyogo College of Medicine, Nishinomiya, 663-8501,
Japan; [2]Laboratory of Developmental Biology and Reproduction, Institute for Advanced Medical
Sciences, Hyogo College of Medicine, Nishinomiya, 663-8501, Japan

Androgens are important factors in spermatogenesis. However, the
biological role of androgen in Sertoli cells is still unclear. In this study,
we analysed mutations and CAG repeats of the androgen receptor gene
in 19 azoospermic and 117 oligozoospermic men. No mutations were
identified, but 9 of 117 oligozoospermic men were found to carry 14 or
15 CAG repeats, as compared to more than 16 CAG repeats in 136 normal
fertile men analysed. Analysis of the androgenic effect on the expression
of transition protein 1 and 2 (TP1, TP2) by co-transfection experiments
showed that androgens regulate the transcription of TP1 via a Dfd-like
molecule and the transcription of TP2 via an androgen-androgen receptor
complex, respectively. We analysed the effect of dihydrotestosterone
(DHT) on protein profiles in the TM4 mouse Sertoli cell line using SELDI-
TOF mass spectrometry. DHT increased the expression of seven proteins
of 4.34, 4.97, 5.68, 5.75, 9.95, 9.98 and 11.30 kDa and decreased six
proteins of 4.94, 4.97, 6.29, 8.57, 12.39 and 19.81 kDa. One of the
decreased molecule (19.80 kDa) is identified to be a translationally
controlled tumor protein. These results show that androgen affects
spermatogenesis by actions on both the germ cells and Sertoli cells.

## Introduction

Androgens are important factors in the development of testis and spermatogenesis (Quigley et
al., 1995; Heinlein and Chang, 2002). Since Sertoli cells have androgen receptors, androgen is
thought to affect spermatogenesis via Sertoli cells (Zirkin, 1993). However, the role of androgen in Sertoli cells is still unclear. The androgen receptor (AR) has also been detected on germ
cells such as spermatogonia and spermatocytes in human. It is therefore possible that androgen
acts directly on germ cells during spermatogenesis. Generally, androgen affects the expression
of target genes via AR. The AR gene, located on the long arm of the X chromosome at Xq11-12,
contains 8 exons (Quigley et al., 1995). The AR gene encodes the N-terminal domain (exon 1),
the DNA binding domain (exon 2 and 3), the hinge region (exon 4) and the ligand binding
domain (exon 5-8) (Fig. 1). A number of defects of the AR gene have been reported leading to
testicular feminization, male infertility and other abnormalities in sexual development. Among
them, some mutations lead to only male infertility (Table 1). In the testis, androgen is secreted
from Leydig cells upon stimulation with luteinizing hormone (LH) to support spermatogenesis

Correspondence: Shinji Komori MD, Department of Obstetrics and Gynecology, Hyogo College of Medicine, 1-1 Mukogawa-cho, Nishinomiya, Hyogo 663-8501, Japan   Telephone:81-798-45-6482
Fax:81-798-46-4163    E-mail: komor615@hyo-med.ac.jp

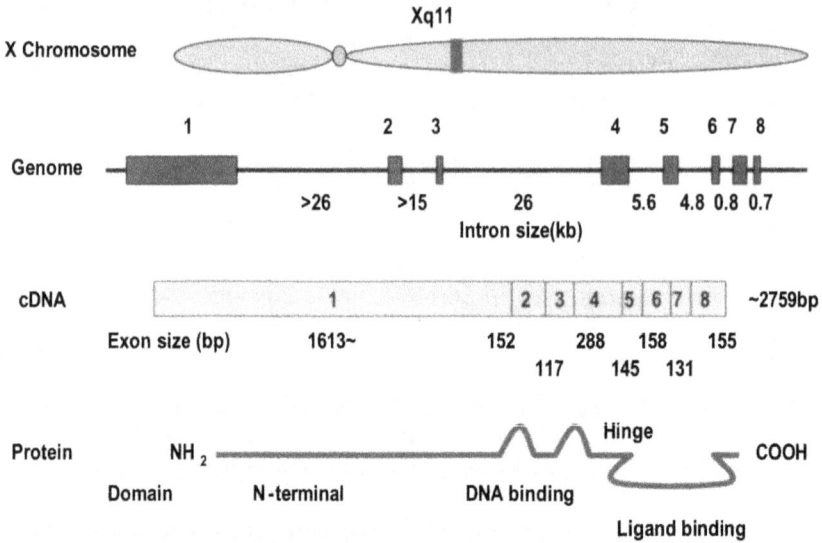

**Figure 1.** Human androgen receptor: The androgen receptor gene is encoded by 8 exons. The scheme indicates the chromosomal location, the genomic organization, the exons and the encoded protein domains of the human androgen receptor.

**Table 1.** Point mutations of the androgen receptor gene in infertile men

| Exon | Amino acid number | Mutation | Patient number | Reports |
|---|---|---|---|---|
| 1 | 390 | Pro(CCG) → Ser(TCG) | 2 | Hiort et al., 2000 |
| 1 | 511 | Val(GTG) → Val(GTA) | 1 | Hiort et al., 1998 |
| 5 | 727 | Asn(AAC) → Lys(AAG) | 1 | Yong et al., 1994 |
| 5 | 756 | Asn(AAT) → Ser(AGT) | 1 | Giwercman et al., 2001 |
| 8 | 886 | Met(ATG) → Val(GTG) | 2 | Ghadessy et al., 1999 |
| 8 | 911 | Val(GTC) → Leu(CTC) | 1 | Knoke et al., 1999 |

by acting on germ cells directly or indirectly through Sertoli cells. Table 1 shows point mutations of the AR gene that caused spermatogenic failure. In Table 2, we list genetic changes of the AR gene associated with androgen insensitivity syndrome (Table 2). Cases 4 and 5 carried no mutations but had short CAG repeats. Short or long CAG repeats within the AR gene have also been reported to be associated with Kennedy disease, prostate cancer and male infertility (Quigley et al., 1995; La Spade, 1991).

In this study, we analyzed: i) mutation and CAG repeats of the AR gene in azoospermic and oligozoospermic men, ii) androgenic effects on the expression of transition proteins (TP1, TP2); and iii) androgenic effects on protein profiles in Sertoli cells

## Analysis of the AR gene in infertile men

Our analysis of the AR gene in 19 azoospermic and 117 oligozoospermic men revealed no mutations (Fig. 2, unpublished data). However, 14 or 15 CAG repeats in 9 out of 117

**Table 2.** Summary of mutations of androgen receptor gene in complete androgen insensitivity syndrome in our clinic

| Patient | Exon | Amino acid number | Mutations | | Binding | Thermostability |
|---|---|---|---|---|---|---|
| 1 | 1 | 258 | Leu | → Pro | Normal | Decreased |
| | 7 | 820 | Gly | → Ala | | |
| 2 | 1 | 194 | Gln | → Arg | | |
| | | 622 | Stop codon | | | |
| 3 | 7 | 855 | Arg | → Cys | NT | NT |
| 4 | 1 | | Gln | 13[a] | Normal | Decreased |
| 5 | 1 | | Gln | 13[a] | Normal | Decreased |
| 6 | 6 | 774 | Arg | → Cys | NT | NT |
| 7 | 2 | 571 | Tyr | → Cys | NT | NT |
| 8 | 5 | 752 | Arg | → Gln | NT | NT |
| 9 | 5 | 752 | Arg | → Cys | NT | NT |
| 10 | 4 | 674 | Leu | → Pro | Normal | Decreased |
| 11 | 6 | 803 | Glu | → Lys | NT | NT |
| 12 | 6 | 803 | Glu | → Lys | NT | NT |

[a]Glutamine repeat number; NT = Not tested

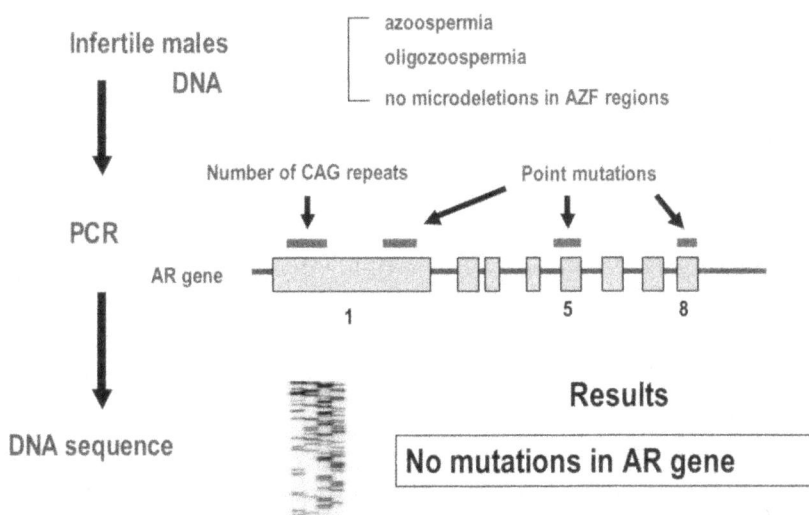

**Figure 2.** Summary of analysis of the androgen receptor gene in infertile males.

oligozoospermic men were observed as compared to more than 16 CAG repeats in 136 normal fertile men (Fig. 3). These results indicated that short CAG repeats within the AR gene may be related to spermatogenic failure (unpublished results).

## Analysis of androgenic effects on the expression of TP1 and TP2

TP1 and TP2 play an important role in condensing the chromosome during spermatogenesis. The deficiency of TP1 and TP2 genes has been shown to result in spermatogenic failure (Yu et al., 2000; Meistrich et al., 2003). To analyse transcriptional regulation of TP1 and TP2 genes by

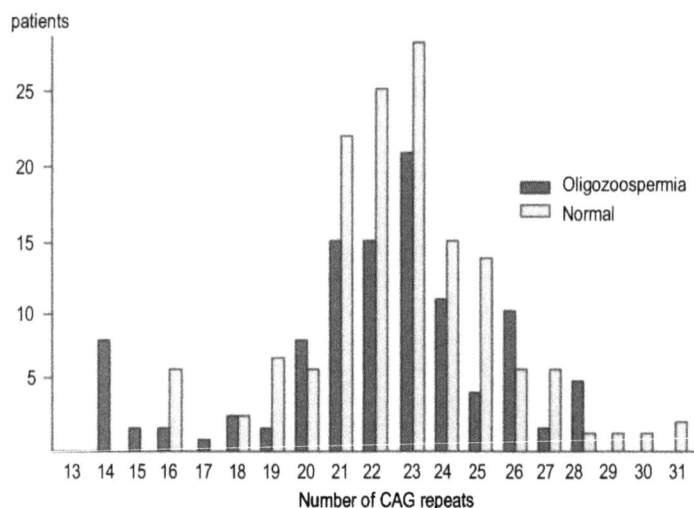

**Figure 3.** Analysis of number of CAG repeats in oligozoospermic males. Gray bar indicates normal fertile males. Black bar indicates oligozoospermic males.

androgen, we constructed a luciferase reporter plasmid under the control of the 5' region of the TP1 and TP2 gene and expression vectors carrying the human AR gene with various CAG repeats (12, 15, 22 or 43 repeats). These plasmids were co-transfected into COS-7 cells by electroporation and 2 days later, luciferase activities were measured with or without dihydrotestosterone (DHT). We found that AR with 22 CAG repeats (wild type) enhanced luciferase activity (unpublished data). On the other hand, increase of luciferase activity was not observed in COS-7 cells transfected with the AR containing short CAG repeats of 12 and 15 or excessively long CAG repeats of 43. The gel mobility shift assays were performed to analyse proteins binding to the 5' region of TP1 and TP2 in COS-7 cells co-transfected with the wild type AR. The TP1 region from -754 to -471 bp and TP2 region from -1218 to -858 bp were identified as the binding site of nuclear proteins. The footprint assays identified these sites as the deformed (Dfd) motif (Lou et al., 1995) and the androgen responsive element-like motif (Quigley et al., 1995), respectively. These results suggest that androgen affects the expression of TP1 via a Dfd-like molecule and the expression of TP2 via androgen-androgen receptor complex in germ cells (unpublished observations).

## Analysis of androgenic effects on protein profiles in Sertoli cells

We analysed the androgenic effect on protein expression in mouse Sertoli cells by culturing TM4 Sertoli cell lines with or without DHT and analysing protein profiles by a protein chip system (Howard et al., 2000; Merchant and Weinberger, 2000; Weinberger et al., 2000; Furuta et al., 2004). We found that DHT treatment increased expression of 4.97 kDa protein at 15 min, 11.30 kDa protein at 24 hrs and 4.34 kDa, 5.68 kDa, 5.75 kDa, 9.95 kDa and 9.98 kDa proteins at 48 hrs (Table 3). On the other hand, expression of 6.29 kDa and 8.57 kDa proteins were decreased at 30 min and expression of 4.94 kDa, 4.97 kDa, 12.39 kDa and 19.81 kDa proteins were decreased at 48 hrs (Table 4). The 19.81 kDa molecule was identified as a translationally controlled tumor protein (TCTP) based on the amino acid sequence (Gross et al., 1989). TCTP has various biological activities including binding to tubulin and $Ca^{2+}$ (Bohm et al., 1989; Guillaume et al., 2001; Tuynder et al., 2002; Cans et al., 2003). Since tubulin is related to mitosis, TCTP may have a role during spermatogenesis.

**Table 3.** Proteins in Sertoli cell line TM4 that were enhanced in expression after DHT treatment

| MW | Tip | pH | Time after DHT |
|---|---|---|---|
| | Increased molecules | | |
| 4970 | Q10 | 5.5 | 15 min |
| 11300 | CM10 | 8.5 | 24 hr |
| 4340 | CM10 | 7.5 | 48 hr |
| 5680 | Q10 | 7.5 | 48 hr |
| 5750 | CM10 | 7.5 | 48 hr |
| 9950 | Q10 | 8.5 | 48 hr |
| 9980 | Q10 | 8.5 | 48 hr |

**Table 4.** Proteins in Sertoli cell line TM4 that were reduced in expression after DHT treatment

| MW | Tip | pH | Time after DHT |
|---|---|---|---|
| | Decreased molecules | | |
| 6290 | Q10 | 7.5 | 30 min |
| 8570 | CM10 | 4.5 | 30 min |
| 4940 | CM10 | 4.5 | 48 hr |
| 4970 | Q10 | 8.5 | 48 hr |
| 12390 | Q10 | 8.5 | 48 hr |
| 19810 | Q10 | 8.5 | 48 hr |

## Conclusion

Our results show that the decrease in the number of CAG repeats of AR is related to defects during spermatogenesis, whereas point mutations of AR are rare in infertile males with spermatogenic failure. Androgen regulates the expression of TP1 and TP2 in germ cells and a number of proteins in Sertoli cells. Thus androgen affects the spermatogenesis by acting on both germ and Sertoli cells.

## Acknowledgement

This work was supported in part by a Grant-in Aid for Scientific Research from the Ministry of Education, Science and Culture (No 16591693 for S Komori and No 14571594 for H Kasumi). We thank Dr Tomohiko Yamasaki and Dr Yonehiro Kanemura for technical support of SELDI-TOF mass spectrometry.

## References

Bohm H, Benndorf R, Gaestel M, Gross B, Nurnberg P, Kraft R, Otto A and Bielka H (1989) The growth-related protein P23 of the Ehrlich ascites tumor: translational control, cloning and primary structure. *Biochemistry International* **19** 277-286

Cans C, Passer BJ, Shalak V, Nancy-Portebois V, Crible V, Amzallag N, Allanic D, Tufino R, Argentini M, Moras D, Fiucci G, Goud B, Mirande M, Amson R and Telerman A (2003) Translationally controlled tumor protein acts as a guanine nucleotide dissociation inhibitor on the translation elongation factor eEF1A *Proceedings of the National Academy of*

*Sciences USA* **100** 13892-13897

Furuta M, Shiraishi T, Okamoto H, Mineta T, Tabuchi K and Shiwa M (2004) Identification of pleiotrophin in conditioned medium secreted from neural stem cells by SELDI-TOF and SELDI-tandem mass spectrometry *Brain Research: Developmental Brain Research* **152** 189-197

Ghadessy FJ, Lim J, Abdullah AA, Panet-Raymond V, Choo CK, Lumbroso R, Tut TG, Gottlieb B, Pinsky L, Trifiro MA and Yong EL (1999) Oligospermic infertility associated with an androgen receptor mutation that disrupts interdomain and coactivator (TIF2) interactions *Journal of Clinical Investigation* **103** 1517-1525

Giwercman YL, Nikoshkov A, Bystrom B, Pousette A, Arver S and Wedell A (2001) A novel mutation (N233K) in the transactivating domain and the N756S mutation in the ligand binding domain of the androgen receptor gene are associated with male infertility *Clinical Endocrinology* **54** 827-834

Gross B, Gaestel M, Bohm H and Bielka H (1989) cDNA sequence coding for a translationally controlled human tumor protein *Nucleic Acids Research* **17** 8367.

Guillaume E, Pineau C, Evrard B, Dupaix A, Moertz E, Sanchez JC, Hochstrasser DF and Jegou B (2001) Cellular distribution of translationally controlled tumor protein in rat and human testes *Proteomics* **1** 880-889

Heinlein CA and Chang C (2002) Androgen receptor (AR) coregulators: an overview *Endocrine Review* **23** 175-200

Hiort O, Holterhus PM, Horter T, Schulze W, Kremke B, Bals-Pratsch M, Sinnecker GH and Kruse K (2000) Significance of mutations in the androgen receptor gene in males with idiopathic infertility *The Journal of Clinical Endocrinology and Metabolism* **85** 2810-2815

Hiort O, Holterhus PM, Horter T, Schulze W, Kremke B, Bals-Pratsch M, Sinnecker GH and Kruse K (1998) Significance of mutations in the androgen receptor gene in males with idiopathic infertility *80th US Endocrine Society Meeting Abstract* p2-38

Howard JC, Heinemann C, Thatcher BJ, Martin B, Gan BS and Reid G (2000) Identification of collagen-binding proteins in *Lactobacillus* spp. with surface-enhanced laser desorption/ionization-time of flight ProteinChip technology *Applied Environmental Microbiology* **66** 4396-4400

Knoke I, Jakubiczka S, Lehnert H and Wieacker P (1999) A new point mutation of the androgen receptor gene in a patient with partial androgen resistance and severe oligozoospermia *Andrologia* **31** 199-201

La Spade AR, Wilson EM, Lubahn DB, Harding AE and Fischbeck KH (1991) Androgen receptor gene mutations in X-linked spinal and bulbar muscular atrophy *Nature* **352** 77-79

Lou L, Bergson C and McGinnis W (1995) Deformed expression in the *Drosophila* central nervous system is controlled by an autoactivated intronic enhancer *Nucleic Acids Research* **23** 3481-3487

Meistrich ML, Mohapatra B, Shirley CR and Zhao M (2003) Roles of transition nuclear proteins in spermiogenesis *Chromosoma* **111** 483-488

Merchant M and Weinberger SR (2000) Recent advancements in surface-enhanced laser desorption/ionization-time of flight-mass spectrometry *Electrophoresis* **21** 1164-1177

Quigley CA, De Bellis A, Marschke KB, El-Awady MK, Wilson EM and French FS (1995) Androgen receptor defects: historical, clinical, and molecular perspectives *Endocrine Review* **16** 271-321

Tuynder M, Susini L, Prieur S, Besse S, Fiucci G, Amson R and Telerman A (2002) Biological models and genes of tumor reversion: cellular reprogramming through tpt1/TCTP and SIAH-1 *Proceedings of the National Academy of Sciences USA* **99** 14976-14981

Weinberger SR, Morris TS and Pawlak M (2000) Recent trends in protein biochip technology *Pharmacogenomics* **1** 395-416

Yong EL, Ng SC, Roy AC, Yun G and Ratnam SS (1994) Pregnancy after hormonal correction of severe spermatogenic defect due to mutation in androgen receptor gene *Lancet* **344** 826-827

Yu YE, Zhang Y, Unni E, Shirley CR, Deng JM, Russell LD, Weil MM, Behringer RR and Meistrich ML (2000) Abnormal spermatogenesis and reduced fertility in transition nuclear protein 1-deficient mice *Proceedings of the National Academy of Sciences USA* **97** 4683-4688

Zirkin BR (1993) Regulation of spermatogenesis in the adult mammal: Gonadotropins and androgens. In *Cell and Molecular Biology of the Testis* Chapter 8 pp 166-188. Eds C Desjardins and LL Ewing. Oxford University Press, Oxford

# Delineating the role of estrogen in regulating epididymal gene expression

Deshpande Shayu[1], Matthew P. Hardy[2] and A. Jagannadha Rao*[1]

[1]Department of Biochemistry, Indian Institute of Science, Bangalore 560012, India; [2]Population Council, Center for Biomedical Research, New York 10021, USA

Maturation of sperm within the epididymis is a pre-requisite for fertilization in mammals. Epididymal function is controlled by a complex array of hormones and growth factors. While testosterone is the primary stimulus for epididymal development and sperm maturation, the importance of estrogen effects on efferent ductules has been increasingly recognized, and points to a need to clarify the role of estrogen receptor-mediated action in the epididymis. Estrogens modulate the expression of genes involved in fluid absorption in the efferent ductules and the epididymis. The present review highlights the role of estrogen in regulation of epididymal gene expression.

## Introduction

At the point of their release from the Sertoli cells, sperm are not fully competent for fertilization and must first undergo a maturational process in the epididymis. The ability to fertilize the female gamete is achieved by an extensive remodeling of the spermatozoa during their passage in the epididymis, when they acquire forward motility, the ability to undergo capacitation and acrosome reaction, all of which are necessary to fuse with the egg plasma membrane (Yanagimachi, 1994). Despite the wealth of information available, much remains to be understood regarding the precise molecular events involved in sperm maturation and the role of the epididymis in this process.

## Epididymal and sperm proteins as potential contraceptive targets

Understanding the process of sperm maturation and the specific events crucial for the fertilizing ability of spermatozoa, would help in identifying potential targets for the development of contraceptives. A multidimensional approach has been used to understand sperm maturation. The first method involves identifying proteins on the sperm membrane that might have been acquired, for instance, eppin, clusterin and human epididymal protein (Kirchhoff et al., 1990; Law and Griswold, 1994), or proteins that are modified during sperm transit in the epididymis such as, PH-20 and galactosyltransferase (Scully et al., 1987; Rutllant and Meyers, 2001). Another strategy identifies key proteins involved in epididymal functioning indirectly and then establishes their bearing on sperm maturation. Contraceptive action may be achieved by impeding the function of these proteins upon binding with specific antibodies (McCauley et al., 2002; O'Rand et al., 2004). Alternatively, through the use of compounds (Ratnasooriya and

---

*Corresponding author
E-mail: ajrao@biochem.iisc.ernet.in

Wadsworth, 1994; Lue *et al.*, 1998) that selectively affect epididymal function or the time required for sperm passage in the epididymis, fertilizing ability can be inhibited. While some of these approaches have shown promise in animal models, the list of acceptable candidate epididymal proteins that might be targeted for reversible contraception in humans remains short. To expand the range of contraceptive choice it is therefore imperative to gain an in-depth understanding of epididymal function.

## Structure of the epididymis

Anatomically, the epididymis can be grossly divided into three segments namely the head (caput), body (corpus) and the tail (cauda). While morphologically these segments are quite distinct in rodents, this segmentation cannot be readily visualized in the primates. In the present study, however, we refer to the epididymis as being divided into the caput, corpus and cauda. The three regions of the epididymis not only exhibit morphological differences, in species such as the rat, but are also functionally distinct, and hence understanding their specific functions would shed light on the formation of the unique epididymal environment.

## Functions of the epididymis

The epididymis has evolved a system for protecting sperm from damage caused by oxidative stress (Aitken, 1999) and microbial attack (Hall *et al.*, 2002). The epididymis is richly endowed with several anti-oxidant enzymes that guard the sperm and the epithelium from free radicals (Aitken, 2002). Besides, a number of sperm proteins of testicular origin are subjected to proteolytic processing in the epididymis (Cuasnicu *et al.*, 2002). The sperm membrane also undergoes lipid remodeling in the epididymis (Jones, 2002). Although the basic structure of the plasma membrane of testicular spermatozoa is similar to the mature spermatozoa, several changes occur at the molecular level, influencing fluidity of the sperm membrane.

As the sperm are transported from the testis to the epididymis, about 74-96% of fluid is reabsorbed in the efferent ductules and the remaining fluid is reabsorbed, mainly in the caput (head) region of the epididymis (Hess *et al.*, 2001). The mechanism of fluid absorption has been elucidated in the studies of the efferent ductules (Ilio and Hess, 1994; Clulow *et al.*, 1998). The epithelia of the efferent ductules and the epididymis express ion exchangers and water channel proteins, enabling them to maintain fluid homeostasis. The movement of water is dependent upon the ionic gradient in the epithelium that is maintained by several ion exchangers such as Na-K ATPase, Na-H exchanger (NHE), carbonic anhydrases (CA) *et cetera* (Lee *et al.*, 2001). The passive movement of water is then facilitated by aquaporins (AQPs), a class of water channel proteins that are present on both the luminal and basal faces of the epithelium. Electrolyte and fluid transport is critical for normal function of the epididymis because it controls the concentrations of luminal components necessary for sperm maturation.

While many of the functions of the epididymis are performed by the entire epididymal duct, there are also region-specific functions that result from localized gene and protein expression patterns. The regional differences are in turn often the result of locally specific responses to hormones and growth factors.

## Regulation of the epididymis by estrogen

The conventional view of estrogen as the female hormone has been modified in light of recent information. Estrogen mediates its actions via the estrogen receptors (ERs), ERα and ERß. An increased interest in the role of estrogens in the male arose from studies demonstrating impaired fertility in mice lacking the receptor for estrogen (Lubahn *et al.*, 1993; Korach, 1994).

## ER localization in the epididymis

There is considerable variation in the expression of ERα and ERß in the male excurrent ducts. Given the variations observed, further analysis of the expression patterns of ERα and ERß at both the mRNA and protein level are warranted and may indicate differences in estrogen-sensitivity across species (Hart, 1990). The present study focused on two animal models, the bonnet monkey (*Macaca radiata*) and the rat. In the baboon epididymis, ERα is detected (Albrecht *et al.*, 2004), but this contrasts with the human and non-human primates such as the marmoset and macaque species (Saunders *et al.*, 2001), where expression levels are low. The mRNA and protein for both ER isoforms were highly expressed in the bonnet monkey epididymides and a distinct nuclear localization was observed for both the receptors in the bonnet monkey (Shayu *et al.*, 2005).

ERß is expressed in the efferent ductules, the entire epididymal tract, prostate and vas deferens of the rat (O'Donnell *et al.*, 2001). High ERα expression is seen in the efferent ductules (Hess *et al.*, 2002). However, while earlier reports showed the presence of ERα mRNA in the rat epididymis, demonstration of the ERα protein by immunohistochemistry could not be established (Hess *et al.*, 1997a). Expression of ERα and ERß protein, assessed by Western blot analysis was abundant in the epididymis of the rat (D. Shayu, unpublished).

## Tools to study the role of estrogen in the male

The earliest insights on the role of estrogen in male came from the observation that exposure of male offspring to the estrogenic compound diethylstilbestrol (DES) during development *in utero*, resulted in a variety of defects such as cryptorchidism, distension of efferent ductules, under-development of the epididymis; and sperm granulomas (Newbold *et al.*, 1985; Fisher *et al.*, 1999; Hess *et al.*, 2002). Thereafter, other estrogenic compounds, including bisphenol-A, genistein, and ethinyl estradiol, were used to study excessive estrogen action on the male reproductive system (Fisher *et al.*, 1999). Conversely, antiestrogenic compounds of the ICI series, such as tamoxifen (ICI 46474), ICI 182780 *et cetera* which bind to ERs were used to inhibit receptor-mediated actions. Antiestrogens and mouse knockout models for ERα (ERαKO) and ERß (ERßKO) have provided essential tools to study estrogen action in the male.

## Effect of ICI 182780 treatment on ER expression in the caput region of the epididymis

We examined the role of estrogen in epididymides obtained from ICI 182780 (ICI) treated bonnet monkey and the rat. Both species were analyzed because, while the rodent serves as a convenient laboratory model, there are limitations in extrapolating data from rodents to higher mammals such as humans. The caput was selected for study, as this region is considered to be estrogen-sensitive (McLachlan *et al.*, 1975). In addition, the caput is most actively involved in protein synthesis and secretion (Cornwall and Hann, 1995).

## Expression of ERα and ERß in the bonnet monkey and rat

ERα mRNA and protein expression were consistently increased in the ICI treated bonnet monkeys (ICI administered via mini-osmotic pumps: 250 µg/day/animal) although serum testosterone concentrations remained unaltered in the 30, 60 and 90 day ICI treated monkeys (Shayu *et al.*, 2005). The expression of ERα and ERß proteins in rat epididymal samples following 8 days of ICI treatment (1 mg/kg B.W.) were significantly lower although steady state mRNA levels for ERα were equivalent in vehicle and ICI treated groups (unpublished observations; Fig. 1). These effects occurred in the absence of any detectable change in the hormonal profile of testosterone (Shayu *et al.*, 2005).

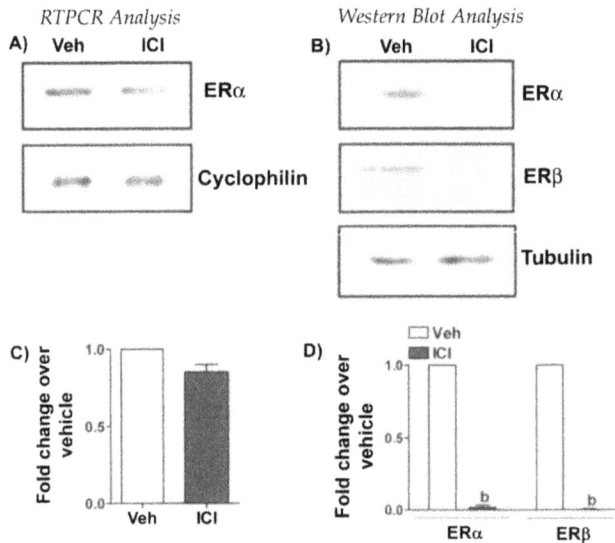

**Fig. 1** Expression of ERα and ERß in the caput of the ICI treated rat. ERα mRNA expression in the vehicle- and ICI-caput was analyzed by semi-quantitative RTPCR with cyclophilin mRNA expression used for normalization of ERα mRNA (panel A). Signal intensities of normalized ERα expression is graphically represented in panel C as fold change over vehicle. Total protein (100 $\mu$g) from the caput region of the vehicle and ICI treated rat epididymis was subjected to SDS-PAGE and Western blotting. The blot was probed using specific antibody for ERα and after stripping, probed for ERß (panel B). The same blot was stripped and probed for the internal control tubulin for normalization of protein loading and a graphical representation of the same is shown in panel D. Data in panels C and D are a mean $\pm$ SEM of three experiments. (b) $P < 0.001$

The results observed in the bonnet monkey epididymis contrasted with our own previous observations and those of others in the rodent, where a drastic decrease in ERα expression was observed following ICI treatment (Oliveira *et al.*, 2003). The increased expression of ERα in the monkey caput after estrogen antagonism may be characteristic of estrogen signaling in primate tissues given that a similar response has been reported in the human endometrium after ICI administration (Dowsett *et al.*, 1995).

## Estrogen mediated regulation of genes involved in fluid absorption

ERαKO males are infertile as a result of abnormal dilation and fluid reabsorption in the rete testis and the efferent ductules (Lubahn *et al.*, 1993). The dilation was a result of fluid build-up in these organs, which in turn causes back-pressure, leading to atrophy of the testes in older animals (Hess *et al.*, 1997b). Expression of CAII and NHE3 decreased in the efferent ductules of both the ERαKO and ICI-treated wild type mice, indicating that these genes involved in fluid absorption, were regulatable by estrogen (Lee *et al.*, 2001; Zhou et al., 2001). Loss of ERα-mediated estrogen action has fewer consequences in the epididymis compared to the efferent ductules. However, morphological abnormalities such as abnormal growth of the initial segment of the ERαKO epididymis and an increase in Periodic Acid Shiff-positive granules in the clear cells of the epididymal epithelium are observed (Hess et al., 2000). We and others have hypothesized that estrogen action is involved in the regulation of epididymal ion exchanger and water channel protein expression levels (Lee et al., 2001; Zhou et al., 2001; Shayu et al., 2005).

In the bonnet monkey, a decrease in AQP1 expression was observed following ICI treatment (Shayu *et al.*, 2005). In rat epididymides also, a sharp reduction in AQP1 and NHE3 mRNA expression occurs after ICI treatment (unpublished observations; Fig. 2). This was associated with decreased AQP1 protein expression. It is to be noted however, that although relatively the expression of ERα receptor following ICI treatment were contrasting in the monkey (increase) and rat (decrease) caput, the expression of AQP1 was down-regulated in both the ICI-treated species.

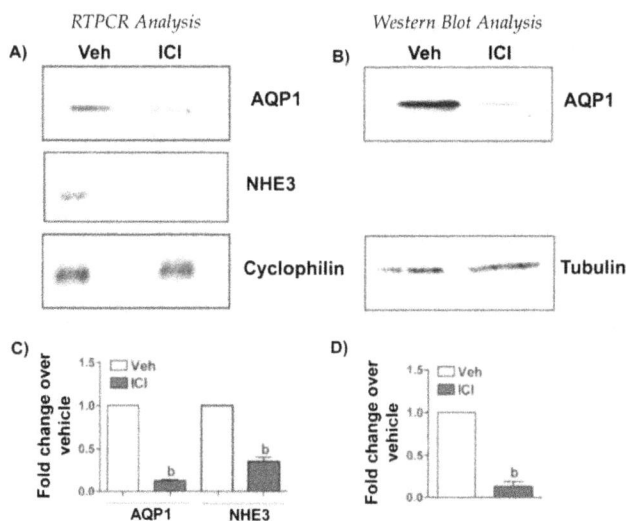

**Fig. 2** Expression of AQP1 and NHE3 mRNA and AQP1 protein levels in the rat vehicle- and ICI-caput regions. The change in mRNA expression of AQP1 and NHE3 following ICI treatment was analyzed by RTPCR in the caput region of the vehicle- and ICI treated rat epididymis (panel A). The expression of AQP1 protein following ICI treatment was also analyzed by Western blot analysis in the caput region using a specific AQP1 antibody (panel B). Panel C represents the densitometric analyses of AQP1 and NHE3 gene expression, normalized to cyclophilin and expressed as fold change relative to vehicle-caput. Panel D depicts histogram of tubulin-normalized AQP1 protein expression and expressed as fold change relative to vehicle-caput. Data in panels C and D are a mean ± SEM of three experiments. (b) *P* < 0.001

### Effects of decreased estrogen signaling on sperm motility

Sperm from the cauda epididymides of ERαKO males were abnormal and the mice were observed to be infertile (Eddy et al., 1996). Transplanted spermatogonia from ERαKO males develop normally in wild-type mouse testes and differentiate into fertile sperm (Mahato et al., 2000). This indicated that the effects on the sperm in ERαKO were not due to altered Sertoli cell function (which can affect germ cell development), but rather due to an impaired function of the excurrent ducts. In our study too, spermatozoa from the 180 day ICI treated monkeys, revealed a precipitous drop in motility and had reduced beat-frequency and progressive motility, which together was indicative of a loss in their fertilization potential (Shayu *et al.*, 2005). Interestingly, these treatment-induced effects occurred in the absence of any obvious alteration in sperm production, and thus ICI treatment probably had an effect on the maturation process and not so much on the sperm production process. Since these effects were observed in the absence of any detectable change in testosterone level, it was reasonable to conclude that the observed defects in motility could be due to blockade of estrogen action.

## Identification of estrogen regulated genes

Given that ERs are expressed abundantly in the excurrent ducts, we asked whether estrogen stimulation plays other roles in the epididymis in addition to fluid absorption. Differential display reverse transcription polymerase chain reaction (DDRTPCR) was carried out comparing the caput regions of vehicle-treated and 30 day or 90 day ICI-treated bonnet monkeys. Among the differentially expressed transcripts identified in the vehicle- and 30 day ICI-caput regions, keratin 19 (K19) was found to be down-regulated in the 30 day ICI-caput (Fig. 3). The trend was confirmed by Northern blot analysis in the monkey caput regions of vehicle and 30 day ICI treated bonnet monkey (Fig. 3). In addition, K19 mRNA expression in the ICI treated rat caput was reduced relative to vehicle control, when assessed by semi-quantitative RTPCR analysis (unpublished observations).

The K19 protein is a member of the intermediate filament family of cytoskeletal proteins. Intermediate filaments such as keratins play an important role as both structural and functional elements of the cytoskeleton. They function as tension-bearing elements to help maintain cell shape and rigidity, and serve to anchor in place several organelles in the cell. In polarized and secretory epithelia such as the intestine, K19 controls polarized localization of apical proteins (Salas *et al.*, 1997). In the epididymal epithelium, which is also secretory (and polarized), the distribution of K19 may be essential in the anchoring of apically placed ion exchanger proteins. Accordingly, the localization of K19 was studied in the caput regions of vehicle- and 30 day ICI treated bonnet monkey using a human K19 antibody (unpublished observations). A distinct localization pattern was observed in the control caput (Fig. 4). Intense staining was present along the apical and lateral regions of the epididymal epithelium. In contrast, staining intensity in the caput region of the 30 day ICI treated monkey was significantly lower (Fig. 4). The staining intensity was negligible in the lateral and basal membranes after ICI administration and was reduced and appeared diffuse in the apical membrane. It is possible for this reason that the loss in K19 expression after ICI treatment leads to aberrant localization of membrane proteins.

Another transcript that was found to be regulated by DDRTPCR was phosphotidylethanolamine N-methyltransferase (PEMT). PEMT catalyzes three sequential methylation reactions converting phosphotidylethanolamine (PE) to phosphatidylcholine (PC) (Vance and Ridgway, 1988). While PEMT expression is mainly found in the liver, its expression has also been observed in the brain, testis and heart (Shields *et al.*, 2001). PC is a major component of mammalian cellular membranes and its appropriate concentration is necessary for the fluidity of the membrane. In addition, oxidation of PC provides an energy source to spermatozoa (Mita and Ueta, 1990). Expression of PEMT was found to be up-regulated in the 90 day ICI-caput (Fig. 5) in the bonnet monkey; however, interestingly ICI treatment in the rat did not elicit a similar response. The role of PEMT in epididymal function needs to be studied further in light of its effects on sperm membrane fluidity and energy metabolism.

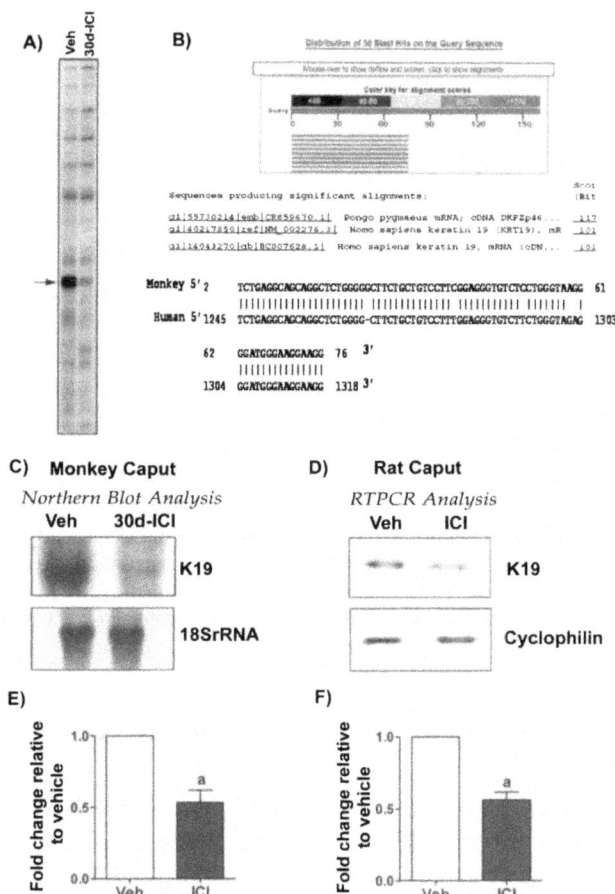

**Fig. 3** DDRTPCR analysis in the caput region of vehicle and 30 day ICI treated bonnet monkeys and validation of expression in the monkey and rat caput. DDRTPCR analysis in the vehicle- and 30 day ICI-caput regions revealed a highly differentially expressed band in the vehicle-caput as shown in panel A. BLAST analysis of this transcript revealed sequence homology with human cytokeratin 19 (K19) (panel B). Sequence of this transcript and its homology with human K19 is also shown. Northern blot analysis (panel C) was performed with 20 μg of RNA from the caput regions of vehicle and 30 day ICI treated monkeys and the K19 transcript was employed as labeled probe. Equality of RNA loading was assessed by stripping and probing the blot with labeled cDNA corresponding to 18S rRNA. RTPCR analysis of K19 was performed in the rat vehicle- and ICI-caput, employing specific primers for rat K19 (panel D). Histograms in panels E-F depict normalized expression of K19 in the ICI-caput as compared to vehicle-caput of the monkey and rat, respectively. Data in panels E and F are a mean ± SEM of three experiments (a) $P < 0.05$

DDRTPCR analysis provided essential clues regarding a much broader role for estrogen in the epididymis. It was therefore felt essential to analyze the 'global' role of estrogen in the epididymis by a high-throughput analysis tool such as microarray, in which profiling was done using the RNA from ERαKO-caput. ERαKO mice were obtained from the laboratory of Prof. Dennis Lubahn, Department of Nutritional Sciences, University of Missouri, Columbia, MO, USA and the epididymides were removed for microarray analysis (unpublished observations). Our analysis revealed several genes to be affected by lack of estrogen action (Fig. 6).

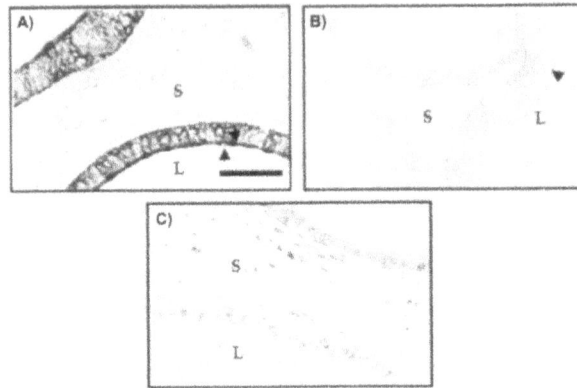

**Fig. 4** Immunolocalization of K19 in the caput region of the vehicle and 30 day ICI treated bonnet monkey. In the vehicle-caput, staining was obtained specifically in the epithelium with no discernable staining in the surrounding stroma (panel A). Staining in the epithelial cells was distinctly observed along the apical and lateral faces of the cell membrane, with intense staining at the apical region (indicated by arrows). The staining pattern was more diffuse in the 30 day ICI-caput region (panel B). Tissue sections, in which addition of primary antibody was omitted, showed no staining and served as negative control (panel C). S and L represent stroma and the epididymal lumen, respectively. Bar in panel A: 100 $\mu$m

## Down-regulation caused by estrogen blockade versus up-regulation

AQP4 and CAII, known to be involved in fluid absorption were down-regulated in ERαKO males relative to wild-type littermates (Fig. 7). CAII is involved in fluid absorption by generating H$^+$ and its decline in the ERαKO would in turn affect Na transport by NHE3 (Lee *et al.*, 2001). The suppression of CAII levels was much greater in ERαKO mice than in rats that received the ICI compound, consistent with results obtained by Lee *et al.* (2001). Androgen stimulation does not affect CAII abundance (Kaunisto *et al.*, 1999), and hence it is likely that estrogen is the key modulator of CAII expression in the caput. AQP4 expression levels were similarly decreased in both ERαKO and ICI-treated males. AQP4 expression in the epididymis has not been reported previously; it has a much higher intrinsic water permeability compared to AQP1 or other AQP isoforms (Verkman and Mitra, 2000). The lack of efferent ductule dilation in the AQP1KO (Zhou *et al.*, 2001) could be due to the compensatory action of AQP4.

Phospholipase A2V (PLA2V) was found to be down-regulated in caput epididymides of ERαKO and ICI-treated males. Localization of PLA2V in the epididymis coincides with the localization of cyclo-oxygenases, which are implicated in regulating fluid secretion (Leung *et al.*, 1998).

The expression of uromodulin (URO) was up-regulated in the ERαKO caput. URO or Tamm-Horsfall protein is a mucoprotein found in urine (Tamm and Horsfall, 1950; Muchmore and Decker, 1985). The glycan moieties of URO compete with fimbriated *E. coli* to bind uroplakin receptors in the convoluted duct of the kidney (Pak *et al.*, 2001), suggesting therefore a function in protecting the urinary tract from bacterial infections. In addition, the major distribution of URO in the thick ascending duct of the kidney suggests a role in ion transport and water permeability. Accordingly, a failure of this role would interfere with NaCl reabsorption, decrease the interstitial osmolality and impair excretion of urine. In this regard, urine concentrating defects are prevalent in patients with mutations in the gene coding for URO (Hart *et al.*, 2002).

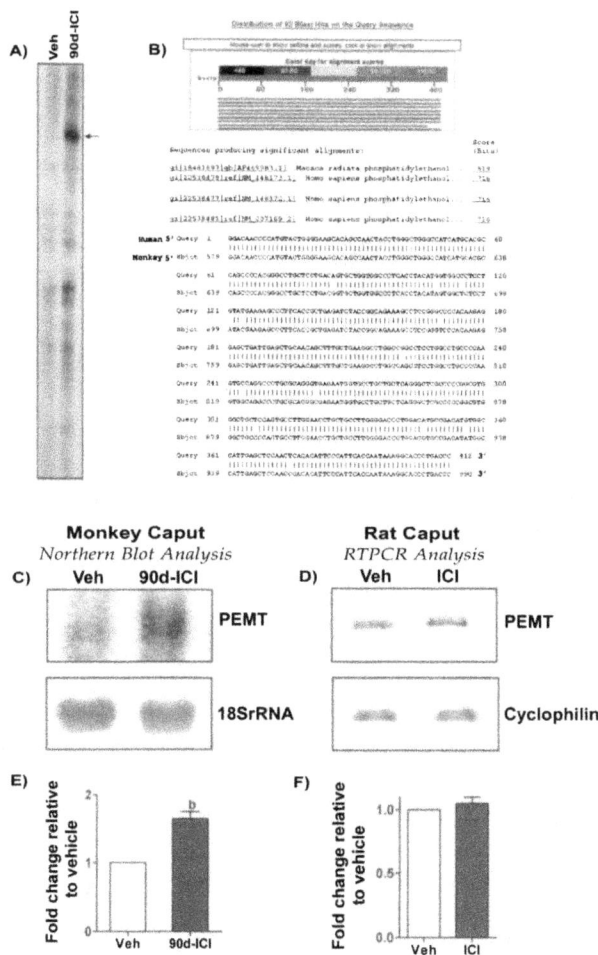

**Fig. 5** Identification of a differentially expressed transcript in 90 day ICI-caput by DDRTPCR analysis and validation of expression in the monkey and rat ICI-caput region. DDRTPCR analysis between the vehicle- and 90 day ICI-caput led to the identification of a transcript, PEMT that was highly expressed in the ICI-caput (panel A). The BLAST hits of the monkey PEMT sequence showed complete homology to human PEMT (panel B). Validation of DDRTPCR was carried out by Northern blot analysis with 20 $\mu$g of RNA from the vehicle- and 90 day ICI-caput regions and the labeled PEMT transcript was used as probe (panel C). The mRNA expression of PEMT in the rat vehicle and ICI-caput regions was analyzed by semi-quantitative RTPCR analysis using specific rat primers (panel D). Histograms in panels E-F show PEMT expression normalized to 18S rRNA or cyclophilin respectively, and plotted as fold change relative to respective vehicle-caput. Data in panels E and F are a mean $\pm$ SEM of three experiments. $P < 0.001$ (b)

The expression of URO in the epididymis has not been reported previously. The presence of URO in the epididymis is interesting since the epididymis and the kidney share common embryonic origin (Rodriguez et al., 2002). Increased URO expression would impede fluid concentration, thereby augmenting the effect offset by reduced ion exchangers in the ERαKO.

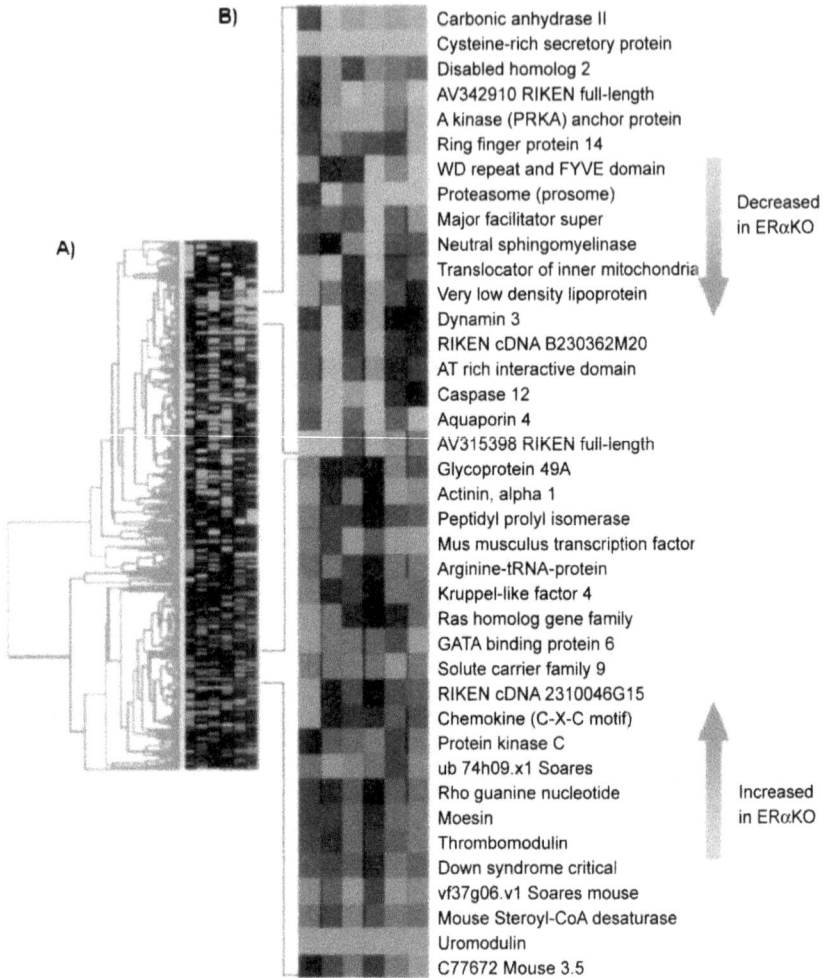

**Fig. 6** Microarray profiling of the genes from the wild-type and ERαKO caput that were differentially regulated are depicted as a cluster (panel A). A portion of the cluster is shown enlarged in panel B. Spots represented in green denote the genes that were down-regulated in the ERαKO-caput while the spots in red demarcate genes up-regulated in the ERαKO.

Microarray analysis revealed a very high expression of URO in the ERαKO caput. While URO expression showed a 3-3.5 fold up-regulation in the ERαKO-caput, interestingly there was no change in its expression in the ICI-caput (unpublished data). Expression of genes such as PEMT and URO that were up-regulated in the ICI treated monkey caput and ERαKO-caput respectively, were not affected by 8-day ICI treatment in the rat, indicating that some changes require a long-term loss of ER mediated signaling. The present findings are the first, identifying probable estrogen regulated genes in the epididymis. In addition to fluid absorption, estrogen may have a role in regulating expression of genes involved in lipid metabolism, maintenance of cell structure, which may or may not be unrelated to fluid absorption.

Reutrakul V, Sangsawan R, Chaichana S and Swerdloff RS (1998) Triptolide: a potential male contraceptive *Journal of Andrology* 19 479-486

Mahato D, Goulding EH, Korach KS and Eddy EM (2000) Spermatogenic cells do not require estrogen receptor-alpha for development or function *Endocrinology* 141 1273-1276

McCauley TC, Kurth BE, Norton EJ, Klotz KL, Westbrook VA, Rao AJ, Herr JC and Diekman AB (2002) Analysis of a human sperm CD52 glycoform in primates: identification of an animal model for immuno-contraceptive vaccine development *Biology of Reproduction* 66 1681-1688

McLachlan JA, Newbold RR and Bullock B (1975) Reproductive tract lesions in male mice exposed prenatally to diethylstilbestrol *Science* 190 991-992

Mita M and Ueta N (1990) Phosphatidylcholine metabolism for energy production in sea urchin spermatozoa *Biochimica et Biophysica Acta* 1047 175-179

Muchmore AV and Decker JM (1985) Uromodulin: a unique 85-kilodalton immunosuppressive glycoprotein isolated from urine of pregnant women *Science* 229 479-481

Newbold RR, Bullock BC and McLachlan JA (1985) Lesions of the rete testis in mice exposed prenatally to diethylstilbestrol *Cancer Research* 45 5145-5150

O'Donnell L, Robertson KM, Jones ME and Simpson ER (2001) Estrogen and spermatogenesis *Endocrine Reviews* 22 289-318

Oliveira CA, Nie R, Carnes K, Franca LR, Prins GS, Saunders PT and Hess RA (2003) The antiestrogen ICI 182,780 decreases the expression of estrogen receptor-alpha but has no effect on estrogen receptor-beta and androgen receptor in rat efferent ductules *Reproductive Biology and Endocrinology* 1 75

O'Rand M G, Widgren EE, Sivashanmugam P, Richardson RT, Hall SH, French FS, VandeVoort CA, Ramachandra SG, Ramesh V and Jagannadha Rao A (2004) Reversible immunocontraception in male monkeys immunized with eppin *Science* 306 1189-1190

Pak J, Pu Y, Zhang ZT, Hasty DL and Wu XR (2001) Tamm-Horsfall protein binds to type 1 fimbriated *Escherichia coli* and prevents *E. coli* from binding to uroplakin Ia and Ib receptors *Journal of Biological Chemistry* 276 9924-9930

Ratnasooriya WD and Wadsworth RM (1994) Tamsulosin, a selective alpha 1-adrenoceptor antagonist, inhibits fertility of male rats *Andrologia* 26 107-110

Rodriguez CM, Kirby JL and Hinton BT (2002) The development of the epididymis. In *The Epididymis:* *from molecules to clinical practice*, edn 1, pp 251-267. Eds B Robaire and B Hinton Kluwer Academic/Plenum publishers, New York

Rutllant J and Meyers SA (2001) Posttranslational processing of PH-20 during epididymal sperm maturation in the horse *Biology of Reproduction* 65 1324-1331

Salas PJ, Rodriguez ML, Viciana AL, Vega-Salas DE and Hauri HP (1997) The apical submembrane cytoskeleton participates in the organization of the apical pole in epithelial cells *Journal of Cell Biology* 137 359-375

Saunders PT, Sharpe RM, Williams K, Macpherson S, Urquart H, Irvine DS and Millar MR (2001) Differential expression of oestrogen receptor alpha and beta proteins in the testes and male reproductive system of human and non-human primates *Molecular Human Reproduction* 7 227-236

Scully NF, Shaper JH and Shur BD (1987) Spatial and temporal expression of cell surface galactosyl-transferase during mouse spermatogenesis and epididymal maturation *Developmental Biology* 124 111-124

Shayu D, Kesava CC, Soundarajan R and Rao AJ (2005) Effects of ICI 182780 on estrogen receptor expression, fluid absorption and sperm motility in the epididymis of the bonnet monkey *Reproductive Biology and Endocrinology* 3 10

Shields DJ, Agellon LB and Vance DE (2001) Structure, expression profile and alternative processing of the human phosphatidylethanolamine N-methyl-transferase (PEMT) gene *Biochimica et Biophysica Acta* 1532 105-114

Tamm I and Horsfall FL, Jr. (1950) Characterization and separation of an inhibitor of viral hemagglutination present in urine. *Proceedings of the Society for Experimental Biology and Medicine* 74 106-108

Vance DE and Ridgway ND (1988) The methylation of phosphatidylethanolamine *Progress in Lipid Research* 27 61-79

Verkman AS and Mitra AK (2000) Structure and function of aquaporin water channels *American Journal of Physiology. Renal Physiology* 278 F13-28

Yanagimachi R (1994) Fertility of mammalian spermatozoa: its development and relativity *Zygote* 2 371-372

Zhou Q, Clarke L, Nie R, Carnes K, Lai LW, Lien YH, Verkman A, Lubahn D, Fisher JS, Katzenellenbogen BS and Hess RA (2001) Estrogen action and male fertility: roles of the sodium/hydrogen exchanger-3 and fluid reabsorption in reproductive tract function *Proceedings of the National Academy of Sciences USA* 98 14132-14137

# Molecular mechanism of oocyte maturation

Samir Bhattacharya[1,2], Dipanjan Basu[1], Navneet AK[2] and Anamika Priyadarshini[1]

[1*]Department of Zoology, School of Life Science, Visva-Bharati, Santiniketan-731235, West Bengal, India;
[2]Indian Institute of Chemical Biology, Raja S. C. Mullick Road, Jadavpur, Kolkata-700032, India

Maturation of vertebrate oocytes is regulated by maturation inducing hormone (MIH), which is progesterone in all vertebrates except in fish, where it is 17α, 20ß dihydroxy progesterone. Once the full growth of the oocytes is achieved, they arrest at prophase of meiosis I. MIH releases oocytes from this arrest. MIH promotes the formation of a dimeric protein kinase complex known as maturation promoting factor (MPF), the regulatory component of which is cyclin B and the catalytic component is cell division cycle (Cdc2) kinase. This complex is activated by phosphorylation at Thr161 but remains inactive due to the inhibitory phosphorylation at Thr14 and Tyr15. MIH stimulates Cdc25, a dual specific phosphatase, that dephosphorylates both Thr14 and Tyr15 and converts pre- or inactive MPF to active MPF. Germinal vesicle break down (GVBD) is the marker of oocyte maturation. In an Indian freshwater perch, *Anabas testudineus*, MIH induced GVBD between 18-20 h. MIH induced oocytes extract in SDS-PAGE showed over-expression of a 30 kDa protein, which is confirmed to be cyclin B by using both monoclonal and polyclonal anti-cyclin B antibodies from various sources. The size of cyclin B in other vertebrates including mammals lies between 46-55 kDa. We have cloned cyclin B gene from perch oocyte and found it to contain the domains required for its function and immunological recognition. We also cloned Cdk1 gene, which is very similar to other vertebrates Cdk1. Perch oocyte Cdc25 is overexpressed prior to GVBD converting inactive MPF to active MPF that affect GVBD. The objective of this overview is to deal with the molecular regulation of MPF activation which causes final maturation of oocytes.

## Introduction

In all animals, immature oocytes become a mature oocyte and then a fertilizable egg. The molecular mechanism(s) of these transitions is of fundamental importance and is critical to the understanding of fertility and reproduction. Oocytes, enter into the process of cell division and get arrested at prophase of first meiosis or meiosis I during their growth period. Meiosis then resumes near or at the end of growth. The prophase I arrested oocyte is described as immature and the process of resumption of meiosis is called meiotic maturation. Meiotic maturation is

---

*Correspondence
E-mail: smrbhattacharya@yahoo.com

characterized by two consecutive M-phases, meiosis I (MI) and meiosis II (MII), without an intervening S-phase, producing haploid gametes. Maturation inducing hormone (MIH) relieves oocytes from this arrest and progesterone has been shown to be MIH in frogs and higher vertebrates, while in fish 17α, 20ß-dihydroxy progesterone is the MIH (Nagahama and Adachi, 1985). MIH regulates the formation of Maturation Promoting Factor (MPF) in germ cells (Masui and Clarke, 1979), whose activity can be attributed to a dimeric protein kinase, cyclin B-cell division cycle (Cdc2) kinase (Nurse, 1990). MPF triggers meiotic maturation events like germinal vesicle breakdown (GVBD) or nuclear envelope breakdown (NEBD) through dispersion of nuclear lamina, chromosome condensation and the formation of metaphase spindles (Lewin, 1990; Peter et al., 1990). Oocyte maturation involves the activation of various signal transduction pathways that converge to activate MPF. This is the most important step for the entry into M-phase of MI and MII. Although the function of MPF in promoting oocyte maturation is ubiquitous, there are species specific differences in the signaling pathways required for MPF activation (Schmitt and Nebreda, 2002). MPF is also shown to phosphorylate and activate elongation factor-1 (EF1) resulting in protein synthesis associated with oocyte maturation in lower vertebrates (Tokumoto et al., 2002).

MPF has been purified from many vertebrates and invertebrates, all of which share a striking commonality in their molecular structures (Yamashita et al., 2000). It is a complex of two proteins, a regulatory component called cyclin B and a catalytic protein kinase Cdc2 (Lohka et al., 1988; Labbe et al., 1989a). The key event to relieve the cell cycle arrest at prophase I in immature oocytes thereby allowing them to enter M- phase is the activation of MPF or cyclin B-Cdc2 kinase. Despite the common molecular structure of MPF in eukaryotes, the mechanisms involved in active MPF formation differ (Kishimoto, 1999; Ye et al., 2003).

MPF is activated by cyclin dependent kinase activating kinase (CAK) that phosphorylates Thr161 of Cdc2 but MPF remains inactive due to two inhibitory phosphorylations at Thr14 and Tyr15 by Myt1/Wee1 kinase; this form of MPF is called pre-MPF. Cdc25, a dual specific phosphatase, dephosphorylates Thr14 and Tyr15 and thus converts inactive pre-MPF to active MPF. Withdrawal of cell cycle arrest at prophase1 in oocytes depends on MPF and Cdc25 activation, the latter remains inactive until activated by Polo like kinase (Plk). Myt 1 inhibition can be effected by Mos which, in turn effects pre-MPF to MPF conversion (Peter et al., 2002a,b). The dynamics of MAPK (Mitogen activated protein kinase) pathway, which is activated downstream of Mos and Plk is very important for G2/M transition (Okano-Uchida et al., 2003). Both of them remain inactive in immature oocytes and are activated at the time of MPF activation during meiotic resumption. The objective of this overview is to describe the crucial events in meiotic resumption, permitting oocytes to cross the most important barrier, to arrive at maturation or G2/M transition.

## Activation of MPF: A heterodimeric cyclin B - Cdc2 kinase

The cyclin B-Cdc2 heterodimer, MPF, has emerged as a key mediator of cell cycle. Meiotic reinitiation by MPF has been studied in the oocytes of a wide range of organisms starting from invertebrates to vertebrates and including human beings (Labbe et al., 1989a,b; Verlhac et al., 1994; Tanaka and Yamashita, 1995; Heikinheimo et al., 1995; Chausson et al., 2004). Myt1 inhibits MPF activation in oocytes while, Cdc25 releases MPF from this inhibition hence the balance of these two regulators during meiotic G2/M phase transition is critical (Fig. 1). G2 phase in prophase I oocytes is maintained by the dominance of Myt1 over Cdc25 thus inhibiting MPF activation. In contrast, Wee1 proteins are not detectable or if present are at very low levels in prophase I oocytes (Okano-Uchida et al., 2003; Kishimoto, 2003). It is therefore important to understand how the balance between Myt1 and Cdc25 is reversed during G2/M transition where MIH is triggering such signaling. By blocking Cdc25 activity with its antibody, the signaling pathway of cyclin B–Cdc2 activation involving Myt1 suppression has been clarified in starfish oocytes. MIH binding to the receptor activates downstream regulators like PI3 kinase and Akt/PKB which are transducers of multiple cellular signals. Akt then phosphorylates and downregulates Myt1 causing the activation of cyclin B-Cdc2 at the meiotic G2/M phase transition (Okumura *et al.*, 2002). In *Xenopus,* activation of the MAPK-p90rsk cascade triggers Myt1 phosphorylation for MPF activation (Palmer *et al.*, 1998).

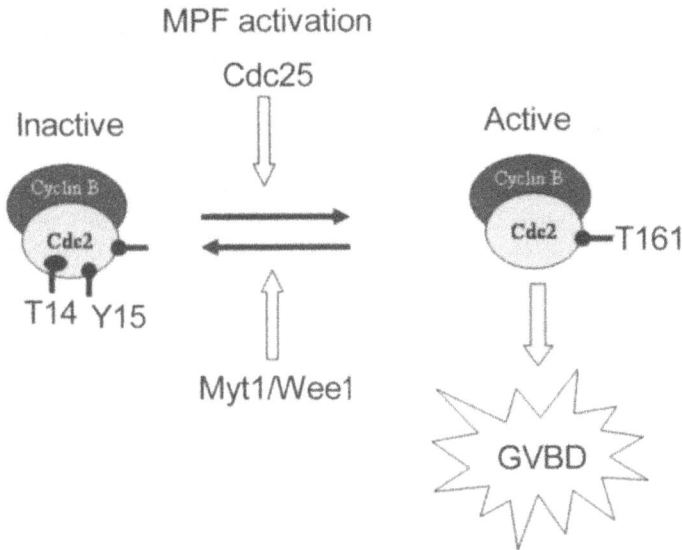

**Fig 1.** Activation and inactivation of Cdc2-cyclin B or maturation promoting factor (MPF). This conversion occurs in the oocytes due to the balance between Myt1 and Cdc25. Thr14 and Tyr15 are phosphorylated by Myt1 rendering MPF inactive while they are dephosphorylated by Cdc25 to activate MPF leading to germinal vesicle breakdown (GVBD).

One of the best studied animal models for MPF activation is *Xenopus*. Purified *Xenopus* MPF is a complex containing p34cdc2, the *Xenopus* homologue of the yeast cdc2/CDC28 cell cycle control gene, and a B-type cyclin (reviewed by Nurse, 1990). The accumulation of cyclin(s), under the induction of MIH permits alterations in the phosphorylation state of p34cdc2 and turns on its protein kinase activity (Gould and Nurse, 1989; Solomon *et al.*, 1990) in *Xenopus* and also in starfish (reviewed by Meijer and Guerrier, 1984). MPF activity appears at the time of GVBD, disappears between the two meiotic divisions, and reappears at second meiotic metaphase. Prior to entry into the M-phase, Cdc2 remains associated with cyclin B and is kept inactive (pre-MPF) through phosphorylation on Tyr15 residue (Gould and Nurse, 1989; Meijer *et al.*, 1991; Norbury *et al.*, 1991; Amon *et al.*, 1992) and in higher vertebrates also on Thr14 (Norbury *et al.*, 1991; Borgne and Meijer, 1996). Activation thus requires the dephosphorylation of Thr14 and Tyr15 by Cdc25. In addition, the complex also requires an activatory phosphorylation of Thr161 (Solomon, 1993). However, amphibian immature oocytes, other than *Xenopus*, contain monomeric Cdc2 but not cyclin B. MPF is formed after hormonal stimulation by binding of the newly produced cyclin B to the pre-existing Cdc2 and is immediately activated through Thr161 phosphorylation (Yoshida *et al.*, 2000). The situation in *Xenopus* is very similar to starfish where the oocyte contains latent amounts of cyclin B-Cdc2 even before meiotic reinitiation occurs in response to 1-methyl adenine which is the maturation inducing hormone in starfish (Kishimoto, 2003).

Dephosphorylation of Cdc2 on Tyr15, an essential step in producing active MPF, appears to be dependent on the concentration of 1-methyl adenine and is also mediated through 26S proteasome (Morinaga et al., 2000). In contrast, pre-MPF is absent in immature bovine and porcine oocytes (Levesque and Sirard, 1996; Kanayama et al., 2002). Studies on mouse, rat and rabbit oocytes, however, demonstrate the existence of pre-formed cyclin B-Cdc2 complex and its potential to effect meiosis reinitiation without new protein synthesis unlike porcine or bovine oocytes where active protein synthesis is absolutely required (Motlik and Kubelka, 1990). Cdc2 kinase and cyclin-B are observed to be synthesized during the first hours of in vitro maturation in bovine oocytes (Wu et al., 1997). MPF activation in G2/M transition also exists in fish oocytes resembling the amphibian model except for Xenopus. Full-grown immature oocytes of goldfish contained a 35 kDa Cdc2. In addition to this protein, a 34 kDa Cdc2 kinase was detected in mature oocytes. It was found that the 34 and 35 kDa Cdc2 proteins are active and inactive forms, respectively of the same protein. The 34 kDa active Cdc2 kinase appeared with the onset of GVBD. However, we have found an unusual Cyclin B in an Indian perch oocytes, which is a 30 kDa protein. When this p30 Cyclin B complexes with p34Cdc2, a significant increase in histone H1 kinase activity of oocytes occurred. Microinjection of this complex to perch oocyte caused GVBD (Fig. 2).

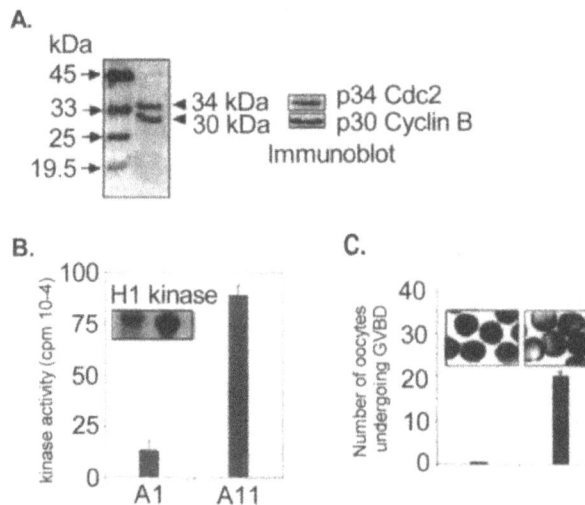

Fig 2. A) Purification of MPF from perch oocytes. MPF was purified by immunoaffinity column chromatography. Cdc2 and cyclin B were separated through 15% SDS-PAGE gel. Unbound eluate from immunoaffinity column is mentioned as A1 whereas the bound MPF as A11. Gel was immunoblotted with mouse monoclonal anti-cyclin B1 and anti-Cdc2 antibodies. Cdc2 is a 34 kDa protein while Cyclin B is 30 kDa. B) A1 and A11 peaks were pooled separately and dialyzed, and volume was reduced by lyophilization; 1.5 μg protein was added to the incubation mixture to examine histone H1 kinase activity, as this is another marker for final maturation. C) Injection of A11 protein (MPF) into the perch oocyte–effected GVBD (Basu et al., 2004).

Cyclin B was absent in immature oocyte extracts and appeared when oocytes underwent GVBD, coinciding with the appearance of the 34 kDa active Cdc2 kinase. Introduction of E. coli produced cyclin B into immature oocyte extracts, which contained the 35 kDa inactive cdc2 kinase but no cyclin B, induced the activation of cdc2 kinase concurrent with the change in apparent molecular mass from 35 to 34 kDa, as found in fish oocytes matured with MIH (Nagahama, 1993). Taken together, it is concluded that MIH induces oocytes to synthesize cyclin B, which in turn activates pre-existing 35 kDa cdc2 kinase. Yet, it cannot be accepted as a generalized model for all teleosts as studies on *Anabas testudineus*, a freshwater perch, have demonstrated the occurrence of pre-MPF in uninduced oocytes and that its conversion to active MPF requires dephosphorylation by Cdc25: a situation very similar to that of *Xenopus* (Basu et al., 2004).

## Oocyte maturation: The release of inhibition

Cdc2, the universal cell cycle regulatory kinase, is held inactive by inhibitory phosphorylation at Thr14 and Tyr15 residues on its association with cyclin B (King et al., 1994). Phosphorylation at Thr161 by the Cdc2 activating kinase (CAK) is necessary for the activation of Cdc2-cyclin B kinase or MPF, whereas, phosphorylations at Thr14 and Try15 by Myt1/Wee1 kinases maintain this complex in an inactive state. In immature *Xenopus* oocytes Thr14 and Tyr15 Myt1 kinase is present but Wee1 kinase is absent. Wee1 kinase could be detected during meiosis II and early embryonic cell phase (Palmer et al., 1998; Murakami and Vande Woude, 1998). The absence of Wee1 ensures S-phase omission at meiosis I (Nakajo et al., 2000). We have recently observed in perch oocytes the presence of Myt1 during meiosis I while the expression of *wee1* gene could be identified during the meiosis I exit and meiosis II and not during MI maturation (Bhattacharya et al., unpublished).

A dual-specific membrane-associated protein kinase, able to phosphorylate Cdc2 on both Thr14 and Tyr15 was initially identified in *Xenopus* and HeLa cell extract (Kornbluth et al., 1994; Atherton-Fessler et al., 1994) and the Thr14 and Tyr15 Myt1 kinase was subsequently cloned (Mueller et al., 1995). In contrast to Wee1, which is localized in the nucleus (McGowan and Russell, 1995), Myt1 is localized in the endoplasmic reticulum and Golgi complex by a membrane-targeting domain on the C-terminal side of the catalytic domain (Li et al., 1997). Specific subcellular localization of protein kinases is of significant importance; indeed the compartmentalization modulated by a nuclear export sequence of cyclin B1 regulates the physiological activity of Cdc2/cyclin B1 (Li et al., 1997; Hagting et al., 1998). Myt1 exhibits more restricted substrate specificity than Wee1, in that it phosphorylates Cdc2 (Cdk1)/cyclin complexes but not Cdk2/cyclin complexes (Booher et al., 1997). This observation strongly suggests that Myt1 specifically regulates G2/M phase transition through the inhibitory phosphorylation of Cdc2. Whereas depletion of Myt1 in *Xenopus* causes nuclear envelope breakdown *in vitro* (Nakajo et al., 2000). Recent studies with a Myt1 ortholog (Wee1.3) isolated from *C. elegans* showed that in co-depletion experiments Wee1.3 is dispensable in the absence of Cdc2 (Burrows et al., 2006). This suggests Cdc2 is a specific target for Myt1 during G2/M in this species. This also implies that factors identified upstream that inhibit Myt1 function may activate Cdc2 by releasing the inhibition imposed by Myt1. In *Xenopus*, activation of Mos, an oocyte specific MAPK kinase, leads to the activation of the protein kinase p90rsk, which in turn can phosphorylate and inhibit the Cdc2 inhibitory kinase Myt1 (Schmitt and Nebreda, 2002). In starfish oocytes, however, Myt1 is inhibited by the protein kinase PKB/Akt. In mouse oocytes, Mos has been proposed to inhibit a MAP phosphatase in addition to activating the MAPK kinase MEK1 (Verlhac et al., 2000). Regulation of Myt1 and Wee1 activities during maturation of oocytes could be regulated by a Mos/MAPK pathway to induce Cdc2 activation (Sagata, 1997). In perch oocytes, the existence of Mos appears to play a role during MPF activated H1 kinase in meiosis

I (Bhattacharya et al., unpublished). There are species-specific variations regarding the role of MAP kinase in oocyte maturation. In mammalian oocytes, MAPK is activated during maturation but oocytes from c-mos knockout mice in which MAPK is not activated, can undergo GVBD, a situation similar to Xenopus although it shows a delayed kinetics in the latter case. Until recently, in Xenopus oocytes MAPK activation was thought to be essential for GVBD as injection of constitutively active MAPK (Haccard et al., 1995) or MEK (Huang et al., 1995) into immature oocytes induced GVBD. However, reports from different laboratories have shown that MAP kinase pathway was not essential for entry into meiosis I (Fisher et al., 1999; Gross et al., 2000). A recent study from Marcel Doree group demonstrated that there exists another pathway through which Myt1 could be inhibited by direct interaction with Mos (Peter et al., 2002a,b). In maturing starfish oocytes, MAPK is activated only after MPF activation and GVBD. In contrast, the Mos-MAPK pathway is not essential for initiating goldfish oocyte maturation although its function as a cyto-static factor during the metaphase II arrest is probable (Kajiura-Kobayashi et al., 2000). Although, there is yet to be a consensus regarding the essentiality of MAP kinase pathway in MPF activation, the involvement of Mos in inhibiting Myt1 activity appears to be more general in vertebrate oocytes. G2/M arrest and consequent cell cycle resumption in both mitosis and meiosis appears to be the result of a fine balance between two mutually opposing factors; one being Wee1/Myt1 kinase activity that keeps Cdc2 phosphorylated at Thr14 and Try15 and the other is the phosphatase activity of Cdc25 that cleaves the inhibitory phosphates. In cells, Cdc2 activation is initiated either by changing the balance in favor of Cdc25 over Wee1/Myt1 or by increasing total Cdc2-cyclin B (Slepchenko and Terasaki, 2003).

## Meiotic maturation: Activation signaling

MPF activation leads to the maturation of oocytes but what the signals are that abruptly activate MPF at the onset of meiotic maturation is still an incomplete story although it is known that the Cdc25 phosphatase family dephosphorylate inhibitory phosphates to achieve MPF activation. In Xenopus oocytes synthesis of Mos begins before MPF activation and GVBD, whereas the situation is different in starfish and mouse oocytes as synthesis of Mos starts after GVBD (Nebreda and Ferby, 2000; Tachibana et al., 2000). How Mos is related to MPF signaling is not yet clear but the role of the Cdc25 family in regulating MPF activity has been worked out in far more detail. Cdc25 are a family of dual specificity protein phosphatases which can dephosphorylate both Thr14 and Tyr15 of cdc2. Genetic studies have detected three different isoforms of Cdc25 in humans: Cdc25A, B and C (Galaktionou and Beach, 1991; Nagata et al., 1991). Of these, Cdc25A is present in G1 phase somatic cells (Perdiguero and Nebreda 2004) and both Cdc25B and Cdc25C, which are 45% identical at the amino acid level (85% identical in the catalytic domains), are present in G2 phase cells and participate in G2/M transition (Lammer et al., 1998). Dephosphorylation of Thr14 and Tyr15 by Cdc25C in late G2 activates the Cdc2-CyclinB complex and triggers the initiation of mitosis. Cdc2-CyclinB complexes in turn are thought to phosphorylate Cdc25C, which further activates Cdc25C, inducing the full activation of Cdc2-CyclinB by forming an autocatalytic feedback loop.

In G2 arrested Xenopus oocytes, Cdc25C is phosphorylated on Ser287 and associated with 14-3-3 proteins. Entry of the oocytes into M-phase of meiosis is triggered by progesterone, which activates a signaling pathway leading to the dephosphorylation of Ser287, probably mediated by the PP1 phosphatase. The activation of Cdc25C during oocyte maturation correlates also with its phosphorylation on multiple sites. These phosphorylations involve several signaling pathways, including Polo kinases and MAP kinases, and might require also the inhibition of the PP2A phosphatase. Finally, Cdc25C is further phosphorylated by its substrate Cdc2-CyclinB, as part of an auto-amplification loop that ensures the high Cdc2-CyclinB activity level required to drive the oocyte through the meiotic cell cycle. In *Xenopus* oocytes, a positive feed-back loop between Cdc2 kinase and its activating phosphatase Cdc25 allows the activation of MPF and entry into the first meiotic division (Karaiskou *et al.*, 1999). New insights into how Cdc25 function is inhibited in prophase I arrested oocytes have recently been emerging. One such regulator is cAMP dependent protein kinase A. In *Xenopus* oocytes, Ruderman and colleagues recently demonstrated that PKA is a relevant candidate for Cdc25 inactivation (Duckworth *et al.*, 2002). In *Xenopus* oocytes PKA negatively regulates Cdc25 by phosphorylating Ser287 (the equivalent of Ser216 in human Cdc25C), which allow binding with 14-3-3 protein. It has been suggested that 14-3-3 binding not only mediates the nuclear exclusion and sequestration of Cdc25 in the cytoplasm but also could directly inhibit Cdc25 function (Oe *et al.*, 2001). Likewise it was shown that in oocyte extracts Plkk1 (Polo like kinase kinase1) or Plx1 (the *Xenopus* homologue of mammalian Plkk1) can activate Cdc25.

However, activation of Plkk1 or Plx1 is shown to be downstream of Cdc2 activity and that Cdc2 induced phosphorylation of Cdc25 is a prerequisite for Cdc25 activation by Plx1 (Karaiskou *et al.*, 1999). The binding of Plk1 to Cdc25 seems to require the priming phosphorylation of Cdc25 by Cdc2 (Elia *et al.*, 2003). Cdc25 activity has also been examined in starfish oocytes using an anti-body to block the signaling pathway that leads to cyclin B-Cdc2 activation. But in starfish, instead of Plk1, Akt has been indicated as the trigger kinase that switches the balance between Myt1 and Cdc25 owing to its capacity to directly phosphorylate and activate Cdc25 (Kishimoto, 2003). An important study, based on knockout experiment, shows that Cdc25B is necessary and sufficient for meiotic resumption in mouse oocytes, even though Cdc25A and Cdc25C are present (Lincoln *et al.*, 2002). Thus Cdc25A and Cdc25C cannot compensate for the function of Cdc25B in meiotic resumption although the absence of Cdc25C can be compensated by Cdc25A and/or Cdc25B (Chen *et al.*, 2001). As fish oocytes does not have a preformed store of cyclin B-Cdc2, Cdc25 activity has not been studied in fish in general. However, we have recently shown that in Indian perch, *Anabas testudineus*, there is an existence of pre-MPF and involvement of Cdc25 (Fig. 3). MIH induces Cdc25 activity prior to the exit of meiosis I to convert pre-MPF to MPF effecting GVBD (Basu *et al.*, 2004). *Anabas* Cdc25C has recently been cloned by us and found to be very close to human putative conserved domains (unpublished data). Our observations in *Anabas* may provide interesting insights to understand the evolution of cell cycle regulation in vertebrates.

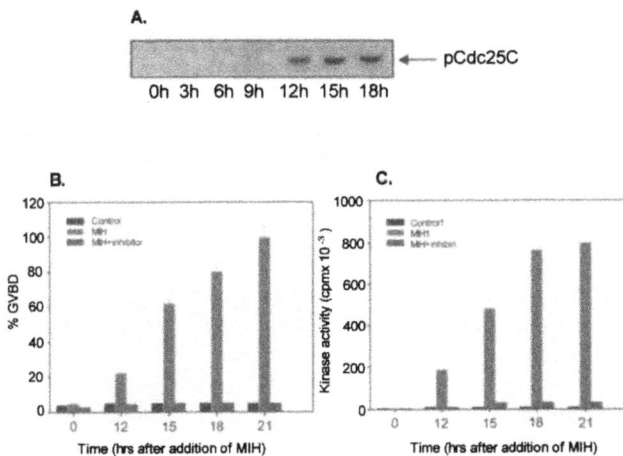

**Fig 3**. Final maturation of perch oocyte closely follows Cdc25 phosphorylation pattern. **A)** Extract from MIH-induced perch oocyte at different time levels was immunoblotted by using rabbit anti-pCdc25C antibody, showing phosphorylation of Cdc25 by Plk1 **B)** MIH-induced GVBD was blocked by tyrosine phosphatase inhibitor, $Na_3VO_4$ which blocks GVBD. **C)** Histone H1 kinase activity of oocyte extract induced by MIH is also ablated by Cdc25C inactivation.

## Conclusive remarks

The meiotic maturation of oocytes is a very unique cycle with reductional meiosis I and equational meiosis II occurring after a single round of S-phase. Omission of S-phase in meiosis I is essential for the generation of haploid germ cells. The most significant feature of oocyte maturation is the extensive networking of feedback signaling starting from the activation of MPF to the downstream of MPF activity. Fig. 4 summarizes the description of MPF activation. However, what happens after MPF activation has not been covered in this review as the mechanism involved there is complex and still not very clear. Whether MPF activation is in the cytoplasm and independent of its nuclear import or it requires nuclear import, is still a shaded area in our understanding. To complete meiotic maturation, a small amount of MPF activity is sufficient to activate the signaling pathways required for oocyte maturation.

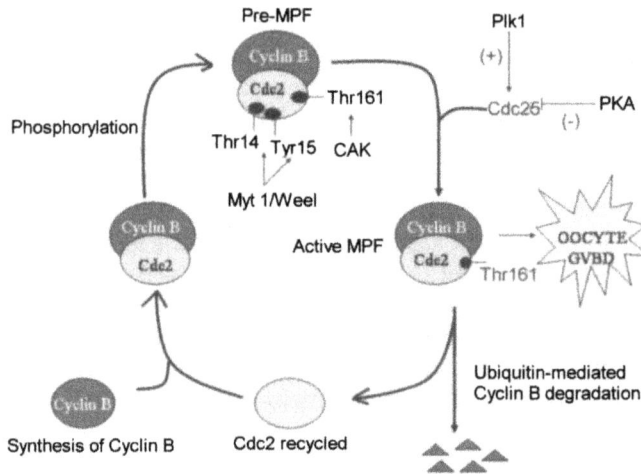

**Fig 4.** Schematic representation of the regulation of oocyte maturation. Formation of MPF occurs due to synthesis of cyclinB as Cdc2 remains in the cell, thereafter both inhibitory phosphorylation at Thr14 and Tyr15 by Myt1/Wee1 and active site phosphorylation at Thr161 occurs, forming pre-MPF or inactive MPF. Removal of inhibitory phosphorylation by Cdc25 dual specific phosphatase converts pre-MPF to active MPF. PKA phosphorylation of Cdc25 made it inactive during pre-MPF whereas Plk1 phosphorylation of Cdc25 activated it. Activated MPF effects oocyte germinal vesicle breakdown (GVBD).

## References

Amon A, Surana U, Muroff I and Nasmyth K (1992) Regulation of p34cdc28 tyrosine phosphorylation is not required for entry into mitosis in *S. cerevisiae Nature* **355** 368-371

Atherton-Fessler S, Liu F, Gabrielli B, Lee MS, Peng CY and Piwnica-Worms H (1994) Cell cycle regulation of the p34cdc2 inhibitory kinases *Molecular Biology of the Cell* **5** 989-1001

Basu D, Navneet AK, Dasgupta S and Bhattacharya S (2004) Cdc2 – cyclin B induced G2 to M transition in perch oocyte is dependent on Cdc25 *Biology of Reproduction* **71** 894-900

Booher RN, Holman PS and Fattaey A (1997) Human Myt1 is a cell cycle-regulated kinase that inhibits Cdc2 but not Cdk2 activity *Journal of Biological Chemistry* **272** 22300-22306

Borgne A and Meijer L (1996) Sequential dephosphorylation of p34cdc2 on Thr-14 and Tyr-15 at the prophase/metaphase transition *Journal of Biological Chemistry* **271** 847– 854

Burrows AE, Sceurman BK, Kosinski ME, Richie CT, Sadler PL, Schumacher JM and Golden A (2006) The *C. elegans* Myt1 ortholog is required for the proper timing of oocyte maturation *Development* **133** 697-709

Chausson F, Paterson LA, Betteley KA, Hannah L, Meijer L and Bentley MG (2004) CDK1/cyclin B regulation during oocyte maturation in two closely related lugworm species, *Arenicola marina* and *Arenicola defodiens* Development, Growth and Differentiation **46** 71–82

Chen MS, Hurov J, White LS, Woodford-Thomas T and Piwnica-Worms H (2001) Absence of apparent phenotype in mice lacking Cdc25C protein phosphatase *Molecular Cellular Biology* **21** 3853–3861

Duckworth BC, Weaver JS and Ruderman JV (2002) G2 arrest in *Xenopus* oocytes depends on phosphorylation of cdc25 by protein kinase A *Proceedings of the National Academy of Sciences USA* **99** 16794–16799

Elia AE, Cantley LC and Yaffe MB (2003) Proteomic screen finds pSer/pThr-binding domain localizing Plk1 to mitotic substrates *Science* **299** 1228-1231

Fisher DL, Brassac T, Galas S and Doree M (1999) Dissociation of MAP kinase activation and MPF activation in hormone-stimulated maturation of *Xenopus* oocytes *Development* **126** 4537–4546

Galaktionov K and Beach D (1991) Specific activation of cdc25 tyrosine phosphatases by B-type cyclins: evidence for multiple roles of mitotic cyclins *Cell* **67** 1181-1194

Gould K and Nurse P (1989) Tyrosine phosphorylation of the fission yeast cdc2 protein kinase regulates entry into mitosis *Nature* **342** 39-45

Gross SD, Schwab MS, Taieb FE, Lewellyn AL, Qian YW and Maller JL (2000) The critical role of the MAP kinase pathway in meiosis II in *Xenopus* oocytes is mediated by p90^Rsk *Current Biology* **10** 430–438

Haccard O, Lewellyn A, Hartley RS, Erikson E and Maller JL (1995) Induction of *Xenopus* oocyte meiotic maturation by MAP kinase *Developmental Biology* **168** 677–682

Hagting A, Karlsson C, Clute P, Jackman M and Pines J (1998) MPF localization is controlled by nuclear export *EMBO Journal* **17** 4127–4138

Heikinheimo O, Lanzendorf SE, Baka SG and Gibbons WE (1995) Cell cycle genes c-mos and cyclin-B 1 are expressed in a specific pattern in human oocytes and preimplantation embryos *Molecular Human Reproduction* **10** 699-707

Huang W, Kessler D and Erikson R (1995) Biochemical and biological analysis of Mek1 phosphorylation site mutants *Molecular Biology of the Cell* **6** 237–245

Kajiura-Kobayashi H, Yoshida N, Sagata N, Yamashita M and Nagahama Y (2000) The Mos/MAPK pathway is involved in metaphase II arrest as a cytostatic factor but is neither necessary nor sufficient for initiating oocyte maturation in goldfish *Development, Genes and Evolution* **210** 416–425

Kanayama N, Miyano T and Lee J (2002) Acquisition of meiotic competence in growing pig oocytes correlates with their ability to activate Cdc2 kinase and MAP kinase *Zygote* **10** 261-270

Karaiskou A, Jessus C, Brassac T and Ozon R (1999) Phosphatase 2A and polo kinase, two antagonistic regulators of cdc25 activation and MPF auto –amplification *Journal of Cell Science* **112** 3747–3756

King RW, Jackson PK and Kirschner MW (1994) Mitosis in transition *Cell* **79** 563-571

Kishimoto T (1999) Activation of MPF at meiosis reinitiation in starfish oocytes *Developmental Biology* **214** 1-8

Kishimoto T (2003) Cell-cycle control during meiotic maturation *Current Opinion in Cell Biology* **15** 654–663

Kornbluth S, Sebastian B, Hunter T and Newport J (1994) Membrane localization of the kinase which phosphorylates p34cdc2 on threonine 14 *Molecular Biology of the Cell* **5** 273–282

Labbe JC, Capony JP, Caput D, Cavadore JC, Derancourt J, Kaghad M, Lelias JM, Picard A and Doree M (1989a) MPF from starfish oocytes at first meiotic metaphase is a heterodimer containing one molecule of cdc2 and one molecule of cyclin B *EMBO Journal* **8** 3053-3058

Labbe JC, Picard A, Peaucellier G, Cavadore JC, Nurse P and Doree M (1989b) Purification of MPF from starfish: identification as the H1 histone kinase p34cdc2 and a possible mechanism for its periodic activation *Cell* **57** 253-263

Lammer C, Wagerer S, Saffrich R, Mertens D, Ansorge W and Hoffmann I (1998) The cdc25B phosphatase is essential for the G2/M phase transition in human cells *Journal of Cell Science* **111** 2445–2453

Levesque JT and Sirard MA (1996) Resumption of meiosis is initiated by the accumulation of cyclin B in bovine oocytes *Biology of Reproduction* **55** 1427-1436

Lewin B (1990) Driving the cell cycle: M phase kinase, its partners, and substrates *Cell* **61** 743-752

Li J, Meyer AN and Donoghue DJ (1997) Nuclear localization of cyclin B1 mediates its biological activity and is regulated by phosphorylation *Proceedings of the National Academy of Sciences USA* **94** 502–507

Lincoln AJ, Wickramasinghe D, Stein P, Schultz RM, Palko ME, De Miguel MP, Tessarollo L and Donovan PJ (2002) Cdc25b phosphatase is required for resumption of meiosis during oocyte maturation *Nature Genetics* **30** 446-449

Lohka MJ, Hayes MK and Maller JL (1988) Purification of maturation-promoting factor, an intracellular regulator of early mitotic events *Proceedings of the National Academy of Sciences USA* **85** 3009-3013

Masui Y and Clarke HJ (1979) Oocyte maturation *International Review of Cytology* **57** 185-282

McGowan CH and Russell P (1995) Cell cycle regulation of human WEE1 *EMBO Journal* **14** 2166-2175

Meijer L and Guerrier P (1984) Maturation and fertilization in starfish oocytes *International Review of Cytology* **86** 129–195

Meijer L, Azzi L and Wang JYJ (1991) Cyclin B targets p34cdc2 for tyrosine phosphorylation *EMBO Journal* **10** 1545-1554

Morinaga C, Izumi K, Sawada H and Sawada M (2000) Activation of maturation-promoting factor and 26S proteasome assembly accelerated by a high concentration of 1-methyladenine in starfish oocytes *Bioscience, Biotechnology and Biochemistry* **64** 268-274

Motlík J and Kubelka M (1990) Cell-cycle aspects of growth and maturation of mammalian oocytes *Molecular Reproduction and Development* **27** 366 – 375

Mueller PR, Coleman TR, Kumagai A and Dunphy WG (1995) Myt1: A membrane-associated inhibitory kinase that phosphorylates Cdc2 on both Threonine-14 and Tyrosine-15 *Science* **270** 86-90

Murakami MS and Vande Woude GF (1998) Analysis of the early embryonic cell cycles of *Xenopus*; regulation of cell cycle length by Xe-wee1 and Mos *Development* **125** 237–248

Nagahama Y (1993) Molecular biology of oocyte maturation in fish In *Perspectives in Comparative Endocrinology* pp 193-198 Eds KG Davey, RE Peter and SS Tobe. Toronto, Canada

Nagahama Y and Adachi S (1985) Identification of maturation-inducing steroid in teleosts, the amgo salmon (*Onycorhynchus rhodurus*) *Developmental Biology* **119** 428-435

Nagata A, Igarashi M, Jinno S, Suto K and Okayama H (1991) An additional homolog of the fission yeast cdc25 gene occurs in humans and is highly expressed in some cancer cells *The New Biologist* **3** 959–967

Nakajo N, Yoshitome S, Iwashita J, Iida M, Uto K, Ueno S, Okamoto K and Sagata N (2000) Absence of Wee1 ensures the meiotic cell cycle in *Xenopus* oocytes *Genes & Development* **14** 328–338

Nebreda AR and Ferby I (2000) Regulation of the meiotic cell cycle in oocytes *Current Opinion in Cell Biology* **12** 666-675

Norbury C, Blow J and Nurse P (1991) Regulatory phosphorylation of the p34cdc2 protein kinase in vertebrates *EMBO Journal* **10** 3321-3329

Nurse P (1990) Universal control mechanism regulating onset of M-phase *Nature* **344** 503-538

Oe T, Nakajo M, Katsuragi Y, Okazaki K and Sagata N (2001) Cytoplasmic occurrence of the Chk1/Cdc25 pathway and regulation of Chk1 in *Xenopus* oocytes *Developmental Biology* **229** 250-261

Okano-Uchida T, Okumura E, Iwashita M, Yoshida H, Tachibana K and Kishimoto T (2003) Distinct regulators for Plk1 activation in starfish meiotic and embryonic cycles *EMBO Journal* **22** 5633-5642

Okumura E, Fukuhara T, Yoshida H, Hanada Si S, Kozutsumi R, Mori M, Tachibana K and Kishimoto T (2002) Akt inhibits Myt1 in the signalling pathway that leads to meiotic G2/M-phase transition *Nature Cell Biology* **4** 111-116

Palmer A, Gavin AC and Nebreda AR (1998) A link between MAP kinase and p34(cdc2)/cyclin B during oocyte maturation: p90(rsk) phosphorylates and inactivates the p34(cdc2) inhibitory kinase Myt1 *EMBO Journal* **17** 5037-5047

Perdiguero E and Nebreda AR (2004) Regulation of Cdc25C activity during G2/M transition *Cell Cycle* **3** 733-737

Peter M, Labbe JC, Doree M and Mandart E (2002a) A new role for Mos in *Xenopus* oocyte maturation: targeting Myt1 independently of MAPK *Development* **129** 2129–2139

Peter M, Le Peuch C, Labbe JC, Meyer AN, Donoghue DJ and Doree M (2002b) Initial activation of cyclin-B1-cdc2 kinase requires phosphorylation of cyclin B1 *EMBO Reports* **3** 551-556

Peter M, Nakagawa J, Doree M, Labbe JC and Nigg EA (1990) *In vitro* disassembly of the nuclear lamina and M phase-specific phosphorylation of lamins by cdc2 kinase *Cell* **61** 591-602

Sagata N (1997) What does Mos do in oocytes and somatic cells? *BioEssays* **19** 13-21

Schmitt A and Nebreda AR (2002) Signaling pathways in oocyte meiotic maturation *Journal of Cell Science* **115** 2457–2459

Slepchenko BM and Terasaki M (2003) Cyclin aggregation and robustness of bio switching *Molecular Biology of the Cell* **14** 4695–4706

Solomon MJ (1993) Activation of the various cyclin/cdc2 protein kinases *Current Opinion in Cell Biology* **5** 180–186

Solomon MJ, Glotzer M, Lee TH, Philippe M and Kirschner MW (1990) Cyclin activation of p34cdc2 *Cell* **63** 1013–1024

Tachibana K, Tanaka D, Isobe T and Kishimoto T (2000) c-Mos forces the mitotic cell cycle to undergo meiosis II to produce haploid gametes *Proceedings of the National Academy of Sciences USA* **97** 14301-14306

Tanaka T and Yamashita M (1995) Pre-MPF is absent in immature oocytes of fishes and amphibians except *Xenopus Development, Growth and Differentiation* **37** 387–393

Tokumoto M, Nagahama Y and Tokumoto T (2002) A major substrate for MPF: cDNA cloning and expression of polypeptide chain elongation factor 1 gamma from goldfish (*Carassius auratus*) *DNA Sequence* **13** 27-31

Verlhac MH, Kubiak JZ, Clarke HJ and Maro B (1994) Microtubule and chromatin behavior follow MAP kinase activity but not MPF activity during meiosis in mouse oocytes *Development* **120** 1017-1025

Verlhac MH, Lefebvre C, Guillaud P, Rassinier P and Maro B (2000) Asymmetric division in mouse oocytes: with or without Mos *Current Biology* **10** 1303–1306

Wu B, Ignotz G, Currie W B and Yang X (1997) Dynamics of maturation-promoting factor and its constituent proteins during *in vitro* maturation of bovine oocytes *Biology of Reproduction* **56** 253-259

Yamashita M, Mita K, Yoshiada N and Kondo T (2000) Molecular mechanisms of the initiation of oocyte maturation: general and species-specific aspects *Progress in Cell Cycle Research* **4** 115-129

Ye J, Flint AP, Luck MR and Campbell KH (2003) Independent activation of MAP kinase and MPF during the initiation of meiotic maturation in pig oocytes *Reproduction* **125** 645-656

**Yoshida N, Mita K and Yamashita M** (2000) The function of the Mos/MAPK pathway during oocyte maturation in the Japanese brown frog *Rana japonica* *Molecular Reproduction and Development* **57** 88-98

# Differential gene expression in transition of primordial to preantral follicles in mouse ovary

Tarala Nandedkar*, Shalmali Dharma, Deepak Modi and Serena D'Souza

*National Institute for Research in Reproductive Health (ICMR), Parel, Mumbai 400 012, India*

In the mammalian ovary, early follicular development is gonadotropin independent. Interaction between the oocyte and granulosa cells possibly plays an important role in transition of primordial to preantral stage. However, the molecular and cellular control of early follicular development and cell-cell interaction is complex and poorly understood. In the present study, we examined gene expression in primordial, primary and preantral follicle by cDNA arrays using Day 2, Day 4 and Day 6 neonatal mouse ovaries that contain the various developmental stages of these follicles, respectively. The results revealed that 30% of the genes were differentially expressed in Day 4 ovaries containing primary follicles as compared to D2 neonatal ovaries. The data were confirmed by the expression of Growth Differentiation Factor-9 in the oocytes of primary and preantral follicles. Also, Stem Cell Factor was localized in the granulosa cells of primary and preantral follicles. Electron microscopic studies of Day 6 ovaries showed projections from granulosa cells and microvilli from oocytes in the follicle during the transition from the primary to preantral stage. Further, initiation of gap junctions were observed at ultrastructure level and corroborated with the expression of specific gap junction protein, connexin 43 in preantral follicles of the ovaries. These results infer that primordial follicles are quiescent while the major activities of cell-cell communication and the production of local paracrine factors, are initiated in primary and preantral follicles of the mouse ovary. These preliminary observations may contribute to the elucidation of molecular and cellular pathways involved in follicle transition.

## Introduction

The basic functional unit in the ovary is the ovarian follicle. About 1-2 million primordial follicles are present in the mammalian ovary at birth. A follicle is composed of somatic cells and an oocyte. The growth of ovarian follicles requires continuous communication between different follicular cell types. The granulosa cells are essential for regulation of growth and maturation of oocytes (Buccione et al., 1990). On the other hand, oocytes help in the proliferation of granulosa cells (Vanderhyden et al., 1990). Thus, the communication of oocytes and granulosa cells is a bidirectional process (Eppig et al., 1997). The two primary cell types in the ovarian follicles are granulosa cells and theca cells, which synthesize hormones as well as autocrine and paracrine factors (Canipari,

---

*Corresponding author
E-mail: cellbioirr@rediffmail.com

2000; Skinner, 2005) that, in turn regulate follicular development (Nandedkar and Raghavan, 1989).

All the primordial follicles are quiescent, only few of these develop into primary follicles. The initiation of the development of a primordial follicle and its transition from the primary to the secondary (preantral) stage is gonadotropin independent (Oktay et al., 1997). Thus, possibly local autocrine & paracrine factors play an important role in the recruitment and transition of primordial follicles to the preantral stage (Gougeon and Busso, 2000).

The squamous granulosa cells in the primordial follicle differentiate into cuboidal form in the primary follicle (Peters et al., 1975; Hirshfield, 1991). These cells then divide to form two layers in secondary (preantral) follicles. A cross talk between oocyte and granulosa cells (Eppig, 2001) through the local paracrine factors regulates follicular development (McNatty et al., 2000). Factors such as Growth Differentiation Factor-9 (GDF9), secreted by the oocyte (Laitinen et al., 1998), and Stem Cell Factor (SCF/ Kit ligand), produced by granulosa cells (Wang and Roy, 2004), have paracrine roles in early folliculogenesis. Granulosa cells in developing follicles produce SCF, which can act on theca and stromal cells and oocytes (Manova et al., 1993; Motro and Bernstein, 1993). On the other hand, c-Kit is highly expressed in oocytes supporting the role of SCF in granulosa cell-oocyte interactions (Motro and Bernstein, 1993).

Immunohistochemical detection of Proliferating Cell Nuclear Antigen (PCNA) (Oktay et al., 1995) or bromodeoxyuridine incorporation in granulosa cells (Gaytan et al., 1996) led to the conclusion that most of the follicles are growing follicles. In the developing ovarian follicle, intercellular communication between the oocyte and the surrounding somatic cells is essential for proper functioning and development of the follicle. This interaction is provided by gap junctions that are present between the follicular cells (Andersen and Albertini, 1976). Gap junctions mediate metabolic co-operation between the oocytes and its somatic companion cells (Heller et al., 1981). Gap junctions form channels, which are composed of six transmembrane protein subunits called connexins (Goodenough, 1976).

Thus morphological changes occur during the transition of primordial to preantral stage with respect to oocyte-granulosa cell interaction, cell-cell communication, development of zona pellucida and differentiation of theca from mesenchymal cells. A number of genes play a crucial role during this period; however, it is not well documented in mammalian ovary. In the present preliminary study, the gene expression in Day 2, Day 4 and Day 6 mouse ovaries containing primordial, primary and preantral follicles, respectively, was elucidated by cDNA arrays. Few genes such as GDF-9, SCF and connexin 43 were validated. Further, morphological changes during early folliculogenesis were studied at the ultrastructural level to understand the interactions between germ and somatic cells along with other details of differentiation of somatic cells during early folliculogenesis.

## Materials and Methods

*Animals*

Swiss mice bred from the Institute's animal colony were maintained at constant light (12 h light: 12 h dark), temperature (24°C) and humidity (60%), and were supplied with food and water *ad libitum*. The Institutional Animal Ethics Committee approved the animal requirement for the present study and experiments were performed in accordance with the guidelines set by the Committee for the Purpose of Control and Supervision of Experiments on Animals, India.

*Ovarian morphology for light and electron microscopy*

Neonatal female Swiss mice were used for the study. Day of birth was considered as Day 0.

Ovaries of Day 2 (n = 10), Day 4 (n = 10), Day 6 (n = 10) and Day 22 (n = 10) mice were dissected out immediately after death, fixed in 4% normal buffered formalin (pH 7.2) for 6 hours, dehydrated in ascending ethanol grades and embedded in paraffin. Sections (5 μm-thick) were mounted on silane-coated slides for histology, immunohistochemistry and *in situ* hybridization.

Ovarian sections were stained by haematoxylin and eosin. The ovarian follicles were classified into primordial, primary and preantral (secondary) follicles based on the shape and number of granulosa cells (Peters and McNatty, 1980).

The Day 6 mouse ovaries were fixed in modified Karnovsky's fluid (Karnovsky, 1965) for 4-6 hrs. The tissues were then post fixed in 1% osmium tetroxide, dehydrated in ascending grades of acetone and embedded in Araldite (Pelco International, USA). Ultrathin sections (60-70 nm) were cut and mounted on uncoated copper grids (200 mesh) using UCT-R ultramicrotome (Leica, Germany). The sections, stained with uranyl acetate and lead citrate, were observed with a Philips Tecnai 12 ( Netherlands) Transmission Electron Microscope at 80 KV. Photographs were captured using Mega View III Sis CCD camera.

*cDNA arrays with ovarian tissues*

Ovaries from Day 2 (n = 60), Day 4 (n = 40) and Day 6 (n = 30) mice were immediately frozen at - 80°C until RNA extraction. Total RNA was extracted from ovarian tissue using Trizol reagent (Life Techonologies Inc., Rockville, MD). Fifty nanograms of total RNA from ovaries were reverse transcribed using a cDNA synthesis system BD SMART PCR cDNA synthesis kit (BD Biosciences, San Jose, CA) according to manufacturer's instructions. The reverse transcribed products were subsequently amplified by LD PCR as per kit instructions.The amplified PCR products were used as probes. The probes were radiolabeled by $^{32}$P-dATP incorporation as reported by Modi et al. (2005).

Hybridization was performed on nylon based mouse cDNA array blots (mouse 1.2 II K) which were commercially obtained (Clontech, Palo Alto,CA). The images were processed and the data was analyzed using the AtlasImage 2.7 software (BD Biosciences).

*Immunohistochemical detection of GDF 9, SCF, PCNA and Connexin 43*

Immunohistochemical detection of GDF 9, SCF, PCNA and Connexin 43 in ovarian tissues was performed with 5-μm sections of paraffin-embedded tissues using antibodies for GDF 9 and SCF (both goat polyclonal), Connexin 43 (rabbit polyclonal) and PCNA (mouse monoclonal). All primary antibodies were used at a dilution of 1:100 mL all were obtained from Santacruz Biotechnology (CA).

*In situ hybridization of GDF 9, SCF and Connexin 43*

To study the distribution of GDF 9, Connexin 43 and SCF transcripts in neonatal mouse ovaries, *in situ* hybridisation analyses was carried out on paraffin sections of Day 21 ovarian sections containing small, medium and large follicles as per the protocol reported earlier (Dharma et al., 2003).

## Results

All results presented below are preliminary.

*Histology of Day 2, Day 4 and Day 6 mouse ovaries (light microscopy)*

The Day 2 mouse ovaries contain a large number of primordial follicles located towards the cortical region. These are compactly placed with hardly any stromal cells in between these follicles. The primordial follicles each contain a distinct oocyte (>15 μm in diameter) with large cytoplasm and a nucleus. The oocyte is surrounded by squamous granulosa cells (Fig. 1a). Towards the medullary region few layers of primary follicles are observed in the Day 4 ovary. A single layer of cuboidal granulosa cells surrounded the oocyte. These primary follicles are about 30–40 μm in size (Fig. 1b). Day 6 ovary has abundant preantral follicles of (~ 50 μm) in size. The granulosa cells multiply forming two layers (Fig.1c). The size of the oocyte does not show any significant increase (20 μm-25 μm in diameter) during early development.

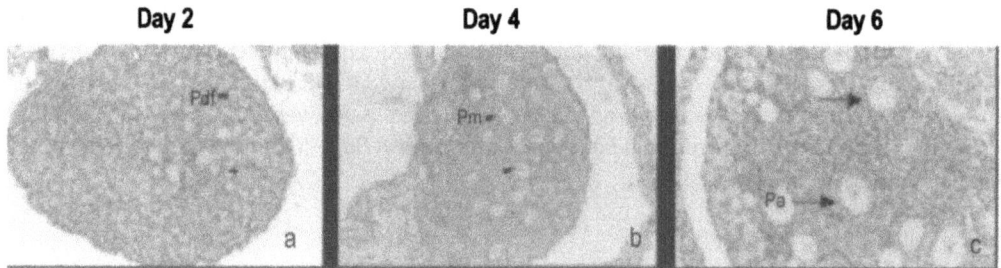

**Fig 1.** Follicular development showing morphology of neonatal mouse ovary. (a) Day 2 ovary: Primordial follicles (Pdf) with oocytes surrounded by flattened squamous granulosa cells; (b) Day 4 ovary: Primary follicles (Pm) with cuboidal granulosa cells (arrow head); (c) Day 6 ovary: Preantral follicles (Pa) with two layers of granulosa cells (line arrow). Magnification 200 X

*Ultrastructure of Day 6 mouse ovaries*

The Day 6 ovaries of mouse contain all three types of follicles; primordial, primary and preantral (secondary) follicles. The primordial follicles were recognized by the presence of squamous shaped granulosa cells surrounding the oocyte without a zona pellucida. The ultrastructure showed abundant mitochondria in the cytoplasm of the oocyte and direct communication between oocyte and granulosa cells by projections from the latter (Fig. 2a). At certain places membrane fusion was observed between the oocyte and squamous granulosa cells (Fig. 2b).The transition of primordial to primary stage revealed a change in the morphology of granulosa cells, which became cuboidal in shape. Oocytes showed microvilli and desmosomes between oocytes and granulosa cells (Fig. 2c). A single layer of cuboidal granulosa cells surrounded the oocyte (Fig. 2d). The presence of basement membrane was observed. Beyond this, primitive theca cells were seen to be differentiated from the stroma mesenchymal cells of the ovary. A layer of zona pellucida was formed covering the oocyte (Fig. 2e). Microvilli from the oocyte enter the zona pellucida (Fig. 2f). In the secondary follicles; two layers of granulosa cells surrounded the oocyte. The oocyte was covered with zona pellucida although not yet fully developed. The basement membrane separated the granulosa cells and theca cells (Fig. 2g). In the secondary follicles the granulosa cells showed rough endoplasmic reticulum (RER), abundant mitochondria and granulated cytoplasm. The nucleus was large with nuclear pores in the nuclear membrane. The gap junctions were distinctly observed between the neighbouring granulosa cells (Fig. 2h).

*cDNA array analysis*

Fig. 3 shows the image of mouse cDNA array hybridized with RNA from Day 2, Day 4 and Day 6 ovaries. As observed, a number of differentially expressed spots were seen. The spots at the diagonally opposite ends were used for aligning the grids. Of the 1200 genes, 398 genes (30%) were found to be differentially expressed during primordial to primary follicle transition (Day 2

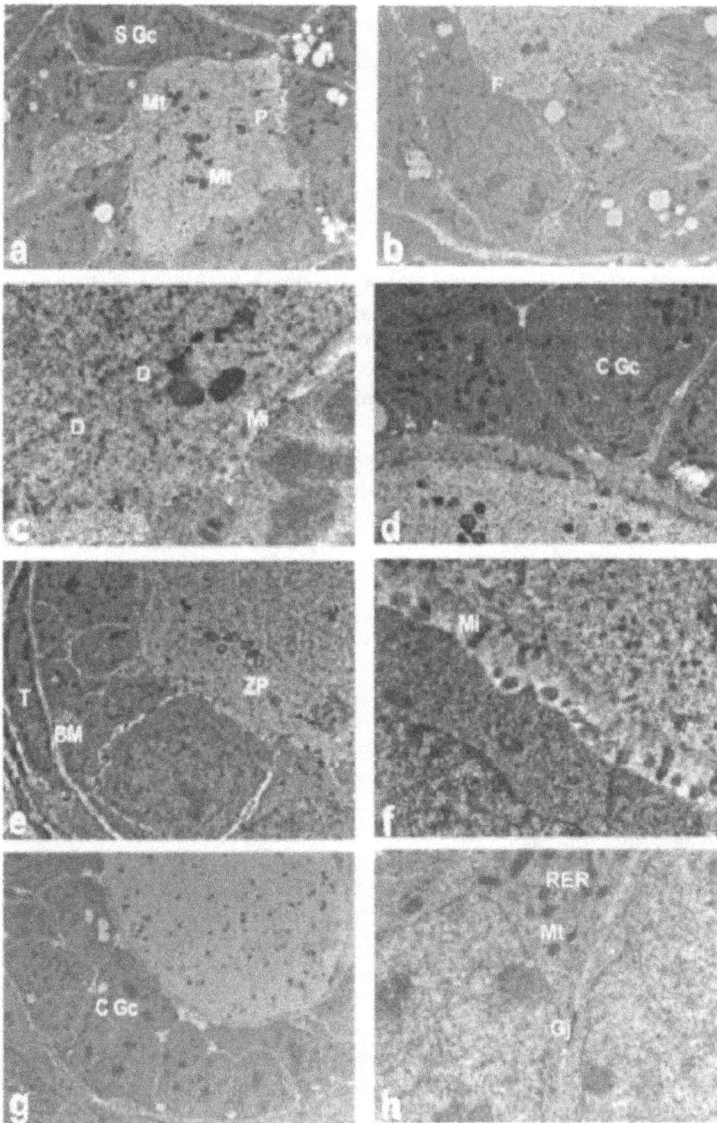

**Fig 2.** Ultrastructure of Day 6 mouse ovary. The three types-primordial, primary and preantral follicles are seen. (a, b) Primordial follicles: (a) oocyte without zona pellucida, abundant mitochondria (Mt) in ooplasm. Squamous granulosa cells (S Gc) surrounding oocyte. Projections (P) of granulosa cells towards oocyte (magnification 4800 X); (b) fusion of plasma membranes (F) (magnification 6800 X). (c,d) primary follicles: (c) desmosomes (D) between the oocyte and granulosa cells, oocyte with microvilli (Mi) entering primitive zona pellucida (ZP) around the oocyte (magnification 18500 X); (d) A single layer of round cuboidal granulosa cells (C Gc) surrounding oocyte (magnification 9300 X); (e) Presence of basement membrane (BM) and primitive theca cells (T) differentiated from stroma (magnification 6800 X); (f) distinct microvilli from the oocyte (magnification 30000 X); (g,h) Preantral follicles: (g) oocyte partially covered with zona pellucida yet not fully developed. Two layers of granulosa cells (magnification 2900 X). (h) The cells showing RER and mitochondria and granulated cytoplasm. Large nucleus with nuclear pores in the nuclear membrane. Distinct gap junction (Gj) between granulosa cells (magnification 15500 X).

| Day 2 | Day 4 | Day 6 |
|---|---|---|

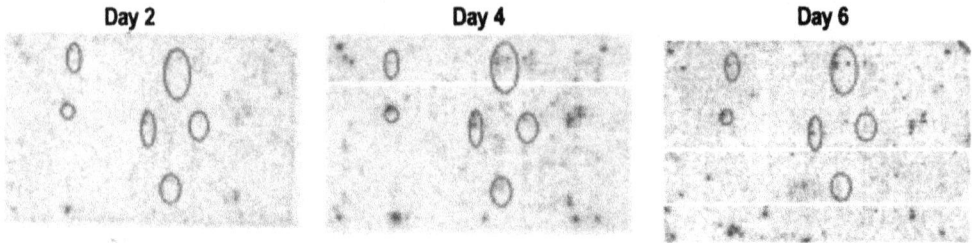

**Figure 3.** Showing the image of mouse cDNA array blot hybridized with RNA from Day 2, Day 4 and Day 6 ovaries. Total RNA was isolated from the Day 2, Day 4 and Day 6 mouse ovaries. $^{32}$P labeled cDNA probes were generated from each RNA samples and hybridized to the BD mouse expression array blots according to User Manual. Blots were exposed to a phosphorImaging screen overnight and scanned.

to Day 4). The genes were categorized according to their functional pathways into several groups. These included immodulators, transcription factors, cytokines, cell cycle regulators neurotransmitters, growth factors and those involved in cell proliferation (Table 1).

**Table 1.** Gene expression profile

| Genes Upregulated | Percentage | Genes Downregulated | Percentage |
|---|---|---|---|
| Day 2 vs Day 4 mouse ovary | | | |
| Cytokines | 16 | Metabolism | 17 |
| Neurotransmitters | 12 | DNA/RNA metabolism | 14 |
| Growth factors | 12 | Protein synthesis | 6 |
| Signal transducers | 10 | Miscellaneous | 5 |
| Immunomodulators | 8 | | |
| Day 4 vs Day 6 mouse ovary | | | |
| Cytokines | 4 | Metabolism | 14 |
| Neurotransmitters | 4 | DNA/RNA metabolism | 3 |
| Growth factors | 4 | Transcription factors | 4 |
| Signal transducers | 2 | Miscellaneous | 5 |
| Immunomodulators | 12 | | |

The genes, which were up regulated, predominantly belonged to cytokine, neurotransmitter, growth factor and signal transducer families. Genes responsible for metabolism showed down regulation on Day 4 as compared to Day 2. During the primary to preantral development (Day 4 to Day 6) only 6.2% of genes showed differential pattern of expression. Interestingly, only a small subset of these overlapped with those seen at Day 2 vs Day 4 transition. These included proteins involved in immunomodulation. Overall, the genes that showed decreased expression during the shift from primordial to primary to secondary stages in the ovary mainly included those involved in protein synthesis and general metabolism.

For the validation of the array results, immunoexpression was undertaken to determine the cellular sites of SCF, GDF-9, Connexin 43 and PCNA in various stages of neonatal mouse ovaries.

*Immunolocalization of GDF 9, SCF, PCNA and Connexin 43*

Immunohistochemical analyses were done to localize the expression of GDF-9, SCF, PCNA and Connexin 43 in Day 4 and Day 6 mouse ovarian follicles. The Day 2 ovaries of mouse contain only primordial follicles and failed to show any staining with antibodies to GDF 9 (Fig. 4a), SCF (Fig. 4f), PCNA (Fig. 4k) and Connexin 43 (Fig. 4p).

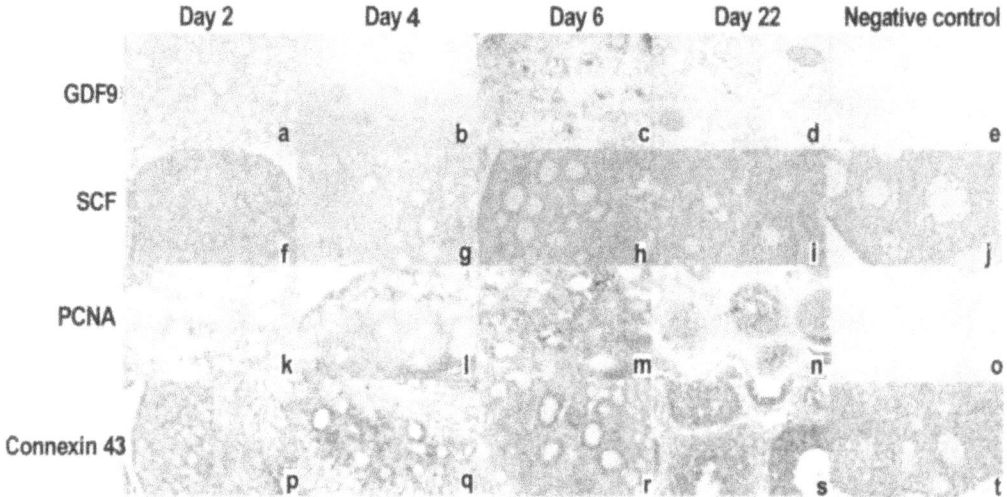

**Fig 4.** Immunostaining for GDF-9, SCF, PCNA and Connexin 43 during follicular development in neonatal mouse ovary. (a) Day 2 ovaries; negative for GDF-9 counterstained with haematoxylin. Intense GDF-9 immunostaining was seen in the oocyte of primary follicles on Day 4 (b) and preantral follicles on Day 6 (c). Intense SCF immunostaining in granulosa cells of preantral follicles on Day 6 (h) but not in primordial follicles (f) and primary follicles (g) was observed. Immunolocalization of PCNA in preantral follicles (m) but not in primordial (k) and primary follicles (l) (No counter staining as PCNA staining is nuclear) was observed. Immunolocalization of Connexin 43 in granulosa cell membrane of primary (q) and preantral (r), but not in primordial follicles (p) was observed. Day 22 ovary (d, i, n, s) used as positive control for respective antibodies. In Day 22 ovary (e, j, o, t), omission of primary antibody yielded no staining, represent the validation of the immunohistochemical localization (counterstained with haematoxylin).

In the Day 4 mouse ovaries along with primordial follicles, development to primary stage was observed. The cytoplasm of the oocyte in the primary follicles demonstrated brown staining for GDF-9 (Fig. 4b) and SCF (Fig. 4g) in granulosa cells but was negative for PCNA (Fig. 4l) and Connexin 43 (Fig. 4q).

In the Day 6 ovaries preantral secondary follicles with two layers of granulosa cells were observed. GDF-9 continued to express in the oocyte (Fig. 4c) and also SCF was present in the granulosa cells of preantral follicles (Fig. 4h). The granulosa cells of the preantral follicles were PCNA positive (Fig. 4m). Connexin 43 was localized in the gap junctions formed between the granulosa cells (Fig. 4r). In the Day 22 ovaries preantral as well as antral follicles were observed. The ooplasm was positive for GDF-9 (Fig. 4d) while cytoplasm of granulosa cells was stained for SCF (Fig. 4i). The nuclei of granulosa cells localized PCNA (Fig. 4n) while Connexin 43 protein was seen in the gap junctions of granulosa cells (Fig. 4s). No staining was observed in the negative controls when preimmune sera were used (Fig. 4e, j, o, t).

*In situ hybridization*

Results revealed that the transcripts of GDF-9, SCF and Connexin 43 were present in preantral/antral follicles (Fig. 5). GDF-9 transcript was detected in the oocytes and in the granulosa cells immediately surrounding the oocyte (Fig. 5a) showing the secretory activity of the oocyte.

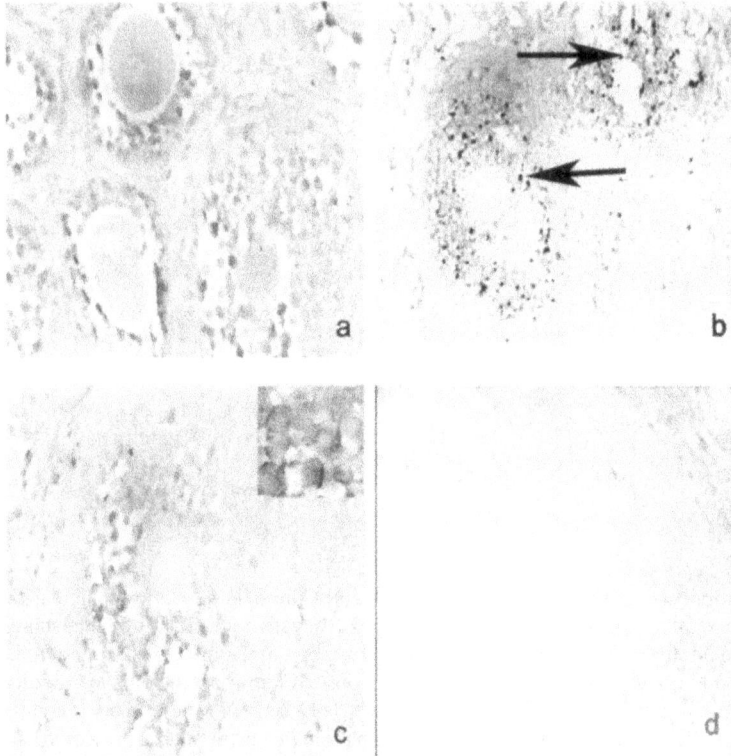

**Figure 5.** *In situ* hybridization studies of GDF-9, SCF and Connexin 43 mRNA in Day 22 mouse ovaries. (a) Moderate to intense staining for GDF-9 mRNA in the nuclei of oocytes and granulosa cells. (b) Expression of SCF (indicated by arrows) in the nuclei of granulosa cells. (c) Expression of Connexin 43 in granulosa cells stained intensely in large follicles of Day 22 ovaries. Inset (magnification 1000 X). (d) Negative control showing no staining (magnification 400 X)

The SCF hybridization signals present over the granulosa cells remained constant in preantral/antral follicles (Fig. 5b). In contrast to the readily detectable Connexin 43 transcript in the cell membrane of granulosa cells in Day 22 mouse ovaries (Fig. 5c), and also in Day 4 and Day 6 ovaries, we were unable to detect it in Day 2 ovary (data not shown).

## Discussion

Dynamic changes occur during folliculogenesis in the follicle structure. The somatic cells proliferate and attain steroidogenic capacity while the oocyte grows and acquires developmental competence. Earlier, the oocyte was believed to be a plausible participant in this complex process but it is now well established that there is an interplay between the oocyte and the surrounding cells that mutually influence the growth and differentiation of both cell lineages (Eppig, 2001). However, it is not clear at what stage of follicular development the communication is initiated. Our preliminary study has demonstrated that distinct molecular and structural alterations occur during the transition of primary to preantral stage in neonatal Day 4 and Day 6 mouse ovaries as revealed by ultrastructure, cDNA arrays and immunolocalization of various proteins.

The granulosa cells of preantral follicles are an actively proliferating population. These cells are joined by desmosomes while gap junctions are absent (Albertini and Andersen, 1974). Recently, a regulatory loop in the ovarian follicle between granulosa cells and the oocyte has been reported (Eppig *et al.*, 2005). Signals originating in the granulosa cells may promote the secretion of paracrine factors by the oocyte and that in turn enhance granulosa cell function as revealed by amino acid uptake by granulosa cells. This transferring of amino acids is likely through gap junctions and is acquired only in the later stage of preantral follicle growth (Eppig *et al.*, 2005). Extending these observations, along with preantral follicle structure by electron microscopy, we further demonstrate the presence of gap junctions between granulosa cells of primary and secondary follicles. This was also confirmed by presence of Connexin 43, a marker for gap junctions, in between granulosa cells on the cell membrane.

Oocyte-granulosa cell communication is essential for oocyte development. FSH prompts proliferation and differentiation of preantral follicles via paracrine factors such as IGF-1 (Adashi *et al.*, 1991) and activin (Miro and Hillier, 1996). In addition, FSH regulates SCF expression in granulosa cells of mouse preantral follicles (Joyce *et al.*, 1999). Further SCF has been reported to promote early oocyte growth *in vitro*. FSH may therefore indirectly regulate oocyte growth in preantral follicles. However, the factors regulating the growth of follicles from the primordial to primary stages are virtually unknown.

In an endeavor to identify the molecular factors that may regulate the transition of primordial follicles to primary and further stages, we performed cDNA array hybridization and scanned the expression of 1200 known gene. A number of genes previously unknown in this process for example activin, FGF and its receptors, estrogen receptor and FSH receptor were found to be differentially expressed (unpublished observations). These results suggested the involvement of multiple complex pathways during follicular transition.

Upregulation of these genes corroborated with the developing and proliferating activities of granulosa cells and metabolically active germ and somatic cells. The presence of mRNA encoding GDF9 and SCF in the oocyte of primary follicles and Connexin 43 in the gap junctions and their translational products supported the array experiments.

The distribution of SCF was evaluated in neonatal mouse ovaries. Corroborating the cDNA array data, an induction of SCF expression was noted in the ovaries of Day 4 animals as compared to Day 2 animals. No difference in SCF expression was evident in the ovaries of Day 4 vs Day 6 animals, further validating the array results. Oocytes of primordial follicles were unstained for SCF, whereas the growing primary and preantral follicles showed distinct cytoplasmic staining in the granulosa cells. Interestingly, the expression of the receptor for SCF (c-kit) is also initiated in the primary follicles. Thus, possible autocrine and paracrine mechanisms are involved in the anti-apoptotic effect of the SCF duet (Hoyer *et al.*, 2005). Additionally, such interactions may also be important for tight binding of the oocyte to the surrounding granulosa cells. Indeed, the microvilli of the oocyte communicate with the projection from granulosa cells through the zona pellucida (Zamboni, 1974). As revealed from the ultrastructural studies, the communication between oocyte and granulosa cells is established during the transition of primary to secondary follicle and the zona pellucida is also formed around the same time.The patches of zona material become confluent and form a continuous layer around the oocyte with microvilli and granulosa cells projection embedded in the matrix as revealed by the ultrastructural observation which agreed with an earlier report (Chiquione, 1960).

It is known that the oocyte expresses various polypeptides such as fibroblast growth factors (Valve *et al.*, 1997), transforming growth factor ß2 (TGFß2) and GDF-9 (McGrath *et al.*, 1995) (the latter two are both members of the TGF ß super family). In the present study, our results have demonstrated that the expression of these factors are initiated as early as Day 4 when the transition of the primordial follicles is initiated. Amongst the members of TGF ß family, GDF-9 has an unusual patterning of 6 cysteines in its mature region compared to the 7 conserved cysteines (Dong et al., 1996). Validating the cDNA array results, by immunolocalization studies, GDF-9 was found to be expressed in the oocytes of the primary follicles in Day 4 mouse ovary but not the primordial follicles of Day 2 ovaries. Furthermore, as evident from the array comparisons of Day 4 and Day 6 ovaries, no change in GDF-9 mRNA was observed and immunohistochemistry data demonstrated a continued expression of GDF-9 in the oocytes of preantral follicles.

One striking feature of the array results was an induction of some proliferation regulating genes particularly cell cycle regulators and growth factors in the Day 4 ovaries as compared to Day 2. PCNA has been used in a number of studies as an index of proliferation of cancer tissues and has also been used to assess the proliferative status of the ovarian follicles (Hall et al., 1990; Kelkar et al., 2003). Corroborating the array results, the granulosa cells of the secondary preantral follicles were PCNA positive as observed by immunohistochemistry in the present study. This also agreed with the increased mitotic activity in these cells observed at light and electron microscopic levels. It is known that the SCF protein observed in the oocytes of primary follicles possibly induces proliferation of granulosa cells through its paracrine action (Nilsson and Skinner, 2004). The over expression of other growth factors in Day 4 and Day 6 ovaries as compared to Day 2 may also contribute to the proliferative status in early follicle growth.

What signals the primordial follicles to transit into the growing phase is an enigma. Our studies have demonstrated the involvement of a number of genes during this process. We are currently clustering our array data in search of a novel pattern that may help us to pin down the early events that regulate this crucial process. We envisage that these observations can contribute significantly towards the elucidation of new molecular pathways involved in follicle transition during folliculogenesis.

## Acknowledgements

The authors thank Dr. C.P. Puri, Director, NIRRH for his support, Dr. S. Mukherjee, Senior Research Officer for her valuable suggestions in the initial stages of the cDNA arrays and the Department of Biotechnology, Government of India for financial assistance. The technical help of Mr. S.T. Ghanekar and the typing assistance of Ms. Supriya Lad are acknowledged.

## References

Adashi EY, Renick CE, Hurwitz A, Ricciarelli E, Hernandez ER, Roberts CT, Leroith D and Rosenfeld R (1991) Insulin-like growth factors: the ovarian connection Human Reproduction 6 1213-1219

Albertini DF and Andersen E (1974) The appearance and structure of intercellular connections during the ontogeny of the rabbit ovarian follicle with particular reference to gap junctions Journal of Cell Biology 63 234-250

Andersen E and Albertini DF (1976) Gap junction between the oocyte and companion follicle cells in the mammalian ovary Journal of Cell Biology 71 680-686

Buccione R, Schroeder AC and Eppig JJ (1990) Interaction between somatic cells and germ cells throughout mammalian oogenesis Biology of Reproduction 43 543-547

Canipari R (2000) Oocyte-granulosa cell interaction Human Reproduction 6 279-289

Chiquione AD (1960) A study of the fine structure of the rabbit primary oocyte Journal of Ultrastructure Research 5 349-363

Dharma SJ, Kelkar RL and Nandedkar TD (2003) Fas and Fas ligand protein and mRNA in normal and atretic mouse ovarian follicles Reproduction 126 783-789

Dong J, Albertini DF, Nishimori K, Kumar TR, Lu N and Matzuk MM (1996) Growth differentiation factor-9 is required during early ovarian folliculogenesis Nature 383 531-535

**Eppig JJ** (2001) Oocyte control of a ovarian follicular development and function in mammals *Reproduction* **122** 829-838

**Eppig JJ, Wigglesworth K, Pendola F and Hirao Y** (1997) Murine oocytes suppress expression of luteining hormone-receptor messenger ribonucleic acid by granulosa cells *Biology of Reproduction* **56** 976-984

**Eppig JJ, Pendola FL, Wigglesworth K and Pendola JK** (2005) Mouse oocyte regulate metabolic cooperativity between granulosa cells and oocyte: Amino acid transport *Biology of Reproduction* **73** 351-357

**Gaytan F, Morales C, Belllido C, Aguilar E and Sanchez-Criado JE** (1996) Proliferative activity in the different ovarian compartment in cycling rats estimated by the 5-bromodeoxyuridine technique *Biology of Reproduction* **57** 1356-1365

**Goodenough DA** (1976) *In vitro* formation of gap junction vesicles *Journal of Cell Biology* **68** 220-231

**Gougeon A and Busso D** (2000) Morphologic and functional determinants of primordial & primary follicles in the monkey ovary *Molecular and Cellular Endocrinology* **163** 33-41

**Hall PA, Levison DA, Woods AL and Eppig JJ** (1990) Proliferating cell nuclear antigen (PCNA) immunolocalization in paraffin sections: an index of cell proliferation with evidence of deregulated expression in some neoplasms *Journal of Pathology* **162** 285-294

**Heller DT, Cahil DM and Schultz RM** (1981) Biochemical studies of mammalian oogenesis: metabolic cooperativity between granulosa cells and growing mouse oocytes *Development Biology* **84** 455-464

**Hirshfield AN** (1991) Development of follicles in the mammalian ovary *International Review of Cytology* **124** 43-101

**Hoyer PE, Byskov AG and Mollgard K** (2005) Stem cell factor and c-Kit in human primordial germ cells and fetal ovaries *Molecular and Cellular Endocrinology* **234** 1-10

**Joyce IM, Pendola FL, Wigglesworth K and Eppig JJ** (1999) Oocyte regulation of Kit ligand expression in mouse ovarian follicles *Developmental Biology* **214** 342-353

**Karnovsky MJ** (1965) A formaldehyde-glutaraldehyde fixative of high osmolality for use in electron microscopy *Journal of Cell Biology* **27** A 137

**Kelkar RL, Dharma SJ and Nandedkar TD** (2003) Expression of Fas and Fas ligand protein and mRNA in mouse oocytes and embryos *Reproduction* **126** 791-799

**Laitinen M, Vuojolainen K, Jaatinen R, Ketola I, Aaltonen J, Lehtonen E, Heikinheimo M and Ritvos O** (1998) A novel growth differentiation factor-9 (GDF-9) related factor is co-expressed with GDF-9 in mouse oocytes during folliculogenesis *Mechanisms of Development* **78** 135-140

**Manova K, Huang EJ, Angeles M, De Leon V, Sanchez S, Pronovost SM, Besmer P and Bachvarova RF** (1993) The expression pattern of the c-kit ligand in gonads of mice supports a role for the c-kit receptor in oocyte growth and in proliferation of spermatogonia *Development Biology* **157** 85-99

**McGrath SA, Esquela AF and Lee SJ** (1995) Oocyte specific expression of growth/differentiation factor -9 Molecular *Endocrinology* **9** 131-136

**McNatty KP, Filder AE, Juengel JL, Quirke LD, Smith PR, Heath DA, Lundy T, O'Connell A and Tisdall DJ** (2000) Control of early ovarian follicular development *Journal of Reproduction and Fertility Suppl* **49** 123-135

**Miro F and Hillier SG** (1996) Modulation of granulosa cell deoxyribonucleic acid synthesis and differentiation by activin *Endocrinology* **137** 464-468

**Modi D, Shah C, Sachdeva G, Gadkar S, Bhartiya D and Puri C** (2005) Ontogeny and cellular localization of SRY transcripts in the human testes and its detection in spermatozoa *Reproduction* **130** 603-613

**Motro B and Bernstein A** (1993) Dynamic changes in ovarian c-kit and steel expression during the estrous reproductive cycle *Developmental Dynamics* **197** 69-79

**Nandedkar TD and Raghavan VP** (1989) Control of ovarian follicular maturation by nonsteroidal parameters: new approaches to female contraception *Advances in Contraceptive Development System* **5** 117-139

**Nilsson EE and Skinner MK** (2004) Kit ligand and basic fibroblast growth factor interactions in the induction of ovarian primordial to primary follicle transition *Molecular and Cellular Endocrinology* **214** 19-25

**Oktay K, Schenken RS and Nelson JF** (1995) Proliferating cell nuclear antigen marks the initiation of follicular growth in the rat *Biology of Reproduction* **53** 295-301

**Oktay K, Briggs D and Gosden RG** (1997) Ontogeny of follicle-stimulating hormone receptor gene expression in isolated human ovarian follicles *Journal of Clinical Endocrinology and Metabolism* **82** 3748-3751

**Peters H and McNatty KP** (1980) *The Ovary: A Correlation of Structure and Function in Mammals* Granada Publishing, London

**Peters H, Byskov AG, Himelstein-Braw R and Faber M** (1975) Follicular growth: the basic event in the mouse and human ovary *Journal of Reproduction and Fertility* **45** 559-566

**Skinner MK** (2005) Regulation of primordial follicle assembly and development *Human Reproduction Update* **11** 461-471

**Valve E, Penttila TI, Paranko J and Harkonen P** (1997) FGF-8 is expressed during specific phases of rodent oocyte and spermatogonium development *Biochemical and Biophysical Research Communication* **232** 173-177

**Vanderhyden BC, Caron PJ, Buccione R and Eppig JJ** (1990) Development pattern of the secretion of cumulus expansion-enabling factor by mouse oocytes and the role of oocytes in promoting granulosa cell differentiation *Developmental Biology* **140** 307-317

**Wang J and Roy SK** (2004) Growth differentiation factor-9 and stem cell factor promote primordial follicle formation in the hamster: modulation by follicle stimulating hormone *Biology of Reproduction* **70** 577-585

**Zamboni L** (1974) Fine morphology of the follicle wall and follicle cell-oocyte association *Biology of Reproduction* **10** 125-149

# Putative human male germ cells from bone marrow stem cells

Nadja Drusenheimer[1], Gerald Wulf[2], Jessica Nolte[1], Jae Ho Lee[1], Arvind Dev[1], Ralf Dressel[3], Jörg Gromoll[4], Jörg Schmidtke[5], Wolfgang Engel[1] and Karim Nayernia[1*]

[1]Institute of Human Genetics, University of Göttingen, D-37073 Göttingen; [2]Department of Haematology and Oncology, University of Göttingen, D-37099 Göttingen; [3]Department of Department of Cellular & Molecular Immunology, University of Göttingen, D-37073 Göttingen; [4]Institute of Reproductive Medicine, University of Münster, D-48129 Münster; [5]Institute of Human Genetics, Medical School of Hannover, D-30625 Hannover

## Abstract

Germ cells must develop along distinct male or female paths to produce the spermatozoa or oocyte required for sexual reproduction. Male germline stem cells maintain spermatogenesis in the postnatal human testis. Here we show that a small population of bone marrow cells is able to transdifferentiate to male germ cell-like cells. We show expression of early germ cell markers (Oct4, Fragilis, Stella and Vasa) and male germ cell specific markers (Dazl, TSPY, Piwil2 and Stra8) in these cells. Our preliminary findings provide direct evidence that human bone marrow cells can differentiate to putative male germ cells and identify bone marrow as a potential source of male germ cells that could sustain sperm production.

## Introduction

Male germ cells are derived from a founder population of primordial germ cells (PGCs) that are set aside early in embryogenesis. PGCs arise from the proximal epiblast, a region of the early embryo that also contributes to the first blood lineages of the embryonic yolk sac (Lawson and Hage, 1994; Zhao and Garbers, 2002). After birth, PGCs differentiate to spermatogonial stem cells in the male which are responsible for maintaining spermatogenesis throughout life by continuous production of daughter cells that differentiate into spermatozoa (de Rooij and Grootegoed, 1998; McLaren, 2000). The common origin of germ cell lineage and blood lineage and the fact that bone marrow stem cells are pluripotent cells which can differentiate to other cell types, prompted us to examine the differentiation potential of human bone marrow (BM) cells to male germ cells. Accumulated evidence suggests that in addition to haematopoietic stem cells (HSCs), bone marrow also harbors endothelial stem cells (ESCs), mesenchymal stem cells (MSCs) and multipotential adult progenitor cells (MAPCs) (Jiang et al., 2002; Kassem, 2004; Ratajczak et al., 2004; Reyes et al., 2002). Adult bone-marrow-derived mesenchymal stem cells are capable of differentiation along several lineages. Recently, it has been discov-

*Corresponding author
E-mail: knayern@gwdg.de

ered that bone marrow grafts to female mice, and possibly humans, can produce new follicles and oocytes in the recipient ovary (Johnson *et al.*, 2005). The authors have also reported that these tissues share genes typical of germ cells and proposed that bone marrow stem cells can migrate and colonize the ovaries to maintain a plentiful stock for reproduction.

In this paper, we present our preliminary findings demonstrating that human MSCs have the potential to express markers characteristic for male germ cell differentiation. This observation heralds new thinking about bone marrow stem cells as a soure for reproductive medicine.

## Materials and methods

### Isolation and culture of human mesenchymal stem cells

Human mesenchymal stem cells (MSCs) were prepared according to previously described methods (Pittenger *et al.*, 1999; Wulf *et al.*, 2004). Following an Internal Review Board's approval, leftover materials from disease-free bone marrow aspirates drawn for diagnostic purposes from male patients of the Haematology Service were obtained. The bone marrow aspirate was mixed with an equal amount of Dulbecco Modified Eagle medium (DMEM, Invitrogen) and centrifuged at 800 g for 10 minutes at 20°C. The cells were then resuspended in DMEM, layered on a Ficoll-Hypaque gradient (density = 1.077 g/cm3; Sigma, Deisenhofen, Germany), and centrifuged again. The low-density mononuclear fraction was collected, washed and resuspended in complete culture medium (DMEM with 10% fetal bovine serum [FBS, Sigma, Deisenhofen, Germany], penicillin/streptomycin [Invitrogen, Karlsruhe, Germnay], and glutamine [Invitrogen, Karlssruhe]) and plated at $2 \times 10^7$ cells/185 mm². The cultures were maintained at 37°C in a humidified atmosphere containing 95% air and 5% $CO_2$ and subcultured prior to confluency.

### Treatment of mesenchymal stem cells

Isolated cells were cultured at 37°C and an atmosphere of 5% carbon dioxide in MesenCult Basal Medium (Cellsystems, St. Katharinen) supplemented with 10% MSC Stimulatory supplements and 50 $\mu$g/ml each penicillin and streptomycin (Invitrogen, Karlsruhe). As this medium prevents differentiation, the cells were cultured during the differentiation analysis in RPMI 1640 (Invitrogen, Karlsruhe) supplemented with 10% FBS and 50 $\mu$g/ml each penicillin and streptomycin. Cell passage was arbitrary, depending on the proliferation state. The cells were detached with Accutase (PAA Laboratories, Cölbe). For induction of differentiation, retinoic acid (RA: Sigma, Deisenhofen, Germany) was added to the RPMI-medium at a final concentration of $10^{-5}$ M or $10^{-6}$ M and the cells were cultured for fifteen days in this medium.

### RNA isolation and RT-PCR analyses

RNA was extracted from cells using the RNeasy-Kit (Qiagen, Hilden, Germany) according to the manufacturer's instructions. RNA from human testis was isolated using the TriReagent (Biomol, Hamburg, Germnay) according to the manufacturer's instruction.

For RT-PCR analyses, 5 $\mu$g of RNA were reverse transcribed into cDNA at 42°C for 50 min in a final volume of 20 $\mu$l containing 200 units of Superscript reverse transcriptase (Invitrogen, Karlsruhe), 0.5 $\mu$g oligo dT Primer, 10 mM DTT and 0.5 mM dNTPs. RT-PCR were carried out with 0.5 or 1 $\mu$l of cDNA, 5 to 30 pmol each of forward and reverse primers and 2 to 5 units of Platinum Taq polymerase (Invitrogen, Karlsruhe) in a final volume of 25 or 50 $\mu$l. The solutions were incubated at 94°C for 4 min and then subjected to 35 cycles of amplification, each consisting of 95°C for 30s (denaturation), 57°C-61°C for 30s-45s (annealing) and 72°C for 60s (primer extension). At the end of the temperature cycles the solutions were incubated at 72°C for 10 min. The PCR products (15 $\mu$l samples) were subjected to electrophoresis on 1.5% (w/v) agarose gels containing 1 $\mu$g/ml ethidium bromide and the amplified fragments were viewed under ultraviolet light and photographed. Glyceraldehyde-3-phosphate dehydrogenase (Gapdh) was used as an internal control. The primers used for RT-PCR analyses are shown in Table 1.

Table 1. *Primers used for the RT-PCR analysis of human mesenchymal stem cells*

| Gene | Forward/reverse | Sequence 5'-3' |
|------|-----------------|----------------|
| c-kit | Forward | CAGACTTAATAGTCCGCGTG |
| | Reverse | TTTGATCATGATGCCCGCCT |
| CyclinA2 | Forward | AGAGGCCGAAGACGAGACGGG |
| | Reverse | GCATAGCAGCAGTGCCCACAA |
| Dazl | Forward | AATCATCCTCCTCCACCACAG |
| | Reverse | GGGCCAGAAAGCCGCTTTAAA |
| Fragilis | Forward | GGGCTCTAGAGAGGAGGCCCC |
| | Reverse | GCAGGGGTTCATGAAGAGGGT |
| Gapdh | Forward | CCAGCAAGAGCACAAGAGGAAGAC |
| | Reverse | AGCACAGGGATACTTTATTAGATG |
| Oct-4 | Forward | GGAGCCGGGCTGGGTTGATCC |
| | Reverse | GGGAGAGCCCAGAGTGGTGAC |
| Piwil2 | Forward | ATGCTTCCATCAGGTAGAGG |
| | Reverse | GATCCATCAAACGCAGTGAC |
| Vasa | Forward | TGAGAATACAAGGACAGGAGCT |
| | Reverse | TCTTCACAAGCTCCCAATCC |

*Immunohistochemistry*

Cells which were treated for 15 days with retinoic acid and untreated cells as controls, were washed with PBS and fixed with 4% ice cold paraformaldehyde for ten minutes. The fixed cells were incubated for 16-20 hours at 4°C with one of the following antibodies in a 1:200 dilution in PBS: anti-Dazl, anti-Piwil2, anti-Stra8 and anti-TSPY. Slides were washed three times with PBS and incubated for 1-2 hours with an anti-rabbit antibody conjugated with Cy3 at 1:500 dilution in PBS. After three further washes with PBS, slides were mounted in 4',6-diamidino,-2-phenylindole mounting solution (Vector Laboratories Inc., Burlingame, CA) and viewed in an Olympus BX 60 fluorescent microscope.

For immunophenotyping of human mesenchymal progenitor cell preparations, mouse fluorochrome-conjugated isotype control antibodies, fluorescein isothiocyanate (FITC) or phycoerythrin (PE)-coupled antibodies against the common leukocyte antigen CD45 (clone HI30, Becton Dickinson), the surface-expressed 5´-ectonucleotidase CD73 (clone 37865X, Becton Dickinson), and the beta 1 integrin CD29 (clone HUTS-21, Becton Dickinson, Heidelberg, Germany) were used following the manufacturers' instructions. Binding of antibodies against the stroma cell surface proteoglycan CD105 (clone 8E11, Biotrends, Cologne, Germany), and the marker for differentiated fibroblasts AS02 (clone AS02, Dianova, Hamburg, Germany) as primary antibodies was detected by anti-mouse IgG conjugate (Sigma, Deisenhofen, Germany). Saturating amounts of antibodies were added to cells for 30 minutes at 4°C, before extensive washing and measurement of 10,000 cells each on a FACScan flow cytometer (Becton Dickinson, Heidelberg, Germany), followed by data analysis with the cell quest (Becton Dickinson, Heidelberg, Germany) or WinMDI programs.

## Results and discussion

We established several MSC cell cultures from human bone marrow. After 3 passages *in vitro*, the MSCs showed homogenous positive surface staining for CD73, CD105 and CD29, which are typically found on mesenchymal stem cell populations: but lacked CD14, CD34, CD45 and AS02 surface expression, which denote hematopoietic or mature fibroblast differentiation, respectively (Fig. 1). To examine the potential of MSCs to differentiate to PGCs and male germ cells, we treated the cells with retinoic acid ($10^{-5}$ M and $10^{-6}$ M, 15 days) and examined expression of PGC and male germ cell specific markers. By RT-PCR analyses, we found that human MSCs were positive for *Oct4, Fragilis, Stella, Vasa, c-kit, cyclin A2* and *Piwil2* (Fig. 2). Oct4 (also known as Oct-3) belongs to the POU (Pit-Oct-Unc) transcription factor family (Scholer et al., 1990). The POU family of transcription factors can activate the expression of their target genes through binding to an octameric sequence motif of an AGTCAAAT consensus sequence (Scholer, 1991). Recent evidence indicates that *Oct4* is almost exclusively expressed in ES cells (Niwa, 2001). During embryonic development, *Oct4* is expressed initially in all blastomeres. Subsequently, its expression becomes restricted to the ICM and is downregulated in the TE and the primitive endoderm. At maturity, *Oct4* expression becomes confined exclusively to the developing germ cells (Pesce and Scholer, 2000). *Fragilis* and *Stella* are involved in initiating germ cell competence and specification and in the demarcation of PGCs from their somatic neighbours (Saitou et al., 2002). It was observed that the expression of *fragilis* first increases in the rim of the epiblast cup, suggesting that one of its roles is to keep together those cells that are predisposed to become PGCs, as they move out of the epiblast during gastrulation. Furthermore, the authors propose an elegant model in which cells at the centre of the community of precursors, which express the highest levels of *fragilis*, are the ones selected to become PGCs and to express high levels of *stella*. Expression of *fragilis* is increased in the migratory PGCs, inducing expression of other germ cell-specific genes such as *stella* (Sato et al., 2002) and the *VASA* homolog (Toyooka et al., 2000). *Vasa* encodes an ATP-dependent RNA helicase which is specific for differentiating germ cells from the late migration stage to the postmeiotic stage. PGCs express *c-Kit* at relatively high levels. In PGCs as well as in hematopoietic cells, this expression is related to a single DNaseI-hypersensitive site (HS2) which is absolutely necessary for its activity (Cairns et al., 2003). We found that human MSCs expressed *Oct4, Fragilis, Stella, Vasa, c-kit, cyclin A2* and *Piwil2*, before and after RA treatment which is an evidence that a population of MSCs shows germ cell characteristics without RA treatment (Fig. 2). Expression of some of these genes was increased after RA treatment which indicates that RA treatment promotes germ cell differentiation of human MSCs.

By immunohistochemical analyses, we found that MSCs show expression of Stra8, Piwil2, Dazl and Tspy (Fig. 3). All these genes are expressed specifically in male germ cells. During mouse embryogenesis, *Stra8* expression is restricted to the male developing gonads, and in adult mice, the expression of *Stra8* is restricted to the premeiotic germ cells (Oulad-Abdelghani et al., 1996). *Piwil2*, which is known in mouse as *Mili*, is expressed in premeiotic male germ cells. Spermatogenesis in the *Mili*-null mice is blocked completely at the early prophase of the first meiosis (Kuramochi-Miyagawa et al., 2001; 2004). In human, *Piwil2* is expressed specifically in testicular premeiotic germ cells (Lee et al., 2006a). Furthermore, we showed previously that *Piwil2* modulates expression of murine spermatogonial stem cell expressed genes (Lee et al., 2006b). DAZL proteins are germ-cell-specific RNA-binding proteins essential for gametogenesis. In humans, loss of the Y chromosomal *DAZ* genes is associated with oligozoospermia or azoospermia. The *DAZ* genes are strong candidates for the AZFc azoospermia factor, one of the most common genetic causes of male infertility (Reijo et al., 1995; Vogt et al., 1996; Ferlin et al., 1999). Tspy, the 'testis-specific protein, Y-encoded', is the product of

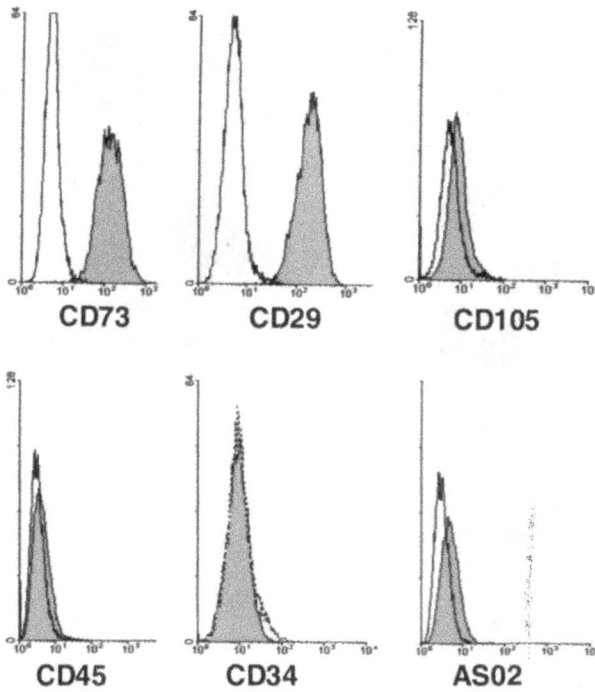

**Fig 1.** The immunophenotype of morphologically homogeneous populations from adherent human bone marrow mesenchymal stem cell cultures after the 3rd passage *in vitro* was analysed by immunofluorescent staining and flowcytometric documentation. Staining profiles of representative samples with 10,000 events each are shown. The markers are represented by shaded histograms, the respective isotype controls fine-lined. The MSCs stained homogeneously strong with markers for mesenchymal progenitors, such as the ectonucleotidase (CD73), the ß1 integrin (CD29) and weakly for the TGF-ß co-receptor endoglin (CD105). The cells were negative for the markers of hematopoetic cells (CD45, CD34), as well as a marker for differentiated fibroblasts (AS02).

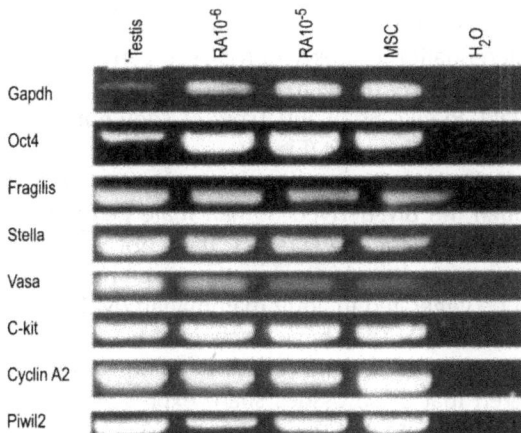

**Fig 2.** RT-PCR analyses of RNA isolated from human mesenchymal stem cells show expression of germ cell specific markers. RNA isolated from human testis served as positive control. $H_2O$, no-template negative control.

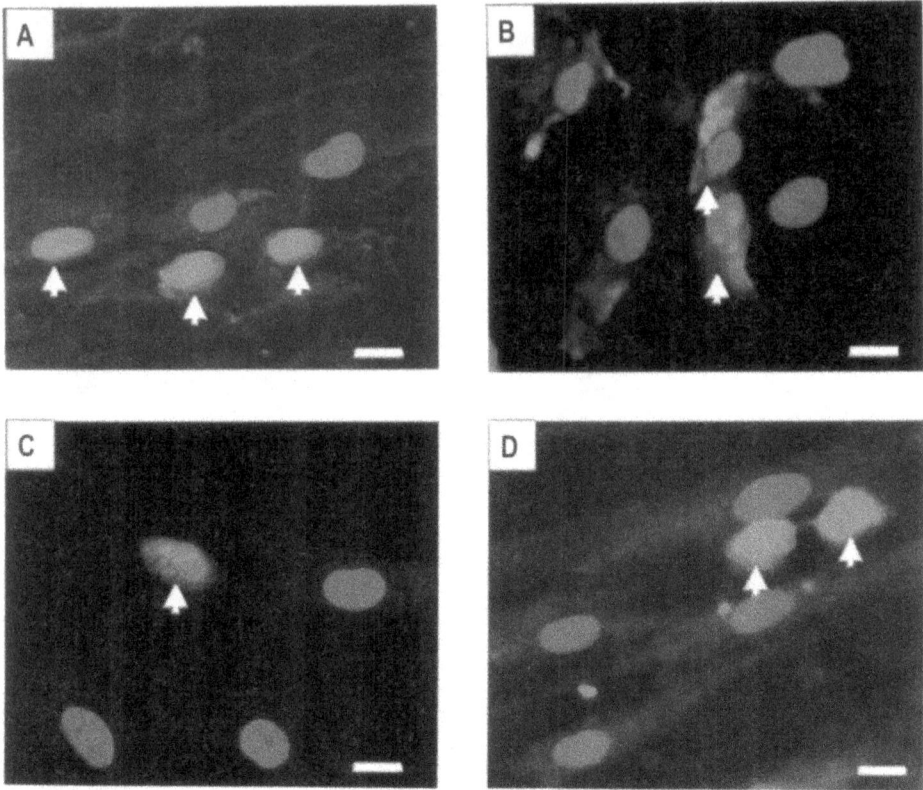

**Fig 3.** Immunohistochemical analyses of human mesenchymal stem cells. Specific antibodies against the male germ cell markers Dazl (A), Stra8 (B), Piwil2 (C) and Tspy (D) (all red) were used. Nuclei are shown in blue, DAPI (4', 6'-diamidino-2-phenylindole) staining. Bar: 10 µm.

a tandem gene cluster on human proximal Yp which is expressed specifically in human spermatogonia (Schnieders et al., 1996). From these results it can be suggested that a population of human mesenchymal stem cells show expression of male germ cell specific markers. To examine the effect of RA on differentiation property of MSCs to germ cells, we treated the cells for 15 days with $10^{-5}$ M RA and $10^{-6}$ M RA, respectively, and performed an indirect immunostaining using antibodies against Dazl, Piwil2, Stra8 and Tspy. As shown in Fig. 4, RA promotes differentiation of MSCs towards male germ cells (Fig. 4).

Our preliminary results indicate that a fraction of bone marrow cells are able to differentiate to male germ cells. Although the BM-derived human male germ cells exhibit expression of germ cell and male germ cell specific markers, we have not yet determined whether these male germ cells can undergo meiosis and form functional spermatozoa. Recently, it was demonstrated that oocytes can be generated in adult mammalian ovaries by putative germ cells in bone marrow (Johnson et al., 2005). They have discovered that bone marrow grafts to female mice, and possibly humans, can produce new follicles and oocytes in the recipient ovary. This team had already discovered that these tissues share expressed genes typical of germ cells (Johnson et al., 2004). They postulate that this situation is one means of replenishing ovarian follicles when numbers in the ovary decline to very few.

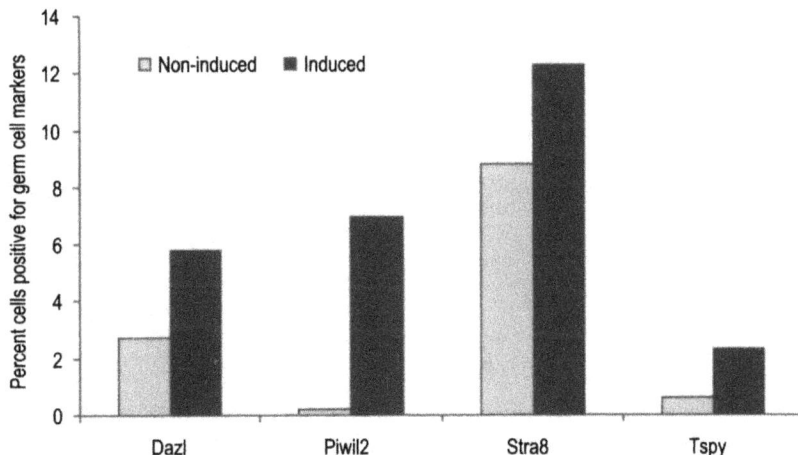

**Fig 4.** Effect of retinoic acid on differentiation of human MSCs to male germ cells. After retinoic acid treatment, an increase in the number of cells which expressed Dazl, Piwil2, Stra8 and Tspy (black) was detected as compared with untreated cells (grey). Percentage of positive cells is shown.

These preliminary results show for the first time that bone marrow stem cells were able to differentiate into a germ cell-like phenotype and may provide a new potential source of male and female germ cells that could be used for production of oocytes and sperm.

## Acknowledgements

The authors thank S. Wolf, B. Sadowski, D. Meyer, Ch. Müller and Leslie Elsner for excellent technical assistance.

## References

Cairns LA, Moroni E, Levantini E, Giorgetti A, Clinger FG, Ronzoni S, Tatangelo L, Tiveron C, DeFelici M, Dolci S, Magli MC, Giglioni B and Ottolenghi S (2003) *Kit* regulatory elements required for expression in developing hematopoietic and germ cell lineages *Blood* **102** 3954–3962

de Rooij DG and Grootegoed JA (1998) Spermatogonial stem cells *Current Opinion in Cell Biology* **10** 694-701

Ferlin A, Moro E, Garolla A and Foresta C (1999) Human male infertility and Y chromosome deletions: role of the AZF-candidate genes DAZ, RBM and DFFRY *Human Reproduction* **14** 1710-1716

Jiang Y, Jahagirdar BN, Reinhardt RL, Schwartz RE, Keene CD, Ortiz-Gonzalez XR, Reyes M, Lenvik T, Lund T, Blackstad M, Du J, Aldrich S, Lisberg A, Low WC, Largaespada DA and Verfaillie CM (2002) Pluripotency of mesenchymal stem cells derived from adult marrow *Nature* **418** 41-49

Johnson J, Canning J, Kaneko T, Pru JK and Tilly JL (2004) Germline stem cells and follicular renewal in the postnatal mammalian ovary *Nature* **428** 145–150

Johnson J, Bagley J, Skaznik-Wikiel M, Lee HJ, Adams GB, Niikura Y, Tschudy KS, Tilly JK, Cortes ML, Forkert R, Spitzer T, Iacomini J, Scadden DT and Tilly JL (2005) Oocyte generation in adult mammalian ovaries by putative germ cells in bone marrow and peripheral blood *Cell* **122** 303-315

Kassem M (2004) Mesenchymal stem cells: biological characteristics and potential clinical applications *Cloning and Stem Cells* **6** 369-374

Kuramochi-Miyagawa S, Kimura T, Yomogida K, Kuroiwa A, Tadokoro Y, Fujita Y, Sato M, Matsuda Y and Nakano T (2001) Two mouse piwi-related genes: *miwi* and *mili Mechanisms of Development* **108** 121-33

Kuramochi-Miyagawa S, Kimura T, Ijiri TW, Isobe T, Asada N, Fujita Y, Ikawa M, Iwai N, Okabe M, Deng W, Lin H Matsuda Y and Nakano T (2004) *Mili,* a mammalian member of *piwi* family gene, is essential for spermatogenesis. *Development* **131** 839-849

Lawson KA and Hage WJ (1994) Clonal analysis of the origin of primordial germ cells in the mouse *Ciba Foundation Symposium* **182** 68-91

Lee JH, Schutte D, Wulf G, Fuzesi L, Radzun HJ, Schweyer S, Engel W and Nayernia K (2006a) Stem cell protein Piwil2 is widely expressed in tumors and inhibits apoptosis through activation of Stat3/Bcl-XL pathway *Human Molecular Genetics* **15** 201-211

Lee JH, Engel W and Nayernia K (2006b) Stem cell protein Piwil2 modulates expression of murine spermatogonial stem cell expressed genes *Molecular Reproduction and Development* **73** 173-179

McLaren A (2000) Germ and somatic cell lineages in the developing gonad *Molecular and Cellular Endocrinology* **163** 3-9

Niwa H (2001) Molecular mechanism to maintain stem cell renewal of ES cells *Cell Structure and Function* **26** 137-148

Oulad-Abdelghani M, Bouillet P, Decimo D, Gansmuller A, Heyberger S, Dolle P, Bronner S, Lutz Y and Chambon P (1996) Characterization of a premeiotic germ cell-specific cytoplasmic protein encoded by Stra8, a novel retinoic acid-response gene *The Journal of Cell Biology* **135** 469-477

Pesce M and Scholer HR (2000) Oct4: control of totipotency and germline determination *Molecular Reproduction and Development* **55** 452-457

Pittenger MF, Mackay AM, Beck SC, Jaiswal RK, Douglas R, Mosca JD, Moorman MA, Simonetti DW, Craig S and Marshak DR (1999) Multi-lineage potential of adult human mesenchymal stem cells *Science* **284** 143-147

Ratajczak MZ, Kucia M, Majka M, Reca R and Ratajcak J (2004) Heterogeneous populations of bone marrow stem cells: are we spotting on the same cells from the different angles? *Folia histochemica et Ccytobiologica* **42** 139-146

Reijo R, Lee TY, Salo P, Alagappan R, Brown LG, Rosenberg M, Rozen S, Jaffe T, Straus D, Hovatta O, de la Chapelle A, Silber S and Page DC (1995) Diverse spermatogenic defects in humans caused by Y chromosome deletions encompassing a novel RNA-binding protein gene *Nature Genetics* **10** 383-393

Reyes M, Dudek A, Jahagirdar B, Koodie L, Marker PH and Verfaillie CM (2002) Origin of endothelial progenitors in human postnatal bone marrow *The Journal of Clinical Investigation* **109** 337-346

Saitou M, Barton SC and Surani MA (2002) A molecular programme for the specification of germ cell fate in mice *Nature* **418** 293-300

Sato M, Kimura T, Kurokawa K, Fujita Y, Abe K, Masuhara M, Tasunaga T, Ryo A, Yamamoto M and Nakano T (2002) Identification of PGC7, a new gene expressed specifically in preimplantation embryos and germ cells *Mechanisms of Development* **113** 91–94

Schnieders F, Dörk T, Arnemann J, Vogel T, Werner M and Schmidtke J (1996) Testis-specific protein, Y-encoded (TSPY) expression in testicular tissues *Human Molecular Genetics* **5** 1801-1807

Scholer HR (1991) Octamania: the POU factors in murine development *Trends in Genetics* **7** 323-329

Scholer HR, Ruppert S, Suzuki N, Chowdhury K and Gruss P (1990) New type of POU domain in germ line-specific protein Oct4 *Nature* **344** 435-439

Toyooka Y, Tsunekawa N, Takahashi Y, Matsui Y, Satoh M and Noce T (2000) Expression and intracellular localization of mouse vasa-homologue protein during germ cell development *Mechanisms of Development* **93** 139–149

Vogt PH, Edelmann A, Kirsch S, Henegariu O, Hirschmann P, Kiesewetter F, Kohn FM, Schill WB, Farah S, Ramos C, Hartmann M, Hartschuh W, Meschede D, Behre HM, Castel A, Nieschlag E, Weidner W, Grone HJ, Jung A, Engel W and Haidl G (1996) Human Y chromosome azoospermia factors (AZF) mapped to different subregions in Yq11 *Human Molecular Genetics* **5** 933-943

Wulf GG, Viereck V, Hemmerlein B, Haase D, Vehmeyer K, Pukrop T, Glass B, Emons G and Trumper L (2004) Mesengenic progenitor cells derived from human placenta *Tissue Engineering* **10** 1136-1147

Zhao GQ and Garbers DL (2002) Male germ cell specification and differentiation *Developmental Cell* **5** 537-547

# Establishment and differentiation of human embryonic stem cell derived germ cells

*Amander T Clark*

Department of Obstetrics, Gynecology and Reproductive Sciences, Program in Development and Stem Cell Biology, University of California at San Francisco, 513 Parnassus Avenue, San Francisco, CA 94143, USA

Germ cells are absolutely essential for fertility. Aberrant germ cell development can result in abnormal gonadal function, incomplete embryogenesis and infertility, or germ cell tumors. Our understanding of the molecular regulation of normal germ cell development in mammals has progressed significantly due to the utility of the mouse as a genetic model system. However, the molecular regulation of human germ cell development is almost completely unknown due to the historical lack of a malleable model. The purpose of this review is to compare the cell-based events leading up to the specification of the germ cell lineage in both mice and humans and to discuss some of the key signaling pathways that have recently been identified, which regulate germ cell specification. In addition, the new cell-based models for differentiating germ cells from both mouse and human embryonic stem cells (ESCs) will be summarized.

## Establishment of the germ cell lineage

There is clear evidence from mouse modeling and gene knockout technology that mammalian germ cells are specified from the epiblast through inductive signaling from the extraembryonic ectoderm and endoderm. This mechanism is distinct from non-mammalian model systems such as *C. elegans*, *Drosophila*, *Xenopus* and *Xebrafish* where germ cell determinants are localized to a distinct region of the oocyte cytoplasm. In mice, the inductive events resulting in establishment of the germ cell lineage from the epiblast requires the appropriate temporal and spatial organization of the extraembryonic ectoderm, endoderm and epiblast. Therefore, for the mouse to be an appropriate model for human germ cell formation, the close apposition of these same cell lineages must also occur. However, the earliest stages of human embryo development are morphologically and structurally distinct from the mouse, particularly at the time of germ cell lineage specification (Figure 1, box). Therefore, humans may have evolved additional mechanisms in order to establish the germ cell lineage, and a human cell based model will be required to test this.

The mouse embryo at implantation (embryonic day (E) 4.5) is composed of three distinct cell lineages: trophectoderm, primitive endoderm and epiblast (Hogan *et al.*, 1994) (Figure 1A; red, yellow and blue respectively). During implantation (E4.5-E5.5), the epiblast cells begin to organize into a simple epithelium surrounding a central cavity called the proamniotic cavity. From E5.5-E6.5 the trophectoderm proliferates forming the extraembryonic ectoderm with a

E-mail: clarka@obgyn.ucsf.edu

central cavity that is continuous with the amniotic cavity. At the midline from E6.0-E6.5, where the epiblast (Figure 1A, blue) tightly interphases with the extraembryonic ectoderm (Figure 1A; red) and endoderm (Figure 1A; yellow), the germ cell lineage is specified (Hogan *et al.*, 1994) (Figure 1A; arrows). This tight opposition is critical for the specification of the first cells of the germ cell lineage which are called primordial germ cells (PGCs).

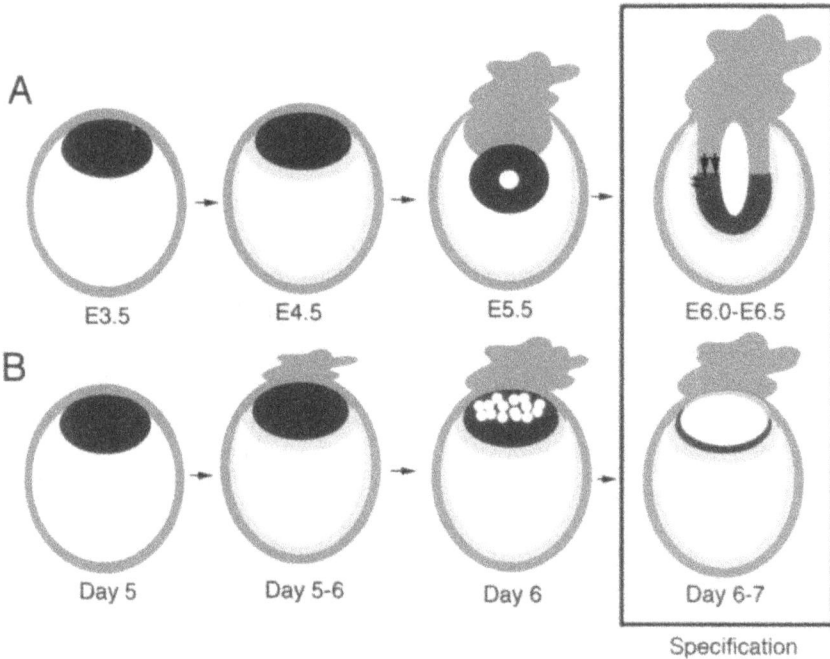

**Fig 1**. Diagrammatic representation of the cell lineages of mouse (A) and human (B) embryos necessary for germ cell specification (box). The mouse blastocyst at E3.5 is composed of ICM (blue) and trophoblast (shown in red). From E3.5-E4.5, the ICM develops into the epiblast (shown in blue) and the primitive endoderm (shown in yellow). From E4.5-E5.5, the epiblast and trophoblast cells proliferate and a central cavity called the proamniotic cavity begins to form within the epiblast cells. The extraembryonic cells including the extraembryonic ectoderm (shown in red) develop in the proximal region whereas the embryonic component (epiblast) develops distally. Together the extraembryonic ectoderm and epiblast form a hollow cylinder around the central cavity (E5.5-E6.0). At the site where the proximal epiblast, extraembryonic ectoderm and primitive endoderm triangulate (arrows), germ cell specification occurs (E6.0-E6.5). The human blastocyst at day 5 is composed of trophoblast (shown in red) and ICM (shown in blue). At implantation (day 5-6), the hypoblast (shown in yellow) differentiates from the ICM. From day 6-7 fluid filled spaces accumulate within the ICM cells creating an amniotic cavity, and a disc like layer of epithelial cells called the epiblast (shown in blue) differentiates from the ICM. The surface of the amniotic cavity is lined by amniotic epithelial cells (shown in green), which are derived from the trophoblast.

In contrast, at the time of implantation in humans (day 5 to 6 after fertilization), the human embryo is composed of two distinct cell lineages (Moore, 1977). These are the trophoblast (equivalent to the trophectoderm in mice), which begins to invade the uterine wall, and the primitive endoderm (hypoblast), which differentiates from the inner cell mass (ICM) (Moore,

1977) (Figure 1B; red and yellow respectively). Further development of the human embryo (day 6), involves the appearance of small fluid-filled spaces between the cells of the ICM (Figure 1B). These spaces coalesce to form a slit-like amniotic cavity (day 6-7) flattening the ICM cells into an epithelial disc called the epiblast (Figure 1B; blue). The amniotic cavity acquires a thin epithelial roof called the amnion, which is hypothesized to be derived from the trophoblast (Figure 1B; green). The floor of the amniotic cavity is formed by the epiblast, with the epiblast cells becoming continuous with the amniotic cells at the periphery (Moore, 1977). Given the vastly different temporal location of these embryonic cell lineages during human embryo development compared to the mouse (Figure 1, box), it is difficult to speculate where the inductive signal will arise in order to stimulate a subpopulation of epiblast cells to form the human germ cell lineage in a similar manner to the mouse. Therefore, due to these temporal and spatial differences in the key inductive cell types, a human cell based model will be necessary to dissect the molecular pathways associated with human germ cell specification and to correctly extrapolate functional information gained from mouse modeling.

## Molecular regulation of germ cell specification from the epiblast in mice

In mice, precursors of the germ cell lineage are first identified in the proximal epiblast of pre to early primitive steak embryos at E6.0-E6.5 (Lawson and Hage, 1994; Tam and Zhou, 1996) (Figure 1). These proximal epiblast cells are not fated to become PGCs until receiving inductive signals of *Bmp2*, *Bmp4* and *Bmp8b* from the extraembryonic ectoderm and primitive endoderm (Winnier *et al.*, 1995; Ying *et al.*, 2000; Ying and Zhao, 2001). Once induced, the PGCs subsequently migrate out of the primitive streak to reside in the extraembryonic mesoderm at the base of the allantois as a tight cluster of approximately 45 cells.

There are a number of loci that mark the specification of PGCs within the proximal epiblast. In particular, *interferon-induced transmembrane protein3* (*Itfm3*) is first induced in the proximal epiblast as a consequence of *Bmp4* signaling from the extraembryonic ectoderm (Saitou *et al.*, 2002). The transcriptional repressor *Blimp1* is also expressed in the proximal epiblast at this stage together with *Itfm3*, however, *Blimp1* is restricted to a single layer of epiblast cells in contact with the extraembryonic ectoderm rather than the broader expression marked by *Itfm3* (Ohinata *et al.*, 2005). *Oct4* is expressed throughout epiblast, however, its expression becomes restricted to the PGC cluster by E7.5 (Scholer *et al.*, 1990a, b; Yeom, 1996). As the newly specified PGCs are about to depart through the primitive streak, they form a tight cluster in the posterior epiblast where they express alkaline phosphatase (*AP*) (Ginsburg *et al.*, 1990). This clustering of PGCs requires cell-to-cell adhesions mediated by E-cadherin (Okamura *et al.*, 2003), which is an essential event in the specification process. Once the PGC cluster has formed at the base of the allantois, these cells continue to express *Itfm3*, *Blimp1*, *Oct4*, *AP* and E-cadherin (Ginsburg *et al.*, 1990; Scholer *et al.*, 1990a, b; Yeom, 1996; Saitou *et al.*, 2002; Okamura *et al.*, 2003; Ohinata *et al.*, 2005). However, now the first definitive germ cell marker called *stella/PGC7/Developmental pluripotency associated 3* (*Dppa3*) is expressed exclusively by the PGCs and not the surrounding somatic cells (Saitou *et al.*, 2002).

Despite this unique and specific expression of *Dppa3* in allantoic PGCs, *Dppa3* also has an interesting temporal expression in the pre-implantation embryo prior to formation of the epiblast. In particular, DPPA3 protein is expressed at high levels by unfertilized and fertilized oocytes (Sato *et al.*, 2002; Payer *et al.*, 2003). Upon fertilization, DPPA3 protein becomes enriched to the two pronuclei and then continues to be expressed throughout the cleavage stages of embryogenesis in both the nucleus and cytoplasm (Sato *et al.*, 2002). At the first lineage decision event to form trophoblast, or ICM, DPPA3 protein is clearly expressed in both cell types (Sato *et al.*, 2002). However, at the second lineage decision event from E3.5-E4.5 of mice where the ICM develops into either the primitive endoderm or epiblast, *DPPA3* is rapidly down regulated and is no longer expressed in the developing embryo only to be re-expressed in allantoic PGCs (Sato *et al.*, 2002). Therefore, unlike any of the other early pluripotency-associated-germ cell markers of PGCs (*ltfm3*, *blimp1* and *Oct4*), *Dppa3* is not expressed in the epiblast, and is only re-expressed in the definitive PGC cluster at E7.25 (Saitou *et al.*, 2002; Sato *et al.*, 2002).

Once specified and expressing *Dppa3*, the PGCs begin to migrate from the extraembryonic mesoderm and into the embryonic endoderm. This migration is mediated by dynamic changes in the expression of the *ltfm* family of genes in both the PGC and the surrounding somatic environment (Tanaka *et al.*, 2005). In particular, *ltfm1*, which is expressed by both the clustered PGCs and the surrounding allantoic mesoderm, is down regulated as the PGCs migrate from the *ltfm1* positive mesodermal environment, to the *ltfm1* negative endodermal environment (Tanaka *et al.*, 2005). A PGC specific conditional knockout of *Oct4* from E7.5 has identified a necessary role for *Oct4* in PGC survival during migration between E9.5 and E10.5 through the hindgut endoderm (Kehler *et al.*, 2004). Furthermore, the PGC-specific genes *nanos3* and *ltfm3*, which are also expressed by the migrating PGCs together with *Dppa3* and *Oct4*, have also been shown to have essential roles in mediating PGC survival (*nanos3*) and homing (*ltfm3*) during migration (Tsuda *et al.*, 2003; Tanaka *et al.*, 2005). As the PGCs enter the gonads at E10.5, the germ cells transition from being called a PGC to a gonocyte. Mouse knockout models for the germ cell specific gene *deleted in azoospermia like (dazl)* have shown variable effects on embryonic germ cell development. However, it is apparent that expression of *dazl* from E11.5 is essential for both male and female germ cell development and that this effect can be significantly modified by genetic background (Ruggiu *et al.*, 1997; Lin and Page, 2005). In the gonad, the gonocytes intimately associate with the surrounding somatic environment and begin to express the ATP dependent germ cell specific RNA helicase, *mouse vasa homologue (mvh)* (Toyooka *et al.*, 2000). From this stage, female germ cells in the ovary begin to enter meiosis I before arresting until recruited into folliculogenesis. Male germ cells in the embryonic testis proliferate before arresting in G0 of the cell cycle.

## Germ cell specification in humans

The germ cell lineage in humans is first identified as a small population of PGCs in the yolk sac of the developing embryo (McKay *et al.*, 1953; Motta *et al.*, 1997). Even though this is the first time that PGCs are actually observed in the human embryo, they are presumably induced much earlier by mechanisms which are yet to be identified. The fact that human PGCs are first localized outside of the embryo (similar to mice), most likely indicates a functional consequence in both species for preventing activation of a somatic cell program that occurs with gastrulation.

The ultrastructural features of human PGCs in the yolk sac endoderm have been determined, and are found to possess large nuclei with dispersed chromatin, prominent nucleoli and very little cytoplasm (Motta *et al.*, 1997). In addition, human PGCs are characterized by expression of the transcription factor OCT4 together with intense cell surface staining of alkaline phosphatase (Gaskell *et al.*, 2004). Remarkably, human embryonic stem cells (hESCs) also share these same ultrastructural features (Sathananthan *et al.*, 2002), and express high levels of both alkaline phosphatase and OCT4 (Thomson *et al.*, 1998).

During the 4th week of human gestation, PGCs migrate from the yolk sac and proliferate, entering the genital ridges during gestational week 6-8. Once in the gonads, the PGCs proliferate rapidly remaining connected by intercellular bridges (Motta et al., 1997; Heyn et al., 1998). In females, the PGCs differentiate into oogonia and begin the process of arresting in prophase I of meiosis during the 12th week of gestation (Motta et al., 1997). In males, PGCs form gonocytes, followed by intermediate spermatogonia and pre-spermatogonia (Gaskell et al., 2004). The pre-spermatogonia are the precursors of the diploid spermatogonial cells of the adult testis.

## Modeling human germ cell formation

Given that the human germ cell lineage is one of the first lineages to be identified at the end of gastrulation, we proposed that hESCs could be used as a model system to examine the initial stages of human PGC establishment given that hESCs had been previously shown to be capable of giving rise to all embryonic lineages as well as the trophoblast (Thomson et al., 1998; Xu et al., 2002; Clark et al., 2004). The first published hESC lines were derived from both fresh and frozen cleavage stage embryos obtained from an *in vitro* fertilization (IVF) clinic with informed consent and Institutional Review Board approval (Thomson et al., 1998). These consented embryos were cultured to the blastocyst stage, before the ICM was isolated, and cultured on a mouse embryonic fibroblast (MEF) feeder layer to derive different lines of hESCs (Thompson et al., 1998). In this first report of hESC derivation, five hESC lines were derived from fourteen independently isolated ICMs, with three hESC lines being cytogenetically XY and two lines cytogenetically XX (Thomson et al., 1998). These hESC lines constitute five of the sixty-seven cell lines listed on the National Institute of Health (NIH) Stem Cell Registry that are eligible for NIH federal funding (http://stemcells.nih.gov/research/registry). However, of the sixty-seven cell lines registered, only twenty-two are available for distribution as of May 2006, and less than 10 have been routinely used in peer-reviewed publications. Given the limited number of hESC lines available for federally funded research, additional cell lines would be useful to add genetic diversity in order to understand genetic modifiers of cell lineage differentiation, or to study the process of hESC derivation, and examine the similarity of hESCs to the ICM from which they are derived.

In order to identify whether germ cells could be differentiated from hESCs we used three cell lines available from the NIH stem cell registry. These cell lines are called WA09 (Wisconsin Alumni Research Foundation, WiCell Research Institute, Madison, WI), UC01 and UC06 (University of California, San Francisco, San Francisco, CA). To begin, we were interested in determining the expression levels of known PGC markers in the undifferentiated hESC population cultured on MEFs (Figure 2B). We found that undifferentiated hESCs expressed markers at both the mRNA and protein level of specified PGCs including the human homologue of mouse *stella/Dppa3* called STELLA RELATED, together with *c-kit* and the evolutionary conserved germ cell molecules Deleted in Azoospermia Like (DAZL), NANOS1 and PUMILIO2 (Jaruzelska et al., 2003; Moore et al., 2003; Clark et al., 2004). The fact that these PGC markers were already expressed by the undifferentiated hESCs could lend itself towards a number of hypotheses including: 1) a subpopulation of the hESC colony is differentiating towards the germ cell lineage; 2) human germ cell specification begins at or before the stage of ICM formation, and 3) the process of hESC derivation selects for a more "germ cell-like" cell type from the outgrowth of ICM cells. Differentiation of a sub population of hESCs within a colony naturally occurs in this model system (Vogel, 2003). This differentiation is traditionally hypothesized to occur at the edges of the colony, with the more pluripotent OCT4 positive cell types localized towards the center. In our work, we found that when hESCs are cultured at high-density and allowed to differentiate, a zone of alpha fetoprotein (AFP) expression is observed at the edges of the colony, particularly in regions where the colonies

begin to merge (Clark et al., 2004). AFP is a classic marker of endoderm, and is not expressed at
the protein level in undifferentiated hESC colonies (Clark et al., 2004).  In comparison, immunohistochemical analysis of undifferentiated hESC colonies cultured under ideal conditions on MEFs
revealed that the germ cell-specific markers DAZL and STELLAR are expressed throughout the
undifferentiated hESC colony at the protein level, and not within subpopulations or zones of
differentiation similar to AFP (Clark et al., 2004). Although expressed in the majority of undifferentiated hESCs, DAZL and STELLAR protein were not expressed within every cell of an undifferentiated hESC colony reflecting the heterogeneous nature of undifferentiated hESCs.  Therefore,
our data suggests that these early PGC markers were not expressed as a consequence of a subpopulation of the hESC colony differentiating towards the germ cell lineage, but instead are expressed as a feature of undifferentiated hESCs.

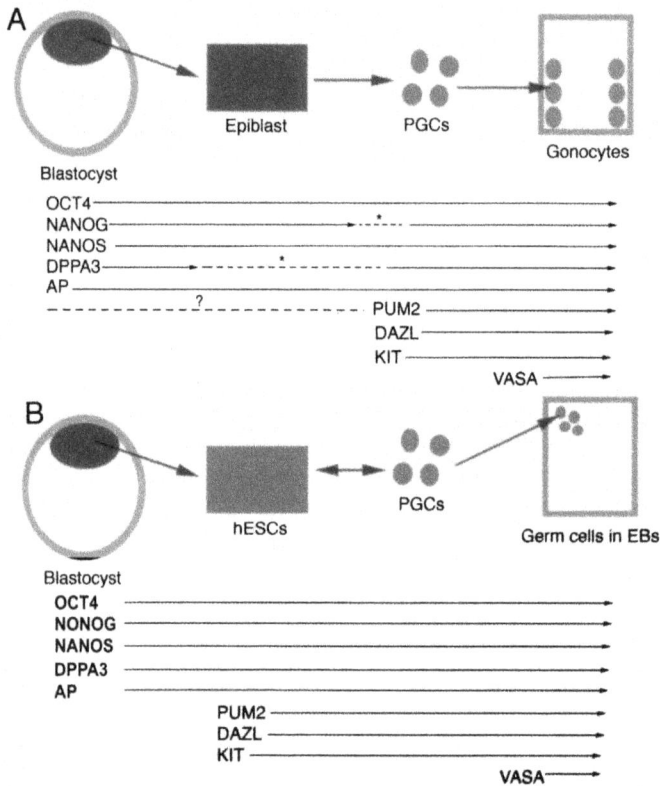

**Fig 2.** Diagrammatic representation of markers that can be used to identify the germ cell
lineage either *in vivo* (A) or from hESCs (B).  The ICM of the blastocyst expresses *OCT4,
NANOG, NANOS, DPPA3 (STELLAR)* and *AP*.  It has not yet been determined whether
human blastocysts express *PUM2* (?).  Marker expression within the human epiblast is not
known, however in the mouse, *Oct4* and *nanog* continue to be expressed in the epiblast,
whereas *Dppa3* expression is turned off until specification of PGCs (*).  *Nanog* is turned off
from the period of specification in the epiblast until resumption of PGC migration (*).
Human PGCs express *OCT4, AP, NANOS, PUM2, DAZL, DPPA3, NANOG* and *Kit*, with
VASA being subsequently expressed once the human germ cells reach the genital ridge.
These hESCs are derived from the ICM of blastocysts (B).  Undifferentiated hESCs express
markers of the ICM and germ cell lineage including *DPPA3, NANOG, OCT4, AP* and
*NANOS* together with the germ cell markers *PUM2, DAZL* and *KIT*.  With differentiation
into embryoid bodies, VASA expression is now detected.

Interestingly, we found that the evolutionary conserved germ cell marker VASA was not expressed at detectable levels at either the mRNA or protein level in the undifferentiated hESC cultures (Clark *et al.*, 2004) (Figure 2B). However, by differentiating hESCs into embryoid bodies (EBs) in suspension culture, we found that VASA, the germ cell-specific marker of the PGC to gonocyte transition was consistently elevated in sub-populations of EBs (Clark *et al.*, 2004). Analysis of these EBs using immunohistochemistry determined that VASA positive cells were generally expressed in clusters, within the EBs, as well as at the edges of EBs. VASA positive cells were not found within every EB analyzed and therefore we believe that this VASA positive population represents a very small percentage of differentiating hESCs (Clark *et al.*, 2004).

Taken together, our work was the first to show that hESCs could be used to differentiate VASA positive germ cells *in vitro*. Furthermore, we were the first to show that hESCs display a number of markers in common with the pre-VASA PGC, and that these markers are expressed throughout undifferentiated hESC colonies. We propose that hESCs have significant potential to be used as a model system to study the earliest molecular events in human germ cell development, in particular those molecular events involved in the developmental progression of germ cells from the PGC stage to the VASA positive gonocyte.

## Embryonic stem cells as a model for germ cell development

Due to their genetic malleability, mouse ESCs have been used successfully to differentiate mouse germ cells *in vitro* (Hubner *et al.*, 2003; Toyooka *et al.*, 2003; Geijsen *et al.*, 2004; Lacham-Kaplan *et al.*, 2006). This creates a unique opportunity to analyze two "equivalent" (where equivalent means derived from the ICM of both species) starting cell types for the molecular regulation of human *versus* mouse germ cell differentiation. In the mouse ESC studies, PGCs (Geijsen *et al.*, 2004), oocytes (Hubner *et al.*, 2003; Lacham-Kaplan *et al.*, 2006) and mature sperm (Toyooka *et al.*, 2003) have all been identified by marker analysis from differentiating mouse ESCs. Male PGCs were isolated by taking advantage of a cell surface marker expressed by both undifferentiated mouse ESCs and migrating PGCs (Geijsen *et al.*, 2004). This marker is called Stage Specific Embryonic Antigen 1 (SSEA1). In order to derive male PGCs, mouse ESCs were cultured as EBs for 5 days before plating onto MEFs and culturing in the presence of retanoic acid (RA) to stimulate both further somatic cell differentiation and PGC self-renewal (Geijsen *et al.*, 2004). Genetic analysis of these SSEA1 positive, RA treated cultures revealed the erasure of DNA methylation at imprinted loci, a hallmark of PGC development. Further culture of these PGCs in RA and secondary isolation with a haploid specific cell surface marker showed that a small proportion of these ESC derived germ cells could spontaneously differentiate into haploid cells (Geijsen *et al.*, 2004). However, Intra Cytoplasmic Sperm Injection (ICSI) of these presumptive haploid cells into mouse oocytes did not result in any live births (Geijsen *et al.*, 2004).

A second study to isolate male germ cells from ESCs took advantage of selectable markers driven by the germ cell specific promoter *mvh* (Toyooka *et al.*, 2003). In these studies, isolation of *mvh* positive cells after 1 day of EB culture was improved by culturing in the presence of a somatic cell line expressing *BMP4* (Toyooka *et al.*, 2003). In order to obtain mature male germ cells, the *mvh*-isolated ESC-derived germ cells were cultured together with E12.5-E13.5 male gonads before transplanting the entire mix under the host testis capsule (Toyooka *et al.*, 2003). Six to eight weeks after transplantation, mature haploid sperm could be observed in the transplants. It will be important to determine whether these haploid ESC-derived male germ cells are capable of fertilization and generation of live young. Especially given that the entire process of generating haploid germ cells *in vitro* (Geijsen *et al.*, 2004), was not capable of achieving viable embryos.

Female germ cells have been derived from mouse ESCs by two independent groups (Hubner et al., 2003; Lacham-Kaplan et al., 2006). The first group used ESCs containing an integrated germ cell-specific *Oct4* promoter transgene driving expression of green fluorescent protein (GFP) (Hubner et al., 2003). GFP positive cells were identified by day 4 of differentiation as adherent cultures, and analysis of these germ cell-specific *Oct4* isolated cells at day 7 revealed that this population expressed the later stage germ cell marker *MVH* (Hubner et al., 2003). With further differentiation, mature follicle-like structures were formed containing an oocyte and surrounding steroid producing somatic cells (Hubner et al., 2003). At later stages of culture, the ESC-derived oocytes expressed the oocyte-specific markers *Gdf9*, *Figα* and the *zona pelucida proteins* as well as markers of meiosis (Hubner et al., 2003). Further culture of these ESC derived oocytes resulted in parthenogenesis and the formation of new blastocyst-like structures (Hubner et al., 2003). It is not known whether these oocyte-like cells could be fertilized and subsequently support the generation of live young. In the most recent study, ovarian-like structures could be identified in differentiating XY ESCs after culturing in the presence of testicular cell (TC) conditioned media (Lacham-Kaplan et al., 2006). TC media was prepared by dissecting the testis of 1 day post partum male mice and culturing for 10 days. TC conditioned media was then collected every three days, and used to condition media to produce germ cells from differentiating EBs (Lacham-Kaplan et al., 2006). Ovarian structures were observed as rapidly as 72 hours after addition of TC conditioned media to day 4-5 EBs (Lacham-Kaplan et al., 2006). It is curious that testis conditioned media derived from XY males resulted in the differentiation of female oocyte-like structures from XY ESCs. It will be interesting to determine what the factors are in TC media that support the preferential formation of female gonadal structures and germ cells from male ESCs.

Taken together, it is clear that mouse ESCs can form germ cells *in vitro* by a variety of methods. However, no *in vitro* derived germ cell to date has been effectively used for fertilization and supporting embryogenesis leading to the birth of live young. This is the quintessential hallmark of a normal, functional germ cell and will most likely be achieved with an understanding of the critical interactions of the germ cell with the surrounding somatic niche cells. In particular, the studies using mouse ESCs to derive germ cell have revealed the importance of a supporting niche for germ cell differentiation to continue from the gonocyte stage. In the mouse ESC studies, this niche was created either spontaneously (Hubner et al., 2003), by co-culture and transplantation into the testis (Toyooka et al., 2003), or by culturing together with either RA (Geijsen et al., 2004) or TC conditioned media (Lacham-Kaplan et al., 2006). In our studies with hESC differentiation towards the germ cell lineage, we also observed the VASA positive presumptive germ cells within clusters rather than dispersed throughout the EBs also indicating regional preference for these cell types and potential formation of a germ cell supporting niche. However, despite this progress, further analysis of somatic-germ cell interactions will undoubtedly be necessary to understand the critical steps for differentiating functional germ cells capable of establishing the next generation. In particular, the control of germ cell self-renewal and meiosis, the establishment of sex-specific imprints and haploid germ cell differentiation all occur in the context of the gonad, and appropriate control of each phenomenon is presumably regulated by somatic cells.

## Conclusions

The initial inductive event to establish the germ cell lineage from the epiblast in mice occurs in the bilaminar embryo before gastrulation and the formation of the three embryonic lineages. The early nature of this event is reflected in the fact that mouse ESCs derived from the ICM of blastocysts express markers traditionally considered to be specific for the germ cell lineage including expression of *Dazl* and *Dppa3* while not expressing protein markers of the somatic cell lineages. Although human embryos develop in a spatially different organization to mouse embryos during the period of germ cell specification, our work has shown that hESCs also express the same conserved PGC markers suggesting that germ cell specification in humans is also one of the first lineages to develop from the epiblast. By differentiating ESCs *in vitro* as either adherent culture or EBs it is now possible to identify later stages of germ cell development in both mice and humans through the expression of VASA. Therefore hESCs, constitute the first malleable model for studying the molecular mechanisms necessary for development of the human germ cell lineage.

## References

Clark AT, Bodnar MS, Fox M, Rodriquez R, Abeyta M, Firpo M and Pera R (2004) Spontaneous differentiation of germ cells from human embryonic stem cells *in vitro Human Molecular Genetics* **13** 727-739

Gaskell TL, Esnal A, Robinson LL, Anderson RA and Saunders PT (2004) Immunohistochemical profiling of germ cells within the human fetal testis: identification of three subpopulations *Biology of Reproduction* **71** 2012-2021

Geijsen N, Horoschak M, Kim K, Grilbnau J, Eggan K and Daley G (2004) Derivation of embryonic germ cells and male gametes from embryonic stem cells *Nature* **427** 148-154

Ginsburg M, Snow MHL and McLaren A (1990) Primordial germ cells in the mouse embryo during gastrulation *Development* **110** 521-528

Heyn R, Makabe S and Motta P (1998) Ultrastructural dynamics of human testicular cords from 6-16 weeks of embryo development. Study by transmission and high resolution scanning electron microscopy *Italian Journal of Anatomy and Embryology = Archivio italiano di anatomia ed embriologia* **103**: 17-29

Hogan B, Beddington R, Costantini F and Lacy E (1994) *Manipulating the mouse embryo: a laboratory manual* Cold Spring Harbor Laboratory Press, Plainview

Hubner K, Fuhrmann G, Christenson L, Kehler J, Reinbold R, De La Fuente R, Wood J, Strauss III J, Boiani M and Scholer H (2003) Derivation of oocytes from mouse embryonic stem cells *Science* **300** 1251-1256

Jaruzelska J, Kotecki M, Kamila K, Spik A and Pera RAR (2003) Conservation of the Pumilio Nanos complex in germ cells from flies to humans *Developmental Genes and Evolution* **213** 120-126

Kehler J, Tolkunova E, Koschorz B, Pesce M, Gentile L, Boiani M, Lomeli H, Nagy A, McLaughlin K, Scholer HR and Tomilin A (2004) OCT4 is required for primordial germ cell survival *European Molecular Biology Organization Reports* **5** 1078-1083

Lacham-Kaplan O, Chy H and Trounson A (2006) Testicular cell conditioned medium supports differentiation of embryonic stem cells into ovarian structures containing oocytes *Stem Cells* **24** 266-273

Lawson KA and Hage WJ (1994) Clonal analysis of the origin of primordial germ cells in the mouse *Ciba Foundation Symposium* **182** 68-91

Lin Y and Page D (2005) Dazl deficiency leads to embryonic arrest of germ cell development in XY C57BL/6 mice *Developmental Biology* **288** 309-316

McKay DG, Hertig AT, Adams EC and Danziger S (1953) Histochemical observations on the germ cells of human embryos *The Anatomical Record* **117** 201-219

Moore FL, Jaruzelska J, Fox MS, Urano J, Firpo MT, Turek PJ, Dorfman DM and Pera RAR (2003) Human Pumilio-2 is expressed in embryonic stem cells and germ cells and interacts with DAZ (Deleted in Azoospermia) and DAZ-Like proteins *Proceedings of the National Academy of Sciences USA* **100** 538-543

Moore KL (1977) *The Developing Human* W.B. Saunders Company, Philadelphia

Motta P, Makabe S and Nottola S (1997) The ultrastructure of human reproduction. The natural history of the female germ cell: origin, migration and differentiation inside the developing ovary *Human Reproduction Update* **3** 281-295

Ohinata Y, Payer B, O'Carroll D, Ancelin K, Ono Y, Sano M, Barton S, Obukhanych T, Nussenzweig M, Tarakhovsky A, Saitou M and Surani MA (2005) Blimp1 is a critical determinant of the germ cell lineage in mice *Nature* **436** 207-213

Okamura D, Kimura T, Nakano T and Matsui Y (2003) Cadherin mediated cell interaction regulates germ cell determination in mice *Development* **130** 6423-6430

Payer B, Saitou M, Barton S, Thresher R, Dixon J, Zahn D, Colledge W, Carlton M, Nakano T and Surani M (2003) Stella is a maternal effect gene required for normal early development in mice *Current Biology* **13** 2110-2117

Ruggiu M, Speed R, Taggart M, McKay SJ, Kilanowski F, Saunders P, Dorin J and Cooke H (1997) The mouse Dazla gene encodes a cytoplasmic protein essential for gametogenesis *Nature* **389** 73-77

Saitou M, Barton SC and Surani MA (2002) A molecular programme for the specification of germ cell fate in mice *Nature* **418** 293-300

Sathananthan H, Pera M and Trounson A (2002) The fine structure of human embryonic stem cells *Reproductive Biomedicine Online* 1:56-61

Sato M, Kimura T, Kurokawa K, Fujita Y, Abe K, Masuhara M, Yasunaga T, Ryo A, Yamamoto M and Nakano T (2002) Identification of PGC7, a new gene expressed specifically in preimplantation embryo's and germ cells *Mechanisms of Development* 113 91-94

Scholer H, Dressler G, Balling R, Rohdewohld H and Gruss P (1990a) Oct-4: a germ line specific transcription factor mapping to the mouse t-complex *European Molecular Biology Organisation Journal* 9 2185-2195

Scholer H, Ruppert S, Suzuki N, Chowdhury K and Gruss P (1990b) New type of POU domain in germ line-specific protein Oct-4 *Nature* 344 435-439

Tam PPL and Zhou SX (1996) The allocation of epiblast cells to ectoderm and germline lineages is influenced by the position of the cells in the gastrulating mouse embryo *Developmental Biology* 178 124-132

Tanaka S, Yamaguchi Y, Tsoi B, Lickert H and Tam P (2005) IFITM/Mil/Fragilis Family Proteins IFITM1 and IFITM3 play distinct roles in mouse primordial germ cell homing and repulsion *Developmental Cell* 9 745-756

Thomson JA, Itskovitz-Eldor J, Shapiro SS, Waknitz MA, Swiergiel JJ, Marshall VS and Jones JM (1998) Embryonic stem cell lines derived from human blastocysts *Science* 282 1145-1147

Toyooka Y, Tsunekawa N, Akasu R and Noce T (2003) Embryonic stem cells can form germ cells *in vitro* Proceedings of the National Academy of Sciences USA 100 11457-11462

Toyooka Y, Tsunekawa N, Takahashi Y, Matsui Y, Satoh M and Noce T (2000) Expression and intracellular localization of mouse Vasa-homologue protein during germ cell development *Mechanisms of Development* 93 139-149

Tsuda M, Sasaoka Y, Kiso M, Abe K, Haraguchi S, Kobayashi S and Saga Y (2003) Conserved role of nanos proteins in germ cell development *Science* 301 1239-1241

Vogel G (2003). 'Stemness' genes still elusive *Science* 302 371

Winnier G, Blessing M, Labosky P and Hogan B (1995) Bone morphogenetic protein 4 is required for mesoderm formation and patterning in the mouse *Genes & Development* 9 2105-2116

Xu R, Chen X, Li D, Li R, Addicks G, Glennon C, Zwaka J and Thompson J (2002) BMP4 initiates human embryonic stem cell differentiation to trophoblast *Nature Biotechnology* 20 1261-1264

Yeom Y (1996) Germ line regulatory element Oct-4 specific for the totipotent cycle of embryonal cells *Development* 122 881-894

Ying Y and Zhao GQ (2001) Cooperation of endoderm-derived BMP2 and extraembryonic ectoderm-derived BMP4 in primordial germ cell generation in the mouse *Developmental Biology* 232 484-492

Ying Y, Liu XM, Marble A, Lawson KA and Zhao GQ (2000) Requirement of Bmp8b for the generation of primordial germ cells in the mouse *Molecular Endocrinology* 14 1053-1063

# Validation of a testis specific serine/threonine kinase [TSSK] family and the substrate of TSSK1 & 2, TSKS, as contraceptive targets

B Xu, Z Hao, KN Jha, L Digilio, C Urekar, YH Kim,
S Pulido, C J Flickinger and JC Herr*

*Center for Research in Contraceptive and Reproductive Health (CRCRH),
Department of Cell Biology, University of Virginia, Charlottesville, VA 22908, USA*

A family of testis specific serine/threonine kinases, TSSK1–4 and SSTK, in addition to the substrate of TSSK1 & 2, TSKS, have been studied during the past several years in our laboratory. This paper will provide a general background on these kinases through review of pertinent literature and then will summarize data from our laboratory germane to evaluating these kinases as candidate targets for future development of small molecule kinase inhibitors that may serve to regulate male fertility. Bio-informatic and structural analyses of human TSSK1–4 and SSTK indicate that these kinases constitute a unique subfamily belonging to the AMPK branch on the human kinome tree. Expression studies showed that all five kinases and the TSKS substrate are testis abundant, if not strictly testis specific, indicating that tissue specific contraceptive targeting is possible. *In situ* hybridization further confirmed that mouse *TSSK2*, *SSTK* and *TSKS* are post-meiotic in their expression patterns, a finding that makes them possible targets of reversible contraceptive intervention by preserving spermatogonia and spermatocytes. Our laboratory detected TSSK2, TSKS and SSTK proteins in mature spermatozoa for the first time. TSKS was localized to the centrioles of human spermatozoa, while TSSK2 was observed in the sperm neck, equatorial segment and mid-piece of the sperm tail, and SSTK was localized in the equatorial segment. The interaction and binding between human TSSK2 and TSKS was confirmed by several methods: this substrate and enzyme interaction offers a particularly interesting opportunity for drug design. *In vitro* kinase assay showed phosphorylation of TSKS by TSSK2. The TSKS phosphopeptide, HGLSPATPIQGCSGPPGS*PEEPPR, was identified by IMAC-LC-FTMS, with serine 285 being phosphorylated (representend by asterisk). These results provide a rationale for high-throughput screening of inhibitors for TSKS phosphorylation and further studies of members of this kinase family as targets for both male contraception and intra-vaginal spermicides.

---

*Corresponding author
E-mail: jch7k@virginia.edu

# General background

*Discovery of the testis-specific serine/threonine kinase [TSSK] family and the substrate of TSSK1 & 2, TSKS*

This family of kinases was discovered relatively recently and its members have generated interest because of their pattern of testicular expression and the testis specific expression of their substrate, TSKS. In a search for new kinases, the first member of the TSSK family, mouse *TSSK1*, was cloned by a PCR based strategy using degenerate oligonucleotide primers annealed to the most conserved motifs within the protein kinase catalytic domain (Bielke *et al.*, 1994). Northern blot analyses of 16 murine tissues revealed a signal band only in the testis. Subsequently, low stringency colony hybridization with the *TSSK1* gene as the probe was used to clone mouse *TSSK2* (Kueng *et al.*, 1997). A yeast two-hybrid screen led to identification of the N-terminal region of a murine molecule that formed complexes with TSSK1 and TSSK2 in the mouse testis. This gene was then named *TSKS* for *TSSK* substrate (Kueng *et al.*, 1997). All three genes were shown to be expressed post-meiotically during spermiogenesis in the mouse. Murine and human *TSSK3* were cloned about the same time by Zuercher *et al.* (2000) and by our laboratory (Visconti *et al.*, 2001) using similar PCR based methods. Human homologues of *TSSK1-3* and *TSKS* were PCR amplified by Hao *et al.* (2004), and *TSSK4* and *SSTK* (small serine/threonine kinase) were discovered by Blast searches of human and mouse genomes (Hao *et al.*, 2004; Chen *et al.*, 2005; Spiridonov *et al.*, 2005). A central transcriptional factor, cAMP responsive element binding protein (CREB), was identified as a TSSK4-interacting protein via a yeast two-hybrid analysis (Chen *et al.*, 2005). Recently, targeted deletion of the *SSTK* gene was achieved in mice, resulting in male sterility (Spiridonov *et al.*, 2005). A defect in DNA condensation in *SSTK* null mutants indicated that SSTK was required for proper post-meiotic chromatin remodeling and male fertility. Interestingly, SSTK expressed in human cell lines specifically interacted with three heat shock proteins, and this association appeared to be required for SSTK kinase activity. The active SSTK phosphorylated several histones *in vitro* (Spiridonov *et al.*, 2005). Variations in the names of the members of the *TSSK* family in the literature can lead to confusion. Here, we name each member based on the human kinome poster that accompanies the article by Manning *et al.* published in *Science* (Manning *et al.*, 2002). Table 1 summarizes this nomenclature for the *TSSK* family.

The precise biological roles of the members of the *TSSK* family and particularly the functions of the substrate of TSSK1 & 2, TSKS, are presently unknown. As we show below, mRNAs of *TSSK2*, *SSTK* and *TSKS* were detected in post-meiotic spermatids by *in situ* hybridization. TSSK2 and TSKS proteins localized to spermatids and persisted in mature spermatozoa. SSTK proteins were observed in the head of the elongating spermatids (Spiridonov *et al.*, 2005) and the equatorial segment of mature spermatozoa. Therefore, the available data on protein localization and patterns of mRNA expression indicate that TSSK kinases and TSKS are transcribed and translated in haploid spermatids and are present in mature spermatozoa, suggesting that they are important to spermiogenesis and possibly may participate in post-testicular events.

*Kinases and spermiogenesis*

Spermiogenesis is the post-meiotic phase of spermatogenesis, in which the haploid spermatids differentiate into mature spermatozoa. Dramatic morphological changes occur during spermiogenesis, including formation of the acrosome and its contents, condensation and reorganization of the chromatin, elongation and species-specific shaping of the cell nucleus and assembly of the flagellum (Sharpe, 1994). These events result from changes in both gene transcription (Hecht, 1988) and translation (Hake *et al.*, 1990) that occur during this developmental period.

Previous studies of the kinases expressed during spermiogenesis are limited. A kinase cascade, ERK/MAPK/MEK, demonstrated to play regulatory roles in meiosis, was proven recently also to play a key role in spermiogenesis and sperm function (Lu *et al.*, 1999; Sun *et al.*, 1999; Inselman and Handel, 2004). This kinase cascade included mitogen-activated protein kinases (MAPK), extracellular signal-regulated protein kinases (ERK), MAPK activating kinase MEK, and other upstream and downstream kinases such as c-mos, rsk, and c-kit. Interestingly, many kinase genes detected in germ line cells including *c-abl*, *c-mos*, *pim-1* and *ras* have testis-specific transcripts, which differ from their forms in somatic cells, suggesting that specificity in transcription or processing of certain kinase genes may occur in haploid male germ cells (Sorrentino *et al.*, 1988; 1991; Iwaoki *et al.*, 1993). Recently, the knockout technique was employed to study kinases in germ cells. The c-kit receptor tyrosine kinase (which probably acts via MAPK) is required for the migration and proliferation of mouse primordial germ cells (Loveland and Schlatt, 1997). Mice devoid of fer protein-tyrosine kinase activity are viable and fertile but display reduced cortactin phosphorylation (Craig *et al.*, 2001). On the other hand, a testicular germ cell-associated serine/threonine kinase, Mak, which is expressed during late prophase I of meiosis and after meiosis (Matsushime *et al.*, 1990; Jinno *et al.*, 1993), is dispensable for sperm formation because no defects in phenotype occurred in *Mak* (-/-) null mice (Shinkai *et al.*, 2002). With regard to selective tissue expression, TESK1 (Toshima *et al.*, 1998), which is structurally similar to LIMK (LIM motif containing protein kinase), is one of the few testis-specific kinases predominantly expressed in testicular germ cells, in addition to the TSSK family. The functions of the majority of kinases found in the testis, however, are not known.

*Phosphorylation is important for capacitation and the acrosome reaction*

After leaving the testis, mammalian spermatozoa undergo maturation during epididymal transit and gain fertilization capacity in the female reproductive tract through the process of capacitation. Capacitation involves molecular changes in both the sperm head and tail that lead to the release of the acrosomal content (acrosome reaction) and the whiplash-like sperm tail motion (hyperactivation). Capacitation and acrosomal reaction can be accomplished *in vitro* using cauda epididymal or ejaculated spermatozoa incubated in defined media (Yanagimachi, 1994).

Protein phosphorylation certainly plays roles in the regulation of capacitation, acrosomal reaction and sperm motility. It could even be argued that reliance of spermatozoa on protein phosphorylation as a means of altering their function is greater than in many other types of cells, since mature spermatozoa are highly differentiated but transcriptionally and translationally inactive (Urner and Sakkas, 2003). However, to date, only the crucial role of the cAMP-PKA kinase pathway in capacitation has been well documented. Briefly, cholesterol efflux leading to an increase in intracellular $Ca^{2+}$, $HCO_3^-$ and $H_2O_2$, results in the activation of membrane-bound adenyl cyclase (AC) to produce cAMP. The cAMP activates protein kinase A (PKA) to phosphorylate proteins on serine (PSP), leading to the tyrosine phosphorylation of a subset of proteins (Visconti et al., 1995a,b; Leclerc et al., 1996; Galantino-Homer et al., 1997; Visconti and Kopf, 1998; Dragileva et al., 1999; Osheroff et al., 1999; Visconti et al., 1999; Dorval et al., 2002). Meanwhile, PKA also activates the $Ca^{2+}$ channel of the outer acrosomal membrane, involving the regulation of intercellular $Ca^{2+}$ during capacitation and acrosomal reaction (Breitbart, 2003).

Recently, additional kinases and phosphorylated proteins have been detected in mammalian spermatozoa. Two members of the A-kinase anchoring protein family, AKAP4 and AKAP3, which are among the most abundant proteins in the sperm flagellum (Fulcher et al., 1995; Miki and Eddy, 1998; Turner et al., 1998; Mandal et al., 1999), undergo capacitation-related phosphorylation on both tyrosine and serine/threonine residues, suggesting that both tyrosine and serine/threonine kinases participate in sperm motility regulation (Ficarro et al., 2003). *Akap4* has recently been deleted in mice (Miki et al., 2002) resulting in spermatozoa that fail to show progressive motility and are infertile. Another fascinating kinase substrate phosphorylated during capacitation is the calcium binding protein, CABYR (Naaby-Hansen et al., 2002). CABYR is also tyrosine and serine/threonine phosphorylated during capacitation, localizes to the fibrous sheath and appears to gain calcium binding ability as a result of phosphorylation (Ficarro et al., 2003). However, at present it is not clear which kinases are responsible for the capacitation dependent phosphorylation of serine/threonine and tyrosine in the AKAPs and CABYR. Calcium/calmodulin-dependent protein kinase IV was detected in human spermatozoa, where it is involved in regulation of motility (Marin-Briggiler et al., 2005). A signalling kinase, glycogen synthase kinase-3 (GSK-3), and two upstream signalling proteins, protein kinase B (PKB; also known as cAkt) and phosphoinositide 3-kinase (PI3-kinase), were also found in spermatozoa. GSK-3 is regulated by serine and tyrosine phosphorylation, and its phosphorylation is correlated with sperm motililty (Somanath et al., 2004). A recent study indicated that PKC, PKA, PTK, PI3K, Akt and the ERK pathways are all involved in the protein phosphorylation associated with the acrosomal reaction, indicating that this process is very complex (Liguori et al., 2005). In our studies described below, we show that TSSK2, TSKS and SSTK are present in mature spermatozoa. Thus, studies of the signal transductional pathway(s) in which the TSSKs are involved may lead to elucidation of kinase networks important for capacitation and the acrosomal reaction.

### Critical evaluation of the TSSK kinase family and TSKS as contraceptive targets

Over the years, alteration in the function of kinases by inhibitors has been applied in a variety of fields. Pharmaceutical companies have gained considerable experience in kinase inhibitor design and consider kinases as "drugable" targets in general (Parang and Sun, 2004). In our laboratory, the TSSK family and the substrate of TSSK1 & 2, TSKS, have been evaluated as candidate contraceptive targets with consideration given to male contraception as well as intra-vaginal spermicidal applications.

*TSSK1-4 and SSTK constitute a testis specific serine/threonine kinase subfamily*

Table 1 summarizes the chromosome localization and NCBI accession number for each member of the *TSSK* family. To date five human *TSSKs* and a *TSSK* pseudogene have been cloned. The pseudogene, designated as *TSSK1*-pseudo, lies 3 kb upstream of *TSSK2* on chromosome 22 and contains a deletion in the kinase domain but is otherwise identical to *TSSK1* on chromosome 5. There are 6 *TSSK* genes in the mouse genome, including a putative *TSSK5* supported by one EST sequence. Mouse and human *TSSK1* and *TSSK2* are intronless genes, implying that they might have been duplicated from other *TSSKs* by a retroviral mechanism during evolution.

TSSK1-4 and SSTK all contain the conserved sequences of serine/threonine kinases without any amino acid substitutions and therefore all five are predicted to be active kinases. TSSK1 and TSSK2 have greatest similarity, and form a clad by virtue of having a kinase domain which is similar to that of SSTK and TSSK3 except for being longer in the carboxyl terminus. TSSK3 and SSTK are among the smallest known kinases and the kinase domain extends from the N to the C terminus. TSSK4, on the other hand, has unique N and C termini. The phylogenetic tree from the human kinome poster (Manning *et al.,* 2002), which depicts the relationships between kinases, indicates that TSSK1-4 and SSTK constitute a unique kinase subfamily. This suggests that specific inhibitors can be developed which would inhibit these kinases by interacting with domains that are present only in the kinases of the TSSK family.

**Table 1.** The mammalian TSSKs in the genome

| Gene name (NCBI accession no.) mouse | Gene name (NCBI accession no.) human | Chromosome mouse | Chromosome human | Substrate |
|---|---|---|---|---|
| *STK22A* (NM_009435) | *TSSK1* (AY028964) | 16 10.4cM | 5q22.2 | TSKS |
| - | *TSSK1*-pseudo | - | 22q11.21 | - |
| *STK22B* (NM_009436) | *TSSK2* (AF362953) | 16 10.4cM | 22q11.21 | TSKS |
| *STK22C* (NM_009436) | *TSSK3* (NM_052841) | 4 D2.3 | 1p35.1 | - |
| *SSTK* (NM_032004) | *SSTK* (AF348007) | 8 C1 | 10p13.11 | Histone |
| *TSSK4* (NM_027673) | *TSSK4* (NM_032037) | 14C1 | 14q11.2 | CREB |
| Putative *TSSK5* (only one EST found) | - | 15D3 | - | - |

Testis-specific expression of TSSK1-3, SSTK and TSKS

Using full-length cDNA probes for each gene, *TSSK1 & 2, TSKS,* were detected only in testis among 8 human tissues, although *TSSK1* displayed two transcripts of 1.7 and 1.3 kb (the 1.3kb band may represent the *TSSK1*-pseudogene mRNA). Similarly, membranes dotted with RNA from 76 human tissues were hybridized to the same *TSSK1, TSSK2* and *TSKS* probes, the hybridization signals were detected only from the testis sites (Hao *et al.,* 2004). In an eight human tissue Northern blot, two *TSSK3* transcripts, approximately 1.2 kb and 1.8 kb, were found only in the testis, suggesting alternatively spliced transcripts (Visconti *et al.,* 2001). An eight mouse tissue Northern blot revealed a single 1.3 kb band of *SSTK* only in the testis (unpublished data). These data indicated that *TSSK1-3, SSTK* and *TSKS* expression were abundant in the testis but not detectable in other tissues, suggesting that the *TSSK* subfamily is testis-specific. Thus tissue specific intervention in the functions of these molecules may be possible.

Real-time PCR analyses of human *TSSK1-3* and *SSTK* in various organs were conducted (Hao *et al.,* 2004). A cDNA panel of 15 human tissues was purchased from Clontech Inc. and single strand cDNA from each was subjected to PCR amplification using primer pairs specific to *TSSK1-3* and *SSTK.* Threshold cycles of *TSSK1-3* and *SSTK* in each tissue were obtained and normalized with *GAPDH* levels. Relative mRNA levels in all tissue samples were expressed as a percentage of RNA found in the testis. The results are in general agreement with the Northern and RNA dot blot analyses, except that *TSSK1* was also found in pancreas (27% of testis level) and *TSSK2* was present in the heart (18%), brain (17%) and placenta (25%) (Hao *et al.,* 2004). The low level of expression of mRNA for *TSSKs* in other tissues besides testis may or may not be significant, since they may be subject to differential translation, and/or the simultaneous presence of substrates may be required for their action.

The relative abundance of the different *TSSKs* mRNA in the human testis were analyzed by real-time PCR. As shown in Fig. 1, *TSSK1* and *TSSK2* transcripts are approximately 10 times more abundant in human testis than those for *TSSK3* and *SSTK* (unpublished data). This observation is in part responsible for subsequently focusing our attention on *TSSK1* and *TSSK2.*

Tissue expression profiles for *TSSK* family members were also analyzed based on the UniGene EST database. Data from this source revealed two main points. 1) In the mouse, *TSSKs* expression are restricted to the testis with only a few exceptions (some low levels of *TSSK2* in colon and *TSSK4* in thymus) but a much broader range of human tissues appear to express *TSSKs.* Similarly, *TSKS* is testis-specific in the mouse (all of the ESTs being from testis) but ESTs for human *TSKS* were noted in placenta, pancreas and brain, *albeit* at a lower incidence than in testis. 2) In human, expression of *TSSK1* and *TSSK2* is more restricted to testis than *TSSK3* and *TSSK4.* Only human *TSSK1* and *TSSK2* were defined as "Expression Restricted to Testis" by the Unigene EST database (that is, ESTs from testis contributed more than half of the EST frequency). In contrast, *TSSK3* was defined as "Expression Restricted to Tongue", and no definition was given to *TSSK4* since it showed no dominant tissue specificity. The disagreement between Northern & dot blotting, real-time PCR and Unigene analyses may be due to the sensitivity of each method. Since *TSSK1* and *TSSK2* are intronless genes, DNA contamination may be responsible for some false data. Our

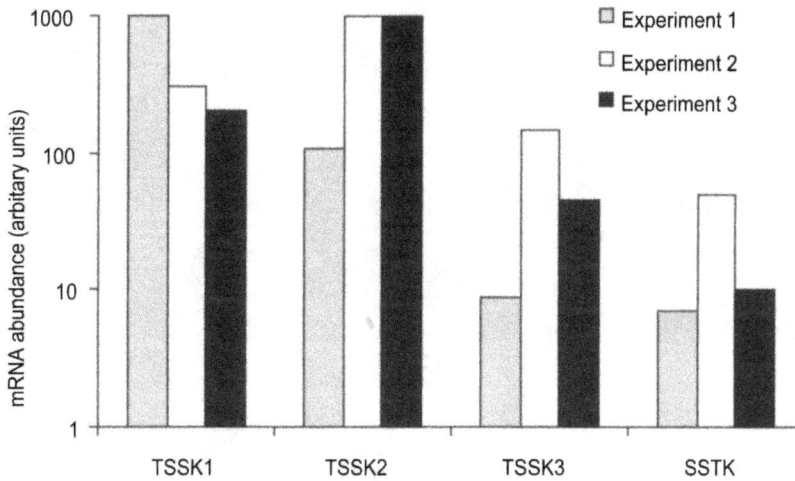

**Fig. 1** Real-time PCR analyses of the relative abundance of *TSSK1-3* and *SSTK* mRNA in the human testis. The abundance of each kinase was normalized with the most abundant transcript in each experiment and multiplied by 1000. Relative abundance was expressed as an arbitrary unit. *TSSK1* and *TSSK2* transcripts are approximately 10 times more abundant in human testis than those for *TSSK3* and *SSTK*.

Northern & dot blotting and real-time PCR data showing predominant testis expression of *TSSK*s are more consistent with the data published by other research groups (Bielke *et al.*, 1994; Kueng *et al.*, 1997; Zuercher *et al.*, 2000; Visconti *et al.*, 2001; Hao *et al.*, 2004; Chen *et al.*, 2005; Spiridonov *et al.*, 2005) than the data extracted from the Unigene database. Moreover, the protein expression of TSSKs should be analyzed in the tissues in which ESTs have been detected to assess the extent to which the mRNAs are translated in different sites.

*Post-meiotic expression of SSTK, TSSK2 and TSKS*

To analyze the onset of *SSTK* mRNA transcription in the procession of testicular cells, *in situ* hybridization of mouse transcripts was carried out using radiolabeled cRNA. Seminiferous tubules hybridized with antisense *SSTK* and viewed in dark field at X100 and X200 magnifications (Fig. 2: panels 1 and 2, respectively) show the presence of *SSTK* transcripts in cells adjacent to the lumens of seminiferous tubules (unpublished observations). Higher magnification (X400) views of seminiferous tubules hybridized with antisense *SSTK* are shown in both bright-field with hematoxylin and eosin staining (Fig. 2: panel 3) and in dark field (panel 4). *SSTK* transcripts are expressed mainly in the post-meiotic spermatids, while labeling of the spermatogonia and the primary spermatocytes is not significantly higher than the background. The control labeling with sense *SSTK*

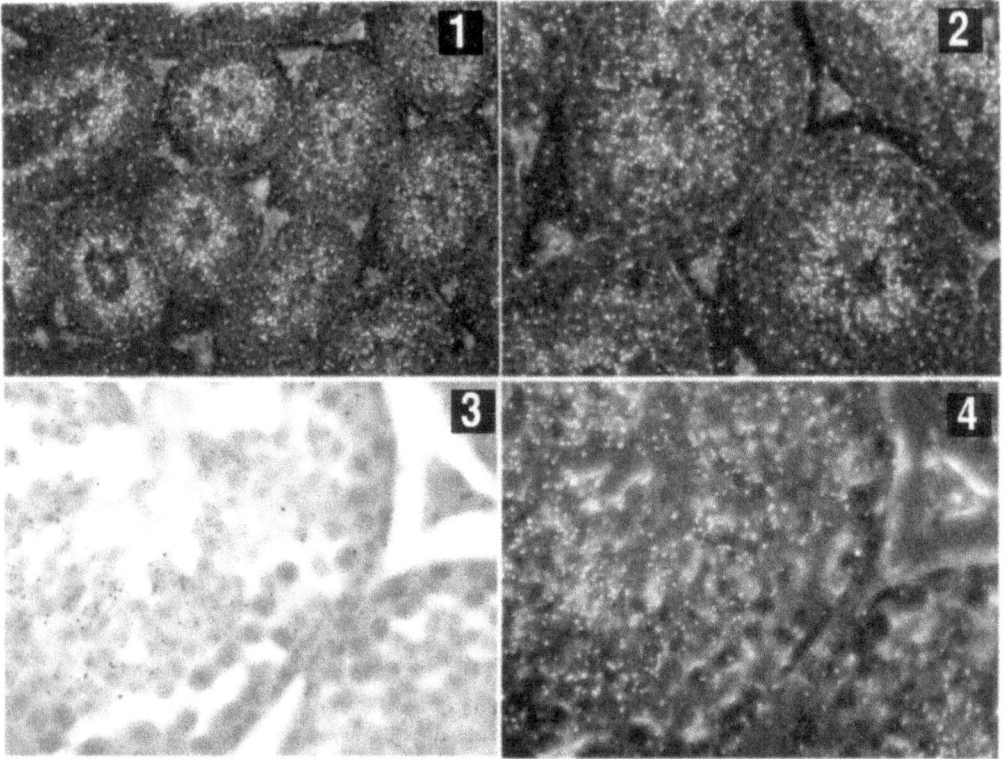

**Fig. 2** Analyses of *SSTK* mRNA expression in adult mouse testis. *In situ* hybridization of *SSTK* transcripts was carried out using radiolabeled mouse *SSTK* cRNA. The low-magnification (X100, X200) views of seminiferous tubules hybridized with antisense *SSTK* (panels 1 and 2) are shown in dark field, the higher magnification (X400) views of seminiferous tubules hybridized with antisense *SSTK* are shown in both bright-field (panel 3), and dark field (panel 4). *SSTK* transcripts are expressed mainly in the post-meiotic spermatids, while labeling of the spermatogonia and the primary spermatocytes is not significantly higher than the background.

revealed no signal above the low background (data not shown). Similar post-meiotic expression patterns were observed for *TSSK2* and *TSKS* mRNA in mouse testis (data not shown). Post-meiotic expression of *SSTK*, *TSSK2* and *TSKS* offers the possibility that a reversible contraceptive agent that targets these proteins might act selectively on spermatids while preserving spermatogonia and spermatocytes.

*Subcellular localization of TSKS, TSSK2 and SSTK in ejaculated human sperm*

Previous studies reported the TSKS protein in testis, but failed to detect it in mature sperm (Kueng *et al.*, 1997; Hao *et al.*, 2004). To study the distribution of TSKS protein further, we generated polyclonal antibodies against human TSKS (unpublished data). The resulting anti-TSKS antibody bound a single 65 kDa band in Western blots of both human testis and sperm proteins (data not shown). Meanwhile, anti-TSSK2 antibody detected a 41 kDa protein in human spermatozoa (Hao et al., 2004). An affinity purified anti-peptide antibody against human and mouse SSTK detected the SSTK protein (30 kDa) in mouse spermatozoa and a 20 kDa possible breakdown product of SSTK in human sperm protein extract by Western analyses (data not shown).

Immunofluorescent studies of the subcellular localization of TSKS, TSSK2 and SSTK were performed on ejaculated human spermatozoa (unpublished results). As shown in Fig. 3, panel 1, with the anti-TSKS antibody, signal was observed predominantly as two very small immunofluorescent dots (green color, indicated by arrows) centrally located at the base of the sperm head. After *in vitro* capacitation and acrosome reaction, this immunofluorescence persisted. A similar pattern of paired dot-like TSKS staining was observed in nearly all of the human spermatozoa studied. Significantly, staining for TSKS protein did not co-localize with the SPANX (human sperm protein associated with the nucleus on the X chromosome) or calreticulin proteins (data not shown), which are markers for the redundant nuclear envelope and calreticulin-containing vesicles, respectively, in the cytoplasmic droplet of the sperm neck (Naaby-Hansen et al., 2001; Westbrook et al., 2001). The staining of two dots in the neck suggested instead that TSKS may be localized to the centrioles of human spermatozoa. Co-staining of γ-tubulin with that produced by the TSKS antibody supports the hypothesis that TSKS is localized to the centrioles (data not shown) and suggests that TSKS plays a role in the formation of the sperm flagellum.

Immunofluorescence localization of TSSK2 in human spermatozoa is shown in Fig. 3, panel 3 (unpublished results). TSSK2 localized to the sperm neck, midpiece of spermatozoa tail, and most intensely, to the equatorial segment. The broader localization of TSSK2 compared to TSKS implies that additional substrates for the enzyme may exist. After *in vitro* capacitation and *in vitro* induced acrosomal reaction, the staining pattern for TSSK2 was unchanged. As reviewed above, this localization suggests that TSSK2 may be involved in phosphorylation events related to capacitation and the acrosome reaction.

Immunofluorescent staining of SSTK in human spermatozoa was performed using affinity purified anti-SSTK antibody. As shown in Fig. 3, panel 5, SSTK is localized to the equatorial segment of human spermatozoa, which might reflect a role for SSTK during fertilization.

## TSSK2 interacts with TSKS

Human TSSK2 open reading frame (ORF) was fused with the GAL4 DNA binding domain in pGBKT7 two-hybrid vector (pGBK-TSSK2) and human TSKS ORF was fused with the GAL4 DNA activation domain in pGADT7 two-hybrid vector (pGAD-TSKS). Yeast host strain AH109 was transformed with this pair of plasmids and an additional two pairs of control plasmids: pGAD, pGBK-TSSK2 (negative control) and pGAD-lgT [SV40 Large T antigen], pGBK-p53 (positive control). Yeast clones harboring each pair of plasmids were cultured in complete drop out medium lacking both leucine and tryptophan to OD$_{600}$

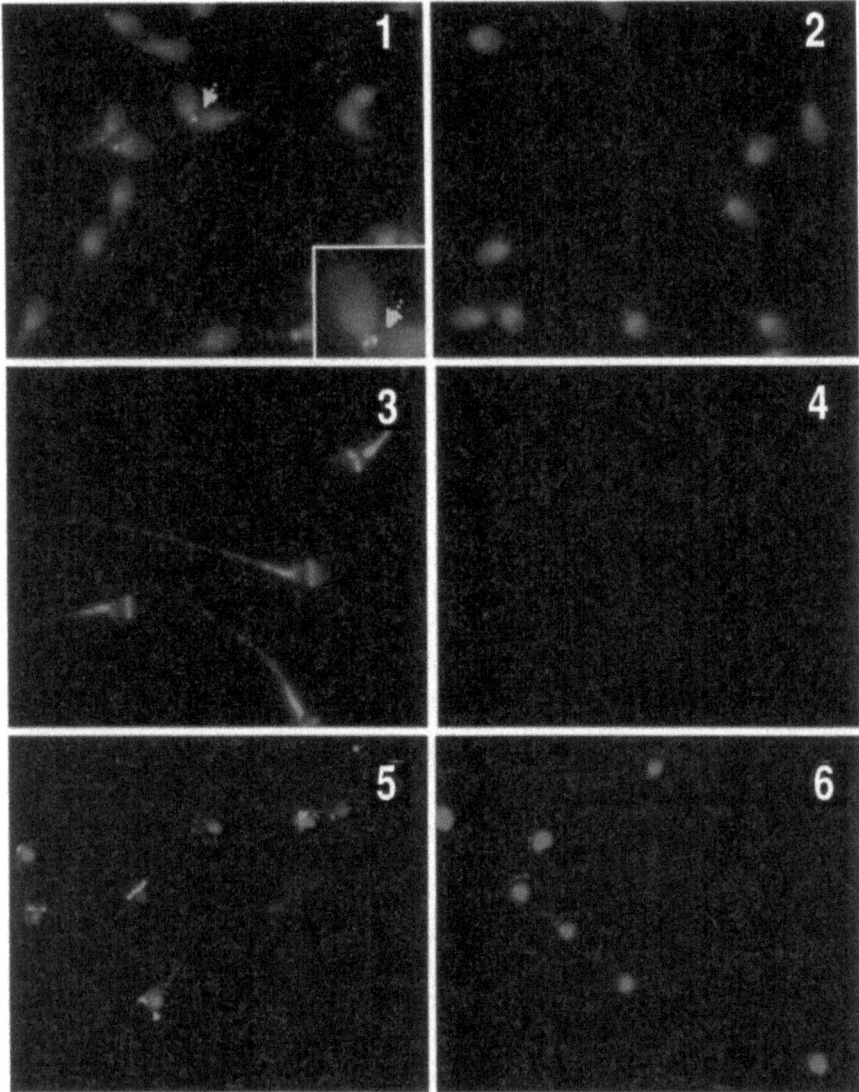

**Fig. 3** Subcellular localization of TSKS, TSSK2 and SSTK in human spermatozoa. Immunofluorescent staining of TSKS (panel 1) and TSSK2 (panel 3) were performed on ejaculated human spermatozoa using polyclonal antibodies. Pre-immune control staining of TSKS (panel 2) and TSSK2 (panel 4) showed low background. Immunofluorescence staining of SSTK (panel 5) on ejaculated human spermatozoa was performed using purified peptide antibody. Secondary antibody staining alone showed very little background (panel 6). Blue color indicates DAPI nuclei staining. As shown above: TSKS was localized to the centrioles of human spermatozoa; TSSK2 was observed in the sperm neck, equatorial segment and mid-piece of the tail; and SSTK was localized in the equatorial segment.

~ 1.0. To quantify the strength of binding interaction, culture supernatants were processed for assaying alpha-galactosidase reporter gene activity using PNP-beta-Gal as substrate. TSKS binding strength with TSSK2 was 12 times stronger than the negative control, while being 1.5 times stronger than the positive control interaction of LgT and p53 (Hao *et al.*, 2004).

In addition, human TSSK2 co-immunoprecipitates with human TSKS in an *in vitro* translation experiment. HA tagged human TSKS and Myc tagged human TSSK were transcribed and co-translated in a rabbit reticulocyte lysate based system from the fusion plasmids pGAD-TSKS and pGBK-TSSK2, respectively. The reaction mix was immunoprecipitated with either rat monoclonal antibody against HA or mouse monoclonal antibody against Myc. Both Myc antibody and HA antibody immunoprecipitated TSSK2/TSKS complexes (Hao *et al.*, 2004). Thus both the immunoprecipitation data and the yeast two-hybrid data showed TSSK2 interacted with TSKS *in vitro*.

To confirm that TSKS and TSSK2 form a complex *in vivo*, co-immunoprecipitation of TSKS/TSSK2 from protein extracts of mouse testis or human spermatozoa was performed using rat anti-TSKS serum or rabbit anti-TSSK2. Both rat anti-TSKS serum and rabbit anti-TSSK2 were capable of co-immunoprecipitating TSSK2/TSKS complexes from either mouse testis or human sperm protein extracts (unpublished). Thus, the formation of TSSK2/TSKS complexes was confirmed both in human and in mouse *in vivo* systems. This substrate and enzyme interaction offers a particularly interesting opportunity for drug design.

*Phosphorylation of TSKS*

To test if TSSK2 is active as a kinase, recombinant TSSK2 was expressed in yeast and *E. coli*. TSSK2 expressed in yeast as a Myc tagged protein was immunoprecipitated with Myc monoclonal antibody and were subjected to an *in vitro* kinase assay. TSSK2 was autophosphorylated. Similarly, bacteria lysates containing recombinant TSSK2 were subjected to an *in vitro* kinase assay in the presence and absence of recombinant TSKS. TSSK2 again autophosphorylated and TSSK2 also phosphorylated TSKS (Hao *et al.*, 2004). Thus we have expressed active TSSK2 in yeast and bacteria and TSSK2 phosphorylated both itself and the TSKS substrate.

Furthermore, TSSK2/TSKS complexes immunoprecipitated with anti-TSKS from mouse testis protein extracts were submitted for phosphopeptide mapping by IMAC-LC-FTMS. $Fe^{3+}$-immobilized metal affinity chromatography (IMAC) enriches digests for peptides containing phospho-amino acids. To decrease nonspecific binding to IMAC, acidic residues were converted to methyl esters before binding. Eluent from the IMAC column was then analyzed in an LC-MS/MS system consisting of a Finnigan LTQ-FT mass spectrometer with a Protana nanospray ion source interfaced to a reversed-phase capillary column. The TSKS IP digest was analyzed using the double play capability of the instrument acquiring full scan mass spectra to determine peptide molecular weights and product ion spectra to determine amino acid sequence in sequential scans. This mode of analysis produces approximately 4000 CAD spectra of ions ranging in abundance over several orders of magnitude. Using FTMS increases the mass accuracy over conventional MS/MS data. The MS data were then analyzed by database searching using the Sequest search algorithm against NCBI nonredundant database. Using this methodology the TSKS phosphopeptide, HGLSPATPIQGCSGPPGS*PEEPPR, was identified, with serine 285 found to be phosphorylated (unpublished data). This result is consistent with our expectation simulated *in silico* from Netphos 2.0 software. Identification of the substrate phosphorylation site at serine 285 will be useful in understanding the kinase substrate interaction and in designing assays for inhibitors.

## Summary

There are several criteria that recommend and support the TSSK family of kinases as suitable contraceptive drug targets. TSSK1-4 and SSTK constitute a unique subfamily of serine/threonine kinases, implying that specific inhibitors can be designed to interact with the unique domains present in this clad. Analyses of mRNA expression for *TSSK1-3* and *SSTK* kinases and the substrate *TSKS* showed that all are predominantly expressed in testis, indicating that tissue specific contraceptive intervention may be possible. *TSSK 1* and *TSSK2* mRNA appear to be 10 fold more abundant in testis than *TSSK 3* and *SSTK*, and *TSSK1* and *TSSK2* are more restricted to testis than *TSSK3* and *TSSK4* according to the Unigene database. *TSSK 1* and *TSSK2* may thus have a higher priority for further study than *TSSK 3* and *TSSK4* and *SSTK*. Mouse *TSSK2, TSKS* and *SSTK* mRNA have the same post-meiotic expression patterns, offering the possibility that contraceptive interference with these gene products may be selective, sparing spermatogonia and spermatocytes. Human TSKS proteins localize in the sperm centrioles, while human TSSK2 proteins are found in the sperm neck, equatorial segment and mid piece of the tail, and human SSTK proteins are present in the equatorial segment. The presence of TSSK2, TSKS and SSTK in mature spermatozoa suggests that the TSSK pathway may play important roles not only in spermiogenesis but also in sperm function. Through yeast two-hybrid interaction and co-immunoprecipitation, human TSKS has been defined as a substrate for TSSK2. This substrate and enzyme interaction offers a particularly attractive opportunity for drug design. TSKS were shown to be phosphorylated in *in vitro* kinase assays and a serine phosphorylation site (Ser285) was detected in mouse TSKS. These phosphorylation data will likely prove useful to establishing high-throughput screening assays for kinase inhibitors.

## Reference

Bielke W, Blaschke RJ, Miescher GC, Zurcher G, Andres AC and Ziemiecki A (1994) Characterization of a novel murine testis-specific serine/threonine kinase *Gene* **139** 235-239

Breitbart H (2003) Signaling pathways in sperm capacitation and acrosome reaction *Cellular and Molecular Biology (Noisy-le-grand)* **49** 321-327

Chen X, Lin G, Wei Y, Hexige S, Niu Y, Liu L,

Yang C and Yu L (2005) TSSK5, a novel member of the testis-specific serine/threonine kinase family, phosphorylates CREB at Ser-133, and stimulates the CRE/CREB responsive pathway *Biochemical and Biophysical Research Communications* **333** 742-749

Craig AW, Zirngibl R, Williams K, Cole LA and Greer PA (2001) Mice devoid of fer protein-tyrosine kinase activity are viable and fertile but display reduced cortactin phosphorylation *Molecular and Cellular Biology* 21 603-613

**Dorval V, Dufour M and Leclerc P** (2002) Regulation of the phosphotyrosine content of human sperm proteins by intracellular Ca²⁺: Role of Ca²⁺-adenosine triphosphatases *Biology of Reproduction* 67 1538-1545

**Dragileva E, Rubinstein S and Breitbart H** (1999) Intracellular Ca²⁺-Mg²⁺-ATPase regulates calcium influx and acrosomal exocytosis in bull and ram spermatozoa *Biology of Reproduction* 61 1226-1234

**Ficarro S, Chertihin O, Westbrook VA, White F, Jayes F, Kalab P, Nartom JA, Shabanowitz J, Herr JC, Hunt D and Visconti PE** (2003) Phosphoproteome analysis of capacitated human sperm. Evidence of tyrosine phosphorylation of AKAP 3 and valosin containing protein/P97 during capacitation *Journal of Biological Chemistry* 278 11579-11589

**Fulcher KD, Mori C, Welch JE, O'Brien DA, Klapper DG and Eddy EM** (1995) Characterization of Fsc1 cDNA for a mouse sperm fibrous sheath component *Biology of Reproduction* 52 41-49

**Galantino-Homer HL, Visconti PE and Kopf GS** (1997) Regulation of protein tyrosine phosphorylation during bovine sperm capacitation by a cyclic adenosine 3'5'-monophosphate-dependent pathway *Biology of Reproduction* 56 707-719

**Hake LE, Alcivar AA and Hecht NB** (1990) Changes in mRNA length accompany translational regulation of the somatic and testis-specific cytochrome c genes during spermatogenesis in the mouse *Development* 110 249-257

**Hao Z, Jha KN, Kim YH, Vemuganti S, Westbrook VA, Chertihin O, Markgraf K, Flickinger CJ, Coppola M, Herr JC and Visconti PE** (2004) Expression analysis of the human testis-specific serine/threonine kinase (TSSK) homologues. A TSSK member is present in the equatorial segment of human sperm *Molecular Human Reproduction* 10 433-444

**Hecht NB** (1988) Post-meiotic gene expression during spermatogenesis *Progress in Clinical and Biological Research* 267 291-313

**Inselman A and Handel MA** (2004) Mitogen-activated protein kinase dynamics during the meiotic G2/MI transition of mouse spermatocytes *Biology of Reproduction* 71 570-578

**Iwaoki Y, Matsuda H, Mutter GL, Watrin F and Wolgemuth DJ** (1993) Differential expression of the proto-oncogenes *c-abl* and *c-mos* in developing mouse germ cells *Experimental Cell Research* 206 212-219

**Jinno A, Tanaka K, Matsushime H, Haneji T and Shibuya M** (1993) Testis-specific mak protein kinase is expressed specifically in the meiotic phase in spermatogenesis and is associated with a 210-kilodalton cellular phosphoprotein *Molecular and Cellular Biology* 13 4146-4156

**Kueng P, Nikolova Z, Djonov V, Hemphill A, Rohrbach V, Boehlen D, Zuercher G, Andres AC and Ziemiecki A** (1997) A novel family of serine/threonine kinases participating in spermiogenesis *Journal of Cell Biology* 139 1851-1859

**Leclerc P, Lamirande E and Gagnon C** (1996) Cyclic adenosine 3'5' monophosphate-dependent regulation of protein tyrosine phosphorylation in relation to human sperm and capacitation *Biology of Reproduction* 55 684-692

**Liguori L, de Lamirande E, Minelli A and Gagnon C** (2005) Various protein kinases regulate human sperm acrosome reaction and the associated phosphorylation of Tyr residues and of the Thr-Glu-Tyr motif *Molecular Human Reproduction* 11 211-221

**Loveland KL and Schlatt S** (1997) Stem cell factor and c-kit in the mammalian testis: lessons originating from Mother Nature's gene knockouts *Journal of Endocrinology* 153 337-344

**Lu Q, Sun QY, Breitbart H and Chen DY** (1999) Expression and phosphorylation of mitogen-activated protein kinases during spermatogenesis and epididymal sperm maturation in mice *Archives of Andrology* 43 55-66

**Mandal A, Naaby-Hansen S, Wolkowicz MJ, Klotz K, Shetty J, Retief JD, Coonrod SA, Kinter M, Sherman N, Cesar F, Flickinger CJ and Herr JC** (1999) FSP95, a testis specific, 95-kilodalton fibrous sheath antigen, that undergoes tyrosine phosphorylation in capacitated human spermatozoa *Biology of Reproduction* 61 1184-1197

**Manning G, Whyte DB, Martinez R, Hunter T and Sudarsanam S** (2002) The protein kinase complement of the human genome *Science* 298 1912-1934

**Marin-Briggiler CI, Jha KN, Chertihin O, Buffone MG, Herr JC, Vazquez-Levin MH and Visconti PE** (2005) Evidence of the presence of calcium/calmodulin-dependent protein kinase IV in human sperm and its involvement in motility regulation *Journal of Cell Science* 118 2013-2022

**Matsushime H, Jinno A, Takagi N and Shibuya M** (1990) A novel mammalian protein kinase gene (*mak*) is highly expressed in testicular germ cells at and after meiosis *Molecular and Cellular Biology* 10 2261-2268

**Miki K and Eddy EM** (1998) Identification of tethering domains for protein kinase A type I alpha regulatory subunits on sperm fibrous sheath protein FSC1 *Journal of Biological Chemistry* **273** 34384-34390

Miki K, Willis WD, Brown PR, Goulding EH, Fulcher KD and Eddy EM (2002) Target disruption of the *Akap4* gene causes defect in sperm flagellum and motility *Developmental Biology* **248** 331-342

Naaby-Hansen S, Wolkowicz MJ, Klotz K, Bush LA, Westbrook VA, Shibahara H, Shetty J, Coonrod SA, Reddi PP, Shannon J, Kinter M, Sherman NE, Fox J, Flickinger CJ and Herr JC (2001) Co-localization of the inositol 1,4,5-trisphosphate receptor and calreticulin in the equatorial segment and in membrane bounded vesicles in the cytoplasmic droplet of human spermatozoa *Molecular Human Reproduction* **7** 923-933

Naaby-Hansen S, Mandal A, Wolkowicz MJ, Sen B, Westbrook VA, Shetty J, Coonrod SA, Klotz KL, Kim YH, Bush LA, Flickinger CJ and Herr JC (2002) CABYR, a novel calcium-binding tyrosine phosphorylation-regulated fibrous sheath protein involved in capacitation *Developmental Biology* **242** 236-254

Osheroff JE, Visconti PE, Valenzuela JP, Travis AJ, Alvarez J and Kopf GS (1999) Regulation of human sperm capacitation by a cholesterol efflux-stimulated signal transduction pathway leading to protein kinase A-mediated up-regulation of protein tyrosine phosphorylation *Molecular Human Reproduction* **5** 1017-1026

Parang K and Sun G (2004) Design strategies for protein kinase inhibitors *Current Opinion in Drug Discovery Development* **7** 617-629

Sharpe RM (1994) Regulation of spermatogenesis. In *The Physiology of Reproduction, vol 1* pp 1363-1434 Eds E Knobil and JD Neill. Raven Press, New York

Shinkai Y, Satoh H, Takeda N, Fukuda M, Chiba E, Kato T, Kuramochi T and Araki Y (2002) A testicular germ cell-associated serine-threonine kinase, MAK, is dispensable for sperm formation *Molecular and Cellular Biology* **22** 3276-3280

Somanath PR, Jack SL and Vijayaraghavan S (2004) Changes in sperm glycogen synthase kinase-3 serine phosphorylation and activity accompany motility initiation and stimulation *Journal of Andrology* **25** 605-617

Sorrentino V, McKinney MD, Giorgi M, Geremia R and Fleissner E (1988) Expression of cellular protooncogenes in the mouse male germ line: a distinctive 24-kilobase *pim-1* transcript is expressed in haploid postmeiotic cells *Proceedings of the National Academy of Sciences USA* **85** 2191-2195

Sorrentino V, Giorgi M, Geremia R, Besmer P and Rossi P (1991) Expression of the *c-kit* proto-oncogene in the murine male germ cells *Oncogene* **6** 149-151

Spiridonov NA, Wong L, Zerfas PM, Starost MF, Pack SD, Paweletz CP and Johnson GR (2005) Identification and characterization of SSTK, a serine/threonine protein kinase essential for male fertility *Molecular and Cellular Biology* **25** 4250-4261

Sun QY, Breitbart H and Schatten H (1999) Role of the MAPK cascade in mammalian germ cells *Reproduction, Fertility and Development* **11** 443-450

Toshima J, Koji T and Mizuno K (1998) Stage-specific expression of testis-specific protein kinase 1 (TESK1) in rat spermatogenic cells *Biochemical and Biophysical Research Communications* **249** 107-112

Turner RM, Johnson LJ, Haig-Ladewig L, Gerton GL and Moss SB (1998) An X-linked gene encodes a major human sperm fibrous sheath protein, hAKAP82. Genomic organization, protein kinase A-RII binding, and distribution of the precursor in the sperm tail *Journal of Biological Chemistry* **273** 32135-32141

Urner F and Sakkas D (2003) Protein phosphorylation in mammalian spermatozoa *Reproduction* **12** 517-526

Visconti PE and Kopf GS (1998) Regulation of protein phosphorylation during sperm capacitation *Biology of Reproduction* **59** 1-6

Visconti PE, Bailey JL, Moore JD, Pan D, Olds-Clarke P and Kopf GS (1995a) Capacitation in mouse spermatozoa I. Correlation between the capacitation state and protein tyrosine phosphorylation *Development* **121** 1129-1137

Visconti PE, Moore GD, Bailey JL, Leclerc P, Connors SA, Pan D, Olds-Clarke P and Kopf GS (1995b) Capacitation of mouse spermatozoa II Protein tyrosine phosphorylation and capacitation are regulated by a cAMP-dependent pathway *Development* **121** 1139-1150

Visconti PE, Galantino-Homer H, Ning X, Moore GD, Valenzuela JP, Jorguez CJ, Alvarez JG and Kopf GS (1999) Cholesterol efflux-mediated signal transduction in mammalian sperm: ß-cyclodextrins initiate transmembrane signaling leading to an increase in protein tyrosine phosphorylation and capacitation *Journal of Biological Chemistry* **274** 3235-3242

Visconti PE, Hao Z, Purdon MA, Stein P, Balsara BR, Testa JR, Herr JC, Moss SB and Kopf GS (2001) Cloning and chromosomal localization of a gene encoding a novel serine/threonine kinase belonging to the subfamily of testis-specific kinases *Genomics* **77** 163-170

**Westbrook VA, Diekman AB, Naaby-Hansen S, Coonrod SA, Klotz KL, Thomas TS, Norton EJ, Flickinger CJ and Herr JC** (2001) Differential nuclear localization of the cancer/testis-associated protein, SPAN-X/CTp11, in transfected cells and in 50% of human spermatozoa *Biology of Reproduction* **64** 345-358

**Yanagimachi R** (1994) Mammalian fertilization. In *The Physiology of Reproduction, second edition* pp189-317 Eds E Knobil and JD Neill. Raven Press, New York

**Zuercher G, Rohrbach V, Andres AC and Ziemiecki A** (2000) A novel member of the testis specific serine kinase family, tssk-3, expressed in the Leydig cells of sexually mature mice *Mechanisms of Development* **93** 175-177

# Role of male reproductive tract CD52 (mrt-CD52) in reproduction

Koji Koyama[1,2], Koichi Ito[1] and Akiko Hasegawa[2]

[1]Department of Obstetrics and Gynecology and [2]Laboratory of Developmental Biology and Reproduction, Institute for Advanced Medical Sciences, Hyogo College of Medicine, 1-1 Mukogawa-cho, Nishinomiya, Hyogo 6638501, Japan

Human CD52 antigen is a highly glycosylated molecule with an unusually small core peptide exclusively expressed on lymphocytes and mature sperm. In the male reproductive tract, it is secreted mainly from the epididymis and inserted into the sperm membrane via the glycosyl-phosphatidyl inositol (GPI) anchor during the passage of the spermatozoa through the epididymis. It has recently been found that the male reproductive tract CD52 (mrt-CD52) is a target antigen of human monoclonal antibody (Mab H6-3C4) obtained from an antisperm antibody-mediated infertile woman. The Mab H6-3C4 shows strong sperm-immobilizing activity with complement and specifically recognizes the N-linked carbohydrate epitope of sperm CD52 but not lymphocyte CD52. Lectin binding assays have revealed the presence of both O-linked as well as N-linked carbohydrate moieties in human mrt-CD52. Mouse monoclonal antibody (1G12) reacting to human mrt-CD52 strongly inhibits penetration of human spermatozoa to the zona denuded hamster oocyte. Mouse CD52 is similar to human CD52 in biological and immunological characteristics. Male and female mice immunized with naturally-occurring mouse mrt-CD52 molecules produce antibodies against the cognate antigen yielding antisera with complement-dependent mouse sperm immobilizing activities.

## Introduction

CD52 is a small glycosylphosphatidylinositol (GPI) anchor protein present in lymphocytes and the epithelial cells of the male reproductive tract (mrt). Originally, CD52 was identified as an antigen recognized by a monoclonal antibody called campath-1 that was produced using human spleen cells as an antigen (Hale et al., 1990; Valentin et al., 1992). The epitope for campath-1 has been shown to consist of the COOH-terminal three amino acids and part of the GPI anchor of CD52. Humanized campath-1 is now used widely as an immunosuppressive drug for organ transplantation and for treatment of leukemia (Buhaescu et al., 2005; Nabhan, 2005).

The structure of human mrt-CD52 is shown in Fig. 1. Human mrt-CD52 contains a unique asparagine N-linked carbohydrate that is not found in other somatic tissues (Fig. 1). In addition, the presence of O-linked carbohydrates linked to serine/threonine residues in human mrt-

E-mail: kkoyama@hyo-med.ac.jp

CD52 has also been suggested (Flori et al., 2005; Hasegawa et al., 2004). The N-linked carbohydrate moiety is immunogenic to generate sperm immobilizing antibody in infertile patients. A human monoclonal antibody (Mab H6-3C4) produced by a hybridoma that was established from an infertile patient's peripheral lymphocyte recognized the N-linked carbohydrate (Hasegawa et al., 2003) and showed strong sperm-immobilizing activity in the presence of complement (Isojima et al., 1987). In this manuscript, we provide an overview of the molecular characteristics and functions of mrt-CD52 in the reproductive process of humans and mice as well.

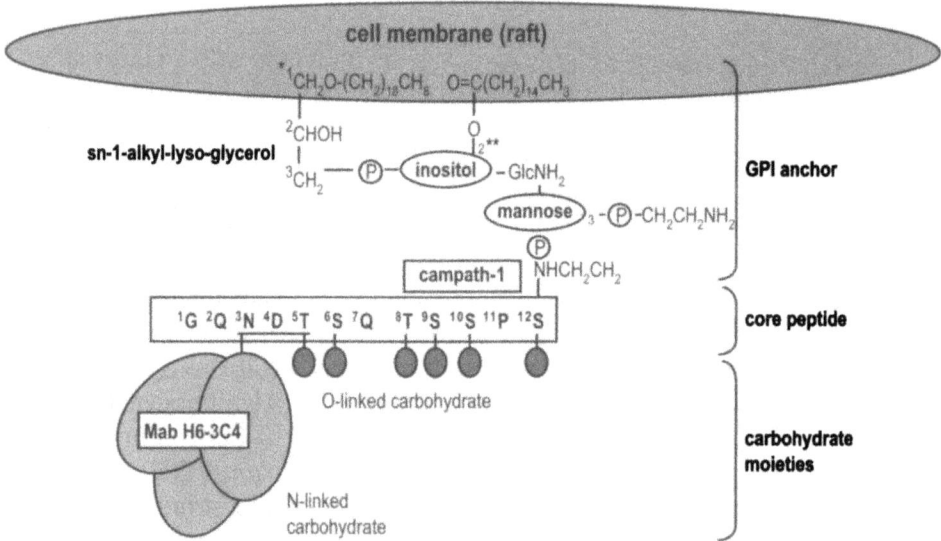

**Fig. 1** Hypothetical structure of human mrt-CD52. The mrt-CD52 molecule is composed of the 12-amino acid core peptide, N-linked and O-linked carbohydrates and a GPI anchor portion inserted in the cell membrane. This drawing is modified from Schröter et al. (1999). The amino acid sequence of the core peptide is shown by the one letter code. Carbohydrate moieties are shown by ovals. Distinctive structures of sn-1-alkyl-lyso-glycerol (single footed) and acylation (mainly palmitoylate) at the 2-position of inositol are shown by a single and double asterisks, respectively.

## Biochemical structure of CD52

The gene encoding CD52 has been isolated from several animal species (Kirchhoff and Schröter, 2001). The deduced amino acid sequence revealed that the amino terminal signal peptide and carboxyl terminal GPI-anchoring peptide are highly conserved, but mature peptide sequences greatly differ among species (Fig. 2). The peptide portion appears to provide a scaffold for the presentation of carbohydrate moieties. The mature peptide sequence contains a consensus sequence, NXT, of the N-linked carbohydrates. Mrt-CD52 is heavily glycosylated forming complex and heterogeneous structures comprising more than 50 different glycoforms. The carbohydrate chains are almost completely sialylated and fucosylated in 10-15% of total mrt-CD52 (Schröter et al., 1999). These features of mrt-CD52 are quite different from those of lymphocyte CD52 which is much smaller in size with low levels of sialylation and no fucosylation (Treumann et al., 1995).

The glycerolipid portion of mrt-CD52 is a sn-1-alkyl-lyso-glycerol type which carries only one fatty acid chain at the 1-position (Fig. 1). The 1-alkyl structure has been reported to be

N-terminal signal peptide     **Mature core peptide**     GPI-anchor signal

**Human**
MKRFLFLLLTISLLVMVQIQTGLS I GQNDTSQTSSPS I ASSNISGGIFLFFVANAIIHLFCFS

**Monkey**
MKRFLFFLLLTISLLVMVQIQTGVT I SQNATSQSSPS I ASSNLSGGGFLFFVANAIIHLFYFS

**Mice (MB7)**
MKSFLLFLTIILLVVIQIQTGSL I GQATTAASGTNKNSTSTKKTPLKS I GASSIIDAGACSFLFFNTLMCLFYLS

**Rat (RB7)**
MNTFLLLLTISLLVVVQIQTGDL I GQNSTAVTTPANKAATTAAATTKAAATTATKTTTAVRKTPGKPPKA I GASSITDVGACTFLFF

**Dog**
MKGFLFLLLTISLLVMIQIQTGVL I GNSTTPRMTTKKVKSATPALSSL I GGGSVLLFLANTLIQLFYLS

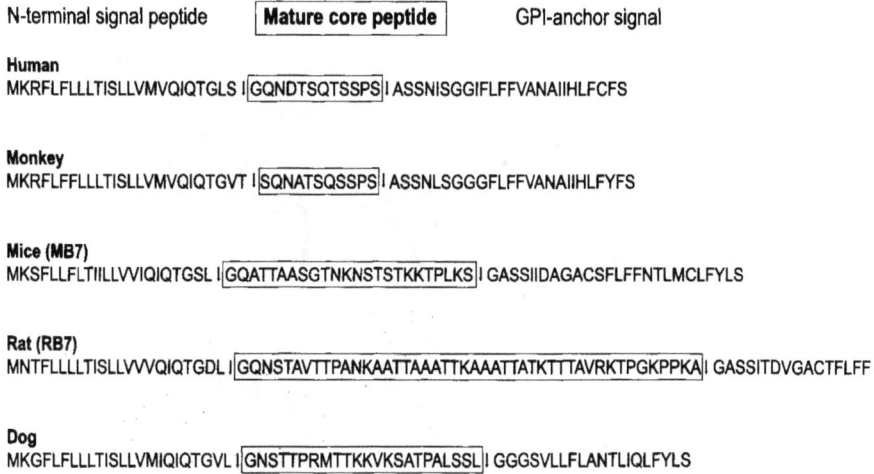

**Fig. 2** Alignment of the deduced amino acid sequences of CD52 from various animal species. Although species homologues of CD52 are highly conserved in the signal peptide and GPI-anchor signal, mature CD52 peptide sequences show low similarity except for the putative N-glycosylation site (NXT) and proline near the COOH-terminus. The potential sites of serine/threonine O-linked carbohydrate moieties are included.

synthesized in male reproductive tracts as a 1-alkyl-2-acetyl-sn-glycero-3-phosphocholine. Such a mono-alkyl structure has not been documented in mammalian species except for male reproductive tracts. The lyso- (single-footed) glycerolipid anchor may play an important role in transporting the mrt-CD52 to spermatozoa from the epithelium in the cauda epididymis. Phospholipase $A_2$, which is present abundantly in seminal plasma, removes the acylation at the 2-position of inositol, although anti-phospholipase $A_2$ inhibitory factors are also detected in the seminal plasma.

### Effects of anti-CD52 antibody on sperm function

Various monoclonal and polyclonal antibodies against mrt-CD52 have been produced in several laboratories. Mab H6-3C4 is a naturally-occurring human monoclonal antibody generated from an infertile patient. Mab H6-3C4 has been suggested to be a pathogenic antibody since it causes strong sperm agglutination and complement–dependent sperm immobilization. Mouse monoclonal antibodies (called as 2C6, 2E5, 2B6, 1G12 and S19) reacting to the mrt-CD52 also have been shown to interfere with sperm motility in the presence of complement (Kameda *et al.*, 1991, 1992; Komori *et al.*, 1997; Diekman *et al.*, 1999, 2000). Immunological reactivities of 2C6, 2E5 and S19 are similar to that of Mab H6-3C4, while 2B6 and 1G12 resemble campath-1. Rabbit antiserum to a synthetic peptide shows human sperm agglutination and complement-dependent sperm immobilization activities. When the antiserum was reacted to swim-up sperm, it stained the whole sperm surface with a patchy pattern. The staining pattern and intensity were not remarkably changed after capacitation. The fluorescence in the apical head region was reduced after acrosome reaction induced by $Ca^{2+}$ inophore (Fig. 3). It has been shown that mouse monoclonal antibodies inhibit binding of human sperm to the zona pellucida in the hemizona assay (Mahony *et al.*, 1991) and penetration of sperm into zona-free hamster eggs (Komori *et al.*, 1997). In our results, blocking effects of mouse monoclonal antibody (1G12) in zona-free hamster egg penetration test were not affected by capacitation or the acrosome reaction (Fig. 4).

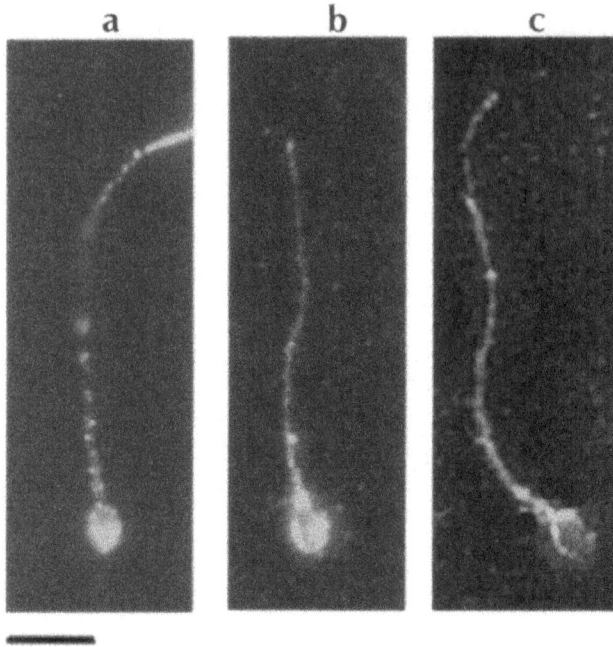

**Fig. 3** Immunofluorescent staining of live human spermatozoa with anti-mrt-CD52 antiserum. Antiserum was produced by immunization with naturally-occurring human mrt-CD52 in human seminal plasma. Swim-up human sperm were prepared by a standard method. To obtain capacitated sperm, swim-up sperm were incubated at 37°C for 3 hr in BWW medium containing 0.4% BSA. To induce acrosome reaction, capacitated sperm were incubated in the medium containing 5 μM $Ca^{2+}$-ionophore, A23187 for 1 hr. Swim-up sperm (a), capacitated sperm (b), and acrosome-reacted sperm (c) were incubated with anti-CD52 antiserum for 1 hr followed by the treatment with FITC- conjugated anti-rabbit IgG. Bar is 10 μm.

## Biological function of CD52

The biological function of mrt-CD52 in mature sperm is not well understood, although it is abundantly expressed on spermatozoa through the maturation process in the epididymis. The highly-negative charges owing to sialylation at the terminal end of carbohydrate chains probably contribute to the prevention of sperm auto-agglutination, immunological attack and adherence to the female reproductive tracts as described by previous reviews (Diekman, 2003).

In the female reproductive tract, sperm are exposed to various immunological and non-specific adhesion molecules. Harmful effects in this environment may be blocked by mrt-CD52 present in a large amount. Staining intensity of spermatozoa by anti-CD52 peptide antibody does not change after acrosome reaction except in the acrosome region (Yeung et al., 2001). The finding that the amount of mrt-CD52 on the spermatozoa does not decrease after the capacitation process suggests that it plays a role in preventing non-specific cell adhesion and premature acrosome reaction.

Considering that Mab H6-3C4 exhibits extremely strong sperm-immobilizing activity with complement via cell membrane damage, it is speculated that mrt-CD52 is a complement suppressive factor to protect spermatozoa. Campath-1 reacting to CD52 positive lymphocyte also causes extremely strong complement-dependent cell lysis (Xia et al., 1993). Functionally-

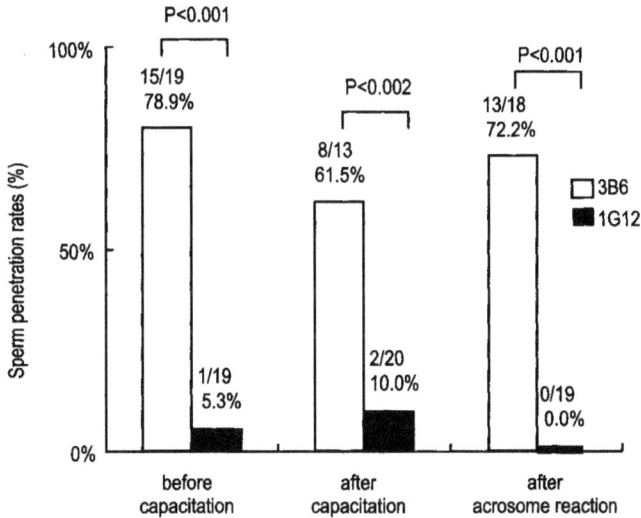

**Fig. 4** Blocking effects of mouse monoclonal antibody (1G12) in zona-free hamster egg penetration test of human sperm before and after capacitation or acrosome reaction. Swim-up human sperm were prepared by a standard method. The capacitated and acrosome-reacted human sperm were obtained as described in the legend to Fig. 3. Swim-up, capacitated or acrosome-reacted human sperm were pretreated with either 1G12 or 3B6 mouse monoclonal antibody for I hr. The pretreated sperm were mixed with zona-free hamster eggs and incubated at 37°C for 18 hr and then the number of eggs penetrated by sperm was assessed after staining with 0.25% acetolacmoid solution.

1G12 (IgMκ): mouse monoclonal antibody recognizing mrt-CD52

3B6 (IgMκ): control mouse antibody with no-binding activity to motile spermatozoa

active complement exists in the female genital tract (Price and Boettcher, 1979) and follicular fluid (Perricone *et al.*, 1992). Recently, complement-regulatory proteins, such as C1-INH, CD55, CD46 and CD59, were found on the surface of spermatozoa (Jiang and Pillai, 1998). CD55 and CD59 have been shown to be GPI anchor proteins (Kirchhoff and Hale, 1996). It suggests that these GPI-anchoring and complement-regulatory proteins including mrt-CD52 protect spermatozoa from complement attack during transportation to the fertilization site.

GPI anchor proteins are known to form a raft in the cell membrane in various organisms. This microdomain structure has been proposed to be a functional domain for signaling and trafficking of cell surface molecules (Simons and Ikonen, 1997). More recently, some mrt-CD52 isoforms bearing the O-linked carbohydrate are suggested to be located in the GM3-rich raft microdomains of human sperm membrane (Ermini *et al.*, 2005). Collectively, mrt-CD52 may be involved in sperm-egg interaction mediated by the raft structure.

## CD52 in different animal species

Molecular cloning of CD52 of different mammalian species including mouse, rat, monkey and dog have revealed that there is low similarity in the mature CD52 peptide sequence except for the putative N-glycosylation site (NXT) and proline near the COOH-terminus (Kirchhoff and Hale, 1996; Kirchhoff and Schröter, 2001) (Fig. 2). The mature peptide sequence of monkey CD52 is 75% (9/12 amino acids) identical to that of human CD52. The CD52 molecule on monkey spermatozoa, like human spermatozoa, shows maturational changes during the passage of spermatozoa in the epididymis and capacitation process (Yeung *et al.*, 2000). The rat CD52 contains the O-linked carbohydrates but no N-linked carbohydrate (Derr *et al.*, 2001), suggesting extreme diversities of CD52 carbohydrate moieties among different animal species.

Campath-1 reacts to monkey (*Macaca fuscicularis*) and human sperm but Mab H6-3C4 recognizes only human spermatozoa (Kameda *et al.*, 1994). Similarly, S19 monoclonal antibody recognizes spermatozoa of chimpanzee (*Pan troglodyte*) but no other non-human primates (McCauley *et al.*, 2002). The carbohydrate epitope recognized by Mab H6-3C4 seems to be restricted to human and chimpanzee spermatozoa. Investigating the differences of the carbohydrate epitopes among species may be helpful for understanding the physiological role of mrt-CD52 molecule in reproduction.

To establish an animal model for the study of mrt-CD52, the immunological and biological characteristics of the mouse homologue of human CD52 has been investigated. The results showed that mouse mrt-CD52 was abundantly produced in the epididymis and vas deferens and bound to the sperm during the passage through the epididymis and vas deferens. Mouse mrt-CD52 also contained the N-linked and O-linked carbohydrates and is negatively charged. The molecular weight ranged from 15 to 20 kDa, suggesting that polymorphic components are included in mouse mrt-CD52. Nasal and systemic immunization with mrt-CD52 molecule showed auto- and iso-antigenicity in male and female mice. The antisera produced from these mice showed a significant complement-dependent sperm-immobilizing activity. The characteristic properties of mouse mrt-CD52 were quite similar to human mrt-CD52.

## Conclusion

Human CD52 antigen is a highly glycosylated, GPI-anchored protein present in lymphocytes and mature spermatozoa. In the male reproductive tract, it is mainly secreted from the epididymis and inserted into the sperm membrane via GPI anchor during their passage through the epididymis. Mouse Mabs to human mrt-CD52 show a strong complement-dependent sperm-immobilizing activity and also block fertilization. N-acetyllactosamine units of the N-linked carbohydrate present in mrt-CD52 have been identified as a pathogenic antigen for infertility by using Mab H6-3C4 generated from an antisperm antibody-mediated infertile woman. Furthermore, mouse mrt-CD52 is also produced in large amounts by the epididymis and vas deferens. It is immunogenic in female mice. This mouse model may be useful to examine the role of mrt-CD52 in reproduction.

## Acknowledgements

This work was supported by the "High-Tech Research Center" Project for Private Universities: matching fund subsidy from MEXT (Ministry of Education, Culture, Sports, Science and Technology), 2004-2008 and by Grant-in-Aid of Science Research (No.15590147) from Japan Society for the Promotion of Science.

## References

Buhaescu I, Segall L, Goldsmith D and Covic A (2005) New immunosuppressive therapies in renal transplantation: monoclonal antibodies *Journal of Nephrology* 18 529-536

Derr P, Yeung CH, Cooper TG and Kirchhoff C (2001) Synthesis and glycosylation of CD52, the major 'matu-ration-associated' antigen of rat spermatozoa, in the cauda epididymidis *Reproduction* 121 435-446

Diekman AB, Norton EJ, Klotz K, Westbrook VA, Shibahara H, Naaby-Hansen S, Flickinger CJ and Herr JC (1999) N-linked glycan of a sperm CD52 glycoform associated with human infertility *FASEB Journal* 13

1303-1313

Diekman AB, Norton EJ, Westbrook VA, Klotz KL, Naaby-Hansen S and Herr JC (2000) Anti-sperm antibodies from infertile patients and their cognate sperm antigens: a review. Identity between SAGA-1, the H6-3C4 antigen, and CD52 *American Journal of Reproductive Immunology* **43** 134-143

Diekman A (2003) Glycoconjugates in sperm function and gamete interactions: how much sugar does it take to sweet-talk the egg? *Cellular and Molecular Life Sciences* **60** 298-308

Ermini L, Secciani F, La Sala GB, Sabatini L, Fineschi D, Hale G and Rosati F (2005) Different glycoforms of the human GPI-anchored antigen CD52 associate differently with lipid microdomains in leukocytes and sperm membranes *Biochemical and Biophysical Research Communications* **16** 1275-1283

Flori F, Giovampaola CD, Focarelli R, Secciani F, La Sala GB, Nicoli A, Hale G and Rosati F (2005) Epitope analysis of immunoglobulins against gp20, a GPI-anchored protein of the human sperm surface homologous to leukocyte antigen CD52 *Tissue Antigens* **66** 209-216

Hale G, Xia MQ, Tighe HP, Dyer MJ and Waldmann H (1990) The CAMPATH-1 antigen (CDw52) *Tissue Antigens* **35** 118-127

Hasegawa A, Fu Y, Tsubamoto H, Tsuji Y, Sawai H, Komori S and Koyama K (2003) Epitope analysis for human sperm-immobilizing monoclonal antibodies, Mab H6-3C4, 1G12 and campath-1 *Molecular Human Reproduction* **68** 337-448

Hasegawa A, Sawai H, Tsubamoto H, Hori M, Isojima S and Koyama K (2004) Possible presence of O-linked carbohydrate in the human male reproductive tract CD52 *Reproductive Immunology* **62** 91-100

Isojima S, Kameda K, Tsuji Y, Shigeta M, Ikeda Y and Koyama K (1987) Establishment and characterization of a human hybridoma secreting monoclonal antibody with high titers of sperm immobilizing and agglutinating activities against human seminal plasma *Journal of Reproductive Immunology* **10** 67-78

Jiang H and Pillai S (1998) Complement regulatory proteins on the sperm surface: relevance to sperm motility *American Journal of Reproductive Immunology* **39** 243-248

Kameda K, Takada Y, Hasegawa A, Tsuji Y, Koyama K and Isojima S (1991) Sperm immobilizing and fertilization-blocking monoclonal antibody 2C6 to human seminal plasma antigen and characterization of the antigen epitope corresponding to the monoclonal antibody *Journal of Reproductive Immunology* **20** 27-41

Kameda K, Tsuji Y, Koyama K and Isojima S (1992) Comparative studies of the antigens recognized by sperm-immobilizing monoclonal antibodies *Biology of Reproduction* **46** 349-357

Kameda K, Fukuda H, Shigeta M, Tsuji Y, Koyama K, Torii R and Isojima S (1994) The effects of patients' sera with sperm-immobilizing antibodies on sperm of the Japanese monkey *Asia-Oceania Journal of Obstetrics and Gynaecology* **20** 433-439

Kirchhoff C and Hale G (1996) Cell-to-cell transfer of glycosylphosphatidylinositol-anchored membrane proteins during sperm maturation *Molecualr Human Reproduction* **2** 177-184

Kirchhoff C and Schröter S (2001) New insights into the origin, structure and role of CD52: a major component of the mammalian sperm glycocalyx *Cells, Tissues, Organs* **168** 93-104

Komori S, Kameda K, Sakata K, Hasegawa A, Toji H, Tsuji Y, Shibahara H, Koyama K and Isojima S (1997) Characterization of fertilization-blocking monoclonal antibody 1G12 with human sperm-immobilizing activity *Clinical and Experimental Immunology* **109** 547-554

Mahony MC, Fulgham DL, Blackmore PF and Alexander NJ (1991) Evaluation of human sperm-zona pellucida tight binding by presence of monoclonal antibodies to sperm antigens *Journal of Reproductive Immunology* **19** 269-285

McCauley TC, Kurth B, Norton E, Klotz K, Westbrook V, Rao A, Herr J and Diekman A (2002) Analysis of a human sperm CD52 glycoform in primates: identification of an animal model for immunocontraceptive vaccine development *Biology of Reproduction* **66** 1681-1688

Nabhan C (2005) The emerging role of alemtuzumab in chronic lymphocytic leukemia *Clinical Lymphoma and Myeloma* **6** 115-121

Perricone R, Pasetto N, De Carolis C, Vaquero E, Piccione E, Baschieri L and Fontana L (1992) Functionally active complement is present in human ovarian follicular fluid and can be activated by seminal plasma *Clinical and Experimental Immunol* **89** 154-157

Price RJ and Boettcher B (1979) The presence of complement in human cervical mucus and its possible relevance to infertility in women with complement-dependent sperm-immobilizing antibodies *Fertility and Sterility* **32** 61-66

Schröter S, Derr, P, Conradt HS, Nimtz M, Hale G and Kirchhoff C (1999) Male-specific modification of human CD52 *Journal of Biological Chemistry* **274** 29862-29873

Simons K and Ikonen E (1997) Functional rafts in cell membranes *Nature* **387** 569-572

Treumann A, Lifely R and Schneider P (1995) Primary structure of CD52 *Journal of Biological Chemistry* **270** 6088-6099

Valentin H, Gelin C, Coulombel L, Zoccola D, Morizet J and Bernard A (1992) The distribution of the CDW52 molecule on blood cells and characterization of its involvement in T cell activation *Transplantation* **54** 97-104

Xia MQ, Hale G, Lifely MR, Ferguson MA, Campbell D, Packman L and Waldmann H (1993) Structure of the CAMPATH-1 antigen, a glycosylphosphatidylinositol-anchored glycoprotein which is an exceptionally good target for complement lysis *The Biochemical Journal* **293** 633-640

Yeung CH, Schroter S, Kirchhoff C and Cooper TG (2000) Maturational changes of the CD52-like epididymal glycoprotein on cynomolgus monkey sperm and their apparent reversal in capacitation conditions *Molecular Reproduction and Development* **57** 280-289

**Yeung C, Perez-Sanchez F, Schröter S, Kirchhoff C and Cooper T** (2001) Changes of the major sperm maturation-associated epididymal protein HE5 (CD52) on human ejaculated spermatozoa during incubation *Molecular Human Reproduction* **7** 617-624

# Regulation of sperm function by protein phosphatase PP1γ2

Srinivasan Vijayaraghavan, Rumela Chakrabarti and Kimberley Myers

Department of Biological Sciences, Kent State University, Kent, Ohio, USA

The intracellular mediators cyclic AMP, calcium and pH regulate sperm function through changes in protein phosphorylation. Protein phosphorylation is the net result of the actions of protein kinases and phosphatases. The protein phosphatase isoform, PP1γ2, with a unique C-terminus extension is highly enriched in spermatozoa and testis. Changes in PP1γ2 catalytic activity, its phosphorylation, and binding to its regulatory proteins change during epididymal maturation. Thus PP1γ2 is a key protein in sperm motility regulation; decreased enzyme activity is associated with increased motility. This review summarizes the current knowledge of this sperm protein phosphatase. The biochemical properties of its regulatory proteins, sds22 and protein 14-3-3, among others, are discussed. Future studies will elucidate sperm signalling pathways involving PP1γ2 and determine if the unique structure of PP1γ2 is critical to normal male gamete development and function. Understanding the role of PP1γ2 will not only contribute to the basic understanding of male gamete functions but also has practical applications in clinical andrology and in the development of male contraceptives.

## Introduction

It is well known that sperm function and metabolism are regulated by the intracellular mediators: cyclic AMP, calcium, and pH (Garbers and Kopf, 1980; Tash and Means, 1983; Vijayaraghavan et al., 1985; Vijayaraghavan and Hoskins, 1985; 1986). However, relatively little is known about how the levels of these mediators are regulated and how they function within spermatozoa. One of the reasons for this limitation is that many of the techniques used to study signalling mechanisms in somatic cells cannot be applied to spermatozoa where there is little or no DNA, RNA, or protein synthesis. An approach that has shed some light on the regulatory mechanisms controlling sperm function is the study of sperm development in the epididymis. Dramatic changes in metabolism, signalling mechanisms, and properties of enzymes occur during acquisition of motility and fertilizing ability of sperm during their passage through the epididymis. It has been shown that the potential for motility already exists in immature testicular and epididymal spermatozoa (Mohri and Yanagimachi, 1980). Motility in intact or demembranated spermatozoa can be induced by changes in the levels of the intracellular mediators or by changes in activities of protein kinases or protein phosphatases (Lindemann,

E-mail: svijayar@kent.edu
Supported by NIH ROHD38520

1978; Vijayaraghavan et al., 1985; 1996).    These same treatments stimulate motility and, under appropriate conditions, also induce hyperactivation in mature spermatozoa (Yanagimachi, 1994). The cytoplasmic effects of the intracellular mediators in immature spermatozoa are held in check and the motility apparatus is consequently inactive. It is not known how this inhibition is removed during sperm development. However, it appears that a shift in levels and patterns of protein phosphorylation accompany motility initiation and development of fertilizing ability of spermatozoa.

Research in sperm protein phosphorylation, until recently, was largely focused on protein kinases and protein kinase A (PKA) in particular (Breitbart, 2003). It is well known that motility stimulation can be affected by cAMP-mediated PKA activation (Hoskins et al., 1975; Lindemann, 1978). Early studies on the role of cAMP in spermatozoa used demembranated sperm models. Reactivation of motility in demembranated spermatozoa requires cAMP. Studies in intact spermatozoa used pharmacological methods to elevate intracellular cAMP levels. In many but not all cases, motility of intact spermatozoa can be induced and stimulated by elevation of cAMP levels (Vijayaraghavan et al., 1985).    These data suggest that cAMP, presumably through PKA activation, increases phosphorylation of protein components required for the expression and maintenance of sperm kinetic activity. The role of cAMP and PKA in sperm function deduced from these studies has now been conclusively verified using genetic approaches. Targeted disruption of sperm adenylyl cyclase and the catalytic subunit of PKA result in infertility in male mice due to lack of sperm motility (Nolan et al., 2004; Livera et al., 2005).

A role for PKA necessarily implies a function for a protein phosphatase since protein phosphorylation is a result of the net activities of protein kinases and phosphatases.   Protein phosphatases can significantly modify and restrict PKA action. If sperm were vigorously motile before demembranation, a requirement for cAMP in the reactivation medium can be bypassed. Inclusion of protein phosphatases in the reactivation media prevents motility initiation (Murofushi et al., 1986). In addition, immotile caput epididymal sperm, and in some cases mature submotile sperm, respond poorly or not at all to cAMP elevation (Vijayaraghavan et al., 1985). These experiments suggest that there exists a mechanism limiting PKA action in immature spermatozoa. It is likely that a protein phosphatase might be involved in the regulation of flagellar motility.

Protein phosphatase 1 (PP1) belongs to a family of protein phosphatases, PPP; which include PP1, PP2A and PP2B (calcineurin) among others.   While the 280-residue core molecule which includes the catalytic domain is conserved in PP1, PP2A, and PP2B, the N- and C-termini of the proteins are divergent (Cohen and Cohen, 1989; Cohen, 2002; Ceulemans and Bollen, 2004). The enzymes differ in their substrate and regulatory properties. The enzyme PP1, which exists in multiple isoforms in diverse organisms, from yeast to mammals, is a highly conserved protein. In mammals there are four catalytic subunit isoforms of PP1, encoded by three genes: PP1α, PP1ß, PP1γ1, and PP1γ2 (Wera and Hemmings, 1995; Cohen, 2002; Ceulemans and Bollen, 2004). There is a high degree of conservation of these isoforms in diverse species. The amino acid sequence of the PP1γ isoform in Xenopus is greater than 90% identical to its human counterpart. The enzymes PP1γ1 and PP1γ2, alternatively spliced variants generated from a single gene (Kitagawa et al., 1990; Varmuza et al., 1999), are identical in all respects except that PP1γ2 has a unique 21-amino-acid carboxy terminus extension. While PP1α, PP1ß and PP1γ1 are ubiquitous, PP1γ2 is predominantly expressed in testis and is highly enriched in spermatozoa (Smith et al., 1996; Vijayaraghavan et al., 1996; Ceulemans and Bollen, 2004).

The enzyme PP1 is known to bind to a large number of targeting and regulatory proteins: to date, more than 50 PP1 binding proteins have been identified (Oliver and Shenolikar, 1998; Ceulemans et al., 2002; Cohen, 2002). Specific cellular and intracellular actions of PP1 are thought to be determined by these regulatory proteins. Some of the regulatory proteins are inhibitors of PP1 catalytic activity; such as PPP1R1 (I1), PPP1R2 (I2), PPP1R11 (I3), NIPP1 and CPI17 among others. These inhibitory proteins are regulated by reversible phosphorylation. The inhibitors, PPP1R1 and its brain form DARP32, inhibit PP1 only when phosphorylated by PKA. On the other hand, the protein PPP1R2, inhibitor I2, inhibits PP1 only in its non-phosphorylated form. Other regulatory proteins serve as targeting proteins that anchor PP1 close to its substrates in the cell (Bollen and Stalmans, 1992). However, why does such a large number of PP1 binding proteins exist? There are far fewer phosphatases than kinases present in cells. While highly specific kinases have few interacting proteins, relatively generic phosphatases may achieve specificity through their regulatory proteins. Thus, subsets of these proteins may have defined functions in regulating PP1 in a cell-specific manner (Bollen and Stalmans, 1992).

While studies with demembranated sperm indirectly suggested a role for a protein phosphatase in sperm function, two reports using protein purification identified the presence of serine/threonine protein phosphatases in extracts from sea urchin and bovine spermatozoa (Tang and Hoskins, 1975; Swarup and Garbers, 1982). The exact nature of the enzyme, however, was not identified. Other reports in the 1980s raised the possibility that the predominant serine/threonine phosphatase, at least in dog spermatozoa, could be the calcium-regulated phosphatase calcineurin (Tash *et al.*, 1988). Surprisingly, there were few subsequent studies on calcineurin in sperm. Calcineurin has not been detected in Western blot analysis of bovine spermatozoa (unpublished data). Calcineurin activity measurements with phosphorylated regulatory subunit RII of PKA (an assay that is likely to be more specific and sensitive than the spectrophotometric measurements used earlier) failed to detect significant enzyme activity in bovine sperm extracts. Other studies have also been unable to detect calcineurin in mouse and bovine spermatozoa using a calmodulin overlay assay and immunocytochemical approaches (Wasco *et al.*, 1989; Moriya *et al.*, 1995). Further studies are required to determine if there is a calcium-regulated phosphatase in spermatozoa.

Enzyme activity and Western blot analyses showed that PP1γ2 is a predominant serine/threonine protein phosphatase in spermatozoa (Smith *et al.*, 1996; Vijayaraghavan *et al.*, 1996). A decline in protein phosphatase activity, likely due to a decrease in catalytic activity of PP1γ2, occurs during epididymal sperm maturation. High protein phosphatase activity is correlated with low motility, whereas low catalytic activity is associated with vigorous motility in bovine and monkey spermatozoa (Smith *et al.*, 1996; Vijayaraghavan *et al.*, 1996). The PP1 inhibitors, okadaic acid (OA) and calyculin A (CA), initiate and stimulate motility of epididymal spermatozoa and promote hyperactivated sperm motility and acrosome reaction (Smith *et al.*, 1996; Vijayaraghavan *et al.*, 1996). As shown in Table 1, PP1 inhibitors initiated motility in immotile caput spermatozoa and stimulated motility in sub-motile caudal spermatozoa.

**Table 1.** Motility induction in bovine epididymal spermatozoa by PP1 inhibitors.

| Section | PP1 inhibitor/conc | | % Motility | Velocity (mm/sec) |
|---|---|---|---|---|
| Caput | Calyculin A | 0 nM | 0 | 0 |
| | | 15 nM | 46 ± 4 | 64 ± 3 |
| | Okadaic acid | 0 μM | 0 | 0 |
| | | 5 μM | 53 ± 3 | 79 ± 2 |
| Caudal | Calyculin A | 0 nM | 20 ± 3 | 48 ± 5 |
| | | 15 nM | 47 ± 3 | 79 ± 5 |
| | Okadaic acid | 0 μM | 78 ± 3 | 81 ± 2 |
| | | 5 μM | 90 ± 2 | 122 ± 4 |

Caudal spermatozoa were rendered sub-motile prior to PP1 inhibition (Vijayaraghavan *et al.*, 1985; Vijayaraghavan and Hoskins, 1986). Spermatozoa were incubated for 10 min at 37°C with the inhibitor before analysis. Data are expressed as mean ± SEM (Vijayaraghavan *et al.*, 1996).

Studies on fowl spermatozoa found that CA and OA stimulated rooster sperm motility (Ashizawa *et al.*, 1995). The enzymes PP1 and PP2A are associated with the axoneme in *Chlamydomonas* (Yang *et al.*, 2000). Taken together these data suggest a key role for the protein phosphatase in flagellar motility in general. This review briefly summarizes our present understanding of PP1γ2 and its regulation in mammalian spermatozoa.

### Presence of PP1γ2 in spermatozoa and its role in motility

Bovine caput and caudal epididymal spermatozoa contain the 39 kDa PP1γ2 but not the 37 kDa PP1γ1 isoform (Vijayaraghavan *et al.*, 1996). As shown in Figure 1, while both PP1γ1 and PP1γ2 are present in testis extracts, epididymal sperm extracts contain only PP1γ2. The Western blot analyses utilized antibodies that specifically recognized the unique PP1γ1 or PP1γ2 C-terminus sequences. Furthermore, the results with the PP1γ2 antibody showed that the enzyme is present in both the soluble and insoluble (pellet) fractions of bovine sperm extracts. The protein is present in spermatozoa from a wide range of mammalian species including humans and non-human primates: immunoreactive PP1γ2 is detected in mouse, rhesus monkey, human (Vijayaraghavan *et al.*, 1996), marmoset, rat, hamster and alpaca spermatozoa (unpublished). The enzyme PP1γ2 is likely to be present in all mammalian spermatozoa. Surprisingly, the carboxyl terminus of PP1γ2, which is not essential for its catalytic activity, is conserved as evidenced by this Western blot analysis.

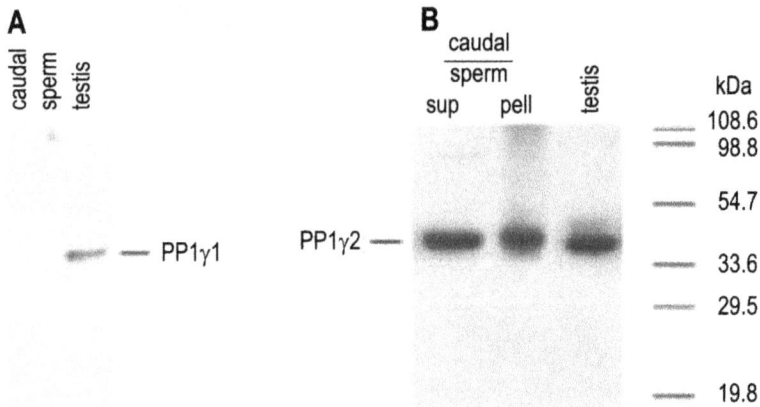

**Fig. 1.** Western blot analyses of sperm and testis extracts with PP1γ1- and PP1γ2-specific antibodies. Extracts were subjected to Western blot analysis with respective antibodies. Bands were visualized using chemiluminescence. **A)** PP1γ1 is present in testis but not sperm extracts. The PP1γ1 antibody (1:1000) is commercially available from Oxford Bio-medical Research (Oxford, MI) and was raised against a peptide corresponding to the C-terminus of the gamma 1 isoform. Per lane 20 μg bovine caudal sperm or mouse testis protein were loaded. **B)** PP1γ2 is present in both soluble (sup) and detergent insoluble (pell) fractions of caudal sperm extracts and in testis extracts. PP1γ2 antibody (1:4000) was raised against a peptide corresponding to the unique 21-amino-acid C-terminus extension of gamma 2 isoform. In each lane 10 μg protein of mouse testis or bovine epididymal sperm extract were loaded.

As shown in Figure 2, soluble extracts of caput epididymal spermatozoa contain significantly higher PP1γ2 activity than caudal epididymal spermatozoa, even though the extracts contain the same amount of immunoreactive PP1γ2. These data suggest that the enzyme may have a higher catalytic activity in immature caput compared to caudal spermatozoa. Protein phosphatases, in general, are regulated by their binding and targeting proteins. Following identification of PP1γ2 we expected that one or more somatic cell protein regulators of PP1 may be present in spermatozoa. Chromatography revealed three distinct pools of PP1γ2 in extracts from caput and caudal epididymal spermatozoa. The proteins associated with these three forms were identified by microsequencing. These findings suggest a sperm-specific role for these PP1γ2 regulatory proteins.

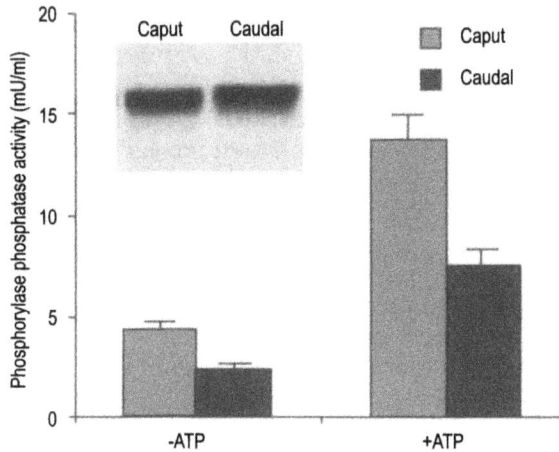

**Fig. 2.** Caput and caudal epididymal sperm extracts display protein phosphatase activity. Activity was measured using a standard assay and is expressed as milliunits per ml of extract. One unit of enzyme is defined as that amount that dephosphorylates 1 μmol of phosphorylasea per minute in the assay. Western blot analysis of the same extracts (inset) with the PP1γ2-specific antibody show that the extracts contain comparable amounts of the enzyme.

## Regulation of PP1γ2 by protein sds22

The protein sds22 was identified in yeast as a PP1 binding protein. Yeast sds22 is named because it is a "suppressor of dism2-mutation" which is a mutation of yeast PP1 that causes mitotic arrest. In yeast, PP1 binding to sds22 inactivates the enzyme against phosphorylasea, a substrate usually used for measuring PP1 catalytic activity (Ohkura and Yanagida, 1991; MacKelvie *et al.*, 1995). However, it is suspected that sds22 may selectively activate yeast PP1 against specific intracellular substrates (Ohkura and Yanagida, 1991). The protein sds22 is a prototypic member of a family of proteins containing leucine-rich repeats (Kobe and Kajava, 2001), a structural motif involved in protein interactions in diverse functions such as signal transduction, cell adhesion, cell development and RNA processing (Kobe and Deisenhofer, 1995; Dinischiotu *et al.*, 1997; Kobe and Kajava, 2001). Human homologues of sds22 were identified using genome and EST data bases (Dinischiotu *et al.*, 1997; Cuelemans *et al.*, 1999). In somatic cells sds22 is a nuclear protein where it appears to be bound to PP1 isoforms α and γ. Partial purification from rat liver showed that sds22-bound PP1 is inactive against a variety of standard substrates used in *in vitro* assays (Dinischiotu *et al.*, 1997). The exact function of sds22 in the nucleus is unknown. Based on its properties in yeast a role for it in somatic cell cycle regulation is suspected.

Micro-sequencing of proteins that co-purified with PP1γ2 from caudal epididymal sperm extracts in immuno-affinity chromatography identified sds22 as a potential PP1γ2 regulatory protein (Huang et al., 2002). The enzyme PP1γ2 bound to sds22 is catalytically inactive against the substrate phosphorylase a (Huang et al., 2002), a situation analogous to yeast and somatic cells. It is not known, if PP1γ2-bound sds22 is preferentially active against other substrates in spermatozoa as in yeast. Intriguingly however, unlike in yeast and somatic cells, a substantial portion of sperm sds22 is cytoplasmic (Huang et al., 2002). It is noteworthy in this regard that PP1γ2 lacks one of the two nuclear localization signals present in other PP1 isoforms and in PP1 in diverse organisms such as yeast and plants. It appears that PP1γ2-sds22 binding is regulated. In immotile caput spermatozoa sds22 is not bound to PP1γ2 and a portion of PP1γ2 is in its free, catalytically active form (Mishra et al., 2003), in sharp contrast to the inactive PP1γ2-sds22 complex isolated from caudal spermatozoa. What regulates sds22 binding to PP1γ2 is not known. Consensus sites for phosphorylation by casein kinase 2, PKA and GSK-3 are present in sds22 (Kwon et al., 1997; Huang et al., 2002). Phosphorylation may alter binding affinity of sds22 for PP1γ2. Alternatively, or in addition, a third protein may be involved in mediating binding interactions between sds22 and PP1γ2 (Mishra et al., 2003). In mammalian spermatozoa, sds22 may play a novel role in regulating PP1γ2 in the cytoplasm. There may exist a cycle of PP1γ2 activation and inactivation due to its binding to and dissociation from sds22 that is a key aspect in the regulation of sperm motility and possibly other sperm functions. Studies are in progress to determine the biochemical mechanisms underlying PP1γ2-sds22 binding and how this binding may regulate sperm function.

### Regulation of PP1γ2 by phosphorylation and binding to protein 14-3-3

The amino acid sequence TPPR at the carboxy-terminus region of PP1γ2 and other PP1 isoforms contains a threonine residue within a consensus site for phosphorylation by cyclin-dependent kinases (cdk). In somatic cells, threonine phosphorylation, which oscillates with the cell cycle, reduces catalytic activity of the enzyme (Brautigan, 1995; Kwon et al., 1997; Liu et al., 1999). Phosphorylation is critical for cell cycle progression and mitosis (Brautigan, 1995). The cdk implicated in this phosphorylation is cdk2. In spermatozoa, a portion of sperm PP1γ2 is also phosphorylated at this threonine residue (amino acid residue 311). The proportion of phosphorylated PP1γ2 is significantly higher in mature caudal than in immature caput epididymal spermatozoa (Huang and Vijayaraghavan, 2004). Phosphorylated PP1γ2 is localized to the posterior region of the sperm head, the region implicated in sperm-egg fusion (Yanagimachi, 1994) and in the principal piece of the sperm tail (Huang and Vijayaraghavan, 2004). It is intriguing that proteins or events involved in mitosis, such as sds22 and cdk-dependant phosphorylation, have unexpected roles in non-dividing cells such as spermatozoa. It is noteworthy that cdk2 knock-out mice are viable; however, both male and female mice lacking cdk2 are sterile due to lack of meiosis (Berthet et al., 2003; Ortega et al., 2003). Phosphorylation, possibly mediated by cdk2, may be a mechanism for regulating PP1γ2 activity for the development of motility and the ability of sperm to fertilize eggs. It is possible that phosphorylation may be a mechanism for anchoring PP1γ2 near its substrates through a bridging molecule such as protein 14-3-3.

Protein 14-3-3 was first discovered as an abundant acidic protein in brain. Its unusual name identifies the column fraction in which it eluted during purification and its migration position in starch gel electrophoresis. Seven different isoforms encoded by seven distinct genes have been identified in mammals: β, γ, ε, η, σ, τ/θ and ζ (Yaffe, 2002). Protein 14-3-3 has subsequently been shown to be highly conserved in all eukaryotic cells: yeast 14-3-3 genes are functionally interchangeable with plant and mammal isoforms (Tzivion and Avruch, 2002). In budding yeast, the protein 14-3-3 gene is essential: cells lacking this protein are not viable (Gelperin *et al.*, 1995). The 14-3-3 isoforms typically associate to form homo- and hetero-dimers *in vivo*. Over one hundred 14-3-3 binding proteins have been identified in somatic cells through affinity chromatography and immunoprecipitation coupled with proteomic analysis (Milne *et al.*, 2002; Jin *et al.*, 2004; Meek *et al.*, 2004; Pozuelo Rubio *et al.*, 2004). Many of the binding partners of protein 14-3-3 are signalling proteins such as protein kinase C, Raf, Bad, PI3-kinase and Cdc25. While binding to non-phosphorylated peptides and proteins *in vitro* has been documented, it is likely that high-affinity 14-3-3 binding *in vivo* is primarily between dimerized 14-3-3 and target proteins containing specific phospho-serine/threonine-containing domains (Shen *et al.*, 2003). Two distinct 14-3-3 protein binding motifs are RS*X* (pS/pT) *XP* (mode 1) and R*XXX* (pS/pT) *XP* (mode 2). While it is known that 14-3-3 is essential for cell function, its exact role is not known. It has been shown to participate in cellular signalling by altering the catalytic activities of bound proteins, changing their sub-cellular localization, affecting binding relationships with other proteins, or protecting the proteins from proteolysis or dephosphorylation (Tzivion and Avruch, 2002; Yaffe, 2002). For example, in somatic cells, 14-3-3 binding to phosphatase Cdc25 protects Cdc25 from dephosphorylation by PP1, thereby preventing premature entry into mitosis (Margolis *et al.*, 2003).

Micro-sequencing of the purified fractions of PP1γ2 in sperm extracts showed that 14-3-3ζ co-eluted with phospho-PP1γ2, while immunoprecipitation and other approaches confirmed that phospho-PP1γ2 binds 14-3-3 in epididymal spermatozoa (Huang *et al.*, 2004). This was the first documentation of a 14-3-3 protein in mature spermatozoa and the first evidence of 14-3-3 binding to PP1γ2. There is also evidence that in addition to PP1γ2, at least three other as yet unidentified sperm phospho-proteins bind to protein 14-3-3ζ (Huang *et al.*, 2004). Thus, it appears that protein 14-3-3 is not only involved in PP1γ2 action but also participates in other signalling pathways in spermatozoa. Protein 14-3-3 is present in spermatozoa isolated from species as diverse as *Xenopus*, turkey, mouse, bovine and human, suggesting an essential role for it in male gamete function (Huang *et al.*, 2004). Studies are underway to examine the role of protein 14-3-3 binding to sperm phospho-proteins in the regulation of sperm function.

### Chloride intracellular channel proteins interact with PP1γ

In addition to 14-3-3ζ, chloride intracellular channel (CLIC) protein 1 (CLIC1) co-eluted with phospho-PP1γ2 in column chromatography. This led to the identification of three CLIC isoforms in mature spermatozoa (Myers *et al.*, 2004). Recombinant versions of all the three CLIC isoforms were shown to bind to endogenous PP1γ2 in pull-down assays (Myers *et al.*, 2004), however, *in vivo* binding relationships have not been demonstrated. The exact role, if any, of CLIC proteins in PP1γ2 regulation or in sperm function is not known.

### Inhibitor I2 and GSK-3

Initial studies on sperm PP1γ2 first focused on identification of heat-stable inhibitors of the enzyme. The first candidate examined was the ubiquitously expressed PKA-regulated inhibitor I1. The established role of PKA in sperm function made inhibitor I1 a logical candidate. Surprisingly, inhibitor I1 activity was undetectable in bovine sperm extracts; however, substantial activity resembling inhibitor I2 was present in heat-stable sperm extracts (Vijayaraghavan *et al.*, 1996; 2000). This activity was thought to be I2-like since inhibition could be reversed by glycogen synthase kinase-3 (GSK-3).

In addition to activation of the PP1-inhibitor I2 complex through phosphorylation of I2 at a threonine residue (DePaoli-Roach, 1984), GSK-3 has several important functions in cellular signal transduction. GSK-3 is regulated both by tyrosine phosphorylation and serine/threonine phosphorylation (Hughes et al., 1993; Wang et al., 1994). In somatic cells, its phosphorylation is mediated by phosphatidyl inositol kinase-3 (PI3-kinase) and protein kinase B (PKB) (Cross et al., 1995). Spermatozoa contain high activity levels of GSK-3. GSK-3 is significantly less phosphorylated (that is, more active) in immotile caput compared to motile caudal epididymal spermatozoa (Vijayaraghavan et al., 1996; 2000). Furthermore, an increase or decrease in motility causes a corresponding increase or decrease in tyrosine and serine phosphorylation of GSK-3 (Somanath et al., 2004). It appears that GSK-3 is inactivated by a combination of tyrosine and serine/threonine phosphorylation. The upstream GSK-3 regulating enzymes, PI3-kinase, PDK1 and PKB, are also present in spermatozoa (Somanath et al., 2004). It is likely that one of the consequences of inactive GSK-3 (for example, in caudal spermatozoa) is the lowering of PP1γ2 activity since inhibitor I2 is likely to be unphosphorylated and thus able to bind to PP1γ2. It therefore follows that low GSK-3 and PP1γ2 activities may be prerequisites for the optimum function of spermatozoa. However, the question of whether an extracellular signal activates sperm GSK-3 remains to be answered.

## Is PP1γ2 indispensable in spermatozoa?

The isoforms PP1γ1 and PP1γ2 are alternatively spliced products of a single gene consisting of 8 exons (Kitagawa et al., 1990; Varmuza et al., 1999). The isoforms result from the inclusion (PP1γ1) or exclusion (PP1γ2) of an intron between exons 7 and 8. This last exon contributes the unique 21-amino-acid extension in PP1γ2. This carboxy terminus sequence does not appear to be essential for catalytic activity, as truncated PP1γ2 lacking the 21-amino-acid C-terminus is able to complement the PP1-deficient yeast cell (Okano et al., 1997). Male mice lacking the PP1γ gene (that is, both PP1γ1 and PP1γ2 isoforms) are sterile due to arrest of spermatogenesis at the spermatid stage (Varmuza et al., 1999; Davies and Varmuza, 2003). Surprisingly, other isoforms of PP1 - PP1α or ß - are able to substitute for PP1γ1/ PP1γ2 in all cellular functions in females and males, except for differentiation of spermatids in the male. The defect in spermatogenesis in PP1γ knockout males could be due to the lack of either PP1γ1 or PP1γ2 or due to the absence of both isoforms. Will mice lacking only one of the isoforms be normal? Can PP1γ1 substitute for PP1γ2 in differentiating cells and spermatozoa? Efforts are underway to produce PP1γ1/PP1γ2 isoform-specific knockout mice to examine the specific requirement for PP1γ2 in male gamete function.

In sum, the enzyme PP1γ2 is a key signalling protein in spermatozoa. The enzyme is regulated in novel ways: by phosphorylation and by proteins identified for the first time in spermatozoa. A schematic representation of PP1γ2 regulatory features is shown in Figure 3. Understanding how regulation of the enzyme is essential in spermatogenesis and in mature sperm function has implications in clinical andrology and in the identification of novel targets for the development of male contraceptives.

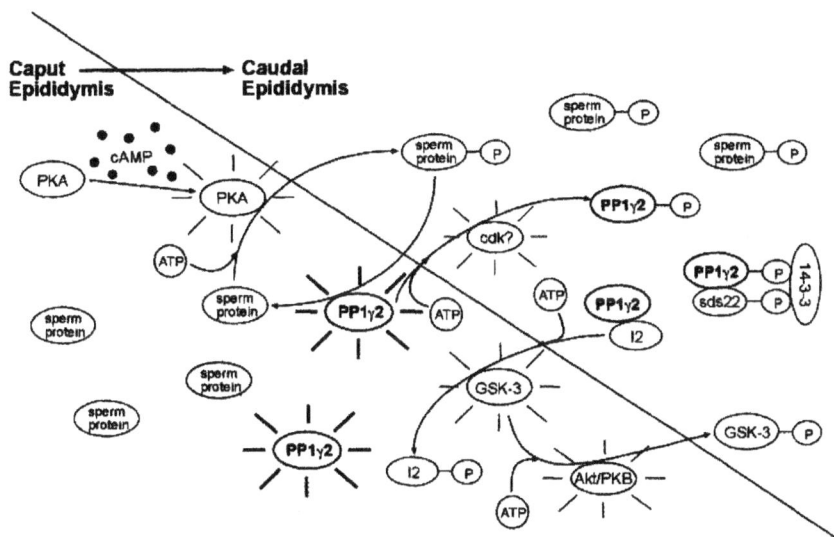

**Fig. 3.** A role for a protein phosphatase in sperm maturation in the epididymis. Activation of protein kinase A or inhibition of protein phosphatases induce motility, suggesting phosphorylation of key proteins increases during epididymal sperm maturation. A decrease in protein phosphatase PP1γ2 activity is thought to occur due to reversible phosphorylation and binding to regulatory proteins. Phosphorylation of PP1γ2 and its binding to sds22 and inhibitor I2 reduce its activity. Low GSK-3 activity causes I2 to be dephosphorylated and bind to PP1γ2. An unknown kinase and binding protein regulate sds22 binding. The cdk responsible for PP1γ2 phosphorylation and the phospho-protein substrates of PP1γ2 are not known. The signals involved in activation of PKA through cAMP or inhibition of PP1γ2 are also unknown.

# References

Ashizawa K, Magome A and Tsuzuki Y (1995) Stimulation of motility and respiration of intact fowl spermatozoa by calyculin A, a specific inhibitor of protein phosphatase-1 and -2A, via a Ca²⁺-dependent mechanism *Journal of Reproduction and Fertility* **105** 109-114

Berthet C, Aleem E, Coppola V, Tessarollo L and Kaldis P (2003) Cdk2 knockout mice are viable *Current Biology* **13** 1775-1785

Bollen M and Stalmans W (1992) The structure, role, and regulation of type 1 protein phosphatases *Critical Reviews in Biochemistry and Molecular Biology* **27** 227-281

Brautigan DL (1995) Flicking the switches: phosphorylation of serine/threonine protein phosphatases *Seminars in Cancer Biology* **6** 211-217

Breitbart H (2003) Signaling pathways in sperm capacitation and acrosome reaction *Cellular and Molecular Biology* **49** 321-327

Ceulemans H and Bollen M (2004) Functional diversity of protein phosphatase 1, a cellular economizer and rest button *Physiological Reviews* **2004** 1-8

Ceulemans H, Van Eynde A, Perez-Callejon E, Beullens M, Stalmans W and Bollen M (1999) Structure and splice products of the human gene encoding sds22, a putative mitotic regulator of protein phosphatase 1 *European Journal of Biochemistry* **262** 36-42

Ceulemans H, Stalmans W and Bollen M (2002) Regulator-driven functional diversification of protein phosphatase-1 in eukaryotic evolution *BioEssays* **24** 371-381

Cohen P (2002) Protein phosphatase 1—targeted in many directions *Journal of Cell Science* **115** 241-256

Cohen P and Cohen PTW (1989) Protein phosphatases come of age *Journal of Biological Chemistry* **264** 21435-21438

Cross DAE, Alessi DR, Cohen P, Andjelkovich M and Hemmings BA (1995) Inhibition of glycogen synthase kinase-3 by insulin mediated by protein kinase B *Nature* **378** 785-789

Davies T and Varmuza S (2003) Development to blastocyst is impaired when intracytoplasmic sperm injection is performed with abnormal sperm from infertile mice harboring a mutation in the protein phosphatase

1cgamma gene *Biology of Reproduction* **68** 1470-1476

**DePaoli-Roach AA** (1984) Synergistic phosphorylation and activation of ATP-Mg-dependent phosphoprotein phosphatase by F A/GSK-3 and casein kinase II (PC0.7) *Journal of Biological Chemistry* **259** 12144-12152

**Dinischiotu A, Beullens M, Stalmans W and Bollen M** (1997) Identification of sds22 as an inhibitory subunit of protein phosphatase-1 in rat liver nuclei *FEBS Letters* **402** 141-144

**Garbers DL and Kopf GS** (1980) The regulation of spermatozoa by calcium and cyclic nucleotides *Advances in Cyclic Nucleotide Research* **13** 251-306

**Gelperin D, Weigle J, Nelson K, Roseboom P, Irie K, Matsumoto K and Lemmon S** (1995) 14-3-3 proteins: potential roles in vesicular transport and Ras signaling in *Saccharomyces cerevisiae Proceedings of National Academy of Sciences USA* **92** 11539-11543

**Hoskins D, Hall M and Munsterman D** (1975) Induction of motility in immature bovine spermatozoa by cyclic AMP phosphodiesterase inhibitors and seminal plasma *Biology of Reproduction* **13** 168-176

**Huang Z and Vijayaraghavan S** (2004) Increased phosphorylation of a distinct subcellular pool of protein phosphatase, PP1gamma2, during epididymal sperm maturation *Biology of Reproduction* **70** 439-447

**Huang Z, Myers K, Khatra B and Vijayaraghavan S** (2004) Protein 14-3-3zeta binds to protein phosphatase PP1gamma2 in bovine epididymal spermatozoa *Biology of Reproduction* **71** 177-184

**Huang Z, Khatra B, Bollen M, Carr DW and Vijayaraghavan S** (2002) Sperm PP1gamma2 is regulated by a homologue of the yeast protein phosphatase binding protein sds22 *Biology of Reproduction* **67** 1936-1942

**Hughes K, Nikolakaki E, Plyte S, Totty NF and Woodgett JR** (1993) Modulation of the glycogen synthase kinase-3 family by tyrosine phosphorylation *European Molecular Biology Organization Journal* **12** 803-808

**Jin J, Smith FD, Start C, Wells CD, Fawcett JP, Kulkarni S, Metalnikov P, O'Donnell P, Taylor P, Taylor L, Zougman A, Woodgett JR, Langeberg LK, Scott JD and Pawson T** (2004) Proteomic, functional and domain-based analysis of *in vivo* 14-3-3 binding proteins involved in cytoskeletal regulation and cellular organization *Current Biology* **14** 1436-1450

**Kitagawa Y, Sasaki K, Shima H, Shibuya M, Sugimura T and Nagao M** (1990) Protein phosphatases possibly involved in rat spermatogenesis *Biochemical and Biophysical Research Communications* **171** 230-235

**Kobe B and Deisenhofer J** (1995) A structural basis of the interactions between leucine-rich repeats and protein ligands *Nature* **374** 183-186

**Kobe B and Kajava AV** (2001) The leucine-rich repeat as a protein recognition motif *Current Opinions in Structural Biology* **11** 725-732

**Kwon YG, Lee SY, Choi Y, Greengard P and Nairn AC** (1997) Cell cycle-dependent phosphorylation of mammalian protein phosphatase 1 by cdc2 kinase *Proceedings of National Academy of Sciences*

*USA* **94** 2168-2173

**Lindemann CB** (1978) A c-AMP-induced increase in the motility of demembranated bull sperm models *Cell* **13** 9-18

**Liu CW, Wang RH, Dohadwala M, Schonthal AH, Villa-Moruzzi E and Berndt N** (1999) Inhibitory phosphorylation of PP1alpha catalytic subunit during the $G_2$/S transition *Journal of Biological Chemistry* **274** 29470-29475

**Livera G, Xie F, Garcia MA, Jaiswal B, Chen J, Law E, Storm DR and Conti M** (2005) Inactivation of the mouse adenylyl cyclase 3 gene disrupts male fertility and spermatozoon function *Molecular Endocrinology* **19** 1277-1290

**MacKelvie SH, Andrews PD and Stark MJ** (1995) The *Saccharomyces cerevisiae* gene sds22 encodes a potential regulator of the mitotic function of yeast type 1 protein phosphatase *Molecular and Cellular Biology* **15** 3777-3785

**Margolis SS, Walsh S, Weiser DC, Yoshida M, Shenolikar S and Kornbluth S** (2003) PP1 control of M phase entry exerted through 14-3-3-regulated Cdc25 dephosphorylation *European Molecular Biology Organisation Journal* **22** 5734-5745

**Meek SE, Lane WS and Piwnica-Worms H** (2004) Comprehensive proteomic analysis of interphase and mitotic 14-3-3-binding proteins *Journal of Biological Chemistry* **279** 32046-32054

**Milne FC, Moorhead G, Pozuelo Rubio M, Wong B, Kulma A, Harthill JE, Villadsen D, Cotelle V and MacKintosh C** (2002) Affinity purification of diverse plant and human 14-3-3-binding partners *Biochemical Society Transactions* **30** 379-381

**Mishra S, Somanath PR, Huang Z and Vijayaraghavan S** (2003) Binding and inactivation of the germ cell-specific protein phosphatase PP1gamma2 by sds22 during epididymal sperm maturation *Biology of Reproduction* **69** 1572-1579

**Mohri H and Yanagimachi R** (1980) Characteristics of motor apparatus in testicular, epididymal and ejaculated spermatozoa *Experimental Cell Research* **127** 191-196

**Moriya M, Fujinaga K, Yazawa M and Katagiri C** (1995) Immunohistochemical localization of the calcium/calmodulin-dependent protein phosphatase, calcineurin, in the mouse testis: Its unique accumulation in spermatid nuclei *Cell and Tissue Research* **281** 273-281

**Murofushi H, Ishiguro K, Takahashi D, Ikeda J and Sakai H** (1986) Regulation of sperm flagellar movement by protein phosphorylation and dephosphorylation *Cell Motility and the Cytoskeleton* **6** 83-88

**Myers K, Somanath PR, Berryman M and Vijayaraghavan S** (2004) Identification of chloride intracellular channel proteins in spermatozoa *Federation of European Biochemical Societies Letters* **566** 136-140

**Nolan MA, Babcock DF, Wennemuth G, Brown W, Burton KA and McKnight GS** (2004) Sperm-specific protein kinase A catalytic subunit Calpha2 orchestrates cAMP signaling for male fertility *Proceedings of Na-*

tional Academy of Sciences USA **101** 13483-13488

Ohkura H and Yanagida M (1991) *S. pombe* gene sds22+ essential for a midmitotic transition encodes a leucine-rich repeat protein that positively modulates protein phosphatase-1 *Cell* **64** 149-157

Okano K, Heng H, Trevisanato S, Tyers M and Varmuza S (1997) Genomic organization and functional analysis of the murine protein phosphatase 1c gamma (Ppp1cc) gene *Genomics* **45** 211-215

Oliver CJ and Shenolikar S (1998) Physiologic importance of protein phosphatase inhibitors *Frontiers in Bioscience* **3** D961-972

Ortega S, Prieto I, Odajima J, Martin A, Dubus P, Sotillo R, Barbero JL, Malumbres M and Barbacid M (2003) Cyclin-dependent kinase 2 is essential for meiosis but not for mitotic cell division in mice *Nature Genetics* **35** 25-31

Pozuelo Rubio M, Geraghty KM, Wong BH, Wood NT, Campbell DG, Morrice N and MacKintosh C (2004) 14-3-3-affinity purification of over 200 human phosphoproteins reveals new links to regulation of cellular metabolism, proliferation, and trafficking *Biochemical Journal* **379** 395-408

Shen YH, Godlewski J, Bronisz A, Zhu J, Comb MJ, Avruch J and Tzivion G (2003) Significance of 14-3-3 self-dimerization for phosphorylation-dependent target binding *Molecular and Biological Cell* **14** 4721-4733

Smith GD, Wolf DP, Trautman KC, da Cruz e Silva EF, Greengard P and Vijayaraghavan S (1996) Primate sperm contain protein phosphatase 1, a biochemical mediator of motility *Biology of Reproduction* **54** 719-727

Somanath PR, Jack SL and Vijayaraghavan S (2004) Changes in sperm glycogen synthase kinase-3 serine phosphorylation and activity accompany motility initiation and stimulation *Journal of Andrology* **25** 605-617

Swarup G and Garbers DL (1982) Phosphoprotein phosphatase activity of sea urchin spermatozoa *Biology of Reproduction* **26** 953-960

Tang FY and Hoskins DD (1975) Phosphoprotein phosphatase of bovine epididymal sperm *Biochemical and Biophysical Research Communications* **62** 328-335

Tash JS and Means AR (1983) Cyclic adenosine 3', 5' monophosphate, calcium and protein phosphorylation in flagellar motility *Biology of Reproduction* **28** 75-104

Tash JS, Krinks M, Patel J, Means RL, Klee CB and Means AR (1988) Identification, characterization, and functional correlation of calmodulin-dependent protein phosphatase in sperm *Journal of Cell Biology* **106** 1625-1633

Tzivion G and Avruch J (2002) 14-3-3 proteins: active cofactors in cellular regulation by serine/threonine phosphorylation *Journal of Biological Chemistry* **277** 3061-3064

Varmuza S, Jurisicova A, Okano K, Hudson J, Boekelheide K and Shipp EB (1999) Spermiogenesis is impaired in mice bearing a targeted mutation in the protein phosphatase 1c gamma gene *Developmental Biology* **205** 98-110

Vijayaraghavan S and Hoskins DD (1985) Forskolin stimulates bovine epididymal sperm motility and cyclic AMP levels *Journal of Cyclic Nucleotide and Protein Phosphorylation Research* **10** 499-510

Vijayaraghavan S and Hoskins DD (1986) Regulation of bovine sperm motility and cyclic adenosine 3', 5'-monophosphate by adenosine and its analogues *Biology of Reproduction* **34** 468-477

Vijayaraghavan S, Critchlow LM and Hoskins DD (1985) Evidence for a role for cellular alkalinization in the cyclic adenosine 3', 5'-monophosphate-mediated initiation of motility in bovine caput spermatozoa *Biology of Reproduction* **32** 489-500

Vijayaraghavan S, Stephens DT, Trautman K, Smith GD, Khatra B, da Cruz e Silva EF and Greengard P (1996) Sperm motility development in the epididymis is associated with decreased glycogen synthase kinase-3 and protein phosphatase 1 activity *Biology of Reproduction* **54** 709-718

Vijayaraghavan S, Mohan J, Gray F, Khatra B and Carr D (2000) A role for phosphoryaltion of glycogen synthase kinase-3alpha in bovine sperm motility regulation *Biology of Reproduction* **62** 1647-1654

Wang Q, Fiol C and DePaoli-Roach P (1994) Glycogen synthase kinase-3-beta is a dual specificity kinase differentially regulated by tyrosine and serine/threonine phosphorylation *Journal of Biological Chemistry* **269** 14566-14574

Wasco WM, Kincaid RL and Orr GA (1989) Identification and characterization of calmodulin-binding proteins in mammalian sperm flagella *Journal of Biological Chemistry* **264** 5104-5111

Wera S and Hemmings BA (1995) Serine/threonine protein phosphatases *Biochemical Journal* **311** 17-29

Yaffe MB (2002) How do 14-3-3 proteins work?—Gatekeeper phosphorylation and the molecular anvil hypothesis *Federation of European Biochemical Societies Letters* **513** 53-57

Yanagimachi R (1994) *Mammalian fertilization*, edn 2. Raven Press Ltd, New York

Yang P, Fox L, Colbran RJ and Sale WS (2000) Protein phosphatases PP1 and PP2A are located in distinct positions in the *Chlamydomonas* flagellar axoneme *Journal of Cell Sciences* **113** 91-102

# The molecular basis of "Curlicue": a sperm motility abnormality linked to the sterility of t haplotype homozygous male mice

SH Pilder[1], J Lu[1], Y Han[1, 2], L Hui[1, 3], SA Samant[1, 4], OO Olugbemiga[1, 5], KW Meyers[6], L Cheng[6] and S Vijayaraghavan[6]

[1]Dept of Anatomy and Cell Biology, School of Medicine, Temple University, 3400 N Broad St, Philadelphia, PA 19140, USA; [2]Dept of Obstetrics and Gynecology, Chinese University of Hong Kong, Hong Kong, New Territory, People's Republic of China; [3]Dept of Obstetrics and Gynecology, University of Pennsylvania Medical Center, Philadelphia, PA 19104, USA; [4]Dept of Medicine, Division of Biological Sciences, University of Chicago, Chicago, IL 60637, USA; [5]Dept of Pathology, Anatomy, and Cell Biology, Meharry Medical College, Nashville, TN 37208, USA; [6]Dept of Biological Sciences, Kent State University, Kent, OH 44242, USA

The t complex, a variant region of chromatin occupying approximately 40-million base pairs of proximal chromosome 17, exists in natural populations of wild mice of the *Mus musculus* species as a family of homologues called t haplotypes (t). Relative to wild-type (+) homologues, all t haplotypes share four large non-overlapping inversions, spanning 95% of the region, leading to intra-inversion recombination suppression in +/t heterozygotes. Non-lethal t homozygous males or complementing recessive lethal t doubly heterozygous males (hereafter both abbreviated "t/t males") are invariably and completely sterile, due to expression of several sperm function abnormalities. One of these traits, "curlicue", describes a condition in which spermatozoa from t/t males fail to reach the site of fertilization *in vivo* because they exhibit a severe loss of vigorous forward motility due to the chronic negative curvature of their flagella. Current data indicate that "curlicue" is the complex phenotypic reflection of the expression of three or more mutations clustered in the distal one-third of the largest and most-distal t complex inversion, *In(17)4*. From proximal to distal, candidates include *Dnahc8*, *Tsga2* and *Tctex5*. Interestingly, new results from high-resolution intra-inversion genetic mapping and protein localization studies suggest that the products of the distal two candidates, *Tsga2* and *Tctex5*, might play synergic roles in the expression of both the "curlicue" motility abnormality and the "stop" sperm-egg interaction aberration, regarded as functionally unrelated traits.

## Genetics of the *t* complex and associated male-specific traits

Mouse (*Mus musculus*) chromosome 17 exhibits polymorphism (Fig. 1) in which approximately 10-20% of natural populations share a variation of its proximal one-third consisting of four non-

E-mail: stephen.pilder@temple.edu

overlapping inversions (Bennett, 1975; Silver and Artzt, 1981; Artzt *et al.*, 1982; Silver, 1989; Hammer *et al.*, 1989; Lyon, 2003). This relatively low frequency deviation, known as the *t* complex, is represented by a family of homologues, the *t* haplotypes (*t*). As a consequence of this variation in chromosome 17 structure, recombination in the *t* complex region is rare in animals heterozygous for a wild-type (+) homologue and a *t* haplotype with concomitant effects on the evolution of regional genes.

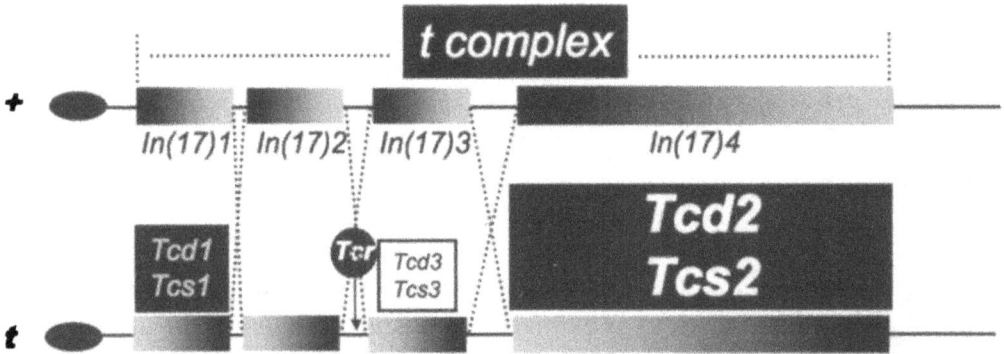

**Fig. 1** The mouse *t* complex. The wild-type (+) and *t* haplotype (*t*) homologues of mouse chromosome 17 are shown (top and bottom, respectively). Positions of *t* complex distorter factors (*tcd1, 2* and *3*) and *t* complex sterility factors (*tcs1, 2* and *3*) are noted above the inversions (shaded rectangles) of the *t* haplotype in which they are located, and the size of each is proportional to the size of the role played by each in either TRD or male sterility. Inversion names (*In[17]1-4*) are shown beneath the + form of each inversion. The central location of the *t* complex responder (*tcr*) in an uninverted region is also shown. Crossed dashed lines indicate the positions of small uninverted regions of chromatin between inversions where rare recombination events between + and *t* would be most likely to occur, and partial *t* haplotypes would be formed.

Heterozygous (+/*t*) males and females are fertile; however, a heterozygous male exhibits meiotic drive or transmission ratio distortion (TRD) in which his progeny receive a positively distorted ratio of his *t* homologue. Results of a series of elegant classical genetic experiments using rare *t* complex recombinants (partial *t* haplotypes) led Lyon (1984) to propose a simple model of the genetic basis of TRD. Essentially, at least three *t* complex distorter (*tcd*) factors, embedded respectively in inversions 1, 3 and 4 (*In[17]1, In[17]3* and *In[17]4* from proximal to distal), act additively on a *t* complex responder factor (*tcr*) located between *In[17]2* and *In[17]3* to cause the transmission of *t* to a distorted ratio of the progeny of a heterozygous male (Fig. 1). Additionally, the ability of individual *tcd* factors to cause the transmission of *t* to high levels varies (Lyon, 1984), with only *tcd2* (in *In[17]4*) capable of driving *t* haplotype transmission to levels greater than 50% in the absence of other *tcds*, although *tcd1* (in *In[17]1*) also shows a substantial effect (Fig. 1).

The primary reason for the low frequency of *t* haplotypes in mouse populations is that *t/t* males are, without deviation, completely sterile. Lyon theorised that *t/t* male sterility may have resulted from natural selection against the propagation of harmful *tcd* factors, with their homozygous expression overwhelming any ameliorative effect of the expression of *tcr* by causing massive sperm dysfunction. This hypothesis was based on results of classical genetic studies (Lyon, 1986, 1992) with partial *t* haplotypes and chromosome 17 homologues containing deletions. These studies demonstrated that: 1) factors causing sterility (*tcs* factors) mapped to the same inversions as the three *tcd* factors; 2) individual *tcs* factors exhibited different abilities to cause sterility when homozygous; 3) the *tcs* factor in *In[17]4* (*tcs2*) was alone in its ability to render males sterile when homozygous and in the complete absence of both of the other sterility factors; 4) homozygosity for *tcs1*, but not *tcs3*, could lead to male sterility in the presence of a single copy of *tcs2*; and 5) deletion of the first inversion in + homologues mimicked the effects of both *tcd1* and *tcs1* when paired with either *t* or appropriate partial *t* haplotypes.

## The functional basis of *t*-associated traits

Lyon's models were compelling in their simplicity, providing a timely impetus to researchers who were attempting to discover the molecular origins of male TRD and sterility linked to the *t* complex. However, in the absence of a viable genetic approach for high-resolution mapping and subsequent identification of the inversion-bound *tcd* and *tcs* factors, investigators employed other approaches to determine the basis of *t*-associated traits.

In an early study (McGrath and Hillman, 1980), investigators determined that spermatozoa from *t/t* males failed to fertilize eggs *in vitro*, even after the egg investments had been removed. Since the experimental mice were outbred, these results were viewed with some skepticism by *t* complex researchers. However, many aspects of this early study were confirmed and extended by experiments using males of a consistent genetic background (Johnson et al., 1995). The accumulated results of these studies indicated that sperm from +/*t* males demonstrated a significant delay in penetrating both the zona pellucida and zona pellucida-free eggs, while sperm from *t/t* males were absolutely unable to complete either step, and exhibited additional deficiencies including a decreased ability to bind to either the zona pellucida or the oolemma. Additional mapping experiments with partial *t* haplotypes demonstrated that genes in both *In[17]1* and *In[17]4* were absolutely essential for the zona binding and oolemma penetration phenotypes exhibited by males carrying one or two *t* haplotypes (Redkar et al., 2002). These studies also supported Lyon's contention that male TRD and sterility may have a similar, if not identical molecular basis.

A long-term study by Olds-Clarke and colleagues concentrated on identifying the underlying physiological basis of poor motility exhibited by spermatozoa from males of *t*-bearing genotypes (Olds-Clarke, 1984; Silver and Olds-Clarke, 1984; Olds-Clarke, 1986, 1989; Lindemann et al., 1990; Olds-Clarke and Wivell, 1992; Olds-Clarke and Johnson, 1993; Pilder et al., 1993). These experiments demonstrated that when cauda epididymal sperm from +/+ males were released into a calcium-containing medium that supports fertilization *in vitro* (IVF medium), they initially moved in a vigorous and progressive fashion with high beat frequency, low amplitude, symmetrical flagellar bends. After about two hours, approximately 30% of these sperm exhibited hyperactivated movement (highly vigorous and nonprogressive motility with low beat frequency, high amplitude, asymmetrical flagellar bends), a characteristic parameter of sperm motility *in vivo*, normally occurring in close proximity to the egg and obligatory for successful fertilization (Ho and Suarez, 2003; Carlson et al., 2003; Marquez and Suarez, 2004). Many of these hyperactivated sperm also exhibited chronic negative curvature of the flagellar midpiece ("fishhook" phenotype) but similar curvature was rarely observed in the principal piece of the flagellum (Fig. 2). Furthermore, if axonemal calcium concentrations were raised above a critical level for a lengthy period of time, sperm tails from +/+ males arrested at one extreme of the beat cycle (Lindemann and Kanous, 1997; Schmitz-Lesich and Lindemann,

2004). These arrested flagella were characterized by the same "fishhook"-like appearance with maximal bending in the flagellar midpiece. Thus, "fishhook" appeared to be a normal flagellar response to increased concentrations of $Ca^{2+}$ in the medium.

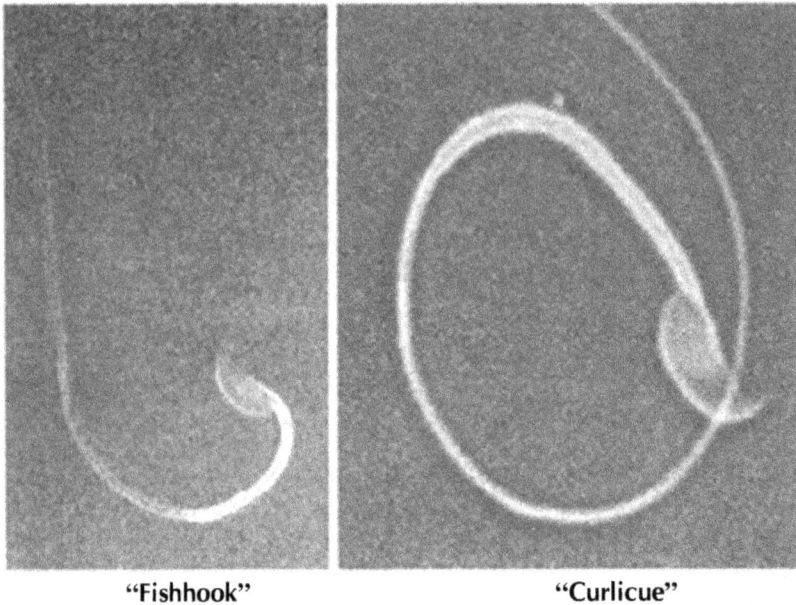

"Fishhook"                    "Curlicue"

**Fig. 2** Flagellar curvature of sperm from $+/t$ ("fishhook") and $t/t$ ("curlicue") males. Both phenotypes demonstrate negative (in the direction opposite to the curvature of the sperm head) midpiece curvature, but only "curlicue" flagella demonstrate chronic negative curvature in the principal piece as well.

Sperm from congenic $+/t$ and $t/t$ males behaved differently from sperm from $+/+$ males in IVF medium. Nearly 60% of the sperm from $+/t$ males hyperactivated within only one hour. However, the "fishhook" tails persisted even during a gradual decline in sperm swimming speed. The principal piece of the tail generally exhibited $+/+$ behavior, although lengthy incubation resulted in transient negative curvature of the principal pieces of the tails of a few sperm as well.

In the case of sperm from $t/t$ males, hyperactivation occurred in over 50% of motile sperm after only a five-minute incubation in IVF medium. Almost as quickly and very abruptly, this highly premature hyperactivation behavior stopped, giving way to a loss of vigor and a precipitous decline in sperm swimming speed, so that almost all motile sperm (> 90%) exhibited a complete loss of regular beat pattern accompanied by a chronic, negative bend of both the flagellar mid-piece and principal piece ("curlicue" phenotype). This trait could best be described as an exaggerated form of $Ca^{2+}$-induced flagellar "arrest" (Fig. 2). Additional *in vivo* research showed that sperm from $t/t$ males failed to reach the site of fertilization (Olds-Clarke, 1986). Concomitant breeding and sperm motility experiments with partial *t* haplotypes demonstrated that the genes responsible for expression of "curlicue" mapped exclusively to *In[17]4* (Pilder et al., 1993).

These studies demonstrated that *t* alleles of genes affecting sperm motility bestowed abnormal calcium sensitivity on the tails of sperm from both $+/t$ and $t/t$ animals and may themselves be calcium-dependent components of the axoneme. Once again, the results supported Lyon's contention that male TRD and sterility phenotypes associated with *t* haplotypes had similar or overlapping molecular origins. However, while these studies made it apparent that the sterility of *t/t* males could result from perturbation of different sperm functions, it was not at all clear whether any of these dissimilar functions shared a common molecular basis.

## The timely discovery of an approach for high-resolution sterility gene mapping within *t* complex inversions

Experiments designed to determine the evolutionary origins of *t* haplotype inversions resulted in the first viable approach for high-resolution localization of intra-inversion genes responsible for *t*-associated male sterility phenotypes (Hammer *et al.*, 1989). The serendipitous discovery of the sterility of males of the *Mus musculus* genetic background heterozygous for a *t* haplotype and a chromosome 17 homologue from the aboriginal mouse species, *Mus spretus* (*s*) indicated that in the *Mus musculus* genetic background, *Mus spretus* alleles would fail to complement the sperm function defects caused by expression of their corresponding *t* alleles. Inversions *In[17]1ˢ*, *In[17]3ˢ* and *In[17]4ˢ* contained the same genetic marker orientation as *In[17]1⁺*, *In[17]3⁺* and *In[17]4⁺* therefore it soon became apparent that inversion-specific *s-+* chromosome 17 homologues could be produced by recombination and utilized to map *t* sterility genes indirectly in males carrying various *s-+/t* chromosome 17 genotypes.

Early studies of *In[17]4* hetero-specific recombinant homologues demonstrated that the ~ 20 Mb *In[17]4* was divisible into three nearly independent hybrid sterility loci, *Hst4*, *Hst6* and *Hst5*, from proximal to distal, with *Hst4* occupying ~ 11 Mb, *Hst6*, ~ 2 Mb and *Hst5*, ~ 7 Mb (Pilder *et al.*, 1991; 1993). Results from the latter study also indicated that spermatozoa from *In[17]4ˢ/t* males exhibited *t/t* levels of "curlicue", and that one or more genes responsible for expression of this phenotype mapped to the small *Hst6* locus (Fig. 3A). In addition, data strongly suggested that the failure of *Mus spretus* alleles to complement *t* defects resulted from the fact that the *Mus spretus* alleles behaved in a nullipotent fashion in the *Mus musculus* genetic background. Thus, expression of the corresponding altered-function *t* alleles would result in *t/t*-like phenotypes (Pilder *et al.*, 1993).

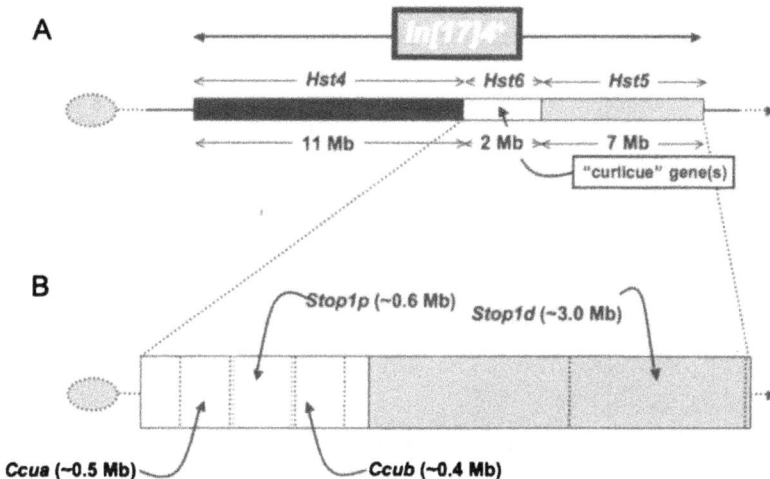

**Fig. 3** A. Low-resolution mapping of a "curlicue" factor(s) in the ~ 2 Mb *Hybrid sterility 6* (*Hst6*) locus was accomplished with primary *In[17]4* hetero-specific recombinant homologues (Pilder *et al.*, 1993). B. The "stop" and "curlicue" phenotypes were partially mapped to high-resolution following the identification of additional *In[17]4* hetero-specific recombinant homologues (Redkar *et al.*, 1998).

Still higher-resolution mapping of "curlicue" as well as the *In[17]4* genes responsible for expression of the aforementioned *t*-associated sperm-oolemma penetration abnormality (later named "stop") was accomplished by measuring the percentage of sperm produced by males heterozygous for a variety of *s-+/t* genotypes that expressed either of these two phenotypes (Redkar *et al.*, 1998; Samant *et al.*, 1999). These experiments suggested that "curlicue" resulted from at least two factors, *Curlicue-a* (*Ccua*), mapping to an interval of less than 0.5 Mb in proximal *Hst6*, and *Curlicue-b* (*Ccub*), in a slightly larger interval in distal *Hst6*. The *In[17]4* component of the "stop" phenotype was concomitantly mapped to the *Stop1* locus, which was thought to be composed of two synergic factors, *Stop1p*, located in a small interval between *Ccua* and *Ccub*, and *Stop1d*, distal to *Ccub* and probably in the *Hst5* locus (Fig. 3B).

## Identification of a *Ccua* candidate gene

The identification of candidate genes for "curlicue" and "stop" factors relied on the premise that their *Mus spretus* alleles were nullipotent or nearly so. Thus, differential display-PCR technology was employed to detect potential candidates for *t* sterility genes in *In[17]4*. In an initial study, mRNA expression in the testis of males homozygous for *Ccua$^s$* was compared to the expression of testis mRNA from *+/+* males (Fossella *et al.*, 2000), resulting in the isolation of an ~15 Kb mRNA expressed abundantly in the testis of *+/+* males but absent from the testis mRNA population of *Ccua$^s$* homozygous animals. Subsequently, this large message was identified as the expressed product of *Dnahc8*, an axonemal dynein heavy chain (axDHC) gene with homology to the *Chlamydomonas reinhardtii* γ-axDHC of the outer row of axonemal dyneins.

The candidacy of *Dnahc8* for *Ccua* was supported by more than the nullipotency of its *Mus spretus* allele and its confirmed location in the center of the small *Ccua* locus (Redkar *et al.*, 1998; Samant *et al.*, 1999; Fossella *et al.*, 2000). Like the *+* allele, *Dnahc8$^t$* was highly expressed in a testis-specific manner but carried seventeen missense mutations scattered throughout its dynein stem and motor units (Samant *et al.*, 2002). Subsequent computational analysis of these mutations indicated that three of them, one near the unique N-terminus of DNAHC8 and two in well-annotated regions of its motor unit, might enhance the calcium sensitivity of the protein while at the same time perturbing the ability of the motor to bind to and/or translocate the adjacent axonemal microtubule doublet (Samant *et al.*, 2005). In addition, unlike other components of axonemal outer row dyneins, including its non-mammalian γ-axDHC orthologues, DNAHC8 was confined to the principal piece of the cauda epididymal sperm tail (Samant *et al.*, 2005). This suggested the possibility that the mid-piece and principal piece motility of mammalian sperm tails might be independently regulated through region-specific paralogous γ-axDHCs with differing dependencies on axonemal calcium concentration. The principal piece-specificity of DNAHC8 also indicated that although DNAHC8$^t$ might play a significant role in the expression of "curlicue", it would most likely play no more than a minor role in TRD since sperm from *+/t* males do not exhibit principal piece flagellar waveform aberrations. However, this result does not rule out potential *Ccub* involvement in TRD.

## *Ccub*: a complex of two haploinsufficient synergic factors

As originally specified (Redkar *et al.*, 1998), sperm from *Ccua$^s$Ccub$^s$/t* males should express *t/t* levels of "curlicue", while sperm from *Ccua$^+$Ccub$^s$/t* males should express high, but significantly reduced levels of "curlicue", relative to *t/t* levels (Fig. 4). However, sperm from a male heterozygous for *t* and an *s-+ In[17]4* recombinant homologue, *S$^{R8}$* (Redkar *et al.*, 1998), theoretically carrying both *Ccua$^s$* and *Ccub$^s$*, failed to exhibit *t/t* levels of "curlicue": suggesting that the map position of *Ccub* required further clarification (Fig. 4; Hui *et al.*, 2006). Thus, two new informative *s-+* recombinant *In[17]4* homologues were derived from another recombinant homologue shown to carry only *Ccub^s* (*S^R2*) as determined in Redkar et al. (1998). The first of these, *S^R12*, incorporating the proximal end of *S^R2*, theoretically contained *Ccub^s* while the second, *S^R14*, produced from the distal part of *S^R2* and overlapping the distal end of *S^R12*, theoretically included neither *Ccua^s* nor *Ccub^s*, as originally determined (Fig. 4; Redkar et al., 1998). Surprisingly, spermatozoa from males heterozygous for *t* and either novel recombinant exhibited negative control (+/t) levels of "curlicue". In addition, breeding experiments demonstrated that *S^R2/t* males were sterile but *S^R2/S^R2* males were completely fertile (data not shown). Together, these new data suggested that: 1) *Ccub* is complex, consisting of two synergic factors, *Ccub1* and *Ccub2*, the first mapping just distal to *Dnahc8* within a genomic interval of ~0.6 Mb and the second mapping to a region ~4-7 Mb distal to *Dnahc8* (Fig. 4); and 2) both *Ccub1^s* and *Ccub2^s* are basically wild-type alleles but are haploinsufficiently expressed in the *Mus musculus* genetic background.

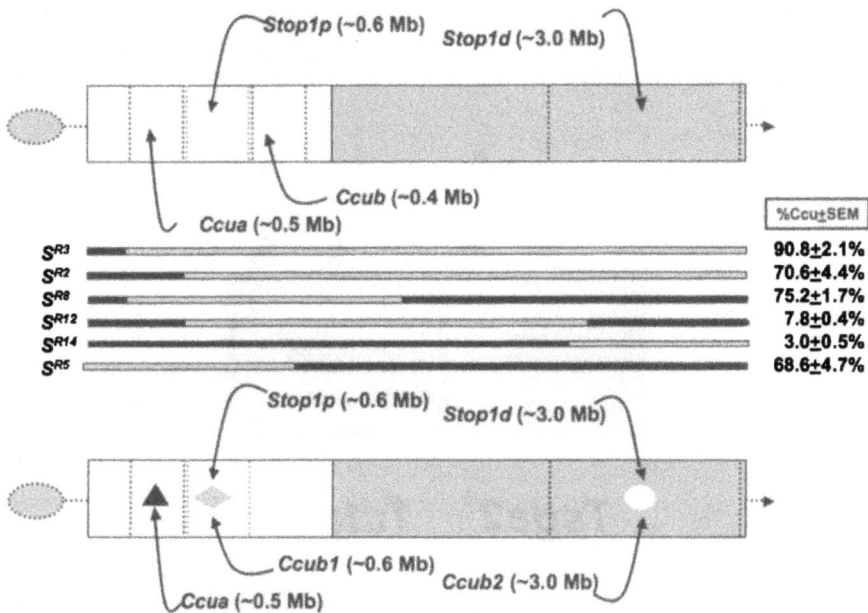

**Fig. 4** Redefinition of the *Ccub* map position. Top, the initial medium-resolution map positions of "stop" and "curlicue" factors as shown in Fig. 3. Middle, rectangular depictions of six *In[17]4* hetero-specific recombinant homologues with the name of each denoted to the left of each rectangle. Black regions represent *Mus musculus* wild-type chromatin, gray regions represent *Mus spretus* chromatin and the position of each rectangular portion is in alignment with the phenotype map shown at the top of the figure. The mean percent "curlicues" + SEM for sperm from each *In[17]4* hetero-specific recombinant homologue/t genotype are denoted to the right of each rectangle. Only sperm from *S^R3/t* males demonstrate the *t/t* level of "curlicues" (>90%). Moderately high levels exhibited by sperm from *S^R2/t*, *S^R8/t* and *S^R5/t* males are not significantly different from each other but are significantly lower than *t/t* levels (and *S^R3/t* levels). This indicates that the latter three genotypes do not carry a complete set of "curlicue" factors. Because sperm neither from *S^R12/t* or *S^R14/t* males exhibit "curlicue" levels significantly different from negative control levels, but are both derived from *S^R2* (known to carry only *Ccub*) and overlap one another, *Ccub* must consist of at least two synergic factors, *Ccub1* and *Ccub2* (whose map positions coincide with *Stop1p* and *Stop1d*, respectively, shown at the bottom). Significant differences between the means for each genotype were determined by the Tukey HSD test after ANOVA. Black triangle, gray diamond and white ellipse represent the map positions of *Dnahc8*, *Tsga2* and *Tctex5*, respectively.

## Identification of *Ccub1* and *Ccub2* candidate genes

Results of differential display-PCR experiments designed to compare the expression of testis mRNA from males homozygous for *Ccua*[s], *Ccub1*[s] and *Ccub2*[s] to the expression of testis mRNA from +/+ males have resulted in the identification of candidate *Ccub1* and *Ccub2* genes, *Tsga2* and *Tctex5*, respectively (data not shown). While the function of *Tsga2* is, at present, unknown, *Tctex5* (also known as *Ppp1r11*) codes for an inhibitory subunit of the sperm-specific protein phosphatase, PP1γ2: an important regulator of both sperm motility activation and sperm capacitation (Vijayaraghavan et al., 1996).

Not only do both genes map to the prescribed *Ccub1* and *Ccub2* loci, but just as our current model of *Ccub1* and *Ccub2* expression suggests, the *Mus spretus* allele of each candidate gene is expressed at no more than 10% of the level of its + and *t* allelic counterparts (Fig. 5). Both genes also demonstrate testis-restricted mRNA expression (Hui et al., 2006; Pilder, unpublished) and computational analyses of their *t* and *Mus spretus* alleles show that while the *Mus spretus* allele of each gene exhibits few, if any, non-synonymous mutations relative to orthologous + alleles, the *t* allele of each demonstrates numerous non-synonymous mutations in otherwise highly conserved, and thus functionally significant, regions of each protein (Fig. 6).

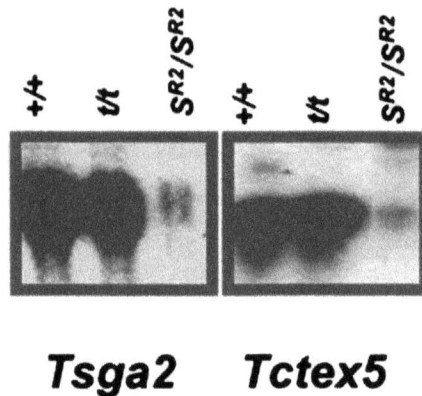

**Tsga2    Tctex5**

**Fig. 5** Northern blot analysis of steady state levels of *Tsga2* (left) and *Tctex5* (right) mRNAs from the testes of +/+, *t/t* and $S^{R2}/S^{R2}$ males.

The candidacy of *Tsga2* and *Tctex5* for *Ccub1* and *Ccub2*, respectively, is further strengthened by Western blots of sperm protein lysates probed with affinity-purified monospecific antisera, clearly demonstrating that two TSGA2 isoforms are present in cauda epididymal mouse sperm, as are two isoforms of TCTEX5 (Fig. 7A). The data concerning TSGA2 are contrary to published studies that determined that TSGA2 is confined to testicular germ cells (Tsuchida et al., 1998; Ju and Huang, 2004); however, neither of these studies had utilized a direct approach to make this determination. Indirect immunofluorescence experiments have confirmed and extended the Western blot results by indicating that both TSGA2 and TCTEX5 are present in cauda epididymal sperm tails and heads (Fig. 7B). Together, these data have provided critical evidence that *Tsga2* and *Tctex5* are viable candidate genes for the synergic flagellar waveform regulators, *Ccub1* and *Ccub2*.

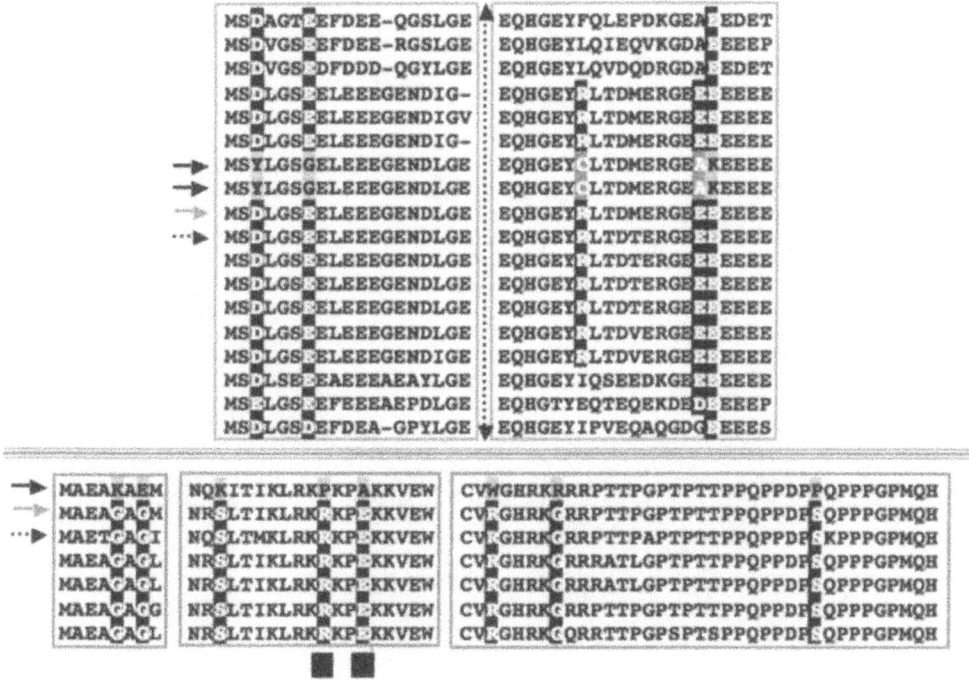

**Fig. 6** Alignments of portions of TSGA2 (top) and TCTEX5 (bottom) orthologous protein sequences demonstrating abundant non-synonymous mutations in t haplotype sequences at highly conserved sites (t haplotypes are indicated to the left of the alignment by solid black arrows; solid gray arrow indicates *Mus spretus* sequence; and dashed black arrow indicates *Mus musculus* sequence). Numerous sites in the highly conserved, acid-rich regions flanking seven MORN repeats in TSGA2 (indicated by vertical dashed arrow) are mutated to non-synonymous amino acids, thus demonstrating significant potential to perturb TSGA2 function. Bottom, numerous residues in the highly conserved N-terminus (left box), PP1-binding site (middle box) and proline-rich C-terminus (right box) of TCTEX5 are replaced by non-synonymous mutations in the t sequence. Black squares indicate two completely conserved residues at the beginning of the PP1-binding site mutated in t haplotypes.

## Do TSGA2 and TCTEX5 live double lives? A speculative conclusion

It is interesting that the newly defined *Ccub1* and *Ccub2* loci are identical to the genomic intervals circumscribing the synergic *In [17]4* sperm-oolemma penetration factors, *Stop1p* and *Stop1d*, respectively (Redkar et al., 1998; Hui et al., 2006; Pilder, unpublished). Both TSGA2 and TCTEX5 are present in cauda epididymal sperm heads as well as tails (as is PP1γ2) therefore it is tempting to speculate that each plays a signalling role common to the different pathways that modulate the calcium dependency of both sperm-oolemma penetration and flagellar waveform transformations. Experiments are currently underway to determine if the heads of capacitated, acrosome-reacted spermatozoa from either +/+ or t/t males continue to exhibit fluorescence in response to probing with either anti-TSGA2 or anti-TCTEX5 antibodies, since a loss of fluorescence might indicate possible involvement of either protein in zona pellucida binding rather than oolemma penetration.

S.H. Pilder et al.

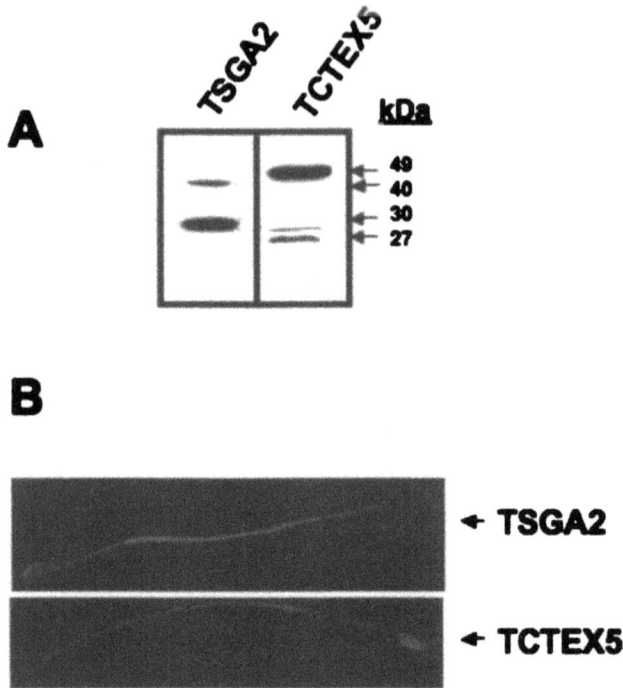

**Fig. 7** A. Western blots of sperm protein lysates probed with affinity-purified antibodies against TSGA2 (left) and TCTEX5 (right). B. TSGA2 (top) and TCTEX5 localises (bottom) to the heads and tails of cauda epididymal spermatozoa by indirect immunofluoresence.

# References

Artzt K, McCormick P and Bennett D (1982) Gene mapping within the T/t complex of the mouse. I. t-lethal genes are nonallelic *Cell* **28** 463-470

Bennett D (1975) The T-Locus of the mouse *Cell* **6** 441-454

Carlson AE, Westenbroek RE, Quill T, Ren D, Clapham DE, Hille B, Garbers DL and Babcock DF (2003) CatSper1 required for evoked $Ca^{2+}$ entry and control of flagellar function in sperm *Proceedings of the National Academy of Sciences USA* **100** 14864-14868

Fossella J, Samant SA, Silver LM, King SM, Vaughan KT, Olds-Clarke P, Johnson KA, Mikami A, Vallee RB and Pilder, SH (2000) An axonemal dynein at the *Hybrid Sterility* 6 locus: implications for t haplotype-specific male sterility and the evolution of species barriers *Mammalian Genome* **11** 8-15

Hammer MF, Schimenti J and Silver LM (1989) Evolution of mouse chromosome 17 and the origin of inversions associated with t haplotypes *Proceedings of the National Academy of Sciences USA* **86** 3261-3265

Ho HC and Suarez SS (2003) Characterization of the intracellular calcium store at the base of the sperm flagellum that regulates hyperactivated motility *Biology of Reproduction* **68** 1590-1596

Hui L, Lu J, Han Y and Pilder SH (2006) The mouse t complex gene Tsga2, encoding polypeptides located in the sperm tail and anterior acrosome, maps to a locus associated with sperm motility and sperm-egg interaction abnormalities *Biology of Reproduction* **74** 633-643

Johnson LR, Pilder SH and Olds-Clarke P (1995) The cellular basis for interaction of sterility factors in the mouse t haplotype *Genetical Research* **66** 189-193

Ju TK and Huang FL (2004) MSAP, the meichroacidin homolog of carp (*Cyprinus carpio*), differs from the rodent counterpart in germline expression and involves flagellar differentiation *Biology of Reproduction* **71** 1419-1429

Lindemann CB and Kanous KS (1997) A model for flagellar motility *International Review of Cytology* **173** 1-72

Lindemann CB, Goltz JS, Kanous KS, Gardner TK and Olds-Clarke P (1990) Evidence for an increased sensitivity to $Ca^{2+}$ in the flagella of sperm from tw32/+ mice *Molecular Reproduction and Development* **26** 69-77

Lyon MF (1984) Transmission ratio distortion in mouse *t*-haplotypes is due to multiple distorter genes acting on a responder locus *Cell* **37** 621-628

Lyon MF (1986) Male sterility of the mouse *t*-complex is due to homozygosity of the distorter genes *Cell* **44** 357-363

Lyon MF (1992) Deletion of mouse *t*-complex distorter-1 produces an effect like the *t*-form of the distorter *Genetical Research* **59** 27-33

Lyon MF (2003) Transmission ratio distortion in mice *Annual Review of Genetics* **37** 393-408

Marquez B and Suarez SS (2004) Different signaling pathways in bovine sperm regulate capacitation and hyperactivation *Biology of Reproduction* **70** 1626-1633

McGrath J and Hillman N (1980) Sterility in mutant (tLx/tLy) male mice. III. *In vitro* fertilization *Journal of Embryology and Experimental Morphology* **59** 49-58

Olds-Clarke P (1984) Genetic analysis of mammalian spermatogenesis: use of the *t* complex in the mouse in studies of spermatogenesis and sperm function *Annals of the New York Academy of Sciences* **438** 406-416

Olds-Clarke P (1986) Motility characteristics of sperm from the uterus and oviducts of female mice after mating to congenic males differing in sperm transport and fertility *Biology of Reproduction* **34** 453-467

Olds-Clarke P (1989) Sperm from tw32/+ mice: capacitation is normal, but hyperactivation is premature and nonhyperactivated sperm are slow *Developmental Biology* **131** 475-482

Olds-Clarke P and Johnson LR (1993) *t* Haplotypes in the mouse compromise sperm flagellar function *Developmental Biology* **155** 14-25

Olds-Clarke P and Wivell W (1992) Impaired transport and fertilization *in vivo* of calcium-treated spermatozoa from +/+ or congenic tw32/+ mice *Biology of Reproduction* **47** 621-628

Pilder SH, Hammer MF and Silver LM (1991) A novel mouse chromosome 17 hybrid sterility locus: implications for the origin of *t* haplotypes *Genetics* **129** 237-246

Pilder SH, Olds-Clarke P, Phillips DM and Silver LM (1993) *Hybrid sterility-6*: a mouse *t* complex locus controlling sperm flagellar assembly and movement *Developmental Biology* **159** 631-642

Redkar AA, Olds-Clarke P, Dugan LM and Pilder SH (1998) High-resolution mapping of sperm function defects in the *t* complex fourth inversion *Mammalian Genome* **9** 825-830

Redkar AA, Si Y, Twine SN, Pilder SH and Olds-Clarke P (2002) Genes in the first and fourth inversions of the mouse *t* complex synergistically mediate sperm capacitation and interactions with the oocyte *Developmental Biology* **226** 267-280

Samant SA, Fossella J, Silver LM and Pilder SH (1999) Mapping and cloning recombinant breakpoints demarcating the *hybrid sterility 6*-specific sperm tail assembly defect *Mammalian Genome* **10** 88-94

Samant SA, Ogunkua O, Hui L, Fossella J and Pilder SH (2002) The *t* complex distorter 2 candidate gene, *Dnahc8*, encodes at least two testis-specific axonemal dynein heavy chains that differ extensively at their amino and carboxyl termini *Developmental Biology* **250** 24-43

Samant SA, Ogunkua OO, Hui L, Lu J, Han Y, Orth JM and Pilder SH (2005) The mouse *t* complex distorter/sterility candidate, *Dnahc8*, expresses a γ-type axonemal dynein heavy chain isoform confined to the principal piece of the sperm tail *Developmental Biology* **285** 57-69

Schmitz-Lesich KA and Lindemann CB (2004) Direct measurement of the passive stiffness of rat sperm and implications to the mechanism of the calcium response *Cell Motility and the Cytoskeleton* **59** 169-179

Silver LM (1989) Gene dosage effects on transmission ratio distortion and fertility in mice that carry t haplotypes *Genetical Research* **54** 221-225

Silver LM and Artzt K (1981) Recombination suppression of mouse *t*-haplotypes due to chromatin mismatching *Nature* **290** 68-70

Silver LM and Olds-Clarke P (1984) Transmission ratio distortion of mouse *t* haplotypes is not a consequence of wild-type sperm degeneration *Developmental Biology* **105** 250-252

Tsuchida J, Nishina Y, Wakabayashi N, Nozaki M, Sakai Y and Nishimune Y (1998) Molecular cloning and characterization of meichroacidin (male meiotic metaphase chromosome-associated acidic protein) *Developmental Biology* **197** 67-76

Vijayaraghavan S, Stephens DT, Trautman K, Smith GD, Khatra B, da Cruz e Silva EF and Greengard P (1996) Sperm motility development in the epididymis is associated with decreased glycogen synthase kinase-3 and protein phosphatase 1 activity *Biology of Reproduction* **54** 709-718

# The role of A-kinase anchoring proteins (AKAPs) in regulating sperm function

DW Carr and AE Hanlon Newell

*Department of Medicine, Oregon Health & Science University and Portland Veterans Affairs Medical Center, Mail Code R&D 8, 3710 SW US Veterans Hospital Road, Portland, OR 97239, USA*

Cyclic AMP (cAMP)-dependent protein kinase (PKA) is a signalling molecule involved in the regulation of many physiological functions including those of cilia and flagella. PKA localizes to specific cellular structures and organelles by binding to AKAP (A-kinase anchoring protein) molecules via interaction with the regulatory subunits (RI and RII) of PKA. AKAPs are capable of forming multi-protein complexes to coordinate the action of several signalling molecules all at a single location. AKAPs also bind to a group of four proteins that share the RII dimerization/docking (R2D2) domain. R2D2 proteins are expressed at high levels in both the testis and spermatozoa and mutants lacking R2D2 proteins exhibit abnormal sperm motility. Thus AKAPs and AKAP associated proteins appear to be key molecules in the biochemical machinery regulating the functions of flagella and cilia.

## PKA and A-kinase anchoring proteins (AKAPs)

PKA is a ubiquitous, broad specificity kinase involved in the regulation of a diverse array of cellular events. The PKA holoenzyme consists of four subunits, two catalytic and two regulatory. The regulatory subunits inhibit the activity of the catalytic subunits. Binding of cAMP to the regulatory subunits promotes the dissociation and activation of the catalytic subunits. Several isoforms of both the catalytic (C$\alpha$, C$\beta$ and C$\gamma$) and regulatory (RI$\alpha$, RI$\beta$b, RII$\alpha$ and RII$\beta$) subunits of PKA have been identified and biochemically characterized (Taylor *et al.*, 1990).

A major advance in signal transduction research in recent years is the understanding that many of the actions of signal transduction molecules are spatially restricted and coordinated by cell-and function-specific targeting of both the enzymes and their substrates. A striking example of such targeted action is PKA anchoring to specific substrates or cellular compartments (reviewed by Wong and Scott, 2004). Protein kinase A anchoring is mediated by binding of the regulatory subunit (R) to an amphipathic helix binding motif located within the A-kinase anchoring proteins (AKAPs) (Carr *et al.*, 1991). Tethering of inactive PKA holoenzymes to specific compartments by AKAPs allows PKA to be immediately available to phosphorylate substrates located specifically near, and perhaps bound to, the AKAP in response to cAMP activation. Numerous AKAPs have been cloned and biochemically characterized. Several AKAPs have been shown to simultaneously bind to PKA and other signal transduction molecules; such as calmodulin, phosphodiesterases, phosphatases and other kinases. Discovery of these interac-

E-mail: carrd@ohsu.edu

tions has led to the model of AKAPs acting as scaffolding proteins that coordinate the actions of several signalling molecules all located within one cellular compartment (Wong and Scott, 2004).

## PKA/AKAPs in spermatozoa

There is ample evidence suggesting that cAMP and the cAMP-dependent protein kinase (PKA) are involved in the regulation of sperm motility (Garbers and Kopf, 1980). If localized PKA action is important in somatic cells, such a mechanism is likely to be even more critical in the highly compartmentalized spermatozoa. Several sperm AKAPs have been identified. AKAP3 (formerly known as AKAP110) is the predominant AKAP detected when an RII overlay assay is used to analyse mammalian sperm lysates (Vijayaraghavan et al., 1999). Mouse tissue Northern analysis has shown that AKAP3 is expressed in testis but not heart, brain, spleen, lung, liver, skeletal muscle and kidney. Immunofluorescent studies have detected AKAP3 in both the principal piece and acrosomal regions of spermatozoa. AKAP3 can be phosphorylated on tyrosine residues and this increase in phosphorylation correlates with an increase in motility (Luconi et al., 2005). AKAP3 has been shown to interact with AKAP4 (formerly known as fibrous sheath component 1 (FSC1) and AKAP82), a protein comprising nearly half of the total proteins found in the fibrous sheath that interacts with both RI and RII (Eddy et al., 2003). Northern analyses and immunofluorescence studies have identified AKAP11 (formerly AKAP220) expression in both human testis and mature spermatozoa (Reinton et al., 2000). AKAP11 also interacts with both RI and RII subunits of PKA and it is known to coordinate the location of PKA and the type 1 protein phosphatase catalytic subunit (PP1c) (Schillace et al., 2001). AKAP 1 (formerly AKAP84) has been shown to anchor PKAIIα as well as c-Myc-binding protein AMY-1 to mitochondria within sperm flagella (Lin et al., 1995; Furusawa et al., 2002). AKAP8 has been shown to be expressed in the mouse epididymis (Giot et al., 2003) and, like AKAP1, binds to AMY-1 (Furusawa et al., 2002): however, AKAP8 directs PKA and AMY-1 to the nuclei, rather than mitochondria. WAVE1 is a member of the Wiskott-Aldrich syndrome family of adaptor proteins and localizes to the mitochondrial sheath following epididymal passage and co-localizes with PKA RII in the mid-piece of the mitochondrial sheath (Rawe et al., 2004). MAP2, also an AKAP, is a microtubule-associated protein that is also expressed in spermatozoa. MAP2 phosphorylation appears to be involved in regulation of either sperm capacitation, acrosome reaction or both (Carr and Acott, 1990).

## AKAP/PDE interaction

As mentioned above, AKAPs often interact with other signalling molecules in addition to PKA, such as phosphodiesterases (PDEs). PDEs modulate the local concentration of cAMP, which in turn regulates the activity of PKA. To function efficiently, these two enzymes need to be located in close proximity to each other. Evidence supporting this concept has been reported for both muscle and Sertoli cells, where PKA and PDE4Ds have been shown to bind simultaneously to AKAPs, forming a 3 protein complex (Dodge et al., 2001; Tasken et al., 2001). Though there are many isoforms of PDE, only a few subtypes have been characterized in spermatozoa, the two major subtypes being PDE1 and PDE4. PDE1 regulates the acrosome reaction while PDE4 regulates motility without affecting the acrosome reaction (Fisch et al., 1998; Fournier et al., 2003; Lefievre et al., 2002; Yan et al., 2001).

Addition of PDE4 specific inhibitors, Rolipram or RS25344, significantly increases the progressive motility of bovine spermatozoa (Bajpai *et al.*, 2006). These data suggest that PDE4 plays a role in regulating motility and suggests a need for co-localization of PDE with PKA/AKAP. Immunocytochemical analysis detects both AKAP3 and PDE4A in the principal piece of bovine spermatozoa. Immunoprecipitation experiments using cells co-transfected with *Akap3* and either *Pde4a5* or *Pde4d* suggest that PDE4A5, but not PDE4D, interacts with AKAP3. Interestingly, the subcellular location and AKAP3 co-localization of PDE4A5 appears to change during sperm maturation. In both caudal and ejaculated spermatozoa, AKAP3 was detected only in the SDS-soluble fraction. However, PDE4A5 is detected primarily in the Triton X-100-soluble fraction of caudal epididymal spermatozoa and primarily in the SDS-soluble fraction of ejaculated spermatozoa. These data are consistent with a model where PDE4A5 migrates closer to AKAP3 at a time that sperm are transitioning from a non-motile state in the cauda to a fully motile state following ejaculation. Thus, AKAP3 appears to bind both PKA and PDE4A and may function as a scaffolding protein in spermatozoa to regulate local cAMP concentrations and modulate sperm functions.

## R2D2 proteins

The use of anchoring inhibitor peptides provides further evidence for the importance of AKAPs in regulating sperm motility. We have found that synthetic peptides encompassing this amphipathic helix-binding domain are very potent competitive inhibitors of PKA/AKAP interaction (Carr *et al.*, 1992). Addition of a stearate moiety to the peptide facilitates membrane permeability. Incubation of stearated Ht31 peptide (S-Ht31) with bovine caudal epididymal spermatozoa inhibits motility in a time- and concentration-dependent manner suggesting that PKA anchoring is required for motility regulation (Vijayaraghavan *et al.*, 1997). A control peptide, S-Ht31-P, identical to S-Ht31 except a proline has been substituted for an isoleucine to prevent amphipathic helix formation, had no effect on motility. Motility inhibition by S-Ht31 is reversible with time, but only if calcium is present in the suspension buffer, suggesting a role for calcium in regulating PKA anchoring. Surprisingly, inhibition of PKA catalytic activity by addition of PKA inhibitors, H-89 or S-PKI, had little effect on basal motility or motility stimulated by agents previously thought to work via PKA activation. We found this surprising because according to our working model of PKA/AKAP interaction, AIPs should mimic the effects of PKA inhibitors. Disrupting the location of PKA should produce the same effect as inhibition of PKA activity. In fact, in every reported case except for spermatozoa, AIPs do mimic PKA inhibitors (Johnson *et al.*, 1994; Rosenmund *et al.*, 1994; Lester *et al.*, 1997; Dodge *et al.*, 1999); for example, addition of either S-Ht31 or H-89 blocks the cAMP inhibition of oxytocin stimulation of PI3 turnover in myometrial cells (Dodge *et al.*, 1999).

These data suggest that proteins interacting with sperm AKAPs regulate motility in a manner that is independent of PKA catalytic activity. Two lines of evidence confirm this hypothesis. First, our measurements of sperm PKA activity show that H-89 is effective at inhibiting the catalytic activity while having little or no effect on sperm motility (Vijayaraghavan *et al.*, 1997). Second, McKnight and colleagues, using spermatozoa from mutant mice lacking RIIα, have shown that the catalytic subunit is no longer located along the flagellum but instead is concentrated in the cytoplasmic droplet, yet these spermatozoa are motile and the mice are fertile (Burton *et al.*, 1999). Thus, the inhibition of motility by AIPs does not occur via the disruption of localized phosphorylation of a PKA substrate. Therefore, the function of AKAPs in spermatozoa appears to be fundamentally different than the function of AKAPs in other cells.

One hypothesis consistent with all of the above data is that sperm AKAPs are interacting with proteins other than RII via the amphipathic helix domain. This would explain how AIPs could be regulating motility in a manner that is independent of both the regulatory and catalytic subunits of PKA. A yeast two-hybrid screen using AKAP3 as bait has identified two proteins in addition to RII that interact with the amphipathic helix region of AKAP3. These two proteins are ropporin (ROPN1) and ropporin-1-like (ROPN1L, formerly ASP) (Carr *et al.*, 2001).

ROPN1 is a binding partner of rhophilin, an effector of the small GTPase, Rho. Northern analysis has shown *Ropn1* to be exclusively expressed in testis and immunofluorescence indicates ROPN1's presence in the principal and end piece of sperm flagella (Fujita *et al.*, 2000). *In vitro* interaction between ROPN1 and AKAP3 has been shown through pull-down assays. In addition, co-localization of ROPN1 and AKAP3 to the spermatozoa flagellum is shown by immunofluorescence (Carr *et al.*, 2001). Of note, ROPN1 localizes to the cytoplasmic droplet, independent of AKAP3 in these same immunostaining studies and AKAP3 appears to localize to the dorsal surface of the acrosome, without ROPN1.

ROPN1L is a relatively uncharacterized protein that is 39% identical to human ROPN1 (Carr *et al.*, 2001). Northern analysis also indicates that *Ropn1L* is highly expressed in the testis. BLAST databases show the presence of ROPN1L in a variety of species of animals, where the amino acid sequence is conserved at greater than 95%. A recent report provides evidence that ROPN1 may be a critical component of the biochemical machinery controlling motility. This research has shown that mutants lacking RSP11, a ROPN1L homolog in the cilia of *Chlamydomonas reinhardtii*, display impaired and sporadic motility (Yang and Yang, 2006).

The N-terminal sequences of ROPN1L and ROPN1 are highly homologous. This similarity is also reflected in the AKAP binding domain contained in the N-terminus of RII and suggests a conserved domain between these proteins. This N-terminus similarity has been found, through database mining and domain alignment, in two additional proteins, Sperm Protein 17 (SP17) and calcium-binding tyrosine phosphorylation-regulated protein (CABYR, formerly FSII). Alignment of these proteins suggests a family of RII homologs that contain the RII dimerization/AKAP docking domain but not the cAMP-binding domain found in the R subunits of PKA (Figure 1). We have termed these proteins RII-like dimerization/docking domain (R2D2) proteins.

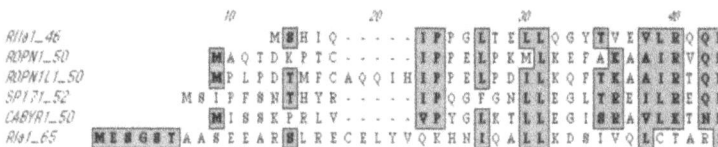

**Fig. 1** Alignment of R2D2 proteins with RI and RII

SP17 was originally thought to be uniquely expressed in spermatozoa and involved in sperm-zona pellucida binding (Richardson *et al.*, 1994). It has been shown that SP17 protein is, indeed, present in the fibrous sheath of the sperm tail as well as the sperm head (Lea *et al.*, 2004). However, recently, RT-PCR has shown that the pattern of SP17 expression is much more broad; including pancreas, lung, kidney, brain, heart, skeletal muscle and aberrant expression in tumor cells (Frayne and Hall, 2002). In addition to the R2D2 domain, SP17 contains two conserved heparin binding motifs, which are believed to contribute to zona pellucida binding, and a C-terminal calmodulin binding motif that has been shown to interact with calmodulin *in vitro* (Lea *et al.*, 2004).

CABYR is a calcium-binding protein that has been shown to localize to the ribs and longitudinal columns of the fibrous sheath of spermatozoa. This protein is tyrosine phosphorylated during the molecular changes in the sperm head and tail that allow for fertilization competence, which encompasses capacitation (Naaby-Hansen *et al.*, 2002). There are six splice variants of human CABYR, and the gene is made up of two contiguous coding regions, CR-A and CR-B, which, though under the same transcriptional control, are separated by a stop codon. It has been shown that CABYR transcripts containing the CR-B region are expressed during spermatogenesis and can assemble into the fibrous sheath of the principle piece of the sperm tail. The CR-A region of CABYR is necessary in isoforms of the protein that can bind calcium (Kim *et al.*, 2005). There are two CABYR splice variants, 281 and 379 that are expressed in a broader array of tissues: 379 in brain, lung, pancreas and testis and 281 in brain, lung and testis. These two variants have been shown to interact with glycogen synthase kinase 3 ß (GSK3ß), an enzyme involved in glycogen metabolism, gene expression, cell proliferation and development (Hsu *et al.*, 2005). Glycogen synthase kinase has also been shown to regulate flagellar assembly in *Chlamydomonas* (Wilson and Lefebvre, 2004).

Each of the R2D2 proteins discussed here have greater sequence identity to RII in the dimerization/docking region (see Figure 1) than RI does to RII, indicating that all 4 R2D2 proteins may bind to AKAPs with higher affinity than RI. Exhaustive database searches and analyses have revealed that RII, RI, ROPN1, ROPN1L, SP17 and CABYR are the only mammalian genes that contain this conserved RII dimerization/docking domain. These genes are highly conserved across many species; including zebrafish, sea urchin, *xenopus* and leishmania.

## R2D2 proteins and primary cilia

As previously mentioned, SP17 and CABYR splice variants have been shown to be expressed in a number of tissues beyond the testis and spermatozoa (Frayne and Hall, 2002; Hsu *et al.*, 2005). Database searches and recent data from our laboratory suggest that ROPN1 and ROPN1L are also expressed at a low level in a variety of tissues. This expression pattern may be explained by the increasingly appreciated presence of cilia on cells in mammalian tissue. Recent work has shown that hair-like projections known as primary cilia are vital to the function of a variety of human organs and are involved in several diseases such as polycystic kidney disease, Bardet-Biedl syndrome and Kartagener syndrome (Vogel, 2005). Cilia are present on almost all human cells including brain neurons and renal epithelial cells and are not only used for cellular movement, but also for sensing and signalling between cells (Pan *et al.*, 2005). We hypothesize that R2D2 proteins are found only in the cilia of most cells and this accounts for their low level expression in tissues other than testis. The recent report that R2D2 proteins are essential for normal functioning of cilia on *Chlamydomonas* suggests they may also play an important role in mammalian cilia and flagella.

## Summary

The importance of PKA/AKAP interaction in regulating a diverse array of cellular functions has been well documented. Previous models have implied that only the regulatory subunit of PKA interacts with the amphipathic helix domain on AKAPs and that this interaction localizes PKA near relevant downstream proteins. However, new data suggests the R2D2 family of proteins, expressed in flagella and cilia, interact with the amphipathic helix region of AKAPs in a manner similar to RII. Thus, the classic model of PKA/AKAP interaction must be broadened to incorporate these structures. In addition, it appears that PDE4A interacts with AKAP3 at times of flagellar motility in sperm, implying AKAP3's involvement as a scaffolding protein for PKA signalling-related proteins in flagella. ROPN1L appears to be important in flagellar motility and CABYR may be essential in flagellar assembly. In addition, because many of these proteins appear to be unique to cells containing cilia and flagella, continued study of these proteins should lead to a better understanding of the proteins responsible for operation and assembly of flagella and cilia in both sperm and other tissues. Characterization of these proteins, therefore, could lead to novel approaches in male contraception, fertility treatment and ciliopathy treatments.

# References

Bajpai M, Fiedler SE, Huang Z, Vijayaraghavan S, Olson GE, Livera G, Conti M and Carr DW (2006) AKAP3 selectively binds PDE4A isoforms in bovine spermatozoa *Biology of Reproduction* **74** 109-118

Burton KA, Treash-Osio B, Muller CH, Dunphy EL and McKnight GS (1999) Deletion of type II alpha regulatory subunit delocalizes protein kinase A in mouse sperm without affecting motility or fertilization *The Journal of Biological Chemistry* **274** 24131-24136

Carr DW and Acott TS (1990) The phosphorylation of a putative sperm microtubule-associated protein 2 (MAP2) is uniquely sensitive to regulation *Biology of Reproduction* **43** 795-805

Carr DW, Stofko-Hahn RE, Fraser ID, Bishop SM, Acott TS, Brennan RG and Scott JD (1991) Interaction of the regulatory subunit (RII) of cAMP-dependent protein kinase with RII-anchoring proteins occurs through an amphipathic helix binding motif *The Journal of Biological Chemistry* **266** 14188-14192

Carr DW, Hausken ZE, Fraser ID, Stofko-Hahn RE and Scott JD (1992) Association of the type II cAMP-dependent protein kinase with a human thyroid RII-anchoring protein. Cloning and characterization of the RII- binding domain *The Journal of Biological Chemistry* **267** 13376-13382

Carr DW, Fujita A, Stentz CL, Liberty GA, Olson GE and Narumiya S (2001) Identification of sperm-specific proteins that interact with A-kinase anchoring proteins in a manner similar to the type II regulatory subunit of PKA *Journal of Biological Chemistry* **276** 17332-17338

Dodge KL, Carr DW and Sanborn BM (1999) Protein kinase A anchoring to the myometrial plasma membrane is required for cyclic adenosine 3',5'-monophosphate regulation of phosphatidylinositide turnover *Endocrinology* **140** 5165-5170

Dodge KL, Khouangsathiene S, Kapiloff MS, Mouton R, Hill EV, Houslay MD, Langeberg LK and Scott JD (2001) mAKAP assembles a protein kinase A/PDE4 phosphodiesterase cAMP signaling module *European Molecular Biology Organisation Journal* **20** 1921-1930

Eddy EM, Toshimori K and O'Brien DA (2003) Fibrous sheath of mammalian spermatozoa *Microscopy Research and Technique* **61** 103-115

Fisch JD, Behr B and Conti M (1998) Enhancement of motility and acrosome reaction in human spermatozoa: differential activation by type-specific phosphodiesterase inhibitors *Human Reproduction* **13** 1248-1254

Fournier V, Leclerc P, Cormier N and Bailey JL (2003) Implication of calmodulin-dependent phosphodiesterase type 1 during bovine sperm capacitation *Journal of Andrology* **24** 104-112

Frayne J and Hall L (2002) A re-evaluation of sperm protein 17 (Sp17) indicates a regulatory role in an A-kinase anchoring protein complex, rather than a unique role in sperm-zona pellucida binding *Reproduction* **124** 767-774

Fujita A, Nakamura K, Kato T, Watanabe N, Ishizaki T, Kimura K, Mizoguchi A and Narumiya S (2000) Ropporin, a sperm-specific binding protein of rhophilin, that is localized in the fibrous sheath of sperm flagella *Journal of Cell Science* **113** 103-112

Furusawa M, Taira T, Iguchi-Ariga SM and Ariga H (2002) AMY-1 interacts with S-AKAP84 and AKAP95 in the cytoplasm and the nucleus, respectively, and inhibits cAMP-dependent protein kinase activity by preventing binding of its catalytic subunit to A-kinase-anchoring protein (AKAP) complex *The Journal of Biological Chemistry* **277** 50885-50892

Garbers DL and Kopf GS (1980) The regulation of spermatozoa by calcium cyclic nucleotides *Advanced Cyclic Nucleotide Research* **13** 251-306

Giot L, Bader JS, Brouwer C, Chaudhuri A, Kuang B, Li Y, Hao YL, Ooi CE, Godwin B, Vitols E, Vijayadamodar G, Pochart P, Machineni H, Welsh M, Kong Y, Zerhusen B, Malcolm R, Varrone Z, Collis A, Minto M, Burgess S, McDaniel L, Stimson E, Spriggs F, Williams J, Neurath K, Ioime N, Agee M, Voss E, Furtak K, Renzulli R, Aanensen N, Carrolla S, Bickelhaupt E, Lazovatsky Y, DaSilva A, Zhong J, Stanyon CA, Finley RL, Jr., White KP, Braverman M, Jarvie T, Gold S, Leach M, Knight J, Shimkets RA, McKenna MP, Chant J and Rothberg JM (2003) A protein interaction map of *Drosophila melanogaster* *Science* **302** 1727-1736

Hsu HC, Lee YL, Cheng TS, Howng SL, Chang LK, Lu PJ and Hong YR (2005) Characterization of two non-testis-specific CABYR variants that bind to GSK3beta with a proline-rich extensin-like domain *Biochemical and Biophysical Research Communications* **329** 1108-1117

Johnson BD, Scheuer T and Catterall WA (1994) Voltage-dependent potentiation of L-type Ca²⁺ channels in skeletal muscle cells requires anchored cAMP-dependent protein kinase *Proceedings of National Academy of Sciences USA* **91** 11492-11496

Kim YH, Jha KN, Mandal A, Vanage G, Farris E, Snow PL, Klotz K, Naaby-Hansen S, Flickinger CJ and Herr JC (2005) Translation and assembly of CABYR coding region B in fibrous sheath and restriction of calcium binding to coding region A *Developmental Biology* **286** 46-56

Lea IA, Widgren EE and O'Rand MG (2004) Association of sperm protein 17 with A-kinase anchoring protein 3 in flagella *Reproductive Biology and Endocrinology* **2** 57

Lefievre L, de Lamirande E and Gagnon C (2002) Presence of cyclic nucleotide phosphodiesterases PDE1A, existing as a stable complex with calmodulin, and PDE3A in human spermatozoa *Biology of Reproduction* **67** 423-430

Lester LB, Langeberg LK and Scott JD (1997) Anchoring of protein kinase A facilitates hormone-mediated insulin secretion *Proceedings of National Academy of Sciences USA* **94** 14942-14947

Lin RY, Moss SB and Rubin CS (1995) Characterization of S-AKAP84, a novel developmentally regulated A-kinase anchor protein of male germ cells *The Journal of Biological Chemistry* **270** 27804-27811

Luconi M, Porazzi I, Ferruzzi P, Marchiani S, Forti G and Baldi E (2005) Tyrosine phosphorylation of the A-kinase anchoring protein 3 (AKAP3) and soluble adenylate cyclase are involved in the increase of human sperm motility by bicarbonate *Biology of Reproduction* **72** 22-32

Naaby-Hansen S, Mandal A, Wolkowicz MJ, Sen B, Westbrook VA, Shetty J, Coonrod SA, Klotz KL, Kim YH, Bush LA, Flickinger CJ and Herr JC (2002) CABYR, a novel calcium-binding tyrosine phosphorylation-regulated fibrous sheath protein involved in capacitation *Developmental Biology* **242** 236-254

Pan J, Wang Q and Snell WJ (2005) Cilium-generated signaling and cilia-related disorders *Laboratory Investigation* **85** 452-463

Rawe VY, Ramalho-Santos J, Payne C, Chemes HE and Schatten G (2004) WAVE1, an A-kinase anchoring protein, during mammalian spermatogenesis *Human Reproduction* **19** 2594-2604

Reinton N, Collas P, Haugen TB, Skalhegg BS, Hansson V, Jahnsen T and Tasken K (2000) Localization of a novel human A-kinase-anchoring protein, hAKAP220, during spermatogenesis *Developmental Biology* **223** 194-204

Richardson RT, Yamasaki N and O'Rand MG (1994) Sequence of a rabbit sperm zona pellucida binding protein and localization during the acrosome reaction *Developmental Biology* **165** 688-701

Rosenmund C, Carr DW, Bergeson SE, Nilaver G, Scott JD and Westbrook GL (1994) Anchoring of protein kinase A is required for modulation of AMPA/kainate receptors on hippocampal neurons *Nature* **368** 853-856

Schillace RV, Voltz JW, Sim AT, Shenolikar S and Scott JD (2001) Multiple interactions within the AKAP220 signaling complex contribute to protein phosphatase 1 regulation *The Journal of Biological Chemistry* **276** 12128-12134

Tasken KA, Collas P, Kemmner WA, Witczak O, Conti M and Tasken K (2001) Phosphodiesterase 4D and protein kinase a type II constitute a signaling unit in the centrosomal area *The Journal of Biological Chemistry* **276** 21999-22002

Taylor SS, Buechler JA and Yonemoto W (1990) cAMP-dependent protein kinase: framework for a diverse family of regulatory enzymes *Annual Review of Biochemistry* **59** 971-1005

Vijayaraghavan S, Goueli SA, Davey MP and Carr DW (1997) Protein kinase A-anchoring inhibitor peptides arrest mammalian sperm motility *The Journal of Biological Chemistry* **272** 4747-4752

Vijayaraghavan S, Liberty GA, Mohan J, Winfrey VP, Olson GE and Carr DW (1999) Isolation and molecular characterization of AKAP110, a novel, sperm- specific protein kinase A-anchoring protein *Molecular Endocrinology* **13** 705-717

Vogel G (2005) News focus: Betting on cilia. *Science* **310** 216-218

Wilson NF and Lefebvre PA (2004) Regulation of flagellar assembly by glycogen synthase kinase 3 in *Chlamydomonas reinhardtii Eukaryotic Cell* **3** 1307-1319

Wong W and Scott JD (2004) AKAP signalling complexes: focal points in space and time *Nature Reviews Molecular Cell Biology* **5** 959-970

Yan C, Zhao AZ, Sonnenburg WK and Beavo JA (2001) Stage and cell-specific expression of calmodulin-dependent phosphodiesterases in mouse testis *Biology of Reproduction* **64** 1746-1754

Yang C and Yang P (2006) The flagellar motility of chlamydomonas pf25 mutant lacking an AKAP-binding protein is overtly sensitive to medium conditions *Molecular Biology of the Cell* **17** 227-238

# Glycodelin: a molecule with multi-functions on spermatozoa

William SB Yeung[1], Kai-Fai Lee[1], Riitta Koistinen[2,3], Hannu Koistinen[3], Markku Seppala[3], PC Ho[1] and Philip CN Chiu[1]

[1] Department of Obstetrics and Gynaecology, University of Hong Kong, Queen Mary Hospital, Pokfulam Road, Hong Kong, China; [2] Department of Obstetrics and Gynaecology and [3] Clinical Chemistry, University Central Hospital, 00029 HUS Helsinki, Finland

Glycodelin is a glycoprotein with three isoforms, namely glycodelin-S, glycodelin-A and glycodelin-F. They have similar protein core but different glycan side chains. Spermatozoa are exposed to these isoforms during their passage to the fertilization site. They first encounter glycodelin-S in the seminal plasma. Data suggest that glycodelin-S suppresses albumin-induced cholesterol loss and maintains the spermatozoa in an uncapacitated state before they enter into the cervical canal where glycodelin-S is removed. This allows albumin in the uterine cavity to initiate capacitation. In the fallopian tube, the spermatozoa are exposed to glycodelin-A and –F produced by the fallopian tube. Glycodelin-A is an endogenous glycoprotein that inhibits the binding of spermatozoa to the zona pellucida. Glycodelin-A may protect the spermatozoa from maternal immune attack by its immunosuppressive activity. Glycodelin-F is the main glycodelin isoform in the follicular fluid. Similar to glycodelin-A, it inhibits spermatozoa-zona pellucida binding. The biological significance of the anti-fertilization activity of glycodelin-A and -F remains to be established. Glycodelin-F also suppresses progesterone-induced acrosome reaction. This is important to prevent premature acrosome reaction when the spermatozoa are swimming through the cumulus mass towards the oocyte and become exposed to progesterone produced by the cumulus cells. In summary, different isoforms of glycodelin act in succession to modulate different aspects of sperm function, and thereby, contribute to the success of fertilization.

### Modulation of sperm functions in the female reproductive tract

Proper functioning of a spermatozoon and an oocyte is crucial to the success of fertilization. After ovulation, the oocyte, the cumulus cell mass and the entrapped follicular fluid are expelled from the ovary and are transported to the oviduct. Spermatozoa in the seminal plasma are deposited into the female vagina during intercourse. In order to fertilize an oocyte, the spermatozoa have to travel a long distance from the cervix through the uterine cavity to the

E-mail: wsbyeung@hkucc.hku.hk

oviduct. Fertilization occurs in the oviduct when the spermatozoa traverse through the cumulus cell mass and reach the oocyte.

It is well known that the reproductive tract modulates sperm functions to prepare the spermatozoa for fertilization. Indeed, the spermatozoa acquire their fertilization capacity in the female reproductive tract by a process known as capacitation, which is an ill-defined process with unclear mechanisms. In addition, the functional parameters of capacitated spermatozoa are modulated by the oviductal cells, cumulus cells and follicular fluid at the site of fertilization. The human oviductal cells may act as a sperm reservoir and maintain the motility and viability of human spermatozoa (Yao et al., 1999a). Evidence suggests that the cumulus oophorus is involved in selecting spermatozoa with normal morphology (Carrell et al., 1993; Hong et al., 2004) and intact acrosome (Yanagimachi, 1994) for fertilization. The influence of the different components on sperm function at the site of fertilization can be antagonistic, suggesting that there is delicate balance among these modulatory activities. Thus human follicular fluid inhibits spermatozoa-zona pellucida binding (Yao et al., 1996) while the cumulus cells reduce this inhibitory effect (Hong et al., 2003). Spermatozoa-oocyte fusion is inhibited by the human oviductal cells but is stimulated by human follicular fluid (Yao et al., 1999b). However, the molecular mechanisms of these interactions are relatively unclear.

## Glycodelin

Glycodelin is a member of the lipocalin protein family. It was previously known as placental protein 14, progesterone-associated endometrial protein, chorionic α2-globulin, placental α2-microglobulin, α-uterine protein and pregnancy-associated α2 globulin. The protein is encoded by a gene on chromosome 9q, band 34 (Van Cong et al., 1991) with seven exons and six introns (Vaisse et al., 1990). Based on the nucleotide sequence, the translated protein has a molecular mass of 18,787 Da (Julkunen et al., 1988). Post-translational glycosylation of the protein produces three known secretory isoforms with molecular sizes ranged from 27-30 kDa (Seppala et al., 2002; Chiu et al., 2003a).

The three isoforms of glycodelin are glycodelin-A (amniotic fluid-derived isoform), glycodelin-S (seminal plasma-derived isoform) and glycodelin-F (follicular fluid-derived isoform). They have the same protein core but with different glycosylation. Prof. A. Dell and co-workers (Dell et al., 1995; Morris et al., 1996) determined the structure of the N-glycans of glycodelin-S and -A using mass spectrometry. They found that the glycans of glycodelin-S were unusual among secreted human glycoproteins in having no sialylated glycans (Morris et al., 1996). Compared with glycodelin-A, glycodelin-S has many more fucose residues. The glycan structure of glycodelin-F is not yet known. However, the lectin binding characteristics and the mobility of glycans in fluorophore-assisted carbohydrate electrophoresis show that the glycosylation of glycodelin-F differs from those of the other two isoforms (Chiu et al., 2003a).

### Interaction of spermatozoa with glycodelin isoforms in the reproductive tract

*In ejaculates*

The epithelial cells of the seminal vesicle in the male synthesize glycodelin-S (Julkunen et al., 1984). There is no synthesis of glycodelin in the testis and epididymis (Koistinen et al., 1997). During ejaculation, glycodelin-S is secreted into the seminal plasma (Julkunen et al., 1984; Koistinen et al., 2000). Its concentration in the seminal plasma is high and may occupy up to 2.5% of the total protein (Bolton et al., 1986). Glycodelin-S is the first glycodelin isoform that interacts with spermatozoa in the reproductive tract.

The binding of glycodelin-S to human spermatozoa is specific and other lipocalins cannot compete with it for binding sites on spermatozoa (Chiu *et al.*, 2005). Gylcodelin-S suppresses capacitation probably via its inhibitory activity on albumin-induced cholesterol efflux from spermatozoa (Chiu *et al.*, 2005). Loss of cholesterol from sperm membrane is known to initiate capacitation (Cross, 1998). This action of glycodelin-S is important in controlling the timing of capacitation as capacitated spermatozoa lose their response to zona pellucida-induced acrosome reaction, a crucial step in fertilization, a few hours after capacitation (Cohen-Dayag *et al.*, 1995). Thus, glycodelin-S ensures that the spermatozoa in the ejaculate are not capacitated before they enter into the female reproductive tract.

In order for capacitation to take place, glycodelin-S is removed during the passage of the spermatozoa through the cervical mucus (Chiu *et al.*, 2005). The low affinity of glycodelin-S to human spermatozoa is consistent with the readiness of removal of sperm-bound glycodelin-S during cervical mucus migration (Chiu *et al.*, 2005). On the other hand, the specific binding, the high concentration of glycodelin-S in seminal plasma and the fast binding kinetics ensure specific and rapid action of glycodelin-S on spermatozoa (Chiu *et al.*, 2005), which is essential to the function of glycodelin-S as the spermatozoa will only be in the seminal plasma briefly before they enter into the cervical canal.

The ability of glycodelin-S in suppressing albumin-induced capacitation has physiological relevance as albumin, though absent in the seminal plasma, is abundant in the uterine fluid (Setchell, 1974). Therefore, the removal of glycodelin-S from spermatozoa in the cervical mucus would allow albumin to initiate capacitation in the uterine cavity. Other molecules in the cervical mucus may also affect capacitation (Tsibris *et al.*, 1982). Glycodelin is present in the cervical mucus (Connor *et al.*, 2000). However, the isoform in the mucus and its action, if any, on sperm function are not known.

Seminal plasma has immunosuppressive activity (Gonzales, 2001). Three observations suggest that glycodelin-S contributes to this activity of seminal plasma. First, immunoadsorption with anti-glycodelin antibody reduces the inhibitory activity of seminal plasma on lymphocyte proliferation (Bolton *et al.*, 1987). Second, the protein backbone of glycodelin-S is identical to that of glycodelin-A (Koistinen *et al.*, 1999), which possesses immunosuppressive activity (Bolton *et al.*, 1987). Third, the protein backbone of glycodelin-A contributes to the immunosuppressive activity of the molecule (Jayachandran *et al.*, 2004).

*In the uterus*

Although the secretory human endometrium is known to produce glycodelin (Julkunen *et al.*, 1986a; Waites *et al.*, 1988), it is only recently concluded that the glycodelin isoform from secretory endometrium, pregnancy decidua, amniotic fluid and pregnancy serum is glycodelin-A based on their similarity in physicochemical analyses (Koistinen *et al.*, 2003). The production of glycodelin in the endometrium shows cyclical variation. Endometrial glycodelin is present in the first 2-3 days of the cycle and is basically absent in the rest of the proliferative phase (Julkunen *et al.*, 1986a). In the secretory endometrium, glycodelin becomes detectable 4 days after the luteinizing hormone surge (LH + 4) and is abundant on LH + 10 (Brown *et al.*, 2000). These observations are in line with several microarray analyses showing a strong up-regulation of endometrial glycodelin mRNA expression in the peri-implantation period when compared to non-receptive endometrium (Kao *et al.*, 2002; Horcajadas *et al.*, 2004). Nearly all pregnancies can be attributed to intercourse before ovulation (Wilcox *et al.*, 1995). The absence of glycodelin-A in the endometrium at this period suggests that spermatozoa do not encounter glycodelin-A in the uterine cavity.

The function of glycodelin-A in the endometrium is likely to be related to feto-maternal defence mechanisms (Clark et al., 1996). This is consistent with the high expression of glycodelin-A in the endometrium in peri-implantation period and during early placentation (Julkunen et al., 1985). There are large granular lymphocytes and natural killer (NK) cells in the decidua (Gurka and Rocklin, 1987; Weetman, 1999). For some unknown reason, the decidual NK cells are less aggressive than the peripheral NK cells (Dosiou and Giudice, 2005). Glycodelin-A has immunosuppressive properties (Bolton et al., 1987). It suppresses proliferation (Rachmilewitz et al., 1999) and enhances apoptosis of T-cells (Mukhopadhyay et al., 2001). The action of glycodelin-A on T-cell activation is synergistically enhanced in the presence of pregnancy zone protein (Skornicka et al., 2005). Glycodelin-A also inhibits peripheral NK cell activity (Okamoto et al., 1991). Its role in reducing the aggressiveness of the decidual NK cells remains to be illustrated.

### In the oviduct

The fallopian tube contains glycodelin mRNA (Saridogan et al., 1997). Glycodelin protein is produced in the fallopian tube throughout the cycle with concentration higher in the secretory phase than in the proliferative phase (Julkunen et al., 1986b). Cultured human tubal epithelial cells secrete glycodelin into the culture medium (Laird et al., 1995). In these studies, the techniques used to study the expression of glycodelin, mRNA analysis or immunological techniques, were incapable of differentiating the various glycodelin isoforms that differ in glycosylation only. With the successive use of anti-glycodelin affinity chromatography and ion-exchange chromatography, we have preliminary data showing that the human oviductal cells produce glycodelin-A and glycodelin-F in about equal proportions. Therefore, the spermatozoa are likely to interact with both glycodelin isoforms in the fallopian tube.

Glycodelin-A was the first glycodelin isoform known to affect sperm function. By comparing the binding of spermatozoa with or without prior glycodelin-A treatment to hemizonae from the same oocyte in a hemizona binding assay, Oehninger and coworkers demonstrated that glycodelin-A dose-dependantly suppressed the binding of spermatozoa to the zona pellucida (Oehninger et al., 1995). Glycodelin-A was the first endogenous glycoprotein found to have such anti-fertilization activity.

As foreign objects, spermatozoa should be immunogenic in the maternal body (Anderson and Tarter, 1982). This is supported by the observation that viable spermatozoa enhance lymphocyte response in vitro (Gutierrez et al., 2003). As glycodelin-A is immunosuppressive, it may protect the oviductal spermatozoa from maternal immune attack by binding to the spermatozoa with its glycan chains and suppressing lymphocyte activity with its protein core, which is important for the immunosuppressive activity of the molecule (Jayachandran et al., 2004).

### In the follicular fluid

During ovulation, follicular fluid is released with the oocyte-cumulus cell mass. They are transported to the oviduct. Our unpublished observations show that follicular fluid contains mainly glycodelin-F. Therefore, oviductal spermatozoa will be exposed to more glycodelin-F as they swim towards the oocyte-cumulus cell mass. Glycodelin-F is the latest identified glycodelin isoform. It was first isolated from human follicular fluid and was named as zona binding inhibitory factor-1 (Yao et al., 1998) because of its inhibitory activity on binding of human spermatozoa to the zona pellucida (Yao et al., 1996). The molecule was subsequently found to be an isoform of glycodelin and was renamed as glycodelin-F in accordance with the glycodelin nomenclature as the major glycodelin isoform in the follicular fluid (Chiu et al., 2003a).

Three observations suggest that the zona binding inhibitory activity of glycodelin may be a physiological phenomenon. First, the activity is present in follicular fluids obtained from both gonadotrophin-stimulated and natural reproductive cycles (Yao et al., 1996; Qiao et al., 1998; Chiu et al., 2002). Second, both glycodelin-A and –F bind to the acrosome region of human spermatozoa, a region in the sperm head that is critical for fertilization (Chiu et al., 2003a). Third, solubilized human zona pellucida proteins displace bound glycodelin-A and -F from isolated sperm membrane (Chiu et al., 2003b; PCN Chiu and WSB Yeung, unpublished data), suggesting that the sperm receptor for glycodelin-A and -F may be part of the zona protein receptor complex (Thaler and Cardullo, 1996), or molecules closely associated with the zona protein receptors. Despite these interesting observations, the physiological significance of the phenomenon remains unknown.

Neither glycodelin-A nor -F affect spontaneous acrosome reaction. However, glycodelin-F suppresses progesterone-induced acrosome reaction while glycodelin-A does not possess such activity (Chiu et al., 2003b). This is in line with the observation that glycodelin-F has an additional high affinity sperm receptor apart from the low affinity receptor that is shared with glycodelin-A (Chiu et al., 2003b). It is likely that this glycodelin-F action on progesterone-induced acrosome reaction occurs *in vivo* as the concentration of glycodelin-F required to elicit the activity is well within the estimated physiological concentration of glycodelin in both the follicular fluid and cumulus cell matrix.

Thus, glycodelin-F may protect spermatozoa from undergoing premature acrosome reaction induced by progesterone secreted from the cumulus cells (Chian et al., 1999) and the luteinised granulosa cells after ovulation. This protective action starts when the spermatozoa are in the oviduct, supported by the observation that human oviductal fluid inhibits progesterone-induced acrosome reaction (Zhu et al., 1994). The spermatozoa are exposed to more progesterone when they migrate towards the oocyte-cumulus mass for fertilization. Glycodelin-F in the follicular fluid provides additional protection against premature acrosome reaction. This protection is important for fertilization as the acrosome reaction reduces the binding affinity of spermatozoa to zona pellucida (Yanagimachi, 1994) and acrosome-reacted spermatozoa has difficulty in penetrating the cumulus mass (Saling, 1989).

## In the cumulus oophorus

The granulosa cells in the ovarian follicle synthesize glycodelin as evidenced by the presence of glycodelin mRNA in luteinised granulosa cells (Tse et al., 2002). In fact, glycodelin immunoreactivity is detectable in the granulosa cells at the late secondary follicle stage (Tse et al., 2002). Instead of synthesizing glycodelin, the cumulus cells specifically take up glycodelin-A and glycodelin-F from the surrounding environment even in the presence of other members of lipocalin family (Tse et al., 2002; PCN Chiu and WSB Yeung, unpublished data). The uptaken glycodelins are converted into a molecule of smaller size. Compared with other glycodelin isoforms, this smaller molecule has a protein core of similar size, suggesting that the molecule has a lower proportion of carbohydrate moieties. The functional significance of this smaller glycodelin molecule on sperm function is being investigated.

Both glycodelin-A and -F inhibit spermatozoa-zona pellucida binding (Oehninger et al., 1995; Chiu et al., 2003a). They have to be removed before fertilization. The observations that the cumulus/corona cells reduce the zona binding inhibitory activity of follicular fluid (Hong et al., 2003) and that the zona binding ability of spermatozoa increase after passing through the cumulus oophorus (Hong *et al.*, 2004) suggest that the removal of glycodelin-A and -F occurs during the migration of the spermatozoa through the cumulus oophorus. How the cumulus cells remove sperm-bound glycodelin is an unanswered question. The cumulus cells may do so as they take up glycodelin from the follicular fluid. However, this depends on fortuitous collision between the spermatozoa and the cumulus cells and is unlikely to remove all the sperm-bound glycodelin within a short period. We are currently testing an alternative possibility; that is, that molecules in the cumulus cell matrix displace the sperm-bound glycodelins.

### Glycosylation and biological activities of glycodelin

The glycodelin isoforms differ only in their glycosylation and have differential effects on sperm function. The differences in glycan structure between glycodelin-A and –S have been used to explain the inability of glycodelin-S in inhibiting spermatozoa-zona pellucida binding (Morris *et al.*, 1996). It is not surprising to find that all known effects of glycodelin on sperm function are glycosylation dependent. Deglycosylation abolishes the binding of glycodelin-S to spermatozoa and the inhibitory activity of the molecule on albumin-induced capacitation (Chiu *et al.*, 2005). Similarly, deglycosylation also abolishes the binding of glycodelin-A and -F and their inhibitory activity on spermatozoa-zona pellucida binding (Chiu *et al.*, 2003b).

The different glycans in the glycodelin isoforms bind to different sperm receptors. Binding kinetic studies demonstrated that the human sperm membrane possesses two glycodelin-F receptors. The low affinity receptor also binds glycodelin-A (Chiu *et al.*, 2003b). Different carbohydrate moieties are involved in the binding of the isoforms to the sperm receptors. Glycodelin-A binding involves mannose, fucose and possibly E-selectin ligand, while that of glycodelin-F involves mannose, fucose and N-acetylglucosamine, but not selectin ligands (Chiu *et al.*, 2004). Glycodelin-S also possesses two receptors with binding properties distinct from that of the other isoforms and glycodelin-A and –F cannot compete with glycodelin-S for its binding sites (Chiu *et al.*, 2005).

Unlike the action on sperm function, the immunosuppressive activity of glycodelin-A is mainly mediated via its protein backbone. Site directed mutagenesis at the glycosylation sites removes the glycan chain but does not affect the anti-proliferative property of glycodelin on the Jurkat T cell line (Jayachandran *et al.*, 2004). However, glycosylation is not totally without effect on the immunosuppressive activities of the molecule. Glycosylation affects the rate of secretion of glycodelin (Jayachandran *et al.*, 2004). It has been suggested that the negatively charged sialic acids of glycodelin-A cause charge repulsion of the carbohydrate side chains, expose the apoptogenic region in the protein core of the molecule to T lymphocytes, and thereby, enhance T-lymphocyte apoptosis (Jayachandran *et al.*, 2004). The lack of sialic acid in glycodelin-S has been used to account for the inability of glycodelin-S in enhancing apoptosis of the T-lymphocytes (Mukhopadhyay *et al.*, 2004).

### Summary

Spermatozoa are exposed to different glycodelin isoforms in the female reproductive tract. The abundance of glycodelin in the reproductive tract and the diverse actions of glycodelin isoforms on spermatozoa are in line with important roles of the gene in regulating sperm function and therefore, fertilization. Glycodelin provides an excellent model for functional glycomics. It demonstrates that post-translational modification in glycosylation is powerful in modulating the biological activity of a single gene. The identification of glycodelin receptors on spermatozoa and the glycans responsible for the biological activities of glycodelin will be useful in designing new drugs for fertility regulation.

## Acknowledgements

This work was supported by the Research Grant Council, Hong Kong (HKU7188/99M, HKU7261/01M, HKU 7408/03M), CRCG, University of Hong Kong, Helsinki University Central Hospital Research Funds, Federation of the Finnish Life and Pension Insurance Companies, the Cancer Society of Finland, the Academy of Finland and the University of Helsinki.

## References

**Anderson DJ and Tarter TH** (1982) Immunosuppressive effects of mouse seminal plasma components *in vivo* and *in vitro Journal of Immunology* **128** 535-539

**Bolton AE, Pinto-Furtado LG, Andrew CE and Chapman MG** (1986) Measurement of the pregnancy-associated proteins, placental protein 14 and pregnancy-associated plasma protein A in human seminal plasma *Clinical Reproduction and Fertility* **4** 233-240

**Bolton AE, Pockley AG, Clough KJ, Mowles EA, Stoker RJ, Westwood OM and Chapman MG** (1987) Identification of placental protein 14 as an immunosuppressive factor in human reproduction *Lancet* **1(8533)** 593-595

**Brown SE, Mandelin E, Oehninger S, Toner JP, Seppala M and Jones HW Jr** (2000) Endometrial glycodelin-A expression in the luteal phase of stimulated ovarian cycles *Fertility and Sterility* **74** 130-133

**Carrell DT, Middleton RG, Peterson CM, Jones KP and Urry RL** (1993) Role of the cumulus in the selection of morphologically normal sperm and induction of the acrosome reaction during human *in vitro* fertilization *Archives of Andrology* **31** 133-137

**Chian RC, Ao A, Clarke HJ, Tulandi T and Tan SL** (1999) Production of steroids from human cumulus cells treated with different concentrations of gonadotropins during culture *in vitro Fertility and Sterility* **71** 61-66

**Chiu PCN, Ho PC, Ng EH and Yeung WSB** (2002) Comparative study of the biological activity of spermatozoa-zona pellucida binding inhibitory factors from human follicular fluid on various sperm function parameters *Molecular Reproduction and Development* **61** 205-212

**Chiu PCN, Koistinen R, Koistinen H, Seppala M, Lee KF and Yeung WSB** (2003a) Zona-binding inhibitory factor-1 from human follicular fluid is an isoform of glycodelin *Biology of Reproduction* **69** 365-372

**Chiu PCN, Koistinen R, Koistinen H, Seppala M, Lee KF and Yeung WSB** (2003b) Binding of zona binding inhibitory factor-1 (ZIF-1) from human follicular fluid on spermatozoa *Journal of Biological Chemistry* **278** 13570-13577

**Chiu PCN, Tsang HY, Koistinen R, Koistinen H, Seppala M, Lee KF and Yeung WSB** (2004) The contribution of D-mannose, L-fucose, N-acetylglucosamine, and selectin residues on the binding of glycodelin isoforms to human spermatozoa *Biology of Reproduction* **70** 1710-1719

**Chiu PCN, Chung MK, Tsang HY, Koistinen R, Koistinen H, Seppala M, Lee KF and Yeung WSB** (2005) Glycodelin-S in human seminal plasma reduces cholesterol efflux and inhibits capacitation of spermatozoa *Journal of Biological Chemistry* **280** 25580-25589

**Clark GF, Oehninger S, Patankar MS, Koistinen R, Dell A, Morris HR, Koistinen H and Seppälä M** (1996) A role for glycoconjugates in human development: the human feto-embryonic defence system hypothesis *Human Reproduction* **11** 467–473

**Cohen-Dayag A, Tur-Kaspa I, Dor J, Mashiach S and Eisenbach M** (1995) Sperm capacitation in humans is transient and correlates with chemotactic responsiveness to follicular factors *Proceedings of National Academy of Sciences USA* **92** 11039-11043

**Connor JP, Brudney A, Ferrer K and Fazleabas AT** (2000) Glycodelin-A expression in the uterine cervix *Gynecological Oncology* **79** 216-219

**Cross NL** (1998) Role of cholesterol in sperm capacitation *Biology of Reproduction* **59** 7-11

**Dell A, Morris HR, Easton RL, Panico M, Patankar M, Oehninger S, Koistinen R, Koistinen H, Seppala M and Clark GF** (1995) Structural analysis of the oligosaccharides derived from glycodelin, a human glycoprotein with potent immunosuppressive and contraceptive activities *Journal of Biological Chemistry* **270** 24116-24126

**Dosiou C and Giudice LC** (2005) Natural killer cells in pregnancy and recurrent pregnancy loss: endocrine and immunologic perspectives *Endocrine Review* **26** 44-62

**Gonzales GF** (2001) Function of seminal vesicles and their role on male fertility *Asian Journal of Andrology* **3** 251-258

**Gurka G and Rocklin RE** (1987) Reproductive immunology *Journal of American Medical Association* **258** 2983-2987

**Gutierrez G, Fitzgerald JS, Pohlmann T, Hoppe I and Markert UR** (2003) Comparative effects of L-tryptophan and 1-methyl-tryptophan on immunoregulation induced by sperm, human pre-implantation embryo and trophoblast supernatants *American Journal of Reproductive Immunology* **50** 309-315

Hong SJ, Tse JY, Ho PC and Yeung WSB (2003) Cumulus cells reduce the spermatozoa-zona binding inhibitory activity of human follicular fluid *Fertility and Sterility* **79** Suppl. **1** 802-807

Hong SJ, Chiu PCN, Lee KF, Tse JM, Ho PC and Yeung WSB (2004) Establishment of a capillary-cumulus model to study the selection of sperm for fertilization by the cumulus oophorus *Human Reproduction* **19** 1562-1569

Horcajadas JA, Riesewijk A, Martin J, Cervero A, Mosselman S, Pellicer A and Simon C (2004) Global gene expression profiling of human endometrial receptivity *Journal of Reproductive Immunology* **63** 41-49

Jayachandran R, Shaila MS and Karande AA (2004) Analysis of the role of oligosaccharides in the apoptotic activity of glycodelin A *Journal of Biological Chemistry* **279** 8585-8591

Julkunen M, Wahlstrom T, Seppala M, Koistinen R, Koskimies A, Stenman UH and Bohn H (1984) Detection and localization of placental protein 14-like protein in human seminal plasma and in the male genital tract *Archives of Andrology* **12** Suppl 59-67

Julkunen M, Rutanen EM, Koskimies A, Ranta T, Bohn H and Seppala M (1985) Distribution of placental protein 14 in tissues and body fluids during pregnancy *British Journal of Obstetrics and Gynaecology* **92** 1145-1151

Julkunen M, Koistinen R, Sjoberg J, Rutanen EM, Wahlstrom T and Seppala M (1986a) Secretory endometrium synthesizes placental protein 14 *Endocrinology* **118** 1782-1786

Julkunen M, Wahlstrom T and Seppala M (1986b) Human fallopian tube contains placental protein 14 *American Journal of Obstetrics and Gynecology* **154** 1076-1079

Julkunen M, Seppala M and Janne OA (1988) Complete amino acid sequence of human placental protein 14: a progesterone-regulated uterine protein homologous to beta-lactoglobulins *Proceedings of National Academy of Sciences USA* **85** 8845-8849

Kao LC, Tulac S, Lobo S, Imani B, Yang JP, Germeyer A, Osteen K, Taylor RN, Lessey BA and Giudice LC (2002) Global gene profiling in human endometrium during the window of implantation *Endocrinology* **143** 2119-2138

Koistinen H, Koistinen R, Kamarainen M, Salo J and Seppala M (1997) Multiple forms of messenger ribonucleic acid encoding glycodelin in male genital tract *Laboratory Investigation* **76** 683-690

Koistinen H, Koistinen R, Seppala M, Burova TV, Choiset Y and Haertle T (1999) Glycodelin and beta-lactoglobulin, lipocalins with a high structural similarity, differ in ligand binding properties *FEBS Letters* **450** 158-162

Koistinen H, Koistinen R, Hyden-Granskog C, Magnus O and Seppala M (2000) Seminal plasma glycodelin and fertilization *in vitro Journal of Andrology* **21** 636-640

Koistinen H, Easton RL, Chiu PCN, Chalabi S, Halttunen M, Dell A, Morris HR, Yeung WSB, Seppala M and Koistinen R (2003) Differences in glycosylation and sperm-egg binding inhibition of pregnancy-related glycodelin *Biology of Reproduction* **69** 1545-1551

Laird SM, Hill CJ, Warren MA, Tuckerman EM and Li TC (1995) The production of placental protein 14 by human uterine tubal epithelial cells in culture *Human Reproduction* **10** 1346-1351

Morris HR, Dell A, Easton RL, Panico M, Koistinen H, Koistinen R, Oehninger S, Patankar MS, Seppala M and Clark GF (1996) Gender-specific glycosylation of human glycodelin affects its contraceptive activity *Journal of Biological Chemistry* **271** 32159-32167

Mukhopadhyay D, Sundereshan S, Rao C and Karande AA (2001) Placental protein 14 induces apoptosis in T cells but not in monocytes *Journal of Biological Chemistry* **276** 28268-28273

Mukhopadhyay D, SundarRaj S, Alok A and Karande AA (2004) Glycodelin A, not glycodelin S, is apoptotically active: Relevance of sialic acid modification *Journal of Biological Chemistry* **279** 8577-8584

Oehninger S, Coddington CC, Hodgen GD and Seppala M (1995) Factors affecting fertilization: endometrial placental protein 14 reduces the capacity of human spermatozoa to bind to the human zona pellucida *Fertility and Sterility* **63** 377-383

Okamoto N, Uchida A, Takakura K, Kariya Y, Kanzaki H, Riittinen L, Koistinen R, Seppala M and Mori T (1991) Suppression by human placental protein 14 of natural killer cell activity *American Journal of Reproductive Immunology* **26** 137-142

Qiao J, Yeung WSB, Yao YQ and Ho PC (1998) The effects of follicular fluid from patients with different indications for IVF treatment on the binding of human spermatozoa to the zona pellucida *Human Reproduction* **13** 128-131

Rachmilewitz J, Riely GJ and Tykocinski ML (1999) Placental protein 14 functions as a direct T-cell inhibitor *Cellular Immunology* **191** 26-33

Saling PM (1989) Mammalian sperm interaction with extracellular matrices of the egg *Oxford Reviews in Reproductive Biology* **11** 339-388

Saridogan E, Djahanbakhch O, Kervancioglu ME, Kahyaoglu F, Shrimanker K and Grudzinskas JG (1997) Placental protein 14 production by human fallopian tube epithelial cells *in vitro Human Reproduction* **12** 1500-1507

Seppala M, Taylor RN, Koistinen H, Koistinen R and Milgrom E (2002) Glycodelin: a major lipocalin protein of the reproductive axis with diverse actions in cell recognition and differentiation *Endocrine Reviews* **23** 401-430

Setchell BP (1974) The contributions of Regnier de Graaf to reproductive biology *European Journal of Obstetrics, Gynecology and Reproductive Biology* **4** 1-13

Skornicka EL, Kiyatkina N, Weber MC, Tykocinski ML and Koo PH (2005) Pregnancy zone protein is a carrier and modulator of placental protein-14 in T-cell growth and cytokine production *Cellular Immunology* **232** 144-156

Thaler CD and Cardullo RA (1996) The initial molecular interaction between mouse sperm and the zona pellucida is a complex binding event *Journal of Biological Chemistry* **271** 23289-23297

Tsibris JC, Thomason JL, Kunigk A, Khan-Dawood FS,

**Kirschner CV and Spellacy WN** (1982) Guaiacol peroxidase levels in human cervical mucus: a possible predictor of ovulation *Contraception* **25** 59-67

**Tse JY, Chiu PCN, Lee KF, Seppala M, Koistinen H, Koistinen R, Yao YQ and Yeung WSB** (2002) The synthesis and fate of glycodelin in human ovary during folliculogenesis *Molecular Human Reproduction* **8** 142-148

**Vaisse C, Atger M, Potier B and Milgrom E** (1990) Human placental protein 14 gene: sequence and characterization of a short duplication *DNA and Cell Biology* **9** 401-413

**Van Cong N, Vaisse C, Gross MS, Slim R, Milgrom E and Bernheim A** (1991) The human placental protein 14 (PP14) gene is localized on chromosome 9q34 *Human Genetics* **86** 515-518

**Waites GT, James RF and Bell SC** (1988) Immunohistological localization of the human endometrial secretory protein pregnancy-associated endometrial alpha 1-globulin, an insulin-like growth factor-binding protein, during the menstrual cycle *Journal of Clinical Endocrinology and Metabolism* **67** 1100-1104

**Weetman AP** (1999) The immunology of pregnancy *Thyroid* **9** 643-646

**Wilcox AJ, Weinberg CR and Baird DD** (1995) Timing of sexual intercourse in relation to ovulation. Effects on the probability of conception, survival of the pregnancy and sex of the baby *New England Journal of Medicine* **333** 1517-1521

**Yanagimachi R** (1994) Fertility of mammalian spermatozoa: its development and relativity *Zygote* **2** 371-372

**Yao YQ, Yeung WSB and Ho PC** (1996) Human follicular fluid inhibits the binding of human spermatozoa to zona pellucida *in vitro Human Reproduction* **11** 2674-2680

**Yao YQ, Chiu PCN, Ip SM, Ho PC and Yeung WSB** (1998) Glycoproteins present in human follicular fluid that inhibit the zona-binding capacity of spermatozoa *Human Reproduction* **13** 2541-2547

**Yao YQ, Ho PC and Yeung WSB** (1999a) Effects of human oviductal cell co-culture on various functional parameters of human spermatozoa *Fertility and Sterility* **71** 232-239

**Yao YQ, Ho PC and Yeung WSB** (1999b) Effects of human follicular fluid on spermatozoa that have been co-cultured with human oviductal cells *Fertility and Sterility* **72** 1079-1084

**Zhu J, Barratt CL, Lippes J, Pacey AA and Cooke ID** (1994) The sequential effects of human cervical mucus, oviductal fluid, and follicular fluid on sperm function *Fertility and Sterility* **61** 1129-1135

# Biological meaning of ubiquitination and DNA fragmentation in human spermatozoa

Monica Muratori[1], Sara Marchiani[1], Luciana Criscuoli[2], Beatrice Fuzzi[2], Lara Tamburino[1], Sara Dabizzi[2], Chiara Pucci[1], Paolo Evangelisti[2], Gianni Forti[1], Ivo Noci[2] and Elisabetta Baldi[1*]

[1]Departments of Clinical Physiopathology, Andrology Unit and [2]Department of Gynaecology, Perinatology and Human Reproduction, University of Florence, viale Pieraccini 6, I-50139 Florence, Italy

The ubiquitin-proteasome is an ubiquitous system mainly devoted to protein degradation. The presence of ubiquitinated proteins in male gametes suggests a role for this system also in reproduction. Available evidence indicate that ubiquitin in spermatozoa may have a role in semen quality control, as ubiquitinated defective spermatozoa in the epididymis are subsequently phagocytosed by epididymal epithelial cells. Moreover, a role both in the regulation of mitochondrial inheritance in mammals (paternal mitochondria are eliminated and their ubiquitination appears to be important for this process) and in sperm-oocyte interaction at fertilization (which is inhibited by an inhibitor of proteasome) have been also suggested. We found that both morphologically normal and abnormal human spermatozoa in semen may be ubiquitinated and that the percentage of ubiquitinated sperm in the ejaculate positively correlates with normal morphology and motility, suggesting that sperm ubiquitination may have a positive role in sperm functions. It remains to be defined if and which patterns of ubiquitination of spermatozoa may distinguish between the different biological functions of this system. In an attempt to answer this question, we set up a method to detect simultaneously ubiquitination and DNA fragmentation by FACScan since the latter parameter is related to a poor quality of semen; in particular, abnormal morphology. We found that DNA fragmented human spermatozoa are also ubiquitinated. Studies are in progress to determine the correlation between the fraction of ubiquitinated-non DNA fragmented spermatozoa and parameters of semen analysis.

## Introduction

The biological impact of the peptide ubiquitin is testified by the fact that in 2004 the Nobel Prize was awarded to the scientists (A. Hershko, A. Ciechanover and I. Rose) who discovered it and disclosed some of its functions. Ubiquitin is a small and highly conserved protein present

E-mail: m.muratori@dfc.unifi.it    e.baldi@dfc.unifi.it

Supported by grants from Italian Ministry of Education and Research (MIUR-COFIN) and the University of Florence.

in apparently all eukaryotic cells which is covalently attached to target proteins which are subsequently degraded by the 26S proteasome. Emerging evidence indicate that ubiquitin conjugation of proteins may also be a signal for other biological functions (for review see Welchman et al., 2005). A few years ago, Sutovsky et al. (2001) postulated a peculiar function for ubiquitin in spermatozoa. They showed evidence supporting the hypothesis that defective spermatozoa are ubiquitinated during epididymal transit for subsequent phagocytosis by epididymal cells. Later on, other roles for sperm ubiquitination have been hypothesized, including the regulation of mitochondrial inheritance in mammals (Thompson et al., 2003) and sperm-oocyte interaction at fertilization (Sakai et al., 2003, 2004; Sutovsky et al., 2004). While the two latter hypotheses imply a positive, direct function for the ubiquitin system in spermatozoa, the former implies a negative one since ubiquitinated sperm are supposed to be damaged and thus ubiquitin represents a signal for their elimination.

## Ejaculated human sperm are ubiquitinated

According to Sutovsky's (2001) hypothesis, if ubiquitinated spermatozoa are found in the ejaculate they should represent cells that have "escaped" the quality control at the epididymal level; that is, they have not been phagocytosed by epididymal epithelial cells for some (still) obscure reasons. In an attempt to elucidate why ubiquitinated sperm are present in the ejaculate, we evaluated sperm ubiquitination in total ejaculates from normo-, astheno-, oligo- and terato-zoospermic subjects by using an anti-ubiquitin antibody from Kamya (USA): the same antibody used by the group of Sutovsky. We found that both morphologically normal and abnormal spermatozoa can be ubiquitinated (Fig. 1) (Muratori et al., 2005), confirming that ubiquitination is not only a marker of defective spermatozoa. To address further this point, we determined the relationship between ubiquitinated sperm and basal parameters of semen analysis. Since the global ejaculate is a complex matrix containing, besides spermatozoa, round cells of spermatogenetic or other origin and epithelial cells from the genital tract, all of which could be ubiquitinated, we have performed the analysis by FACScan considering only the flame shape region characteristic of spermatozoa (Muratori et al., 2003). Recently, we have demonstrated that this region also contains round bodies of uncertain origin termed by us M540 bodies (Muratori et al., 2004). M540 bodies are virtually devoid of chromatin material and are also ubiquitinated (Muratori et al., 2005). Hence, ubiquitinated M540 bodies were subtracted in the sperm ubiquitination analyses. When the percentage of ubiquitinated spermatozoa was related to basal parameters of semen analysis, a positive correlation with a good quality of semen was found (Muratori et al., 2005). In particular, the parameter that shows the highest correlation with sperm ubiquitination is normal morphology (Fig. 2). This result implies that sperm ubiquitination may also be important for post-ejaculation sperm functions, such as, capacitation, hyperactivation and fertilization. In addition, sperm ubiquitination may represent a good predictor of fertilization ability of human spermatozoa.

## Are ubiquitinated human spermatozoa also DNA fragmented?

Although our results about a positive correlation between sperm ubiquitination and normal morphology points to a "good" role of ubiquitination in sperm functions, it must be stressed that among the global ubiquitinated spermatozoa, morphologically abnormal spermatozoa are also present (Sutovsky et al., 2001; Muratori et al., 2005). In such a situation, sperm ubiquitination

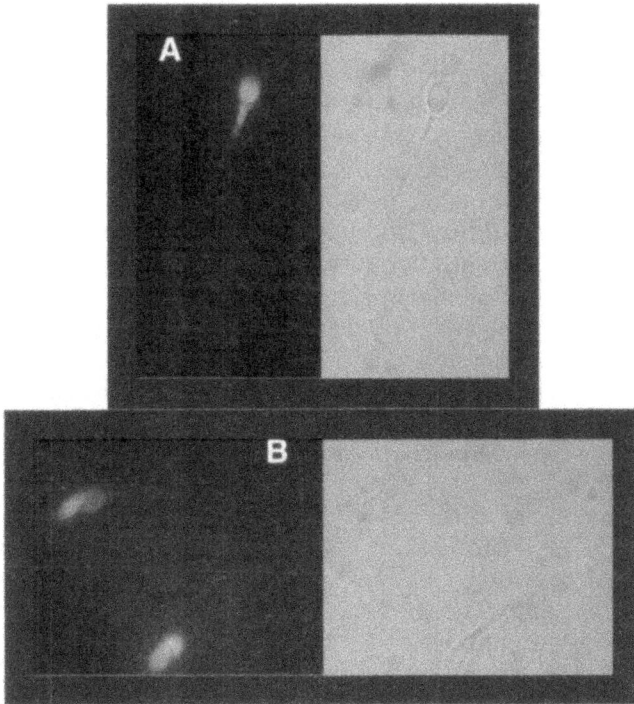

**Fig. 1** Micrographs of semen samples from a normozoospermic (upper panels) and an oligoasthenoteratozoospermic man by fluorescence microscopy after double staining with anti-ubiquitin antibody (green) and propidium iodide (nuclei lebelling, red). The image has been obtained by overlapping the two fluorescence signals. Bright images corresponding to the same field are shown in the right panels. Reproduced from Muratori *et al.* (2005) with permission.

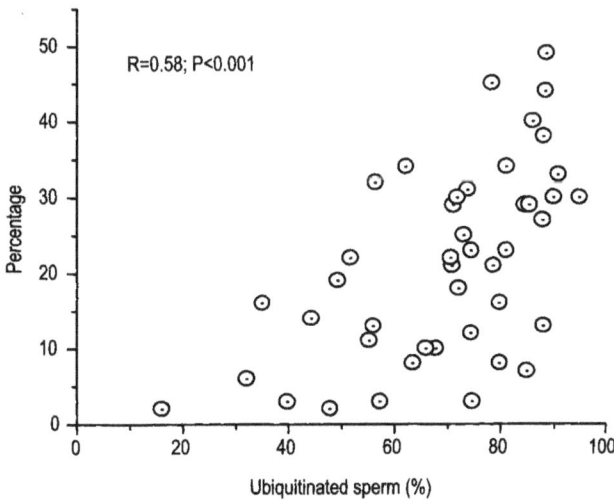

**Fig. 2** Scatter plot between the percentage of ubiquitinated sperm from 45 semen samples and percentage of normal morphology. Reproduced from Muratori *et al.* (2005) with permission.

may not be the ideal parameter as a predictor of fertilizing ability of spermatozoa, since it includes a heterogeneous population of spermatozoa. Previous data from the group of Sutovsky (Sutovsky et al., 2002) indicated that approximately 30% ubiquitinated bovine ejaculated sperm are TUNEL-positive (and thus DNA fragmented) and that these cells show variable morphological defects: even apparently normal spermatozoa were both ubiquitinated and TUNEL positive. It is well known that sperm DNA fragmentation is an anomaly frequently found in sub-fertile or infertile patients (for review see Muratori et al., 2006) that negatively correlates with a poor quality of semen (Lopes et al., 1998; Muratori et al., 2000). We reasoned that if DNA fragmented human spermatozoa also shows ubiquitination, it would be possible to isolate (and thus subtract) ubiquitinated and DNA fragmented spermatozoa from the total percentage of ubiquitinated spermatozoa to obtain a more homogenous fraction. Presumably, such a fraction should show higher correlations with a good quality of semen. To test whether ubiquitinated human spermatozoa also show DNA fragmentation, we developed an assay to detect simultaneously the 2 parameters by using anti-ubiquitin antibody and TUNEL. As shown in Fig. 3, both DNA fragmented and DNA intact spermatozoa show ubiquitination. In particular, Fig. 3 A shows micrographs obtained by fluorescence microscopy with a DNA-fragmented ubiquitinated spermatozoon (upper panels) and a DNA-intact ubiquitinated one (lower panels). Fig. 3B reports typical dot plots obtained with this assay, showing the negative control (absence of primary antibody for ubiquitin and of TdT enzyme for TUNEL analysis, left panel) and the corresponding test sample (right panel).

Virtually, all DNA fragmented spermatozoa are also ubiquitinated (UR quadrants). Non-DNA fragmented, ubiquitin positive sperm are in the UL quadrant. In this quadrant also ubiquitinated M540 bodies are also present. Hence their amount was determined in a different aliquot of the same sample and manually subtracted from the percentage of the UL quadrant population shown in Fig. 3B. Preliminary experiments, conducted in 8 semen samples obtained from patients undergoing semen analyses for couple infertility demonstrate that the fraction of ubiquitinated non DNA fragmented spermatozoa after subtraction of M540 bodies is highly correlated with progressive motility ($r = 0.74$ $p < 0.05$) and normal morphology ($r = 0.75$, $p < 0.05$) (unpublished data).

## Conclusion: "good" and "bad" ubiquitin in spermatozoa

As stated above, ubiquitination may have a different function(s) in reproduction. All these functions have the apparent purpose of favouring the mechanism of conception: from the elimination of damaged spermatozoa from the ejaculate (Sutovsky et al., 2001) to the modulation of the fertilization process with both a role in sperm-oocyte interaction (Sakai et al., 2003) and in the elimination of mitochondria of paternal origin (a well known phenomenon occurring in mammals) (Thompson et al., 2003). Our data (Muratori et al., 2005) demonstrating that sperm ubiquitination is related to good semen quality and, in particular, with normal morphology indicate that spermatozoa need to be ubiquitinated for successful fertilization. However, in the total fraction of ubiquitinated spermatozoa, DNA fragmented spermatozoa are also included, as demonstrated here (Fig. 3). Overall, our data demonstrate that different elements present in the ejaculate are ubiquitinated: DNA fragmented and DNA intact spermatozoa and M540 bodies. While ubiquitinated bodies are related to a poor quality of semen (Muratori et al., 2005), the total fraction of ubiquitinated sperm is related to a good quality (Muratori et al., 2005) and, according to our preliminary experiments conducted in a limited number of samples, the fraction of DNA intact ubiquitinated spermatozoa appears to correlate better. Studies are in progress to evaluate whether ubiquitination in DNA intact and DNA fragmented spermatozoa is associated with a different sperm structure or different proteins.

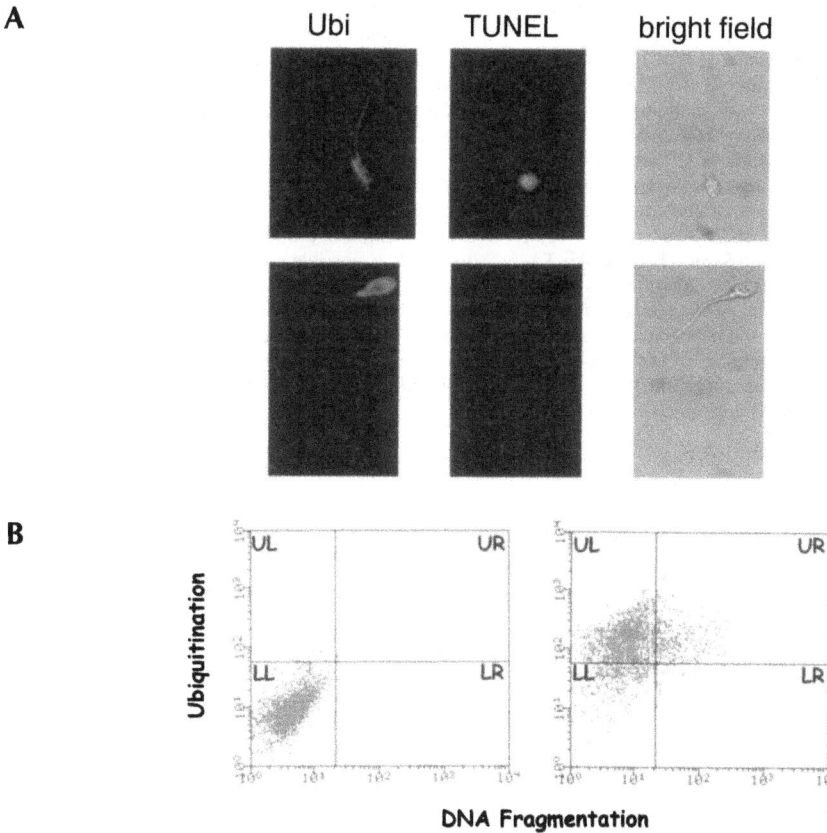

**Fig. 3** Panel A: Micrographs of semen samples showing a spermatozoon positive for both ubiquitin (Ubi) and TUNEL labelling (upper micrographs) and a spermatozoon positive for ubiquitin and negative for TUNEL (lower micrographs). Bright images corresponding to the same field are shown in the right panels. Panel B: Dot plots showing negative sperm sample (negative control for ubiquitin labelling obtained in the absence of the primary antibody and for TUNEL in the absence of TdT enzyme, left dot plot) and the corresponding test sample (obtained with double staining for ubiquitin and TUNEL, right dot plot). UL = fraction of ubiquitinated non DNA fragmented spermatozoa.

# References

**Lopes S, Sun JG, Jurisicova A, Meriano J and Casper RF** (1998) Sperm deoxyribonucleic acid fragmentation is increased in poor-quality semen samples and correlates with failed fertilization in intracytoplasmic sperm injection *Fertility and Sterility* **69** 528-532

**Muratori M, Piomboni P, Baldi E, Filimberti E, Pecchioli P, Moretti E, Gambera L, Baccetti B, Biagiotti R, Forti G and Maggi M** (2000) Functional and ultrastructural features of DNA-fragmented human sperm *Journal of Andrology* **21** 903-912

**Muratori M, Maggi M, Spinelli S, Filimberti E, Forti G and Baldi E** (2003) Spontaneous DNA fragmentation in swim-up selected human spermatozoa during long term incubation *Journal of Andrology* **24** 253-262

**Muratori M, Porazzi I, Luconi M, Marchiani S, Forti G and Baldi E** (2004) AnnexinV binding and merocyanine staining fail to detect human sperm capacitation *Journal of Andrology* **25** 797-810

**Muratori M, Marchiani S, Forti G and Baldi E** (2005) Sperm ubiquitination positively correlates to normal morphology in human semen *Human Reproduction* **20** 1035-1043

**Muratori M, Marchiani S, Maggi M, Forti G and Baldi E** (2006) Origin and biological significance of DNA fragmentation in human spermatozoa *Frontiers in Bioscence* **11** 1491-1499

**Sakai N, Sawada H and Yokosawa H** (2003) Extracellular ubiquitin system implicated in fertilization of the ascidian, *Halocynthia roretzi*: isolation and characterization *Developmental Biology* **264** 299-307

**Sakai N, Sawada MT and Sawada H** (2004) Non-traditional roles of ubiquitin-proteasome system in fertilization and gametogenesis *International Journal of Biochemistry and Cell Biology* **36** 776-784

**Sutovsky P, Moreno R, Ramalho-Santos J, Dominko T, Thompson WE and Schatten G** (2001) A putative, ubiquitin-dependent mechanism for the recognition and elimination of defective spermatozoa in the mammalian epididymis *Journal Cell Science* **114** 1665-1675

**Sutovsky P, Neuber E and Schatten G** (2002) Ubiquitin-dependent sperm quality control mechanism recognizes spermatozoa with DNA defects as revealed by dual ubiquitin-TUNEL assay *Molecular Reproduction and Development* **61** 406-413

**Sutovsky P, Manandhar G, McCauley TC, Caamano JN, Sutovsky M, Thompson WE and Day BN** (2004) Proteasomal interference prevents zona pellucida penetration and fertilization in mammals *Biology of Reproduction* **71** 1625-1637

**Thompson WE, Ramalho-Santos J and Sutovsky P** (2003) Ubiquitination of prohibitin in mammalian sperm mitochondria: possible roles in the regulation of mitochondrial inheritance and sperm quality control *Biology of Reproduction* **69** 254-260

**Welchman RL, Gordon C and Mayer RJ** (2005) Ubiquitin and ubiquitin-like proteins as multifunctional signals *Nature Review Molecular Cellular Biology* **6** 599-609

# A testis specific auto-antigen TSA70 belongs to Odf2/Cenexin family

V Khole and M Wakle

*Department of Gamete Immunobiology, National Institute for Research in Reproductive Health, J.M. Street, Parel, Mumbai- 400012, India*

The occlusion of testicular outflow following vasectomy leads to autoimmunity which is characterised by the production of antisperm antibodies. Although post-vasectomy autoimmune response has been reported in several species, very little is known about the sperm auto-antigens that are targeted. Using a vasectomised mouse, a number of monoclonal antibodies were generated with the aim of characterising the targeted sperm specific antigens. All the monoclonal antibodies were found to react with testicular proteins. One of the antibodies, D5E5 was then used for immunochemical characterisation of its cognate antigen that was found to be a testis specific autoantigen of ~70 kDa, termed TSA70. TSA70 is expressed postmeiotically in a stage specific pattern during spermiogenesis. TSA70 was observed to be conserved across the species as seen by its presence on rat, bull, marmoset and human spermatozoa. On mouse spermatozoa it was localised at the tip of acrosome and sperm tail as seen by indirect immunofluorescence. Following capacitation it was seen to spread all over the acrosome. However, the localisation on the acrosomal tip persisted even after acrosome reaction, which suggests that the antigen is likely to play a physiological role post acrosome reaction. Solubilisation with Triton X100, revealed the acrosomal component of TSA70 to be a matrix protein, whereas its counterpart on the tail had both a soluble as well as a particulate form. *In vitro* studies showed that the monoclonal antibody significantly reduced progressive motility of mouse spermatozoa. Preliminary sequence analysis showed that TSA70 belongs to the Odf2/ Cenexin family. These characterisation studies suggest that TSA70 is a conserved, testis specific sperm autoantigen with a definitive physiological role in reproduction.

## Introduction

Provision of safe and sustained effective fertility control for the world population is urgently needed (Aitken, 2002). Contraceptive choices for men are currently limited to condoms and vasectomy (Holden, 2002; Nass and Strauss, 2004), whereas, several different choices and approaches are available for contraception in women. A vaccine approach for contraception

E-mail: kholevv@icmr.org.in

would definitely be a valuable addition to the existing armamentarium of different approaches used for family planning. Vaccine based on sperm antigen is very promising and therefore, in spite of no real breakthrough, are being actively pursued. The targets have been sperm antigens coming either from the testis or the epididymis. Realising the feasibility of this approach and the need for additional methods for contraception, the National Institute for Child Health and Human Development (NICHD) convened a workshop to identify novel strategies involving testicular and epididymal antigens for developing a male contraceptive in the 21st century (DePaolo et al., 2000).

Different approaches, such as biochemical approaches, 2D proteomics and neonatal tolerisation (Ensrud and Hamilton,1991; Khole et al., 2000), have been exploited by investigators to identify sperm molecules. In addition to this, immunoinfertile sera (Linnet and Hjort, 1977) and vasectomy induced auto-antibodies (Handley et al., 1988; Nakamura et al.,1994; Verdier et al., 2002) have also been successfully used for the identification of functionally important sperm proteins.

The process of spermatogenesis commences after the acquisition of fully functional immune system. At puberty, when immune competence is already established, differentiating germ cells commence a new programme that leads to the formation of mature spermatozoa. During this process an array of new surface molecules are expressed on the differentiating germ cells which do not belong to the family of those considered as 'self' by the immune system. Under physiological conditions both the blood-testis and blood-epididymis barrier prevent undesirable autoimmunity by sequestring these sperm auto antigens from immune cells. Any change via trauma and/or alterations in this immunological barrier will lead to production of antisperm antibodies (ASA). Antisperm antibodies are one of the main causes of immune infertility. ASA may impair sperm function at various stages of reproduction such as the transport of the spermatozoa in the female genital tract (Bronson et al., 1984), sperm capacitation or the acrosome reaction (Lansford et al., 1990) and sperm–zona binding or sperm–oolemma fusion (Bronson et al., 1982; Alexander, 1984). ASA have also been reported to be associated with inflammation, cryptorchidism, varicocele and surgical interventions in the genital tract (Gubin et al., 1998). The rates of incidence of ASA reported in literature among infertile couples lies in the range of 9-36%. Although the antibody mediated effect on sperm function depends on several factors such as titre, class of the immunoglobulin and genital or systemic localisation, it is the inciting sperm antigen which plays a crucial role. The detection of ASA as such does not have clear prognostic value; since several fertile men (~ 19%) and as many as 43% fertile women show the presence of ASA in their sera (Chiu and Chamley, 2002). This high incidence of ASA in fertile population indicates that not all ASA cause infertility and/or are functionally relevant. The existence of qualitative differences in the antigenic profile of the fertile and infertile spermatozoa have been demonstrated by comparing the immunoreactivity between these two populations (Paradisi et al., 1996). It has also been proposed that spermatozoa from an infertile individual differ substantially at the immunogenic level from that of fertile individual (Paradisi et al., 1996). For a better understanding of the mechanism involved in immunological infertility, it is imperative to identify and characterise the sperm antigen capable of eliciting the production of ASA that are functionally relevant in the impairment of reproductive function. Though a plethora of sperm molecules have been identified only a few have been shown to have a definitive role in the process of fertilisation. Therefore, sperm antigens which are not only potentially immunogenic but are also cell specific and moreover, biologically relevant need to be identified and characterised.

Occlusion of testicular outflow by performing vasectomy for male contraception results in appearance of ASA. The immunogenicity of sperm antigens is indicated by the high incidence of ASA in the sera and accessory gland fluids of vasectomised men (Aitken et al., 1988).

Moreover, the presence of such antibodies is correlated with the persistence of infertility in vasovasostomised individuals even though the vas was found to be patent. Though the dynamics of ASA following vasectomy have been observed in various animal models, only a few sperm auto-antigens such as protamine (Samuel *et al.*, 1978), Fertilisation Antigen-1 (FA1: Naz and Zhu, 1998), human Nuclear Auto Antigenic Sperm Protein (NASP: Batova *et al.*, 2000), a novel Asparginase Like Protein (ALP; Bush *et al.*, 2002), a number of Outer Dense Fiber molecules (ODF; Flickinger *et al.*, 2001) and 97 kDa fox Sperm Protein 13 (fSP13: Verdier *et al.*, 2005) have been precisely identified. Because of the low titres of the sperm auto antibodies and the polyclonality of the serum, it has been extremely difficult to determine the nature and the number of sperm auto antigens involved in vasectomy induced autoimmunity. Therefore, we decided to raise monoclonal antibodies using a vasectomised mouse. Nakamura and co-workers (Nakamura *et al.*, 1994) reported identification and characterisation of a 67 kDa sperm autoantigen using a similar approach. Using a vasectomised mouse model, we have generated a panel of monoclonal antibodies (mAbs).

### Humoral immune response to sperm auto antigens following vasectomy

All the mice responded to vasectomy by the production of ASA as detected by ELISA though some individual variations were observed in their titre (data not shown). On the other hand sham operated animals did not show any rise in antisperm antibodies when compared to the preimmune sera. The ASA titre showed a typical pattern of immune response with an early increase in ASA titre on day 20 post vasectomy followed by a near plateau by day 50, with a sudden drop by day 60 post vasectomy (Handley *et al.*, 1990). In sham operated animals the titre was almost similar as compared to the pre-surgery sera till day 60 post vasectomy. Following vasectomy, degeneration of spermatogenesis was observed in several of the seminiferous tubules and these alterations were focal in nature. Of the several likely factors as seen in the present study, the immunological mechanisms seem to be the most likely factor that could bring about these alterations post vasectomy (Bigazzi *et al.*, 1976; Alexander and Anderson,1979).

### Testicular autoantigens are major targets following vasectomy in a murine model

The ASA following vasectomy are known to target either testicular or epididymal proteins. Although it was reported that antibodies to sperm and testicular autoantigens are a hallmark of vasectomy induced autoimmunity in a murine system (Ishakia and Alexander, 1984), Turner and his group have shown that vasectomy affects the synthesis and secretion of proteins in rat caput epididymis (Turner *et al.*, 1999, 2000).

Ishakia and Alexander (1984) have generated five mAbs using vasectomised mouse. Four of the mAbs were sperm specific and identified sperm surface autoantigens appearing on the spermatid cells. Based on these observations it was suggested that in case of mouse, post vasectomy antibodies are predominantly to autoantigens of testicular origin. Our data also shows that the antigens identified by all the mAbs were also of testicular origin (data not shown). It is likely that in a murine system testicular damage is more obvious due to the lower distensibility of the murine epididymis as suggested by Weiske (2001). In this study, splenocytes from a vasectomised mouse producing antisperm antibodies were successfully fused with mouse myeloma cells to produce mAbs. One of the mAb, D5E5, was chosen for further characterisation based on its high titre. This antibody identified a testis specific autoantigen of ~70 kDa hereafter refered to as TSA70.

## TSA70 is a testis specific sperm auto antigen of ~ 70 kDa

In Western blot, the mAb D5E5 identified multiple bands in the region of ~ 65-80 kDa with the testicular protein and a single band of ~ 70 kDa with testicular as well as epididymal sperm protein (Fig.1). In the present study, it was also noted that TSA70 is specific to testis as no reactivity was seen in Western blots with any other accessory reproductive tissues including epididymis or somatic tissues (Fig. 1). This confirms that the protein is of testicular origin and is produced in the testis and not acquired by spermatozoa during the epididymal maturation process. The multiple bands identified in the testis could be due to an unprocessed immature form of the protein occuring during germ cell differentiation. Similar observations have been made for a major mouse FS protein: mRNA for which has been cloned and characterised in two different laboratories (Carrera *et al.*, 1994; Fulcher *et al.*, 1995). It has been shown that the translation product, a precursor of a calculated molecular mass of about 93 kDa, is processed just before FS assembly to a 73 kDa protein (Carrera *et al.*, 1994). The same has been reported for the protein sp56 which is an acrosomal matrix protein. It has been shown that the form of sp56 in pachytene spermatocytes and spermatids during acrosomal biogenesis has a higher molecular weight (~ 67 kDa) than for spermatozoa and the size differences were apparently due to alterations in the carbohydrate side chains (Kim *et al.*, 2001). Processing of a testicular antigen, 2B1, has been reported in the epididymis (Jones *et al.*, 1990). A precursor glycoprotein of ~ 60 kDa first appears postmeiotically on the surface of stage VI to VIII round spermatid. During elongation of the spermatids, 2B1 is excluded from the sperm head domain and is sequestered onto the tail. As spermatozoa pass through the caput epididymis. 2B1 is cleaved endolytically at arginine residue 312 to form heteromeric glycoproteins of ~ 40 kDa and ~ 19 kDa.

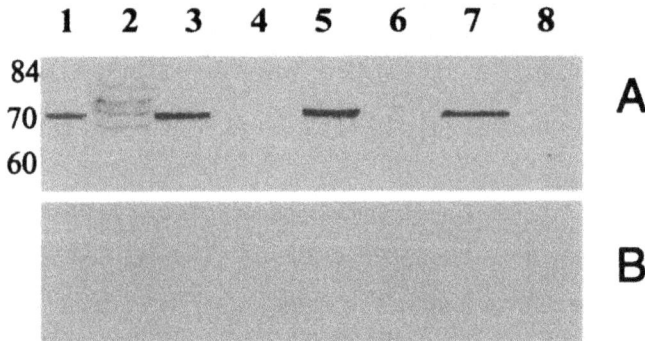

**Fig. 1** Immunoblot analysis with sperm and tissue proteins from the mouse testis and three regions of the epididymis (1A). The mAb identified multiple bands in the region of ~ 65-80 kDa with testicular tissue protein (lane 2) but a single band at ~ 70 kDa with sperm protein from testis (lane 1), while a band of slightly lesser molecular weight identified with the sperm protein from caput (lane 3), corpus (lane 5) and cauda epididymis (lane 7). No reactivity was seen with the epididymal tissue protein from caput (lane 4), corpus (lane 6) and cauda (lane 8). No reactivity was observed for the myeloma culture supernatant (1B) with any of the sperm and/or tissue proteins. This shows the specificity of the mAb and its reactivity for the ~ 70 kDa protein.

## TSA70 is expressed post meiotically in stage specific pattern

Immunohistochemistry results (Fig. 2) show that the source of the antigen is the testicular germ cells (elongating spermatids). The antigen is post meiotic and makes its appearance much

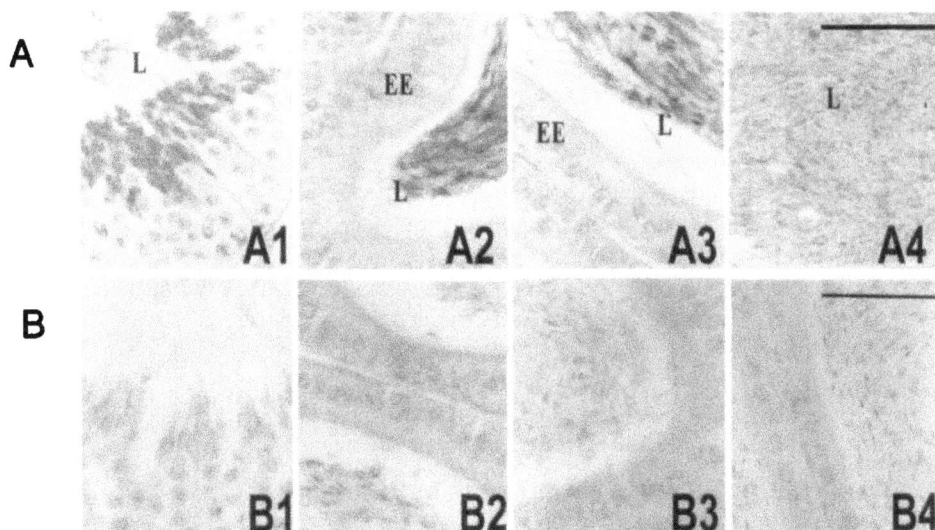

**Fig. 2** The localisation of antigen in mouse testis and epididymis (2A). In mouse testis (A1) the antigen is localised postmeiotically starting from elongating spermatids. Reactivity was also seen on the spermatozoa in the lumen (L). In the epididymis (A2-A4) the reactivity was seen on the spermatozoa in the lumen (L) of caput (A2), corpus (A3) and cauda (A4) but no reactivity was observed in the epididymal epithelium (EE). Absence of reactivity with the myeloma culture supernatant was observed for all the tissues as seen in Fig 2B. B1-B4 are the myeloma controls corresponding to A1-A4. Scale bar = 100 μm

after the development of the immune system hence is not recognised as "self". The multiple bands identified on Western blot with testicular protein could be attributed to the expression of the antigen on spermatids at different stages during spermiogenesis. It could be that the precursor form of the protein is expressed first in the step 8 spermatids and is eventually processed as it is acquired by the spermatozoa. The mAb identifies the antigen in a stage specific manner. Immunohistochemical localisation reveals that the expression of the antigenic determinant on the sperm head initiates in step 8 spermatids (rat and mouse) and is then acquired by the spermatozoa while the antigenic determinant on the tail region starts its expression from step 16 spermatids in rat and step 15 in mouse. The expression of the antigenic determinant on the head is coincident with the acrosome phase of spermiogenesis when the acrosome no longer grows but only conforms to the changing shape of the spermatid nucleus. The final third of the spermiogenesis is primarily concerned with the assembly of accessory components to the axoneme of the tail (Oko, 1998). Since fibrous sheath (FS) and outer dense fibre (ODFs) make up a major portion of spermatozoa, it indicates that most of the protein synthesis during this time period is directed towards the assembly of these tail components (Oko, 1988). This suggests that TSA70 could be a cytoskeletal component of the sperm.

### TSA70 is localised on identical domains on mouse testicular and epididymal sperm

Immunofluorescence studies have demonstrated that the protein was localised on two different domains represented by the tip of the acrosome as well as on the spermatozoa tail (principal piece and end piece). The domain specificity pattern is the same for spermatozoa from testis as well as from the caput, corpus and cauda epididymis. The localisation of antigenic determinants on both the acrosome and principal piece of the sperm tail suggest the  possibility of common epitopes between the two domains. A similar pattern of localisation has been reported for TPX-1(testis specific protein 1), a member of Cysteine Rich Secretory Protein (CRISP) family. The protein has been cloned from the rat testicular library using an ODF specific antiserum. The immunochemical characterisation of TPX-1 revealed that the protein exists as 25 and 27 kDa forms in two different components of rat spermatids: the ODFs and the acrosome. Concurrent with sperm head formation, TPX-1 protein is found to be incorporated into the developing spermatid tail and specifically the ODFs. The authors have suggested that TPX-1 may have functional significance in the process of sperm head development and tail function (O'Bryan et al., 2001). TPX-1 has also been shown to be present on human sperm acrosome as an intracellular antigen (Busso et al., 2005). This may also be applicable to the protein TSA70 and is under investigation.

### TSA70 is conserved across the species

TSA70 is observed to be conserved across species by indirect immunoflurescence (Fig. 3). Although the antigen is conserved across the species, its domain specificity varies from species to species.   It is interesting to note that evolutionarily close species (that is, mouse and rat) show identical pattern. It was observed that in bull (*Bos taurus*), marmoset (*Callithrix jaccus*) and human

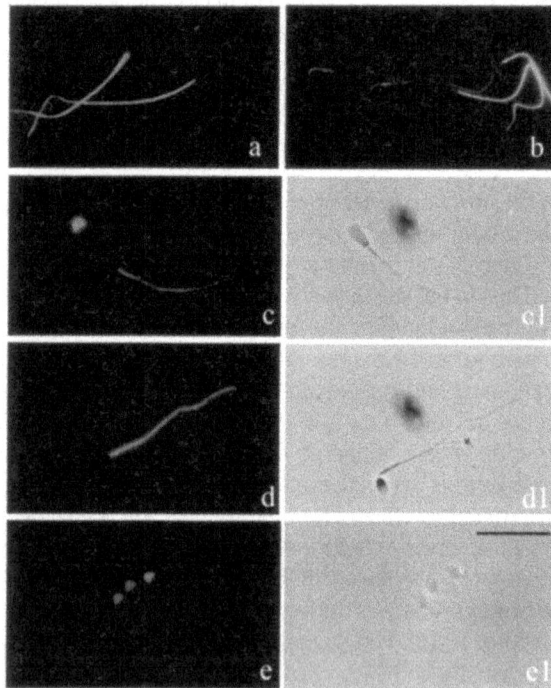

**Fig. 3** The protein is conserved across the species as seen by the indirect immunofluorescence localisation studies on rat, bull, marmoset and human sperm (3A). Rat spermatozoa (b) showed identical domain localisation as that of mouse spermatozoa (a). In phylogenetically divergent species the localisation was different. In bull sperm (c), domains on both the acrosome and principal piece were identified; in marmoset sperm, (d) immunofluorescence was detected only on the tail while in human sperm (e) the localisation was restricted to the acrosomal region. c1-e1 are corresponding phase contrast images. Scale bar = 100 μm.

spermatozoa TSA70 showed a species specific pattern. In case of the bull, TSA70 was localised on both the acrosome and tail. By contrast, in marmoset spermatozoa it was seen only on the sperm tail whereas in human sperm it was restricted to the acrosome. Western Blot analysis showed species specific variation in the molecular weights of the conserved antigen (data not shown). This could have occurred during the process of evolution and the difference in the molecular size may be due to differential post-translational processing of the protein in each species.

## TSA70 relocates following capacitation

The localisation of the antigen on the tip of the acrosome by the mAb D5E5 led us to probe into its role in the acrosome reaction (AR). Indirect immunofluorescence studies revealed that the fluorescence was observed over the acrosome in the capacitated mouse spermatozoa (Fig. 4B), which was lost following AR (Fig. 4C). Both phenomena did not affect the fluorescence on the sperm tail (principal piece and end piece). However, the fluorescence at the tip of the acrosome persisted even after the AR. From these observations one could infer that the antigen at the tip of the acrosome may have a putative role in secondary binding or anchorage to the oocyte. The fluorescence on the acrosome after capacitation may suggest that our mAb recognises some hidden epitopes on the acrosome that are exposed or unmasked following capacitation due to a change in membrane fluidity. This relocation of sperm surface molecules during capacitation and the acrosome reaction has been reported earlier for other sperm proteins like 2B1 (Jones et al., 1990), PH20 (Myles and Primakoff, 1984), MC31 (Saxena and Toshimori, 2004) and TPX-1 (Busso et al., 2005). After the acrosome reaction, PH20, a surface antigen of guinea pig spermatozoa migrates from the postacrosomal region to a new location on the inner acrosomal membrane. The 2B1 molecule shifts from the sperm tail to sperm head after capacitation, while MC31 relocates from the principal piece to the sperm head after capacitation and to the equatorial segment after the acrosome reaction. TPX-1 on human sperm is observed to relocate from the acrosomal cap region to the equatorial segment following the AR. In our study, the shift in location of the antigen could be attributed to the migration of the molecule or the destabilisation of the membrane which may unmask certain hidden epitopes to which mAb D5E5 was found to react. The exocytosis phenomenon during the AR itself explains the loss of these antigens after the AR, which is very clear by the loss of the fluorescent labeling on the acrosome after the AR. The persistence of the antigen at the tip of the acrosome even after the AR suggests that the antigen has a physiological /functional role during post-acrosome reaction, which needs to be elucidated further.

## TSA70 is an acrosomal matrix protein

The retention of TSA70 at the acrosomal tip even after the AR reflects its tight association with the sperm head which in turn indicates its particulate nature. It has been suggested that the position and solubility of a specific acrosomal protein may govern its function during the course of the AR and thereafter (Kim et al., 2001). A component of acrosomal matrix is predicted to be associated with the sperm head for a longer period of time than would a soluble protein (Kim et al., 2001). Acrosomal proteins could be either soluble proteins or acrosomal matrix proteins based on their solubility following Triton X100 extraction under conditions that block proteolysis (Huang et al., 1985). Western blot analysis of rat sperm proteins from isolated heads and tails (Fig. 5) showed an identical band with a molecular weight of ~ 70 kDa in each. However, in the head the mAb  reacted only with the particulate fraction while on the tail it reacted

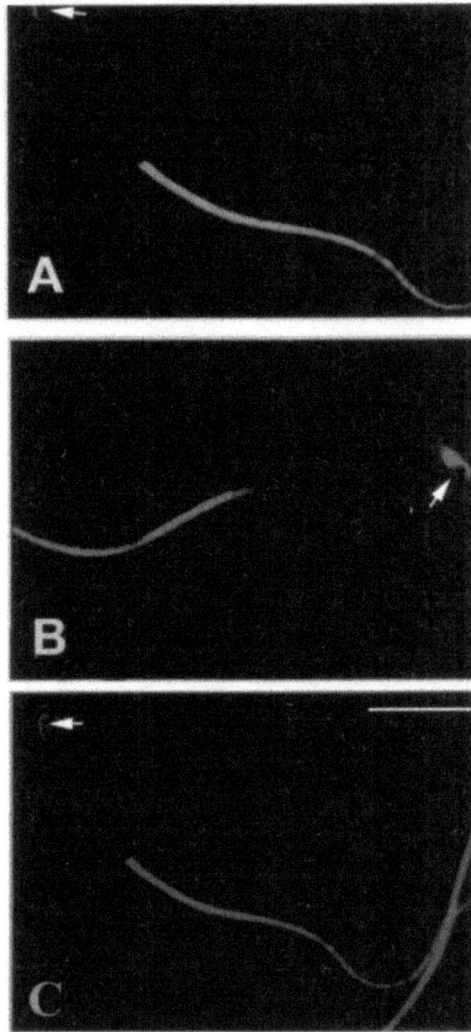

**Fig. 4** Effect of capacitation and acrosome reaction on localization of TSA70 in mouse spermatozoa. Panel (A) shows the immunofluorescent localisation on the normal mouse spermatozoa. Following capacitation the fluorescent localisation from the tip of acrosome shifted to the acrosomal region (B). After acrosome reaction the fluorescence on the acrosomal region was lost but it persisted at the tip of acrosome (C). Localisation on the sperm tail was not seen to be altered by any of the phenomena. Scale bar = 100 μm.

with both the soluble as well as the particulate fractions. These results indicate that on the head TSA70 is an acrosomal matrix (particulate) protein and on the tail it has both soluble as well as particulate forms. The fluorescence observed over the acrosome following capacitation may represent the soluble form of the protein which is made available only after capacitation as a result of destabilisation of the plasma membrane.

216

116

86.5

48.3

32.2

25.6

17.5    **1   2   3   4   5**

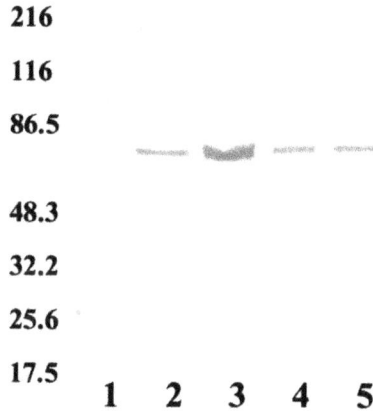

**Fig. 5** TSA70 is an acrosomal matrix protein. Western blotting of particulate and soluble fractions of isolated head and tail proteins from rat spermatozoa was performed. The mAb did not stain proteins from the soluble fraction of heads (lane 1) but stain proteins from particulate fraction of heads (lane 2) and both the soluble and the particulate fractions of tails (lane 3 & 4).The molecular weight of the protein identified in all the lanes was identical (~ 70 kDa). The reactivity with the total rat caudal sperm protein is shown in lane 5.

## TSA70 is essential for mouse sperm progressive motility

The localisation of the antigen on the tail is very important from the point of view of its function and therefore studies on the effect of the mAb on motility were undertaken. When the effect of the mAb on the motility was assessed *in vitro*, a sharp and time dependent decline in the forward progressive motility as compared to the total sperm motility was observed (data not shown). Forward progressive motility is an essential requirement for the successful unassisted fertilisation and since the antibody against the cognate antigen affects the same, it could be hypothesised that the auto-antigen has a functional role in sperm motility. Although the mAb has an effect on sperm motility, no apparent effect on viability was seen.

## TSA70 Is a cytoskeletal protein

When immunogold labeling was performed on rat sperm, gold particles were localised on both FS and ODF suggesting a common epitope between these two flagellar components (Fig. 6A), which has been shown earlier by various investigators (Tres and Kierszenbaum,1996; Escalier et al., 1997; Oko, 1988). These two cytoskeletal components have been shown to have immunological and biochemical similarities (Oko, 1988). In the developing spermatid head (Fig. 6B), the gold particles were observed in the perinuclear theca (PT). These observations further extend the possibility of a certain epitope common to all the three cytoskeletal structures unique to spermatozoa. Though the sperm head cytoskeletal structures are composed of unique proteins in the spermatogenic cells, they may be related to sperm flagellar proteins. The PT, FS and ODF are the highly specialised, unique cytoskeletal components of the spermatozoa which appear to have no counterpart in somatic cells (Miranda-Vizuete et al., 2003). The presence of TSA70 in all these components as seen by the electron micrography studies confirms that it is a sperm cytoskeletal protein.

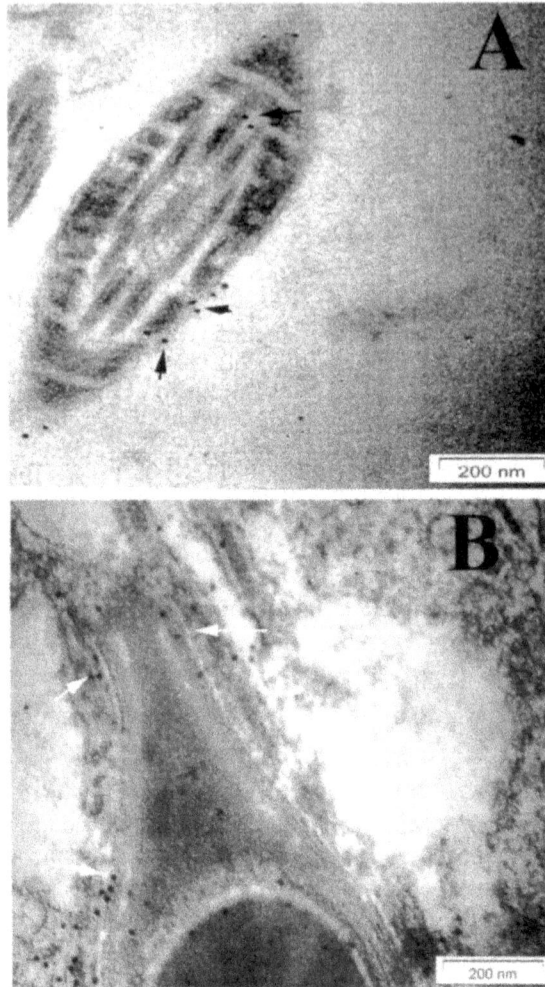

**Fig. 6** Immunogold labeling showing the cytoskeletal nature of the protein. In rat caudal spermatozoa (A), the gold particles were observed on the outer dense fibers (ODF) and fibrous sheath (FS). In rat testicular sections (B), the gold particles were observed on the perinuclear theca of the developing spermatid head. These observations together indicate sharing of epitopes between these cytoskeletal components of sperm. Magnification = x 49,000.

## TSA70 belongs to the Odf2/Cenexin family

The amino acid sequence data revealed that TSA70 shows homology to Odf2/Cenexin. The two proteins, Odf2 and Cenexin, are 91% identical with each other and belong to the same family of proteins. Looking at the characterisation studies carried out so far, we can speculate that TSA70 could be a novel form of cytoskeletal protein (sperm auto-antigen) that belongs to the Odf2/Cenexin family. The presence of TSA70 in the three cytoskeletal components unique to spermatozoa underlies the importance of this protein in sperm function.

The PT is a condensed layer of selected cytoplasmic protein that is sandwiched between the nuclear envelope and the inner acrosomal membrane apically and between the nuclear enve-lope and the plasma membrane caudally. It is assumed to play a pivotal role in nuclear shaping, acrosomal-nuclear docking and sperm-oocyte interactions (Aul and Oko, 2002; Sutovsky et al., 2003). ODF surrounds the axoneme of the whole sperm flagellum while FS defines the extent of principal piece. Though the functional role is not clearly elucidated, ODF is known to provide tensile strength and protect the sperm from shearing forces during their transit through both the epididymis and female genital tract. FS, on the other hand, tethers glycolytic enzymes that provide energy during hyper activation. In addition, this accessory structure serves as a scaffold and organising center for multiple signaling and metabolic cascades that are critical for normal flagellar function; such as the degree of flexibility, plane of flagellar motion and the shape of flagellar beat (Eddy et al., 2003).

The function of the ODFs and FS is not fully elucidated though; the structural integrity of ODF as well as FS is believed to be associated with the sperm motility and male fertility. The extensive homology between TSA70 and Odf2/Cenexin proteins indicates an important role for TSA70 in sperm function, which needs to be elucidated.

## Conclusion

The mammalian spermatozoon is a highly differentiated and polarised cell. Different sub-cellular organelles characterise each region of the spermatozoon. The cytoskeleton exerts a direct effect on the function of sperm by influencing the distribution of sub-cellular organelles and plasma mem-brane molecules. Each domain on the spermatozoa is destined to perform its physiological role at an exact time during the course of fertilisation. That the spermatozoon should behave in this manner is especially important since it is a cell of very limited biosynthetic potential, lacking the machinery to synthesise new molecules for subsequent intracellular routing and insertion into the appropriate membrane domain. Novel cytoskeletal elements, as defined by their ability to resist extraction from spermatozoa by detergent and/or high salt, have been identified but their function remains unclear. Information should be obtained to understand the cytoskeletal influences on the polarisation of sperm components and ultimately on sperm function. It will therefore be interesting to investigate the role of TSA70, a new member of the sperm cytoskeletal protein family. This information may provide us with an insight into its important role in sperm function which could be used for diagnosis of infertility as well as fertility regulation.

## Acknowledgements

We are grateful to Dr C P Puri, Director, National Institute of Research in Reproductive Health for his continued interest and encouragement and Dr. Sarena D'Souza for help with EM studies. Ms Monali Wakle is grateful to Indian Council of Medical Research (ICMR) for a Senior Re-search Fellowship.

## References

**Aitken RJ, Parslow JM, Hargreave TB and Hendry WF** (1988) Influence of antisperm antibodies on human sperm function *British Journal of Urology* **62** 367-373

**Aitken RJ** (2002) Immunocontraceptive vaccines for hu-man use *Journal of Reproductive Immunology* **57** 273-287

**Alexander NJ** (1984) Antibodies to human spermatozoa impede sperm penetration of cervical mucus of ham-ster eggs *Fertility and Sterility* **41** 433-439

**Alexander NJ and Anderson DJ** (1979) Vasectomy: con-sequences of autoimmunity to sperm antigens *Fertility and Sterility* **32** 253-260

Aul RB and Oko RJ (2002) The major subacrosomal occupant of bull spermatozoa is a novel histone H2B variant associated with the forming acrosome during spermiogenesis *Developmental Biology* **242** 376-387

Batova IN, Richardson RT, Widgren EE and O'Rand MG (2000) Analysis of the autoimmune epitopes on human testicular NASP using recombinant and synthetic peptides *Clinical and Experimental Immunology* **121** 201-209

Bigazzi PE, Kosuda LL, Harnick LL, Brown RC and Rose NR (1976) Antibodies to testicular antigens in vasectomised rabbits *Clinical Immunology Immunopathology* **5** 182-194

Bronson RA, Cooper GW and Rosenfeld DL (1982) Sperm specific isoantibodies and autoantibodies inhibit the binding of human sperm to human zona pellucida *Fertility and Sterility* **38** 724-729

Bronson RA, Cooper GW and Rosenfeld DL (1984) Sperm antibodies: their role in infertility *Fertility and Sterility* **42** 171-183

Bush LA, Herr JC, Wolkowicz M, Sherman NE, Shore A and Flickinger CJ (2002) Novel asparginase–like protein is a sperm autoantigen in rats *Molecular Reproduction and Development* **62** 233 -247

Busso D, Cohen DJ, Hayashi M, Kasahara M and Cuasnicu PS (2005) Human testicular protein TPX1/CRISP-2: localization in spermatozoa, fate after capacitation and relevance for gamete interaction *Molecular Human Reproduction* **4** 299-305

Carrera A, Gerton GL and Moss SB (1994) The major fibrous sheath polypeptide of mouse sperm: structural and functional similarities to the A-kinase anchoring proteins *Developmental Biology* **165** 272-284

Chiu WW and Chamley LW (2002) Use of antisperm antibodies in differential display Western blotting to identify sperm proteins important in fertility *Human Reproduction* **17** 984-989

DePaolo LV, Hinton BT and Braun RE (2000) Male contraception: views to the 21st century, Bethesda, MD, USA, 9-10 September 1999 *Trends in Endocrinology and Metabolism* **11** 66-69

Eddy EM, Toshimori K and O'Brien DA (2003) Fibrous sheath of mammalian spermatozoa *Microscopy Research and Technique* **61** 103-115

Escalier D, Gallo JM and Schrevel J (1997) Immunochemical characterization of a human sperm fibrous sheath protein, its developmental expression pattern, and morphogenetic relationships with actin *The Journal of Histochemistry and Cytochemistry* **45** 909-922

Ensrud KM and Hamilton DW (1991) Use of neonatal tolerization and chemical immunosuppression for the production of monoclonal antibodies to maturation specific sperm surface molecule *Journal of Andrology* **12** 305-314

Flickinger CJ, Rao J, Bush LA, Sherman NE, Oko RJ, Jayes FC and Herr JC (2001) Outer dense fiber protein are dominant post obstruction autoantigen in adult Lewis rats *Biology of Reproduction* **64** 1451-1459

Fulcher KD, Mori C, Welch JE, O'Brien DA, Klapper DG and Eddy EM (1995) Characterization of Fsc1 cDNA for a mouse sperm fibrous sheath component *Biology of Reproduction* **52** 41-49

Gubin DA, Dmochowski R and Kutteh WH (1998) Multivariant analysis of men from infertile couples with and without antisperm antibodies *American Journal of Reproductive Immunology* **39** 157-160

Handley HH Jr, Flickinger CJ and Herr JC (1988) Post vasectomy sperm autoimmunogens in the Lewis rat *Biology of Reproduction* **39** 1239-1250

Handley HH Jr, Flickinger CJ and Herr JC (1990) Biphasic production of antisperm autoantibodies follow vasectomy of the Lewis rat *Journal of Reproductive Immunology* **17** 53-67

Holden C (2002) Research on contraception still in the doldrums *Science* **296** 2172-2173

Huang TT Jr, Hardy D, Yanagimachi H, Teuscher C, Tung K, Wild G and Yanagimachi R (1985) pH and protease control of acrosomal content stasis and release during the guinea pig sperm acrosome reaction *Biology of Reproduction* **32** 451-462

Ishakia M and Alexander NJ (1984) Vasectomy induced autoimmunity: antisperm and antinuclear autoimmune monoclonal antibodies *American Journal of Reproductive Immunology* **5** 117-124

Jones R, Shalgi R, Hoyland J and Phillips DM (1990) Topographical rearrangement of plasma membrane antigen during capacitation of rat spermatozoa *in vitro Developmental Biology* **139** 349-362

Khole V, Joshi S and Singh S (2000) Identification of epididymis-specific antigens using neonatal tolerization *American Journal of Reproductive Immunology* **44** 350-358

Kim KS, Moon CC and Gerton GL (2001) Mouse sperm protein sp56 is a component of the acrosomal matrix *Biology of Reproduction* **64** 36-43

Lansford B, Haas GG Jr, DeBault LE and Wolf DP (1990) Effect of sperm-associated antibodies on the acrosomal status of human sperm *Journal of Andrology* **11** 532-538

Linnet L and Hjort T (1977) Sperm agglutinins in seminal plasma and serum after vasectomy: correlation between immunological and clinical findings *Clinical Experimental Immunology* **30** 413-420

Miranda-Vizuete A, Tsang K, Yu Y, Jimenez A, Pelto Huikko M, Flickinger CJ, Sutovsky P and Oko R (2003) Cloning and developmental analysis of murid spermatid-specific thioredoxin-2 (SPTRX-2), a novel sperm fibrous sheath protein and autoantigen *Journal of Biological Chemistry* **278** 44874-44885

Myles DG and Primakoff P (1984) Localized surface antigens of guinea pig sperm migrate to new regions prior to fertilization *Journal of Cell Biology* **99** 1634-1641

Nakamura S, Tsuji Y and Koyama K (1994) Identification and characterization of a sperm peptide antigen recognized by a monoclonal antisperm antibody derived from a vasectomized mouse *Biochemica Biophysica Research Communication* **205** 1503-1509

Nass SJ and Strauss JF 3rd (2004) Strategies to facilitate the development of new contraceptives *Nature Reviews- Drug Discovery* **3** 885-890

Naz RK and Zhu X (1998) Recombinant Fertilization

Antigen-1 causes a contraceptive effect in actively immunized mice *Biology of Reproduction* **59** 1095-1100

O'Bryan MK, Sebire K, Meinhardt A, Edgar K, Keah HH, Hearn MT and De Kretser DM (2001) Tpx-1 is a component of the outer dense fibers and acrosome of rat spermatozoa *Molecular Reproduction and Development* **58** 116-125

Oko R (1988) Comparative analysis of proteins from the fibrous sheath and outer dense fibers of rat spermatozoa *Biology of Reproduction* **39** 169-182

Oko R (1998) Occurrence and formation of cytoskeletal proteins in mammalian spermatozoa *Andrologia* **30** 193-206

Paradisi R, Bellavia E, Pession A, Venturoli S, Bach V and Flamigni C (1996) Characterization of human sperm antigens reacting with antisperm antibodies from autologous sera and seminal plasma: comparison among infertile subpopulations *International Journal of Andrology* **19** 345-352

Samuel T, Linnet L and Rumke P (1978) Post vasectomy autoimmunity to protamine in relation to the formation of granulomas and sperm agglutinating antibodies *Clinical and Experimental Immunology* **33** 261 – 269

Saxena DK and Toshimori K (2004) Molecular modifications of MC31/CE9, a sperm surface molecule, during sperm capacitation and the acrosome reaction in the rat: Is MC31/CE9 required for fertilization? *Biology of Reproduction* **70** 993-1000

Sutovsky P, Mananndhar G, Wu A and Oko R (2003) Interactions of sperm perinuclear theca with the oocyte: Implications for oocyte activation, antipolyspermy defense, and assisted reproduction *Microscopy Research and Technique* **61** 362-378

Tres LL and Kierszenbaum AL (1996) Sak57, an acidic keratin initially present in the spermatid manchette before becoming a component of paraaxonemal structures of the developing tail *Molecular Reproduction and Development* **44** 395-407

TurnerTT, Riley TA, Mruk DD and Cheng CY (1999) Obstruction of the vas deferens alters  protein secretion by the rat caput epididymal epithelium *in vivo Journal of Andrology* **20** 289-297

Turner TT, Riley TA, Vagnetti M, Flickinger CJ, Caldwell JA and Hunt DF (2000) Postvasectomy alterations in protein synthesis and secretion in the rat caput epididymis are not repaired after vasovasostomy *Journal of Andrology* **21** 276-290

Verdier Y, Chaffaux S and Boue F (2002) Identification of post vasectomy sperm auto–antigens in fox (*Vulpes vulpes*) by two-dimensional gel electrophoresis and Western blotting *Journal of Reproductive Immunology* **54** 65-80

Verdier Y, Guillaume F, Nelly R, Zoltan K, Tamas J and Boué F (2005) Identification of a new testis specific sperm antigen localized on principal piece of the spermatozoa tail in the fox (*Vulpes vulpes*) *Biology of Reproduction* **72** 502-508

Weiske WH (2001) Vasectomy *Andrologia* **33** 125-134

# Modulators of spermatogenic cell survival

## Chandrima Shaha

*Cell Death and Differentiation Research, National Institute of Immunology, Aruna Asaf Ali Marg, New Delhi 110067, India*

Apoptosis is a process of cell suicide, the mechanisms of which are encoded in the chromosomes of all nucleated cells. Apoptosis occurs spontaneously throughout mammalian spermatogenesis for the development of normal mature spermatozoa and for the elimination of excess or abnormal germ cells: a critical prerequisite for functional spermatogenesis under physiological conditions. Any deregulation of the apoptotic process during spermatogenesis would lead to defective sperm formation and while increased apoptosis could potentially lead to infertility, decreased cell death could do the same by disrupting testicular homeostasis due to accumulation of cells. Male germ cell apoptosis occurs through two major pathways, involving either mitochondria (intrinsic) or cell surface death receptors (extrinsic). The mitochondrial pathway of apoptosis involves the Bcl-2 group of proteins and different members of this group are involved in diverse situations. The cell death receptor pathway involves members of the TNF receptor superfamily. The stimuli for germ cell apoptosis are internal cues that control proper homeostasis of the testicular tissue or external agents including testicular toxins, heat stress and chemotherapeutic agents. In addition, an imbalance of hormones can lead to the apoptosis of germ cells. The pathway of apoptosis adopted by the germ cells depends on the type of stimuli they receive. This review discusses the recent advances made in the understanding of the mechanisms of germ cell apoptosis that may provide clues to the control of male fertility or treating germ cell tumors and other testis associated pathological conditions.

## Introduction

Cell death, a tightly controlled, finely orchestrated event, may be described either as apoptosis or nonapoptotic cell death, traditionally called 'necrosis'. Apoptosis is a process of cell suicide, the mechanisms of which are encoded in the chromosomes of all nucleated cells. Much progress has been made in understanding the regulation of apoptosis in various extragonadal cell systems (Hengartner, 2000). This mode of cell death constitutes a tightly regulated series of energy-dependent molecular and biochemical events orchestrated by a genetic program. The process is defined by the morphologic appearance of the dying cell which includes cytoplasmic blebbing, chromatin condensation, nuclear fragmentation and cell shrinkage. Biochemical fea-

E-mail: cshaha@nii.res.in

tures associated with apoptosis include high molecular weight DNA fragmentation and the formation of an oligonucleosomal ladder, externalization of phosphatidylserine that is normally confined to inner surface of the plasma membrane and proteolytic cleavage of a number of intracellular substrates (Hengartner, 2000). A large number of studies have been published on germ cell
apoptosis, and it appears that multiple pathways of apoptosis exist in germ cells depending on the
type of stimuli received. In this review, the studies that provide a definitive link to a particular
death pathway associated with a given stimulus are included and those describing only morphologic features of apoptosis are not discussed.

## The process of spermatogenesis

Spermatogenesis is a dynamic process in which stem spermatogonia, through a series of events,
become mature spermatozoa. Apoptosis occurs spontaneously throughout mammalian spermatogenesis for the development of normal mature spermatozoa, and for the elimination of excess or
abnormal germ cells (Sinha Hikim et al., 2003). Stem spermatogonia undergo mitosis to produce
additional stem cells and differentiating spermatogonia, which undergo mitotic divisions to form
primary spermatocytes. The spermatocytes enter a lengthy meiotic phase as preleptotene spermatocytes and proceed through two cell divisions (meiosis I and II) to give rise to haploid spermatids.
These in turn undergo a complex process of morphological and functional differentiation collectively known as spermiogenesis resulting in the production of mature spermatozoa. Not all germ
cells achieve maturity and cell death by apoptosis appears to be a constant feature of normal
spermatogenesis in a variety of mammalian species to maintain proper germ cell numbers. In adult
rat this loss is incurred mostly during spermatogonial development (up to 75%) and to a lesser
extent during maturation divisions of spermatocytes and spermatid development. Therefore, any
deregulation of the apoptotic process during spermatogenesis would lead to defective sperm formation. While increased apoptosis could potentially lead to infertility, decreased cell death could
do the same by disrupting testicular homeostasis due to accumulation of cells.

## Checkpoints of cell-death in the testis

The first wave of spermatogenesis is initiated when gonocytes differentiate into spermatogonia.
While some spermatogonia become self-renewing spermatogonial stem cells, most differentiate
into spermatocytes, and at approximately day 10 after birth in mice and at puberty in man, initiate
meiosis. The first wave of spermatogenesis is accompanied by extensive germ cell apoptosis. In
the adult, spermatogenesis is characterized by continuous germ cell maturation that includes mitotic proliferation of spermatogonia, meiotic division of spermatocytes, differentiation of spermatids and finally release of spermatozoa into the tubular lumen. To achieve precise homeostasis of
each germ cell type in the adult, germ cell renewal, proliferation, export and apoptosis must be
finely balanced. During the entire process of spermatogenesis, apoptosis occurs at the various
stages where it is used as a means to control the overproliferating population. This appears to occur
at the cost of substantial germ cell wastage because it is estimated that up to 75% of potential
spermatozoa degenerate in the testes of adult mammals. Apoptotic germ cells are either sloughed
off into the tubular lumen or phagocytosed by Sertoli cells. During release of the mature spermatozoon into the tubule lumen, a residual body is formed from the excess cytoplasm which is
phagocytosed by Sertoli cells. These residual bodies may be removed by a nuclear-independent
apoptotic process since several markers of apoptosis have been detected in cytoplasmic bodies
within Sertoli cells of the rat testis (Rey, 2003).

## Pathways of apoptosis

The signalling events leading to apoptosis can be divided into two major pathways, involving either mitochondria (intrinsic) or death receptors (extrinsic) (Fig 1). Typically, the death receptor pathway is mediated by cell surface receptors that are engaged by soluble ligands leading to their activation through adaptor proteins. The mitochondrial pathway on the other hand is activated by intrinsic mechanisms and do not involve cell death receptors (Hengartner, 2000). The death receptor pathway mediated activation can use the mitochondrial pathway to amplify apoptotic signals. Recent advances have led to the identification of four major functional groups of molecules involved in triggering and affecting the apoptotic process. These are the caspases, the adaptor proteins which control the activation of initiator caspases, members of the tumor necrosis factor (TNF) receptor (TNF-R) super family and members of the Bcl-2 family of proteins. A group of cysteine proteases called caspases, is essential for programmed cell death in a variety of species. At least 14 caspases have been identified in mammals. These enzymes recognize tetrapeptide motifs and cleave their substrates on the carboxyl side of an aspartate residue. Individual caspases have distinct substrate specificities that are determined by the pattern of amino acids upstream of the cleavage site. Cells that are exposed to extracellular death ligands, TNF and Fas ligand, undergo death-receptor-mediated apoptosis through the extrinsic death pathway. These ligands are bound by trimeric cell surface death receptors TNF-R for TNF and Fas for Fas Ligand (FasL) that recruit adaptor proteins, including Fas associated death domain (FADD) and TNF receptor associated death domain (TRADD), and the receptor-interacting protein RIP (Green, 2005).

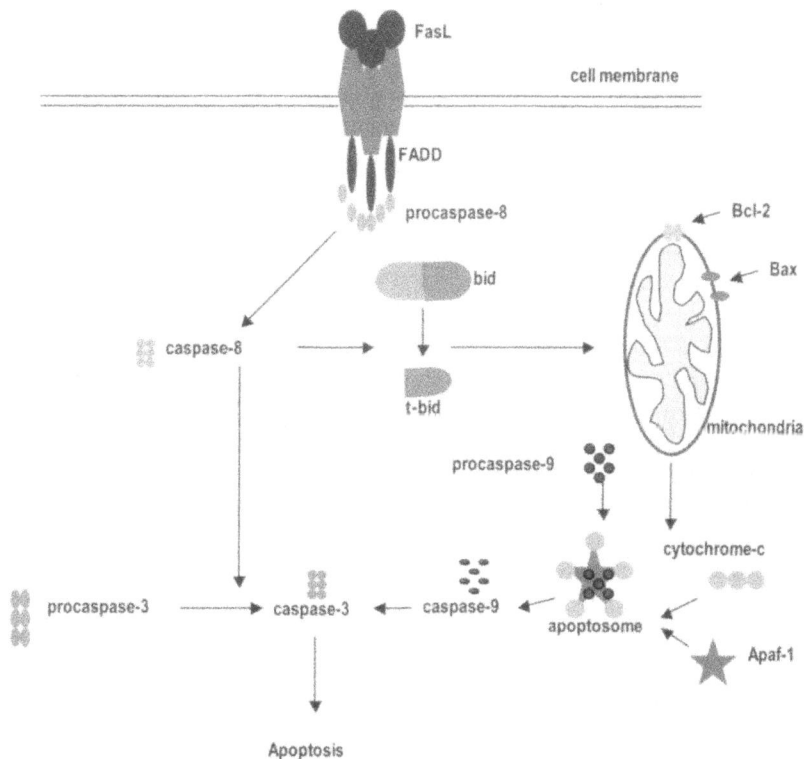

**Fig. 1.** Pathways of apoptosis. Figure shows some members of the extrinsic (death receptor mediated) and intrinsic (mitochondrial) pathways of apoptosis.

## Death receptor pathway in germ cell death

FasL is a 281 amino acid long type II transmembrane molecule belonging to the large TNF family of proteins which bind to and activate members of the TNF receptor family (Janssen et al., 2003). FasL and its corresponding receptor, Fas (CD95/APO-1), both interact as oligomers and the activated Fas receptor complex initiates a pro-apoptotic death signal in the receptor-bearing cell (Curtin and Cotter, 2003). As mentioned above, the existence of Fas and FasL in the testicular cells is undisputed but there is controversy regarding their localization. Many studies report localization of FasL in the Sertoli cells and Fas receptor on the germ cells (Koji et al., 2001), while other studies report localization of FasL in the germ cells as well (D'Alessio et al., 2001). Studies from our laboratories show that under stimulation with estradiol, both FasL and Fas are localized to different generations of germ cells (Nair and Shaha, 2003).

*Fas and FasL deficient models:* Despite the apparent requirement of Fas and FasL for damage-induced germ cell apoptosis, *lpr* mice containing a spontaneous loss of function mutation in the Fas gene are fertile with normal spermatogenesis (Lee et al., 1999). These studies suggest that this is due to a "salvage pathway" that restores expression of Fas in the testes of these mutants. While similar levels of Fas protein and germ cell apoptosis are present in the testes of both control and *lpr* mice, control mice expressed significantly higher amounts of Fas in all other organs tested. Thus, it appears that in the testes, either the *lpr* mutation has no effect due to tissue specific regulation, or a unique system is present to restore proper levels of Fas. The FasL mutant *gld* mice that lack a functional Fas-signaling pathway have a small but significant increase in testis weight and numbers of spermatid heads per testis compared with wild-type mice. In addition, *gld* mice have a small increase in the spontaneous incidence of germ cell apoptosis, as indicated by characteristic DNA fragmentation (Richburg and Nanez, 2003).

*Fas/FasL and toxins:* Different toxins influence testicular Fas and FasL, for example, ethane-1,2-dimethanesulfonate (EDS) is a cytotoxic alkylating agent that is selective toxin for adult rat Leydig cells. Exposure to the toxin causes Leydig cell ablation resulting in a rapid decrease in testosterone concentrations followed by a characteristic pattern of germ cell loss (Bartlett et al., 1986). In such treatments, there is an increase in testicular Fas with an accompanying elevation of germ cell apoptosis (Nandi et al., 1999) suggesting that Fas is involved in EDS induced cell death. In contrast to this report, a decline in the expression of FasL and Fas in the pachytene spermatocytes and spermatids have been reported after treatment with EDS (Woolveridge et al., 1999). It is therefore possible that this treatment which creates withdrawal of testosterone support, alters germ cell Fas in a germ cell specific manner. While Leydig cells provide testosterone supply to germ cells, Sertoli cells provide structural support. In addition, paracrine factors from Sertoli cells also help germ cells to survive. Sertoli cells are targets of phthalates, ubiquitous environmental toxicants, capable of producing testicular atrophy in laboratory animals. Phthalate exposure results in a rapid induction of testicular germ cell apoptosis spermatocytes being the most sensitive population (Akingbemi et al., 2004; Salazar et al., 2004). In addition to environmental pollutants, ionizing radiation also shows detrimental effects on germ cells. The tumor suppressor protein p53 has been identified as a major proapoptotic factor regulating the testicular response to x-irradiation. After exposure to 5-Gy x-irradiation, Fas mRNA expression does not change at pretreatment levels in p53 knock-out mice. However, Fas mRNA increases in a time-dependent manner in wild-type mice following exposure to 5-Gy radiation, indicating that radiation-induced Fas expression is p53 dependent (Embree-Ku et al., 2002). Additional reports show an increase in Fas expression following exposure to radiation resulting in cell death (Lee et al., 1999). Among other compounds that affect testes, it is well known that ethanol exposure disrupts the hypothalamic-pituitary-gonadal axis and adversely affects the secretory function of Sertoli cells. Fas ligand levels are increased within the testes of rats chronically administered ethanol (Zhu *et al.*, 2000). Adult Wistar rats fed with ethanol for 12 weeks show marked Sertoli cell vacuolization, germ cell degeneration and upregulation of FasL (Eid *et al.*, 2002). Therefore, ethanol induced deregulation of germ cell survival appears to be mediated through the Fas/FasL pathway. Busulfan is an alkylating drug now used almost exclusively to treat chronic myeloid leukaemia. It is known to cause germ cell depletion in rodents, however, the pathway of apoptosis appears to be not mediated by the Fas receptor. Busulfan induced apoptosis appears to be mediated by loss of c-kit/SCF signaling (Choi *et al.*, 2004). The summary of the data presented above show the stimulation of the Fas/FasL pathway by a variety of stimuli, indicating that this system is available for control of germ cell apoptosis under a variety of toxic insults.

*Fas/FasL and temperature:* The scrotal location of the testes allows the germ cells to survive at a lower temperature than that of the main body. It is known that germ cell apoptosis occurs following exposure to elevated temperatures. Studies using the *gld* and *lprcg* mice which harbor loss of function mutations in FasL and Fas, respectively, show that heat-induced germ cell apoptosis is not blocked in these mice. Therefore, evidence that the Fas signaling system is not required for heat-induced germ cell apoptosis in the testis is provided by these studies (Vera *et al.*, 2004). In contrast, additional studies show that the expression levels of Fas and p53 increased significantly from 1 to 3 days after heat exposure suggesting that germ cell apoptosis induced by heat exposure is possibly mediated by the Fas/FasL system (Miura *et al.*, 2002). Cryptorchidism, a condition in which the testes are exposed to body temperature rather than scrotal temperature, thereby creating a condition of increased heat stress, is a frequent male sexual disorder in mammals which affects the histology of the tunica propria, interstitial tissue, blood vessels, seminiferous epithelium and testis functioning (Bernal-Manas *et al.*, 2005). P53-dependent apoptosis appears responsible for the initial phase of germ cell loss in experimental cryptorchidism based on a 3-day delay of apoptosis in P53$^{-/-}$ mice. P53$^{-/-}$, *lpr/lpr* (a spontaneous mutation in the Fas gene, which causes autoimmune disease) double-mutant mice with unilateral cryptorchidism show testicular weight reduction and germ cell apoptosis where the Fas production increased in the time frame of p53-independent apoptosis in the experimental cryptorchid testis of wild-type mice. These results suggest that Fas is involved in testicular germ cell apoptosis, and that Fas-dependent apoptosis is related to the p53-independent phase of germ cell apoptosis in the cryptorchid testis (Yin *et al.*, 2002). Taken together, these results demonstrate that the mitochondria and possibly also endoplasmic reticulum-dependent pathways are the key apoptotic pathways for heat-induced germ cell death in the testis (Sinha Hikim *et al.*, 2003). In rats with experimental autoimmune orchities, germ cell death occurs through an apoptotic mechanism preceding germ cell sloughing. Immunohistochemical data suggest that the Fas/FasL system mediates germ cell apoptosis in an autocrine and/or paracrine way during experimental autoimmune orchities (Theas *et al.*, 2003). In ischemia-reperfusion testes, increased Fas-positive germ cells and increased FasL expression in Sertoli cells as compared to normal testes are known where most of the apoptotic cells are confined to seminiferous epithelial stages of XI and XII (Koji *et al.*, 2001). Some reports suggest a simultaneous upregulation of Bax protein and FasL during ischemia-reperfusion (Lysiak *et al.*, 2000). In normal human males, less than 10% of spermatozoa express Fas but this number is much higher in males with defects in sperm number, motility and morphology (Sakkas *et al.*, 1999). Disruption of FasL expression in rat germ cells was found to result in increased cell survival, further demonstrating that the Fas pathway regulates male germ cell apoptosis (Lee *et al.*, 1999). Similarly, human testicular tissue treated with an antibody able to bind to FasL and prevent interaction with Fas resulted in an inhibition of germ cell apoptosis (Pentikainen *et al.*, 1999). Conversely, mouse germ cells treated with an antibody to activate Fas showed a higher induction of apoptosis and decreased survival as compared to control cells (Lee et al., 1999). There-fore, various testicular conditions like orchities, cryptorchidism and ischemia reperfusion leads to activation of the Fas/FasL pathway. This serves as a clue to potential mechanisms that can trigger the cell death receptor pathway.

*Fas/FasL and hormones:* Testosterone is absolutely essential for male germ cell survival, however, the amount of testosterone that is required is very critical. Excess testosterone can cause death, for example, after injection of testosterone undecanoate, both the apoptotic signal in germ cells and the expression of Fas/FasL in the testis increases correlatively in a time-dependent manner, reaching a maximum on day 30 (Zhou et al., 2001). In contrast, testoster-one withdrawal stimulates caspase activity and produces DNA fragmentation in Sertoli cells, with only a weak effect on DNA fragmentation and caspase activity in germ cells. This sug-gests that acute apoptosis of germ cells in the adult human testis occurs in a caspase-indepen-dent way and is controlled by Sertoli cells via an as yet undetermined mechanism (Tesarik et al., 2002). Evidence has accumulated over several decades now that estrogen is essential for spermatogenesis and that intratesticular concentrations of estrogen are very high (Hess, 2003). It has also been realized that estrogen-like chemicals present in the environment adversely affects male reproductive function. These chemicals include industrial pollutants, like bisphenol-A and polychlorinated biphenols, and pesticides like DDT, methoxychlor or chlorodecone. The extent of exposure to these chemicals on members of a population differs: occupations in agriculture, petrochemicals or the construction industry entails higher exposure. Since estrogen receptors are present in the pituitary and spermatogenic cells (Korach, 1994), estrogen-like chemicals can act as agonists or antagonists for the hormone and interfere with spermatogen-esis. The significance of estrogens in the male is of great interest due to the demonstration of impaired fertility in mice lacking estrogen receptor-$\alpha$ (Eddy et al., 1996) and the discovery of a second estrogen receptor-$\beta$, which is expressed in most of the spermatogenic cells (Kuiper et al., 1996). Therefore, agents able to mimic estrogens can potentially alter the action of the hormone on spermatogenic cells leading to functional impairment of the male gamete. Reports of lowered fertility rates as a consequence of exposure to agents with estrogenic activity termed as endocrine disruptors are well documented in wild life populations (Enmark et al., 1997). Taken together, the effects of estrogen on testicular function provide a conceptual basis to examine the speculative link between increased exposure to environmental estrogens and reduced fertility.

We used a diethylstilbestrol (DES) induced spermatogenic cell apoptosis model (Nair and Shaha, 2003) as this was ideal to study the mechanism of estrogen induced spermatogenic cell death. This is because DES can mimic estrogen action and has also been widely used as a model estrogen to study the effects on the neonatal male rat reproductive tract (Williams et al., 2000; Goyal et al., 2001). This study demonstrated that haploid spermatogenic cells undergo apoptosis in response to DES, which is in contrast to toxin induced cell death models where the diploid population of spermatocytes are most affected (Eid et al., 2002). This result was not surprising because the two regulators of the cell death pathway, Fas and FasL, were over ex-pressed in spermatids after DES exposure (Fig. 2). Interestingly, the largest number of apoptotic spermatogenic cells were visible in stage VII of the seminiferous epithelial cycle, a stage were both Fas and FasL expression were most pronounced and not in stage V or XII where FasL and Fas expression respectively were not prominent. Importantly, the observation that the number of apoptotic cells is highest in seminiferous tubule stages where both Fas/FasL expression was high reinforced the idea that these two proteins expressed in spermatogenic cells are the main modulators of death. Understandibly, in the presence of death stimuli, germ cells could alter the expression of death proteins to orchestrate their own death. In response to estrogen, stage

VII of the cycle of the seminiferous epithelium appears to be most sensitive. One other important finding that comes out of this study is that the activation of caspase-8 is in consonance with the fact that engagement of Fas with FasL prompts the formation of the death inducing signalling complex composed by the adaptor molecules like FADD and procaspase-8, followed by the release of active caspase-8. Our data clearly shows that the mitochondria, are involved in DES induced death as shown by the translocation of Bax from the cytosol to the mitochondria causing the release of cytochrome-c and loss of mitochondrial membrane potential. Therefore, the death signal in this case was amplified through the mitochondrial pathway. In view of the above data, it could be inferred that normal testicular homeostasis may involve the Fas/FasL system to maintain proper spermatogenic cell number, which may be under estrogen regulation. Interpreting the above results we propose that the spermatogenic cell in response to estrogen activates the death receptor mediated apoptotic pathway, which in turn can operate through a mitochondria dependent and a mitochondria independent mechanism to bring about cell death. This provides a failsafe means for elimination of injured cells that could contribute genetic damage to the next generation.

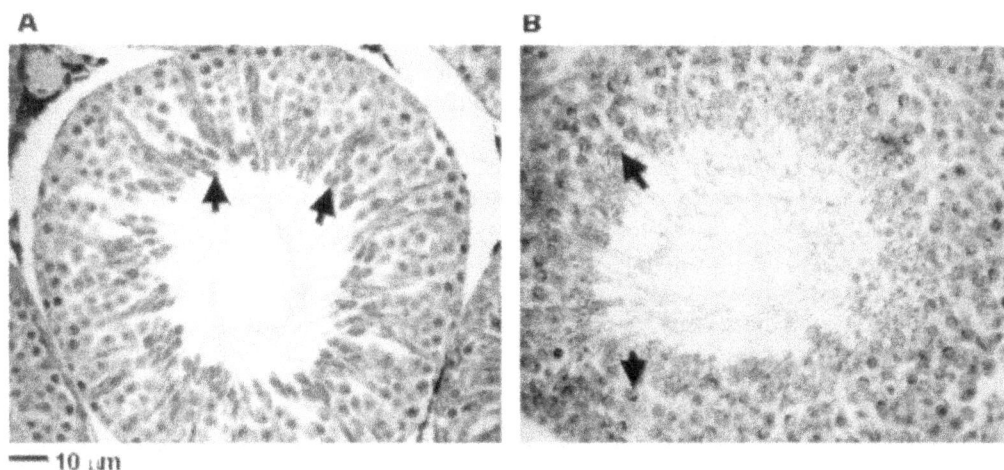

— 10 µm

**Fig. 2.** FasL and Fas staining in the seminiferous tubule. A. Anti-FasL antibody reactivity in the seminiferous tubule showing elongated spermatids staining for FasL. B. Anti-Fas antibody showing reactivity to round spermatids demonstrating the existence of Fas receptor. Arrows indicate respective antibody reactivity.

## The mitochondrial pathway of germ cell death

Bcl-2 family proteins are central regulators of apoptosis that occur through the mitochondrial pathway. These proteins contain up to four conserved Bcl-2 homology (BH) domains, BH1, BH2, BH3 and BH4, which are recognized by their amino acid sequence similarity. Both anti- and pro-apoptotic Bcl-2 family proteins have been identified. As in other cell types, Bcl-2 family proteins appear to be the major regulators of the mitochondrial death pathway in the male germline, largely by controlling the release of cytochrome-c from mitochondria. This leads to the activation of an Apaf-1/caspase-9 complex, resulting in further downstream caspase activation and substrate cleavage. Many Bcl-2 family proteins are capable of physically interacting with each other, forming a complex network of homo- and heterodimers, and these physical interactions sometimes play important roles in the opposing effects of pro- and anti-apoptotic members of the family. The pro-apoptotic members of the Bcl-2 family can be broadly classified into two groups. One group, including Bax, Bak and Bok, exhibit cytotoxic effects independently of their ability to bind other Bcl-2 family proteins; including Bcl-2 and other cytoprotective members of the family such as Bcl-xL, Bcl-w, Bfl-1 and Mcl-1. The second group of pro-apoptotic Bcl-2 family proteins varies widely in their amino acid sequences, often containing only a single region of similarity, specifically, the BH3 domain. These "BH3-only" proteins appear to possess no intrinsic or autonomous cytodestructive activity and instead operate as trans-dominant inhibitors of the survival proteins. Their antagonism of proteins such as Bcl-2 and Bcl-X$_L$ depends on binding via their BH3 domains to a hydrophobic pocket on target anti-apoptotic proteins.

*Bcl proteins and toxins:* Bcl-x exists in two isoforms, the anti-apoptotic form Bcl-xL, and the proapoptotic form Bcl-xS. The critical balance between the two appears to be important for cell survival, however, it is still not clear how exactly the vital balance is maintained. The effects of chronic exposure of 4-tert-octylphenol on the testicular development of prepubertal male rats show that expression of Bcl-xL mRNA was significantly decreased in the treated groups, whereas the expressions of Bcl-2 and Bax mRNA were not significantly changed (Kim et al., 2004). TNF-alpha promotes cell survival in the rat seminiferous epithelium and this prosurvival effect can be blocked by infliximab, a TNF-alpha antagonist. Bcl-xL is upregulated in mitochondrial membranes by TNF-alpha: possibly the mechanism of increased cell survival induced by TNF-alpha (Suominen et al., 2004). Gene expression studied by microarray in rat testis samples after treatment of rats with di(2-ethylhexyl) phthalate (DEHP) shows an activation of Fas/FasL cascade and Apaf-1/caspase-9 cascade while Bcl-2 decreases (Kijima et al., 2004). Short-term exposure (43°C for 15 min) of the rat testis to mild heat results within 6 h in stage- and cell-specific activation of germ cell apoptosis which is preceded by a redistribution of Bax from a cytoplasmic to paranuclear localization in heat-susceptible germ cells. The relocation of Bax is accompanied by cytosolic translocation of cytochrome c and is associated with activation of the initiator caspase 9 and the executioner caspases-3, -6, and -7 and cleavage of poly(ADP) ribose polymerase. Taken together, these results demonstrate that the mitochondria and possibly also endoplasmic reticulum-dependent pathways are the key apoptotic pathways for heat-induced germ cell death in the testis (Sinha Hikim et al., 2003).

Epidemiological surveys and animal experimental studies suggest that exposure to 2-bromopropane (2-BP) could result in reproductive and hematopoietic disorders. In animal studies, downregulation of Bcl-2 after the first or second injection of 2-BP and upregulation of Bax after the first treatment contributed to the initiation of primary apoptosis of spermatogonia. These results indicate that 2-BP resulted in apoptotic death of testicular germ cells and that this process involves the Bcl-2 family genes (Yu et al., 2001). P53 overexpression leads to impairment of spermatogenesis in the rat testes (Fujisawa et al., 2001).

Using an *in vitro* spermatogenic cell apoptosis model, we showed the possible role of Ca$^{2+}$ in regulating Bcl-xS and Bcl-xL expression. A metabolite of the common industrial solvent *n*-hexane, 2,5-hexanedione, caused a significant increase in reactive oxygen species followed by an enhancement of intracellular Ca$^{2+}$ through the T-type Ca$^{2+}$ channels (Fig. 3). Consequent to the above changes, expression of Bcl-xS increased with a concomitant drop in Bcl-xL expression thus altering the ratio of the two proteins. Impediment of Ca$^{2+}$ influx by using a T-type Ca$^{2+}$ channel blocker, pimozide, resulted in a decrease in Bcl-xS and an increase in Bcl-xL expression. This caused prevention of mitochondrial potential loss, reduction of caspase-3 activity, inhibition of DNA fragmentation and an increase in cell survival. Alternatively, Ca$^{2+}$ ionophores caused an increase of Bcl-xS encoding isoform over the Bcl-xL encoding isoform (Mishra et al., 2006).

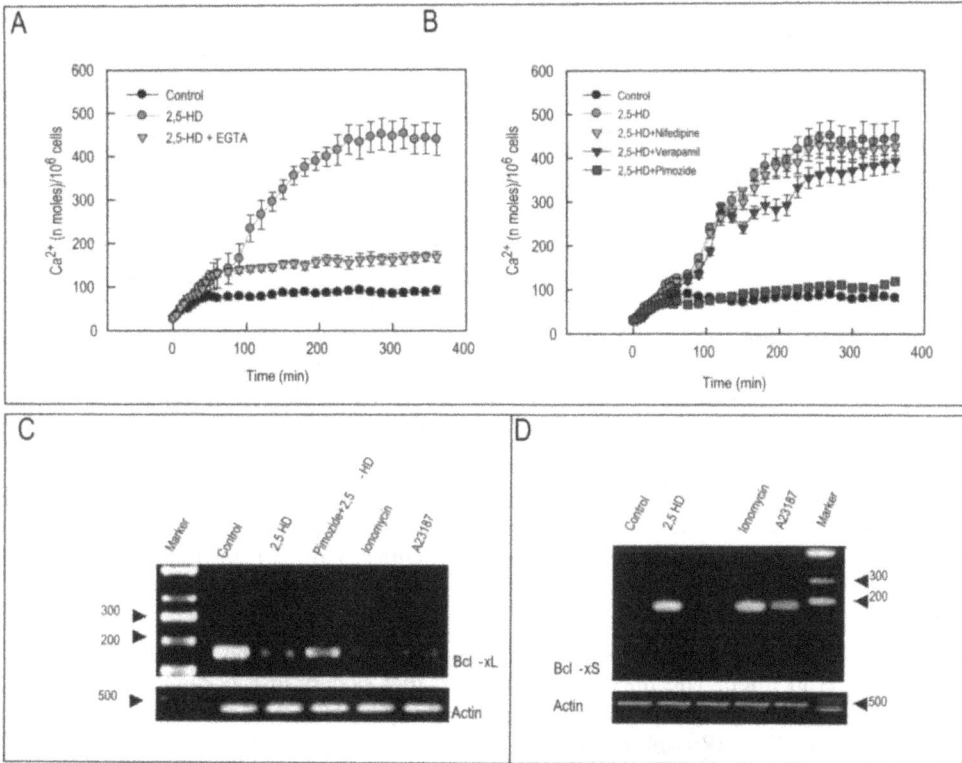

**Fig. 3.** Intracellular Ca$^{2+}$ changes in response to 2,5-HD. A. Treatment with 2,5-HD induces an increase in Ca$^{2+}$ concentration. Cellular Ca$^{2+}$ was monitored after labeling the cells with fluorescent probe fluo-3AM (5 μM) containing 1 μM pluronic acid F-127 and 0.25 mM sulfinpyrazone. Note that presence of EGTA was able to reduce Ca$^{2+}$ influx. Data represent the mean ± S.E. (n = 3). B. Graph demonstrates the effects of inhibition of Ca$^{2+}$ channels. Note the ability of pimozide (20 μM), a T-type Ca$^{2+}$ channel blocker to restrict entry of Ca$^{2+}$ as compared to inhibitors of L-type Ca$^{2+}$ channels, nifedipin (10 μM) and verapamil (20 μM) that failed to stop Ca$^{2+}$ entry. Data represent the mean ± S.E. (n = 3). C and D. Detection of Bcl-x mRNA. C. Note the reduction of Bcl-xL expression after 2,5-HD exposure in comparison to the groups where pimozide (20 mM) was present or the control group. Presence of ionomycin (1 mM) and A23187 (100 nM) shows a decrease in BcL-xL expression. D. Detection of Bcl-xS mRNA by RT-PCR. Note the increase of Bcl-xS expression after 2,5-HD exposure in comparison to the groups where pimozide was present or the control group. Presence of ionomycin (1 mM) and A23187 (100 nM) shows an increase in BcL-xS expression.

*Bcl deficient models:* Bcl-w is a death-protecting member of the Bcl-2 family of apoptosis-regulating proteins. Mice that are mutant for Bcl-w display progressive and nearly complete testicular degeneration (Russell *et al.*, 2001). Mutations in *Bcl-w* result in male sterility, increased germ cell apoptosis and reduced testes size (Ross *et al.*, 1998). Furthermore, adult *Bcl-w$^{-/-}$* mice were unable to produce any mature sperm (Ross *et al.*, 1998). Transgenic (TG) mice overexpressing Bcl-w driven by a chicken beta-actin promoter develop normally but are infertile. The adult transgenic testes display disrupted spermatogenesis with various severities ranging from thin seminiferous epithelium containing less germ cells to Sertoli cell-only appearance. Data suggest that regulated spatial and temporal expression of Bcl-w is required for normal testicular development and spermatogenesis, and overexpression of Bcl-w inhibits germ cell cycle entry and/or cell cycle progression leading to disrupted spermatogenesis (Yan et al., 2003) Other Bcl-2 family members, the anti-apoptotic proteins Bcl-2 and Bcl-xL, and the pro-apoptotic Bax, do appear to play a role during this early period of apoptosis. Mice overexpressing Bcl-2 or Bcl-xL exhibit male sterility characterized by defects in spermatogenesis, apparently due to suppression of the early apoptotic wave. During the early wave, a high level of Bax expression is normally seen in germ cells of the testes (Rodriguez et al., 1997; Jahnukainen et al., 2004). Bax-deficient mice display similar phenotypes to those overexpressing Bcl-2 or Bcl-xL including an increase in premeiotic germ cells (Knudson et al., 1995; Rodriguez et al., 1997). While Bcl-2-deficient mice display normal spermatogenesis, those expressing lower levels of Bcl-xL demonstrate increased germ cell death similar to Bax overexpression (Knudson et al., 1995). Thus, it appears that the early apoptotic wave in mice required for the formation of mature sperm is dependent on the proper balance of anti-apoptotic genes such as Bcl-xL, and pro-apoptotic genes such as Bax. The lack of such balance in Bax knockout mice likely helps to explain the resulting increase in early germ cell number. Interestingly, mice deficient for Bax exhibited an eventual increase in cell death, possibly due to the excessive number of spermatogonia unable to be supported by the adjacent Sertoli cells (Knudson et al., 1995). This resulted in a block in the maturity of spermatocytes and cell death leading to the loss of almost all spermatocytes and mature sperm (Russell et al., 2001).

Bax is a multidomain, proapoptotic member of the Bcl-2 family that is required for normal spermatogenesis in mice. The massive hyperplasia that occurs in Bax-deficient mice subsequently results in Bax independent cell death that may be triggered by overcrowding of the seminiferous epithelium (Russell et al., 2002). Bcl6 protein was detected in testicular germ cells, mainly spermatocytes, of normal mice but its physiological role is largely unknown. The number of spermatozoa in the cauda epididymis of Bcl6-deficient (Bcl6$^{-/-}$) mice is lower than that of Bcl6$^{+/+}$ mice. Numerous apoptotic spermatocytes at the metaphase I stage with induction of Bax protein in adult Bcl6$^{-/-}$ testes are seen. The incidence of apoptosis in heterozygous Bcl6$^{+/-}$ mice is also higher than that of Bcl6$^{+/+}$ mice. Treatment of testes of adult Bcl6$^{+/+}$ mice with a mild hyperthermia results in germ cell apoptosis predominantly in metaphase-I spermatocytes with induction of Bax protein. Thus, Bcl6 possibly has a role as a stabilizer in protecting spermatocytes from apoptosis induced by stressors (Kojima et al., 2001). The BH3-only proteins of the Bcl-2 family initiate apoptosis through the activation of Bax-like relatives. The hematopoietic compartments of bik($^{-/-}$)bim($^{-/-}$) and bim($^{-/-}$) mice were indistinguishable. However, although testes develop normally in mice lacking either Bik or Bim, adult bik($^{-/-}$)bim($^{-/-}$) males were infertile, with reduced testicular cellularity and no spermatozoa. The testes of young bik($^{-/-}$)bim($^{-/-}$) males, like those lacking Bax, exhibited increased numbers of spermatogonia and spermatocytes, although loss of Bik plus Bim blocked spermatogenesis somewhat later than Bax deficiency. The initial excess of early germ cells suggests that spermatogenesis fails because supporting Sertoli cells are overwhelmed. Thus, Bik and Bim share, upstream of Bax, the role of eliminating supernumerary germ cells during the first wave of spermatogenesis, a process vital for normal testicular development (Coultas et al., 2005).

*Bcl proteins and hormones:* Our studies show that exposure to 17β-estradiol induces FasL expression and subsequent ligation of this ligand with Fas increases nitric oxide formation through the activation of inducible nitric oxide synthase (Mishra and Shaha 2005). Superoxide is generated as a byproduct of estradiol metabolism through a Fas/FasL independent pathway. Ketoconazole, a cytochrome P450 monooxygenase inhibitor, reduces superoxide formation while anti-FasL antibodies are unable to do so. The peroxynitrites formed as reactive products of superoxide and nitric oxide lead to the induction of a transient mitochondrial hyperpolarization resulting in the release of cytochrome-c from the organelle. Scavenging of the nitrites with an intracellular peroxynitrite scavenger reduces hyperpolarization and cell death. Interestingly, prevention of nitric oxide generation to reduce nitrite formation leads to the inhibition of hyperpolarization as well as caspase-3 activation, however, inhibition of caspase-8 cleavage was unable to prevent hyperpolarization but reduced caspase-3 activity indicating the existence of a dual pathway. Loss of cellular glutathione as a result of estradiol exposure increases the susceptibility of the cells to damage by peroxynitrites. Taken together, this study established the ability of spermatogenic cells to respond to estrogens independent of testicular somatic cells and demonstrated the active participation of two pathways of caspase-3 activation, one involving the mitochondria and the other being mitochondria independent.

## Conclusions and perspectives

The Sertoli cells are largely responsible for orchestrating the development of germ cells through sequential phases of mitosis, meiosis and differentiation. It is possible that apoptosis of germ cells serves as a mechanism to reduce the germ cell population to a level which the Sertoli cell can support. It is clear that apoptosis of testicular germ cells is a critical prerequisite for functional spermatogenesis under physiological conditions. Alterations in the physiologic regulation of apoptosis are now recognized to be associated in the pathogenesis of various human diseases (Thompson, 1995). Toxicants that injure or disrupt the functions of Sertoli cells can effectively reduce their supportive capacity and result in an increased elimination of germ cells via apoptosis. The Fas-signaling system is proposed as a mechanism by which Sertoli cells can actively initiate the efficient removal of select germ cells. The regulation of germ cell apoptosis is likely influenced by alterations in the cell-specific expression of both Fas and FasL. Increased rates of germ cell apoptosis have been observed in testicular biopsies from human patients suffering from certain pathological conditions (Hadziselimovic et al., 1998) or infertility (Feng et al., 1999). Chemical-induced disruption of germ cell apoptosis regulatory mechanisms could participate in the pathogenesis of germ cell over-proliferation or deletion. The final common consequence for either of these conditions is infertility. Interestingly, there are differences between different ethnic groups in the rates of spontaneous apoptosis in humans with rates discovered to be higher in Chinese than in Caucasian men (Sinha Hikim and Swerdloff, 1999). These differences suggest the possibility that varying human populations may differ in their susceptibility to chemical-induced infertility based on differences in the sensitivity of their germ cells to undergo apoptosis. These observations are likely to further fuel the widening interest in the mechanisms regulating apoptosis in the testis.

This review has attempted to highlight the recent development in the field of apoptosis in general and the role of this mode of cell death in spermatogenesis (Fig. 4). Although germ cell apoptosis can also be triggered by various non-hormonal regulatory stimuli, including testicular toxins, heat stress and chemotherapeutic agents, the mechanisms by which these hormonal and non-hormonal factors regulate germ cell apoptosis are not well understood. Studies using genetically altered mice either overexpressing or harbouring a null mutation of specific genes indicate that germ cell apoptosis like that of other cell systems, is regulated by multiple genes that either inhibit or promote cell death. The challenge is now to identify the intracellular regulators of germ cell apoptosis and to elucidate their regulation by a variety of death signals. The mechanisms by which various pro-apoptotic and anti-apoptotic genes control spermatogenesis will provide insight into the molecular components of the death machinery within the

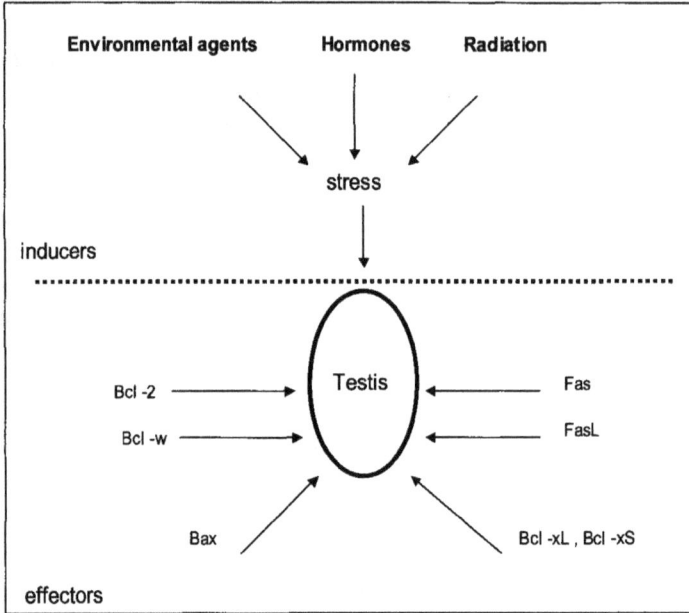

**Fig. 4.** Model of germ cell apoptosis. A summary diagram showing the inducers and effectors of germ cell apoptosis.

testis that determine whether germ cells grow, divide or die. This information is likely to be applicable to the assessment and management of male factor infertility as well as to more targeted approaches to male contraception. In addition, tools for the management of testicular tumours which result due to anomalous apoptosis of cells could be developed from the knowledge of apoptotic molecular pathways operative in the testis.

## Acknowledgement

Parts of the work described from the author's laboratory was supported by grants to the National Institute of Immunology from the Department of Biotechnology and a grant from Indo-US Collaboration on 'Contraceptive and Reproductive Health Research (CRHR)', USA.

## References

Akingbemi BT, Ge R, Klinefelter GR, Zirkin BR and Hardy MP (2004) Phthalate-induced Leydig cell hyperplasia is associated with multiple endocrine disturbances *Proceedings of the National Academy of Sciences USA* **101** 775-780

Bartlett JM, Kerr JB and Sharpe RM (1986) The effect of selective destruction and regeneration of rat Leydig cells on the intratesticular distribution of testosterone and morphology of the seminiferous epithelium *Journal of Andrology* **7** 240-253

Bernal-Manas CM, Morales E, Pastor LM, Pinart E, Bonet S, Rosa PL, Dolors BM, Zuasti A, Ferrer C and Canteras M (2005) Proliferation and apoptosis of spermatogonia in postpuberal boar (*Sus domesticus*) testes with spontaneous unilateral and bilateral abdominal cryptorchidism *Acta Histochemistry* **107** 365-372

Choi YJ, Ok DW, Kwon DN, Chung JI, Kim HC, Yeo SM, Kim T, Seo HG and Kim JH (2004) Murine male germ cell apoptosis induced by busulfan treatment correlates with loss of c-kit-expression in a Fas/FasL- and p53-independent manner *FEBS Letters* **575** 41-51

Coultas L, Bouillet P, Loveland KL, Meachem S, Perlman

H, Adams JM and Strasser A (2005) Concomitant loss of proapoptotic BH3-only Bcl-2 antagonists Bik and Bim arrests spermatogenesis *EMBO Journal* 24 3963-3973

Curtin JF and Cotter TG (2003) Live and let die: regulatory mechanisms in Fas-mediated apoptosis *Cell Signaling* 15 983-992

D'Alessio A, Riccioli A, Lauretti P, Padula F, Muciaccia B, De Cesaris P, Filippini A, Nagata S and Ziparo E (2001) Testicular FasL is expressed by sperm cells *Proceedings of the National Academy of Sciences U S A* 98 3316-3321

Eddy EM, Washburn TF, Bunch DO, Goulding EH, Gladen BC, Lubahn DB and Korach KS (1996) Targeted disruption of the estrogen receptor gene in male mice causes alteration of spermatogenesis and infertility *Endocrinology* 137 4796-4805

Eid NA, Shibata MA, Ito Y, Kusakabe K, Hammad H and Otsuki Y (2002) Involvement of Fas system and active caspases in apoptotic signalling in testicular germ cells of ethanol-treated rats *International Journal of Andrology* 25 159-167

Embree-Ku M, Venturini D and Boekelheide K (2002) Fas is involved in the p53-dependent apoptotic response to ionizing radiation in mouse testis *Biology of Reproduction* 66 1456-1461

Enmark E, Pelto-Huikko M, Grandien K, Lagercrantz S, Lagercrantz J, Fried G, Nordenskjold M and Gustafsson JA (1997) Human estrogen receptor beta-gene structure, chromosomal localization, and expression pattern *Journal of Clinical Endocrinology and Metabolism* 82 4258-4265

Feng HL, Sandlow JI, Sparks AE, Sandra A and Zheng LJ (1999) Decreased expression of the c-kit receptor is associated with increased apoptosis in subfertile human testes *Fertility and Sterility* 71 85-89

Fujisawa M, Shirakawa T, Fujioka H, Gotoh A, Okada H, Arakawa S and Kamidono S (2001) Adenovirus-mediated p53 gene transfer to rat testis impairs spermatogenesis *Archives of Andrology* 46 223-231

Goyal HO, Braden TD, Mansour M, Williams CS, Kamaleldin A and Srivastava KK (2001) Diethylstilbestrol-treated adult rats with altered epididymal sperm numbers and sperm motility parameters, but without alterations in sperm production and sperm morphology *Biology of Reproduction* 64 927-934

Green DR (2005) Apoptotic pathways: ten minutes to dead *Cell* 121 671-674

Hadziselimovic F, Geneto R and Emmons LR (1998) Increased apoptosis in the contralateral testes of patients with testicular torsion as a factor for infertility *Journal of Urology* 160 1158-1160

Hengartner MO (2000) The biochemistry of apoptosis *Nature* 407 770-776

Hess RA (2003) Estrogen in the adult male reproductive tract: A review *Reproductive Biology and Endocrinology* 1 1-52

Jahnukainen K, Chrysis D, Hou M, Parvinen M, Eksborg S and Soder O (2004) Increased apoptosis occurring during the first wave of spermatogenesis is stage-specific and primarily affects midpachytene spermatocytes in the rat testis *Biology of Reproduction* 70 290-296

Janssen O, Qian J, Linkermann A and Kabelitz D (2003) CD95 ligand—death factor and costimulatory molecule? *Cell Death and Differentiation* 10 1215-1225

Kijima K, Toyosawa K, Yasuba M, Matsuoka N, Adachi T, Komiyama M and Mori C (2004) Gene expression analysis of the rat testis after treatment with di(2-ethylhexyl) phthalate using cDNA microarray and real-time RT-PCR *Toxicology and Applied Pharmacology* 200 103-110

Kim SK, Lee HJ, Yang H, Kim HS and Yoon YD (2004) Prepubertal exposure to 4-tert-octylphenol induces apoptosis of testicular germ cells in adult rat *Archives of Andrology* 50 427-441

Knudson CM, Tung KS, Tourtellotte WG, Brown GA and Korsmeyer SJ (1995) Bax-deficient mice with lymphoid hyperplasia and male germ cell death *Science* 270 96-99

Koji T, Hishikawa Y, Ando H, Nakanishi Y and Kobayashi N (2001) Expression of Fas and Fas ligand in normal and ischemia-reperfusion testes: involvement of the Fas system in the induction of germ cell apoptosis in the damaged mouse testis *Biology of Reproduction* 64 946-954

Kojima S, Hatano M, Okada S, Fukuda T, Toyama Y, Yuasa S, Ito H and Tokuhisa T (2001) Testicular germ cell apoptosis in Bcl6-deficient mice *Development* 128 57-65

Korach KS (1994) Insights from the study of animals lacking functional estrogen receptor *Science* 266 1524-1527

Kuiper GG, Enmark E, Pelto-Huikko M, Nilsson S and Gustafsson JA (1996) Cloning of a novel receptor expressed in rat prostate and ovary *Proceedings of the National Academy of Sciences USA* 93 5925-5930

Lee J, Richburg JH, Shipp EB, Meistrich ML and Boekelheide K (1999) The Fas system, a regulator of testicular germ cell apoptosis, is differentially up-regulated in Sertoli cell versus germ cell injury of the testis *Endocrinology* 140 852-858

Lysiak JJ, Turner SD and Turner TT (2000) Molecular pathway of germ cell apoptosis following ischemia/reperfusion of the rat testis *Biology of Reproduction* 63 1465-1472

Mishra DP and Shaha C (2005) Estrogen-induced spermatogenic cell apoptosis occurs via the mitochondrial pathway: role of superoxide and nitric oxide *Journal of Biological Chemistry* 280 6181-6196

Mishra DP, Pal R and Shaha C (2006) Changes in cytosolic $Ca^{2+}$ levels regulate Bcl-xS and Bcl-xL expression in spermatogenic cells during apoptotic death *Journal of Biological Chemistry* 281 2133-2143

Miura M, Sasagawa I, Suzuki Y, Nakada T and Fujii J (2002) Apoptosis and expression of apoptosis-related genes in the mouse testis following heat exposure *Fertility and Sterility* 77 787-793

Nair R and Shaha C (2003) Diethylstilbestrol induces rat spermatogenic cell apoptosis *in vivo* through increased expression of spermatogenic cell Fas/FasL system *Jour-

*nal of Biological Chemistry* **278** 6470-6481

Nandi S, Banerjee PP and Zirkin BR (1999) Germ cell apoptosis in the testes of Sprague Dawley rats following testosterone withdrawal by ethane 1,2-dimethanesulfonate administration: relationship to Fas? *Biology of Reproduction* **61** 70-75

Pentikainen V, Erkkila K and Dunkel L (1999) Fas regulates germ cell apoptosis in the human testis *in vitro* *American Journal of Physiology* **276** E310-E316

Rey R (2003) Regulation of spermatogenesis *Endocrinology and Development* **5** 38-55

Richburg JH and Nanez A (2003) Fas- or FasL-deficient mice display an increased sensitivity to nitrobenzene-induced testicular germ cell apoptosis *Toxicology Letters* **139** 1-10

Rodriguez I, Ody C, Araki K, Garcia I and Vassalli P (1997) An early and massive wave of germinal cell apoptosis is required for the development of functional spermatogenesis *EMBO Journal* **16** 2262-2270

Ross AJ, Waymire KG, Moss JE, Parlow AF, Skinner MK, Russell LD and MacGregor GR (1998) Testicular degeneration in Bclw-deficient mice *Nature Genetics* **18** 251-256

Russell LD, Chiarini-Garcia H, Korsmeyer SJ and Knudson CM (2002) Bax-dependent spermatogonia apoptosis is required for testicular development and spermatogenesis *Biology of Reproduction* **66** 950-958

Russell LD, Warren J, Debeljuk L, Richardson LL, Mahar PL, Waymire KG, Amy SP, Ross AJ and MacGregor GR (2001) Spermatogenesis in Bclw-deficient mice *Biology of Reproduction* **65** 318-332

Sakkas D, Mariethoz E and St John JC (1999) Abnormal sperm parameters in humans are indicative of an abortive apoptotic mechanism linked to the Fas-mediated pathway *Experimental Cell Research* **251** 350-355

Salazar V, Castillo C, Ariznavarreta C, Campon R and Tresguerres JA (2004) Effect of oral intake of dibutyl phthalate on reproductive parameters of Long Evans rats and pre-pubertal development of their offspring *Toxicology* **205** 131-137

Sinha Hikim AP and Swerdloff RS (1999) Hormonal and genetic control of germ cell apoptosis in the testis *Reviews in Reproduction* **4** 38-47

Sinha Hikim AP, Lue Y, Diaz-Romero M, Yen PH, Wang C and Swerdloff RS (2003) Deciphering the pathways of germ cell apoptosis in the testis *Journal of Steroid Biochemistry and Molecular Biology* **85** 175-182

Suominen JS, Wang Y, Kaipia A and Toppari J (2004) Tumor necrosis factor-alpha (TNF-alpha) promotes cell survival during spermatogenesis, and this effect can be blocked by infliximab, a TNF-alpha antagonist *European Journal of Endocrinology* **151** 629-640

Tesarik J, Martinez F, Rienzi L, Iacobelli M, Ubaldi F, Mendoza C and Greco E (2002) *In-vitro* effects of FSH and testosterone withdrawal on caspase activation and DNA fragmentation in different cell types of human seminiferous epithelium *Human Reproduction* **17** 1811-1819

Theas S, Rival C and Lustig L (2003) Germ cell apoptosis in autoimmune orchitis: involvement of the Fas-FasL system *American Journal of Reproductive Immunology* **50** 166-176

Thompson CB (1995) Apoptosis in the pathogenesis and treatment of disease *Science* **267** 1456-1462

Vera Y, Diaz-Romero M, Rodriguez S, Lue Y, Wang C, Swerdloff RS and Sinha Hikim AP (2004) Mitochondria-dependent pathway is involved in heat-induced male germ cell death: lessons from mutant mice *Biology of Reproduction* **70** 1534-1540

Williams K, Saunders PT, Atanassova N, Fisher JS, Turner KJ, Millar MR, McKinnell C and Sharpe RM (2000) Induction of progesterone receptor immunoexpression in stromal tissue throughout the male reproductive tract after neonatal oestrogen treatment of rats *Molecular and Cellular Endocrinology* **164** 117-131

Woolveridge I, Boer-Brouwer M, Taylor MF, Teerds KJ, Wu FC and Morris ID (1999) Apoptosis in the rat spermatogenic epithelium following androgen withdrawal: changes in apoptosis-related genes *Biology of Reproduction* **60** 461-470

Yan W, Huang JX, Lax AS, Pelliniemi L, Salminen E, Poutanen M and Toppari J (2003) Overexpression of Bcl-w in the testis disrupts spermatogenesis: revelation of a role of Bcl-w in male germ cell cycle control *Molecular Endocrinology* **17** 1868-1879

Yin Y, Stahl BC, DeWolf WC and Morgentaler A (2002) P53 and Fas are sequential mechanisms of testicular germ cell apoptosis *Journal of Andrology* **23** 64-70

Yu X, Kubota H, Wang R, Saegusa J, Ogawa Y, Ichihara G, Takeuchi Y and Hisanaga N (2001) Involvement of Bcl-2 family genes and Fas signaling system in primary and secondary male germ cell apoptosis induced by 2-bromopropane in rat *Toxicology Applied Pharmacology* **174** 35-48

Zhou XC, Wei P, Hu ZY, Gao F, Zhou RJ and Liu YX (2001) Role of Fas/FasL genes in azoospermia or oligozoospermia induced by testosterone undecanoate in rhesus monkey *Acta Pharmacology Sinica* **22** 1028-1033

Zhu Q, Meisinger J, Emanuele NV, Emanuele MA, LaPaglia N and Van Thiel DH (2000) Ethanol exposure enhances apoptosis within the testes *Alcohol Clinical Experimental Research* **24** 1550-1556

# Features that affect secretion and assembly of zona pellucida glycoproteins during mammalian oogenesis

Luca Jovine[1, 2, 5], Huayu Qi[1, 3], Zev Williams[1, 4], Eveline S. Litscher[1] and Paul M. Wassarman[1, 5]

[1] Brookdale Dept. Molecular, Cell and Developmental Biology, Mount Sinai School of Medicine, One Gustave L. Levy Place, New York, NY 10029-6574, USA; [2] Karolinska Institute, Department of Biosciences and Nutrition, Center for Structural Biochemistry, S-141 57 Huddinge, Sweden; [3] Present address: Dept. Cardiovascular Research, Harvard Medical School, 320 Longwood Avenue, Boston, MA 02115, USA; [4] Present address: Brigham and Women's Hospital, Harvard Medical School, 75 Francis St., Boston, MA 02115, USA

For sperm to fertilize eggs, they must bind to and penetrate the zona pellucida (ZP) that surrounds the plasma membrane of all mammalian eggs. The ZP first appears during oocyte growth and increases in thickness as oocytes increase in diameter. The ZP is an extracellular matrix composed of long, crosslinked filaments. In mice, three glycoproteins, called mZP1-3, are synthesised and secreted by growing oocytes and assembled into a thick ($\sim 6.5$ $\mu$m) extracellular coat over a 2-3 week period. Recently, we identified several regions of nascent ZP glycoproteins that affect their secretion and incorporation into the ZP (assembly) by growing oocytes. Among these are the ZP domain, the consensus furin cleavage site (CFCS) and the C-terminal propeptide (CTP) with its transmembrane domain (TMD), external hydrophobic patch (EHP), charged patch (CP), conserved cysteine (Cys) residue, and short cytoplasmic tail (CT). Particularly important is the ZP domain, a $\sim 260$ amino acid region with 8 conserved Cys residues that is common to a variety of extracellular proteins of diverse functions found in a wide range of multicellular eukaryotes. Our results show that the ZP domain functions as a polymerisation module and that its N-terminal half, including 4 conserved Cys residues, is largely responsible for this role. Additionally, two conserved hydrophobic sequences, one within the ZP domain (internal hydrophobic patch; IHP) and another within the CTP (EHP), apparently regulate polymerisation of nascent ZP glycoproteins. Collectively, our findings suggest a general mechanism for assembly of all ZP domain proteins based on coupling between proteolytic processing and polymerisation.

E-mail: paul.wassarman@mssm.edu or luca.jovine@biosci.ki.se
[5] To whom correspondence should be directed

## Introduction

The zona pellucida (ZP) is a thick extracellular coat that surrounds the plasma membrane of all mammalian eggs (Gwatkin, 1977; Yanagimachi, 1994; Wassarman et al., 2001) (Fig. 1). It plays significant roles during oogenesis (for example, development of Graafian follicles), fertilisation (for example, species-specific fertilisation and prevention of polyspermy) and preimplantation development (for example, prevention of dissociation of cleavage-stage blastomeres). The ZP appears during oocyte growth, while oocytes are arrested in first meiotic prophase and in creases in thickness as oocytes increase in diameter.

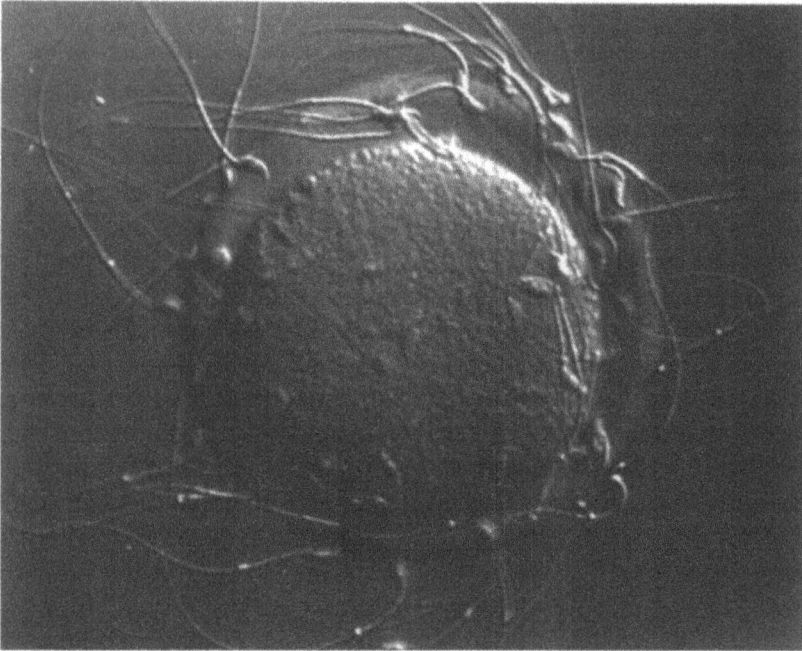

**Fig. 1** Binding of free-swimming mouse sperm to the zona pellucida (ZP) of an ovulated mouse egg. The light micrograph was taken using Nomarski differential interference contrast optics.

In mice, three ZP glycoproteins, called mZP1 ($\sim$ 200 kDa Mr), mZP2 ($\sim$ 120 kDa Mr) and mZP3 ($\sim$ 83 kDa Mr), are synthesised and secreted by growing oocytes and assembled into long, crosslinked filaments (Wassarman, 1988; Wassarman and Mortillo, 1991; Jovine et al., 2002a). Evidence suggests that all three ZP glycoproteins have a structural role and that, in addition, mZP3 and mZP2 serve as primary and secondary sperm receptors, respectively, during fertilisation (Wassarman, 1999; Wassarman et al., 2001). Mouse ZP polypeptides are encoded by single-copy genes (mZP1: 12 exons, chromosome 19; mZP2: 18 exons, chromosome 7; mZP3: 8 exons, chromosome 5). Each ZP glycoprotein consists of a unique polypeptide that is heterogeneously glycosylated with both complex-type asparagine- (N-) and serine/threonine- (O-) linked oligosaccharides. A similar constellation of glycoproteins is found in the vitelline envelopes (VE) of fish, amphibian and bird eggs (Jovine et al., 2002a). Female mice that are homozygous nulls for either mZP2 or mZP3 produce eggs that lack a ZP and are infertile (Liu et al., 1996; Rankin et al., 1996; 2001), whereas heterozygous nulls for mZP3 are fertile but their eggs have a thin ZP (Wassarman et al., 1997). The evidence suggests that mZP2 and mZP3 are structural glycoproteins essential for both assembly of a ZP and fertility.

ZP glycoproteins have been characterised from a wide variety of mammals; including rodents, domesticated animals, marsupials; and primates (Jovine et al., 2002a; Spargo and Hope, 2003). While molecular weights (Mrs) differ for ZP glycoproteins from different species, due in large part to differential glycosylation and other modifications, apparently all ZP consist of only a few glycoproteins whose polypeptides are related to those of mZP1-3. The primary structures of ZP2- and ZP3-related ZP glycoproteins from different mammals are relatively well conserved (~65-98% identity), whereas ZP1-related glycoproteins are conserved to a lesser degree (~40% identity). It is also apparent that ZP1-3 have regions of polypeptide in common, suggesting that these regions may be derived from a common ancestral gene (Spargo and Hope, 2003). It should be noted that there are additional ZP glycoproteins present in some mammals (for example, ZPB; Lefievre et al., 2004): all containing a ZP domain and highly related to ZP1-3.

Nascent mouse ZP polypeptides have several features in common (Figs. 2 and 3). They have an N-terminal signal peptide (SP), a ZP domain close to the C-terminus, a consensus furin cleavage site (CFCS), an external hydrophobic patch (EHP), a C-terminal transmembrane domain (TMD), and a short cytoplasmic tail (CT). mZP1 also contains a trefoil (P) domain just upstream of its ZP domain. The EHP, TMD and CT of ZP polypeptides are downstream of the CFCS and are part of the C-terminal propeptide (CTP) that is removed from ZP glycoproteins prior to their incorporation into the ZP. The SP targets nascent ZP glycoproteins to the secretory pathway, the TMD is used to anchor the glycoproteins in secretory vesicles and plasma membrane, and the CFCS is used to release the glycoproteins into the extracellular space.

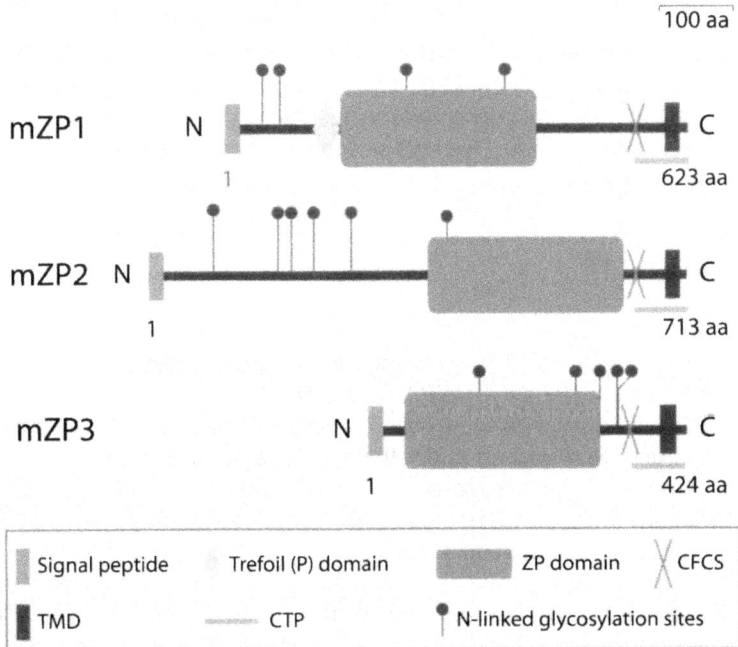

**Fig. 2** Schematic representation of the overall architecture of mouse ZP glycoproteins ZP1, ZP2 and ZP3. The polypeptide of each ZP glycoprotein is drawn to scale, with the N and C termini indicated. Key features of the polypeptide, including the N-terminal signal peptide (*green*), trefoil (P) domain (*yellow*), ZP domain (*red*), CFCS (*X*), TMD (*black*) and C-terminal propeptide (CTP) region (*blue bar*) are depicted. N-linked glycosylation sites supported by experimental evidence are shown. The number of amino acids in the polypeptide of each ZP glycoprotein is indicated.

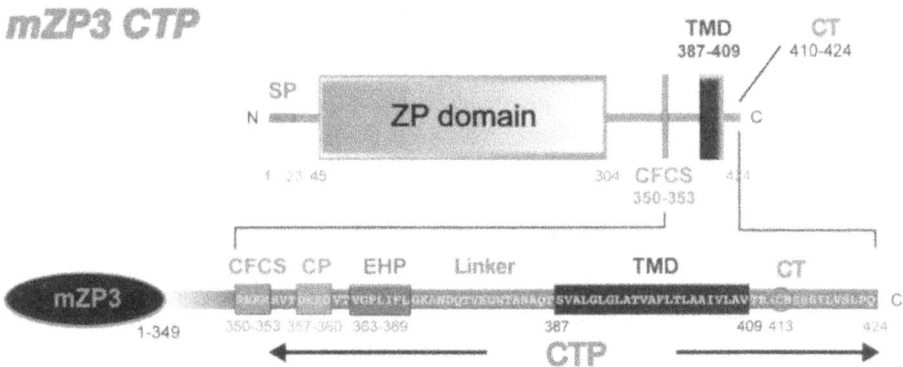

**Fig. 3** Sequence of the C-terminal propeptide (CTP) of mZP3. Polypeptide boundaries are marked by gray bars, with the signal peptide (SP) in red; the ZP domain, CFCS, charged peptide (CP), EHP and TMD are depicted as pink, orange, green, cyan and blue rectangles, respectively; conserved $Cys_{413}$ is circled in red.

Recently, we investigated the effects of different regions of ZP polypeptides on secretion and assembly of nascent ZP glycoproteins. Much of our research has focused on the ZP domain (Bork and Sander, 1992): a conserved domain common to hundreds of proteins of diverse functions from a wide variety of tissues in mammals, birds, amphibia, fish, flies, worms, tunicates and molluscs (Jovine et al., 2005). Mutations in genes encoding ZP domain proteins can result in severe human pathologies. In addition, we have examined the effects of other regions of ZP polypeptides, such as the CFCS, EHP, TMD and other features of the CTP [for example, a charged patch, CP, and a single conserved Cys residue (Cys431 in mZP3)], on secretion and assembly of ZP glycoproteins. Here, we review some aspects of this research and discuss the implications of our results.

## Results

*Overview of experimental approaches*

In some experiments, to examine ZP glycoprotein expression, secretion and assembly, cDNAs encoding epitope-tagged mZP2 (Myc-mZP2) and mZP3 (Flag-mZP3) were microinjected into the germinal vesicle (nucleus) of growing mouse oocytes (Qi et al., 2002; Jovine et al., 2002b; 2004). cDNA constructs were placed under the control of a SV40 promoter that is known to drive expression of reporter genes in isolated mouse oocytes cultured *in vitro* (Chalifour et al., 1986; 1987). Deletions and point mutations were generated in mZP2 and mZP3 cDNAs by overlap extension PCR and site-directed mutagenesis. To examine expression of Myc-mZP2 and Flag-mZP3, laser scanning confocal microscopy (LSCM) was used with fixed (formaldehyde), permeabilised (Triton X-100) and non-permeabilised injected oocytes. To isolate ZP, injected oocytes were exposed to NP-40 and pipetted up and down through a glass pipette with a tip diameter smaller than that of oocytes. Isolated ZP were then fixed with formaldehyde. In other experiments, embryonal carcinoma (EC), Chinese hamster ovary (CHO) or human embryonic kidney (293T) cells were stably or transiently transfected and expression and secretion of recombinant proteins were detected by Western immunoblotting using monoclonal or polyclonal antibodies directed against Myc, Flag, mZP2 or mZP3 (Williams and Wassarman, 2001; Jovine et al., 2002b; 2004; unpublished results).

*Nascent ZP glycoproteins are packaged into large secretory vesicles and incorporated into the innermost layer of the ZP*

Ultrastructural studies of growing mouse oocytes revealed that the Golgi changes from flattened stacks of lamellae, with few, if any, vacuoles or granuoles, in early stages of oocyte growth to extensive arrays of swollen stacked lamellae, with many large vacuoles, in late stages of growth (Wassarman and Josefowicz, 1978). These changes, as well as those of the endoplasmic reticulum (ER), suggest that growing mouse oocytes become actively engaged in trafficking of various glycoproteins, including those of the ZP.

Epitope-tagged mZP2 and mZP3 are expressed and secreted by growing mouse oocytes after microinjection with the corresponding cDNAs, with recombinant protein representing ~ 5% of total nascent ZP protein (Qi et al., 2002). To localize Myc-mZP2 and Flag-mZP3, microinjected oocytes were immunolabeled with either anti-Myc or anti-Flag, respectively, and examined by LSCM. With fixed, non-permeabilized oocytes, Myc-mZP2 and Flag-mZP3 were concentrated primarily around the plasma membrane/ZP region of the oocytes with no significant intracellular signal detected. With fixed, permeabilized oocytes, Myc-mZP2 and Flag-mZP3 were also detected intracellularly in relatively large membranous vesicles. Using a monoclonal antibody directed against vesicle associated membrane protein (VAMP) (Sudhof, 1995; Lin and Scheller, 2000) and antibodies directed against Myc- or Flag-tagged recombinant ZP glycoproteins, confirmed that nascent ZP glycoproteins and VAMP colocalised to secretory vesicles that originated from the *trans*-Golgi (Fig. 4A). It was noted that, in general, the vast majority of the vesicles were doughnut shaped, having an empty cavity in the middle (lumen), and were quite large, ~2.3 ± 0.32 μm in diameter, as compared to secretory vesicles of somatic cells

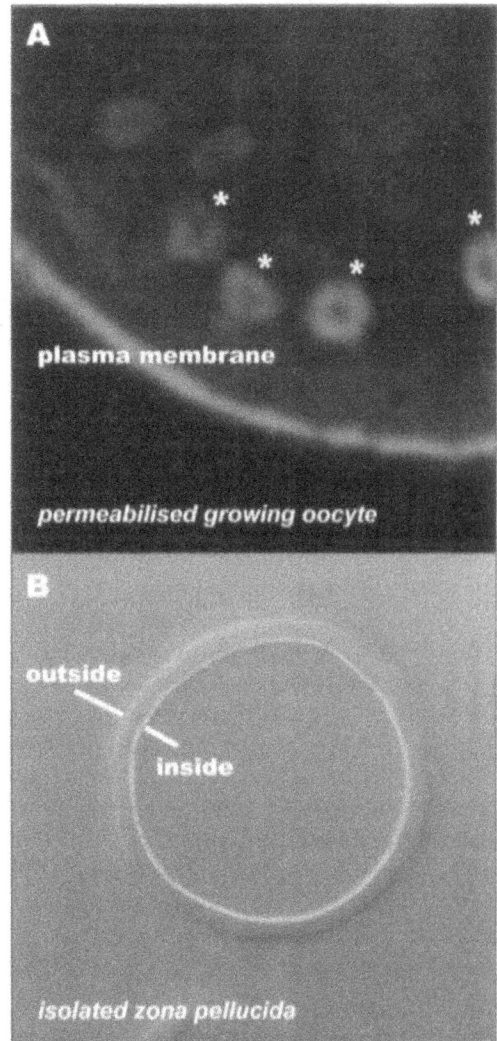

**Fig. 4** Immunolabeling of growing oocyte secretory vesicles and the innermost region of the ZP. Panel A (top): Oocytes from 14 day-old mice were microinjected with Myc-mZP2, cultured for ~1 day, stained with anti-Myc followed by Texas Red-conjugated goat anti-mouse IgG secondary antibody and subjected to LSCM. A fluorescent image is presented. Note the presence of Myc-mZP2 in large, doughnut-shaped vesicles located in the cortical region of the oocyte (indicated by asterisks). Panel B (bottom): Oocytes from 13 day-old mice were microinjected with Flag-mZP3, cultured for ~1 day and ZP were isolated in the presence of NP-40. Isolated ZP were stained with anti-Flag followed by FITC-conjugated secondary antibody and subjected to LSCM. A fluorescent image is presented. Note the presence of Flag-mZP3 only in the innermost region of the oocyte.

($\sim$ 0.1-0.2 $\mu$m in diameter). These secretory vesicles fuse with oocyte plasma membrane and release nascent ZP glycoprotein into the extracellular space to be incorporated into the thickening ZP.

Since Myc-mZP2 and Flag-mZP3 localised to the plasma membrane/ZP region of non-permeabilised oocytes, we examined whether epitope-tagged, recombinant ZP glycoproteins were incorporated into the thickening ZP. Oocytes were microinjected with either Myc-mZP2 or Flag-mZP3, cultured for either $\sim$ 1 or $\sim$ 2 days, and ZP were isolated in the presence of 1% NP-40 (in PBS/PVP-40) followed by thorough washing in PBS/PVP-40. Isolated ZP were fixed, immunolabeled either with anti-Myc or anti-Flag, and examined by LSCM. In both cases, immunofluorescence was observed solely on the inner surface of isolated ZP (Fig. 4B). Furthermore, longer culture times resulted in an increase in the intensity of the fluorescent signal; for example, Myc-mZP2 and Flag-mZP3 fluorescence intensities (luminosity measurements) increased $\sim$ 1.5-fold when culture time was doubled from $\sim$ 1 to $\sim$ 2 days. Collectively, these results suggest that nascent recombinant mZP2 and mZP3 are incorporated into the thickening ZP exclusively at its innermost surface. This implies that the ZP thickens from the inside to the outside.

*Removal of the CTP is required for secretion of ZP glycoproteins by growing oocytes*

ZP glycoproteins are processed at a CFCS (Fig. 3) by a member of the furin family of serine proteases prior to incorporation into the ZP (Litscher et al., 1999; Williams and Wassarman, 2001; Kiefer and Saling, 2002). The processing results in removal of the CTP from ZP glycoproteins. To examine the requirement for cleavage at the CFCS, a mutated form of Flag-mZP3-Myc, designated Δ-Flag-mZP3-Myc, was used to assess secretion and incorporation of nascent ZP glycoprotein into the ZP (Qi et al., 2002). In Δ-Flag-mZP3-Myc, the CFCS was mutated from Arg-Asn-Arg-Arg to Arg-Asn-Gly-Glu. Results obtained with mammalian cells transfected with this mutant construct suggest that the mutation abolished the cleavage site for furin-like enzymes and resulted in accumulation of unprocessed ZP glycoproteins in the ER of transfected cells (Williams and Wassarman, 2001). Secretion of mutated recombinant mZP3 by the cells, as compared to wild-type mZP3, was reduced as much as $\sim$ 20-fold.

To examine the effect of mutation of the CFCS in growing oocytes, isolated oocytes were microinjected with either Flag-mZP3-Myc or Δ–Flag-mZP3-Myc, cultured for $\sim$ 1 day, and immunolabeled with either anti-Myc or anti-Flag. When probed with anti-Flag and subjected to LSCM, oocytes microinjected with either construct displayed immunolabeling along the plasma membrane/ZP region of the oocytes; on the other hand, very little signal was detected along this region when oocytes were probed with anti-Myc. It was noted that there was a large increase in vesicular staining in oocytes microinjected with Δ–Flag-mZP3-Myc, as compared to Flag-mZP3-Myc, suggesting that mutation of the CFCS significantly slowed down trafficking of mZP3 through the secretory pathway. Furthermore, immunolabeling was observed along the inner surface of isolated ZP following microinjection of Δ-Flag-mZP3-Myc into growing oocytes, whereas when probed with anti-Myc no immunolabeling was detected. Collectively, these results suggest that mutation of the CFCS caused an enhanced retention of mZP3 within oocytes, although some mutated mZP3 escaped intracellular retention and was incorporated into the ZP. Like wild-type mZP3, the latter was proteolytically processed at its C-terminus, possibly by a protease other than furin or by furin at a significantly reduced rate.

*Mutation of a conserved Cys in the CT of ZP glycoprotein precursors does not affect secretion or assembly*

Aside from 2-4 basic juxtamembrane amino acids commonly found in type I transmembrane proteins (Boyd and Beckwith, 1990), a single Cys residue is the only feature conserved in the short CT of ZP glycoproteins (Fig. 3). Since the other Cys residues are present as intramolecular disulfides in the extracellular portion of ZP glycoproteins (Boja *et al.*, 2003; Yonezawa and Nakano, 2003) and since Cys residues can be important for assembly of transmembrane protein complexes (Locker and Griffiths, 1999; Rozanov *et al.*, 2001), we assessed whether mutation of Cys413 of mZP3 would affect secretion and incorporation of the glycoprotein into the ZP. We found that mutant Cys413Ser protein was secreted and incorporated by microinjected oocytes to the same extent as wild-type mZP3 and, in addition, that secretion of the mutant protein was unaffected in transfected cell lines (Jovine *et al.*, 2004). This suggests that Cys413 is not re-quired for either secretion or incorporation of ZP glycoproteins into the ZP.

*The TMD of ZP glycoprotein precursors is not involved in specific interactions but is required for processing at the CFCS*

Truncation of ZP polypeptides before the TMD does not prevent packaging of mZP2 and mZP3 into oocyte secretory vesicles whereas it completely prevents their incorporation into the ZP (Jovine *et al.*, 2002b). However, unlike the situation with wild-type recombinant ZP proteins, the lumen of secretory vesicles is filled with truncated nascent ZP glycoprotein (that is, secre-tory vesicles did not display a doughnut-shaped immunofluorescent signal). Furthermore, ZP proteins truncated just before the TMD are secreted by transfected mammalian cells as effi-ciently as wild-type recombinant ZP proteins (Jovine *et al.*, 2002b).

Assembly of extracellular complexes can be mediated by specific interactions between their TMDs (Harrison, 1996). To assess whether this was the case for ZP glycoproteins, the TMD of mZP3 (aa 387-409) (Fig. 3) was replaced by the single-spanning, C-terminal TMD of human CD7 (aa 178-201), which is very different in sequence and is not involved in specific interac-tions (Schanberg *et al.*, 1991). The resulting construct was efficiently secreted by transfected cells and was incorporated into the ZP of microinjected oocytes as well as wild-type recombi-nant mZP3 (Jovine *et al.*, 2004). These results suggest that the conserved TMD is not involved in specific interactions, but, because it is required for assembly (Jovine *et al.*, 2002b), it ensures proper localisation and/or topological orientation of nascent ZP proteins so that incorporation into the ZP can take place.

To determine whether truncation of mZP3 before its TMD affected proteolytic processing, a truncated construct with a C-terminal Flag-tag was produced (mZP3$_{373}$-Flag). Immunoblotting of cells transfected with this construct revealed that the Flag-tag was retained in the secreted protein, indicating that cleavage at the CFCS did not take place (Jovine *et al.*, 2004). These observations suggest that neither the TMD nor the short CT are required for secretion, but that the TMD is required for cleavage at the CFCS and incorporation of ZP glycoproteins into the ZP.

*Polymerisation of ZP glycoproteins into filaments depends on the ZP domain*

It was proposed in 1995 that the ZP domain might play a role in polymerisation of ZP domain-containing proteins into filaments or matrices (Killick *et al.*, 1995; Legan *et al.*, 1997). Subse-quently, development of an assay to follow incorporation of epitope-tagged recombinant ZP glycoproteins into the ZP of mouse oocytes permitted an experimental approach to investigate the role of the ZP domain (Qi *et al.*, 2002). Using such an assay, it was found that the ZP domain, together with the N-terminal SP and CTP, are both necessary and sufficient for incorpo-ration of nascent protein into the ZP (Jovine *et al.*, 2002b). Furthermore, electron microscope analyses of human Tamm-Horsfall Protein (THP)/uromodulin, a ZP domain-containing protein (Serafini-Cessi *et al.*, 2003), revealed that proteolytic digestion of non-ZP domain sequences did not disrupt the organisation of THP filaments (Jovine *et al.*, 2002b). The filaments apparently consist of two protofilaments wound around each other to form a right-handed double helix (axial repeat ~120 Å; diameter 90-140 Å) (Fig. 5). These results demonstrate that the ZP domain of extracellular proteins functions as a polymerisation module and suggest that other functions of ZP domain proteins can be ascribed to sequences N- and C-terminal to the ZP domain.

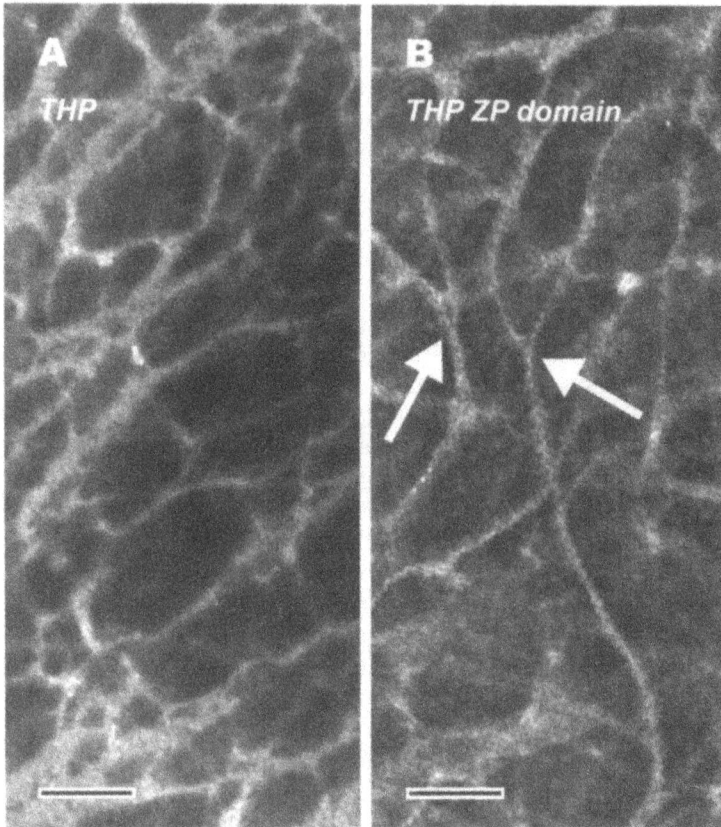

**Fig. 5** Filaments of THP and THP-ZPD. Electron microscopic analyses of negatively stained samples of native THP (panel A, left) and the isolated ZP domain of THP (THP-ZPD) (panel B, right). It should be noted that both samples consist of filamentous structures, with THP-ZPD apparently consisting of two protofilaments (*arrows*), wound around each other to form a right-handed double helix (axial repeat of ~120 Å and a diameter of 90-140 Å). Bars represent 0.1 μm.

Several lines of evidence suggest that the ZP domain consists of two subdomains (ZP-N and ZP-C) (Fig. 6). The first subdomain (ZP-N) includes conserved Cys 1-4 and can be found in mammalian proteins such as PLAC1, Oosp1, Papillote and CG16798 (Cocchia *et al.*, 2000; Hemberger *et al.*, 2000; Yan *et al.*, 2001; Jazwinska and Affolter, 2004; Bokel *et al.*, 2005). By performing a search of all genome sequences currently available through Ensembl with a specifically developed Hidden Markov Model (HMM) profile, we identified three other proteins sharing homology with ZP-N only (Jovine *et al.*, 2006). In view of the fact that no sequence containing

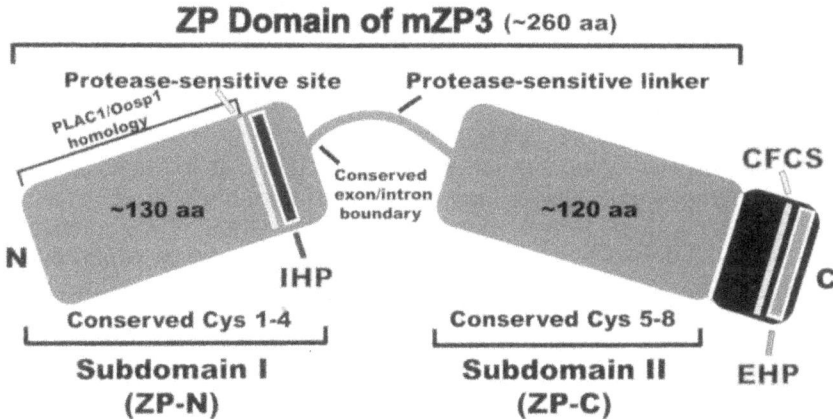

## ZP Domain of mZP3 (~260 aa)

**Fig. 6** The ZP domain consists of two subdomains. Disulfide bond connectivity, limited proteolytic digestion, homology to PLAC1/Oosp1 proteins, conservation of exon-intron boundaries, identification of glycosylation sites, and the relative location of the IHP (*blue*) and EHP (*green*) sequences are all consistent with the view that two subdomains interact to form the entire ZP domain (*red*). Although not apparent from their primary structure and disulfide connectivity, it is possible that the two subdomains share a similar three-dimensional structure.

only the second subdomain (ZP-C) of the ZP domain has been identified thus far, the conservation of ZP-N suggests that it might be responsible for polymerisation of proteins containing a ZP domain.

To test this hypothesis, recombinant ZP-N was expressed in *E. coli* as a fusion to maltose binding protein (MBP-ZP-N) (Jovine *et al.*, 2006). Considerable amounts of the protein were expressed in a soluble form when the corresponding DNA was transformed into a highly engineered *E. coli* strain that facilitates formation of disulfide bonds. Mass spectrometric analysis of affinity-purified MBP-ZP-N suggested that its ZP domain Cys residues form disulfides with the same connectivity as in full-length ZP domain proteins. Although the 53 kDa fusion protein appeared to be soluble by ultracentrifugation at 100,000 x g, it eluted in the void volume of a 300 kDa Mr cut-off size-exclusion column. Analysis of purified MBP-ZP-N by electron microscopy revealed that the protein assembled into long filaments apparently consisting of at least two protofilaments. Localisation of MBP-ZP-N to the filaments was confirmed by immunogold labeling using a primary monoclonal antibody directed against MBP. Finally, Western blot analyses under non-reducing conditions revealed, in addition to monomeric MBP-ZP-N, a ladder of bands migrating at Mrs equal to 2n x 53 kDa (n = 1, 2, ...). These bands apparently correspond to fragments of MBP-ZP-N filaments that were not completely depolymerised due to incomplete protein denaturation in the absence of reducing agents. Collectively, these results suggest that ZP-N is able to polymerize into filaments on its own and, in agreement with studies of full-length ZP domain proteins, a homodimer of ZP-N apparently constitutes the filament building block.

*A sequence between the CFCS and TMD is required for secretion of truncated ZP glycoproteins*

ZP protein constructs truncated just before the TMD are efficiently secreted by transfected cells (Jovine *et al.*, 2002b). On the other hand, polypeptides truncated immediately after the CFCS are retained in the ER and not secreted (Williams and Wassarman, 2001; Jovine et al., 2002b). These findings suggest that elements required for secretion are located between the CFCS and TMD.

Sequence alignments of the C-terminal propeptides of ZP1-3 homologues reveal that the CFCS is followed by a short stretch rich in charged amino acids, designated as a "charged patch" (CP) (Fig. 3). Since the CP sequence is relatively conserved in mammalian ZP3 homologues, we assessed whether this motif plays a role in protein secretion (Jovine et al., 2004). The CP of mZP3 was mutated to Ala-Ala-Ala-Ala in the context of both a full-length protein (ZP3-Flag-ΔCP) and a construct truncated before the TMD (ZP3-Flag-370-ΔCP). Recombinant ZP glycoproteins encoded by both mutant constructs were secreted at levels comparable to wild-type ZP glycoproteins, suggesting that the CP is not required for secretion of mZP3 (Jovine et al., 2004).

A second short conserved motif, consisting of an almost invariant GP sequence immediately followed by 4-5 hydrophobic amino acids, is found C-terminal to the CP. This element, designated as an "external hydrophobic patch" (EHP) (Fig. 3), is connected to the TMD by a linker that is neither conserved in sequence or length nor required for secretion of truncated ZP glycoproteins (Jovine et al. 2004). Interestingly, fish homologues of ZP1 and ZP3, that are synthesised by the liver and travel in the blood to the oocyte vitelline envelope (VE) (Hyllner et al., 2001), lack a TMD and end just after the EHP. The CTPs of these proteins, that are missing from assembled VE proteins (Sugiyama et al., 1999), are essentially equivalent to construct ZP3-FLAG-370 (Jovine et al., 2004). Similarly, precursors of avian ZP1 homologues, also secreted by the liver (Bausek et al., 2000; Sasanami et al., 2003), terminate with the EHP. Finally, EHPs were identified in sequences of precursors of other ZP domain proteins, both with and without C-terminal TMDs (Jovine et al., 2004).

To determine whether the EHP plays a role in secretion, a mZP3 construct truncated immediately before the EHP was produced (ZP3-Flag-362). Mammalian cells transfected with this construct failed to secrete recombinant mZP3 and identical results were obtained using a corresponding deletion mutant of mZP2 (Jovine et al., 2004). Although the EHP is required for secretion of constructs lacking a TMD, it is dispensable when the TMD is present: when the entire EHP was replaced by a single Arg residue in the context of full-length mZP3 (ZP3-Flag-ΔEHP), secretion was comparable to that of wild-type recombinant protein (Jovine et al., 2004). The latter finding contrasts with a report that substitution of the EHP with a Flag-tag abolishes secretion of an enhanced green fluorescent protein (EGFP)-mZP3 fusion construct (Zhao et al., 2003). Analyses of the effect of single amino acid changes in the EHP of truncated construct ZP3-Flag-370 were also carried out (Jovine et al., 2004). The ZP3-Flag-370-Gly$_{364}$Ala mutant is equivalent to construct ZP3-Flag-362 in that it was not secreted, whereas mutants Pro$_{365}$Ala and Phe$_{368}$Ser were secreted at 1-2% of the levels of wild-type recombinant mZP3. The severity of these mutations correlates with the degree of conservation of the corresponding amino acids. Overall, these results suggest that ZP glycoproteins must contain either an EHP or a TMD to be secreted.

*Assembly of ZP glycoproteins requires the presence of an EHP and IHP*

Two short hydrophobic motifs are conserved in ZP domain proteins (Figs. 3 and 6). One, described above, is termed an external hydrophobic patch (EHP) and is present in the CTP between the CFCS and TMD (Zhao et al., 2003; Jovine et al., 2004). The other, called an internal hydrophobic patch (IHP), is present in the ZP domain itself following conserved Cys residues 1-4 (Jovine et al., 2004). The motifs, which in several cases (for example, in mZP1 and mZP2) are highly similar in sequence, are both predicted to form ß-strands. The relative locations of the EHP and IHP are consistent with the proposal that the ZP domain consists of two subdomains (Patra et al., 2000; Yonezawa and Nakano, 2003), with each subdomain connected to a C-terminal hydrophobic patch by a short, protease sensitive linker (Jovine et al., 2004).

Results of LSCM of growing oocytes microinjected with ZP3-Flag-353 were consistent with cell transfection experiments described above; that is, mZP3 was not detected in secretory vesicles or the ZP. Similar analyses were carried out using oocytes microinjected with ZP3-Flag-ΔCP and ZP3-Flag-ΔEHP mutant constructs (Jovine et al., 2004). In the former case, mutant recombinant mZP3 was incorporated into the ZP to the same extent as wild-type mZP3, suggesting that the CP does not play a role in assembly. On the other hand, deletion of the EHP from full-length mZP3 did not prevent packaging of mutant protein into vesicles but completely abolished its incorporation into the ZP. It can be concluded that both the EHP and TMD must be present for incorporation of nascent ZP glycoprotein into the ZP.

While, under some conditions, modification of the EHP can apparently impair secretion of ZP proteins (see above; Zhao et al., 2003), we found that deletion of the EHP does not affect secretion of ZP glycoproteins but prevents their incorporation into the ZP. Similarly, mutation of the IHP also prevents incorporation of full-length ZP glycoproteins into the ZP without affecting secretion. On the other hand, secretion is severely inhibited when either the IHP or EHP is mutated in the context of a ZP protein truncated before the TMD (Jovine et al., 2004). These results suggest that the EHP and IHP are functionally related to each other and, together with the CFCS and TMD, control incorporation of nascent ZP glycoproteins into the ZP.

### A general mechanism for assembly of ZP proteins into filaments

The observations described above led us to propose a mechanism for activation of polymerisation of ZP proteins (Jovine et al., 2004; 2005) (Fig. 7). The proposal is based upon the loss of the EHP when the CTP of ZP protein precursors is removed by proteolytic cleavage at the CFCS prior to polymerisation of ZP proteins. Presumably, presence of both EHP and IHP within ZP protein precursors prevents the ZP domain from participating in polymerisation within the cell. This mechanism probably applies to ZP domain proteins in general because it relies on sequence elements (EHP and IHP) and events (coupling between proteolytic processing and polymerisation) conserved in all ZP domain proteins. In this context, cleavage of inhibitory sequences from protein precursors with concomitant exposure of polymerisation elements has been shown to regulate assembly of several other types of proteins (Taylor et al., 1997; Bourne et al., 2000; Handford et al., 2000; Mosesson et al., 2001; Gamblin et al., 2003).

### Summary

Several conclusions about the secretion and assembly of nascent ZP glycoproteins by growing mouse oocytes can be drawn from the results of experiments described above. (1) ZP glycoproteins are packaged into large secretory vesicles in growing oocytes and, after secretion, are assembled into the innermost region of the ZP, which therefore grows from the inside to the outside. (2) The CFCS of ZP glycoproteins must be cleaved for normal levels of secretion by growing oocytes; when the CFCS is not cleaved nascent ZP glycoproteins accumulate in the ER. (3) The CP and conserved Cys residue in the CTP of ZP glycoproteins are not required for either secretion or assembly. (4) The presence of either an EHP or a TMD is required for ZP glycoprotein secretion by growing oocytes. (5) The presence of both an EHP and a TMD is required for assembly of mammalian ZP glycoproteins through the ZP domain. (6) The TMD of ZP glycoproteins can be replaced by an unrelated TMD without altering its functions. (7) The

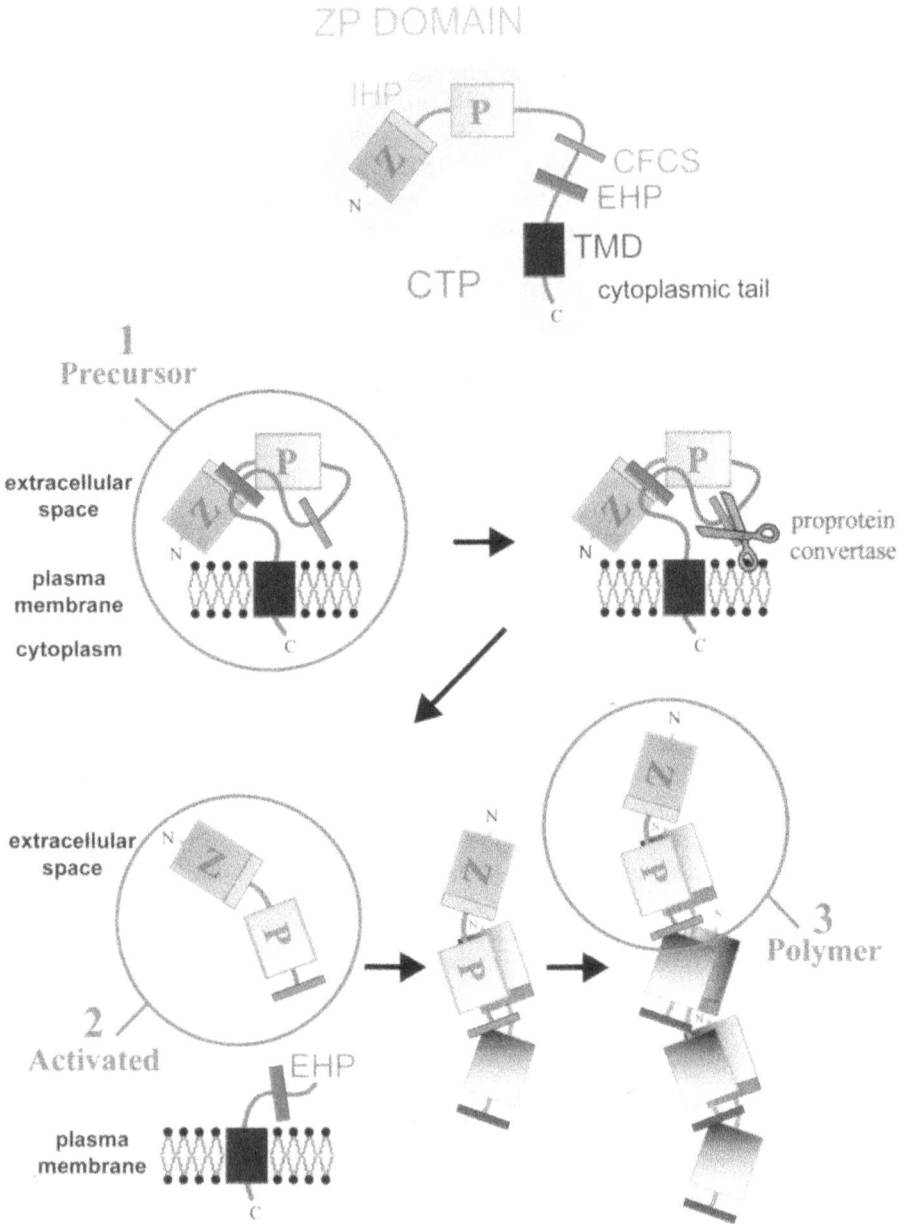

Fig. 7 A general mechanism for assembly of ZP domain proteins. In all ZP domain precursors, the ZP domain is followed by a C-terminal propeptide (CTP) that contains a basic cleavage site (such as a CFCS), and EHP, and, in most cases, a TMD or GPI-anchor (top panel). Precursors do not polymerize within the cell either as a result of direct interaction between the EHP and IHP or because they adopt an inactive conformation dependent on the presence of both patches (middle left panel). C-terminal processing at the CFCS by a proprotein convertase (middle right panel) would lead to dissociation of mature proteins from the EHP (bottom left panel), activating them for assembly into filaments and matrices (bottom right panel).

presence of both the EHP in the CTP and the IHP in the ZP domain prevents ZP glycoprotein assembly inside growing oocytes. (8) The ZP domain of ZP glycoproteins consists of two conserved subdomains and functions as a polymerisation module due to its N-terminal subdomain.

The conclusions drawn above reflect the essential roles played by polypeptides of ZP glycoproteins in the formation of a ZP around growing oocytes during mammalian oogenesis. In particular, they point out a critical role for the ZP domain as a structural element in the assembly of ZP filaments. They also suggest a general mechanism for assembly of all ZP domain proteins based on coupling between proteolytic processing and polymerisation. ZP domain proteins represent a large and important class of molecules that will continue to grow in size and should be of great interest to investigators in several areas of research for many years to come.

## Acknowledgements

We thank Dr. Satish Gupta for his kind invitation to present some of our research at the International Congress on Gamete Biology in New Delhi, India in February 2006. Our research was supported in part by National Institutes of Health grant HD35105 (PMW) and by a Human Frontier Science Program long-term fellowship (LJ).

## References

Bausek N, Waclawek M, Schneider, WJ and Wohlrab F (2000) The major chicken egg envelope protein ZP1 is different from ZPB and is synthesized in the liver *Journal of Biological Chemistry* **275** 28866-28872

Boja ES, Hoodbhoy T, Fales HM and Dean J (2003) Structural characterization of native mouse zona pellucida proteins using mass spectrometry *Journal of Biological Chemistry* **278** 34189-34202

Bokel C, Prokop A and Brown NH (2005) Papillote and Piopio: *Drosophila* ZP-domain proteins required for cell adhesion to the apical extracellular matrix and microtubule organization *Journal of Cell Science* **118** 633-642

Bork P and Sander C (1992) A large domain common to sperm receptors (ZP2 and ZP3) and TGF-ß type III receptor *FEBS Letters* **300** 237-240

Bourne Y, Watson MH, Arvai AS, Bernstein SL, Reed SI and Tainer JA (2000) Crystal structure and mutational analysis of the *Saccharomyces cerevisiae* cell cycle regulatory protein Cks1: implications for domain swapping, anion binding and protein interactions *Structure Folding & Design* **8** 841-850

Boyd D and Beckwith J (1990) The role of charged amino acids in the localization of secreted and membrane proteins *Cell* **62** 1031-1033

Chalifour LE, Wirak DO, Wassarman PM and DePamphilis ML (1986) Expression of simian virus early and late genes in mouse oocytes and embryos *Journal of Virology* **59** 619-627

Chalifour LE, Wirak DO, Hansen U, Wassarman PM and DePamphilis ML (1987) *Cis*- and *trans*-acting sequences required for the expression of simian virus 40 genes in mouse oocytes *Genes & Development* **1** 1096-1106

Cocchia M, Huber R, Pantano S, Chen EY, Ma P, Forabosco A, Ko MS and Schlessinger D (2000) *PLAC1*, an Xq26 gene with placenta-specific expression *Genomics* **68** 305-312

Gamblin TC, Chen F, Zambrano A, Abraha A, Lagalwar S, Guillozet AL, Lu M, Fu Y, Garcia-Sierra F, LaPointe M, Miller R, Berry RW, Binder LI and Cryns VL (2003) Caspase cleavage of tau: linking amyloid and neurofibrillary tangles in Alzheimer's disease *Proceedings of the National Academy of Sciences USA* **100** 10032-10037

Gwatkin RBL (1977) *Fertilization: Mechanisms in Man and Mammals* Plenum Press, NY

Handford PA, Downing AK, Reinhardt DP and Sakai LY (2000) Fibrillin: from domain structure to supramolecular assembly *Matrix Biology* **19** 457-470

Harrison PT (1996) Protein:protein interactions in the lipid bilayer *Molecular Membrane Biology* **13** 67-79

Hemberger M, Himmelbauer H, Ruschmann J, Zeitz C and Fundele R (2000) cDNA subtraction cloning reveals novel genes whose temporal and spatial expression indicates association with trophoblast invasion *Developmental Biology* **222** 158-169

Hyllner SJ, Westerlund L, Olsson PE and Schopen A (2001) Cloning of rainbow trout egg envelope proteins: members of a unique group of structural proteins *Biology of Reproduction* **64** 805-811

Jazwinska A and Affolter M (2004) A family of genes encoding zona pellucida (ZP) domain proteins is expressed in various epithelial tissues during *Drosophila* embryogenesis *Gene Expression Patterns* **4** 413-421

Jovine L, Litscher ES and Wassarman PM (2002a) Egg zona pellucida, egg vitelline envelope, and related extracellular glycoproteins *Advances in Developmen-*

tal *Biology and Biochemistry* **12** 31-53

**Jovine L, Qi H, Williams Z, Litscher E and Wassarman PM** (2002b) The ZP domain is a conserved module for polymerization of extracellular proteins *Nature Cell Biology* **4** 457-461

**Jovine L, Qi H, Williams Z, Litscher ES and Wassarman PM** (2004) A duplicated motif controls assembly of zona pellucida domain proteins *Proceedings of the National Academy of Sciences USA* **101** 5922-5927

**Jovine L, Darie CC, Litscher ES and Wassarman PM** (2005) Zona pellucida domain proteins *Annual Review of Biochemistry* **74** 83-114

**Jovine L, Janssen WG, Litscher ES and Wassarman PM** (2006) The PLAC1-homology region of the ZP domain is sufficient for protein polymerization *BMC Biochemistry* **7** 11

**Kiefer SM and Saling P** (2002) Proteolytic processing of human zona pellucida proteins *Biology of Reproduction* **66** 407-414

**Killick R, Legan PK, Malenczak C and Richardson GP** (1995) Molecular cloning of chick ß-tectorin, an extracellular matrix molecule of the inner ear *Journal of Cell Biology* **129** 535-547

**Lefievre L, Conner SJ, Salpekar A, Olufowobi O, Ashton P, Pavlovic B, Lenton W, Afnan M, Brewis IA, Monk M, Hughes DC and Barratt CL** (2004) Four zona pellucida glycoproteins are expressed in the human *Human Reproduction* **19** 1580-1586

**Legan PK, Rau A, Keen JN and Richardson GP** (1997) The mouse tectorins. Modular matrix proteins of the inner ear homologous to components of the sperm-egg adhesion system *Journal of Biological Chemistry* **272** 8791-8801

**Lin RC and Scheller RH** (2000) Mechanisms of synaptic vesicle exocytosis *Annual Review of Cell and Developmental Biology* **16** 19-49

**Litscher ES, Qi H and Wassarman PM** (1999) Mouse zona pellucida glycoproteins mZP2 and mZP3 undergo carboxy-terminal proteolytic processing in growing oocytes *Biochemistry* **38** 12280-12287

**Liu C, Litscher ES, Mortillo S, Sakai Y, Kinloch RA, Stewart CL and Wassarman PM** (1996) Targeted disruption of the mZP3 gene results in production of eggs lacking a zona pellucida and infertility in female mice *Proceedings of the National Academy of Sciences USA* **93** 5431-5436

**Locker JK and Griffiths G** (1999) An unconventional role for cytoplasmic disulfide bonds in vaccinia virus proteins *Journal of Cell Biology* **144** 267-279

**Mosesson MW, Siebenlist KR and Meh DA** (2001) The structure and biological features of fibrinogen and fibrin *Annals of the New York Academy of Sciences* **936** 11-30

**Patra AK, Gahlay GK, Reddy BV, Gupta SK and Panda AK** (2000) Refolding, structural transition and spermatozoa-binding of recombinant bonnet monkey (*Macaca radiata*) zona pellucida glycoprotein-C expressed in *Escherichia coli* *European Journal of Biochemistry* **267** 7075-7081

**Qi H, Williams Z and Wassarman PM** (2002) Secretion and assembly of zona pellucida glycoproteins by growing mouse oocytes microinjected with epitope-tagged cDNAs for mZP2 and mZP3 *Molecular Biology of the Cell* **13** 530-541

**Rankin T, Familiari M, Lee E, Ginsburg A, Dwyer N, Blanchette-Mackie J, Drago J, Westphal H and Dean J** (1996) Mice homzygous for an insertional mutation in the Zp3 gene lack a zona pellucida and are infertile *Development* **122** 2903-2910

**Rankin TL, O'Brien M, Lee E, Wigglesworth K, Eppig J and Dean J** (2001) Defective zonae pellucidae in Zp2-null mice disrupt folliculogenesis, fertility, and development *Development* **128** 1119-1126

**Rozanov DV, Deryugina EI, Ratnikov BI, Monosov EZ, Marchenko GN, Quigle JP and Strongin AY** (2001) Mutation analysis of membrane type-1 matrix metalloproteinase (MT1-MMP). The role of the cytoplasmic tail Cys(574), the active site Glu(240), and furin cleavage motifs in oligomerization, processing, and self-proteolysis of MT1-MMP expressed in breast carcinoma cells *Journal of Biological Chemistry* **276** 25705-25714

**Sasanami T, Pan J and Mori M** (2003) Expression of perivitelline membrane glycoprotein ZP1 in the liver of Japanese quail (*Coturnix japonica*) after *in vivo* treatment with diethylstilbestrol *Journal of Steroid Biochemistry and Molecular Biology* **84** 109-116

**Schanberg LE, Fleenor DE, Kurtzberg J, Haynes BF and Kaufman RE** (1991) Isolation and characterization of the genomic human CD7 gene: structural similarity with the murine Thy-1 gene *Proceedings of the National Academy of Sciences USA* **88** 603-607

**Serafini-Cessi F, Malagolini N and Cavallone D** (2003) Tamm-Horsfall glycoprotein: biology and clinical relevance *American Journal of Kidney Diseases* **42** 658-676

**Spargo SC and Hope RM** (2003) Evolution and nomenclature of the zona pellucida gene family *Biology of Reproduction* **68** 358-362

**Sudhof TC** (1995) The synaptic vesicle cycle: a cascade of protein-protein interactions *Nature* **375** 645-653

**Sugiyama H, Murata K, Iuchi I, Nomura K and Yamagami K** (1999) Formation of mature egg envelope subunit proteins from their precursors (choriogenins) in the fish, *Oryzias latipes*: loss of partial C-terminal sequences of the choriogenins *Journal of Biochemistry* **125** 469-475

**Taylor KM, Trimby AR and Campbell AK** (1997) Mutation of recombinant complement component C9 reveals the significance of the N-terminal region for polymerization *Immunology* **91** 20-27

**Wassarman PM** (1988) Zona pellucida glycoproteins *Annual Review of Biochemistry* **57** 415-442

**Wassarman PM** (1999) Mammalian fertilization: molecular aspects of gamete adhesion, exocytosis, and fusion *Cell* **96** 175-183

**Wassarman PM and Josefowicz WJ** (1978) Oocyte development in the mouse: an ultrastructural comparison of oocytes isolated at various stages of growth and meiotic competence *Journal of Morphology* **156** 209-236

**Wassarman PM and Mortillo S** (1991) Structure of the mouse egg extracellular coat, the zona pellucida *International Review of Cytology* **130** 85-110

**Wassarman PM, Qi H and Litscher ES** (1997) Mutant female mice carrying a single *mZP3* allele produce eggs with a thin zona pellucida, but reproduce normally *Proceedings of the Royal Society of London Series B Biological Sciences* **264** 323-328

**Wassarman PM, Jovine L and Litscher ES** (2001) A profile of fertilization in mammals *Nature Cell Biology* **3** E59-E64

**Williams Z and Wassarman PM** (2001) Secretion of mouse ZP3, the sperm receptor, requires cleavage of its polypeptide at a consensus furin cleavage-site *Biochemistry* **40** 929-937

**Yan C, Pendola FL, Jacob R, Lau AL, Eppig JJ and Matzuk MM** (2001) *Oosp1* encodes a novel mouse oocyte-secreted protein *Genesis* **31** 105-110

**Yanagimachi R** (1994) Mammalian fertilization In: *The Physiology of Reproduction* pp. 189-317 Eds E Knobil and JD Neill. Raven Press, NY

**Yonezawa N and Nakano M** (2003) Identification of the carboxyl termini of porcine zona pellucida proteins using mass spectrometry *Biochemical and Biophysical Research Communications* **307** 877-882

**Zhao M, Gold L, Dorward H, Liang LF, Hoodbhoy T, Boja E, Fales HM and Dean J** (2003) Mutation of a conserved hydrophobic patch prevents incorporation of ZP3 into the zona pellucida surrounding mouse eggs *Molecular and Cellular Biology* **23** 8982-8991

# Structural and functional attributes of zona pellucida glycoproteins

Satish K. Gupta*, Sanchita Chakravarty, Suraj K., Pankaj Bansal, Anasua Ganguly, Manish K. Jain and Beena Bhandari

*Gamete Antigen Laboratory, National Institute of Immunology, Aruna Asaf Ali Marg, New Delhi-110 067, India*

A translucent matrix termed the zona pellucida (ZP) surrounds the mammalian oocyte. It plays a critical role in fertilization by acting as a "docking site" for binding of spermatozoa followed by induction of the acrosome reaction in the zona bound sperm. Recent analyses of the genes of the human oocyte revealed that the ZP matrix is composed of four glycoproteins, designated as ZP1, ZP2, ZP3 and ZP4, instead of 3 found in the mouse ZP. Comparison of the deduced amino acid (aa) sequences of the human ZP glycoproteins with those from various species, revealed that these are evolutionarily conserved. Phylogenetic analysis revealed that ZP1 and ZP4 may be related as these have the highest sequence identity at the aa level within a given species. Each zona protein has a signal sequence driving these proteins to the endoplasmic reticulum, a ~260 aa long 'ZP domain' comprising of 8-10 conserved cysteine residues, a C-terminal, hydrophobic transmembrane-like region and a short cytoplasmic tail. In order to understand the structure-function relationship of human ZP glycoproteins, our lab has cloned and expressed ZP2, ZP3 and ZP4 proteins both in *E. coli* as well as baculovirus expression systems. Simultaneously, our group has been able to amplify the cDNA encoding human ZP1. Employing baculovirus-expressed recombinant ZP glycoproteins; our group has provided evidence for the first time that in human, in addition to ZP3, ZP4 is also able to induce acrosomal exocytosis in the capacitated spermatozoa. ZP3 mediated induction of the acrosome reaction can be inhibited by pertussis toxin suggesting the involvement of $G_i$ protein in downstream signaling in contrast to ZP4, which follows a $G_i$ protein independent pathway. Hence, elucidation of the role of individual ZP glycoproteins in humans will provide a better insight into the gamete interaction culminating in fertilization.

## Introduction

In mammals, fertilization is a well-orchestrated event culminating in the union of two highly specialized haploid cells; that is, the spermatozoa and the oocyte. A comprehensive under-

---

*Corresponding author
E-mail: skgupta@nii.res.in

standing of the molecular mechanisms of human gamete interaction will facilitate improving *in vitro* fertilization success rates and may aid in developing novel contraceptive strategies aiming to block at pre-fertilization stages. The spermatozoa first encounters the zona pellucida (ZP): a translucent, extra-cellular matrix surrounding the egg. The ZP plays a crucial role in fertilization by serving as a species-selective substrate for sperm binding as well as an agonist for regulated exocytosis of the spermatozoon's acrosomal vesicle. On reaching the perivitelline space, the spermatozoon fuses with the egg plasma membrane. This fusion event gives rise to the cortical reaction which produces changes in the ZP matrix helping in the block to polyspermy.

The ZP is the extra-cellular translucent coat of sulfated glycoproteinaceous matrix synthesized and secreted by the oocyte during follicular development. The size, rigidity as well as the thickness of the ZP varies from species to species. It can vary in thickness from < 2 μm in marsupials to 27 μm in cows (Dunbar and Wolgemuth, 1984). Scanning electron micrographs (SEM) of zonae from several species demonstrate that its surface ultrastructure consists of an extensive fibrous network interspersed with numerous pores of varying sizes. These pores appear larger at the outer surface of the zona than inner surface (Wassarman, 1987; Wassarman and Litscher, 1995). The spongy outer surface of zona with larger pores may facilitate sperm penetrability as human ZP with a more compact and smoother outer surface has been shown to be less penetrable (Familiari *et al.*, 1988; 1992). With the help of cross-linking experiments and localization with specific monoclonal antibodies (MAbs), it has been demonstrated that the mouse ZP matrix consists of 2-3 μm long interconnected filaments each having a 14-15 nm structural repeat, considered to be a ZP2-ZP3 heterodimer cross linked by ZP1 (El-Mestrah *et al.*, 2002).

In the present review, we deliberate upon the structural and biochemical attributes of the ZP glycoproteins and analyze their functional significance in the intricate process of fertilization.

## Structural attributes of ZP glycoproteins

*i) Biochemical characterization of ZP*

Various ZP glycoproteins that constitute the ZP matrix have been characterized from several species by gel electrophoresis. In mouse, the ZP is composed of three families of glycoproteins designated as ZP1 (180-200 kDa), ZP2 (120-240 kDa) and ZP3 (83 kDa), based on their mobility on a SDS-PAGE (sodium dodecyl sulfate-polyacrylamide gel electrophoresis) run under non-reducing conditions (Bleil and Wassarman, 1980). Porcine ZP, under non-reducing conditions, resolved into ZP1 (80-90 kDa) and ZP3 (55 kDa; Yurewicz *et al.*, 1987); whereas under reducing conditions, it resolved into 4 bands designated as ZP1 (82 kDa), ZP2 (61 kDa), ZP3 (55 kDa) and ZP4 (21 kDa). The bands corresponding to ZP2 and ZP4 have been shown to originate from ZP1 by proteolytic cleavage and are connected by disulfide bonds (Hedrick and Wardrip, 1987; Yurewicz *et al.*, 1987). The porcine ZP1 is homologous to mouse ZP2 (Taya *et al.*, 1995). Chemical deglycosylation or digestion of porcine ZP3 with enzymes such as endo-ß-galactosidase revealed that biochemically, it is composed of two distinct glycoproteins designated as ZP3α (equivalent to mouse ZP1) and ZP3ß (equivalent to mouse ZP3: Hedrick and Wardrip, 1987; Yurewicz *et al.*, 1987).

The characterization of native ZP glycoproteins from human oocytes by various groups has revealed heterogeneity and variability in their mobility on SDS-PAGE (Shabanowitz and O'Rand, 1988; Bercegeay *et al.*, 1995; Gupta *et al.*, 1998; Bauskin *et al.*, 1999). The human ZP, under reducing conditions, resolved into 3 bands corresponding to ZP1 (90-110 kDa), ZP2 (64-78 kDa) and ZP3 (57-73 kDa; Shabanowitz and O'Rand, 1988). The human ZP has been further characterized by two-dimensional electrophoresis followed by Western blot employing antibodies generated against the respective homologous recombinant proteins of bonnet monkey

(*Macaca radiata*: Gupta *et al.*, 1998). The studies revealed that human ZP1 is comprised of two chains of 63-58 kDa and 55-43 kDa, each consisting of multiple isomers. The human ZP2 was observed to be comprised of a major 65 kDa component and a minor ~ 96 kDa component. The 65 kDa component displayed a higher degree of charged isomers in comparison with the 96 kDa component. The human ZP3 was observed as a broad band in the range of 65-58 kDa. Using antibodies against synthetic peptides, it has been further documented that ZP2 is comprised of 90-110 kDa and ZP3 of 53-60 kDa (Bauskin *et al.*, 1999). Recent studies, however, revealed that human ZP is composed of four glycoproteins (Lefievre *et al.*, 2004, Conner *et al.*, 2005), resulting in their redesignation as ZP1, ZP2, ZP3 and ZP4 (ZP4 previously designated as ZP1/ZPB in non-human primates and humans).

Though the ZP proteins from various species have a very similar polypeptide core, there are differences in their relative order of migration on SDS-PAGE. It is observed because of the differential post-translational modifications including glycosylation, which modify the effective molecular weight of the ZP polypeptides and hence, their migration (Wassarman, 1999). Analysis of ZP glycoproteins by two-dimensional electrophoresis also revealed that each family of ZP glycoproteins is comprised of multiple isomers. It may be due to the variations in glycosylations of both N-(Asparagine) and O-(Serine/Threonine) linked oligosaccharides and sulfation that yield extensive charge heterogeneity in the ZP proteins as a result of which they exist as several isoelectric species (Wassarman and Litscher, 1995). Some of the discrepancies in the apparent molecular weights assigned to various ZP glycoproteins may also be contributed by the different nomenclature used by various investigators. In the present article, we have used the nomenclature of ZP1, ZP2, ZP3 and ZP4 as described for human ZP glycoproteins (Conner *et al.*, 2005).

*ii) Genomic organization of ZP glycoproteins*

Studies pertaining to the genomic organization of the *Zp* genes have revealed by and large a similar intron/exon structure for each gene family. However, the length of the chromosomal region comprising the locus of these genes varies from as small as 6.5 kilobases (kb) in mouse *Zp1* to as big as 18.3 kb in human *Zp3* (Chamberlin and Dean, 1990; Epifano *et al.*, 1995). The mouse *Zp1* is composed of 12 exons ranging in size from 82-364 base pairs (bp) and encodes for a 623 aa long polypeptide. The exon size in mouse *Zp2* ranges from 45-190 bp with a mRNA transcript of 2201 nucleotides (nt) encoding for a polypeptide of 713 aa while the mouse *Zp3* gene consists of 8 exons spanning 92-338 bp transcribed into 1302 nt mRNA encoding for a 424 aa polypeptide (Kinloch *et al.*, 1988; Liang and Dean, 1993).

The human *Zp1* is composed of 12 exons encoding a polypeptide of 638 aa residues. The human *Zp2* has 19 exons (one more than in mouse) transcribed into a 2235 nt long mRNA transcript, which codes for a 745 aa long polypeptide. The human *Zp3* is comprised of 8 exons, the transcript of which encodes for a 424 aa polypeptide. The exon map for human *Zp4* spans 11 kb comprised of 12 exons and the transcript encodes for a 540 aa long polypeptide (Harris *et al.*, 1994).

The cDNA clones of ZP glycoproteins from different species have also been characterized (Harris *et al.*, 1994; Spargo and Hope, 2003). Comparison of the nt and deduced aa sequences of the various ZP glycoproteins reveal certain common features among them. These are: i) short 5' and 3' untranslated regions; ii) N-terminal hydrophobic signal peptide sequence that coerces the protein into the secretory pathway and gets cleaved off from the mature protein; iii) potential N- and O- linked glycosylation sites; iv) a C-terminus hydrophobic transmembrane-like domain that plays a role in the intracellular trafficking of the proteins; v) a potential tetra basic furin cleavage site, RXR/KR, upstream of the transmembrane-like domain; and vi) a ZP signature domain.

A schematic representation depicting the various structural features of the human ZP glyco-proteins is presented in Fig. 1. All four human zona proteins have a signal peptide ranging from 21-38 aa at the N-terminus. The signal peptide is cleaved in the mature protein by signal peptidase present in the oocyte. All the four proteins also have a 'ZP domain'. The position of 'ZP domain' within the individual ZP protein varies. In ZP1, it is comprised of 279-548 aa, ZP2 from 372-637 aa, ZP3 from 45-304 aa and ZP4 from 188-460 aa (Fig. 1). In ZP3, the 'ZP domain' lies more towards the N-terminus. All four proteins have a consensus furin cleavage site (CFCS). In addition to the above features, human ZP1 (234-274 aa) and ZP4 (141-183 aa) also have a Trefoil motif or a P-domain, which is absent in ZP2 as well as ZP3. It is a cysteine rich domain of approximately 45 aa found in some extracellular eukaryotic proteins like spasmolytic polypep-tide (SP), intestinal trefoil factor (ITF) et cetera (Bork, 1993). The domain was originally iden-tified from mucosal tissues, where it may have a regulatory or structural role and has also been implicated as a growth factor in other tissues.

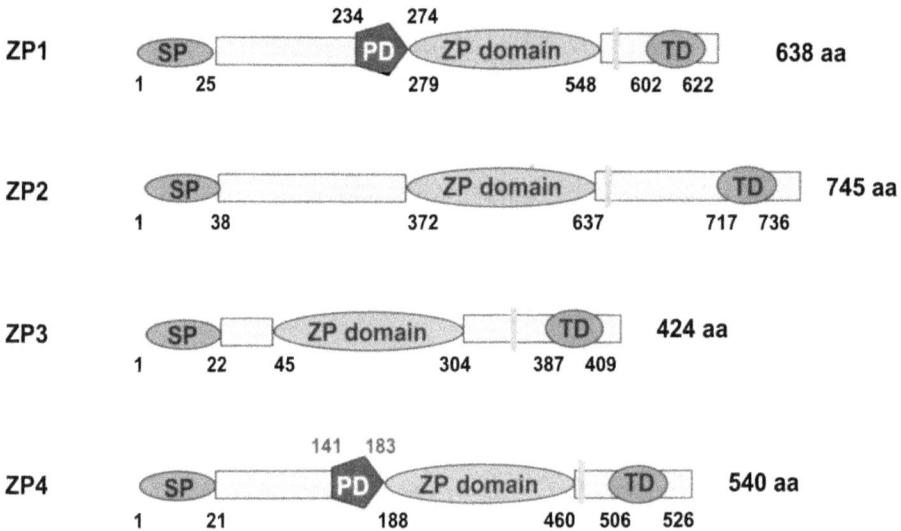

**Figure 1: Schematic representation of the structural attributes of human ZP glycoproteins.** The structural organization and the predicted domains of the different zona proteins are depicted where the amino acid (aa) residues spanning each domain are specified (not drawn to scale). Signal peptide (SP) is shown in orange, Trefoil domain (PD) in blue, 'ZP domain' in pink, consensus furin cleavage site (CFCS) as a light blue bar and transmembrane-like do-main (TD) in green. The 'ZP domain' in all the four human zona proteins consists of approxi-mately 8 conserved Cys residues and precedes the CFCS where the zona proteins are pre-dicted to be cleaved prior to secretion and subsequent incorporation in the zona matrix. The trefoil domain is present only in human ZP1 and ZP4.

(a)  Significance of the 'ZP domain'

All members belonging to the ZP family share a sequence motif referred to as 'ZP domain'. A ZP domain consists of ~ 260 aa with 8 conserved cysteine (Cys) residues as well as conserved polar and hydrophobic patterns which probably aid the proteins to assume their three dimensional con-formation (Bork and Sander, 1992). The 'ZP domain' is found in a large number of eukaryotic extracellular proteins with diverse biological functions like the ZP proteins, Tamm-Horsfall protein (THP; Uromodulin), transforming growth factor ß receptor III, pancreatic secretory granule glyco-protein (GP-2), α- and ß-tectorins, NompA (no-mechanoreceptor-potential-A), Dumpy and cuticulin-1, *Drosophila* genes *miniature* and *dusky et cetera* (Wassarman *et al.*, 2001; Roch *et al.*, 2003).

Identification of 'ZP domain' within various proteins involved in the formation of filaments or matrices led to the proposal that this motif is responsible for polymerization of the proteins into filaments of similar supramolecular structure and hence their assembly (Jovine et al., 2002).

(b) Relevance of the furin cleavage site for secretion of ZP glycoproteins and their assembly in the ZP matrix

The furin cleavage site is said to be involved in the secretion of the mature ZP glycoproteins. Nascent mouse ZP3 from transfected cells was not secreted when the furin cleavage site was modified by site-directed mutagenesis or by using a specific inhibitor of furin-like enzymes (Williams and Wassarman, 2001). This observation was further supported by another study wherein the secretion and assembly of Myc and Flag epitope tagged mouse ZP2 and ZP3, microinjected into the germinal vesicles (nucleus) of oocytes from juvenile mice, was investigated. It was observed that the excision of the C-terminal region of the glycoproteins was essential for their assembly into the ZP matrix (Qi et al., 2002). These findings point towards the importance of the furin cleavage site as in the absence of which the mature protein is not secreted. However, another study examining the involvement of the furin cleavage site in the secretion of mouse ZP3 in mouse fibroblasts and its incorporation into the ZP of the mouse oocyte indicates that the cleavage at the furin cleavage site is not an absolute necessity for ZP3 secretion, intracellular trafficking or its incorporation into the zona matrix (Zhao et al., 2002).

*iii) Conservation and evolutionary changes in ZP glycoproteins during evolution*

It seems that the overall backbone structure of the ZP proteins from different species is conserved. It may be due to the conserved nature of the Cys residues present in various zona proteins. The ZP2 sequence from various species has 20 conserved Cys residues with 1 or 2 non-conserved ones. Similarly, ZP3 family has 14 conserved Cys residues. The ZP1 sequences, however, seem to be more divergent than the ZP2 or ZP3 groups. A considerable homology has been shown to exist between the vitelline envelope (VE) glycoproteins present in fish oocytes and mammalian ZP glycoproteins. The three VE glycoproteins present in the anuran *Xenopus laevis*, designated as gp37, gp69 and gp41, are homologous to the three mammalian ZP glycoproteins, ZP1, ZP2 and ZP3 respectively (Hedrick, 1996). The aa sequence of ZP2 of a marsupial, the brushtail possum (*Trichosurus vulpecula*), shares high sequence similarity with the respective *Xenopus* protein (Haines et al., 1999). Even the avian ZP glycoproteins share a considerable degree of homology with their respective mammalian, fish and amphibian counterparts (Takeuchi et al., 1999; Bausek et al., 2000). Phylogenetic analysis demonstrated a hierarchy of relatedness of *Zp* genes and these can be grouped into 4 sub-families (Fig. 2). Discovery of chicken *Zp1* gene (Bausek et al., 2000) suggested that gene duplication predating the divergence of fish and amphibians may have given rise to two paralogous group of genes, *Zp1* and *Zp4* (Spargo and Hope, 2003). The phylogenetic tree shown in Fig. 2 supports the above notion as ZP1 and ZP4 sequences of various species are related to each other more closely than ZP2 and ZP3. All these observations indicate that probably the egg envelope proteins arose from a common ancestor and diversified over a period of time during evolution to give them species-specific characteristics (Fig. 2).

*iv) Sequence similarity of ZP glycoproteins within a given species*

Apart from the homology observed within a given ZP protein from different species, a considerable amount of homology is also seen between the aa sequence of ZP1 and ZP2 of different species indicating the involvement of a common ancestral gene. A domain of mouse ZP1

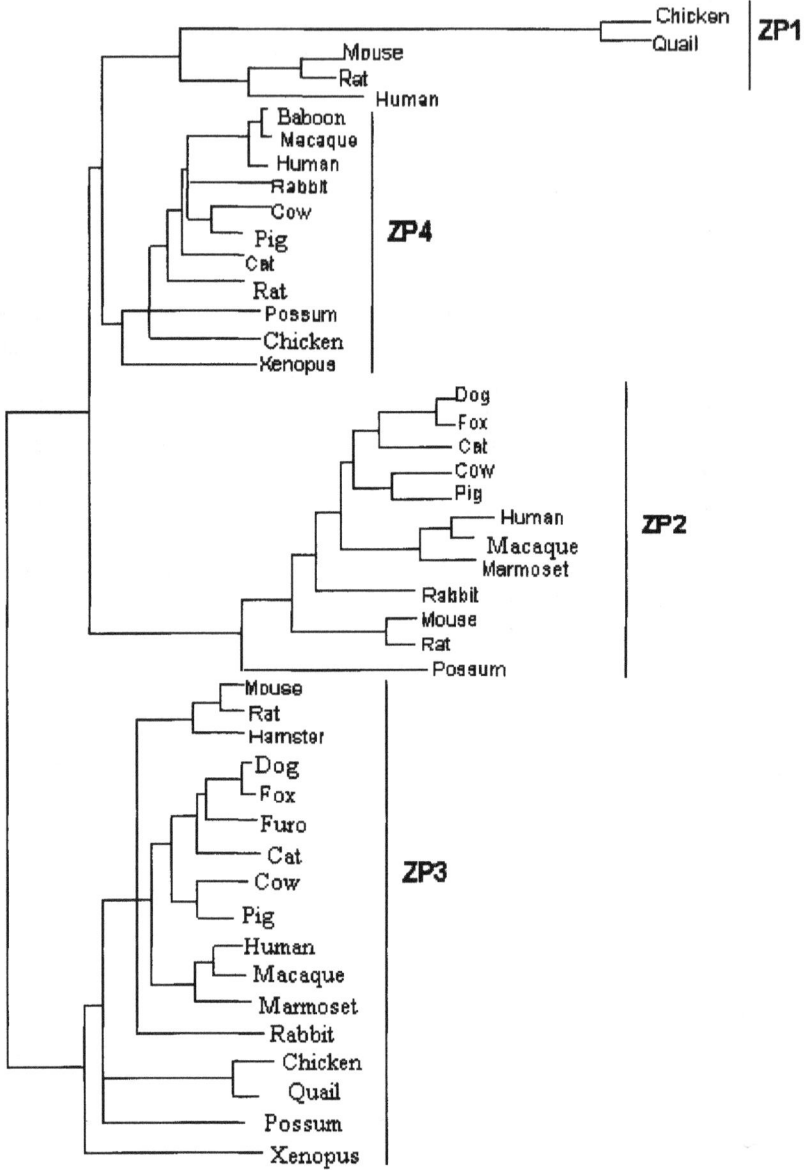

**Figure 2: Evolutionary relationship of ZP glycoproteins between different species.** Using ClustalW, a phylogenetic tree of different ZP glycoproteins from various species has been generated and the different clads show the evolutionary divergence amongst the various ZP glycoproteins. The ordering of the nodes describes how lineages have diverged over the course of evolution and the relative branch lengths indicate rates of evolution along a particular branch. The following sequences have been used for the generation of the phylogenetic tree: chicken ZP1 (NP_990014), quail ZP1 (BAB47585), mouse ZP1 (NP_033606), rat ZP1 (NP_445961), human ZP1 (P60852), dog ZP2 (NP_001003304), fox ZP2 (AAT37676), cat ZP2 (NP_0001009875), cow ZP2 (NP_776398), pig ZP2 (P42099), human ZP2 (AAA61335), Macaque ZP2 (AAP13259), marmoset ZP2

**Figure 2 Contd.** (CAA71740), rabbit ZP2 (P48829), mouse ZP2 (NP_035905), rat ZP2 (O55767), possum ZP2 (AAC28737), mouse ZP3 (P10761), rat ZP3 (P97708), hamster ZP3 (P23491), dog ZP3 (P48831), fox ZP3 (AAT37677), furo ZP3 (AAU14267), cat ZP3 (P48832), cow ZP3 (NP_776399), pig ZP3 (P42098), human ZP3 (AAA61336), Macaque ZP3 (P53785), marmoset ZP3 (P53786), rabbit ZP3 (P48833), chicken ZP3 (AAV35193), quail ZP3 (BAB86301), possum ZP3 (AAC28736), *Xenopus* ZP3 (AAB39079), baboon ZP4 (AAP13260), Macaque ZP4 (AAP13262), human ZP4 (AAA74391), rabbit ZP4 (Q00193), cow ZP4 (Q9BH11), pig ZP4 (Q07287), cat ZP4 (P48834), rat ZP4 (NP_758833), possum ZP4 (AAF73042), chicken ZP4 (NP_990210) and *Xenopus* ZP4 (AAA91465).

represented by aa residues 268-623 is found to have 47% similarity (32% identity) with mouse ZP2 (aa residues 363-713) and is encoded by 8 exons in both cases (Epifano *et al.*, 1995). Probably, mouse ZP1 as well as mouse ZP2 might have originated from the same ancestral gene that has been duplicated and re-utilized by exon shuffling. In addition to the sequence conservation between the above two genes, the two proteins also share conserved positions of the 10 Cys residues in their aa sequences suggesting a similarity in the structural aspects of the two proteins. In humans, comparison of the deduced aa sequence of ZP2 and ZP4 revealed 47% identity (Table 1). A high (40%) sequence identity at the aa level between human ZP2 and ZP4 was also observed.

**Table 1.** Percent identity between the deduced amino acid sequences of mature polypeptide of human ZP proteins

|  | *Percent identity at amino acid level* | | | |
|  | ZP1 | ZP2 | ZP3 | ZP4 |
| --- | --- | --- | --- | --- |
| ZP1 | - | 33 | 23 | 47 |
| ZP2 | 33 | - | 20 | 40 |
| ZP3 | 23 | 20 | - | 21 |
| ZP4 | 47 | 40 | 21 | - |

## Functional attributes of ZP glycoproteins

Attempts have been made by various investigators to delineate the functional relevance of ZP glycoproteins at various stages of fertilization. These studies have been conducted in various animal models either by employing proteins purified from native source or those obtained by recombinant DNA technology.

*i) Delineating the role of ZP glycoproteins during fertilization in mouse model*

The binding characteristics of mouse ZP glycoproteins and their ability to induce acrosomal exocytosis are summarized in Table 2. Different studies have demonstrated that in mouse, ZP3 is responsible for the initial binding of the sperm to the oocyte and induction of acrosomal exocytosis, thereby assigning it the primary sperm receptor function (Bleil and Wassarman, 1980; 1983; Mortillo and Wassarman, 1991; Beebe *et al.*, 1992). Mouse ZP3 purified from ovulated eggs but not from fertilized eggs/one cell embryos binds to the capacitated sperm. However, ZP3 does not bind to acrosome-reacted sperm. Approximately, 60% of the ZP3 binding sites are present on the acrosomal cap region whereas 40% of the sites reside in the post-acrosomal region (Bleil and Wassarman, 1986; Mortillo and Wassarman, 1991). However, subsequent studies suggest that only 20% of the ZP3 binding sites are located on the acrosomal cap and the remaining sites are present on the post-acrosomal region (Kerr *et al.*, 2002). ZP3 has been shown to induce the acrosome reaction when incubated with the capacitated mouse sperm (Bleil and Wassarman, 1986; Kinloch et al., 1991; Beebe et al., 1992). The role of ZP3 in binding sperm has been further confirmed by observations that showed that mice with homozygous mutation in the ZP3 gene failed to conceive on mating with males of proven fertility (Liu et al., 1996; Rankin et al., 1996).

There is considerable evidence to suggest that carbohydrate moieties, in particular O-linked oligosaccharides present on ZP3, play an important role in its binding to spermatozoa (Florman et al., 1984; Florman and Wassarman, 1985). Subsequently, it was demonstrated that terminal galactose residues in $\alpha$ or ß linkages as well as N-acetylglucosamine in ß linkage of O-linked oligosaccharides are critical (Wassarman and Litscher, 1995). However, mice deficient in glycosyl transferase, which amends terminal galactose in $\alpha$ linkage, are fully fertile (Thall et al., 1995) suggesting that galactose in either ß linkage or N-acetylglucosamine or both of them may be the critical moieties. It has also been suggested by site-directed mutagenesis studies, wherein the mutation of serine (Ser) residues $Ser_{332}$ and $Ser_{334}$ of mouse ZP3 completely abrogated its sperm receptor activity (Chen et al., 1998), that these aa residues may carry O-linked oligosaccharides that are recognized by the receptors present on the acrosome-intact spermatozoa. However, mass spectrometric studies of mouse ZP3 showed that neither $Ser_{332}$ nor $Ser_{334}$ are occupied by O-linked oligosaccharide side chains (Boja et al., 2003). Using exon-swapping studies in the mouse ZP3, it has been recently documented that spermatozoa recognize and bind to a region of ZP3 polypeptide immediately downstream of its ZP domain that is encoded by mouse ZP3 exon-7 (Williams et al., 2006).

Following the acrosome reaction, ZP2 acts as the secondary sperm receptor and helps in the binding of the acrosome-reacted spermatozoa to the oocyte (Bleil et al., 1988). ZP2 binds exclusively to the acrosome-reacted spermatozoa. ZP2, in mouse model, also plays an important role in blocking polyspermy following fertilization. Subsequent to the cortical reaction, ZP2 undergoes limited proteolytic cleavage resulting in the formation of small molecular weight fragments, which do not dissociate but remain non-covalently bound (Wassarman, 1987). It results in an increase in the non-covalent interactions that makes ZP resistant to proteolytic cleavage and acts as a protective barrier for polyspermy. Concomitantly, it is also associated with the loss of ZP3 activity as a primary sperm receptor. The crucial role of ZP2 during fertilization and development of follicles is further supported by the investigations using ZP2 null mice (Rankin et al., 2001). In these animals, a thin zona matrix composed of ZP1 and ZP3 was formed in the early follicles but was not sustained in the pre-ovulatory follicles. The presence of abnormal zona did not affect initial folliculogenesis but a significant decrease in the number of antral stage follicles was observed. Moreover, no 2-cell embryos were detected when these mice were mated with normal males.

In the mouse model, ZP1 has been implicated to have a role in maintaining the structural integrity of the zona matrix by cross-linking the ZP2-ZP3 filaments (Greve and Wassarman, 1985). The ovaries from ZP1 null mice have perturbed folliculogenesis and on mating with males, fewer 2-cell embryos were produced and the litter size was smaller than that obtained on mating of normal females (Rankin et al., 1999).

*ii) Lessons learnt from porcine and rabbit models*

Contrary to the mouse model, initial studies in the pig model suggested that ZP3$\alpha$ (equivalent to mouse ZP1) is the putative sperm receptor (Yurewicz et al., 1993). However, subsequently, it was demonstrated that high molecular weight hetero-complexes of ZP3$\alpha$ (~ ZP1) with ZP3ß (~ ZP3) bind with very high affinity to boar sperm membrane vesicles suggesting that both the glycoproteins are involved in the sperm binding process (Yurewicz et al., 1998). In the rabbit model, rec55 (homologue of mouse ZP1, expressed in baculovirus) has been shown to bind to the spermatozoa in a dose-dependent manner (Prasad et al., 1996). Both rec45 (homologue of mouse ZP3) and rec55 components of rabbit ZP have been shown to bind to recombinant Sp17 (sperm autoantigen), suggesting the involvement of a molecular mechanism similar to that found in the porcine system (Yamasaki *et al.*, 1995). These findings strongly suggest that the functional role of individual ZP glycoproteins during fertilization, as determined in the mouse model, may not be the same in other species. Hence, in order to delineate the functional significance of individual ZP glycoproteins in a given species, the same have to be investigated in an unbiased manner, rather than in the context of the deductions made from the mouse model.

### iii) Role of ZP glycoproteins during fertilization in primates

#### (a) Bonnet Monkey

The major focus of our group is to delineate the role of ZP glycoproteins during fertilization in primates. To begin with, the cDNA encoding bonnet monkey (*Macaca radiata*) ZP2 (Jethanandani *et al.*, 1998), ZP3 (Kolluri *et al.*, 1995) and ZP4 (Gupta *et al.*, 1997) were cloned and sequenced. Comparison of the deduced aa sequences of bonnet monkey zona proteins revealed a very high (>92%) sequence identity with the respective human homologues. The baculovirus-expressed bonnet monkey ZP3 binds to the head region of capacitated spermatozoa and failed to bind to the acrosome-reacted spermatozoa (Gahlay and Gupta, 2003). The baculovirus-expressed bonnet monkey recombinant ZP3, when incubated with capacitated spermatozoa, led to a significant increase in the induction of the acrosome reaction (Gahlay and Gupta, 2003). On the other hand, *E. coli*-expressed recombinant ZP3 failed to induce any significant increase in the acrosome reaction, thereby suggesting that the glycosylation of ZP3 is critical for induction of the acrosome reaction. Interestingly, bonnet monkey recombinant ZP4 expressed both in *E. coli* (Govind *et al.*, 2001) as well as baculovirus (unpublished observations) expression systems bind to the head of capacitated spermatozoa as revealed by immunofluorescence studies. In acrosome-reacted spermatozoa, the binding of *E. coli*-expressed protein was located in equatorial segment, post-acrosome domain and mid-piece region (Govind *et al.*, 2001). The binding of the baculovirus-expressed recombinant ZP4 was observed on the tip of the inner acrosomal membrane, equatorial segment and the mid-piece region of acrosome-reacted spermatozoa (unpublished observations). Competitive inhibition studies or antibody-mediated inhibition studies revealed that the binding of recombinant bonnet monkey ZP4 was specific to capacitated spermatozoa (Govind *et al.*, 2001; Gahlay *et al.*, 2002). These studies suggest that in primates, ZP4 may also have a role during fertilization.

#### (b) Humans

To delineate the functional significance of ZP glycoproteins during fertilization in humans, initial studies employed either intact or solubilized zonae (Cross *et al.*, 1988; Franken *et al.*, 1996; Oehninger, 2003). Intact as well as acid-disaggregated ZP induces acrosomal exocytosis in human spermatozoa (Cross *et al.*, 1988). Further, employing solubilized human zona, a dose-dependent increase in the induction of the acrosome reaction was observed (Franken *et al.*, 1996). Solubilized human zona not only induces the acrosome reaction in the capacitated human spermatozoa but also brings about an effective inhibition of the human sperm-egg interaction *in vitro* (Lee *et al.*, 1992; Franken *et al.*, 1996).

A critical appraisal of the functional attributes of the individual human ZP glycoproteins has been hampered due to the paucity of availability of a large number of human oocytes owing to ethical considerations. Further, the protocols to purify individual zona proteins that constitute

ZP matrix also need to be standardized. To circumvent this, various investigators have obtained human ZP glycoproteins by using recombinant DNA technology (van Duin et al., 1994; Chapman et al., 1998; Harris et al., 1999; Tsubamoto et al., 1999; Chakravarty et al., 2005). The glycosylated recombinant human ZP3 expressed in CHO cells induces the acrosome reaction in human sperm and promotes fusion of human sperm with zona free hamster oocytes (van Duin et al., 1994). The polypeptide backbone of ZP3 expressed in E. coli is also competent to induce acrosomal exocytosis (Chapman et al., 1998). These observations raise further controversies regarding the critical requirement of the carbohydrate moieties of ZP3 for its biological activity.

To delineate the functional relevance of the individual human zona glycoproteins, our group has cloned and expressed these in a baculovirus expression system. The analyses by SDS-PAGE and Western blot of the purified recombinant human ZP2, ZP3 and ZP4 revealed apparent molecular weights of ~110, ~65 and ~70-75 kDa, respectively (Chakravarty et al., 2005). The expressed recombinant proteins had both N- and O-linked glycosylations. Major oligosaccharides were represented by the binding of lectins, concanavalin A (specific for mannose α 1-3 or mannose α 1-6 residues) and Jacalin (specific for α-O-glycosides of Gal or GalNAc moieties). Both recombinant ZP3 and ZP4 bind to the capacitated spermatozoa whereas recombinant ZP2 failed to bind to the capacitated spermatozoa (unpublished results). Recombinant ZP2 binds only to the acrosome-reacted spermatozoa. When capacitated sperm were incubated with recombinant ZP3, a significant increase in the acrosomal exocytosis was observed (Chakravarty et al., 2005). Interestingly, recombinant ZP4 also led to a significant increase in the acrosomal exocytosis. However, E. coli-expressed recombinant ZP3 and ZP4 failed to induce acrosomal exocytosis. Recombinant ZP3 failed to induce acrosomal exocytosis in the presence of pertussis toxin, an inhibitor of the $G_i$ pathway, suggesting that induction of acrosomal exocytosis by ZP3 involves activation of $G_i$ complex. On the contrary, pertussis toxin failed to inhibit acrosomal exocytosis mediated by recombinant ZP4, suggesting that its action is not dependent on the activation of $G_i$ pathway (Chakravarty et al., 2005). These studies suggest that in humans, ZP4 in addition to ZP3 also has an important role in the binding of spermatozoa to the oocyte and subsequent induction of acrosomal exocytosis (Table 2).

**Table 2.** Summary of the binding characteristics of zona proteins and their ability to induce acrosomal exocytosis in mice and humans

| Zona protein | Species | Binding of zona protein to spermatozoa | | Induction of acrosomal exocytosis |
| | | Capacitated | Acrosome-reacted | |
| --- | --- | --- | --- | --- |
| ZP1 | Mouse | negative | negative | negative |
| | Human | ND | ND | ND |
| ZP2 | Mouse | negative | positive | negative |
| | Human | negative | positive | negative |
| ZP3 | Mouse | positive | negative | positive |
| | Human | positive[b] | positive[b] | positive |
| ZP4 | Mouse[a] | NA | NA | NA |
| | Human | positive[b] | positive[b] | positive |

ND: Not Determined; NA: Not Applicable
a: The ZP4 protein has not been demonstrated in the mouse ZP
b: unpublished observations

## Role of supramolecular structure of the ZP matrix during fertilization

During fertilization, spermatozoa bind to the ZP matrix rather than the constituent individual glycoproteins. Nonetheless, studies with the individual glycoproteins either purified from native source or obtained by recombinant DNA technology, have provided useful information on their role during fertilization. The supramolecular structure that is formed by the interaction among the four glycoproteins that constitute the human ZP matrix is very critical. In mice, ZP3 is essential for the formation of the ZP matrix (Liu *et al.*, 1996; Rankin *et al.*, 1998). In ZP2 null mice, the width of the ZP1/ZP3 matrix is considerably thinner than the ZP2/ZP3 matrix in ZP1 null mice (Rankin *et al.*, 1999; Rankin *et al.*, 2001). It may be due to the limited amount of ZP1 that is made by the mouse oocyte. Mapping of disulfide bonds by micro scale mass spectrometry revealed that there are two main isoforms of the 'Zona Domain'; type I represented by ZP3 and one or more type II represented by ZP1, ZP2 and ZP4 (Boja *et al.*, 2005). It appears that formation of the ZP matrix requires both isoforms of 'Zona Domain'. The importance of the supramolecular structure of the ZP matrix is reiterated by the observation that genetically engineered mice, where the formation of the ZP matrix is compromised (ZP3 null mice), fail to fertilize (Rankin *et al.*, 1996; Liu *et al.*, 1996).

Genetically altered mice in which human ZP1 replaces endogenous mouse ZP1 (human ZP1 rescue mice) are fertile. In contrast to the one/two–cell embryos obtained from normal mice that failed to bind fresh sperm, two-cell embryos obtained from human ZP2 rescue mice by *in vitro* fertilization continue to bind sperm (Dean, 2004). In human ZP2 rescue mice, subsequent to cortical reaction, human ZP2 is not cleaved. These observations suggest that the cleavage status of ZP2 regulates the three-dimensional structure of the zona matrix, rendering it permissive (intact ZP2) or non-permissive (cleaved ZP2) for sperm binding (Dean, 2004). The three dimensional ZP matrix structure to which spermatozoa bind and may involve protein, carbohydrate or both have not been determined and will require further investigations.

## Concluding comments

Characterization of ZP glycoproteins from various species has revealed that there is a considerable degree of conservation in the basic structure of these proteins. The functional role of each of the ZP glycoproteins during the complex process of fertilization as determined using the mouse model, in which ZP3 acts as a primary sperm receptor, ZP2 as a putative secondary receptor and ZP1 as a stabilizer of the zona matrix may not be correct for the ZP glycoproteins of other species. Recent observations suggest that in other animal models including human, more than one zona protein may be involved in binding of the spermatozoa to the oocyte and also in the induction of acrosomal exocytosis. The physiological significance of the observations that human ZP is composed of four glycoproteins, in contrast to mouse ZP that is composed of three glycoproteins, needs to be ascertained. In future, it will be pertinent to resolve the nature and importance of the supramolecular structure, resulting from the interaction of the various zona glycoproteins, in the complex process of fertilization.

## Acknowledgements

The authors gratefully acknowledge the financial support provided by the Department of Biotechnology, Government of India either independently or under the Indo-US Joint Program on Contraceptive and Reproductive Health Research and the Indian Council of Medical Research, Government of India for carrying out these studies. SC is recipient of Senior Research Fellowship, Council of Scientific and Industrial Research, Government of India. The views expressed by the authors do not necessarily reflect the views of the funding agencies.

# References

Bausek N, Waclawek M, Schneider WJ and Wohlrab F (2000) The major chicken egg envelope protein ZP1 is different from ZPB and is synthesized in the liver *Journal of Biological Chemistry* 275 28866-28872

Bauskin AR, Franken DR, Eberspaecher U and Donner P (1999) Characterization of human zona pellucida glycoproteins *Molecular Human Reproduction* 5 534-540

Beebe S, Leyton L, Burks D, Ishikawa M, Fuerst T, Dean J and Saling P (1992) Recombinant mouse ZP3 inhibits sperm binding and induces the acrosome reaction *Developmental Biology* 151 48-54

Bercegeay S, Jean M, Lucas H and Barriere P (1995) Composition of human zona pellucida as revealed by SDS-PAGE after silver staining *Molecular Reproduction and Development* 41 355-359

Bleil JD and Wassarman PM (1980) Mammalian sperm-egg interaction: Identification of a glycoprotein in mouse zonae pellucidae possessing receptor activity for sperm *Cell* 20 873-882

Bleil JD and Wassarman PM (1983) Sperm-egg interactions in the mouse: sequence of events and induction of acrosome reaction by a zona pellucida glycoprotein *Developmental Biology* 95 317-324

Bleil JD and Wassarman PM (1986) Autoradiographic visualization of the mouse egg's sperm receptor bound to sperm *Journal of Cell Biology* 102 1363-1371

Bleil JD, Greve JM and Wassarman PM (1988) Identification of a secondary sperm receptor in the mouse egg zona pellucida: role in maintenance of binding of acrosome-reacted sperm to eggs *Developmental Biology* 128 376-385

Boja ES, Hoodbhoy T, Fales HM and Dean J (2003) Structural characterization of native mouse zona pellucida proteins using mass spectrometry *Journal of Biological Chemistry* 278 34189-34202

Boja ES, Hoodbhoy T, Garfield M and Fales HM (2005) Structural conservation of mouse and rat zona pellucida glycoproteins. Probing the native rat zona pellucida proteome by mass spectrometry *Biochemistry* 44 16445-16460

Bork P (1993) A trefoil domain in the major rabbit zona pellucida protein *Protein Science* 2 669-670

Bork P and Sander C (1992) A large domain common to sperm receptors (ZP2 and ZP3) and TGF-beta type III receptor *FEBS Letters* 300 237-240

Chakravarty S, Suraj K and Gupta SK (2005) Baculovirus expressed recombinant human zona pellucida glycoprotein-B induces acrosomal exocytosis in capacitated spermatozoa in addition to zona pellucida glycoprotein-C *Molecular Human Reproduction* 11 365-372

Chamberlin ME and Dean J (1990) Human homolog of the mouse sperm receptor *Proceedings of the National Academy of Sciences USA* 87 6014-6018

Chapman NR, Kessopoulou E, Andrews P, Hornby D and Barratt CR (1998) The polypeptide backbone of recombinant human zona pellucida glycoprotein-3 initiates acrosomal exocytosis in human spermatozoa *in vitro* *Biochemical Journal* 330 839-845

Chen J, Litscher ES and Wassarman PM (1998) Inactivation of the mouse sperm receptor, mZP3, by site-directed mutagenesis of individual serine residues located at the combining site for sperm *Proceedings of the National Academy of Sciences USA* 95 6193-6197

Conner SJ, Lefievre L, Hughes DC and Barratt CL (2005) Cracking the egg: increased complexity in the zona pellucida *Human Reproduction* 20 1148-1152

Cross NL, Morales P, Overstreet JW and Hanson FW (1988) Induction of acrosome reactions by the human zona pellucida *Biology of Reproduction* 38 235-244

Dean J (2004) Reassessing the molecular biology of sperm-egg recognition with mouse genetics *Bioessays* 26 29-38

Dunbar BS and Wolgemuth DJ (1984) Structure and function of the mammalian zona pellucida, a unique extracellular matrix *Modern Cell Biology, New York* 3 77-111

El-Mestrah M, Castle PE, Borossa G and Kan FW (2002) Subcellular distribution of ZP1, ZP2, and ZP3 glycoproteins during folliculogenesis and demonstration of their topographical disposition within the zona matrix of mouse ovarian oocytes *Biology of Reproduction* 66 866-876

Epifano O, Liang LF and Dean J (1995) Mouse Zp1 encodes a zona pellucida protein homologous to egg envelope proteins in mammals and fish *Journal of Biological Chemistry* 270 27254-27258

Familiari G, Nottola SA, Micara G, Aragona C and Motta PM (1988) Is the sperm-binding capability of the zona pellucida linked to its surface structure? A scanning electron microscopic study of human *in vitro* fertilization *Journal of In Vitro Fertilization Embryo Transfer* 5 134-143

Familiari G, Nottola SA, Macchiarelli G, Micara G, Aragona C and Motta PM (1992) Human zona pellucida during *in vitro* fertilization: an ultrastructural study using saponin, ruthenium red, and osmium-thiocarbohydrazide *Molecular Reproduction and Development* 32 51-61

Florman HM and Wassarman PM (1985) O-linked oligosaccharides of mouse egg ZP3 account for its sperm receptor activity *Cell* 41 313-324

Florman HM, Bechtol KB and Wassarman PM (1984) Enzymatic dissection of the functions of the mouse egg's receptor for sperm *Developmental Biology* 106 243-255

Franken DR, Henkel R, Kaskar K and Habenicht UF (1996) Defining bioassay conditions to evaluate sperm/zona interaction: inhibition of zona binding mediated by solubilized human zona pellucida *Journal of Assisted Reproduction and Genetics* 13 329-332

Gahlay GK and Gupta SK (2003) Glycosylation of zona pellucida glycoprotein-3 is required for inducing acrosomal exocytosis in the bonnet monkey *Cellular and Molecular Biology (Noisy-le-grand)* 49 389-397

Gahlay GK, Srivastava N, Govind CK and Gupta SK (2002) Primate recombinant zona pellucida proteins expressed in *Escherichia coli* bind to spermatozoa *Journal of Reproductive Immunology* **53** 67-77

Govind CK, Gahlay GK, Choudhury S and Gupta SK (2001) Purified and refolded recombinant bonnet monkey (*Macaca radiata*) zona pellucida glycoprotein-B expressed in *Escherichia coli* binds to spermatozoa *Biology of Reproduction* **64** 1147-1152

Greve JM and Wassarman PM (1985) Mouse egg extracellular coat is a matrix of interconnected filaments possessing a structural repeat *Journal of Molecular Biology* **181** 253-264

Gupta SK, Sharma M, Behera AK, Bisht R and Kaul R (1997) Sequence of complementary deoxyribonucleic acid encoding bonnet monkey (*Macaca radiata*) zona pellucida glycoprotein-ZP1 and its high-level expression in *Escherichia coli Biology of Reproduction* **57** 532-538

Gupta SK, Yurewicz EC, Sacco AG, Kaul R, Jethanandani P and Govind CK (1998) Human zona pellucida glycoproteins: characterization using antibodies against recombinant non-human primate ZP1, ZP2 and ZP3 *Molecular Human Reproduction* **4** 1058-1064

Haines BP, Rathjen PD, Hope RM, Whyatt LM, Holland MK and Breed WG (1999) Isolation and characterization of a cDNA encoding a zona pellucida protein (ZPB) from the marsupial *Trichosurus vulpecula* (brushtail possum) *Molecular Reproduction and Development* **52** 174-182

Harris JD, Hibler DW, Fontenot GK, Hsu KT, Yurewicz EC and Sacco AG (1994) Cloning and characterization of zona pellucida genes and cDNAs from a variety of mammalian species: the ZPA, ZPB and ZPC gene families *DNA Sequence* **4** 361-393

Harris JD, Seid CA, Fontenot GK and Liu HF (1999) Expression and purification of recombinant human zona pellucida proteins *Protein Expression and Purification* **16** 298-307

Hedrick JL (1996) Comparative structural and antigenic properties of zona pellucida glycoproteins *Journal of Reproduction and Fertility Supplement* **50** 9-17

Hedrick JL and Wardrip NJ (1987) On the macromolecular composition of the zona pellucida from porcine oocytes *Developmental Biology* **121** 478-488

Jethanandani P, Santhanam R and Gupta SK (1998) Molecular cloning and expression in *Escherichia coli* of cDNA encoding bonnet monkey (*Macaca radiata*) zona pellucida glycoprotein-ZP2 *Molecular Reproduction and Development* **50** 229-239

Jovine L, Qi H, Williams Z, Litscher E and Wassarman PM (2002) The ZP domain is a conserved module for polymerization of extracellular proteins *Nature Cell Biology* **4** 457-461

Kerr CL, Hanna WF, Shaper JH and Wright WW (2002) Characterization of zona pellucida glycoprotein 3 (ZP3) and ZP2 binding sites on acrosome-intact mouse sperm *Biology of Reproduction* **66** 1585-1595

Kinloch RA, Roller RJ, Fimiani CM, Wassarman DA and Wassarman PM (1988) Primary structure of the mouse sperm receptor polypeptide determined by genomic cloning *Proceedings of the National Academy of Sciences USA* **85** 6409-6413

Kinloch RA, Mortillo S, Stewart CL and Wassarman PM (1991) Embryonal carcinoma cells transfected with ZP3 genes differentially glycosylate similar polypeptides and secrete active mouse sperm receptor *Journal of Cell Biology* **115** 655-664

Kolluri SK, Kaul R, Banerjee K and Gupta SK (1995) Nucleotide sequence of cDNA encoding bonnet monkey (*Macaca radiata*) zona pellucida glycoprotein-ZP3 *Reproduction Fertility and Development* **7** 1209-1212

Lee MA, Check JH and Kopf GS (1992) A guanine nucleotide-binding regulatory protein in human sperm mediates acrosomal exocytosis induced by the human zona pellucida *Molecular Reproduction and Development* **31** 78-86

Lefievre L, Conner SJ, Salpekar A, Olufowobi O, Ashton P, Pavlovic B, Lenton W, Afnan M, Brewis IA, Monk M, Hughes DC and Barratt CL (2004) Four zona pellucida glycoproteins are expressed in the human *Human Reproduction* **19** 1580-1586

Liang LF and Dean J (1993) Conservation of mammalian secondary sperm receptor genes enables the promoter of the human gene to function in mouse oocytes *Developmental Biology* **156** 399-408

Liu C, Litscher ES, Mortillo S, Sakai Y, Kinloch RA, Stewart CL and Wassarman PM (1996) Targeted disruption of the mZP3 gene results in production of eggs lacking a zona pellucida and infertility in female mice *Proceedings of the National Academy of Sciences USA* **93** 5431-5436

Mortillo S and Wassarman PM (1991) Differential binding of gold-labeled zona pellucida glycoproteins mZP2 and mZP3 to mouse sperm membrane compartments *Development* **113** 141-149

Oehninger S (2003) Biochemical and functional characterization of the human zona pellucida *Reproductive Biomedicine Online* **7** 641-648

Prasad SV, Wilkins B, Skinner SM and Dunbar BS (1996) Evaluating zona pellucida structure and function using antibodies to rabbit 55 kDa ZP protein expressed in baculovirus expression system *Molecular Reproduction and Development* **43** 519-529

Qi H, Williams Z and Wassarman PM (2002) Secretion and assembly of zona pellucida glycoproteins by growing mouse oocytes microinjected with epitope-tagged cDNAs for mZP2 and mZP3 *Molecular Biology of the Cell* **13** 530-541

Rankin T, Familari M, Lee E, Ginsberg A, Dwyer N, Blanchette-Mackie J, Drago J, Westphal H and Dean J (1996) Mice homozygous for an insertional mutation in the Zp3 gene lack a zona pellucida and are infertile *Development* **122** 2903-2910

Rankin TL, Tong ZB, Castle PE, Lee E, Gore-Langton R, Nelson LM and Dean J (1998) Human ZP3 restores fertility in Zp3 null mice without affecting order-specific sperm binding *Development* **125** 2415-2424

Rankin T, Talbot P, Lee E and Dean J (1999) Abnormal zonae pellucidae in mice lacking ZP1 result in early embryonic loss *Development* **126** 3847-3855

Rankin TL, O'Brien M, Lee E, Wigglesworth K, Eppig J

and Dean J (2001) Defective zonae pellucidae in Zp2-null mice disrupt folliculogenesis, fertility and development *Development* **128** 1119-1126

Roch F, Alonso CR and Akam M (2003) Drosophila miniature and dusky encode ZP proteins required for cytoskeletal reorganization during wing morphogenesis *Journal of Cell Science* **116** 1199-1207

Shabanowitz RB and O'Rand MG (1988) Characterization of the human zona pellucida from fertilized and unfertilized eggs *Journal of Reproduction and Fertility* **82** 151-161

Spargo SC and Hope RM (2003) Evolution and nomenclature of the zona pellucida gene family *Biology of Reproduction* **68** 358-362

Takeuchi Y, Nishimura K, Aoki N, Adachi T, Sato C, Kitajima K and Matsuda T (1999) A 42-kDa glycoprotein from chicken egg-envelope, an avian homolog of the ZPC family glycoproteins in mammalian zona pellucida. Its first identification, cDNA cloning and granulosa cell-specific expression *European Journal of Biochemistry* **260** 736-742

Taya T, Yamasaki N, Tsubamoto H, Hasegawa A and Koyama K (1995) Cloning of a cDNA coding for porcine zona pellucida glycoprotein ZP1 and its genomic organization *Biochemical and Biophysical Research Communications* **207** 790-799

Thall AD, Maly P and Lowe JB (1995) Oocyte Gal alpha 1,3Gal epitopes implicated in sperm adhesion to the zona pellucida glycoprotein ZP3 are not required for fertilization in the mouse *Journal of Biological Chemistry* **270** 21437-21440

Tsubamoto H, Hasegawa A, Nakata Y, Naito S, Yamasaki N and Koyama K (1999) Expression of recombinant human zona pellucida protein 2 and its binding capacity to spermatozoa *Biology of Reproduction* **61** 1649-1654

van Duin M, Polman J, De Breet IT, van Ginneken K, Bunschoten H, Grootenhuis A, Brindle J and Aitken RJ (1994) Recombinant human zona pellucida protein ZP3 produced by Chinese hamster ovary cells induces the human sperm acrosome reaction and promotes sperm-egg interaction *Biology of Reproduction* **51** 607-617

Wassarman PM (1987) Early events in mammalian fertilization *Annual Review of Cell Biology* **3** 109-142

Wassarman PM (1999) Mammalian fertilization: molecular aspects of gamete adhesion, exocytosis, and fusion *Cell* **96** 175-183

Wassarman PM and Litscher ES (1995) Sperm-egg recognition mechanisms in mammals *Current Topics in Developmental Biology* **30** 1-19

Wassarman PM, Jovine L and Litscher ES (2001) A profile of fertilization in mammals *Nature Cell Biology* **3** E59-64

Williams Z and Wassarman PM (2001) Secretion of mouse ZP3, the sperm receptor, requires cleavage of its polypeptide at a consensus furin cleavage-site *Biochemistry* **40** 929-937

Williams Z, Litscher ES, Jovine L and Wassarman PM (2006) Polypeptide encoded by mouse ZP3 exon-7 is necessary and sufficient for binding of mouse sperm in vitro *Journal of Cellular Physiology* **207** 30-39

Yamasaki N, Richardson RT and O'Rand MG (1995) Expression of the rabbit sperm protein Sp17 in COS cells and interaction of recombinant Sp17 with the rabbit zona pellucida *Molecular Reproduction and Development* **40** 48-55

Yurewicz EC, Sacco AG and Subramanian MG (1987) Structural characterization of the Mr = 55,000 antigen (ZP3) of porcine oocyte zona pellucida. Purification and characterization of alpha- and beta-glycoproteins following digestion of lactosaminoglycan with endo-beta-galactosidase *Journal of Biological Chemistry* **262** 564-571

Yurewicz EC, Pack BA, Armant DR and Sacco AG (1993) Porcine zona pellucida ZP3 alpha glycoprotein mediates binding of the biotin-labeled M(r) 55,000 family (ZP3) to boar sperm membrane vesicles *Molecular Reproduction and Development* **36** 382-389

Yurewicz EC, Sacco AG, Gupta SK, Xu N and Gage DA (1998) Hetero-oligomerization-dependent binding of pig oocyte zona pellucida glycoproteins ZPB and ZPC to boar sperm membrane vesicles *Journal of Biological Chemistry* **273** 7488-7494

Zhao M, Gold L, Ginsberg AM, Liang LF and Dean J (2002) Conserved furin cleavage site not essential for secretion and integration of ZP3 into the extracellular egg coat of transgenic mice *Molecular and Cellular Biology* **22** 3111-3120

# Structural significance of *N*-glycans of the zona pellucida on species-selective recognition of spermatozoa between pig and cattle

N Yonezawa[1,2], S Kanai[1] and M Nakano[1,2]

[1]*Graduate School of Science and Technology, and* [2]*Department of Chemistry, Faculty of Science, Chiba University, 1-33, Yayoi-cho, Inage-ku, Chiba 263-8522, Japan*

The zona pellucida that surrounds the mammalian oocyte plays a role in species-selective sperm-egg interactions. In the pig and cattle, the zona pellucida consists of ZPA, ZPB and ZPC. Sperm binding activity of porcine zona glycoproteins is conferred by tri- and tetra-antennary complex-type chains and in cattle it is conferred by a high-mannose-type chain of five mannose residues. Non-reducing terminal residues of these *N*-linked chains, ß-galactosyl residues in pig and α-mannosyl residues in cattle, are involved in the binding of zona glycoproteins to respective spermatozoa. The major *N*-linked chains of recombinant porcine ZPB expressed using the baculovirus-Sf9 cell expression system are pauci- and high-mannose-type chains that are different in structure to the major neutral *N*-linked chains of the porcine zona but similar to those of the bovine zona. The mixture of porcine ZPB/ZPC co-expressed in Sf9 cells binds to bovine sperm but not to porcine sperm, indicating an essential role of the *N*-linked chains in species-selective recognition of sperm in pig and cattle. Asn to Asp mutations at either of two of the *N*-glycosylation sites of ZPB, residue 203 or 220, significantly reduce the sperm-binding activity of the ZPB/ZPC mixture, while a similar mutation at Asn333 has no effect on binding. These results coincide with our previous report that tri- and tetra-antennary complex-type chains are localized at Asn220 in native porcine ZPB and suggest that the *N*-glycans located in the N-terminal half of the ZP domain of porcine ZPB are involved in sperm-zona binding.

## Introduction

The zona pellucida, a transparent envelope of the mammalian oocyte, plays a role in species-selective sperm-egg interactions (Wassarman and Litscher, 2001). Porcine and bovine zona glycoprotein components are called ZPA, ZPB and ZPC in the order of the size of the cDNAs that encode the polypeptides (Table 1; Harris *et al.*, 1994). In pig, the components are also described as ZP1, ZP3α and ZP3ß, respectively (Yurewicz *et al.*, 1987). Mouse zona pellucida consists of ZP1, ZP2 and ZP3 in the order of the apparent molecular mass of the proteins on SDS-polyacrylamide gel electrophoresis under non-reducing conditions (Wassarman, 1988). In human, it was found that ZP1 and ZPB are distinct gene products and the zona pellucida consists of four glycoproteins ZP1, ZP2, ZP3 and ZP4 (Lefievre *et al.*, 2004). According to this nomenclature, porcine and

---

E-mail: nyoneza@faculty.chiba-u.jp

bovine ZPA, ZPB and ZPC are described as ZP2, ZP4 and ZP3, respectively (Table 1). According to the recently proposed nomenclature, ZP1, ZP2, ZP3 and ZP4 are also described as ZPB1, ZPA, ZPC and ZPB2, respectively (Spargo and Hope, 2003). We describe porcine and bovine zona glycoproteins as ZPA, ZPB and ZPC in this manuscript.

**Table 1.** Zona pellucida components of pig, cattle and mouse

| Species | Nomenclatures used in this paper (apparent molecular mass) | | |
|---|---|---|---|
| Pig | ZPA (90 kDa) | ZPB (55 kDa) | ZPC (55 kDa) |
| Cattle | ZPA (78 kDa) | ZPB (78 kDa) | ZPC (78 kDa) |
| Mouse | ZP2 (120 kDa) | ZP1 (200 kDa) | ZP3 (83 kDa) |
| | Recently proposed nomenclatures | | |
| | ZP2 | ZP1 | ZP4 | ZP3 |
| | ZPA | ZPB1 | ZPB2 | ZPC |

Carbohydrate contents differ between respective components from several species, while considerable homology is observed between the amino acid sequences of respective components. Particularly, most of the positions of Cys residues are almost the same between species, suggesting that the higher-order structures of the zona proteins are fairly conserved during evolution (Jovine et al., 2005). The zona glycoproteins are synthesized as transmembrane proteins. The putative transmembrane region is cleaved by furin or furin-like protease to yield the C-terminus of the mature protein (Fig. 1).

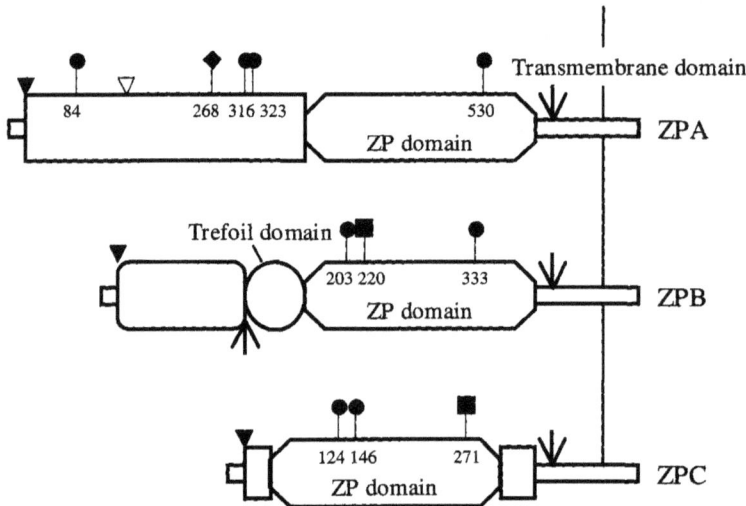

**Fig. 1** Schematic representation of three main components of porcine zona glycoproteins. Porcine and bovine zona components have the same domain structures except that mature porcine ZPB lacks N-terminal region by a putative processing at the site indicated by an upward arrow. The cleavage sites of signal peptides are indicated by inverted closed triangles. Proproteins are processed at the site indicated by downward arrows by furin or furin-like enzymes followed by carboxypeptidases. ZPA is further processed upon fertilization by an unidentified enzyme at the site indicated by an inverted open triangle. N-glycosylation sites are indicated by closed circles, closed squares and closed diamonds that are linked to polypeptides by bars. High-mannose-type chains are localized at the site of ZPA indicated by a closed diamond. Tri- and tetra-antennary complex-type chains are localized at the sites of ZPB and ZPC indicated by closed squares. Di-antennary chains are predominant at all the sites.

The non-reducing terminal residues of carbohydrate chains of mouse ZP3, such as α-Gal (Bleil and Wassarman, 1988), ß-GlcNAc (Miller *et al.*, 1992), α-Fuc in the context of Lewis[x] (Johnston *et al.*, 1998), α-Man (Cornwall *et al.*, 1991) and ß-Gal (Mori *et al.*, 1997), are thought to mediate mouse sperm binding. O-linked chains on mouse ZP3 have been proposed to be sperm ligand carbohydrate chains (Florman and Wassarman, 1985). Subsequently, the sperm ligand O-linked chains are shown to be linked to Ser332 and Ser334 (Chen *et al.*, 1998). Nevertheless, a recent structural analysis using mass spectrometry does not show evidence for the glycosylation at Ser332 and Ser334 (Boja *et al.*, 2003). Recent structural analysis of mouse zona glycoproteins has shown that both O- and N-linked chains lack Lewis[x] type structures (Easton *et al.*, 2000). The *in vivo* studies so far performed using transgenic mice lacking each glycosyltransferase gene do not support these proposed sperm-binding sites on the mouse zona pellucida (Thall *et al.*, 1995; Lu and Shur, 1997; Shi *et al.*, 2004). Alternatively, it is proposed that mouse sperm recognise the supramolecular structure of the zona matrix, but not the carbo-hydrate structures, based on the data obtained from the mice rescued using human counterpart of mouse ZP2 (Rankin *et al.*, 2003; Hoodbhoy and Dean, 2004). Therefore, the essential role of the carbohydrate moiety in mouse sperm-egg binding is now being debated (Dell *et al.*, 2003; Hoodbhoy and Dean, 2004). Recent studies do not completely deny the *in vitro* studies on the involvement of carbohydrate chains in mouse sperm-egg binding. Given that *in vitro* experi-ments examining mouse sperm-egg binding suggest the involvement of multiple receptors on spermatozoa or ligands on the zona pellucida (Thaler and Cardullo, 1996), the redundant sperm-binding sites might compensate for the lack of some of the sperm-binding sugar residues, therefore, the transgenic mice may be fertile.

The *in vitro* studies on the involvement of carbohydrate chains of porcine and bovine zona glycoproteins in sperm-egg binding have been performed according to the mouse model. One of our present concerns is whether or not carbohydrate chains are essential for sperm-egg bind-ing in pig and cattle. We would like to discuss this subject reviewing our studies.

## Glycoproteins of porcine and bovine zonae pellucidae

The porcine zona pellucida is about 16 μm in width and contains about 30 ng of glycoproteins. The bovine zona pellucida is of similar size to porcine zona and contains about 20 ng of glycoproteins. The yield of zona glycoprotein mixture from one ovary is ten times higher in pig than in cattle (Noguchi *et al.*, 1994). Mature porcine ZPB lacks a N-terminal region probably because of processing with an unidentified enzyme, while the mature bovine ZPB has the N-terminal region (Fig. 1). The C-termini of bovine zona glycoproteins and porcine ZPA have not been determined yet. The C-termini of porcine ZPB and ZPC are Ala462 and Ser332, respec-tively (Yonezawa and Nakano, 2003). The molar ratio of mouse ZP1:ZP2:ZP3 is estimated as 1:4:4 (Epifano *et al.*, 1995), which is similar to that of porcine zona pellucida but quite different from that of bovine zona pellucida (Table 2). In mouse, ZP2 and ZP3 form filamentous equimo-lar complexes and ZP1 cross-links the filaments (Greve and Wassarman, 1985). It is supposed that in pig, ZPB and ZPC form equimolar complexes and ZPA crosslinks the complexes. Bovine zona architecture seems to be quite different from that of porcine zona pellucida.

## Structures of carbohydrate chains from porcine zona glycoproteins

The structures of N-linked carbohydrate chains from porcine zona protein mixture and from porcine ZPB/ZPC mixture have been determined for the first time in mammals (Mori *et al.*, 1991; Noguchi *et al.*, 1992; Noguchi and Nakano, 1992; Nakano *et al.*, 1996). The N-linked

**Table 2.** Molecular masses, residue numbers and stoichiometries of the protein moieties of porcine and bovine zona glycoproteins

|  | ZPA | | Pig ZPB | | ZPC |
|---|---|---|---|---|---|
| Molecular mass (Da) | 67,700 | | 36,024 | | 34,356 |
| Residue number | 605[a] | | 326 | | 310 |
| Weight ratio | 1 | : | 3 | : | 3 |
| Molar ratio | 1 | : | 5.6 | : | 5.9 |
|  | ZPA | | Bovine ZPB | | ZPC |
| Molecular mass (Da) | 67,578 | | 48,617 | | 35,450 |
| Residue number | 602[a] | | 440[a] | | 317[a] |
| Weight ratio | 1.21 | : | 1 | : | 1.45 |
| Molar ratio | 1 | : | 1.2 | : | 2.3 |

All Cys residues are assumed to form disulfide linkages.
[a]C-termini of these components are assumed to be furin cleavage site.

chains comprise neutral and acidic chains in the molar ratio of about 1:3. Both neutral chains and acidic chains comprise di-, tri- and tetra-antennary complex-type chains with a Fuc residue at the innermost GlcNAc. The structures of di-antennary acidic chains have been analyzed in detail (Noguchi and Nakano, 1992). These chains are classified into four groups. The first group represents the sialylated chains without the sulfated N-acetyllactosamine repeats, whereas the other three groups have chains of various lengths that are different in the number of monosulfated N-acetyllactosamine unit. Fucosylated N-acetyllactosamine also exists in the non-reducing portion of acidic chains (Mori et al., 1998).

The structures of eight neutral O-linked chains (Hirano et al., 1993) and 26 acidic O-linked chains have been determined (Hokke et al., 1993; 1994). Major structures are core 1 type with N-acetyllactosamine repeats without branches. Acidic chains are different in the number of monosulfated N-acetyllactosamine units as observed in acidic N-linked chains.

## Involvement of carbohydrate chains from porcine zona pellucida in sperm-egg binding

It has been reported that O-linked chains obtained from porcine zona glycoprotein mixture by alkaline-borohydrate treatment inhibit sperm-egg binding, whereas N-linked chain mixture obtained by digestion with N-glycanase does not (Yurewicz et al., 1991). Thereafter, we reported that N-linked chains obtained from porcine ZPB/ZPC mixture inhibit sperm-egg binding and the neutral chain mixture has the inhibitory activity but the acidic chain mixture does not (Noguchi et al., 1992). Thus, there are conflicting reports on the sperm-binding activity of carbohydrate chains from porcine zona pellucida.

Tri- and tetra-antennary neutral chains (Fig. 2) show an inhibitory activity stronger than that of di-antennary neutral chains (Kudo et al., 1998). Most of these oligosaccharides have ß-galactosyl residues at the non-reducing ends. The removal of the non-reducing terminal ß-galactosyl residues from either the tri- and tetra-antennary chains or endo-ß-galactosidase-digested glycoproteins significantly reduces their inhibition of sperm-egg binding, indicating that the ß-galactosyl residues at the non-reducing ends are involved in porcine sperm-egg binding (Yonezawa et al., 2005a).

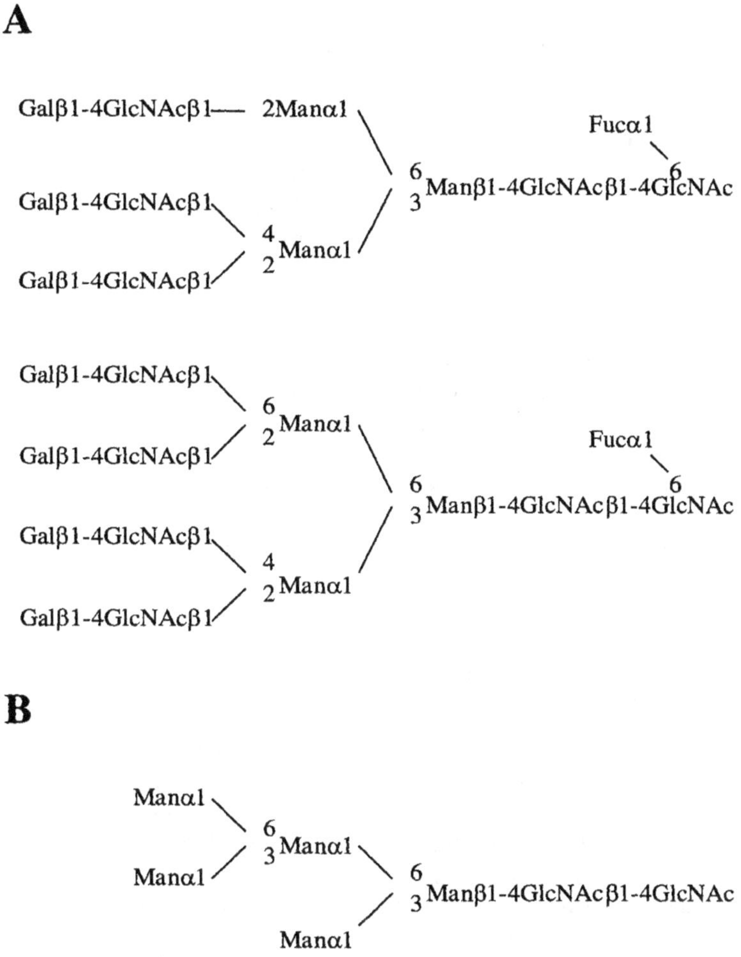

## A

```
Galβ1-4GlcNAcβ1——2Manα1 \
                            \
                             6
                             3 Manβ1-4GlcNAcβ1-4GlcNAc
Galβ1-4GlcNAcβ1 \          /        Fucα1
                 4        /             \
                 2 Manα1 /               6
Galβ1-4GlcNAcβ1 /
```

```
Galβ1-4GlcNAcβ1 \
                 6
                 2 Manα1 \
Galβ1-4GlcNAcβ1 /         \
                           6              Fucα1
                           3 Manβ1-4GlcNAcβ1-4GlcNAc
Galβ1-4GlcNAcβ1 \         /                  \
                 4       /                     6
                 2 Manα1
Galβ1-4GlcNAcβ1 /
```

## B

```
Manα1 \
       6
       3 Manα1 \
Manα1 /         \
                 6
                 3 Manβ1-4GlcNAcβ1-4GlcNAc
Manα1 /
```

**Fig 2.** *N*-linked neutral carbohydrate chains active in sperm binding. (A) In pig, the tri- and tetra-antennary complex-type chains, in which Fuc is attached to the innermost GlcNAc residue, have sperm-binding activity. (B) In cattle, the high-mannose-type chain that possesses five Man residues has sperm-binding activity.

### Identification of sperm binding site in porcine ZPB

ZPB and ZPC can be separated from each other by reverse-phase HPLC after removal of heterogenous *N*-acetyllactosamine repeats in the non-reducing terminal region of carbohydrate chains by digestion with endo-ß-galactosidase (Yurewicz *et al.*, 1987; Yonezawa *et al.*, 1995a). The digested ZPB retains inhibitory activity for sperm-egg binding similar to that of solubilized zona glycoprotein mixture in a competition assay, whereas ZPC does not (Berger *et al.*, 1989; Sacco *et al.*, 1989; Yurewicz *et al.*, 1993; Yonezawa *et al.*, 1995b). The *N*-linked chains of endo-ß-galactosidase-digested ZPB do not necessarily have the original structures. However, a large part of the neutral *N*-linked chains of the ZPB/ZPC mixture do not have *N*-acetyllactosamine repeats and therefore are not susceptible to endo-ß-galactosidase (Noguchi *et al.*, 1992). The elimination of N-linked chains from endo-ß-galactosidase-digested ZPB by digestion with N-glycanase markedly reduces its inhibitory effect on sperm-zona binding, whereas the elimination of O-linked chains by alkali treatment barely reduces the inhibitory effect (Yonezawa et al., 1995b).

The N-terminal fragment of ZPB (residues 137-247), which contains two N-linked chains at Asn203 and Asn220 (see Fig. 1), significantly inhibits sperm-egg binding. However, the fragment (residues 325-341) that has one N-linked chain at Asn333 and the fragment (residues 248-324) that has two or three O-linked chains do not inhibit sperm-egg binding (Yonezawa et al., 1997). In conclusion, these studies show that N-linked chains located at the N-terminal region of ZPB are involved in porcine sperm-egg binding.

Porcine ZPB purified by reverse-phase HPLC is actually contaminated with a trace amount of ZPC and heterocomplex formation of ZPB and ZPC is essential for sperm-binding activity of the glycoproteins (Yurewicz et al., 1998). Neither pure ZPB nor pure ZPC shows sperm-binding activity. Since the complex shows sperm-binding activity in the presence of a trace amount of ZPC, ZPB might be predominantly involved in the activity: but it is not yet clarified whether one of ZPB and ZPC or both in the heterocomplex directly bind to sperm.

It is possible that the N-terminal fragment of ZPB in our studies contains ZPC fragment and therefore shows sperm-binding activity. Thus, it is necessary to re-examine the sperm-binding site in the ZPB/ZPC complex taking the influences of enzymatic digestion on the heterocomplex formation into consideration.

### Localization of N-linked carbohydrate chains in porcine ZPA, ZPB and ZPC

The glycopeptide (residues 197-211) containing Asn203 and the glycopeptide (residues 217-236 and 241-247) containing Asn220 are separated after digestion of the N-terminal fragment (residues 137-247) of porcine ZPB with chymotrypsin. Structural analysis of the neutral fraction released from the glycopeptides indicates that tri- and tetra-antennary chains are localized at Asn220 (Fig. 1: Kudo et al., 1998). ZPC also has three N-linked chains at Asn124, Asn146 and Asn271 (Fig. 1). Tri- and tetra-antennary neutral chains are localized at Asn271 (Yonezawa et al., 1999). Thus, tri- and tetra-antennary chains are localized in the N-terminal half of the ZP domain in ZPB, while in ZPC the chains are localized in the C-terminal half of the ZP domain (Nakano and Yonezawa, 2001). We speculate that the localization of tri- and tetra-antennary chains in the ZP domain is related to sperm-binding activity. ZPA has six potential N-glycosylation sites. Asn268, Asn316, Asn323 and Asn530 are glycosylated and the proteolytic peptide containing Asn84 and Asn93 has only one N-linked chain (von Witzendorff et al., 2005). Asn84 is not detected by Edman degradation but Asn93 is detected (Hasegawa et al., 1994), suggesting that Asn84 is glycosylated but Asn93 is not (Fig. 1). Similar to ZPB and ZPC, di-antennary complex-type chains with a Fuc are predominant in ZPA. The ZP domain of ZPA has only one N-glycosylation site at Asn530 and tri- and tetra-antennary chains may be localized outside of the ZP domain such as Asn84 and Asn316, although localization of tri- and tetra-antennary chains is not fully determined. A remarkable difference in the structures of N-linked chains between ZPA and ZPB/ZPC is that ZPA has high-mannose-type chain containing five Man residues probably located at Asn268 (von Witzendorff et al., 2005). High- mannose-type chains are not detected in ZPB/ZPC mixture (Noguchi et al., 1992).

### Structures of carbohydrate chains from bovine zona glycoproteins

Only the structures of N-linked chains are reported (Katsumata et al., 1996). The N-linked chains from bovine ovarian egg zona are composed of neutral (23%) and acidic (77%) chains. Almost all the acidic chains are neutralized by sialidase digestion, indicating that sialylation is predominant in bovine zona pellucida, while in porcine zona pellucida sulfation of N-acetyllactosamine unit is more dominant (Noguchi and Nakano, 1992). The major neutral chain consists of one structure, high-mannose-type chain, $Man_5GlcNAc_2$ (Fig. 2) and the acidic chains are di-, tri- and tetra-antennary, fucosylated complex-type chains that have N-acetyllactosamine repeats in the non-reducing regions.

## Involvement of carbohydrate chains from bovine zona pellucida in sperm-egg binding

Neutral N-linked chains released from bovine zona glycoproteins and $Man_5GlcNAc_2$ commercially purchased inhibit sperm-egg binding and *in vitro* fertilization (Amari et al., 2001). When non-reducing terminal α-mannosyl residues are eliminated from the bovine zona glycoproteins by α-mannosidase digestion, the inhibitory activity for sperm-egg binding is reduced. These indicate that the α-mannosyl residues play an essential role in bovine sperm-egg binding.

N-glycosylation sites of bovine ZPB and ZPC are not determined yet. N-glycosylation sites of bovine ZPA are Asn83, Asn191 and Asn527 (Ikeda et al., 2002). Neutral chain, $Man_5GlcNAc_2$, is found at Asn83 and Asn191, but there is very little of this active chain at Asn527 in the ZP domain of ZPA.

The number of sperm binding to fertilized eggs is about one-half of that of sperm binding to ovarian eggs. The amount of $Man_5GlcNAc_2$ remains unchanged after *in vitro* fertilization, however, the amount of the acidic chains decreases to 32 mol/100 mol in the fertilized egg zonae (Katsumata et al., 1996). Endo-ß-galactosidase digestion of bovine zona mixture does not reduce the inhibitory activity for sperm-egg binding, suggesting that the acidic chains possessing N-acetyllactosamine repeats are not involved in sperm-egg binding (Yonezawa et al., 2001). Thus, it is not yet elucidated why the sperm binding activity of bovine zona pellucida is reduced during fertilization.

## Sperm binding active components of bovine zona pellucida

ZPB has the highest sperm binding activity among bovine zona glycoproteins, while ZPA and ZPC also have weak but significant sperm-binding activities (Yonezawa et al., 2001). ZPB is not completely purified by reverse-phase HPLC after digestion with endo-ß-galactosidase and the ZPB fraction contains ZPC. Since porcine ZPB without a trace amount of ZPC does not show sperm-binding activity, it is probable that in cattle also heterocomplexes of ZPB and ZPC show sperm-binding activity but ZPB or ZPC alone does not.

## Are carbohydrate moieties of porcine and bovine zona glycoproteins significant in sperm-egg binding?

Since the carbohydrate chains released from zona glycoproteins show only weak inhibitory activity for sperm-egg binding compared to the activities of native zona glycoproteins in pig and cattle, it is necessary to examine the effects of modification of carbohydrate moieties of zona glycoproteins by glycosidase digestion on the sperm-zona binding as mentioned above. Recently, we established the expression system of recombinant porcine zona glycoproteins using baculovirus-Sf9 cells (Yonezawa et al., 2005b). Carbohydrate structures of the recombinants are expected to be quite different from those of native porcine zona glycoproteins. The major structures of N-linked chains of recombinant ZPB (rZPB) are estimated by MALDI-TOF MS analysis and are confirmed using several lectins. Since rZPA and rZPC show the same patterns of lectin staining as rZPB, the structures of N-linked chains of recombinant zona glycoproteins may be roughly the same, that is, pauci- and high-mannose-type chains with or without a Fuc at the innermost GlcNAc. Fortunately, these are similar in structure to the major neutral N-linked chain of native bovine zona glycoproteins. The rZPB and rZPC can be co-expressed by infecting Sf9 cells with the mixture of respective recombinant baculoviruses. Beads coated with rZPB/rZPC bind bovine sperm but not porcine sperm. The rZPB/rZPC mixture does not significantly inhibit the binding of porcine sperm to plastic wells coated with porcine zona mixture. In contrast, rZPB/rZPC significantly inhibits the binding of bovine sperm to bovine zona. Digestion of rZPB/rZPC with α-mannosidase almost abolishes its inhibitory activity for bovine sperm-zona binding, demonstrating the essential role of non-reducing terminal α-mannosyl residues of rZPB/rZPC in its sperm-binding activity. The rZPB/rZPC mixture as well as solubilized, native bovine zona bind to the acrosomal region of bovine spermatozoa as shown by indirect immunofluorescence detection of zona glycoproteins (Fig. 3). These three different assays show consistent results that porcine rZPB/rZPC bind to bovine sperm but not to porcine sperm. These results strongly suggest that the carbohydrate moieties of the zona glycoproteins are essential for the species-selective recognition of bovine and porcine spermatozoa.

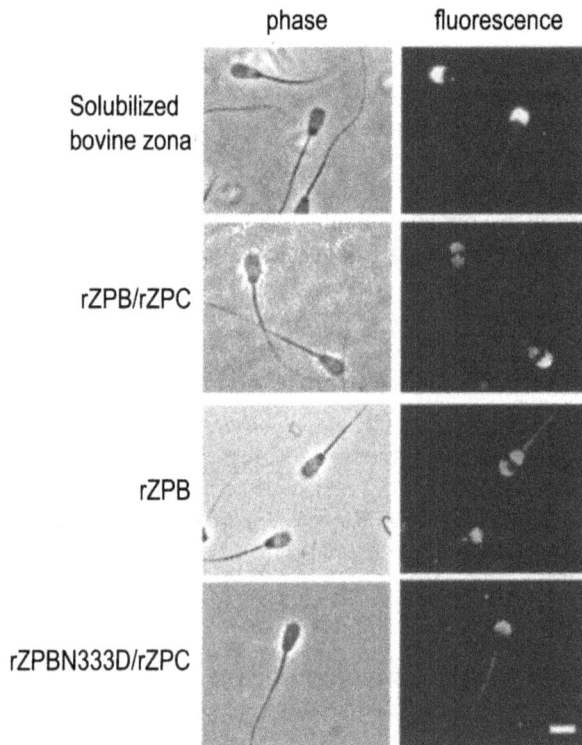

**Fig. 3** Binding of recombinant zona glycoproteins to spermatozoa examined by indirect immunofluorescence staining. Suspensions of bovine sperm were incubated with solubilized bovine zona, a mixture of recombinant porcine ZPB and ZPC (rZPB/rZPC), rZPB or a mixture of rZPB in which Asn333 is mutated to Asp and rZPC (rZPBN333D/rZPC). The proteins that bound to spermatozoa were detected using a cocktail of anti-porcine ZPA, anti-porcine ZPB and anti-porcine ZPC antibodies as the primary antibodies, and fluorescein-conjugated goat anti-rabbit IgG antibody as the secondary antibody. The sperm were observed using fluorescence microscopy. Scale bar represent 10 μm. Phase, phase-contrast image; fluorescence, fluorescence image.

The rZPB shows a sperm-binding activity much weaker than that of rZPB/rZPC in the first two assays. Binding of rZPB to the postacrosomal region of bovine spermatozoa is much stronger than that of solubilized bovine zona (Fig. 3). Therefore, nonphysiological binding of rZPB might increase in the absence of rZPC. We have not yet examined whether or not rZPB and rZPC interact with each other when they are co-expressed in Sf9 cells. However, it is strongly suggested that rZPB and rZPC form heterocomplexes in a similar manner to native ZPB and ZPC, since only rZPB/rZPC shows physiologically significant sperm-binding activity.

Based on the studies on mouse sperm-egg binding, the hypothesis has been established that sperm-zona recognition is mediated by specific interactions between certain carbohydrate chains of the zona and corresponding complementary carbohydrate-binding proteins on the sperm plasma membranes (Wassarman and Litscher, 2001). According to this hypothesis, the neutral N-linked chains of zona glycoproteins which are different in structure between pig and cattle are good candidates for species-specific sperm ligand. Actually, ß-galactosyl residues at the neutral complex-type chain and α-mannosyl residues at the non-reducing ends of the high-mannose-type chain are involved in porcine and bovine sperm-egg binding, respectively. However, binding of sperm to the intact zona is not species-specific but at most species-selective between pig and cattle. Porcine sperm can tightly bind to bovine eggs (Sinowatz et al., 2003). The number of bovine sperm which bind to one immature porcine egg is about half of that to one immature bovine egg (Takahashi et al., unpublished). The involvement of the acidic N-linked chains and O-linked chains in sperm binding is not yet investigated well both in pig and cattle. Since the binding of mouse spermatozoa to zona is multivalent and complex, the interaction between the non-reducing terminal residues of neutral chains of porcine and bovine zonae and respective spermatozoa may explain only a part of the sperm-zona recognition mechanisms.

### Re-examination of the involvement of N-linked chains of porcine ZPB in sperm binding using rZPB mutants

Three separate Asn to Asp site-directed mutations in the polypeptide sequence of porcine ZPB eliminate N-linked glycosylation at specific sites on rZPB (Yonezawa et al., 2005b). These mutations, which are located at Asn203, Asn222 and Asn333 of ZPB, result in mutant rZPBs, referred to as rZPBN203D, rZPBN220D and rZPBN333D, respectively.

Each of the rZPB mutants is co-expressed with rZPC in Sf9 cells. The rZPBN333D/rZPC mixture inhibits the binding of bovine spermatozoa to bovine zona as well as solubilized bovine zona. The rZPBN333D/rZPC mixture binds to the acrosomal region of bovine spermatozoa (Fig. 3). On the other hand, rZPBN220D/rZPC does not inhibit, significantly, the sperm-zona binding. The rZPBN203D/rZPC mixture inhibits the sperm-zona binding significantly but the inhibitory effect is much weaker than that of rZPB/rZPC. These results coincide with our previous report that the fragment (residues 325-341) of porcine ZPB that is N-glycosylated at Asn333 does not have sperm-binding activity (Yonezawa et al., 1997). rZPBN333D does not show sperm-binding activity by itself but shows the activity by co-expression with rZPC, indicating the N-glycans linked to Asn333 are not necessary for heterocomplex formation and sperm binding of rZPB/rZPC. These results also coincide with our previous results that the N-terminal fragment containing two N-linked chains at Asn203 and Asn220 exhibits sperm-binding activity (Yonezawa et al., 1997) and the tri- and tetra-antennary chains are localized at Asn220 of porcine ZPB (Kudo et al., 1998). Taken together, these results suggest that Asn220 is the most important of the three N-glycosylation sites of ZPB for the sperm-binding activity of ZPB/ZPC. However, we cannot rule out the possibility that the loss of sperm-binding activity caused by the Asn to Asp mutation in rZPB is caused by a change in tertiary structure rather than a loss of glycosylation. The effects of the amino acid substitutions on the interaction between rZPB and rZPC are important issues that will be addressed in future experiments.

## Conclusion

Our recent studies indicate that bovine spermatozoa recognise α-mannosyl residues at the non-reducing ends of the carbohydrate chains linked to the recombinant porcine zona polypeptides as well as those linked to the native bovine zona polypeptides and that porcine sperm do not recognise α-mannosyl residues on the recombinant porcine zona polypeptides *in vitro* (Fig. 4). In pig and cattle, ZPB and ZPC form a framework on which the carbohydrate chains active in sperm binding can exhibit high avidity for spermatozoa and the specificity of sperm binding may be based on the structures of carbohydrate chains. However, the candidate sperm proteins that bind specifically to α-mannosyl residues in cattle and to ß-galactosyl residues in pig have not yet been identified. In order to prove the significance of the carbohydrate chains in sperm binding, the sperm proteins need to be found and characterized as already have been tried in mouse.

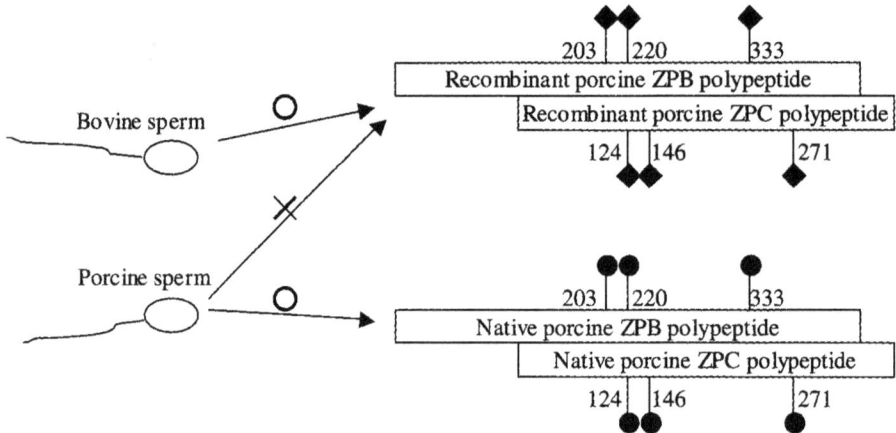

**Fig. 4** Schematic representation of the specificity in recognition between sperm and zona glycoproteins. Bovine sperm bind to the heterocomplex of recombinant porcine ZPB and ZPC that have pauci- and high-mannose-type chains (closed diamonds with bars). Porcine sperm bind to the heterocomplex of native porcine ZPB and ZPC that have complex-type chains (closed circles with bars) but not to the heterocomplex of recombinant ZPB and ZPC.

## References

Amari S, Yonezawa N, Mitsui S, Katsumata T, Hamano S, Kuwayama M, Hashimoto Y, Suzuki A, Takeda Y and Nakano M (2001) Essential role of the nonreducing terminal α-mannosyl residues of the N-linked carbohydrate chain of bovine zona pellucida glycoproteins in sperm-egg binding *Molecular Reproduction and Development* **59** 221-226

Berger T, Davis A, Wardrip NJ and Hedrick JL (1989) Sperm binding to the pig zona pellucida and inhibition of binding by solubilized components of the zona pellucida *Journal of Reproduction and Fertility* **86** 559-565

Bleil JD and Wassarman PM (1988) Galactose at the nonreducing terminus of O-linked oligosaccharides of mouse egg zona pellucida glycoprotein ZP3 is essential for the glycoprotein's sperm receptor activity *Proceedings of the National Academy of Sciences USA* **85** 6778-6782

Boja ES, Hoodbhoy T, Fales HM and Dean J (2003) Structural characterization of native mouse zona pellucida proteins using mass spectrometry *Journal of Biological Chemistry* **278** 34189-34202

Chen J, Litscher ES and Wassarman PM (1998) Inactivation of the mouse sperm receptor, mZP3, by site-directed mutagenesis of individual serine residues located at the combining site for sperm *Proceedings of the National Academy of Sciences USA* **95** 6193-6197

Cornwall GA, Tulsiani DRP and Orgebin-Crist MC (1991) Inhibition of the mouse sperm surface α-D-mannosidase inhibits sperm-egg binding in vitro *Biology of Reproduction* **44** 913-921

Dell A, Chalabi S, Easton RL, Haslam SM, Sutton-Smith M, Patankar MS, Lattanzio F, Panico M, Morris HR and Clark GF (2003) Murine and human zona pellucida 3 derived from mouse eggs express identical O-glycans *Proceedings of the National Academy of Sciences USA* **100** 15631-15636

Easton RL, Patankar MS, Lattanzio FA, Leaven TH, Morris HR, Clark GF and Dell A (2000) Structural analysis of murine zona pellucida glycans: Evidence for the expression of core 2-type O-glycans and the Sda antigen *Journal of Biological Chemistry* **275** 7731-7742

Epifano O, Liang LF, Familari M, Moos MCJr and Dean J (1995) Coordinate expression of the three zona pellucida genes during mouse oogenesis *Development* **121** 1947-1956

Florman HM and Wassarman PM (1985) O-linked oligosaccharides of mouse egg ZP3 account for its sperm receptor activity *Cell* **41** 313-324

Greve JM and Wassarman PM (1985) Mouse egg extracellular coat is a matrix of interconnected filaments possessing a structural repeat *Journal of Molecular Biology* **181** 253-264

Harris JD, Hibler DW, Fontenot GK, Hsu KT, Yurewicz ED and Sacco AG (1994) Cloning and characterization of zona pellucida genes and cDNA from a variety of mammalian species: the ZPA, ZPB and ZPC gene families *DNA Sequence-The Journal of Sequencing and Mapping* **4** 361-393

Hasegawa A, Koyama K, Okazaki Y, Sugimoto M and Isojima S (1994) Amino acid sequence of a porcine zona pellucida glycoprotein ZP4 determined by peptide mapping and cDNA cloning *Journal of Reproduction and Fertility* **100** 245-255

Hirano T, Takasaki S, Hedrick JL, Wardrip NJ, Amano J and Kobata A (1993) O-Linked neutral sugar chains of porcine zona pellucida glycoproteins *European Journal of Biochemistry* **214** 763-769

Hokke CH, Damm JBL, Kamerling JP and Vliegenthart JFG (1993) Structure of three acidic O-linked carbohydrate chains of porcine zona pellucida glycoproteins *FEBS Letters* **329** 29-34

Hokke CH, Damm JBL, Penninkhof B, Aitken RJ, Kamerling JP and Vliegenthart JFG (1994) Structure of the O-linked carbohydrate chains of porcine zona pellucida glycoproteins *European Journal of Biochemistry* **221** 491-512

Hoodbhoy T and Dean J (2004) Insights into the molecular basis of sperm-egg recognition in mammals *Reproduction* **127** 417-422

Ikeda K, Yonezawa N, Naoi K, Katsumata T, Hamano S and Nakano M (2002) Localization of N-linked carbohydrate chains in glycoprotein ZPA of the bovine egg zona pellucida *European Journal of Biochemistry* **269** 4257-4266

Johnston DS, Wright WW, Shaper JH, Hokke CH, Van den Eijnden DH and Joziasse DH (1998) Murine sperm-zona binding, a fucosyl residue is required for a high affinity sperm-binding ligand *Journal of Biological Chemistry* **273** 1888-1895

Jovine L, Darie CC, Litscher ES and Wassarman PM (2005) Zona pellucida domain proteins *Annual Review of Biochemistry* **74** 83-114

Katsumata T, Noguchi S, Yonezawa N, Tanokura M and Nakano M (1996) Structural characterization of the N-linked carbohydrate chains of the zona pellucida glycoproteins from bovine ovarian and fertilized eggs *European Journal of Biochemistry* **240** 448-453

Kudo K, Yonezawa N, Katsumata T, Aoki H and Nakano M (1998) Localization of carbohydrate chains of pig sperm ligand in the glycoprotein ZPB of egg zona pellucida *European Journal of Biochemistry* **252** 492-499

Lefievre L, Conner SJ, Salpekar A, Olufowobi O, Ashton P, Pavlovic B, Lenton W, Afnan M, Brewis IA, Monk M, Hughes DC and Barratt CL (2004) Four zona pellucida glycoproteins are expressed in the human *Human Reproduction* **19** 1580-1586

Lu Q and Shur BD (1997) Sperm from beta 1,4-galactosyltransferase-null mice are refractory to ZP3-induced acrosome reactions and penetrate the zona pellucida poorly *Development* **124** 4121-4131

Miller DJ, Macek MB and Shur BD (1992) Complementarity between sperm surface ß-1,4-galactosyltransferase and egg-coat ZP3 mediates sperm-egg binding *Nature* **357** 589-593

Mori E, Takasaki S, Hedrick JL, Wardrip NJ, Mori T and Kobata A (1991) Neutral oligosaccharide structures linked to asparagines of porcine zona pellucida glycoproteins *Biochemistry* **30** 2078-2087

Mori E, Mori T and Takasaki S (1997) Binding of mouse sperm to beta-galactose residues on egg zona pellucida and asialofetuin-coupled beads *Biochemical and Biophysical Research Communications* **238** 95-99

Mori E, Hedrick JL, Wardrip NJ, Mori T and Takasaki S (1998) Occurrence of reducing terminal N-acetylglucosamine 3-sulfate and fucosylated outer chains in acidic N-glycans of porcine zona pellucida glycoproteins *Glycoconjugate Journal* **15** 447-456

Nakano M and Yonezawa N (2001) Localization of sperm ligand carbohydrate chains in pig zona pellucida glycoproteins *Cells Tissues Organs* **168** 65-75

Nakano M, Yonezawa N, Hatanaka Y and Noguchi S (1996) Structure and function of the N-linked carbohydrate chains of pig zona pellucida glycoproteins *Journal of Reproduction and Fertility Supplement* **50** 25-34

Noguchi S and Nakano M (1992) Structure of the acidic N-linked carbohydrate chains of the 55-kDa glycoprotein family (PZP3) from porcine zona pellucida *European Journal of Biochemistry* **209** 883-894

Noguchi S, Hatanaka Y, Tobita T and Nakano M (1992) Structural analysis of the N-linked carbohydrate chains of the 55-kDa glycoprotein family (PZP3) from porcine zona pellucida *European Journal of Biochemistry* **204** 1089-1100

N. Yonezawa et al.

Noguchi S, Yonezawa N, Katsumata T, Hashizume K, Kuwayama M, Hamano S, Watanabe S and Nakano M (1994) Characterization of the zona pellucida glycoproteins from bovine ovarian and fertilized eggs *Biochimica et Biophysica Acta* **1201** 7-14

Rankin TL, Coleman JS, Epifano O, Hoodbhoy T, Turner SG, Castle PE, Lee E, Gore-Langton R and Dean J (2003) Fertility and taxon-specific sperm binding persist after replacement of mouse sperm receptors with human homologs *Developmental Cell* **5** 33-43

Sacco AG, Yurewicz EC, Subramanian MG and Matzat PD (1989) Porcine zona pellucida: association of sperm receptor activity with the alpha-glycoprotein component of the Mr = 55,000 family *Biology of Reproduction* **41** 523-532

Shi S, Williams SA, Seppo A, Kurniawan H, Chen W, Ye Z, Marth JD and Stanley P (2004) Inactivation of the *Mgat1* gene in oocytes impairs oogenesis, but embryos lacking complex and hybrid *N*-glycans develop and implant *Molecular and Cellular Biology* **24** 9920-9929

Sinowatz F, Wessa E, Neumuller C and Palma G (2003) On the species specificity of sperm binding and sperm penetration of the zona pellucida *Reproduction of Domestic Animals* **38** 141-146

Spargo SC and Hope RM (2003) Evolution and nomenclature of the zona pellucida gene family *Biology of Reproduction* **68** 358-362

Thaler CD and Cardullo RA (1996) The initial molecular interaction between mouse sperm and the zona pellucida is a complex binding event *Journal of Biological Chemistry* **271** 23289-23297

Thall AD, Maly P and Lowe JB (1995) Oocyte Galα1,3Gal epitopes implicated in sperm adhesion to the zona pellucida glycoprotein ZP3 are not required for fertilization in the mouse *Journal of Biological Chemistry* **270** 21437-21440

von Witzendorff D, Ekhlasi-Hundrieser M, Dostalova Z, Resch M, Rath D, Michelmann HW and Topfer-Petersen E (2005) Analysis of *N*-linked glycans of porcine zona pellucida glycoprotein ZPA by MALDI-TOF MS: a contribution to understanding zona pellucida structure *Glycobiology* **15** 475-488

Wassarman PM (1988) Zona pellucida glycoproteins *Annual Review of Biochemistry* **57** 415-442

Wassarman PM and Litscher ES (2001) Towards the molecular basis of sperm and egg interaction during mammalian fertilization *Cells Tissues Organs* **168** 36-45

Yonezawa N and Nakano M (2003) Identification of the carboxyl termini of porcine zona pellucida glycoproteins ZPB and ZPC *Biochemical and Biophysical Research Communications* **307** 877-882

Yonezawa N, Hatanaka Y, Takeyama H and Nakano M (1995a) Binding of pig sperm receptor in the zona pellucida to the boar sperm acrosome *Journal of Reproduction and Fertility* **103** 1-8

Yonezawa N, Aoki H, Hatanaka Y and Nakano M (1995b) Involvement of *N*-linked carbohydrate chains of pig zona pellucida in sperm-egg binding *European Journal of Biochemistry* **233** 35-41

Yonezawa N, Mitsui S, Kudo K and Nakano M (1997) Identification of an *N*-glycosylated region of pig zona pellucida glycoprotein ZPB that is involved in sperm binding *European Journal of Biochemistry* **248** 86-92

Yonezawa N, Fukui N, Kudo K and Nakano M (1999) Localization of neutral *N*-linked carbohydrate chains in pig zona pellucida glycoprotein ZPC *European Journal of Biochemistry* **260** 57-63

Yonezawa N, Fukui N, Kuno M, Shinoda M, Goko S, Mitsui S and Nakano M (2001) Molecular cloning of bovine zona pellucida glycoproteins ZPA and ZPB and analysis for sperm-binding component of the zona *European Journal of Biochemistry* **268** 3587-3594

Yonezawa N, Amari S, Takahashi K, Ikeda K, Imai FL, Kanai S, Kikuchi K and Nakano M (2005a) Participation of the nonreducing terminal beta-galactosyl residues of the neutral *N*-linked carbohydrate chains of porcine zona pellucida glycoproteins in sperm-egg binding *Molecular Reproduction and Development* **70** 222-227

Yonezawa N, Kudo K, Terauchi H, Kanai S, Yoda N, Tanokura M, Ito K, Miura K, Katsumata T and Nakano M (2005a) Recombinant porcine zona pellucida glycoproteins expressed in Sf9 cells bind to bovine sperm but not to porcine sperm *Journal of Biological Chemistry* **280** 20189-20196

Yurewicz EC, Sacco AG and Subramanian MG (1987) Structural characterization of the Mr = 55,000 antigen (ZP3) of porcine oocyte zona pellucida *Journal of Biological Chemistry* **262** 564-571

Yurewicz EC, Pack BA and Sacco AG (1991) Isolation, composition, and biological activity of sugar chains of porcine oocyte zona pellucida 55K glycoproteins *Molecular Reproduction and Development* **30** 126-134

Yurewicz EC, Pack BA, Armant DR and Sacco AG (1993) Porcine zona pellucida ZP3 alpha glycoprotein mediates binding of the biotin-labeled M(r) 55,000 family (ZP3) to boar sperm membrane vesicles *Molecular Reproduction and Development* **36** 382-389

Yurewicz EC, Sacco AG, Gupta SK, Xu N and Gage DA (1998) Hetero-oligomerization-dependent binding of pig oocyte zona pellucida glycoproteins ZPB and ZPC to boar sperm membrane vesicles *Journal of Biological Chemistry* **273** 7488-7494

# Contribution of zona proteins to oocyte growth

*Akiko Hasegawa[1] and Koji Koyama[1,2]*

[1]Laboratory of Developmental Biology and Reproduction, Institute for Advanced Medical Sciences, Hyogo College of Medicine; [2]Department of Obstetrics and Gynecology, Hyogo College of Medicine, 1-1, Mukogawa-cho, Nishinomiya 663-8501, Japan

The zona pellucida is an extracellular matrix surrounding growing oocytes, ovulated eggs and preimplantation embryos. In mice, the zona pellucida is composed of three glycoproteins designated as ZPA, ZPB and ZPC. *ZpA* and *ZpC* knockout mice are infertile due to incomplete folliculogenesis. In this article, the role of ZPA in the growth and survival of oocytes and the mechanism by which it is incorporated into zona structure is described. Oocyte-granulosa cell complexes (OGCs) were injected with short interfering RNAs (siRNAs) of mouse *ZpA* and cultured in α-MEM for 5 days. The survival rate of these oocytes and the diameter of surviving oocytes were significantly lower as compared to the control oocytes. The zona pellucida was also abnormal in morphology. These observations suggest that mouse ZPA protein contributes to oocyte growth as well as zona pellucida formation. In addition, a minigene construct carrying pig *ZpA* was transferred into mice to examine whether the gene is expressed and the protein product incorporated into the mouse zona pellucida. Pig ZPA was synthesized and secreted by the mouse oocytes but not used to assemble a chimeric zona pellucida. These results suggest that there are different requirements for zona assembly in pigs and mice.

## Introduction

The zona pellucida (ZP) surrounding mammalian oocyte is an extracellular matrix constituted of three or four glycoproteins. In mice, three glycoproteins designated as ZP1, ZP2 and ZP3 are encoded by *Zp1/ZpB*, *Zp2/ZpA* and *Zp3/ZpC* genes, respectively (Harris et al., 1994). In humans, four glycoproteins termed as ZP1, ZP2, ZP3 and ZP4 are encoded by *Zp1/ZpB1*, *Zp2/ZpA*, *Zp3/ZpC* and *Zp4/ZpB2* genes, respectively (Lefievre et al., 2004). Currently the proposed terminology in mice and humans is shown in Table 1 with bold characters. The terminology does not directly correspond to pig ZP proteins and genes. In pigs, ZPB/ZP3α, ZPA/ZP1 and ZPC/ZP3ß proteins are encoded by *ZpB*, *ZpA*, and *ZpC* genes, respectively. In this article, we use the terminology as shown at the bottom of Table 1.

The ZP plays a role in recognition of spermatozoa, induction of the acrosome reaction and prevention of polyspermy. It mediates relatively species-specific sperm-egg recognition and acrosome-reaction induction that are critical for the penetration of spermatozoa into the ZP matrix. Following fertilization, the ZP prevents the penetration of additional sperm and protects the preimplantation embryo passing down the oviduct (Wassarman, 1987; Yanagimachi, 1994).

E-mail: kkoyama@hyo-med.ac.jp

**Table 1.** Zona pellucida proteins and coding genes in three mammalian species.

| | | | | | |
|---|---|---|---|---|---|
| Mouse | protein | **ZP1** | **ZP2** | **ZP3** | |
| | gene | **Zp1**/ZpB | **Zp2**/ZpA | **Zp3**/ZpC | |
| Human | protein | **ZP1** | **ZP2** | **ZP3** | **ZP4** |
| | gene | **Zp1**/ZpB1 | **Zp2**/ZpA | **Zp3**/ZpC | **Zp4**/ZpB2 |
| Pig | protein | ZPB/ZP3α | ZPA/ZP1 | ZPC/ZP3ß | |
| | gene | ZpB | ZpA | ZpC | |
| Terminology | protein | ZPB | ZPA | ZPC | |
| in this article | gene | ZpB | ZpA | ZpC | |

Bold characters represent currently proposed terminology

In mice, ZPC is responsible for primary sperm-binding to the ZP and induction of the acrosome reaction. ZPA plays a role as a secondary sperm receptor that maintains the binding of the acrosome-reacted spermatozoa to the zona pellucida (Bleil et al., 1988; Wassarman, 1999). In pigs, however, combination of ZPB and ZPC contribute to sperm binding to the ZP (Yurewicz et al., 1998). It has also been reported that pig ZPA together with 55K protein (containing ZPC and ZPB) effectively induce the acrosome reaction (Berger et al., 1989). The biochemical structure of pig zona proteins also differs from mice and humans. The primary sequences of ZP components of pigs are similar to those of mouse and human but the charge heterogeneity owing to carbohydrate chains is much higher than in mice and humans (Hedrick and Wardrip, 1986; Koyama et al., 1991). Pig ZPA, but not the mouse counterpart, is partially cleaved into the two components (pig ZP2 and ZP4) by proteolysis before fertilization (Hatanaka et al., 1992; Hasegawa et al., 1994).

In this article, we describe the effects of mouse ZPA protein on oocyte growth *in vitro* as analyzed using a RNA interference technique. We then present molecular mechanisms of the expression of pig ZPA and its incorporation into mouse zona pellucida as analyzed using transgenic mice.

### Inhibitory effects of short interfering RNA (siRNA) of ZPA on oocyte growth *in-vitro*

In this study, RNA interference technique (Fire et al., 1998) was used to examine whether ZPA protein contributes to the oocyte growth and follicular development (unpublished observations). The procedure is shown in Fig. 1. Oocyte-granulosa cell complexes (OGCs) were collected from juvenile mouse ovaries by mechanical dissection and collagenase treatment, injected with two sets of custom-prepared siRNAs of mouse ZPA [Dharmacon RNA Technologies http://www.dharmacon.com/] at a concentration of $40 \mu M$ through the granulosa cell layers and cultivated in α-MEM [Sigma M4526] for 5 days. Granulosa cells were removed by pipetting and assessed for the survival rate of oocytes, the diameter of surviving oocytes and the morphology of the ZP.

The survival rate of the oocytes injected with ZPA siRNA was significantly lower than that of the control oocytes (Table 2). The diameter of siRNA-injected oocytes was also significantly smaller than that of the control group. RT-PCR analysis confirmed that transcription of the *ZpA* gene was suppressed in ZPA siRNA-injected oocytes with no changes in transcription of *ZpB*, *ZpC* and *Gapdh* genes (data not shown). Morphologically, siRNA injection led to detachment of the ZP from the oocyte (Fig. 2b), while control oocytes showed ZP tightly attached to the oocyte (Fig. 2a).

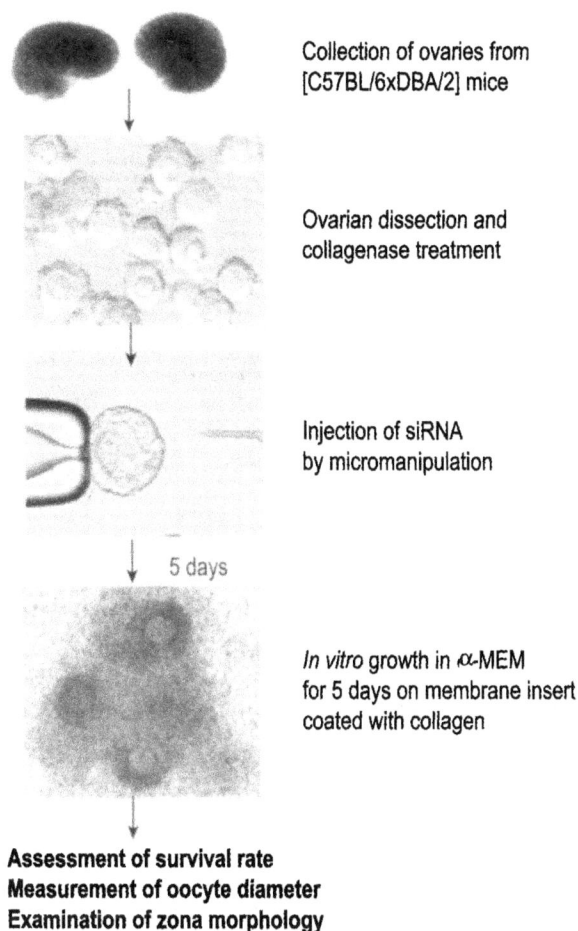

Collection of ovaries from [C57BL/6xDBA/2] mice

Ovarian dissection and collagenase treatment

Injection of siRNA by micromanipulation

5 days

*In vitro* growth in α-MEM for 5 days on membrane insert coated with collagen

**Assessment of survival rate**
**Measurement of oocyte diameter**
**Examination of zona morphology**

**Fig. 1** Analysis of the role of ZPA in the growth of oocytes using siRNA.
Oocyte-granulosa cell complexes (OGCs) were collected from juvenile mice and treated with collagenase to remove theca cells. Oocytes were injected with siRNA and cultivated for 5 days in α-MEM. Granulosa cells were removed by pipetting and the survival rate, diameter of oocytes and zona morphology were analyzed.

**Table 2.** Effects of siRNA injection on oocyte growth *in vitro*

|  | Collected OGCs | Surviving oocytes (%) | Diameters (μm) (mean ± SD) |
|---|---|---|---|
| Control (Vehicle only) | 71 | 63 (88.7%)[a] | 62.15 ± 4.52 [c] |
| siRNA-ZPA | 70 | 26 (37.1%)[b] | 58.93 ± 4.60 [d] |

$P < 0.01$: t-test (a vs b, c vs d)

It has been reported that *ZpA* or *ZpC* knockout mice are infertile due to defective ovarian follicle formation (Liu *et al.*, 1996; Rankin *et al.*, 1996, 2001). The mice failed to form ZP

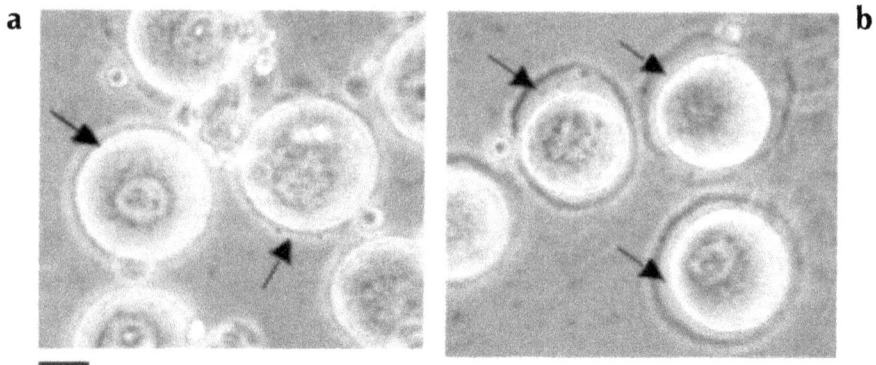

**Fig. 2** Effect of siRNA on oocyte growth and zona formation
The ZP in vehicle-injected oocytes showed normal morphology (a), while siRNA-injected oocytes formed abnormal ZP which were detached from the oocyte cytoplasm (b). Arrows show ZP formed around the oocytes. Bar is 25 μm.

despite the presence of other components. In the present study, we showed that knockdown of ZPA protein by siRNA impaired normal zona formation. This abnormal ZP probably did not support the oocyte-granulosa cell communications that are essential for oocyte growth. Similar experiments were done using antisense oligonucleotides (Tong et al., 1995).

## Production of pig ZPA incorporated transgenic mouse

To examine whether the zona can be assembled in mice using pig ZPA, a transgene was constructed with a partial pig *ZpA* cDNA of exon 1-6, pig genomic *ZpA* DNA encompassing exon 7-15 and a partial pig *ZpA* cDNA of exon 16-18 followed by a His-tag sequence (Tsubamoto et al., 1999). The 1.4-kb mouse *ZpA* promoter segment (Liang et al., 1990) was kindly provided by Dr. J. Dean, Laboratory of Cellular and Developmental Biology, National Institutes of Health, USA. It was placed upstream of the transgene. Immunohistochemical analyses using a pig ZPA-specific monoclonal antibody (MAb-5H4) detected pig ZPA inside the oocyte cytoplasm but not the ZP region defined by staining with a Schiff's base (Fig. 3). These results suggest that pig ZPA protein was produced by the transgene but not used to form a chimeric ZP in mice.

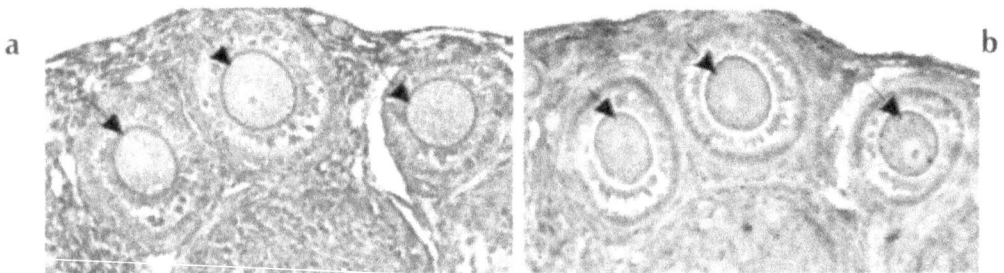

**Fig. 3** Immunohistochemical detection of pig ZPA in the ovary of a transgenic mouse
Paraffin-embedded ovarian sections were treated with MAb-5H4 coupled to an ABC enhancement. (a): The ZP was defined by staining with a Schiff's base (as shown by arrows). (b): Immunohistochemical staining of the oocyte showing strong positive signals of pig ZPA protein (as shown by arrows). Bar is 50 μm.

To clarify whether pig ZPA is secreted from the oocyte, unfixed ovarian oocytes were used for detection of pig ZPA protein because the antibody does not penetrate into unfixed cells. Immunocytochemical analysis using MAb-5H4 detected pig ZPA in ZP before germinal vesicle breakdown (Fig. 4a) but not after (Fig. 4b). An ovulated oocyte from the transgenic mice did not show a positive reaction with MAb-5H4 (Fig. 4c).

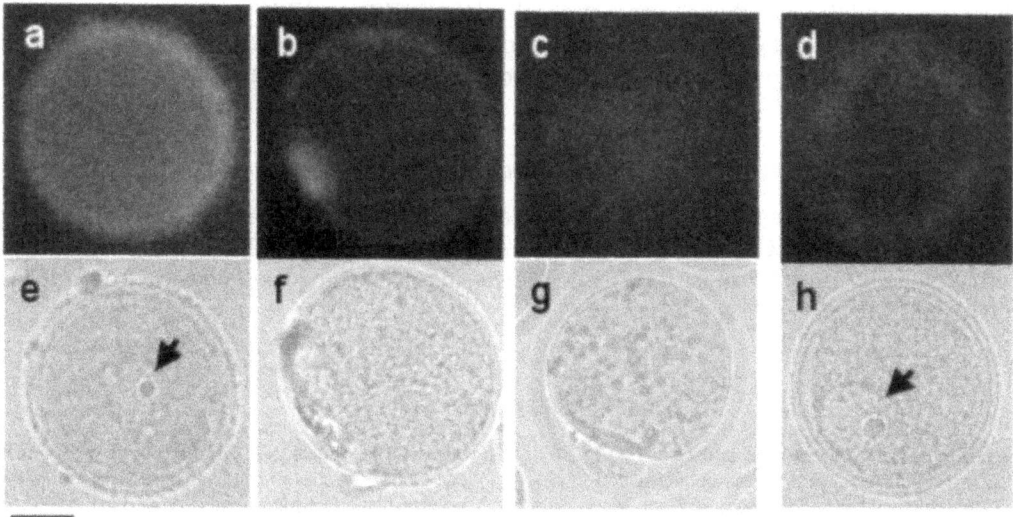

**Fig. 4** Immunofluorescent staining of pig ZPA with MAb-5H4 in transgenic mouse oocytes (a): Ovarian oocyte containing germinal vesicle showed positive reaction to MAb-5H4 in the zona pellucida. (b): Reduced staining in ovarian oocyte after germinal vesicle breakdown. (c): No staining in ovulated eggs. (d): Control ovarian oocyte. Arrows indicate germinal vesicles. Bar is 25 $\mu$m. (e) (f) (g) (h) are light-field observations of the oocytes same as in (a) (b) (c) (d).

These results indicate that pig ZPA is produced in mouse immature oocyte and temporarily stays within the ZP but was not used to assemble mouse ZP. Similar observations were reported demonstrating that mouse ZPA is present on the plasma membrane of the egg from the ZPC-knockout mice but did not deposit around the oocyte (Qi and Wassarman, 1999). It was shown that ZPA was secreted. Increased fluidity and changes in the oocyte plasma membrane composition following oocyte maturation may be responsible for the diffusion of ZPA.

Recently, hydrophobic segments of the mouse ZP domain in ZPA, ZPB and ZPC proteins have been reported to be important for assembly of these zona proteins in zona matrix (Jovine et al., 2002; 2004). It is suggested that a hydrophobic amino-acid sequence of PGPLVLV (483-489) of mouse ZPA, located in the ZP domain, is crucial for protein secretion. The corresponding sequence is replaced by PGPLTLT in pig ZPA. The change of the two sites of hydrophobic amino acid (Valine) in mice to hydroxyl amino acid (Threonine) in pigs may be also crucial in the failure to form chimeric ZP of mice and pigs.

It has been shown that human ZPA and ZPC can form a chimeric ZP in mice (Rankin et al., 1998; 2003). The amino acid sequences of pig and human ZPA are 54% and 60% identical to that of mouse ZPA, respectively. The difference (6%) in the amino acid sequence cannot explain why pig ZPA did not assemble with mouse zona proteins, since *Xenopus laevis* zona proteins are assembled with mouse zona proteins in spite of 39-48% amino acid similarity (Doren et al., 1999). Absence or replacement of amino acid(s) at critical site(s) may result in the failure of pig ZPA assembly in mouse zona. Further work is necessary to understand ZP assembly.

# Conclusion

In this article, we showed that the interference with ZPA-messenger RNA function affected zona formation and oocyte growth. This is consistent with the report that ZPA-knockout mice have decreased numbers of ovarian follicles and are infertile. In addition, we showed that the pig ZPA protein did not assemble into mouse zona protein using a transgenic mouse technique. These findings differ from the reports showing that human ZPA and ZPC can form a chimeric ZP matrix in mice. It is not clear why pig ZPA is not incorporated into mouse ZP. Further investigations are necessary to clarify species differences in zona matrix formation.

# Acknowledgements

This work was supported by Grant-in-Aid of Scientific Research (No. 14370536) from Ministry of Education, Science, Culture, Sports and Technology (MEXT), 2002-2004 and by the "High-Tech Research Center" Project for Private Universities: matching fund subsidy from MEXT, 2004-2008.

# References

Berger T, Davis A, Wardrip NJ and Hedrick JL (1989) Sperm binding to the pig zona pellucida and inhibition of binding by solubilized components of the zona pellucida *Journal of Reproduction and Fertility* **86** 559-565

Bleil JD, Greve JM and Wassarman PM (1988) Identification of a secondary sperm receptor in the mouse egg zona pellucida: role in maintenance of binding of acrosome-reacted sperm to eggs *Delopmental Biology* **128** 376-385

Doren S, Landsberger N, Dwyer N, Gold L, Blanchette-Mackie J and Dean J (1999) Incorporation of mouse zona pellucida proteins into the envelope of *Xenopus laevis* oocytes *Development Genes Evolution* **209** 330-339

Fire A, Xu S, Montgomery MK, Kostas SA, Driver SE and Mello CC (1998) Potent and specific genetic interference by double-stranded RNA in *Caenorhabditis elegans Nature* **391** 806-811

Harris JD, Hibler DW, Fontenot GK, Hsu KT, Yurewicz EC and Sacco AG (1994) Cloning and characterization of zona pellucida genes and cDNAs from a variety of mammalian species: the ZPA, ZPB and ZPC gene families *DNA Sequence* **4** 361-393

Hasegawa A, Koyama K, Okazaki Y, Sugimoto M and Isojima S (1994) Amino acid sequence of a porcine zona pellucida glycoprotein ZP4 determined by peptide mapping and cDNA cloning *Journal of Reproduction and Fertility* **100** 245-255

Hatanaka Y, Nagai T, Tobita T and Nakano M (1992) Changes in the properties and composition of zona pellucida of pigs during fertilization *in vitro Journal of Reproduction and Fertility* **95** 431-440

Hedrick JL and Wardrip NJ (1986) Isolation of the zona pellucida and purification of its glycoprotein families from pig oocytes *Analytical Biochemistry* **157** 63-70

Jovine L, Qi H, Williams Z, Litscher E and Wassarman PM (2002) The ZP domain is a conserved module for polymerization of extracellular proteins *Nature Cell Biology* **4** 457-461

Jovine L, Qi H, Williams Z, Litscher ES and Wassarman PM (2004) A duplicated motif controls assembly of zona pellucida domain proteins *Proceedings National Academy of Sciences USA* **101** 5922-5927

Koyama K, Hasegawa A, Inoue M and Isojima S (1991) Blocking of human sperm-zona interaction by monoclonal antibodies to a glycoprotein family (ZP4) of porcine zona pellucida *Biology of Reproduction* **45** 727-735

Lefievre L, Conner SJ, Salpekar A, Olufowobi O, Ashton P, Pavlovic B, Lenton W, Afnan M, Brewis IA, Monk A, Hughes DC and Barratt CL (2004) Four zona pellucida glycoproteins are expressed in the human *Human Reproduction* **9** 1580-1586

Liang LF, Chamow SM and Dean J (1990) Oocyte-specific expression of mouse Zp-2: developmental regulation of the zona pellucida genes *Molecular Cell Biology* **10** 1507-1515

Liu C, Litscher ES, Mortillo S, Sakai Y, Kinloch RA, Stewart CL and Wassarman PM (1996) Targeted disruption of the mZP3 gene results in production of eggs lacking a zona pellucida and infertility in female mice *Proceedings National Academy of Sciences USA.* **93** 5431-5436

Qi H and Wassarman PM (1999) Secretion of zona pellucida glycoprotein mZP2 by growing oocytes from mZP3$^{+/+}$ and mZP3$^{-/-}$ mice *Developmental Genetics* **25** 95-102

Rankin T, Familari M, Lee E, Ginsberg A, Dwyer N, Blanchette-Mackie J, Drago J, Westphal H and Dean J (1996) Mice homozygous for an insertional mutation in the Zp3 gene lack a zona pellucida and are infertile

*Development* **122** 2903-2910

Rankin TL, Tong ZB, Castle PE, Lee E, Gore-Langton R, Nelson LM and Dean J (1998) Human ZP3 restores fertility in Zp3 null mice without affecting order-specific sperm binding *Development* **125** 2415-2424

Rankin TL, O'Brien M, Lee E, Wigglesworth K, Eppig J and Dean J (2001) Defective zonae pellucidae in Zp2-null mice disrupt folliculogenesis, fertility and development *Development* **128** 1119-1126

Rankin TL, Coleman JS, Epifano O, Hoodbhoy T, Turner SG, Castle PE, Lee E, Gore-Langton R and Dean J (2003) Fertility and taxon-specific sperm binding persist after replacement of mouse sperm receptors with human homologs *Developmental Cell* **5** 33-43

Tong ZB, Nelson LM and Dean J (1995) Inhibition of zona pellucida gene expression by antisense oligonucleotides injected into mouse oocytes *The Journal of Biological Chemistry* **270** 849-853

Tsubamoto H, Yamasaki N, Hasegewa A and Koyama K (1999) Expression of a recombinant porcine zona pellucida glycoprotein ZP1 in mammalian cells *Protein Expression and Purification* **17** 8-15

Wassarman PM (1987) The biology and chemistry of fertilization *Science* **235** 553-560

Wassarman PM (1999) *Mammalian fertilization*: molecular aspects of gamete adhesion, exocytosis, and fusion *Cell* **96** 175-183

Yanagimachi R (1994) Mammalian fertilization. In *The physiology of Reproduction*, edn 5, pp 189-317 Eds E Knobil, JD Neil. Raven Press, New York

Yurewicz EC, Sacco AG, Gupta SK, Xu N and Gage DA (1998) Hetero-oligomerization-dependent binding of pig oocyte zona pellucida glycoproteins ZPB and ZPC to boar sperm membrane vesicles *The Journal of Biological Chemistry* **273** 7488-7494

# Understanding the physiology of pre-fertilisation events in the human spermatozoa – a necessary prerequisite to developing rational therapy

S J Conner[1,2], L Lefièvre[1,2], J Kirkman-Brown[1,2], F Michelangeli[3], C Jimenez-Gonzalez[3], G S M Machado-Oliveira[3], K L Pixton[4], I A Brewis[5], C L R Barratt[1,2+] and S J Publicover[3]

[1]Reproductive Biology and Genetics Group, Division of Reproductive and Child Health, University of Birmingham, Edgbaston, Birmingham B15 2TT, UK, [2]Assisted Conception Unit, Birmingham Women's Hospital, Birmingham B15 2TG, UK, [3]School of Biosciences, University of Birmingham, Edgbaston, Birmingham, B15 2TT, UK, [4]School of Biological Sciences, The University of Reading, Whiteknights, PO Box 228, Reading, RG6 6AJ, UK, [5]Department of Medical Biochemistry and Immunology, Henry Wellcome Building, Cardiff University, Cardiff, CF14 4XN, UK

Sperm dysfunction is the single most common defined cause of infertility. One in 15 men is sub-fertile and the condition is increasing in frequency. However, the diagnosis is poor and, excluding assisted conception, there is no treatment. The reason for this is our limited understanding of the biochemical, molecular and genetic functions of the spermatozoon. The underlying premise of our research programme is to establish a rudimentary understanding of the processes necessary for successful fertilisation. In this manuscript, we detail advances in our understanding of calcium signalling in the cell and outline genetic and proteomic technologies that are being used to improve the diagnosis of the condition.

## Introduction

The first part of this review discusses the premise that there is clear need to significantly improve our understanding of the cellular basis of normal sperm function. This knowledge is fundamental for two key developments in male fertility: first, to provide the basis for effective diagnostic tools and, secondly, to facilitate the study of the physiology of abnormal/dysfunctional cells, which is central to developing rational, non-Assisted Reproductive Technology (ART) therapy. Our primary focus is detailing the recent progress determining the nature and operation of the calcium signalling network – the calcium toolkit- in the spermatozoon (see Jimenez-Gonzalez et al., 2006). This charts the rapid progress in this area but also identifies noticeable gaps in knowledge. Using this as a base, we examine strategies to determine the dysfunction of spermatozoa, in particular the use of proteomics.

Epidemiological data suggests that 1 out of 7 couples are classed as sub-fertile (Templeton et al., 1990). Sperm dysfunction is the single most common cause of infertility and affects ap-

[+]Corresponding author: Dr. C. L. R. Barratt, Room 330, Institute of Biomedical Research, Medical School, Division of Reproductive and Child Health, University of Birmingham, Birmingham B15 2TG United Kingdom
E-mail: c.l.barratt@bham.ac.uk

proximately 1 out of 15 men (HFEA, 2005). Studies using semen assessment as the criterion for sub-fertility (sperm concentration < $20X10^6$ cells/ml) show that 1 out of 5 eighteen year olds are classed as sub-fertile (Andersen et al., 2000). This is a high proportion of the population compared with other common diseases such as diabetes (2.8% of the population: Wild et al., 2004). Thus, male sub-fertility is a very significant global problem. Recent reports suggest that its prevalence is increasing (Sharpe and Irvine, 2004).

## Improving the diagnosis of male fertility

There is a clear requirement to produce new and robust tests of sperm function to diagnose male infertility. The value of traditional semen parameters (concentration, motility and morphology) has been debated for almost 60 years and, perhaps not surprisingly, the debate continues (see Bjorndahl and Barratt, 2005). Suffice it to say, if appropriate quality control (QC) measures are in place to ensure the assessment is valid, traditional semen parameters do provide some degree of prognostic and diagnostic information for the infertile couple (Tomlinson et al., 1999). However, even with rigorous QC procedures in place, it is only at the lower ranges of the spectrum that these parameters are most useful and, even then they can only be used as guidance for couples, and do not represent absolute values.

Traditional semen analysis is therefore only a limited first line tool in the diagnosis of male infertility. Consequently, andrologists have focussed on developing simple, robust and effective tests of sperm function. Yet despite the plethora of potential assays available, results have been very disappointing (ESHRE, 1996; Muller, 2000). In fact, assessing the current data, it is difficult to see that there are any more effective methods to assess sperm function than the simplest (and oldest) sperm function assay - penetration of spermatozoa into human cervical mucus (or artificial substitutes) (Aitken et al., 1992; Ivic et al., 2002). Our attempts to short cut the necessity to develop a detailed understanding of sperm function has led to a number of false dawns and inappropriate treatments; for example, the indiscriminate use of pentoxyfylline.

Although male infertility is primarily manifested as abnormal semen parameters (concentration, motility and morphology), there is a significant incidence of 'hidden' male infertility; that is, a man may be infertile even though he has a "normal semen profile" [unexplained infertility]. Importantly, total fertilisation failure at IVF is a relatively common occurrence in such men [approximately 5-25% of cases: Tournaye et al., 2002] and whilst the detailed molecular mechanisms involved are unknown, there are subtle changes in sperm motility (Barratt et al., 1989), impaired development of hyperactivation (Mackenna et al., 1993) and defects in sperm binding to the zona pellucida/ induction of the acrosome reaction (Liu and Baker, 2000).

## Development of effective drug-based non-ART therapy

With such an important health issue as male infertility, we would expect a number of rational and effective treatments to be available. However, there are no drug treatments to enhance sperm function that have been shown to be effective in randomised controlled trials (Kamischke and Nieschlag, 2002). Thus, remarkably, the only treatment option for the sub-fertile man is in vitro fertilisation or intra cytoplasmic sperm injection (IVF/ICSI – collectively termed assisted conception). Assisted conception is very expensive, invasive, has limited success, a number of side effects, is not widely available and poses significant concerns about the long term health of children (Maher et al., 2003; review Hansen et al., 2005). Yet, the number of patients treated with assisted conception is continuously increasing. In the USA, for example, there was a 78% increase in ART treatments from 1996-2002 (CDC, 2005) and currently, up to 4% of births are a result of ART (Andersen et al., 2005). Put simply, this means that, as a result of our ignorance of the causes of sperm dysfunction, we are currently subjecting an increasing proportion of women to inappropriate invasive therapy to treat their partners.

In summary, we need to understand in cellular, biochemical and molecular terms how spermatozoa works in order to address both of the above (Barratt and Publicover, 2001). As the functioning of a spermatozoon is critically dependent upon the tight regulation of calcium which, when disrupted, results in fertilisation failure (Krausz *et al.*, 1996) our research is focused on this area.

## Understanding the calcium toolkit as a basic function of the cell

*Functional importance of sperm [Ca²⁺]ᵢ signalling*

Although there is little doubt that spermatozoa use a range of cell messengers (for example, cAMP), data accumulated over the last few years have shown that $[Ca^{2+}]_i$ also plays a major role in all important sperm functions that occur after ejaculation (reviewed in Jimenez-Gonzalez *et al.*, 2006). However, in contrast to somatic cells (Berridge, 2005), an understanding of $Ca^{2+}$-signalling in the sperm cell (despite its great importance) is only now developing (Fig. 1). The simple pump-leak model is initially attractive for spermatozoa. Due to the small size of the cell, diffusion is unlikely to be a limiting factor and it is reasonable to assume that $Ca^{2+}$-influx from the extracellular compartment can effect a rapid rise in $[Ca^{2+}]_i$ in any part of the cytoplasm. Although evidence for expression of mechanisms for $Ca^{2+}$ store mobilisation in mammalian spermatozoa has been available for some time (Kuroda *et al.*, 1999), direct demonstration of store mobilisation proved difficult, possibly because the store was labile or only partially filled (Publicover and Barratt, 1999). However, investigations on the induction of the acrosome reaction by zona strongly implicated a mechanism involving mobilisation of a $Ca^{2+}$ store (probably the acrosome – see below) and activation of capacitative $Ca^{2+}$ influx (Blackmore, 1993; O'Toole *et al.*, 2000; Evans and Florman, 2002). More recently, it has become apparent that a second, separately regulated store may exist, which functions primarily to regulate flagellar beat (Ho and Suarez, 2003; Harper *et al.*, 2004). Thus sperm may well possess a relatively complex $Ca^{2+}$-signalling apparatus including pump-leak and multiple stores.

## Ca²⁺ flux at the plasmalemma (ion channels)

*i) Voltage-operated Ca²⁺ channels*

Voltage operated $Ca^{2+}$ channels (VOCCs) similar to those of somatic cells have been detected in mature and immature sperm cells (Table 1) (Arnoult *et al.*, 1996; Jimenez-Gonzalez *et al.*, 2006). Application of electrophysiological techniques to spermatozoa is still extremely difficult (Gu *et al.*, 2004), but patch clamping of immature male germ cells (rodent and human) has shown consistently that these cells display typical low-voltage activated, fast-inactivating (T-type) currents (Arnoult *et al.*, 1996; Jagannathan *et al.*, 2002). No high-voltage-activated currents have been observed using this method. Electrophysiological recordings from male germ cells of knockout mice showed that the loss of the T-type channel α1G (CaV3.1) had little effect on the currents, suggesting that α1H (CaV3.2) is the main functional VOCC in wild type male germ cells (Stamboulian *et al.*, 2004). Most pharmacological analyses of these channels shows a characteristic T-channel profile, though there is an unusually high sensitivity to dihydropyridines (Arnoult *et al.*, 1998), which may have been misinterpreted as evidence for participation of L-type VOCCs in some studies. Wennemuth *et al.* (2000) reported that the $Ca^{2+}$ currents in mouse spermatogenic cells were sensitive to the application of ω-conotoxin GVIA, a blocker of N-type $Ca^{2+}$ channels. However, a residual ω-conotoxin GVIA-insensitive current was present, suggesting that there are at least two components to the voltage-operated

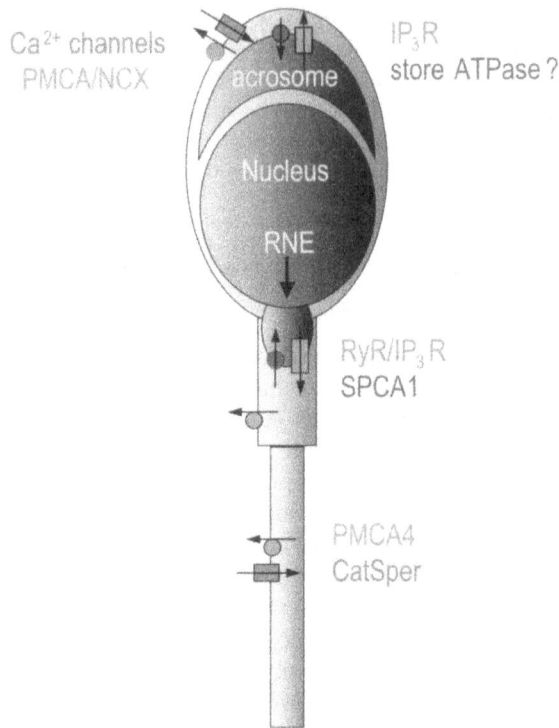

**Fig 1.** Two-store model for $Ca^{2+}$-signalling in human spermatozoa: Plasma membrane $Ca^{2+}$ ATPases (PMCA) and $Na^{+}$-$Ca^{2+}$ exchangers (NCX) are shown in green; $Ca^{2+}$ channels in the plasma membrane are shown in red; ATPases responsible for filling intracellular stores are shown in blue and channels for mobilisation of stored $Ca^{2+}$ are shown in orange. Identified or putative components of the $Ca^{2+}$-signalling toolkit are labelled (using the same colour coding) adjacent to their localisation (adapted from Jimenez-Gonzalez et al., 2006).

**Table 1.** Summary of studies which have illustrated expression of voltage operated $Ca^{2+}$ channels in male germ cells (human).

| | | | |
|---|---|---|---|
| Spermatozoa | In situ RT-PCR | CaV1.2 detected | Goodwin et al., 2000 |
| Spermatozoa | RT-PCR RNase protection assays | CaV3.1, CaV3.2, CaV3.3, CaV1.2, CaV2.3, CaV2.2 | Park et al., 2003 |
| Spermatozoa | Immunogold transmission EM, Immunostaining (confocal microscope) | CaV3.1, CaV3.2, CaV3.3, CaV1.2, CaV2.3-detection/ regional localization | Trevino et al., 2004 |
| Sperm RNA | RT-PCR | CaV3.1 not detected | Jacob and Benoff, 2000 |
| Testis/male germ cells | PCR | CaV3.1 and CaV3.2 full sequence | Jagannathan et al., 2002 |
| Testis | In situ hybridization | CaV3.1 and CaV3.2 in germ cells and somatic cells | Jagannathan et al., 2002 |
| Multi-tissue northern blot panel | Northern blot | CaV3.1 detected (small amounts), CaV3.3 not detected | Monteil et al., 2000a, b |
| Testis | RT-PCR | CaV3.2 detected (489bp); CaV3.1 and CaV3.3 not detected | Son et al., 2000 |

Column 1 shows the tissue/cells studied, columns 2 and 3 summarise the methods used and the main findings and column 4 gives the relevant references (Adapted from Jimenez-Gonzalez et al., 2006).

$Ca^{2+}$ currents in these cells. Functional expression of T channels in mature spermatozoa is yet to be confirmed, though photometric measurements of the responses of murine sperm to solubilised zona pellucida show a large, brief (200 ms) $[Ca^{2+}]_i$ transient consistent with T-channel activation (Arnoult et al., 1999; see below).

Transcripts for α 1C [Cav1.2], α1B [Cav2.2], α1E [Cav2.3], α1G [Cav3.1], α1H [Cav3.2] and α1I [Cav3.3] $Ca^{2+}$ channels are present in motile human spermatozoa (Park et al., 2003). Immunostaining experiments using specific antibodies showed that in human spermatozoa, α1H [Cav3.2] is localised to the principal piece of the tail and the back of the sperm head and α1I [Cav3.3] is restricted to the midpiece (Trevino et al., 2004). High voltage activated channels show similar localization in mouse and human sperm (Trevino et al., 1998; 2004; Westenbroek and Babcock, 1999; Wennemuth et al., 2000). L type channels have been detected in all studies, using RT-PCR and/or immunostainng techniques (Goodwin et al., 2000). However, the sensitivity of sperm T-type channels to drugs previously considered L-channel specific (see above) means that pharmacological data used to identify these channels must be treated cautiously.

*ii) Store-operated channels*

In both non-excitable and excitable cells, mobilisation of intracellular $Ca^{2+}$ stores is believed to activate $Ca^{2+}$-permeable ion channels (store-operated channels - SOCs) in the plasmalemma-capacitative $Ca^{2+}$ entry (Putney, 1990). Pharmacological mobilization of stored $Ca^{2+}$ (using thapsigargin or cyclopiazonic acid) activates $Ca^{2+}$ influx in non-capacitated human sperm (Blackmore, 1993). This influx produces a sustained rise in intracellular $Ca^{2+}$, which is believed to lead to the acrosome reaction in mammalian and non-mammalian spermatozoa (O'Toole et al., 2000; Gonzalez-Martinez et al., 2001; Hirohashi and Vacquier, 2003) and regulates chemotactic behaviour in ascidian spermatozoa (Yoshida et al., 2003). TRP2 may be a necessary component of the channel in mouse sperm responsible for the sustained rise in intracellular $Ca^{2+}$ (triggered by ZP3) that leads to acrosome reaction, which could be regulated either by the depletion of $Ca^{2+}$ from internal stores (O'Toole et al., 2000) or through receptor activation (Harteneck et al., 2000).

*iii) Cyclic nucleotide-gated channels and CatSpers*

Cyclic nucleotide signalling is pivotal in the functioning of all spermatozoa. In invertebrate cells cGMP is of importance in regulation of motility and the acrosome reaction (Kaupp et al., 2003). In sperm of mammals (including humans) levels of cGMP are much lower and cAMP appears to be of greater importance (Ain et al., 1999; Leflevre et al., 2000). Manipulation of cGMP or cAMP levels in mouse spermatozoa induced a transient elevation of $[Ca^{2+}]_i$ lasting 20-60 s (cGMP being significantly more effective), an effect apparently due to $Ca^{2+}$-influx through cyclic nucleotide-gated channels (Kobori et al., 2000). A cyclic nucleotide gated (CNG) $Ca^{2+}$-permeable channel is expressed in spermatozoa, occurring in the principal piece of the flagellum, and is more sensitive to cGMP than cAMP (Weyand et al., 1994; Wiesner et al., 1998). Olfactory receptors, which activate a cyclic nucleotide-mediated $Ca^{2+}$ influx and control chemotactic activity, have recently been described in human sperm (Spehr et al., 2003; see below).

CatSpers are a novel family of ion channels, expressed exclusively in spermatozoa. Four different subunits have been identified: CatSper1 (Ren et al., 2001), CatSper2 (Quill et al., 2001) and CatSpers -3 and -4 (Lobley et al., 2003). In mature cells CatSper2 protein is localized to the sperm flagellum (Quill et al., 2001) and CatSper1 to the principal piece (Ren et al., 2001), suggesting that CatSper channels may be involved in regulating sperm motility. Preliminary studies suggest that CatSper expression is reduced in human spermatozoa with poor motility (Nikpoor et al., 2004). Studies with CatSper knockout mice showed that in mutant spermatozoa lacking Catsper1 motility was severely decreased and, as a consequence, the sperm could not fertilize (Ren et al., 2001). Furthermore, in CatSper1$^{-/-}$ cells there was no increase in flagellar $[Ca^{2+}]_i$ upon application of cell-permeant cAMP/cGMP. This observation suggested that the channel might be cAMP-gated. Subsequently, Carlson et al. (2003) and Quill et al. (2003) confirmed the importance of CatSper 1 and CatSper 2 in sperm motility (more specifically in the hyperactivated movement required for zona penetration). It was also shown that CatSper1 is required for depolarisation-evoked $Ca^{2+}$ entry, suggesting that CatSper1 functions as a voltage-gated $Ca^{2+}$ channel (facilitated by cyclic nucleotides) which regulates hyperactivated motility.

## Ca$^{2+}$ clearance mechanisms in sperm

In most cells $Ca^{2+}$ clearance is undertaken primarily by ATP requiring $Ca^{2+}$ pumps ($Ca^{2+}$-ATPases) or $Na^+$-$Ca^{2+}$ exchangers, which extrude $Ca^{2+}$ either out of the cell or into intracellular $Ca^{2+}$ stores (Michelangeli et al., 2005). Analysis of $Ca^{2+}$ clearance in mouse spermatozoa suggests that both $Ca^{2+}$ pumps and $Ca^{2+}$ exchangers are important contributors to $Ca^{2+}$ clearance in mammalian sperm, although the relative importance of each is yet to be determined (reviewed in Jimenez-Gonzalez et al., 2006).

## Ca$^{2+}$ pumps

To date, three types of ATP-utilising $Ca^{2+}$ pumps have been identified: the plasma membrane $Ca^{2+}$ ATPase (PMCA); the Sarcoplasmic-Endoplasmic $Ca^{2+}$ ATPase (SERCA); and the Secretory Pathway $Ca^{2+}$ ATPase (SPCA) (Michelangeli et al., 2005; Toyoshima and Inesi, 2004).

### PMCA in spermatozoa

PMCA protein is present in rat spermatids and mouse spermatozoa (Berrios et al., 1998; Wennemuth et al., 2003). PMCA4 is the main isoform expressed in spermatozoa (Okunade et al., 2004) and is primarily confined to the principal piece of the flagellum (Wennemuth et al., 2003; Okunade et al., 2004; Schuh et al., 2004). Mice null for PMCA4 showed normal spermatogenesis and mating behavior but the male mice were infertile (Okunade et al., 2004; Schuh et al., 2004). The sperm failed to respond to conditions that induce hyperactivated motility and eventually became non-motile (Okunade et al., 2004). Measurement of $[Ca^{2+}]_i$ showed that, after 60 min incubation in capacitating medium, resting $[Ca^{2+}]_i$ was increased from 157 to 370 nM in PMCA4-deficient sperm (Schuh et al., 2004). This effect could be mimicked using the PMCA inhibitor 5-(and -6)-carboxyeosin diacetate succinimidyl ester on wild-type mice. A similar failure of hyperactivated motility was observed.

### SERCA in spermatozoa

The presence and activity of SERCA in mature spermatozoa is controversial. Wennemuth et al. (2003) suggested that $Ca^{2+}$ clearance in mature spermatozoa is unlikely to involve SERCA. In our laboratory, using an anti-SERCA antibody (which recognised all known mammalian SERCA isoforms), no cross-reactivity was detected in Western blots using human sperm (Harper et al., 2005). Furthermore, thapsigargin induced $Ca^{2+}$-mobilization and disruption of $Ca^{2+}$-signalling in sperm in the 1-10 µM range only, which is far higher than concentrations used to specifically inhibit SERCA (Wictome et al., 1992).

In somatic cells, the secretory pathway $Ca^{2+}$ ATPase (SPCA) are found located on the Golgi apparatus or secretory vesicles (Wootton *et al.*, 2004) and are believed to control the levels of both $Ca^{2+}$ and $Mn^{2+}$ within the Golgi in order to regulate its function (Michelangeli *et al.*, 2005). We have shown that rat germ cells (spermatids) express the mRNA for SPCA1 (Wootton *et al.*, 2004). SPCA1 is also present in mature human spermatozoa and is localised to the anterior mid-piece, extending into the rear of the head (Harper *et al.*, 2005), perhaps reflecting expression in the putative $Ca^{2+}$ store of the redundant nuclear envelope (RNE) (Ho and Suarez, 2003).

In summary, there is strong evidence to indicate that sperm express both PMCA and SPCA and that these $Ca^{2+}$ pumps play a major role in controlling spermatozoa $Ca^{2+}$ homeostasis. The role for SERCA in mature sperm is more tenuous but there appears to be some evidence that it may play a role during spermatogenesis.

## The $Na^+$-$Ca^{2+}$ exchanger

There are two groups (families) of $Na^+/Ca^{2+}$ exchanger: $Na^+/Ca^{2+}$ exchanger (NCX) and $K^+$-dependent $Na^+/Ca^{2+}$ exchangers (NCKX). These exchangers appear to contribute significantly to $Ca^{2+}$ clearance in mouse sperm (Wennemuth *et al.*, 2003) and NCKX expressed in the tail of sea urchin sperm plays an important role in $[Ca^{2+}]_i$ homeostasis (Su and Vacquier, 2002).

## Mobilisation of stored $Ca^{2+}$

*Inositol 1,4,5-triphosphate receptors in spermatozoa*

The inositol 1,4,5-triphophate-sensitive $Ca^{2+}$ channel (commonly referred to as the IP3 receptor or IP3R) has been studied extensively in a variety of cell types including sperm (Jimenez-Gonzalez *et al.*, 2006). Several studies have shown that sperm from a variety of mammals, including humans, contain proteins which cross-react with a number of different IP3R-specific antibodies (Kuroda *et al.*, 1999; Minelli *et al.*, 2000; Ho and Suarez, 2003). Use of isoform-specific IP3R antibodies also showed that IP3R1 was predominantly located to the anterior acrosome (Kuroda *et al.*, 1999; Ho and Suarez, 2003). Interestingly, the IP3R labelling was lost or reduced once the sperm had undergone the acrosome reaction further confirming the IP3R location as being the outer acrosomal membrane (Kuroda *et al.*, 1999).

Kuroda *et al.* (1999) showed that the posterior region of the head, the midpiece and part of the tail was specifically labelled with an IP3R3-specific antibody in human sperm, while the presence of IP3R2 was not detected at all. In contrast, in bull sperm an IP3R1-specific antibody stained the RNE (Ho and Suarez, 2003). Anti-IP3R staining also localises to this region in approximately half of human spermatozoa, though the most intense staining was observed in the acrosomal region (>90% of cells: Naaby-Hansen *et al.*, 2001).

In order to assess whether $Ca^{2+}$ mobilization from IP3R containing $Ca^{2+}$ stores contribute to sperm function, thimerosal an IP3R activator has been used. Herrick *et al.* (2005) showed that thimerosal was able to induce the acrosome reaction, confirming a role for IP3R in sperm physiology. It is clear that two IP3R containing $Ca^{2+}$ stores are present within spermatozoa, one in the acrosome and the other, a much smaller $Ca^{2+}$ store, located within the RNE. To further support the notion that both these compartments are bona fide $Ca^{2+}$ stores, studies by Naaby-Hansen *et al.* (2001) and by Ho and Suarez (2003), showed that both regions also contained calreticulin (a low affinity, high capacity $Ca^{2+}$ buffering protein) that is always associated with IP3R containing $Ca^{2+}$ stores in somatic cells. Interestingly, $Ca^{2+}$ stored in the RNE of human spermatozoa is apparently mobilised upon progesterone stimulation by $Ca^{2+}$-induced $Ca^{2+}$ release, in an IP3-independent manner (Harper et al., 2004; see below).

## Ryanodine receptors (RyR) in spermatozoa

Initial evidence for the presence of RyR in sperm came from a study by Giannini et al. (1995), who employed both in situ hybridization methods using ribo-probes specific for the different RyR isoforms and immuno-histochemical methods using RyR isoform specific antibodies. From the analysis of mouse testis sections it was concluded that germ cells such as spermatocytes and spermatids expressed both RyR1 and RyR3 but not RyR2 (Giannini et al., 1995). A number of later studies showed that developing mouse spermatocytes and spermatids expressed both RyR1 and RyR3 (Trevino et al., 1998; Chiarella et al., 2004). Furthermore, Trevino et al. (1998) showed that only RyR3 could be detected in mature spermatozoa and that similar staining patterns were observed in both intact and acrosome-reacted spermatozoa, indicating that the localisation of the RyR3 was unlikely to be on the acrosome (Trevino et al. 1998). We have also shown that human spermatozoa can be specifically labelled with a fluorescent analogue of ryanodine (BODIPY-FL-X- ryanodine). This labelling, which appears to be mainly focussed to the rear of the sperm head that is, around the head and mid-piece junction, with only low levels of labelling around the acrosome (Harper et al., 2004), co-localises with SPCA1 and with the oscillations of $[Ca^{2+}]_i$ that occur in response to progesterone stimulation.

A number of pharmacological based studies have also indicated the presence and functionality of RyR in mature spermatozoa. Minelli et al. (2000) using digitonin permeabilised bovine sperm showed that both caffeine and ryanodine decreased $45Ca^{2+}$ accumulation within the spermatozoa in a similar manner to IP3, indicating the activation of $Ca^{2+}$ efflux channel with RyR-like properties. In intact human spermatozoa, we have also demonstrated that the progesterone-induced intracellular $[Ca^{2+}]$ oscillations are IP3-independent but could be modified by ryanodine, with low doses increasing the frequency of these intracellular $[Ca^{2+}]$ oscillations and higher doses reducing the frequency (Harper et al., 2004). In addition, the RyR inhibitor tetracaine could abolish these $Ca^{2+}$ oscillations all together (Harper et al., 2004). Spermatozoa contain a functional RyR that not only plays an important physiological role in regulating agonist-induced $Ca^{2+}$ changes but also in sperm development.

## Actions of $Ca^{2+}$-mobilising agonists: Progesterone as an example

In human spermatozoa the $[Ca^{2+}]_i$ response to progesterone has been studied in great detail. In fact, human spermatozoa appear to be unusually sensitive to progesterone (Kirkman-Brown et al., 2002). When stimulated with 3 $\mu$M progesterone, believed to be representative of concentrations present in the vicinity of the oocyte-cumulus, human spermatozoa generate a biphasic $[Ca^{2+}]_i$ response consisting of a transient (lasting 1-2 min) followed by a sustained elevation. We have suggested previously that the response to progesterone may be similar to that induced by ZP, activating a VOCC (though probably not T-type: Blackmore and Eisoldt, 1999) and possibly converging with the ZP-activated pathway on activation of store-operated influx (Barratt and Publicover, 2001). Extracellular La3 + can completely inhibit the response to progesterone (Blackmore et al., 1990), confirming the importance of membrane $Ca^{2+}$ channels, and a late component of the initial $[Ca^{2+}]_i$ transient (that is particularly sensitive to occlusion by prior progesterone stimulation; Harper et al., 2003) is sensitive to nifedipine (Kirkman-Brown et al., 2003). However, the balance of evidence from studies which have specifically attempted to demonstrate a role for VOCCs in response to progesterone does not support this model (Blackmore and Eisoldt, 1999; Garcia and Meizel, 1999). Stimulation of sperm from PLCδ4-knockout mice with 50-100 $\mu$M progesterone generates a response of reduced amplitude and greatly reduced duration compared to that of wild-type cells (Fukami *et al.*, 2003), consistent with a requirement for emptying of an IP3-sensitive store, though the high doses required to evoke large responses in murine sperm may be acting by a different pathway to that normally studied in human spermatozoa, which saturates at approximately 300 nM progesterone (Baldi *et al.*, 1991; Harper *et al.*, 2003). Attempts to demonstrate pharmacologically that the sustained elevation of $[Ca^{2+}]_i$ is due to activation of store-operated channels have produced equivocal data (Harper *et al.*, 2004; 2005). Recently, it has been shown that progesterone also activates repeated $[Ca^{2+}]_i$ oscillations in human spermatozoa, which are the result of store mobilisation (Harper *et al.*, 2004; Kirkman-Brown *et al.*, 2004). If progesterone is applied as a gradient (to represent more closely the stimulus encountered as a spermatozoon approaches the oocyte) then the initial $[Ca^{2+}]_i$ transient, a characteristic of all previous studies, does not occur, but $[Ca^{2+}]_i$ oscillations occur in a large proportion of cells (Harper *et al.*, 2004). Though IP3Rs have been localised to this area of the spermatozoa, the $[Ca^{2+}]_i$ oscillations are resistant to pharmacological treatments designed to inhibit PLC or IP3Rs, suggesting that IP3 generation is not required for their generation (Harper *et al.*, 2004). Instead, $Ca^{2+}$ influx induced by progesterone apparently activates a ryanodine (like) receptor located in the sperm neck/midpiece (probably on the RNE) leading to repetitive bursts of $Ca^{2+}$-induced $Ca^{2+}$ release. Recent work in our laboratory (Machado-Oliveira, unpublished data) shows that nitric oxide (NO) can also induce mobilisation of stored $Ca^{2+}$, apparently from the same store. Intriguingly, NO treatment interacts with the effects of very low dose (20 nM) progesterone: apparently acting synergistically at the ryanodine-like receptor (Fig. 2). Re-uptake of $Ca^{2+}$ during oscillations is thapsigargin-insensitive and apparently is dependent (at least in part) on activity of SPCA1 (Harper *et al.*, 2005).

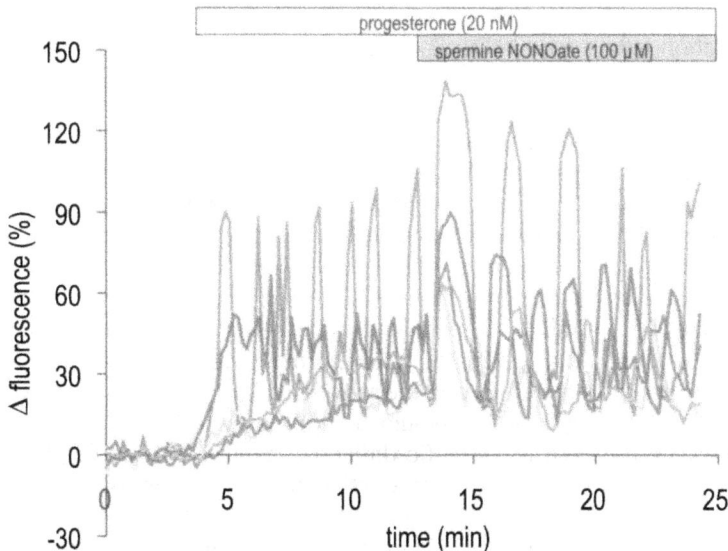

**Fig 2**. Interacting effects of progesterone and nitric oxide on sperm $Ca^{2+}$ signalling: Responses of 5 cells are shown, each in a different colour. Application of 20 nM progesterone fails to induce the biphasic response seen with mM concentrations, but nearly all cells show a slow rise in $[Ca^{2+}]_i$, which sometimes has superimposed oscillations (green trace). However, when 100 $\mu$M spermine NONOate is then applied, all cells generate a significant transient, which is followed by oscillations in many cells.

## Summary - the Ca²⁺ signalling 'toolkit' in sperm

On the basis of the data summarised above, a complex model for sperm $Ca^{2+}$-homeostasis involving several types of $Ca^{2+}$ permeable channel in the plasma membrane and at least two stores is appropriate (Fig. 1). Furthermore, it is clear that these toolkit components are distributed to allow localisation of $[Ca^{2+}]_i$ signals. As well as a range of VOCCs, which are clearly localised to sperm regions, the CatSpers, which are essential for activation of hyperactivated mobility, are expressed specifically in the principal piece of the sperm tail, as is PMCA4. It appears that the acrosome functions as an IP3-releasable store, activated by agonists linked to PLC. Recent studies suggest mobilisation of acrosomal $Ca^{2+}$ is intimately involved in activation of acrosome reaction (de Blas et al., 2002; Herrick et al., 2005). A separate store, probably the RNE, exists in the neck region of the sperm and plays a key role in regulation of flagellar beat mode (see above). Future work must address a number of questions to do with interrelatedness of these mechanisms.

## Methods to determine the pathology of sperm dysfunction

*Mouse models for male infertility: The role of knock out mice*

In order to understand the cause of the dysfunction, in particular, the regulation of calcium, it may be instructive to compare human phenotypes with two specific examples of knock out mice: CatSper1 and CatSper2 (Ren et al., 2001; Carlson et al., 2003; Quill et al., 2003) and PLCδ4 (Fukami et al., 2001; 2003). In the case of CatSper, the mutant sperm are capable of undergoing capacitation and acrosome reaction but not hyperactivation. These results support the hypothesis that $Ca^{2+}$ entry channels are crucial for the control of hyperactivation of motility. Consequently, we could predict abnormalities in the expression of CatSper in men with unexplained fertilisation failure who fail to undergo hyperactivation in response to biological agonist. In the PLCδ4 mutant animals, null males produced fewer and smaller litters *in vivo* and, *in vitro*, fewer eggs were fertilised. The majority of sperm from the PLCδ4-null mice could not fuse with zona-free eggs but fertilisation was achieved by ICSI (Fukami et al., 2001; 2003). The defect in PLCδ4 knock out animals relate to abnormalities in the influx of calcium, abnormal mobilization of calcium stores and impaired spatio-temporal dynamics (Fukami et al., 2003). There are distinct similarities between the pathology (and putative signalling mechanisms) of these mice and those in men with defective zona-induced acrosome reactions, however no detailed studies have been performed.

With the increasing number of knock out mice being generated with a male infertility phenotype (Matzuk and Lamb, 2002; *www.germonline.org*; Wiederkehr et al., 2004; *www.reprogenomics.jax.org*) (Fig. 3) and the relative ease of screening men, we should expect that the causes of sperm dysfunction would be well known. However, although knock out animals are very useful there are specific difficulties translating findings in mice to men (i) there is significant redundancy in the reproductive process; (ii) the pathology whilst similar is not the same and very detailed studies are needed on both the mouse and man to determine the real differences; and (iii) fertilisation in humans has a number of very specific differences to that in mice (see Barratt and Publicover, 2001). Consequently, successful examples of identifying gene defects in sub-fertile men by screening for genes knocked out in mice are rare (Miyamoto et al., 2003). The usual case is that no mutations are found and much effort has been wasted; for example, the examination of men with globozoospermia for mutations in casein kinase IIα (encoded by Csnk2α2 gene) (Pirrello et al., 2005).

Thus alternative (complementary) strategies to determine the defects in men with sperm dysfunction are required. Simplistically, differences between normal and dysfunctional cells

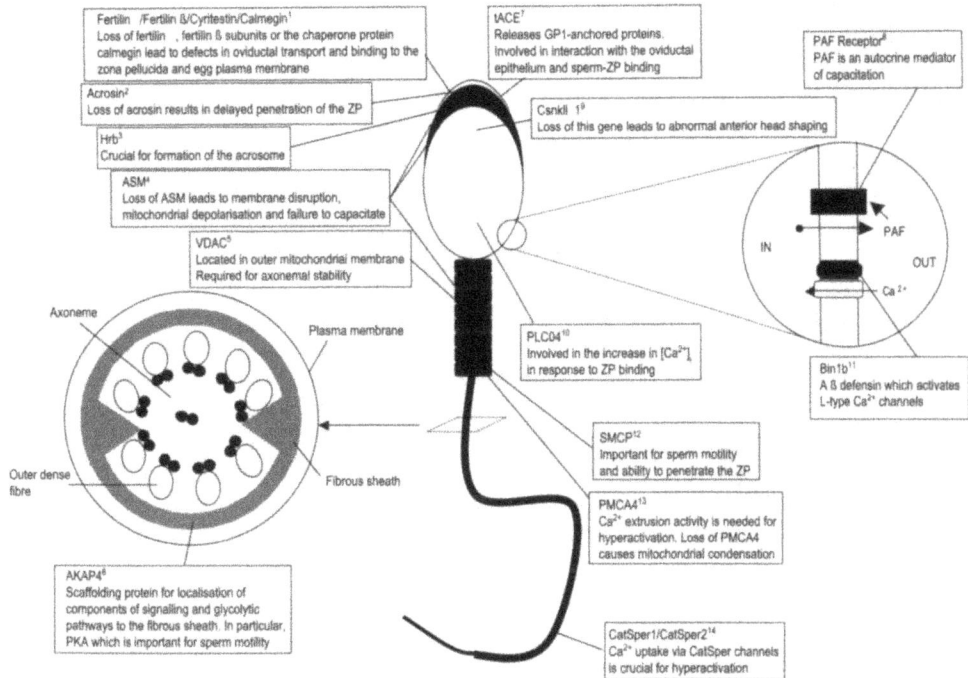

**Fig 3.** A schematic diagram of a spermatozoon illustrating the site of expression and effect of gene knock out experiments in the mouse: particular emphasis is placed on the process of sperm capacitation, transport in the female tract and interaction with the zona pellucida. [1]Ikawa *et al.*, 1997; Shamsadin *et al.*, 1999; Cho *et al.*, 1998; 2000; Nishimura *et al.*, 2001, [2]Baba *et al.*, 1994, [3]Kang-Decker *et al.*, 2001, [4]Butler *et al.*, 2002, [5]Sampson *et al.*, 2001, [6]Miki *et al.*, 2002, [7]Hagaman *et al.*, 1998; Kondoh *et al.*, 2005, [8]Koizumi *et al.*, 2003, [9]Escalier *et al.*, 2003, [10]Fukami *et al.*, 2001, [11]Zhou *et al.*, 2005, [12]Nayernia *et al.*, 2003, [13]Okunade *et al.*, 2004; Schuh *et al.*, 2004, [14]Carlson *et al.*, 2003; Quill *et al.*, 2003; Ren *et al.*, 2001 (adapted from Conner and Barratt, 2006).

can be examined using transcript or proteomic profiling. Whether sperm have functional mRNA is open to debate and thus a transcriptome approach may be limited (Ostermeier *et al.*, 2002). However, spermatozoa, as they are transcriptionally inactive, are ideal cells for proteomic analysis to examine normal cell functions and changes associated with defined correlates of fertilisation success (Ainsworth, 2005).

### Proteomics for understanding the sperm cell and diagnosis of sperm dysfunction

*The sperm proteome*

Comprehensive and systematic identification and quantification of proteins expressed in cells and tissues are providing important and fascinating insights into the dynamics of cell function. There has been a wealth of detailed proteomic studies to identify molecular signatures of disease states; for example, phospho-protein networks in cancer cells (Irish *et al.*, 2004).

Although spermatozoa is ideal to study from a proteomic perspective as it is transcriptionally inactive, there have been relatively few studies examining the proteome of human spermatozoa (Ainsworth, 2005). Studies have used sperm antibodies in an attempt to detect potential sperm targets for male contraception. This is a logical approach because antisperm antibodies are associated with sterility, *albeit* in a very limited number of cases. Unfortunately, however, this rational approach has met with limited success with very few robust candidate proteins being identified (review Bohring and Krause, 2003).

A small number of studies have attempted initial characterisation of the sperm plasma membrane (Shetty et al., 2001). Further studies have examined specific processes, for example calcium-binding proteins and proteins that are tyrosine phosphorylated (Naaby Hansen et al., 2002; Ficarro et al., 2003). Interestingly, it is over 10 years since the discovery of tyrosine phosphorylation as a putative marker of capacitation yet the role of the proteins and their sequence of activation is very sketchy and, with the exception of the AKAPs (AKAP3 and AKAP4), only a small number of candidate proteins have been identified (see Naz and Rajesh, 2004) and we are still a long way from obtaining even a minimal 'picture' of events.

In our laboratory, we have been using proteomic strategies to identify defects in sperm function responsible for failures in fertilisation (Lefievre et al., 2003; 2004; Pixton et al., 2004). Specifically we are interested in identifying differences in sperm protein expression between control (fertile) men and patients with spermatozoa that failed to fertilise oocytes *in vitro*. Our initial studies have focused on a 2D gel-based approach and developing a series of fertile controls (with several ejaculates) in order to determine, if any differences observed in the patient samples are real. Initial results are interesting. To our surprise, there was relatively little intra-donor and inter-donor variation (1.4% and 1.8% of the total number of spots identified, respectively) (see Pixton et al., 2004). However, differences between gels do occur and when accounting for this, we have categorized one man (Pixton et al., 2004), where we have identified 20 differences from the control that we are confident represent true differences.

We expanded this study to include a further five men (six patients in total, designated A-F), all of whom were normozoospermic but had complete failure of fertilisation. 2D proteomic analysis revealed several consistent differences in protein expression levels within the group compared to normozoospermic controls (Table 2). Given the vast scope for errors occurring during spermatogenesis, capacitation and fertilisation this is a promising finding that indicates that such proteomic studies have the potential to identify important biomarkers for fertilisation success.

However, our approach was limited as it only employed 2-D PAGE. A combination of proteomic approaches is required; for example, Ostrowski and colleagues (Ostrowski et al., 2002) only identified 38 potential proteins using 2-D PAGE. Whilst they use only colloidal Coomassie staining instead of more sensitive silver staining for the sequencing gel this does raise concerns about this approach. A number of proteins were not resolved (for example, dynein heavy chains which have a large molecular mass) and complementary approaches were needed to provide a detailed picture. One dimensional gels identified another 110 proteins. A second approach involved isolated axonemes, followed by digestion and analysis directly by LC/MS/MS or multi dimensional LC/MS/MS. This led to the identification of a further 66 proteins. In conclusion, more than one proteomic strategy needs to be employed as all of these workflows identify proteins not observed by any of the other approaches.

Indeed in studies we have performed on the human zona pellucida (ZP), we had to try several different methods to definitively identify ZP1 as a fourth zona protein. The most successful approach used direct trypsin digestion on solubilised human zonae followed by MS/MS (Lefievre et al., 2004).

A further difficulty in comparing sperm samples (for example, patient *versus* control) is exactly how to do this. Our initial studies have used gel-based software to compare different gels which is appropriate for absolute differences (Pixton et al., 2004). What is really required is comparison between the samples run and analyzed simultaneously (that is, internal controls).

**Table 2.** Summary table of proteins found to have undergone at least a 4-fold change in expression levels as, determined by 2-D electrophoresis, in at least 2 out of the 6 patients assessed.

| Protein | Expression profile | No. of peptides identified | Accession number |
|---|---|---|---|
| Phosphoprotein phosphatase 1-gamma catalytic chain | Reduced in patients B, C and F | 4 | NP_002701 |
| Isocitrate dehydrogenase (NAD) alpha chain precursor | Reduced in patients C, E and F | 4 | NP_005521 |
| Glutathione-S-transferase Mu 3 | Reduced in patients A, C, E and F | 10 | P21266 |
| Secretory actin binding protein | Increased in patients A, B, E and F | 18 | P12273 |
| Lysozyme-like acrosomal sperm-specific secretory protein ALLP17 | Reduced in patients D and E | 8 | AAK01478 |
| Clusterin | Reduced in patients C, E and F | 8 | P10909 |
| Lactate dehydrogenase (testis-specific) | Reduced in patients B, C, E and F | 12 | AAD14939 |
| Voltage dependent anion channel 2 | Reduced in patients D and F | 10 | NP_003366 |
| Semenogelin I | Increased in patients C, D and E | 9 | CAA05213 |
| Semenogelin II | Reduced in patients C and D | 4 | CAA87637 |

Semen samples were collected from patients recruited following failed fertilisation at IVF at the Assisted Conception Unit, Birmingham Women's Hospital and also from research donors who were normozoospermic in accordance with WHO guidelines (WHO, 1999) and of proven fertility. Semen was collected by masturbation after 2 to 3 days abstinence and allowed to liquefy for 30 min at 37°C (5% $CO_2$ in air). Samples were subjected to 2-D electrophoresis and tandem mass spectrometry as described in Pixton *et al.* (2004). Protein spots were selected for analysis on the basis of a >4-fold increase or decrease in one gel compared to another using Phoretix 2D v5.1 software, NonLinear Dynamics Ltd., Newcastle.

Several proteomic technologies to do this exist. Interestingly, Baker and colleagues (Baker *et al.*, 2005) analysed post-translational modifications occurring during sperm maturation in the rat using 2-D fluorescence DIGE (difference gel electrophoresis). This used sample labelling with different cyanine dyes, sample multiplexing and electrophoresis in the same gel and resolution of the different cyanine-labelled samples. Eight unambiguous proteins were identified and one (ß subunit of the mitochondrial F1ATPase) was shown to be serine phosphorylated as sperm transit the epididymis. Potentially this approach can identify subtle (20%) differences in protein expression and clearly represents the best way forward for analysis of protein expression in gels (Lilley and Friedman, 2004).

## Conclusion

There is an urgent need to develop a more detailed understanding of the physiological, biochemical and molecular functioning of the human spermatozoa. We can use this knowledge as a platform to improve the diagnosis of male infertility and importantly to develop potential non-ART based therapies. The tools at our disposal have never been more sophisticated and it is likely that rapid progress will be made in this area within the next 5 years. Perhaps then we will see a decrease in the use of inappropriate ART treatment.

## Acknowledgements

This work was supported by funding from MRC, BBSRC, The Wellcome Trust, Lord Dowding Fund for Humane Research, Genosis (Ltd) and Fonds de recherche en santé du Québec. Approval was obtained from the Local Research Ethics Committee (LREC #5570) for proteomics studies on failed–fertilisation patients. The authors wish to thank all past members of the group who have contributed to our research. We would also like to acknowledge the staff in the Assisted Conception Unit for assistance with semen samples and providing human eggs. The authors would particularly like to thank all patients and donors who took part in our research and who continue to do so.

## Disclaimer

Part of this review contains previous arguments and adaptations from manuscripts produced by our group in the last 12 months in particular: Barratt and Kirkman-Brown, 2006; Harper and Publicover, 2005; Jimenez-Gonzalez et al., 2006; Conner and Barratt, 2006; Conner et al., 2005; Harper et al., 2005.

## References

Ain R, Uma Devi K, Shivaji S and Seshagiri PB (1999) Pentoxifylline-stimulated capacitation and acrosome reaction in hamster spermatozoa: involvement of intracellular signalling molecules Molecular Human Reproduction 5 618-626

Ainsworth C (2005) Cell biology: the secret life of sperm Nature 436 770-771

Aitken RJ, Bowie H, Buckingham D, Harkiss D, Richardson DW and West KM (1992) Sperm penetration into a hyaluronic acid polymer as a means of monitoring functional competence Journal of Andrology 13 44-54

Andersen AG, Jensen TK, Carlsen E, Jorgensen N, Andersson AM, Krarup T, Keiding N and Skakkebaek NE (2000) High frequency of sub-optimal semen quality in an unselected population of young men Human Reproduction 15 366-372

Andersen AN, Gianaroli L, Felberbaum R, de Mouzon J, Nygren KG; The European IVF-monitoring programme (EIM), European Society of Human Reproduction and Embryology (ESHRE) (2005) Assisted reproductive technology in Europe, 2001. Results generated from European registers by ESHRE Human Reproduction 20 1158-1176

Arnoult C, Kazam IG, Visconti PE, Kopf GS, Villaz M and Florman HM (1999) Control of the low voltage-activated calcium channel of mouse sperm by egg ZP3 and by membrane hyperpolarization during capacitation Proceedings of the National Academy of Sciences USA 96 6757-6762

Arnoult C, Cardullo RA, Lemos JR and Florman HM (1996) Activation of mouse sperm T-type Ca$^{2+}$ channels by adhesion to the egg zona pellucida Proceedings of the National Academy of Sciences USA 93 13004-13009

Arnoult C, Villaz M and Florman HM (1998) Pharmacological properties of the T-type Ca$^{2+}$ current of mouse

spermatogenic cells Molecular Pharmacology 53 1104-1111

Baba T, Azuma S, Kashiwabara S and Toyoda Y (1994) Sperm from mice carrying a targeted mutation of the acrosin gene can penetrate the oocyte zona pellucida and effect fertilization Journal of Biological Chemistry 269 31845-31849

Baker MA, Witherdin R, Hetherington L, Cunningham-Smith K and Aitken RJ (2005) Identification of post-translational modifications that occur during sperm maturation using difference in two-dimensional gel electrophoresis Proteomics 5 1003-1012

Baldi E, Casano R, Falsetti C, Krausz C, Maggi M and Forti G (1991) Intracellular calcium accumulation and responsiveness to progesterone in capacitating human spermatozoa Journal of Andrology 12 323-330

Barratt CL and Publicover SJ (2001) Interaction between sperm and zona pellucida in male fertility Lancet 358 1660-1662

Barratt CL and Kirkman-Brown J (2006) Man-made versus female-made environment—will the real capacitation please stand up? Human Reproduction Update 12 1-2

Barratt CLR, Osborn J, Harrison PE, Monks N, Lenton EA and Cooke ID (1989) The hypo-osmotic swelling test and the sperm mucus penetration test in determining fertilisation of the human oocyte Human Reproduction 4 430-434

Berridge MJ (2005) Unlocking the secrets of cell signalling Annual Review of Physiology 67 1-21

Berrios J, Osses N, Opazo C, Arenas G, Mercado L, Benos DJ and Reyes JG (1998) Intracellular Ca$^{2+}$ homeostasis in rat round spermatids Biology of the Cell 90 391-398

Bjorndahl L and Barratt CL (2005) Semen analysis: setting standards for the measurement of sperm numbers Journal of Andrology 26 11

Blackmore PF (1993) Thapsigargin elevates and potentiates the ability of progesterone to increase intracellular free calcium in human sperm: possible role of perinuclear calcium *Cell Calcium* **14** 53-60

Blackmore PF and Eisoldt S (1999) The neoglycoprotein mannose-bovine serum albumin, but not progesterone, activates T-type calcium channels in human spermatozoa *Molecular Human Reproduction* **5** 498-506

Blackmore PF, Beebe SJ, Danforth DR and Alexander N (1990) Progesterone and 17 alpha-hydroxy-progesterone. Novel stimulators of calcium influx in human sperm *Journal of Biological Chemistry* **265** 1376-1380

Bohring C and Krause W (2003) Immune infertility: Towards a better understanding of sperm (auto)-immunity. The value of proteomic analysis *Human Reproduction* **18** 915-924

Butler A, He X, Gordon RE, Wu HS, Gatt S and Schuchman EH (2002) Reproductive pathology and sperm physiology in acid sphingomyelinase-deficient mice *American Journal of Pathology* **161** 1061-1075

Carlson AE, Westenbroek RE, Quill T, Ren D, Clapham DE, Hille B, Garbers DL and Babcock OF (2003) CatSper1 required for evoked Ca²⁺ entry and control of flagellar function in sperm *Proceedings of the National Academy of Sciences USA* **100** 14864-14868

CDC (2005) www.cdc.gov

Chiarella P, Puglisi R, Sorrentino V, Boitani C and Stefanini M (2004) Ryanodine receptors are expressed and functionally active in mouse spermatogenic cells and their inhibition interferes with spermatogonial differentiation *Journal of Cell Science* **117** 4127-4134

Cho C, Bunch DO, Faure JE, Goulding EH, Eddy EM, Primakoff P and Myles DG (1998) Fertilization defects in sperm from mice lacking fertilin beta *Science* **281** 1857-1859

Cho C, Ge H, Branciforte D, Primakoff P and Myles DG (2000) Analysis of mouse fertilin in wild-type and fertilin beta (-/-) sperm: evidence for C-terminal modification alpha/beta dimerization and lack of essential role of fertilin alpha in sperm-egg fusion *Developmental Biology* **222** 289-295

Conner SJ and Barratt CLR (2006) Genomic and proteomic approaches to defining sperm production and function. In *The Sperm Cell- Production, Maturation, Fertilisation, Regeneration* pp 49-71 Eds C De Jonge and CLRBarratt. Cambridge University Press, Cambridge, UK

Conner SJ, Lefievre L, Hughes DC and Barratt CL (2005) Cracking the egg: increased complexity in the zona pellucida *Human Reproduction* **20** 1148-1152

de Blas G, Michaut M, Trevino CL, Tomes CN, Yunes R, Darszon A and Mayorga LS (2002) The intra-acrosomal calcium pool plays a direct role in acrosomal exocytosis *Journal of Biological Chemistry* **277** 49326-49331

Escalier D, Silvius D and Xu X (2003) Spermatogenesis of mice lacking CK2alpha': failure of germ cell survival and characteristic modifications of the spermatid nucleus *Molecular Reproduction and Development* **66** 190-201

ESHRE (European Society of Human Reproduction and Embryology) Andrology Special Interest Group (1996) Consensus workshop on advanced diagnostic andrology techniques *Human Reproduction* **11** 1463-1479

Evans JP and Florman HM (2002) The state of the union: the cell biology of fertilization *Nature Cell Biology* Supplement s57-s63

Ficarro S, Chertihin O, Westbrook VA, White F, Jayes F, Kalab P, Marto JA, Shabanowitz J, Herr JC, Hunt DF and Visconti PE (2003) Phosphoproteome analysis of capacitated human sperm: Evidence of tyrosine phosphorylation of a kinase-anchoring protein 3 and valosin-containing protein/p97 during capacitation *Journal of Biological Chemistry* **278** 11579-11589

Fukami K, Nakao K, Inoue T, Kataoka Y, Kurokawa M, Fissore RA, Nakamura K, Katsuki M, Mikoshiba K, Yoshida N and Takenawa T (2001) Requirement of phospholipase Cdelta4 for the zona pellucida-induced acrosome reaction *Science* **292** 920-923

Fukami K, Yoshida M, Inoue T, Kurokawa M, Fissore RA, Yoshida N, Mikoshiba K and Takenawa T (2003) Phospholipase Cδ4 is required for Ca²⁺ mobilization essential for acrosome reaction in sperm *Journal of Cell Biology* **161** 79-88

Garcia MA and Meizel S (1999) Progesterone-mediated calcium influx and acrosome reaction of human spermatozoa: pharmacological investigation of T-type calcium channels *Biology of Reproduction* **60** 102-109

Giannini G, Conti A, Mammarella S, Scrobogna M and Sorrentino V (1995) The ryanodine receptor/calcium channel genes are widely and differentially expressed in murine brain and peripheral tissues *Journal of Cell Biology* **128** 893-904

Gonzalez-Martinez MT, Galindo BE, De La Torre L, Zapata O, Rodriguez E, Florman HM and Darszon A (2001) A sustained increase in intracellular Ca²⁺ is required for the acrosome reaction in sea urchin sperm *Developmental Biology* **236** 220-229

Goodwin LO, Karabinus DS, Pergolizzi RG and Benoff S (2000) L-type voltage-dependent calcium channel alpha-1C subunit mRNA is present in ejaculated human spermatozoa *Molecular Human Reproduction* **6** 127-136

Gu Y, Kirkman-Brown JC, Korchev Y, Barratt CL and Publicover SJ (2004) Multi-state, 4-aminopyridine-sensitive ion channels in human spermatozoa *Developmental Biology* **274** 308-317

Hagaman JR, Moyer JS, Bachman ES, Sibony M, Magyar PL, Welch JE, Smithies O, Krege JH and O'Brian DA (1998) Angiotensin-converting enzyme and male fertility *Proceedings of the National Academy of Sciences USA* **95** 2552-2557

Hansen M, Bower C, Milne E, de Klerk N and Kurinczuk JJ (2005) Assisted reproductive technologies and the risk of birth defects-a systematic review *Human Reproduction* **20** 328-338

Harper CV and Publicover SJ (2005) Reassessing the role of progesterone in fertilization—compartmentalized calcium signalling in human spermatozoa? *Human Reproduction* **20** 2675-2680

Harper CV, Kirkman-Brown JC, Barratt CL and Publicover SJ (2003) Encoding of progesterone stimulus intensity by intracellular [Ca$^{2+}$] ([Ca$^{2+}$]i) in human spermatozoa *Biochemical Journal* **372** 407-417

Harper CV, Barratt CL and Publicover SJ (2004) Stimulation of human spermatozoa with progesterone gradients to simulate approach to the oocyte. Induction of [Ca$^{2+}$]$_i$ oscillations and cyclical transitions in flagellar beating *Journal of Biological Chemistry* **279** 46315-46325

Harper C, Wootton L, Michelangeli F, Lefievre L, Barratt C and Publicover S (2005) Secretory pathway Ca$^{2+}$-ATPase (SPCA1) Ca$^{2+}$ pumps, not SERCAs, regulate complex [Ca$^{2+}$]i signals in human spermatozoa *Journal of Cell Science* **118** 1673-1185

Harteneck C, Plant TD and Schultz G (2000) From worm to man: three subfamilies of TRP channels *Trends in Neurosciences* **23** 159-166

Herrick SB, Schweissinger DL, Kim SW, Bayan KR, Mann S and Cardullo RA (2005) The acrosomal vesicle of mouse sperm is a calcium store *Journal of Cell Physiology* **202** 663-671

HFEA (2005) www.hfea.gov.uk

Hirohashi N and Vacquier D (2003) Store-operated calcium channels trigger exocytosis of the sea urchin sperm acrosomal vesicle *Biochemical and Biophysical Research Communications* **304** 285-292

Ho HC and Suarez SS (2003) Characterization of the intracellular calcium store at the base of the sperm flagellum that regulates hyperactivated motility *Biology of Reproduction* **68** 1590-1596

Ikawa M, Wada I, Kominami K, Watanabe D, Toshimori K, Nishimune Y and Okabe M (1997) The putative chaperone calmegin is required for sperm fertility *Nature* **387** 607-611

Irish JM, Hovland R, Krutzik PO, Perez OD, Bruserud O, Gjertsen BT and Nolan GP (2004) Single cell profiling of potentiated phospho-protein networks in cancer cells *Cell* **118** 217-228

Ivic A, Onyeaka H, Girling A, Brewis IA, Ola B, Hammadieh N, Papaioannou S and Barratt CL (2002) Critical evaluation of methylcellulose as an alternative medium in sperm migration tests *Human Reproduction* **17** 143-149

Jacob A and Benoff S (2000) Full length low voltage-activated ('T-type' Ca$^{2+}$ channel α1G mRNA is not detected in mammalian testis and sperm *Journal of Andrology* **56** 48

Jagannathan S, Punt EL, Gu Y, Arnoult C, Sakkas D, Barratt CL and Publicover SJ (2002) Identification and localization of T-type voltage-operated calcium channel subunits in human male germ cells. Expression of multiple isoforms *Journal of Biological Chemistry* **277** 8449-8456

Jimenez-Gonzalez C, Michelangeli F, Harper CV, Barratt CLR and Publicover SJ (2006) Calcium signalling in human spermatozoa: a specialized 'toolkit' of channels, transporters and stores *Human Reproduction Update* **3** 253-267

Kamischke A and Nieschlag E (2002) Diagnosis and treatment of male infertility. In *Assisted Reproductive Technology-Accomplishments and New Horizons* pp 231-254 Eds CJ De Jonge and CLR Barratt. Cambridge University Press, Cambridge, UK

Kang-Decker N, Mantchev GT, Juneja SC, McNiven MA and van Deursen JM (2001) Lack of acrosome formation in Hrb-deficient mice *Science* **294** 1531-1533

Kaupp UB, Solzin J, Hildebrand E, Brown JE, Helbig A, Hagen V, Beyermann M, Pampaloni F and Weyand I (2003) The signal flow and motor response controlling chemotaxis of sea urchin sperm *Nature Cell Biology* **5** 109-117

Kirkman-Brown JC, Punt EL, Barratt CL and Publicover SJ (2002) Zona pellucida and progesterone-induced Ca$^{2+}$ signalling and acrosome reaction in human spermatozoa *Journal of Andrology* **23** 306-315

Kirkman-Brown JC, Barratt CL and Publicover SJ (2003) Nifedipine reveals the existence of two discrete components of the progesterone-induced [Ca$^{2+}$]$_i$ transient in human spermatozoa *Developmental Biology* **259** 71-82

Kirkman-Brown JC, Barratt CL and Publicover SJ (2004) Slow calcium oscillations in human spermatozoa *Biochemical Journal* **378** 827-832

Kobori H, Miyazaki S and Kuwabara Y (2000) Characterization of intracellular Ca$^{2+}$ increase in response to progesterone and cyclic nucleotides in mouse spermatozoa *Biology of Reproduction* **63** 113-120

Koizumi H, Yamaguchi N, Hattori M, Ishikawa TO, Aoki J, Taketo MM, Inoue K and Arai H (2003) Targeted disruption of intracellular type I platelet activating factor-acetylhydrolase catalytic subunits causes severe impairment in spermatogenesis *Journal of Biological Chemistry* **278** 12489-12494

Kondoh G, Tojo H, Nakatani Y, Komazawa N, Murata G, Yamagata K, Maeda Y, Kinoshita T, Okabe M, Taguchi R and Takeda J (2005) Angiotensin-converting enzyme is a GPI-anchored protein releasing factor crucial for fertilization *Nature Medicine* **11** 160-166

Krausz C, Bonaccorsi L, Maggio P, Luconi M, Criscuoli L, Fuzzi B, Pellegrini S, Forti G and Baldi E (1996) Two functional assays of sperm responsiveness to progesterone and their predictive values in in-vitro fertilization *Human Reproduction* **11** 1661-1667

Kuroda Y, Kaneko S, Yoshimura Y, Nozawa S and Mikoshiba K (1999) Are there inositol 1,4,5-triphosphate (IP3) receptors in human sperm? *Life Sciences* **65** 135-143

Lefièvre L, De Lamirande E and Gagnon C (2000) The cyclic GMP-specific phosphodiesterase inhibitor, sildenafil, stimulates human sperm motility and capacitation but not acrosome reaction *Journal of Andrology* **21** 929-937

Lefièvre L, Barratt CL, Harper CV, Conner SJ, Flesch FM, Deeks E, Moseley FL, Pixton KL, Brewis IA and Publicover SJ (2003) Physiological and proteomic approaches to studying prefertilization events in the human *Reproductive Biomedicine Online* **7** 419-427

Lefièvre L, Conner SJ, Salpekar A, Olufowobi O, Ashton P, Pavlovic B, Lenton W, Afnan M, Brewis IA, Monk M, Hughes DC and Barratt CL. (2004) Four zona pellucida glycoproteins are expressed in the human *Human Reproduction* 19 1580-1586

Lilley KS and Friedman DB (2004) All about DIGE: quantification technology for differential-display 2D-gel proteomics *Expert Review of Proteomics* 1 401-409

Liu DY and Baker HW (2000) Defective sperm-zona pellucida interaction: a major cause of failure of fertilization in clinical in-vitro fertilization *Human Reproduction* 15 702-708

Lobley A, Pierron V, Reynolds L, Allen L and Michalovich D (2003) Identification of human and mouse CatSper3 and CatSper4 genes: characterisation of a common interaction domain and evidence for expression in testis *Reproductive Biology and Endocrinology* 1 53

Matzuk MM and Lamb DJ (2002) Genetic dissection of mammalian fertility pathways *Nature Cell Biology* 4 (Suppl) 41-49

Mackenna A, Barratt CL, Kessopoulou E and Cooke I (1993) The contribution of a hidden male factor to unexplained infertility *Fertility and Sterility* 59 405-411

Maher ER, Afnan M and Barratt CL (2003) Epigenetic risks related to assisted reproductive technologies: epigenetics, imprinting, ART and icebergs? *Human Reproduction* 18 2508-2511

Michelangeli F, Ogunbayo OA and Wootton LL (2005) A plethora of interacting organellar Ca$^{2+}$ stores Current *Opinion in Cell Biology* 17 135-140

Miki K, Willis WD, Brown PR, Goulding EH, Fulcher KD and Eddy EM (2002) Targeted disruption of the Akap4 gene causes defects in sperm flagellum and motility *Developmental Biology* 248 331-342

Minelli A, Allegrucci C, Rosati R and Mezzasoma I (2000) Molecular and binding characteristics of IP3 receptors in bovine spermatozoa *Molecular Reproduction and Development* 56 527-533

Miyamoto T, Hasuike S, Yogev L, Maduro MR, Ishikawa M, Westphal H and Lamb DJ (2003) Azoospermia in patients heterozygous for a mutation in SYCP3 *Lancet* 362 1714-1719

Monteil A, Chemin J, Bourinet E, Mennessier G, Lory P and Nargeot J (2000a) Molecular and functional properties of the human alpha (1G) subunit that forms T-type calcium channels *Journal of Biological Chemistry* 275 6090-6100

Monteil A, Chemin J, Leuranguer V, Altier C, Mennessier G, Bourinet E, Lory P and Nargeot J (2000b) Specific properties of T-type calcium channels generated by the human alpha 1I subunit *Journal of Biological Chemistry* 275 16530-16535

Muller CH (2000) Rationale, interpretation, validation, and uses of sperm function tests *Journal of Andrology* 21 10-30

Naaby-Hansen S, Mandal A, Wolkowicz MJ, Sen B, Westbrook VA, Shetty J, Coonrod SA, Klotz KL, Kim YH, Bush LA, Flickinger CJ and Herr JC (2002) CABYR a novel calcium-binding tyrosine phosphorylation-regulated fibrous sheath protein involved in capacitation *Developmental Biology* 242 236-254

Naaby-Hansen S, Wolkowicz MJ, Klotz K, Bush LA, Westbrook VA, Shibahara H, Shetty J, Coonrod SA, Reddi PP, Shannon J, Kinter M, Sherman NE, Fox J, Flickinger CJ and Herr JC (2001) Co-localization of the inositol 1,4,5-triphosphate receptor and calreticulin in the equatorial segment and in membrane bounded vesicles in the cytoplasmic droplet of human spermatozoa *Molecular Human Reproduction* 10 923-933

Nayernia K, Drabent B, Adham IM, Moschner M, Wolf S, Meinhardt A and Engel W (2003) Male mice lacking three germ cell expressed genes are fertile *Biology of Reproduction* 69 1973-1978

Naz RK and Rajesh PB (2004) Role of tyrosine phosphorylation in sperm capacitation/acrosome reaction *Reproductive Biology and Endocrinology* 2 75

Nikpoor P, Mowla SJ, Movahedin M, Ziaee SA and Tiraihi T (2004) CatSper gene expression in postnatal development of mouse testis and in sub-fertile men with deficient sperm motility *Human Reproduction* 19 124-128

Nishimura H, Cho C, Branciforte DR, Myles DG and Primakoff P (2001) Analysis of loss of adhesive function in sperm lacking cyritestin or fertilin beta *Developmental Biology* 233 204-213

O'Toole CMB, Arnoult C, Darszon A, Steinhardt RA and Florman HM (2000) Ca$^{2+}$ entry through store-operated channels in mouse sperm is initiated by egg ZP3 and drives the acrosome reaction *Molecular Biology of the Cell* 11 1571-1584

Okunade GW, Miller ML, Pyne GJ, Sutliff RL, O'Connor KT, Neumann JC, Andringa A, Miller DA, Prasad V, Doetschman T, Paul RJ and Shull GE (2004) Targeted ablation of plasma membrane Ca$^{2+}$-ATPase (PMCA) 1 and 4 indicates a major housekeeping function for PMCA1 and a critical role in hyperactivated sperm motility and male fertility for PMCA4 *Journal of Biological Chemistry* 279 33742-33750

Ostermeier GC, Dix DJ, Miller D, Khatri P and Krawetz SA (2002) Spermatozoal RNA profiles of normal fertile men *Lancet* 360 772-777

Ostrowski LE, Blackburn K, Radde KM, Moyer MB, Schlatzer DM, Moseley A and Boucher RC (2002) A proteomic analysis of human cilia: identification of novel components *Molecular & Cellular Proteomics* 1 451-465

Park JY, Ahn HJ, Gu JG, Lee KH, Kim JS, Kang HW and Lee JH (2003) Molecular identification of Ca$^{2+}$ channels in human sperm *Experimental Molecular Medicine* 35 285-292

Pirrello O, Machey N, Schmidt F, Terriou P, Menezo Y and Viville S (2005) Search for mutations involved in human globozoospermia *Human Reproduction* 20 1314-1318

Pixton KL, Deeks ED, Flesch FM, Moseley FL, Bjorndahl L, Ashton PR, Barratt CL and Brewis IA (2004) Sperm proteome mapping of a patient who experienced failed fertilization at IVF reveals altered expression of at least 20 proteins compared with fertile donors: case report *Human Reproduction* 19 1438-1447

Publicover SJ and Barratt C (1999) Voltage-operated Ca$^{2+}$

channels and the acrosome reaction: which channels are present and what do they do? *Human Reproduction* **14** 873-879

Putney JW Jr (1990) Receptor-regulated calcium entry *Pharmacology & Therapeutics* **48** 427-434

Quill TA, Ren D, Clapham DE and Garbers DL (2001) A voltage-gated ion channel expressed specifically in spermatozoa *Proceedings of the National Academy of Sciences USA* **98** 12527-12531

Quill TA, Sugden SA, Rossi KL, Doolittle LK, Hammer RE and Garbers DL (2003) Hyperactivated sperm motility driven by CatSper2 is required for fertilization *Proceedings of the National Academy of Sciences USA* **100** 14869-14874

Ren D, Navarro B, Perez G, Jackson AC, Hsu S, Shi Q, Tilly JL and Clapham DE (2001) A sperm ion channel required for sperm motility and male fertility *Nature* **413** 603-609

Sampson MJ, Decker WK, Beaudet AL, Ruiteneek W, Armstrong D, Hicks MJ and Craigen WJ (2001) Immotile sperm and infertility in mice lacking mitochondrial voltage-dependent anion channel type 3 *Journal of Biological Chemistry* **276** 39206-39212

Schuh K, Cartwright EJ, Jankevics E, Bundschu K, Liebermann J, Williams JC, Armesilla AL, Emerson M, Oceandy D, Knobeloch KP and Neyses L (2004) Plasma membrane Ca²⁺ ATPase 4 is required for sperm motility and male fertility *Journal of Biological Chemistry* **279** 28220-28226

Shamsadin R, Adham IM, Nayernia K, Heinlein UA, Oberwinker H and Engel W (1999) Male mice deficient for germ-cell cyritestin are infertile *Biology of Reproduction* **61** 1445-1451

Sharpe RM and Irvine DS (2004) How strong is the evidence of a link between environmental chemicals and adverse effects on human reproductive health? *British Medical Journal* **328** 447-451

Shetty J, Diekman AB, Jayes FC, Sherman NE, Naaby-Hansen S, Flickinger CJ and Herr JC (2001) Differential extraction and enrichment of human sperm surface proteins in a proteome: identification of immunocontraceptive candidates *Electrophoresis* **22** 3053-3066

Son WY, Lee JH, Lee JH and Han CT (2000) Acrosome reaction of human spermatozoa is mainly mediated by α1H T-type calcium channels *Molecular Human Reproduction* **6** 893-897

Spehr M, Gisselmann G, Poplawski A, Riffell JA, Wetzel CH, Zimmer RK and Hatt H (2003) Identification of a testicular odorant receptor mediating human sperm chemotaxis *Science* **299** 2054-2058

Stamboulian S, Kim D, Shin HS, Ronjat M, De Waard M and Arnoult C (2004) Biophysical and pharmacological characterization of spermatogenic T-type calcium current in mice lacking the Ca$_v$3.1 (alpha1G) calcium channel: Ca$_v$3.2 (alpha1H) is the main functional calcium channel in wild-type spermatogenic cells *Journal of Cellular Physiology* **200** 116-124

Su YH and Vacquier VD (2002) A flagellar K⁺-dependent Na⁺/Ca²⁺ exchanger keeps Ca²⁺ low in sea urchin spermatozoa *Proceedings of the National Academy of Sciences USA* **99** 6743-6748

Templeton A, Fraser C and Thompson B (1990) The epidemiology of infertility in Aberdeen *British Medical Journal* **301** 148-152

Tomlinson MJ, Kessopoulou E and Barratt CL (1999) The diagnostic and prognostic value of traditional semen parameters *Journal of Andrology* **20** 588-593

Tournaye H, Verheyen G, Albano C, Camus M, Van Landuyt L, Devroey P and Van Steirteghem A (2002) Intracytoplasmic sperm injection versus in vitro fertilization: a randomized controlled trial and a meta-analysis of the literature *Fertility and Sterility* **78** 1030-1037

Toyoshima C and Inesi G (2004) Structural basis of ion pumping by Ca²⁺-ATPase of the sarcoplasmic reticulum *Annual Review of Biochemistry* **73** 269-292

Trevino CL, Santi CM, Beltran C, Hernandez-Cruz A, Darszon A and Lomeli H (1998) Localisation of inositol trisphosphate and ryanodine receptors during mouse spermatogenesis: possible functional implications *Zygote* **6** 159-172

Trevino CL, Felix R, Castellano LE, Gutierrez C, Rodriguez D, Pacheco J, Lopez-Gonzalez I, Gomora JC, Tsutsumi V, Hernandez-Cruz A, Fiordelisio T, Scaling AL and Darszon A (2004) Expression and differential cell distribution of low-threshold Ca²⁺ channels in mammalian male germ cells and sperm *Federation of European Biochemical Societies Letters* **563** 87-92

Wennemuth G, Westenbroek RE, Xu T, Hille B and Babcock DF (2000) CaV2.2 and CaV2.3 (N- and R-type) Ca²⁺ channels in depolarization-evoked entry of Ca²⁺ into mouse sperm *Journal of Biological Chemistry* **275** 21210-21217

Wennemuth G, Babcock DF and Hille B (2003) Calcium clearance mechanisms of mouse sperm *The Journal of General Physiology* **122** 115-128

Westenbroek RE and Babcock DF (1999) Discrete regional distributions suggest diverse functional roles of calcium channel alpha1 subunits in sperm *Developmental Biology* **207** 457-469

Weyand I, Godde M, Frings S, Weiner J, Muller F, Altenhofen W, Hatt H and Kaupp UB (1994) Cloning and functional expression of a cyclic-nucleotide-gated channel from mammalian sperm *Nature* **368** 859-863

Wictome M, Henderson I, Lee AG and East JM (1992) Mechanism of inhibition of the calcium pump of sarcoplasmic reticulum by thapsigargin *Biochemical Journal* **283** 525-529

Wiederkehr C, Basavaraj R, Sarrauste de Menthiere C, Koch R, Schlecht U, Hermida L, Masdoua B, Ishii R, Cassen V, Yamamoto M, Lane C, Cherry M, Lamb N and Primig M (2004) Database model and specification of GermOnline Release 2.0, a cross-species community annotation knowledge base on germ cell differentiation *Bioinformatics* **20** 808-811

Wiesner B, Weiner J, Middendorff R, Hagen V, Kaupp UB and Weyand I (1998) Cyclic nucleotide-gated channels on the flagellum control Ca²⁺ entry into sperm *The Journal of Cell Biology* **142** 473-484

**Wild S, Roglic G, Green A, Sicree R and King H** (2004) Global prevalence of diabetes: estimates for the year 2000 and projections for 2030 *Diabetes Care* **27** 1047-1053

**Wootton LL, Argent CC, Wheatley M and Michelangeli F** (2004) The expression, activity and localisation of the secretory pathway $Ca^{2+}$-ATPase (SPCA1) in different mammalian tissues *Biochimica et Biophysica Acta* **1664** 189-197

**World Health Organization (WHO)** (1999) WHO Laboratory Manual for the Examination of Human Semen and Sperm-Cervical Mucus Interaction Cambridge University Press, Cambridge, United Kingdom

**Yoshida M, Ishikawa M, Izumi H, De Santis R and Morisawa M** (2003) Store-operated calcium channel regulates the chemotactic behavior of ascidian sperm *Proceedings of the National Academy of Sciences USA* **100** 149-154

**Zhou Q, Shima JE, Nie R, Friel PJ and Griswold MD** (2005) Androgen-regulated transcripts in the neonatal mouse testis as determined through microarray analysis *Biology of Reproduction* **72** 1010-1019

# Biology of sperm capacitation: evidence for multiple signalling pathways

*Daulat R P Tulsiani[1], Hai-Tao Zeng[2] and Aida Abou-Haila[3]*

[1]Departments of Obstetrics & Gynecology and Cell & Developmental Biology, Vanderbilt University School of Medicine, Nashville, TN 37232, USA; [2]Research Center of Molecular Biology, Xiang-Ya School of Medicine, Central South University, Changsha, Hunan 410078, People's Republic of China; [3]UFR Biomédicale, Université René Descartes Paris V, 45 rue des Saints-Pères, Paris, France

The endpoint of *in vitro/in vivo* capacitation is the ability of sperm surface receptors to bind to their complementary ligands on zona pellucida, the extracellular glycocalyx that surrounds the egg, and undergo the $Ca^{2+}$-dependent signal transduction. The net result is the fenestration and fusion of the sperm plasma membrane and the underlying outer acrosomal membrane at multiple sites and exocytosis of acrosomal contents. The hydrolytic action of glycohydrolases and proteinases, released at the site of sperm-zona binding, along with the enhanced thrust generated by the hyperactivated flagellar motility of the bound spermatozoon, are important factors that regulate the fertilization process. This report discusses the physiological significance of calmodulin, a 17 kDa $Ca^{2+}$ sensor protein, in sperm function. The *in vitro* experimental approaches described in this article provide evidence strongly suggesting that calmodulin plays an important role in the priming (that is, capacitation) of mouse spermatozoa as well as in the agonist-induced acrosome reaction. In addition, we have used several calmodulin antagonists in an attempt to characterize further the morphological and biochemical changes associated with sperm capacitation. Data presented in this report suggest that calmodulin antagonists prevent capacitation by interfering with multiple regulatory pathways and do so either with or without adverse effects on sperm motility and protein tyrosine phosphorylation of sperm components.

### Introduction

Mammalian fertilization is the result of a complex set of molecular events which enable the capacitated spermatozoon to recognize and bind to the egg's extracellular coat, the zona pellucida, undergo acrosomal exocytosis and fertilize the egg (Yanagimachi, 1994; Tulsiani *et al.*, 1997). In order to successfully fertilize an egg, mammalian spermatozoa undergo a series of biochemical and functional changes during: i) development in the testis (spermatogenesis); ii) maturation in the epididymis, a process termed epididymal maturation; and iii) capacitation in the female genital tract. Although the biochemical and functional modifications during capacitation may vary from species to species, the net result of the multifaceted process in all species

---

E-mail: daulat.tulsiani@vanderbilt.edu

examined is sperm hyperactivity and their ability to traverse the vestments that surround the ovulated egg (Yanagimachi, 1994; Tulsiani et al., 1998). These vestments include the matrix of the cumulus oophorus (cumulus cells) and the zona pellucida (Tulsiani et al., 1998).

Mammalian spermatozoa contain two parts: i) the head with the acrosomal (anterior head) and post-acrosomal (posterior head) regions; and ii) flagellum comprising of the middle, principal, and end piece (Tulsiani and Abou-Haila, 2001). Whereas, the carbohydrate-binding molecules (receptors) responsible for binding to complementary sugar residues on the zona pellucida (ligands) and initiating a signal transduction cascade prior to the induction of the acrosome reaction are localized on the surface of the anterior head region of capacitated spermatozoa in many species, the hyperactivated motility is a result of molecular changes on the flagellum (Olds-Clarke, 1990; Tulsiani and Abou-Haila, 2001). Thus, sperm capacitation is the net result of changes in membrane properties and multiple enzyme activities on the head and flagellum that activate the cell signalling pathway.

Much of the knowledge on sperm capacitation has come from the pioneering work by Austin (1951) and Chang (1951). Over five decades ago, the two investigators independently reported functional changes on spermatozoa recovered from the female genital tract after mating or the use of epididymal spermatozoa preincubated with oviductal secretions (Yanagimachi and Chang, 1963). The precise site of capacitation may vary in different species; however, several studies suggest that capacitation is most efficient when spermatozoa pass through the uterus and oviduct. The secretions collected from the oviduct of estrous females have been demonstrated to be most efficient in rendering the functional changes in spermatozoa (Yanagimachi, 1994).

The preparatory modifications associated with capacitation include removal of seminal plasma proteins/glycoproteins adsorbed to the surface of ejaculated spermatozoa and modifications/ reorganization of sperm surface molecules (Yanagimachi, 1994). The net change during capacitation is a combined effect of multiple alterations in sperm plasma membrane proteins/glycoproteins and sterols (mainly cholesterol) that modifies the ion channels in the plasmalemma of spermatozoa. The net result of these changes is plasma membrane fluidity and activation of intracellular second messengers. Combined, these modifications result in sperm hyperactivity, their ability to bind to the zona pellucida and their responsiveness to undergo the acrosome reaction (Cross, 1998). The current knowledge on the dynamics and physiology of capacitation has been extensively discussed in two reviews (Cross, 1998; Flesch and Gadella, 2000) and will not be repeated here. The major focus of this article will be to summarize recent advances on the capacitation-associated changes in spermatozoa. Our intention is also to discuss the physiological significance of calmodulin, a 17 kDa acidic protein, in sperm function and also to present evidence for the occurrence of at least two pathways that regulate capacitation.

### *In vivo* capacitation: deposition and transport of spermatozoa through the female genital tract

At coitus, millions of spermatozoa are deposited into the female genital tract. A vast majority of the deposited cells are eliminated; however, only a small percentage of sperm rapidly enters the highly folded mucus-filled cervix that serves to: (1) prevent entry of seminal plasma into the uterus; (2) exclude morphologically abnormal spermatozoa as well as potentially infectious microbes; and (3) store spermatozoa (Yanagimachi, 1994). Once in the cervix, spermatozoa move by complex passive and active processes towards the uterus and oviduct, the likely sites of *in vivo* capacitation in many species. Spermatozoa that enter the distal oviductal region (ampulla), the site of fertilization, are confronted with ovulated eggs surrounded by cumulus complex (Tulsiani *et al.*, 1998). The complex is thought to be dispersed by hyaluronidase, an enzyme present within the sperm acrosome (Meyer and Rosenberger, 1999) as well as on the sperm surface (Lin *et al.*, 1994), allowing capacitated spermatozoa passage through the dispersed cumulus complex.

## *In vitro* capacitation

Ejaculated or cauda epididymal spermatozoa can also be capacitated *in vitro* by incubating in a chemically-defined medium supplemented with bovine serum albumin (Davis, 1981) or methyl-ß-cyclodextrin (Zeng and Tulsiani, 2003), energy substances (such as glucose and pyruvate) and reagents used in Krebs-Ringer bicarbonate medium. Albumin, the major protein in the female genital tract secretions or methyl-ß-cyclodextrin is thought to facilitate capacitation by efflux of sterols, mainly cholesterol, from capacitating spermatozoa. A recent report (Wu *et al.*, 2001) presented evidence suggesting that albumin also has a role as an acceptor of phospholipids, including ether phospholipid 1-O-alkyl-2-acetyl-sn-glyceryl-3-phosphocholine or platelet activating factor (PAF). Mouse spermatozoa, incubated in a medium that favors capacitation, released PAF and expressed its receptor on the plasma membrane. Inclusion of PAF, in the capacitation medium, stimulated sperm motility and fertilizing ability of spermatozoa (Wu *et al.*, 2001). Furthermore, sperm cells from PAF receptor KO mice displayed significantly reduced rate of capacitation and *in vitro* fertilization. Combined, these data are consistent with the suggestion that capacitating sperm cells release PAF and lose membrane sterols that promote capacitation. The loss of cholesterol is thought to destabilize the sperm plasma membrane bilayer making the membrane permeable and fusogenic (Yanagimachi, 1994). Recent *in vitro* evidence suggests that there is a gradual loss of cholesterol during capacitation; however, the acrosomal responsiveness does not develop for some time, a result consistent with the author's suggestion that cholesterol loss precedes capacitation (Cross, 1998).

## Initiation of capacitation

As capacitation proceeds, a number of biochemical and morphological changes occur on spermatozoa. The known changes include: (1) efflux of cholesterol; (2) increased adenylyl cyclase activity and increased levels of cAMP; (3) protein tyrosine phosphorylation of a subset of sperm components; (4) elevated intrasperm pH; (5) $Ca^{2+}$ influx; (6) loss of sperm surface molecules; (7) modification/alteration of the sperm plasma membrane; (8) changes in the lectin-binding pattern; and (9) membrane priming. However, with the exception of cholesterol efflux (Cross, 1998), the sequence of other modifications and their significance in sperm capacitation remains evasive.

In addition to the efflux of cholesterol, another important change that initiates capacitation is the composition and distribution of sperm plasma membrane phospholipids. The loss of cholesterol decreases cholesterol/phospholipid ratio that could account for the capacitation-associated changes in membrane fluidity and distribution of sperm surface proteins/glycoproteins (Yanagimachi, 1994). Experimental evidence suggests that there is little or no change in the total amount of phospholipids; however, the distribution of each lipid changes in the outer and inner leaflet of the sperm plasma membrane (Yanagimachi, 1994). Moreover, evidence suggests methylation of phospholipid [that is, conversion of phosphatidyl ethanolamine to the phosphatidyl choline] and conversion of intracellular phosphatidic acid to cardiolipin, an anionic lipid that is inserted into the outer leaflet of the sperm plasma membrane. Thus, the plasma membrane of the capacitating/capacitated spermatozoa undergoes a variety of biochemical changes and reorientation.

Another important cell surface alteration in the capacitating/capacitated spermatozoa is a change in the lectin-binding patterns, a result consistent with the suggestion that the cell surface glycan chains are altered in capacitating/capacitated spermatozoa (Ahuja, 1985). One possible explanation could be that uncapacitated sperm cells lose surface-coating molecules that expose new glycoproteins with different glycan chains. A second possibility could be that the existing sperm surface glycoconjugates are modified *in vivo* by glycosyltransferase activities reported by us to be present in the female reproductive tract secretions (Tulsiani et al., 1996). Since the levels of these synthetic enzymes are regulated during estrous cycle, it is reasonable to assume that they have a role in modifying the existing sperm surface glycans. A third possible explanation for the changes in the lectin binding properties *in vivo* could be the association of an oviductal glycoprotein on the surface of spermatozoa (King and Killian, 1994; Abe et al., 1995). Finally, the acrosomal contents present within the acrosome of uncapacitated spermatozoa could become accessible on the surface over the acrosome as capacitation proceeds (Abou-Haila and Tulsiani, 2003). Many intra-acrosomal glycohydrolases are glycoproteins (Tulsiani et al., 1998). Their exposure to the surface of capacitating spermatozoa will be expected to alter their lectin-binding properties.

The principal goal of a majority of investigators engaged in this area of research has been to find *in vitro* conditions that enhance the ability of spermatozoa to fertilize an egg and not to elucidate the mechanism underlying capacitation-associated sperm priming. The general assumption has been that when the spermatozoon successfully fertilizes an egg *in vitro*, it signals capacitation; however, failure to fertilize the egg should not be considered as unsuccessful capacitation since other factors may influence fertilization (Yanagimachi, 1994). Since uncapacitated spermatozoa do not undergo the agonist-induced acrosome reaction, their responsiveness to agonists (zona pellucida, neoglycoproteins, and progesterone *et cetera*) is generally considered an acceptable endpoint for *in vitro* capacitation (Cross, 1998).

### Capacitating/capacitated spermatozoa undergo physiological priming prior to the interaction of opposite gametes

The sperm acrosome is filled with a host of glycohydrolases and proteinases (Tulsiani et al., 1998). The enzymes are thought to aid in the dispersion and digestion of the vestments surrounding the egg. These vestments include cumulus cells along with the innermost layer of cells, the corona radiata, and the zona pellucida (Tulsiani et al., 1998). For an acrosomal enzyme to be functional in the dispersion of cumulus cells, it has to be accessible on the surface of capacitated (acrosome-intact) spermatozoa. Interestingly, there are numerous published reports suggesting that several intra-acrosomal antigens redistribute/translocate on the surface of capacitating/capacitated spermatozoa (for review, see Tulsiani and Abou-Haila, 2004). Our immunocytochemical and biochemical approaches have provided evidence strongly suggesting that the mouse spermatozoa, incubated in a medium that favors capacitation, undergo membrane changes in a time-dependent manner. The net result is the progressive accessibility of acrosomal glycohydrolases on the surface of capacitating/capacitated spermatozoa (Abou-Haila and Tulsiani, 2003; Tulsiani and Abou-Haila, 2004).

To determine whether intra-acrosomal glycohydrolases are suitable markers to examine the capacitation-associated sperm membrane priming, we used indirect immunofluorescent (IIF) microscopy. In these studies, we used affinity purified antibodies against two glycohydrolases that cross-reacted with the intra-acrosomal enzymes only when uncapacitated sperm cells were permeabilized. Incubation of spermatozoa in a medium that favors *in vitro* capacitation induced membrane priming that allowed the antibodies to cross-react with the acrosomal enzymes in capacitating (acrosome-intact) spermatozoa without permeabilization. This was revealed by the appearance of several distinct fluorescent patterns, including an initial immunopositive lining over the acrosome cap to an intense immunopositive reaction throughout the acrosome (Fig. 1H, I). These early immunopositive patterns were followed by the appearance of intense fluorescent spots (droplets) (Fig. 1J) that seem to establish contact with the plasma membrane in a time-dependent manner (Fig. 1K, L). In some spermatozoa that appeared to have a swollen acrosome, an immunopositive reaction (stage I) was first seen at the borders of the anterior acrosome (Fig. 2A, D). As capacitation proceeds, other stages were identified by the presence of a large and intense immunofluorescent spot (Fig. 2B, E) or many small spots (Fig. 2C, F), respectively, inside the acrosomal region. Figures 1 and 2 present results of immunostaining using anti-ß-D-galactosidase IgG. Similar fluorescent patterns were observed with the antibody to ß-D-glucuronidase (data not included).

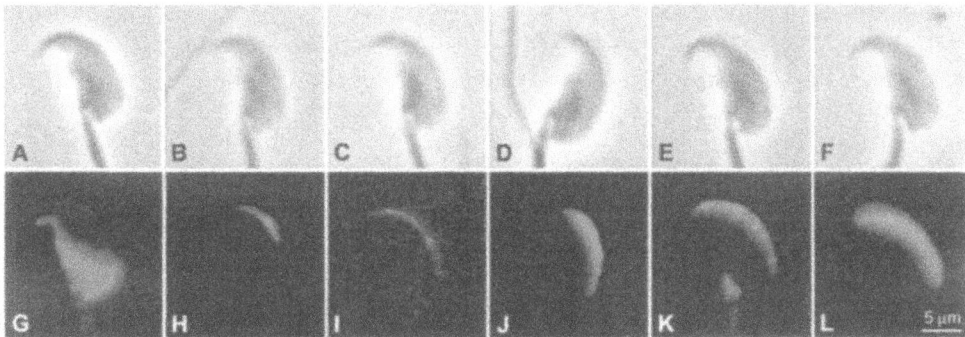

**Fig. 1** Immunolocalization of ß-D-galactosidase in capacitating mouse spermatozoa. Matched phase-contrast (A-F) and indirect immunofluorescence patterns (G-L) after sperm cells were incubated in a medium that favors capacitation, fixed without permeabilization and immunostained using anti-ß-D-galactosidase IgG as primary antibody. The distinct fluorescent patterns were grouped into three stages: stage I: H-I; stage II: j; stage III: K,L. The different images were photographed after 30 min incubation. Reproduced from Abou-Haila and Tulsiani (2003) with permission.

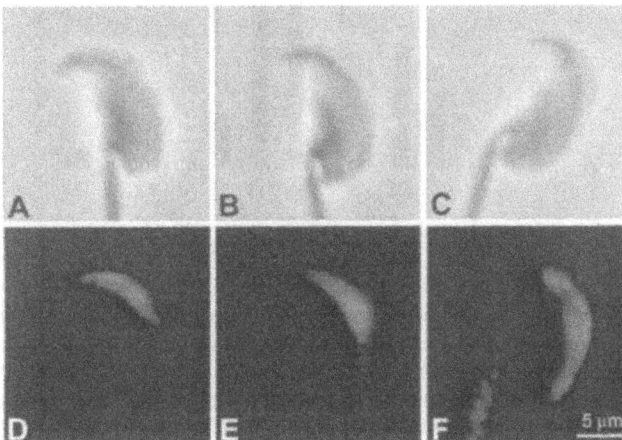

**Fig. 2** Immunolocalization of ß-D-galactosidase in capacitating mouse spermatozoa. Matched phase-contrast (A-C) and indirect immunofluorescence (D-F) of fixed non-permeabilized sperm cells treated with anti-ß-D-galactosidase IgG as primary antibody. Note the pattern of the swollen acrosome in the anterior region of the head and the immunofluorescent reaction which is heterogeneously distributed at the border of the acrosome in D and F. The immunopositive reaction in D-F corresponds to stages I-III, respectively. Reproduced from Abou-Haila and Tulsiani (2003) with permission.

The various immunopositive patterns seen in Figs. 1 and 2 were grouped into three progressive stages (stage I, stage II, and stage III). The number of spermatozoa in each stage was calculated and plotted as a function of the incubation time. Data presented in Fig. 3 demonstrate that the percentage of spermatozoa in each stage changed with incubation time (Fig. 3A,B left). For instance, the percentage of spermatozoa in stage I increased from ~20% to ~30% between 15 and 30 min incubation; however, it decreased to ~5% after 60 min (Fig. 3A). In contrast, the percentage of immunopositive spermatozoa in stages II and III alone (Fig. 3A) or combined (Fig. 3B) increased between 30 and 60 min incubation. The time course for the loss of stage I coincided with the gain of immunopositive sperm in stages II and III (Fig. 3B, left), implying that spermatozoa in stage I undergo progressive membrane priming to become intensely immunopositive cells seen in stages II and III.

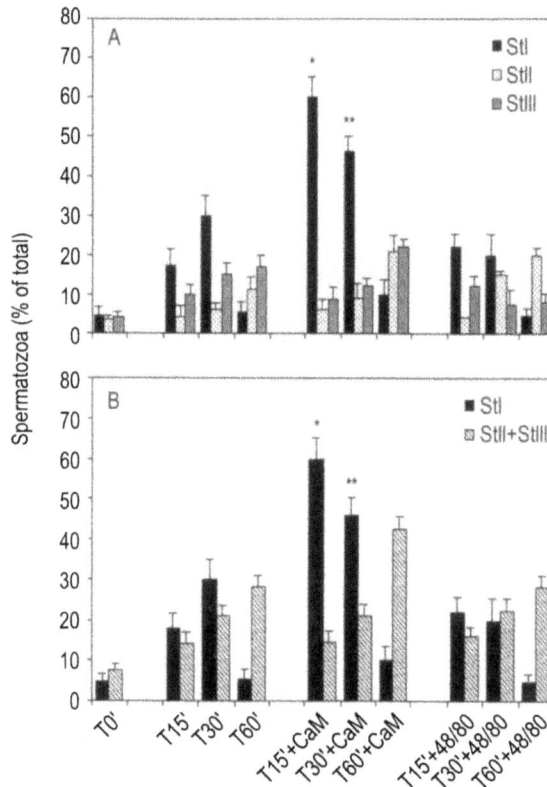

Fig. 3 Quantification of the sperm priming as revealed by the indirect immunofluorescence patterns using anti-ß-D-galactosidase as in Figs. 1 and 2. Spermatozoa were fixed in 4% paraformaldehyde before (T0') and after incubation of 15 (T15'), 30 (T30') or 60 min (T60') in the absence (left), presence of 10 $\mu$M calmodulin (middle) or presence of 2 $\mu$M compound 48/80 (right) in the incubation medium. The fixed non-permeabilized cells were incubated with primary antibody and FITC-labeled secondary antibodies. The percentage of immunopositive spermatozoa in stages I-III are presented in (A), or grouped in stage I and stages II and III combined (B). Data represent the mean ± SD of two independent experiments done in triplicate. Groups that are statistically different are marked with asterisks: *$P<0.005$; **$P<0.05$. Reproduced from Abou-Haila and Tulsiani (2003) with permission.

In a recent report (Kim *et al.*, 2001), it was reported that the mouse sperm component Sp56, a molecule initially immunolocalized on the surface of capacitated spermatozoa (Bookbinder *et al.*, 1995), was actually present in the acrosomal matrix (Foster *et al.*, 1997). The molecule was absent from the sperm surface unless the outer acrosomal membrane and plasma membrane have begun fusing or have ruptured (Kim *et al.*, 2001). To explain how the Sp56 protein could have been mistaken as a cell surface molecule, the investigators hypothesized that sperm capacitation represents transitional state whereby the membranes are modified by destabilization or initial fusion events (Kim *et al.*, 2001). These fusion phases could allow the intra-acrosomal matrix protein Sp56 to become exposed to the extracellular milieu and, under conditions used for immunolocalization, the molecule appeared to have a surface localization. The proposed explanation is consistent with our data strongly suggesting the surface exposure of intra-acrosomal glycohydrolases on capacitating spermatozoa in a time-dependent manner (see above).

### Is the surface exposure of acrosomal contents functionally significant?

The fact that sperm cells recognize and bind to the zona pellucida only after capacitation suggests major changes on the sperm surface during this process. It is plausible that some of the exposed acrosomal contents, in addition to the carbohydrate-binding molecules (receptors) on the sperm plasma membrane (for reviews, see Tulsiani and Abou-Haila, 2001), participate in sperm-zona (egg) interaction. The growing list of acrosomal molecules suggested to have a role in adhesion to the zona pellucida includes acrosomal matrix component (Foster *et al.*, 1997; Kim *et al.*, 2001), proacrosin/acrosin (Tesarik *et al.*, 1990), hyaluronidase (Meyer and Rosenberger, 1999), glycohydrolases (Abou-Haila and Tulsiani, 2003), an acrosomal protein Sp10 (Coonrod *et al.*, 1996) and perhaps arylsulfatase A (Yang and Srivastava, 1974; Tantibhedhyangkul *et al.*, 2002). The reported presence of several intra-acrosomal molecules on the surface of capacitated spermatozoa is consistent with our suggestion that the capacitation-associated exposure of these molecules is functionally significant.

### Role of calmodulin in sperm capacitation and the agonist-induced acrosome reaction

It may be recalled that sperm capacitation and induction of the acrosome reaction are $Ca^{2+}$-dependent processes. Since $Ca^{2+}$/calmodulin plays a significant role in several cell signaling pathways and membrane fusion events, we used a pharmacological approach to examine the functional significance of calmodulin in capacitation and triggering the agonist-induced acrosome reaction (Bendahmane *et al.*, 2001). Inclusion of calmodulin antagonists (calmodulin-binding domain, compound 48/80, ophiobolin A, calmidazolium, W5, W7 and W13), either in the *in vitro* capacitation medium or after sperm capacitation, blocked the zona pellucida-/neoglycoprotein-induced acrosome reaction. Purified calmodulin (CaM) largely reversed the acrosome reaction blocking effects of antagonists during capacitation. These data allowed us to suggest that calmodulin regulates these events by modulating sperm membrane components (Bendahmane *et al.*, 2001).

It is important to mention that calmodulin antagonists used in our study are reported to have an adverse effect on multiple key molecules of the cell signalling pathway. For instance, W5, W7, W13 and ophiobolin A are reported to effect $Ca^{2+}$/calmodulin-dependent phosphodiesterase; calmodulin-binding domain inhibits activation of calmodulin-dependent protein kinase II; calmidazolium inhibits calmodulin-dependent $Ca^{2+}$ATPase that releases $Ca^{2+}$; and compound

48/80 has been reported to inhibit phosphodiesterase C, ADP-ribosylation as well as activation of G-proteins. This has allowed us to suggest that the antagonists prevent capacitation by interfering with multiple signaling molecules (Bendahmane et al., 2001).

### Calmodulin accelerates the priming of sperm membranes during capacitation

To examine the possibility that calmodulin may be involved in progressive sperm membrane priming, the cauda epididymal spermatozoa were capacitated in the absence or presence of 10 $\mu$M purified calmodulin or 2 $\mu$M compound 48/80, a water soluble CaM antagonist. The presence of calmodulin in the capacitation medium did not alter the overall rate of sperm capacitation; however, its presence significantly accelerated initial stages of membrane priming (Fig. 3A, B, middle). Inclusion of the calmodulin antagonist in the capacitation medium prevented CaM-induced acceleration of the initial stages of the capacitation-associated membrane priming (Fig. 3A, B, right).

### Common components and potential similarities between calcium-dependent sperm capacitation and early events of Ca²⁺-triggered membrane fusion in somatic cells and viruses

Accumulated evidence has helped to consolidate the view that, as in regulated secretory events in somatic cells, calcium is required for capacitation (DasGupta et al., 1993) as well as for the induction of the acrosome reaction (Florman et al., 1992). Calcium exerts its effect on sperm functions through calmodulin and other $Ca^{2+}$-binding proteins that undergo conformation changes upon interaction with the divalent cation. Calmodulin is known to regulate many signaling pathways in somatic cells by modulating the activities of enzymes, proteins and ion pumps. Sperm cells contain high levels of calmodulin in the head and flagellum regions, the localizations consistent with its reported role in sperm function (Bendahmane et al., 2001).

In addition to calmodulin, the sperm acrosome contains synaptotagmins, a family of transmembrane proteins suggested to have a regulatory role in membrane fusion events (Calakos and Scheller, 1996; Shiavo et al., 1998). There is a certain degree of functional and topological convergence between synaptotagmin 1 and calmodulin. First, both molecules are involved in $Ca^{2+}$-dependent regulation of phospholipid binding at the base of SNARE (soluble N-ethylmaleimide-sensitive factor attachment protein receptor) complex composed of the vesicle protein synaptobrevin (VAMP2) and two plasma membrane partners, Syntaxin I and SNAP-25 (Coorssen et al., 2003). Second, certain clostrial toxins have been reported to target the two processes. For instance, BoNT/A cleaves a C-terminal fragment of SNAP-25 implicated in $Ca^{2+}$ sensoring and interaction with synaptotagmin: it also cleaves VAMP at peptide bond Q76-F77 precisely separating the C-terminal calmodulin/phospholipid-binding domain (VAMP 77-90) from N-terminal superhelical region. Thus, it is reasonable to argue that the two calcium sensor proteins interact with SNAREs, a large family whose members have been found on either the target membrane, the t-SNAREs or the transport vesicles that shuttle between pairs of communicating membranes and the v-SNAREs (Quetglass et al., 2002). This implies that SNARE proteins play a vital role in assembling complexes that form a bridge between fusing membranes.

A formed transport vesicle (v-SNARE) is likely to inspect many potential targets before it finds a complementary target membrane (t-SNARE). This crucial step is thought to be controlled by Rab proteins, a family of small GTPase, that enable v-SNARE and t-SNARE to find a correct match. After the vesicle (v-SNARE) encounters the correct target membrane, the physical attachment of v-SNARE and t-SNARE (tethering) occurs. The attachment event is thought to be the earliest known event that precedes the formation of a trans-SNARE complex (Gerona *et al.*, 2000; Waters and Hughson, 2000). The vesicles remain bound allowing the Rab proteins to exchange GTP for GDP that enables tight adhesion (docking) of the vesicle to the target membrane prior to their fusion. This implies that the docking of the transport vesicle to the target membrane and their subsequent fusion are two separate events. For instance, it is possible to prevent fusion while permitting docking by keeping cytoplasmic concentration of $Ca^{2+}$ low. This results in an accumulation of vesicles attached to (but not fused to) the target membrane and may represent the endpoint of sperm capacitation before sperm-egg binding and influx of additional $Ca^{2+}$ which signals membrane fusion and acrosomal exocytosis (Florman *et al.*, 1992).

It should be noted that the priming of membranes and their actual fusion during secretory and endocytotic pathways can be broken down into several steps. First, there is formation of vesicles from the donor membrane. Second, the vesicles are transported to their destination. Third, tethering and docking of vesicles with the target plasma membrane occurs. Finally, the vesicles fuse with the plasma membrane. Does the priming of membranes during sperm capacitation represent the first three steps of secretory pathway? As stated above, the endpoint of capacitation, a process unique to the male gamete, is the sperm activation and the agonist-induced fusion of the outer acrosomal membrane and the plasma membrane. Thus, it is reasonable to argue that sperm capacitation and the acrosomal reaction utilize many of the molecules that regulate the membrane fusion among eukaryotes. The common molecules in these pathways include N-ethylmaleimide-sensitive factor (NSF) suggested to have an important role in intracellular fusion, soluble NSF attachment proteins [SNAPS] and a large family of SNARE proteins (Mayer, 2001; Chamberlain and Gould, 2002). At the functional level, SNARE proteins take part in bringing two membranes closer to each other. Other molecules, such as NSF, SNAPs and large size molecules (> 250 kDa) are thought to participate in the transport of cargo proteins during tethering, an event representing the earliest known step in membrane targeting and fusion (Waters and Hughson, 2000). Interestingly, all these molecules, in addition to $Ca^{2+}$ sensor proteins (calmodulin and synaptotagmin), and Rab3A have been identified in the sperm acrosome (for review, see Ramalho-Santos *et al.*, 2002).

This brief discussion seems to suggest potential similarities between sperm capacitation and early stages of $Ca^{2+}$-triggered membrane fusion in somatic cells and viruses. For instance, there is a contact and merger of the two phospholipid bilayers of fusing membranes, hemifusion of the bilayer at the site of membrane contact followed by fusion pore formation (Zimmerberg and Chermordik, 1999). This implies that the sperm capacitation, like early events of membrane fusion, represents progressive membrane priming and not an all-or-nothing change. While the direct evidence in support of the proposed similarities between sperm capacitation and early events of membrane fusion is still lacking, the time-dependent appearance of several immunopositive patterns, including the appearance of droplets (vesicle-like spots) in the acrosome of capacitating spermatozoa (Figs. 1 and 2) favors our argument. The reported presence of several other intra-acrosomal antigens on the surface of capacitating/capacitated spermatozoa (see above) is consistent with our suggestion.

Although relatively little is known about the mechanism(s) underlying physiological priming of spermatozoa, the endpoint of this process is responsiveness of cells to undergo the acrosome reaction (Cross, 1998). Thus, it is reasonable to suggest that the outer acrosomal membrane and the plasma membrane, the two membranes that will ultimately fuse during the acrosome reaction, realign in capacitating/capacitated spermatozoa in a manner similar to the step-wise preparation of fusing membranes. First, the two membranes come together by a progressive evagination of the outer acrosomal membrane (Fig. 4). Second, complementary molecules on the fusing membranes allow their close contact (tethering) and tight adhesion when facing leaflets of the two lipid bilayers intermingle (docking). Finally, pores develop in distinctive phases (destabilization), some of which may have features typical of membrane channels. The dynamic aspects of the fusion pores, thought to be the priming steps prior to the membrane fusion among eukaryotes and among stations of secretory pathways (Jahn and Grubmüller, 2002; Müller et al., 2002), may occur in capacitating spermatozoa (see Fig. 4 in Tulsiani and Abou-Haila, 2004). These fusion pores could allow intra-acrosomal contents and/ or antibodies to diffuse through these pores revealing intra-acrosomal contents on the surface of capacitating/capacitated spermatozoa.

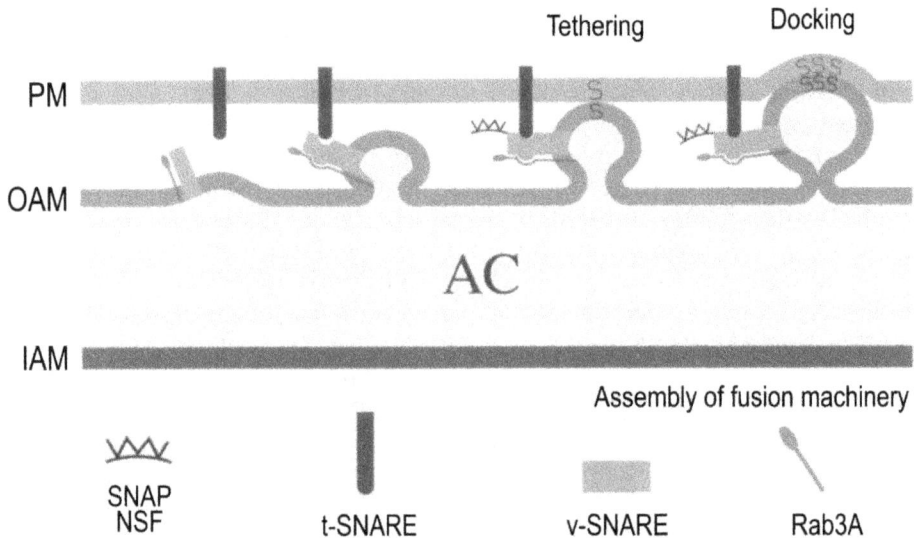

**Fig. 4** A hypothetic model (based on immunofluorescence patterns seen in Figs 1 and 2) illustrating putative membrane changes that may allow surface exposure of the intra-acrosomal glycohydrolase (and perhaps other acrosomal contents) or allow antibodies to react with intra-acrosomal glycohydrolases. The figure illustrates how the outer acrosomal membrane evaginates, forming a vesicle that enlarges and comes close to the plasma membrane, making initial contact (tethering) using complementary SNARE proteins on the two membranes. The Rab protein(s) is likely responsible for checking that the match between v- and t-SNAREs is correct. Following the initial contact, there is a tight adhesion (docking) of the formed vesicle with the target plasma membrane. The initial contact and subsequent adhesion results in partial destabilization and destabilization of the membranes due to the formation of hemifusion and fusion pores, respectively. These pores could allow diffusion of intra-acrosomal glycohydrolases and/or antibody to interact without sperm permeabilization. AC, acrosomal contents; PM, plasma membrane; OAM and IAM, outer and inner acrosomal membranes. Reproduced from Tulsiani and Abou-Haila (2004) with permission.

## Capacitation-associated sperm protein tyrosine phosphorylation: effects of calmodulin antagonists and purified calmodulin

It has been demonstrated that the cauda epididymal spermatozoa incubated in a defined medium supplemented with bovine serum albumin or methyl-ß-cyclodextrin undergo capacitation-associated increase in protein tyrosine phosphorylation of a subset of sperm molecules (Visconti et al., 1995; Visconti and Kopf, 1998). Since calmodulin antagonists prevent or significantly inhibit in vitro capacitation, we used six calmodulin antagonists in an attempt to characterize further the role of calmodulin in capacitation-associated protein tyrosine phosphorylation. The cauda epididymal spermatozoa were incubated in the medium supplemented with 3 mg bovine serum albumin/ml or 1 mM methyl-ß-cyclodextrin in the absence or presence of six calmodulin antagonists individually, and the *in vitro* capacitation was carried out by incubation for 60 or 90 minutes at 37°C under 5% $CO_2$ in air. Interestingly, three of the antagonists (compound 48/80, W13 and calmodulin-binding domain) had no effect on protein tyrosine phosphorylation or sperm motility (Fig. 5, Table 1). In contrast, the other three calmodulin antagonists inhibited protein tyrosine phosphorylation of all sperm components (apparent molecular weights 42, 56, 66, 82 and 95 kDa) and adversely affected their motility without altering their viability (Fig. 6, Table 1). Inclusion of CaM (10 mM) during *in vitro* capacitation significantly increased tyrosine phosphorylation of 82 kDa and 95 kDa components (Fig. 7).

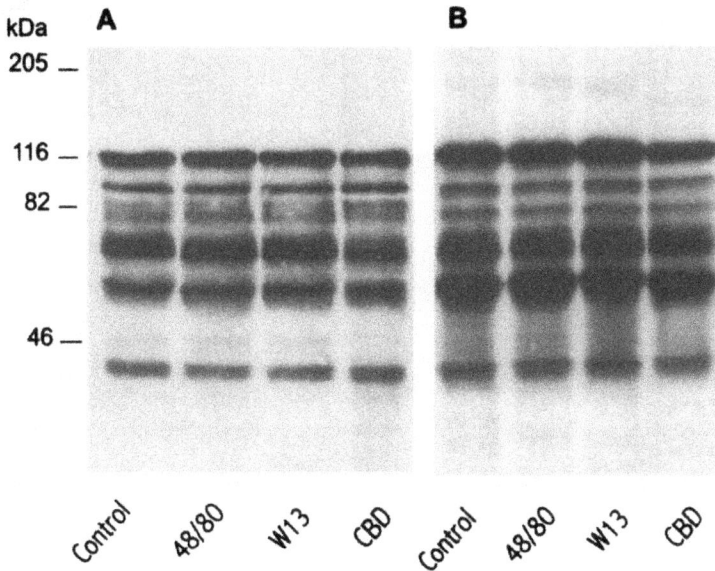

**Fig. 5** Effects of CaM antagonists on protein tyrosine phosphorylation of the mouse sperm components. Spermatozoa were incubated at 37°C for 90 min in enriched Krebs-Ringer bicarbonate medium supplemented with 3 mg BSA/ml (A) or 1 mM methyl-ß-cyclodextrin (B) in the absence (control) or presence of different concentrations of calmodulin antagonists (compound 48/80, 2 $\mu$M; W13, 1.58 $\mu$M; calmodulin-binding domain, 0.16 $\mu$M). Protein tyrosine phosphorylation was revealed as described (Zeng and Tulsiani 2003). A slight variation in phosphorylated proteins around 80 kDa in this figure (compare A versus B) is probably due to different running conditions. Reproduced from Zeng and Tulsiani (2003) with permission.

How does calmodulin accelerate early stages of sperm priming (Fig. 3) as well as influence capacitation-associated protein tyrosine phosphorylation (Fig. 7)? The calcium sensor protein is reported to stimulate adenylyl cyclase (Gross *et al.*, 1987; Kopf and Vacquier, 1984), a sperm plasma membrane enzyme responsible for the synthesis of cAMP. Thus, it seems likely that the three antagonists that inhibit capacitation-associated protein tyrosine phosphorylation and

**Table 1** Effect of calmodulin antagonists on viability, motility and protein tyrosine phosphorylation of mouse spermatozoa.

| Addition* | Concentration ($\mu$M) | Viability[†] (%) | Motility[†] (%) | Protein tyrosine phosphorylation |
|---|---|---|---|---|
| None (control) | | 51.3 ± 3.1 | 45.7 ± 5.1[‡] | + + + +[§] |
| Ophiobolin A | 20 | 50.3 ± 4.0 | 1.0 ± 1.0[¶] | ±[¶,**] |
| W7 | 25 | 50.7 ± 2.1 | 1.3 ± 0.6[¶] | +[¶] |
| Calmidazolium | 2.5 | 48.3 ± 3.5 | 23.3 ± 4.7[¶] | + +[¶] |
| Compound 48/80 | 2 | 51.0 ± 3.6 | 45.0 ± 6.0[‡] | + + + +[§] |
| W13 | 1.58 | 49.7 ± 4.5 | 41.3 ± 5.7[‡] | + + + +[§] |
| Calmodulin-binding domain | 0.16 | 52.7 ± 3.8 | 44.0 ± 4.6[‡] | + + + +[§] |

*The cauda epididymal spermatozoa were capacitated by incubating with or without the indicated concentration of CaM antagonists at 37°C for 90 minutes under 5% $CO_2$ in air. Following this incubation, aliquots were checked for sperm viability, sperm motility and sperm protein tyrosine phosphorylation. Reproduced from Zeng and Tulsiani (2003) with permission.

[†]Values are the mean of three independent assays ± s.d.
[‡]Most of the motile spermatozoa (>95%) displayed hyperactivated motility.
[§]Normal levels of protein tyrosine phosphorylation (see Fig. 5).
[¶]Significantly different from the control group.
**Traces

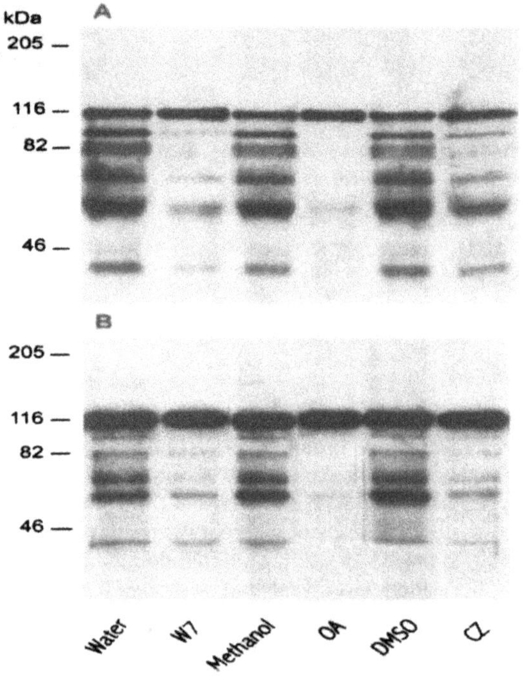

**Fig. 6** Effects of calmodulin antagonists on protein tyrosine phosphorylation of the mouse sperm components. Spermatozoa were incubated for 90 min at 37°C in enriched Krebs-Ringer bicarbonate medium supplemented with 3 mg BSA/ml (A) or 1 mM methyl-ß-cyclodextrin (B) in the absence or presence of different concentrations of calmodulin antagonists (W7, 25 $\mu$M; ophiobolin A, 20 $\mu$M; calmidazolium, 2.5 $\mu$M). The lanes with solvent alone (water, methanol and DMSO) are controls for each antagonist. Tyrosine phosphorylation of the sperm components was revealed as above. Reproduced from Zeng and Tulsiani (2003) with permission.

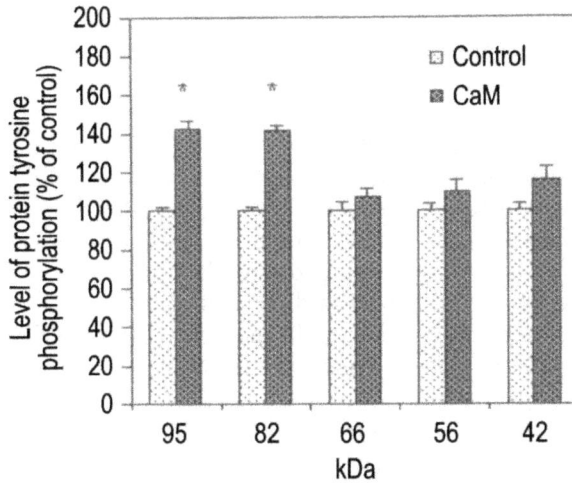

**Fig. 7** Effect of calmodulin on protein tyrosine phosphorylation of mouse sperm components. Cauda epididymal spermatozoa were incubated in enriched Krebs-Ringer bicarbonate medium supplemented with 3 mg/ml BSA in the absence or presence of 10 $\mu$M calmodulin. After 90 min at 37°C under 5% $CO_2$ in air, sperm cells were pelleted, washed, extracted, resolved by SDS-PAGE, and protein tyrosine phosphorylated components were revealed as described (Zeng and Tulsiani 2003). Bands from the exposed X-ray film were scanned and the intensities of control and calmodulin-treated groups quantified. The asterisk (*) indicates a significant difference ($P < 0.05$) between the control and experimental groups. Reproduced from Zeng and Tulsiani (2003) with permission.

sperm motility do so by their effects on calmodulin-dependent adenylyl cyclase activity. The inhibition of this enzyme is expected to start a chain reaction by reducing levels of sperm cAMP and inhibiting the proposed cross-talk between cAMP and tyrosine kinase second messenger systems (Visconti and Kopf, 1998), thereby inhibiting capacitation-associated alterations. As a corollary, the inclusion of purified calmodulin in the capacitation medium will stimulate adenylyl cyclase activity, which will lead to increased levels of cAMP. The increased cAMP levels, suggested to act upstream on phosphorylation (Visconti and Kopf, 1998), might be important in enhancing capacitation-associated sperm membrane priming and reported increase in protein tyrosine phosphorylation of 82 kDa and 95 kDa sperm molecules (Fig. 7). Whether the 95 kDa component is a unique hexokinase (Kalab et al., 1994), a testis-specific 95 kilodalton fibrous sheath antigen (Mandal et al., 1999), a molecule which has both ZP3-binding and a tyrosine kinase activity (Leyton and Saling, 1989) or an unidentified molecule is not yet known.

## How does calmodulin promote capacitation?

Does calmodulin enter the spermatozoa and trigger the intracellular signaling cascade that promotes capacitation? Inclusion of calmodulin in the culture medium has been reported to promote DNA synthesis and cell proliferation in human leukemia lymphocytes (Crocker et al., 1988). These data strongly suggest that the acidic protein can influence mitosis through an extracellular mechanism. Furthermore, extracellular calmodulin inhibits monocyte tumor necrosis factor release and augments neutrophil elastase release (Houston et al., 1997). These data, in conjunction with the reported occurrence of receptor(s) on the monocytic cell line

(Houston *et al.*, 1997), suggest that calmodulin possesses an extracellular signalling role in addition to its intracellular regulatory function. Although a calcium-dependent calmodulin-binding protein has been reported on bovine spermatozoa (Noland *et al.*, 1985), it is not known whether purified calmodulin functions via this calcium-dependent binding protein or mouse spermatozoa possess surface receptor(s).

Since all antagonists inhibited/prevented capacitation [as judged by the poor response of spermatozoa to undergo the agonist-induced acrosome reaction (see Table 1 in Zeng and Tulsiani, 2003)], it is reasonable to assume that their blocking effects are due to the inhibition of calmodulin in the sperm head. However, why only W7, ophiobolin A and calmidazolium adversely affected sperm motility and protein tyrosine phosphorylation (see above) is not known at the present time. One possibility is that these three antagonists might not elicit their effects through calmodulin alone but have other molecular targets such as FSP95, a fibrous-sheath protein (Mandal *et al.*, 1999). Alternatively, perhaps only some of the reagents are able to enter the sperm tail and block calmodulin.

The differential effects of the two sets of calmodulin antagonists on capacitation-associated protein tyrosine phosphorylation and sperm motility raise many interesting questions on the inter-relationship between sperm cholesterol, protein tyrosine phosphorylation and sperm capacitation. Cross (1998) has reported that when human spermatozoa are incubated in capacitating medium *in vitro*, there is a gradual loss of cholesterol; however, the acrosomal responsiveness does not develop for some time, a result consistent with the author's suggestion that cholesterol loss precedes capacitation. It will be interesting to determine if calmodulin antagonists act upstream or downstream to the cholesterol efflux. New data will further our understanding of molecules that mediate capacitation.

Our data provide evidence suggesting the occurrence of at least two pathways that regulate sperm motility, protein tyrosine phosphorylation and sperm capacitation. The first event might adversely affect sperm motility before inhibiting protein tyrosine phosphorylation or *vice versa*; the second possible event could be inhibition/prevention of capacitation without affecting sperm motility or protein tyrosine phosphorylation (see Fig. 5 in Zeng and Tulsiani, 2003). These data suggest a close interrelationship between protein tyrosine phosphorylation of a subset of sperm components and motility. The proposed relationship is consistent with the experimental evidence from many investigators, suggesting that protein tyrosine phosphorylation plays an important role in the control of the hyperactivated motility (Vijayaraghavan *et al.*, 1997; Si and Okuno, 1999; Holt and Harrison, 2002).

In conclusion, we have attempted to discuss capacitation-associated membrane priming of spermatozoa and the involvement of various sperm components that modulate capacitation and potential cross-talk between the outer acrosomal membrane and the plasma membrane during the assembly of fusion machinery. We have highlighted common components and potential similarities between sperm capacitation and early events of $Ca^{2+}$-triggered membrane fusion among eukaryotes and viruses. In addition, we have used a pharmacological approach to examine the interrelationship between protein tyrosine phosphorylation and sperm motility. The biochemical approaches provide evidence strongly suggesting the occurrence of multiple pathways that regulate capacitation-associated sperm motility and protein tyrosine phosphorylation.

## Acknowledgments

The secretarial assistance of Loreita Little is gratefully acknowledged. We are indebted to Lynne Black for critically reading the manuscript. The work reported in this article was supported in part by grants HD25869 and HD34041 from the National Institute of Child Health and Human Development.

# References

**Abe H, Sendai Y, Satoh T and Hoshi H** (1995) Bovine oviduct-specific glycoprotein: a potent factor for maintenance of viability and motility of bovine spermatozoa *in vitro Molecular Reproduction & Development* **42** 226-232

**Abou-Haila A and Tulsiani DRP** (2003) Evidence for the capacitation-associated membrane priming of mouse spermatozoa *Histochemistry & Cell Biology* **119** 179-187

**Ahuja K** (1985) Lectin-coated agarose beads in the investigation of sperm capacitation in the hamster *Developmental Biology* **104** 131-142

**Austin CR** (1951) Observation on the penetration of the sperm into the mammalian ovum *Australian Journal of Science & Research* **4** 581-596

**Bendahmane M, Lynch C and Tulsiani DRP** (2001) Calmodulin signals capacitation and triggers the agonist-induced acrosome reaction in mouse spermatozoa *Archives of Biochemistry and Biophysics* **390** 1-8

**Bookbinder LH, Cheng A and Bleil JD** (1995) Tissue- and species-specific expression of Sp56, a mouse sperm fertilization protein *Science* **269** 86-89

**Calakos N and Scheller RH** (1996) Synaptic vesicle biogenesis, docking and fusion: a molecular description *Physiological Review* **76** 1-29

**Chamberlain LH and Gould GW** (2002) The vesicle and target-SNARE proteins that mediate Glut 4 vesicle fusion are localized in detergent insoluble lipid rafts present on distinct intracellular membranes *Journal of Biological Chemistry* **277** 49750-49754

**Chang MC** (1951) Fertilizing capacity of spermatozoa deposited into the fallopian tubes *Nature* **168** 697-698

**Coonrod SA, Herr JC and Westhusin ME** (1996) Inhibition of bovine fertilization *in vitro* by antibodies to Sp10 *Journal of Reproduction and Fertility* **107** 287-297

**Coorssen JS, Blank PS, Albertorio F, Bezrukov L, Kolosova I, Chen X, Backlund PS Jr and Zimmerberg J** (2003) Regulated secretion: SNARE density, vesicle fusion and calcium dependence *Journal of Cell Science* **116** 2087-2097

**Crocker G, Dawson RA, Barton CH and MacNeil S** (1988) An extracellular role for calmodulin-like activity in cell proliferation *Biochemical Journal* **253** 877-884

**Cross NL** (1998) Role of cholesterol in sperm capacitation *Biology of Reproduction* **59** 7-11

**DasGupta S, Mills CL and Fraser LR** (1993) $Ca^{2+}$-related changes in the capacitation state of human spermatozoa assessed by a chlorotetracycline fluorescence assay *Journal of Reproduction and Fertility* **99** 135-143

**Davis BK** (1981) Timing of fertilization in mammals: sperm cholesterol/phospholipid ratio as a determinant of the capacitation interval *Proceedings of National Academy of Sciences USA* **78** 7560-7564

**Flesch FM and Gadella BM** (2000) Dynamics of the mammalian sperm plasma membrane in the process of fertilization *Biochimica Biophysica Acta* **1469** 197-235

**Florman HM, Corron ME, Kim TDH and Babcock DF** (1992) Activation of voltage-dependent calcium channels of mammalian sperm is required for zona-induced acrosomal exocytosis *Developmental Biology* **152** 304-314

**Foster JA, Friday BB, Maulit MT, Blobel C, Winfrey VP, Olson GE, Kim K-S and Gerton GL** (1997) AM67, a secretory component of the guinea pig sperm acrosomal matrix, is related to mouse sperm protein Sp56 and the complement component 4-binding proteins *Journal of Biological Chemistry* **272** 12714-12722

**Gerona RR, Larsen EC, Kowalchyk JA and Martin TF** (2000) The C-terminus of SNAP25 is essential for $Ca^{2+}$-dependent binding of synaptotagmin to SNARE complexes *Journal of Cell Biology* **275** 6328-6336

**Gross MK, Toscano DG and Toscano WA** (1987) Calmodulin-mediated adenylate cyclase from mammalian spermatozoa *Journal of Biological Chemistry* **256** 7590-7596

**Holt WV and Harrison RAP** (2002) Bicarbonate stimulation of boar sperm motility via a protein kinase A-dependent pathway: between cell and between ejaculate differences are not due to deficiencies in protein kinase A activation *Journal of Andrology* **23** 557-565

**Houston DS, Carson CW and Esmon CT** (1997) Endothelial cells and extracellular calmodulin inhibit monocyte tumor necrosis factor release and augment neutrophil elastase release *Journal of Biological Chemistry* **272** 11778-11785

**Jahn R and Grubmüller H** (2002) Membrane fusion *Current Opinions in Cell Biology* **14** 488-497

**Kalab P, Visconti PE, Leclere P and Kopf GS** (1994) p95, the major phosphotyrosine containing protein in mouse spermatozoa is a hexokinase with unique properties *Journal of Biological Chemistry* **269** 3810-3817

**Kim K-S, Cha MC and Gerton GL** (2001) Mouse sperm protein Sp56 is a component of the acrosomal matrix *Biology of Reproduction* **64** 36-43

**King RS and Killian GL** (1994) Purification of bovine estrus-associated protein and localization of binding on sperm *Biology of Reproduction* **51** 34-42

**Kopf GS and Vacquir VD** (1984) Characterization of calmodulin stimulated adenylate cyclase from abalone spermatozoa *Journal of Biological Chemistry* **259** 7590-7596

**Leyton L and Saling P** (1989) p95 kd sperm proteins bind ZP3 and serve as tyrosine kinase substrates in response to zona binding *Cell* **57** 1123-1130

**Lin Y, Mahan K, Lathrop W, Myles D and Primakoff P** (1994) A hyaluronidase activity of the sperm plasma membrane PH-20 enables sperm to penetrate the cumulus cell layer surrounding the egg *Journal of Cell Biology* **125** 1157-1163

**Mandal A, Naaby-Hensen S, Wolkowicz MJ, Klotz K, Shetty J, Relief JD, Coonrod SA, Kinler M, Sherman N, Cesar F, Flickinger CJ and Herr JC** (1999) FSP95, a testis-specific 95 kilodalton fibrous sheath antigen that undergoes tyrosine phosphorylation in capacitated spermatozoa *Biology of Reproduction* **61** 1184-1197

**Mayer A** (2001) What drives membrane fusion in eukaryocytes? *Trends in Biochemical Science* **26** 717-723

**Meyer SA and Rosenberger AE** (1999) A plasma membrane-associated hyaluronidase is localized to the posterior acrosome region of stallion sperm and is associated with sperm function *Biology of Reproduction* **61** 444-451

**Müller O, Bayer MJ, Peters C, Anderson JS, Mann M and Mayer A** (2002) The Vtc proteins in vacuole fusion: coupling NSF activity to Vo trans-complex formation *EMBO Journal* **21** 259-269

**Noland TD, Van Eldik J, Garbers DL and Burgess WH** (1985) Distribution of calmodulin and calmo-dulin-binding proteins in membranes from bovine epididymal spermatozoa *Gamete Research* **11** 297-303

**Olds-Clarke P** (1990) Variation in the quality of sperm motility and its relationship to sperm motility. In *Fertilization in Mammals* pp 91-99 Eds B Barister, J Cummins and ERS Roldan. Serono Symposium, Norwell, MA

**Quetglass S, Iborra C, Sasakawa N, Haro LD, Kumakura K, Sato K, LeVeque C and Seager M** (2002) Calmodulin and lipid binding to synaptobrevin regulates calcium-dependent exocytosis *EMBO Journal* **21** 3970-3979

**Ramalho-Santos J, Schatten G and Moreno RD** (2002) Control of membrane fusion during spermatogenesis and the acrosome reaction *Biology of Reproduction* **67** 1043-1051

**Shiavo G, Osborne SL and Sgouros JG** (1998) Synaptotagmins: more isoforms than function *Biochemical Biophysical Research Communication* **248** 1-8

**Si Y and Okuno M** (1999) Role of tyrosine phosphorylation of flagellar proteins in hamster sperm hyper-activation *Biology of Reproduction* **61** 240-246

**Tantibhedhyangkul J, Weerachatyanukul W, Carmona E, Xu H, Anupriwan A, Michand D and Tanphaichitr N** (2002) Role of sperm surface aryl sulfatase A in mouse sperm-zona pellucida binding *Biology of Reproduction* **67** 212-219

**Tesarik J, Drahorad J, Testart J and Mendoza H** (1990) Acrosin activation follows its surface exposure and precedes membrane fusion in human sperm acrosome reaction *Development* **110** 391-400

**Tulsiani DRP, Chayko CA, Orgebin-Crist M-C and Araki Y** (1996) Temporal surge of glycosyltransferase activities in the genital tract of the hamster during the estrous cycle *Biology of Reproduction* **54** 1032-1037

**Tulsiani DRP, Yoshida-Komiya H and Araki Y** (1997) Mammalian fertilization: A carbohydrate-mediated event *Biology of Reproduction* **57** 487-494

**Tulsiani DRP, Abou-Haila A, Loeser CR and Pereira BMJ** (1998) The biological and functional significance of the sperm acrosome and acrosomal enzymes in mammalian fertilization *Experimental Cell Research* **240** 151-164

**Tulsiani DRP and Abou-Haila A** (2001) Mammalian sperm molecules that are potentially important in interaction with female genital tract and egg vestments *Zygote* **9** 51-59

**Tulsiani DRP and Abou-Haila A** (2004) Is sperm capacitation analogous to early phases of $Ca^{2+}$-triggered membrane fusion in somatic cells and viruses? *BioEssays* **26** 281-290

**Vijayaraghavan S, Trautman KD, Goueli SA and Carr DW** (1997) A tyrosine phosphorylated 55-kilodalton motility-associated bovine sperm protein is regulated by cyclic adenosine 3', 5'-monophosphate and calcium *Biology of Reproduction* **59** 1-6

**Visconti PE, Bailey JL, Moore GD, Olds-Clarke P and Kopf GS** (1995) Capacitation of mouse spermatozoa 1. Correlation between the capacitation state and protein phosphorylation *Development* **121** 1129-1137

**Visconti PE and Kopf GS** (1998) Regulation of protein phosphorylation during capacitation *Biology of Reproduction* **59** 1-6

**Waters MG and Hughson FM** (2000) Membrane tethering and fusion in the secretory and endocytotic pathways *Traffic* **1** 588-597

**Wu C, Stojanov T, Chami O, Ishii S, Shimuzu T, Li C and O'Neill C** (2001) Evidence for the autocrine induction of capacitation of mammalian spermatozoa *Journal of Biological Chemistry* **276** 26962-26968

**Yanagimachi R and Chang MC** (1963) Sperm ascent through the oviduct of hamster and rabbit in relation to the time of ovulation *Journal of Reproduction and Fertility* **6** 281-282

**Yanagimachi R** (1994) Mammalian fertilization. In *Physiology of Reproduction* pp 189-317 Eds E Knobil and JD Neil. Raven Press, New York

**Yang C-H and Srivastava PN** (1974) Purification and properties of aryl sulfatases from rabbit sperm acrosome *Proceedings of Society for Experimental Biology and Medicine* **145** 721-725

**Zeng H-T and Tulsiani DRP** (2003) Calmodulin antagonists differentially affect capacitation-associated protein tyrosine phosphorylation of mouse sperm components *Journal of Cell Science* **116** 1981-1989

**Zimmerberg J and Chermodik LV** (1999) Membrane fusion *Advances in Drug Delivery Review* **38** 197-205

# Molecular mechanisms of sperm capacitation: progesterone-induced secondary calcium oscillations reflect the attainment of a capacitated state

R J Aitken* and E A McLaughlin

*ARC Centre of Excellence in Biotechnology and Development, Discipline of Biological Sciences, University of Newcastle, Callaghan, NSW 2708, Australia*

Progesterone has an extragenomic action on human spermatozoa characterised by the rapid induction of a calcium transient followed by a plateau phase during which $[Ca^{2+}]_i$ remains significantly above baseline. By imaging the calcium responses generated in individual cells, we have demonstrated that during this plateau phase, spermatozoa exhibit a series of asynchronous secondary calcium oscillations. The incidence of such oscillations was dependent upon sperm capacitation and showed significant inter-individual variation. The oscillations were dependent upon the influx of extracellular calcium via mechanisms that were insensitive to inhibitors of L-type voltage operated calcium channels (nifedipine, verapamil, diltiazem), G-proteins (pertussis toxin) or the GABA (A) receptor (bicuculline). However, treatment with an inhibitor of the GABA-associated chloride channel (picrotoxin) significantly suppressed the incidence of secondary calcium oscillations in pentoxifylline-treated cells, as did two inhibitors of T-type calcium channels (pimozide and amiloride). We hypothesise that the sub-population of spermatozoa exhibiting secondary calcium oscillations are characterised by a hyperpolarized plasma membrane that sets T-type channels in a closed but activation-competent state. The secondary calcium oscillations created via these channels do not induce acrosomal exocytosis *per se* but may prime the cells so that this event is rapidly triggered when the spermatozoa make contact with the zona pellucida.

## Introduction

Our future ability to manage both the high incidence of male infertility (Hull et al., 1985) and the unmet needs in male contraception (Lyttle and Kopf, 2003) will depend upon a quantum increase in our understanding of the molecular mechanisms that regulate human sperm function. A central concept in sperm cell biology is that freshly ejaculated human spermatozoa are immature and functionally incompetent. As they ascend the female reproductive tract they undergo a series of maturational changes known collectively as 'capacitation' and, as a consequence of which, they gain the ability to recognize the oocyte and initiate the cascade of cell-

E-mail: jaitken@mail.newcastle.edu.au
*To whom correspondence should be addressed:

cell interactions that culminate in fertilization (Austin, 1951; Chang, 1951; Yanagimachi, 1994; Aitken, 1997; Urner and Sakkas, 2003). The biochemical basis of sperm capacitation is known to involve a complex array of post-translational changes, including efflux of cholesterol from the plasma membrane and high levels of tyrosine phosphorylation induced by a capacitation-dependent increase in the intracellular concentrations of cAMP (White and Aitken, 1989; Visconti *et al.*, 1995; Aitken *et al.*, 1995; 1998a). Capacitated spermatozoa are essentially primed cells that will respond to an influx of extracellular calcium by undergoing the acrosome reaction. The latter comprises an exocytotic event associated with the release of proteolytic enzymes that facilitate sperm passage through the zona pellucida and, simultaneously, a remodelling of the sperm surface such that this cell becomes competent to recognise and fuse with the vitelline membrane of the oocyte (Yanagimachi, 1994).

The calcium influx that induces the acrosome reaction may be triggered by two constituents of the cumulus oocyte complex, progesterone (Osman *et al.*, 1989; Blackmore *et al.*, 1990) and ZP3 (Arnoult *et al.*, 1996), possibly acting in sequence (Roldan *et al.*, 1994). The ability of progesterone to stimulate a calcium influx into human spermatozoa involves an extragenomic mechanism activated by the binding of this steroid to cell surface receptors that are yet to be fully characterized. In the human testes, mRNA for both genomic and extragenomic progesterone receptors are certainly present (E. McLaughlin, unpublished). With respect to the mature gamete, evidence has been presented for a truncated version of the B-isoform of the genomic receptor (Shah *et al.*, 2005) as well as for non-genomic progesterone receptors with distant homologies to cytochrome *b*5 (Losel *et al.*, 2005; J. Chick, unpublished). In fish spermatozoa, yet another form of progesterone receptor has been identified that possesses the 7-pass trans-membrane structure typical of G-protein coupled receptors (Thomas *et al.*, 2005). The mechanisms by which such potential receptors might mediate progesterone-induced calcium transients in human spermatozoa are still unresolved.

Spectrofluorometric monitoring of the calcium responses to progesterone in heterogeneous populations of spermatozoa has revealed a rapid peak during which cytosolic $Ca^{2+}$ reaches around 0.5 µM, followed by a plateau phase when the intracellular level of this cation remains significantly above baseline (Blackmore *et al.*, 1990, 1991; Aitken *et al.*, 1996a; Kirkman-Brown *et al.*, 2000). The calcium peak is characterized by an influx of calcium from the extracellular space via mechanisms involving depolarization of the human sperm membrane potential, driven largely by a sodium influx and, in the early phase of the response, by a chloride efflux (Foresta *et al.*, 1993; Patrat *et al.*, 2002). This depolarization effect is not causally involved in the induction of the initial calcium transient by progesterone since inhibitors of voltage-operated calcium channels including verapamil, diltiazem and nifedipine (at doses that are adequate to antagonize this class of channel) do not disrupt the initial influx of calcium triggered by progesterone (Aitken *et al.*, 1996a; Morales *et al.*, 2000; Bonaccorsi *et al.*, 2001; Patrat *et al.*, 2002; Guzman-Grenfell and Gonzalez-Martinez, 2004).

An important feature of the extragenomic progesterone response is that it is essentially ubiquitous, more than 90% of normal human sperm populations exhibiting a clear-cut calcium response on exposure to this steroid (Aitken *et al.*, 1996a). In contrast, the biological response to the calcium transient induced by this steroid is not universal and involves a small and variable subgroup of cells, amounting to little more than 20% of the total sperm population (Harper *et al.*, 2003). The size of this biologically reactive subpopulation of spermatozoa is heavily influenced by the capacitation status of these cells, as reflected in their level of tyrosine phosphorylation (Aitken *et al.*, 1996b). However, the way in which capacitation supports the extragenomic action of progesterone is unknown. Resolution of this problem is essential if we are to understand how the extragenomic action of this steroid can influence sperm function. It has been suggested that the tyrosine phosphorylation associated with capacitation stimulates a secondary calcium increase in response to progesterone and that it is this second transient that is responsible for acrosomal exocytosis (Tesarik *et al.*, 1996). In this study, we have carefully examined the secondary calcium oscillations induced by progesterone, using single cell imaging techniques, and investigated their role in eliciting a biological response to this steroid.

## Materials and Methods

This study was based upon human semen samples donated by a panel of healthy donors after 2 or 3 days abstinence. All donors used in this research were normozoospermic according to the criteria set out by the World Health Organization (1992). The methods used to recruit the donors and the research undertaken in this study were approved by the Human Fertilization and Embryology Authority of the United Kingdom and the Human Ethics Committees of the University of Newcastle, NSW and the University of Edinburgh, UK. All samples were produced into sterile containers and left for at least 30 min to liquefy before processing.

The spermatozoa were isolated by discontinuous Percoll gradient centrifugation using a simple 2-step design incorporating 44% and 88% Percoll respectively. Isotonic (100%) Percoll was created by supplementing 20 ml of 10 X concentrated Earle's medium (Gibco, Paisley, UK) with 3 ml of 20% Albuminar (Armour Pharmaceutical Company, Eastbourne, UK), 6 mg sodium pyruvate, 0.74 ml of a sodium lactate syrup, 2 ml of a penicillin/streptomycin preparation (10,000 IU/ml penicillin and 1000 µg/ml streptomycin; Gibco) and adding 180 ml of Percoll (Pharmacia, Uppsala, Sweden). The medium used to dilute the isotonic Percoll was HEPES - buffered Biggers Whitten and Whittingham medium (BWW) supplemented with 20 mM HEPES and 0.3% Albuminar (Biggers *et al.*, 1971). Semen (1 to 3 ml) was layered on the top of each gradient and centrifuged at 500 g for 20 min. Purified populations of highly motile spermatozoa were subsequently recovered from the base of the 88% Percoll fraction, washed with a 7 ml volume of medium BWW, centrifuged at 500 g for 5 min and resuspended at a concentration of $2 \times 10^7$ cells/ml.

For certain experiments requiring low levels of extracellular calcium, the isotonic Percoll was prepared using 10 X concentrated calcium - depleted BWW, in which sodium chloride was used in place of calcium chloride. This isotonic Percoll preparation was subsequently diluted with 1 X calcium - depleted BWW to create the gradients needed for sperm preparation. The same medium was used to wash the spermatozoa recovered from the high density fraction and prepare them for intracellular calcium analysis. This buffer contained $2.0 \pm 0.5 \mu M$ calcium as measured by atomic adsorption analysis.

### Sperm-oocyte fusion

In order to monitor the ability of human spermatozoa to exhibit a functional acrosome reaction accompanied by the concomitant generation of a fusogenic equatorial segment, zona - free hamster ova were used in a heterologous *in vitro* fertilization assay (Aitken, 1986). For this assay, the isolated spermatozoa were diluted 1:1 with the relevant test solution and incubated for a period of 3 h at 37°C, in an atmosphere of 5% $CO_2$ in air. As a positive control 2.5 µM A23187 (Calbiochem Novabiochem, Nottingham, UK) was used, formulated as the free acid and maintained as a 100 mM stock solution in dimethylsulfoxide (DMSO): this preparation consistently gave 100% oocyte penetration. Progesterone (Sigma Chemical company, St Louis,

MO) was prepared daily as a 50 mM stock in DMSO and diluted with medium BWW to give a final concentration of 5 µM. At the end of the incubation period, the spermatozoa were pelleted by centrifugation at 500 g, resuspended in fresh medium BWW and distributed as 50 µl droplets under liquid paraffin.

Zona-free hamster oocytes were prepared as described (Aitken, 1986; Aitken et al., 1996b) and dispersed into the droplets at 5 oocytes/drop and about 15 oocytes/sample. After a further 3 h incubation at 37°C, the oocytes are recovered from the droplets, washed free of loosely adherent spermatozoa, compressed to a depth of about 30 µm under a 22 X 22 mm coverslip on a glass slide and assessed for the presence of decondensing sperm heads with an attached or closely associated tail, by phase contrast microscopy. The number of spermatozoa penetrating each egg was assessed and expressed in terms of the percentage of oocytes penetrated and as the mean number of spermatozoa penetrating each oocyte (total number of penetrations /total number of oocytes).

*Image analysis of calcium transients*

Intracellular calcium levels were monitored using the dual excitation dye, fura 2. The acetoxymethyl ester of fura 2 (Calbiochem-Novabiochem) was prepared as a 200 µM stock in DMSO and frozen in aliquots at -20°C. On the day of analysis, this probe was diluted 1 in 50 with the sperm suspension to give a final concentration of 4 µM fura 2. Loading of the dye was achieved by a 30 min incubation in the dark at 37°C, under 5% $CO_2$ in air. Once loaded, the cells were washed through 1 ml of 50% Percoll at 500 g for 5 min. The supernatant was then removed and the pellet washed again with a 2 ml volume of fresh BWW medium, centrifuged at 500 g for 5 min and finally resuspended at approximately 30 X $10^6$ spermatozoa/ml in fresh BWW. A 50 µl droplet of the fura-loaded sperm suspension was pipetted onto a quartz coverslip coated with 0.3% poly-L-lysine in an Applied Imaging (Tyne and Wear, UK) incubation chamber at 37°C. Once a majority of the spermatozoa had attached to the cover slip, the remainder were removed by washing out the chamber with fresh medium BWW. Imaging of the calcium transients in the attached cells was achieved using an Applied Imaging Magiscan system incorporating the Tardis calcium imaging software (Tyne and Wear, UK), calibrated for use with human spermatozoa. The image capture sequence was initiated by recording, averaging and storing 8 background frames from microscopic fields that did not contain spermatozoa, using excitation wavelengths of 340 and 380 nm and an emission wavelength of 510 nm. The information contained in these frames was subsequently used to correct for any grey level offset in the camera and residual background fluorescence in the ambient medium. Images of the spermatozoa were subsequently captured at 340 and 380 nm, averaged, background-subtracted, ratioed and displayed as a pseudo-coloured image in real time; the entire cycle from initial excitation to display of the ratioed image took 2 secs. Calibration was performed as previously described (Aitken et al., 1996a; Anderson et al., 1992) using 1 µM ionomycin and 10 mM EGTA to generate maximum and minimum fluorescence values and a Kd for fura 2 at 37°C of 224 nM. The results for the image analysis experiments are presented as individual traces or expressed in terms of either the percentage of imaged cells giving a response (defined as a rise in intracellular calcium more than twice the baseline value) or the mean ± SE intracellular calcium concentration for the sperm population.

*Inhibitors*

A variety of different inhibitors were used in an attempt to disrupt the calcium primary and secondary transients induced by progesterone. Inhibition of capacitation was achieved using strategies (bicarbonate withdrawal, exposure to 0.05% 2 mercaptoethanol, replacement of glucose with 2-deoxyglucose) that have previously been shown to disrupt the ability of human spermatozoa to attain a capacitated state (Aitken *et al.*, 1995;1998a,b). The calcium channel antagonists verapamil (1 and 10 µM), nifedipine (0.25 and 0.5 µM) and diltiazem (1 and 10 µM) were used at concentrations that have previously been shown to inhibit the acrosome reaction in mammalian spermatozoa (Fraser and McIntyre, 1989; Florman *et al.*, 1992; Shi and Roldan, 1995; Aitken *et al.*, 1996b). Pertussis toxin was employed at a concentration (1 µg/ml) compatible with the preservation of sperm viability and in excess of the dose needed to block zona-induced acrosome reactions in these cells (Tesarik *et al.*, 1993). The significance of the GABA (A) receptor and associated chloride channel was assessed using doses of bicuculline (10 µM) and picrotoxin (200 µM) that have been shown to suppress sperm function and are more than sufficient to antagonize GABA (A) receptor activation in other tissues (Baldi *et al.*, 1991; Calogero *et al.*, 1999; Kuroda *et al.*, 1999; Whyment *et al.*, 2004). Pimozide (10 µM) and amiloride (100 µM) were used to antagonize the activation of T-type calcium channels at doses that were compatible with sperm viability and were able to suppress the zona–pellucida induced acrosome reaction in murine spermatozoa (Arnoult *et al.*, 1996).

*Statistics*

The data were analysed by analysis of variance (ANOVA) using the Superanova program (Abacus Concepts, Berkeley, CA) on an Apple Macintosh computer. Percentage values were subjected to angular transformation prior to ANOVA analysis (Fisher and Yates, 1963). Paired comparisons were conducted with Wilcoxon signed rank test, while differences in the incidence of certain patterns of response were assessed by Chi-squared analysis.

# Results

*Induction of calcium oscillations in human sperm by progesterone*

Addition of 5 µM progesterone to populations of human spermatozoa induced a calcium transient in the acrosomal domain of the sperm head as reported previously (Aitken *et al.*, 1996b). Fig. 1 illustrates the typical pattern of progesterone-induced intracellular calcium ($[Ca^{2+}]_i$) change in a population of human spermatozoa, characterized by an immediate synchronized calcium peak during which calcium levels rise to around 0.4-1.0 µM, followed by a 'plateau phase' during which calcium levels remain significantly elevated above baseline (unpublished data). Image analysis of individual cells treated with progesterone revealed that the elevated $[Ca^{2+}]_i$ observed during the plateau phase were the product of asynchronous calcium oscillations occurring at different frequencies and amplitudes in individual cells (Fig. 2). Analysis of 652 cells from 19 independent samples subjected to a conventional capacitation protocol (3-h incubation in medium BWW) revealed that a majority of the spermatozoa (73.9 ± 4.1%) only exhibited a single, immediate calcium transient in response to progesterone (Fig. 2A). However, a further 20.23 ± 3.1% of cells exhibited at least one secondary calcium oscillation, during which $[Ca^{2+}]_i$ rose to at least twice the baseline value, following the primary, progesterone-induced calcium transient (Fig. 2 B, C, D). In 4 independent samples a mean of 2.2 ± 0.3 secondary oscillations were observed during the 500 sec recording period. The oscillations typically lasted for around 50 sec, although more prolonged secondary elevations in intracellular calcium were occasionally observed (Fig. 2D), and reached amplitudes of approximately the same as the initial progesterone induced peak. Analysis of the percentage of spermatozoa capable of showing secondary calcium oscillations in individual samples revealed a wide range of activities, giving 10th-90th percentile values of 1% - 41% in 19 samples capacitated in BWW for 3 h.

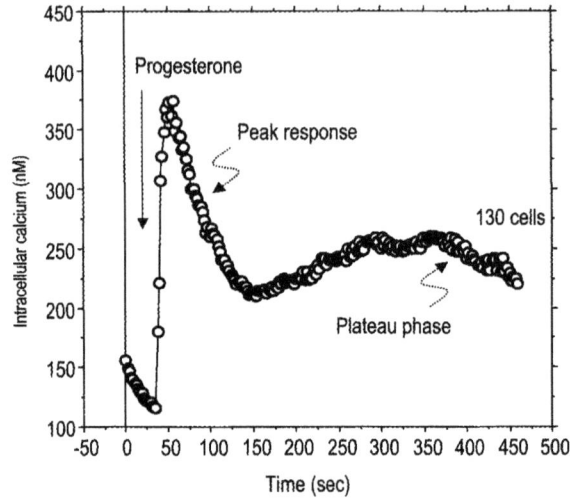

**Fig. 1** Induction of a calcium transient in human spermatozoa with progesterone. Following a rapid calcium transient, a plateau phase is reached during which calcium levels remain significantly elevated above baseline.

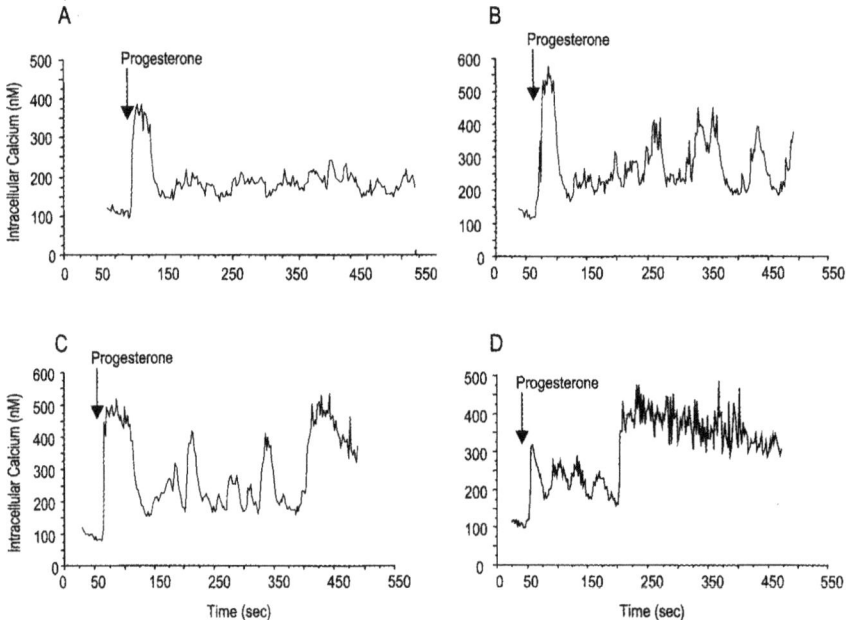

**Fig. 2** Single cell imaging of the calcium profiles induced by progesterone. (A) Following an initial rapid transient that is synchronized across all of the cells in a population, a majority of spermatozoa maintain a steady-state calcium level that is just above baseline. However, a minority of cells will exhibit secondary intracellular calcium transients oscillating at rates and magnitudes that show considerable cell-to-cell variation (B-D).

*Secondary calcium oscillations do not require presence of progesterone*

Further experiments established that once progesterone had stimulated the initial transient, this steroid did not have to be present for the secondary oscillations to occur. Thus, if human spermatozoa were exposed to 5 µM progesterone and then, after 200 sec, the preparation was flushed with fresh medium to remove the steroid, secondary calcium transients could still be detected 30 min later (Fig. 3). Of 116 imaged spermatozoa from 4 independent samples, 30.1% of the spermatozoa were found to exhibit secondary calcium transients and in 81% of these cells calcium oscillations continued to be observed during the ensuing 250 sec (Fig. 3). Re-addition of progesterone during the plateau period did not stimulate a secondary response indicating that once the initial progesterone-induced transient had occurred, the signal transduction machinery responsible for mediating the primary calcium influx had been completely down-regulated (Fig. 3).

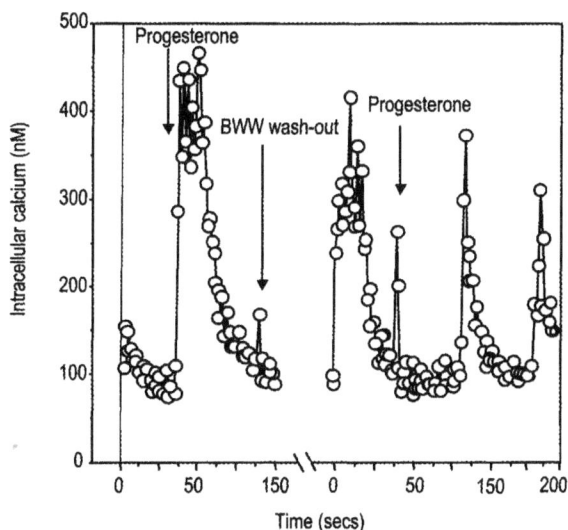

**Fig. 3** Secondary calcium oscillations do not depend upon the continued presence of progesterone and cannot be elicited by progesterone after the primary transient has occurred. In this experiment, the spermatozoa were washed free of progesterone immediately after the primary transient had been induced. After a 30 min lapse the changes in $[Ca^{2+}]_i$ were monitored. Spontaneous calcium oscillations continued in these cells, whereas progesterone could no longer elicit a response. Representative trace from an experiment in which 116 cells were imaged.

*Secondary oscillations involve extracellular calcium*

In previous studies it has been established that the primary calcium transient induced by progesterone involves an influx of extracellular calcium (Aitken *et al.*, 1996a). In order to determine whether the secondary calcium transients also involved the entry of extracellular calcium or the episodic release of calcium from intracellular calcium stores, $[Ca^{2+}]_i$ was monitored in spermatozoa incubated in BWW medium lacking added calcium (unpublished data). If spermatozoa were isolated and incubated in medium lacking significant extracellular calcium, the $[Ca^{2+}]_i$ fell to extremely low levels and no primary transients or secondary oscillations were observed on addition of progesterone (Fig. 4A).

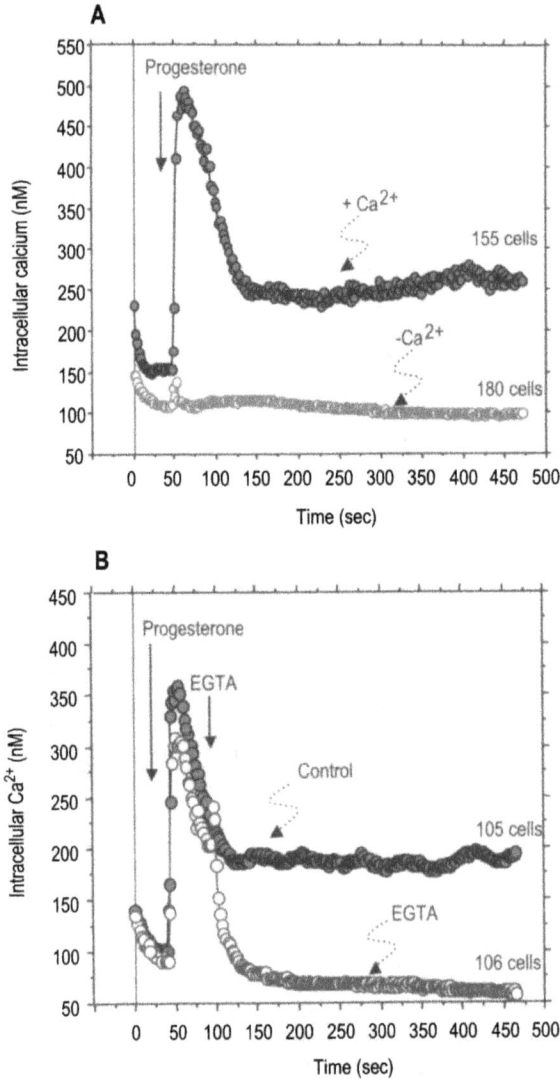

**Fig. 4** Impact of extracellular calcium on calcium oscillations in human spermatozoa. (A) in the absence of added extracellular calcium neither the primary progesterone-induced calcium transient nor the secondary calcium oscillations were observed. (B) If EGTA (3 mM) was added to the spermatozoa after the completion of the initial calcium transient, then secondary oscillations failed to occur.

The importance of extracellular calcium in the generation of secondary calcium oscillations was also indicated in an experiment in which calcium was removed from the incubation immediately after the induction of the primary transient with progesterone by the addition of EGTA (Fig. 4B). This treatment completely suppressed the occurrence of secondary calcium oscillations (unpublished data). Thus whereas 20% of 105 control cells exhibited at least one secondary calcium transient after the initial response, only 1 of 106 EGTA treated cells exhibited a secondary calcium wave; a difference that was highly statistically significant ($P < 0.001$). So, if internal calcium stores are ultimately involved in the mediation of secondary calcium transients, they have to be filled via an influx of extracellular calcium.

### Sperm capacitation and calcium oscillations

The phosphodiesterase inhibitor, pentoxifylline, has previously been shown to promote sperm capacitation, significantly enhancing the responses of human spermatozoa to progesterone, in terms of both the acrosome reaction and sperm-oocyte fusion (Kay et al., 1994; Aitken et al., 1996b). In order to determine whether the promotion of sperm capacitation had a significant impact on the patterns of calcium oscillation seen in response to progesterone, human spermatozoa were incubated for 3 h in 3 mM pentoxifylline prior to steroid exposure. In these experiments, pentoxifylline significantly enhanced the proportion of cells exhibiting secondary calcium oscillations following the administration of progesterone. Thus, in a population of 1176 cells from 34 individual samples, the mean percentage of spermatozoa exhibiting secondary calcium oscillations was 32% compared with 20% in 652 imaged control cells ($P < 0.02$). However, the size of the oscillating sperm population showed considerable inter-individual variation, giving 10th and 90th percentile values of 12 and 61% respectively ($n = 34$) (unpublished data).

The pentoxifylline data suggested that the capacity of human spermatozoa to exhibit secondary calcium oscillations was associated with the capacitation status of these cells. In order to examine the strength of this association, a variety of treatments were assessed that are known to suppress sperm capacitation including treatment with membrane permeant thiols (2-mercaptoethanol), exposure to 2–deoxyglucose or removal of bicarbonate.

*2-Mercaptoethanol*: Addition of 0.05% 2-mercaptoethanol to populations of human spermatozoa during capacitation did not prevent the induction of a primary calcium transient by progesterone; 93.2% of 147 cells imaged following treatment with this reductant exhibited a primary calcium transient compared with 97.1% of 90 control spermatozoa (Fig. 5A, C). However, the element of the calcium response that was dramatically different in the 2-mercaptoethanol treated samples, was the level of intracellular calcium achieved during the plateau phase ($146.5 \pm 3.9$ nM), which was significantly lower ($P < 0.001$) than that observed in control ($229.1 \pm 9.0$ nM) and pentoxifylline ($210.9 \pm 10.1$ nM) treated samples (Fig. 5B, C, D). This failure to sustain elevated intracellular calcium levels during the plateau phase was associated with a significant inhibition ($P < 0.001$) in the percentage of cells exhibiting secondary calcium oscillations (15%) compared with cells treated with pentoxifylline (38%) or control spermatozoa (22%) in a data set comprising 307 imaged cells from 3 independent donors (unpublished data).

*2-Deoxyglucose*: A second treatment that has been shown to suppress tyrosine phosphorylation in human spermatozoa, involves the use of media in which glucose has been replaced with an equimolar amount of 2-deoxyglucose (Aitken et al., 1998a). As long as spermatozoa are simultaneously supplied with exogenous lactate (25 mM), they are able to sustain their viability and motility in the presence of this compound. However, spermatozoa incubated in the presence of 2-deoxyglucose cannot undergo tyrosine phosphorylation during sperm capacitation and exhibit a consequential loss of functional responsiveness to progesterone (Aitken et al., 1998a). In order to determine whether the failure to respond to progesterone under these conditions was associated with a change in the ability of the spermatozoa to exhibit secondary calcium oscillations, progesterone-induced calcium transients were monitored in cells incubated for 3 h in medium BWW supplemented with 2-deoxyglucose. In a sample of 359 imaged cells from 3 independent donors, treatment with 2-deoxyglucose significantly ($P < 0.001$) reduced the percentage of spermatozoa exhibiting secondary calcium transients (6%) compared with controls in glucose containing BWW, with (41.6%) or without (16.6%) 3 mM pentoxifylline supplementation (unpublished data).

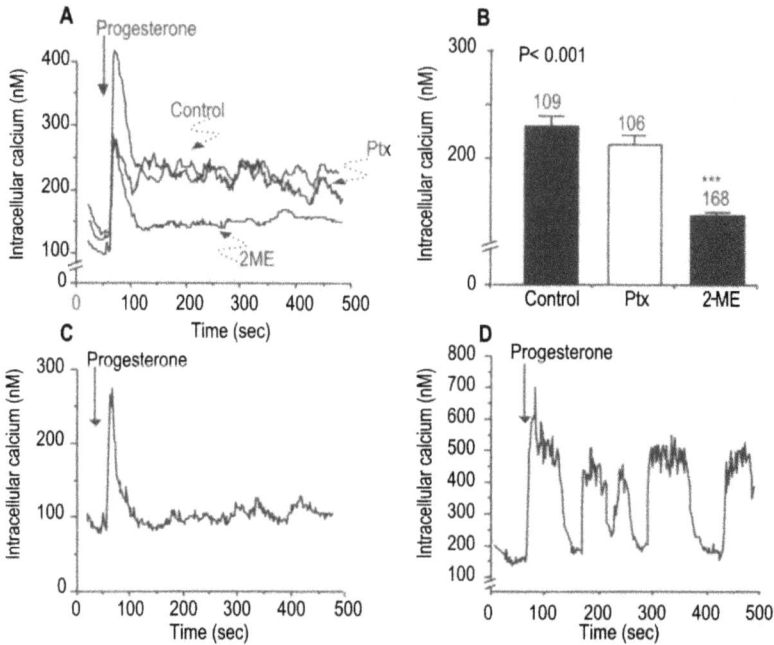

**Fig. 5** Impact of 2-mercaptoethanol on the calcium responses generated in human spermatozoa in response to the extragenomic action of progesterone. (A) The mean calcium levels recorded in one of three independent experiments illustrating the suppressive effect of 2-mercaptoethanol, particularly during the plateau phase of the progesterone response; number of cells imaged in this experiment was 55, 28 and 49 for the 2-mercaptoethanol, pentoxifylline and control incubations, respectively. (B) Mean levels of calcium observed during the plateau phase, 200 sec after the administration of progesterone; ***$P < 0.001$ for differences between the 2-mecaptoethanol treated cells and control or pentoxifylline-treated samples; numbers at head of each column gives total number of cells imaged. (C) An example of a 2-mercaptoethanol treated cell showing a primary calcium transient but no secondary oscillations. (D) Clear secondary calcium oscillations generated in representative cells from a pentoxifylline-treated sample.

*Bicarbonate.* Another treatment that is known to suppress capacitation in human spermatozoa is the removal of $HCO_3^-$ (Aitken et al., 1998a). Under these conditions, human spermatozoa do not phosphorylate tyrosine and cannot express their capacity for fertilization. When spermatozoa were incubated in $HCO_3^-$ free medium they significantly lost their ability to respond to progesterone in terms of both the primary progesterone–induced calcium transient and the secondary oscillations. Thus, only 28.3% of 219 cells exhibited a primary calcium transient compared with 81% of 161 control cells. Addition of pentoxifylline (3.0 mM) and 5 mM dbcAMP to the $HCO_3^-$-free medium did not improve this situation (only 8.8% of 68 cells responding). However, buffering the medium to pH 8.6 with 20 mM TAPS completely restored both the ability of human spermatozoa to exhibit both a primary calcium transient and secondary calcium oscillations (data not shown).

### Calcium channel antagonists

Since the extragenomic action of progesterone is associated with a rapid depolarization of the sperm plasma membrane (Patrat et al., 2002), it was of interest to determine whether the secondary calcium oscillations were associated with the opening of voltage operated calcium channels (VOCC).

*Diltiazem*: In order to evaluate the impact of the calcium channel blocker, diltiazem, on the responses of human spermatozoa to progesterone stimulation, cells were primed with 3.0 mM pentoxifylline for 3 h, washed and then stimulated with either progesterone alone or progesterone plus diltiazem at 1 μM and 10 μM doses. ANOVA analysis demonstrated a highly significant interaction between treatment and the pattern of calcium oscillations ($P < 0.001$). As the dose of diltiazem increased the percentage of spermatozoa exhibiting a single primary transient declined. In contrast, the percentage of cells exhibiting secondary calcium oscillations was not suppressed by the presence of diltiazem but increased from $25.0 \pm 5.3\%$ to $44.0 \pm 6.5\%$ (Fig. 6A).

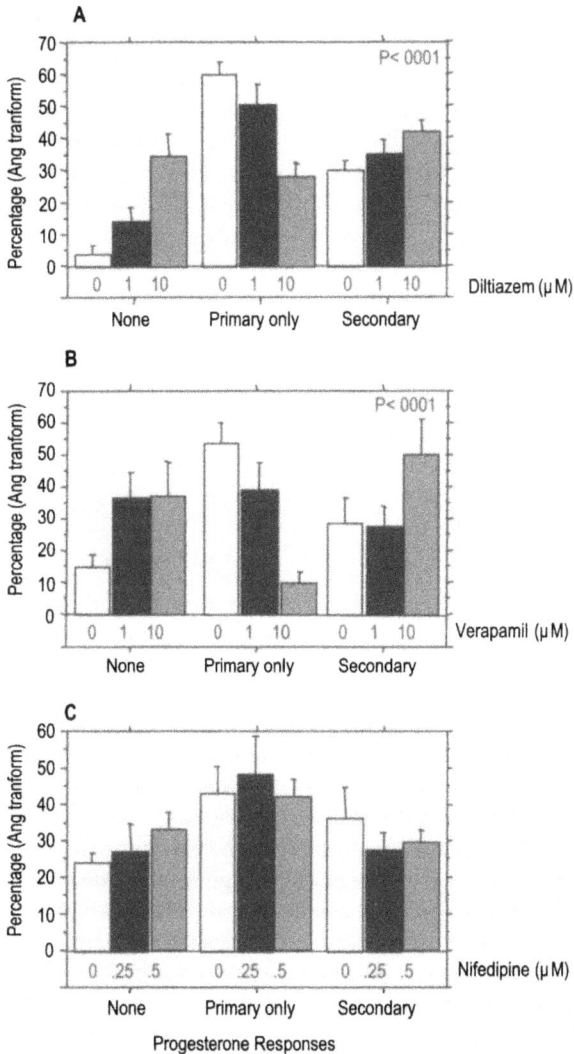

**Fig. 6** Impact of voltage-operated calcium channel blockers on the calcium transients generated in populations of human spermatozoa. All cells were primed with 3.0 mM pentoxifylline for 3 h and then treated with 5 μM progesterone along with inhibitor. (A) diltiazem; (B) verapamil; (C) nifedipine. Percentage data were angular (Ang) transformed prior to ANOVA analysis; significance of differences between treatments is given in panels A and B.

Moreover, the mean levels of $[Ca^{2+}]_i$ achieved during the plateau phase of the progesterone response, when the secondary calcium oscillations were occurring, was unaffected by diltiazem treatment (data not shown).

*Verapamil:* Using an identical experimental paradigm, spermatozoa were pre-incubated with 3.0 mM pentoxifylline and then stimulated with either progesterone alone or progesterone plus verapamil at doses of 1 μM and 10 μM (unpublished data). This experiment generated a similar response to that observed with diltiazem. ANOVA analysis again revealed a highly significant interaction between the pattern of calcium oscillation and treatment with the calcium channel antagonist (Fig. 6B; $P < 0.001$). Verapamil treatment significantly decreased the percentage of spermatozoa exhibiting a single transient but increased the proportion exhibiting secondary calcium transients, rising from $27.8 \pm 3.0\%$ in the controls to $54.7 \pm 14.4\%$ in the cells treated with 10 μM verapamil. This antagonist also had no significant influence on the $[Ca^{2+}]_i$ levels achieved during the plateau phase of the progesterone response (data not shown).

*Nifedipine:* Under identical conditions, nifedipine, at doses that are effective at disrupting VOCC (0.25 and 0.5 μM) had no significant effect on the incidence of secondary calcium oscillations or plateau phase $[Ca^{2+}]_i$ in cells primed with 3.0 mM pentoxifylline and treated with progesterone (Fig. 6 C).

### Pertussis toxin

Since the biological responsiveness of human spermatozoa to physiological agonists such as the zona pellucida can be primed by progesterone and inhibited with pertussis toxin (Tesarik et al., 1993), the latter was assessed for its impact on secondary calcium oscillations in pentoxifylline-primed cells, following progesterone exposure. As illustrated in Fig. 7A, pertussis toxin had no significant effect on the percentage of spermatozoa exhibiting primary or secondary calcium transients in response to progesterone. This G-protein inhibitor also had no effect on the levels of $[Ca^{2+}]_i$ achieved during the plateau phase of the response (data not shown).

### *Bicuculline and Picrotoxin*

Since the functional responses of human spermatozoa to progesterone can be suppressed by an antagonist of GABA (A) receptor (bicuculline) and a blocker of GABA (A) receptor-coupled chloride channel (picrotoxin), the impact of these reagents on the primary and secondary calcium transients induced by progesterone was assessed (Kuroda et al., 1999; Calogero et al., 1999). At a dose (10 μM) that is more than sufficient to antagonize the GABA (A) receptor in other tissues (Whyment et al., 2004), bicuculline affected neither the primary calcium transient induced by progesterone nor the proportion of cells generating secondary calcium oscillations, regardless of whether the inhibitor was added immediately before or immediately after the progesterone (Fig. 7B). Moreover, bicuculline had no impact on the levels of $[Ca^{2+}]_i$ achieved during the plateau phase of the progesterone-induced response (data not shown).

A slightly different picture emerged with picrotoxin. Using a dose of this compound (200 μM) that is maximally efficient in the suppression of GABA (A) receptor activation in other tissues (Wang et al., 2003), no significant impact was observed on the mean levels of $[Ca^{2+}]_i$ achieved during the primary transient or secondary plateau phases of the progesterone-induced response (data not shown). However, a significant ($P < 0.01$) interaction was observed between picrotoxin treatment and progesterone response pattern by ANOVA analysis (Fig. 7C). This change was reflected by an increase in the percentage of non-responsive cells in the presence of picrotoxin accompanied by a decline in the percentage of cells exhibiting second-

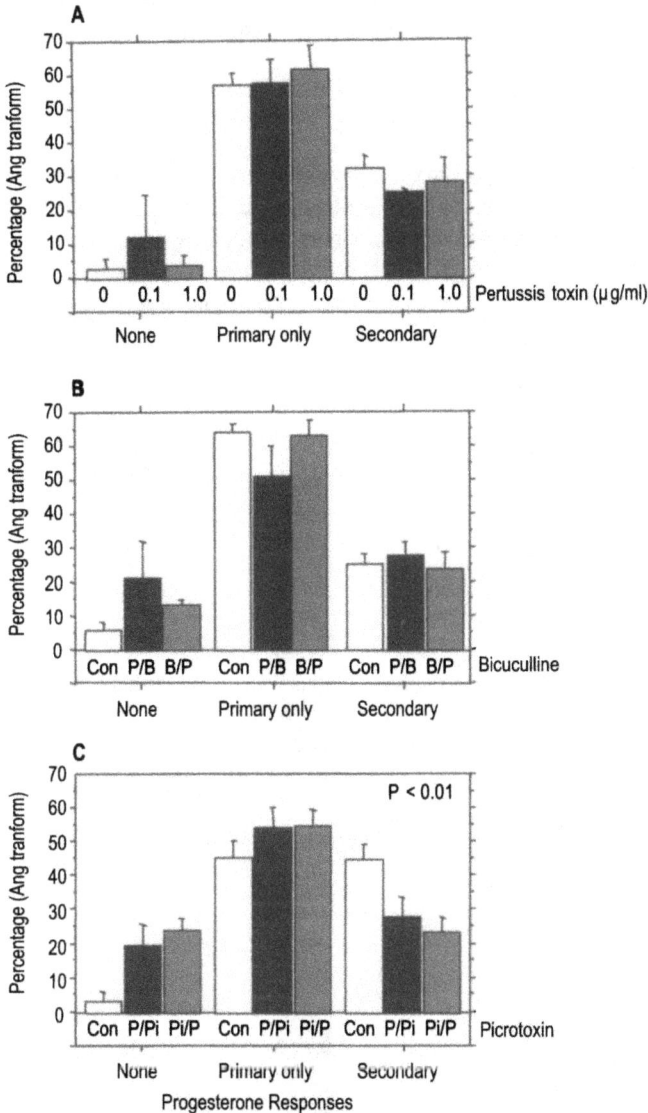

**Fig. 7**. Impact of inhibitors of G-protein activation and GABA signaling pathways on responses of human spermatozoa to the extragenomic action of progesterone. All cells were pre-treated with pentoxifylline for 3 h prior to progesterone stimulation. (A) dose-dependent study of pertussis toxin; (B) bicuculline. Progesterone (5 μM) was added along with bicuculline (P/B) or after 250 sec pretreatment with this antagonist (B/P). (C) picrotoxin. Progesterone (5 μM) was added along with picrotoxin (P/Pi) or after 250 sec pretreatment with this antagonist (Pi/P). Percentage data were angular transformed prior to ANOVA analysis; significance of differences between treatments is given in panel C.

ary responses. When the data across the 3 replicate experiments were pooled, 54/105 (51%) imaged cells treated with pentoxifylline alone exhibited secondary calcium transients following progesterone treatment, compared with 22/100 (22%) and 15/94 (16%) when the picrotoxin was administered either immediately after, or immediately before, progesterone exposure. These differences in the frequencies of secondary calcium responses were statistically different ($P <$ 0.001; chi-square analysis).

### Pimozide and amiloride

Since male germ cells possess T-type calcium channels that are activated on contact with the zona pellucida and inhibited by pimozide and amiloride (Arnoult *et al.*, 1996), the ability of these reagents to inhibit the secondary calcium transients induced by progesterone was assessed. Addition of 10 µM pimozide synchronously with progesterone to pentoxifylline stimulated cells, resulted in a significant interaction between the presence of this reagent and the pattern of calcium response as revealed by ANOVA ($P < 0.01$; Fig. 8A).

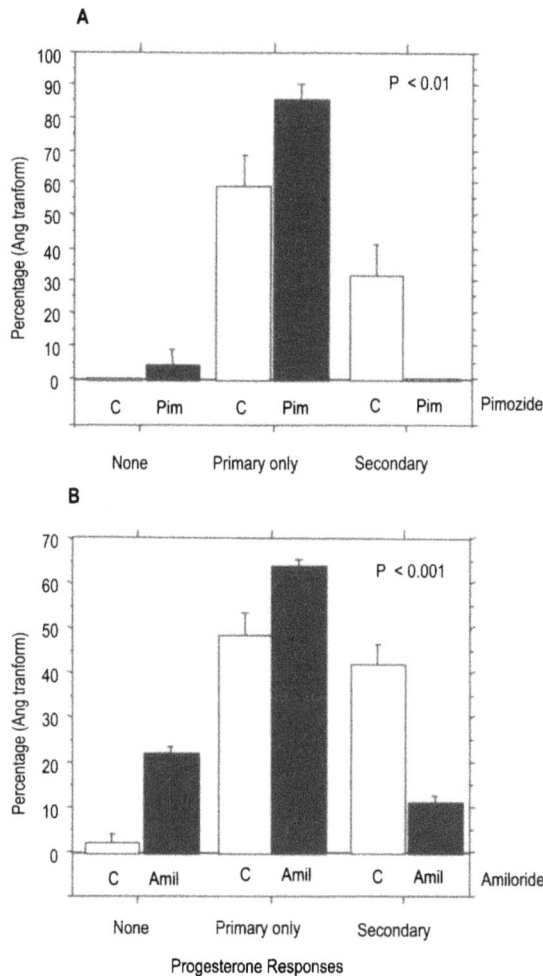

**Fig. 8** Impact of T-type calcium channel inhibitors on the extragenomic action of progesterone on human spermatozoa. All cells were pretreated with pentoxifylline for 3 h prior to progesterone stimulation. (A) 10 µM pimozide (Pim) was added to the spermatozoa along with 5 µM progesterone. (B) 100 µM amiloride (Amil) was added to the spermatozoa along with 5 µM progesterone. Percentage data were angular transformed prior to ANOVA analysis; significance of differences between treatments is given in panels A and B.

Out of a total of 72 control cells imaged in this experiment, 24 (33%) exhibited secondary calcium oscillations, compared with 0 out of 84 cells treated with pimozide ($P < 0.001$; chi-square). Similarly, administration of amiloride at the point of progesterone stimulation, resulted in a highly significant interaction between the presence of reagent and $[Ca^{2+}]_i$ ($P < 0.001$; Fig. 8B). Of 151 control cells treated with pentoxifylline and progesterone, 63 (42%) exhibited secondary calcium oscillations, compared with 8 (7%) out of 117 cells treated with progesterone and amiloride ($P < 0001$; chi square).

### Sperm-oocyte fusion

To determine whether the suppression of secondary calcium oscillations with pimozide or amiloride disrupted the functional responses of human spermatozoa to progesterone stimulation, assessments of sperm–oocyte fusion were undertaken. Verapamil was included in this analysis as an example of a calcium channel antagonist that did not suppress calcium transients. The results of this analysis are presented in Fig. 9. They indicate that the presence of these reagents had no significant impact on sperm motility. However, a moderate suppression of sperm-oocyte fusion was observed with amiloride and pimozide that, in the case of the former, was statistically significant ($P < 0.05$; Wilcoxon sign rank test).

## Discussion

The results of this study reveal a high order of complexity in the response of human spermatozoa to the extragenomic actions of progesterone. Not only does this steroid induce an immediate calcium transient, as previously reported, but this phenomenon is followed by secondary calcium oscillations, the incidence of which depends heavily on the capacitation status of the spermatozoa. Thus the frequency of secondary calcium oscillations were significantly enhanced following preincubation with the phosphodiesterase inhibitor, pentoxifylline, which is known to stimulate the capacitation of human spermatozoa via the elevation of intracellular cyclic AMP. Conversely, treatments that suppress capacitation, such as bicarbonate withdrawal, exposure to membrane permeant thiols or incubation in the presence of 2-deoxyglucose, were all highly effective in disrupting the secondary calcium oscillations induced by progesterone. These findings are in keeping with reports by others (Kirkman-Brown *et al.*, 2004) who also observed secondary calcium oscillations in human spermatozoa treated with progesterone, the incidence of which was promoted by extending the duration of sperm capacitation.

A characteristic feature of these secondary calcium oscillations is that they only occur in a subpopulation of spermatozoa. This is in contrast to the primary progesterone response, which occurs in the vast majority of cells and is not dependent on, or indicative of, capacitation status. The proportion of cells behaving in this manner shows significant inter-individual variation but approximates to the region of 20% of the capacitated human sperm populations. Since around 20% of such sperm populations also acrosome react on exposure to progesterone (Harper *et al.*, 2003), a reasonable hypothesis would be that the subpopulation of cells showing secondary calcium oscillations, represents the same subpopulation that has been primed by capacitation to acrosome react on exposure to biological stimuli such as progesterone. If this is the case, then resolving the nature, control and biological significance of these secondary calcium oscillations will have profound implications for our understanding of male fertility and defective sperm function.

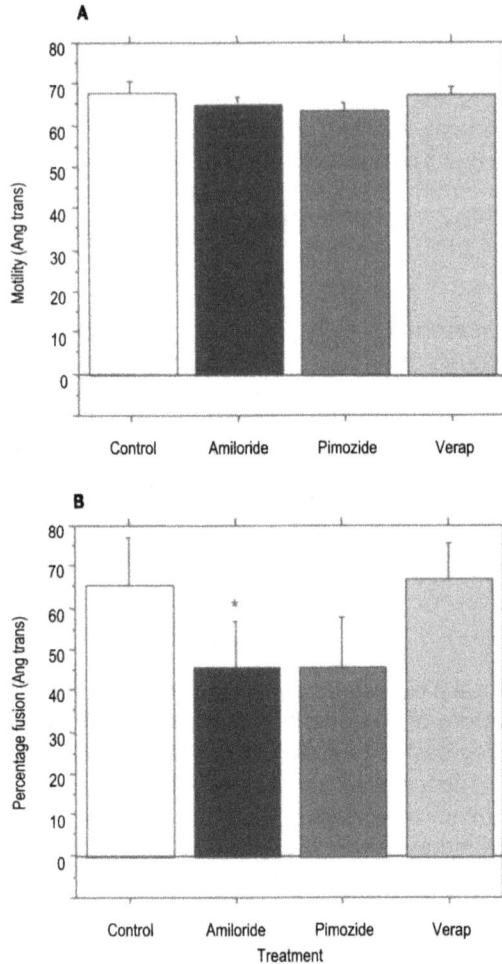

**Fig. 9** Impact of T-type calcium channel inhibitors on the fertilizing ability of human spermatozoa. Sperm function was assessed in a heterologous *in vitro* fertilization assay employing zona-free hamster oocytes. This assay detects the presence of acrosome reacted human spermatozoa that are competent to fuse with the vitelline membrane of the oocyte. All cells were pretreated with pentoxifylline for 3 h prior to progesterone stimulation. A, motility; B, percentage of oocytes penetrated. Inhibitors used were amiloride (100 μM), pimozide (10 μM) and verapamil (1 μM). *$P < 0.05$ for difference from control incubation by Wilcoxon signed rank test.

In term of their nature, analysis of the changes in $[Ca^{2+}]_i$ following calcium depletion with EGTA either prior to, or immediately after, the initial progesterone-induced calcium transient, revealed that these secondary calcium oscillations involve the entry of extracellular calcium through channels in the sperm plasma membrane. Once the primary calcium transient has occurred, the secondary calcium oscillations can proceed in the presence of very low levels (~5 μM) of extracellular calcium but not in the complete absence of this cation (Harper *et al.*, 2004). Thus some influx of extracellular calcium is necessary to sustain the secondary calcium oscillations induced by progesterone. It is possible that this influx is necessary to replenish intracellular calcium stores (Rossato *et al.*, 2001) and it is the cyclical emptying and refilling of these stores that creates the calcium oscillations. Depletion of these stores may trigger the capacitative influx of extracellular calcium through store-operated calcium channels in the plasma membrane. This potential for capacitative calcium entry is known to be tyrosine phosphorylation dependent and may contribute significantly to the attainment of a capacitated state (Rossato *et al.*, 2001; Dorval *et al.*, 2003).

Detailed analysis of the biochemical pathways responsible for regulating calcium influx revealed that voltage-operated L-type calcium channels are not involved in shaping the spermatozoon's response to progesterone stimulation. The presence of L-type channel blockers such as verapamil, diltiazem or nifedipine (at doses that are known to disrupt such channels in other tissues) did not influence the secondary calcium responses to progesterone stimulation. Although higher doses of nifedipine (10 µM) have been shown to impact upon the incidence of secondary calcium oscillations in progesterone-treated human spermatozoa (Kirkman-Brown *et al.*, 2004), at this dose, nifedipine lacks specificity and can also impact upon T-type channel activity (see below).

The induction of acrosomal exocytosis by the zona glycoprotein ZP3 involves the activation of G-proteins that are sensitive to inhibition with pertussis toxin (Schuffner *et al.*, 2002). In contrast, the calcium influxes activated by prostaglandins E1 and E2 in human spermatozoa are not sensitive to pertussis toxin disruption (Shimizu *et al.*, 1998). The results obtained in this study indicate that the secondary calcium oscillations induced by progesterone are, like the prostaglandin-induced transients, insensitive to pertussis toxin inhibition and therefore not regulated by G-proteins. Similarly, the primary calcium transient induced by progesterone has been shown to be resistant to inhibition with this toxin (Foresta *et al.*, 1993). Furthermore, the influx of extracellular calcium associated with secondary calcium oscillations did not appear to involve the activation of GABA (A) receptors since an inhibitor of this receptor, bicuculline, had no significant effect on the incidence of spermatozoa exhibiting primary or secondary calcium transients in this study. However, picrotoxin, an inhibitor of the GABA (A) receptor chloride channel, did have a significant effect on the incidence of secondary calcium oscillations following progesterone exposure, without disrupting the primary calcium transient (Turner *et al.*, 1994). The chloride channel must therefore play some role in creating the conditions necessary for the appearance of secondary calcium oscillations. The nature of this contribution probably reflects the importance of progesterone-induced chloride efflux in achieving depolarization of the sperm plasma membrane, as discussed below.

T-type calcium channels appear to be the main route by which extracellular calcium enters spermatozoa during the secondary calcium oscillations elicited by progesterone. Thus the T-type channel inhibitors, pimozide and amiloride, were both effective in reducing the incidence of cells exhibiting secondary calcium oscillations, without influencing the primary calcium transient (Fig. 8). This type of calcium channel has previously been demonstrated in both mouse and rat spermatozoa and is known to be involved in mediating the calcium influx associated with the induction of the acrosome reaction on the surface of the zona pellucida (Arnoult *et al.*, 1996). The existence of T-type channels on the surface of human spermatozoa has also been confirmed and implicated in the calcium influx and acrosomal exocytosis induced by neoglycoprotein mimics of the zona pellucida (Blackmore and Eisoldt, 1999). However, these channels are not involved in the primary calcium transient induced by progesterone (Blackmore and Eisoldt, 1999).

The setting of T-type channels by progesterone is consistent with the ability of this steroid to stimulate hyperpolarization, following transient depolarization, in subsets of human spermatozoa stimulated with progesterone (Patrat *et al.*, 2002). The initial depolarization response to progesterone is driven largely by a sodium influx, although a minor contribution is made by the chloride efflux mediated by the picrotoxin-sensitive channel associated with the GABA receptor. As a consequence of the sodium influx/chloride efflux elicited by progesterone, a depolarization is induced that takes the plasma membrane potential of human spermatozoa from around $-50$ mV to $+15$ mV in a subpopulation of cells comprising less than 30% of the entire population. Within this subpopulation membrane depolarization is followed by hyperpolarization to around $-60$ mV (Patrat et al., 2002). This is the degree of polarization needed to maintain T-type channels in the sperm plasma membrane in a closed but activation-competent state (Arnoult et al., 1996). In biological terms, this hyperpolarization-dependent setting of T-type channels by progesterone in a sperm subpopulation would account for the priming effect of progesterone on the subsequent ability of these cells to acrosome react on the zona surface. A role for progesterone in effecting the final stages of sperm capacitation is also in keeping with data indicating that progesterone can enhance the capacitation status of murine spermatozoa and that this capacitation-inducing effect is completely negated when T-type channels are blocked with $NiCl_2$ (Senuma et al., 2001).

The secondary calcium oscillations induced in a subpopulation of cells by progesterone are thus reflective of the attainment of this primed, fully capacitated state in which hyperpolarization, as a consequence of sodium influx and chloride efflux, leads to the activation of T-type channels. Similar spontaneous calcium spikes have previously been observed during neuronal development in Xenopus, and these transients are also dependent on the entry of extracellular calcium through T-type channels (Gu et al., 1994). When spermatozoa that have been primed in this way bind to the zona pellucida, the T-type channels open and elicit a second calcium-entry pathway involving transient receptor potential (Trp) proteins, thereby producing the sustained increase in intracellular calcium needed to precipitate acrosomal exocytosis at the zona surface (Jungnickel et al., 2001).

Progesterone is a weak biological stimulus for the acrosome reaction. Although this steroid induces a detectable change in sensitive bioassays such as the hamster oocyte penetration assay (Aitken et al., 1996b; Francavilla et al., 2002), the concomitant rate of acrosomal exocytosis is low to undetectable, depending on the assay conditions employed (Harper et al., 2004; Harper and Publicover, 2005). Harper et al. (2004) could find no positive relationship between the subpopulation of spermatozoa exhibiting an acrosome reaction in response to progesterone and the presence of secondary calcium transients. Indeed, fewer of the cells exhibiting calcium oscillations underwent an acrosome reaction than the non-oscillating cells. Such results reinforce the notion that many of the acrosome reactions precipitated by progesterone and detected with conventional histochemical methods are dysfunctional. After all, if large numbers of spermatozoa really did acrosome react on contact with progesterone, they would no longer be able to bind to the zona pellucida and fertilize the oocyte. However, the hamster oocyte penetration test is such a sensitive bioassay that it can detect a biological response to progesterone exposure, in terms of the presence of spermatozoa have not only undergone a physiological acrosome reaction but also generated a fusogenic equatorial segment (Aitken et al., 1996b; Francavilla et al., 2002). Suppression of the secondary calcium transients with T-type channel antagonists gave a moderate suppression of this biological response. Our interpretation of these results is that the generation of secondary calcium transients in response to progesterone is a reflection of the capacitation status of the spermatozoa. While a small proportion of highly capacitated (possibly over-capacitated) cells may respond to these calcium signals by undergoing a physiological acrosome reaction (in terms of mechanism, if not timing), a vast majority of oscillating cells are simply primed by this response so that they can rapidly undergo the acrosome reaction on binding to the zona pellucida (Harper et al., 2004; Harper and Publicover, 2005). Furthermore, progesterone-induced secondary calcium oscillations may also promote fertilisation by inducing changes in the flagellar beat pattern compatible with hyperactivation (Jaiswal et *al.*, 1999; Harper and Publicover, 2005). In this light, progesterone can be seen as an integral part of the global oocyte-cumulus signaling complex that orchestrates a series of contemporaneous behavioural changes in the spermatozoa that combine to facilitate efficient fertilisation.

## Acknowledgements

The authors gratefully acknowledge the technical assistance of Wendy Knox and Diana Harkiss in the conduct of these studies. This study was partially supported by funds from the ARC Centre of Excellence in Biotechnology and Development.

## References

Aitken RJ (1986) The zona-free hamster oocyte penetration test and the diagnosis of male fertility *International Journal of Andrology Supplement* **6** 1-19

Aitken RJ (1997) Molecular mechanisms regulating human sperm function *Molecular Human Reproduction* **3** 169-173

Aitken RJ, Paterson M, Fisher H, Buckingham DW and van Duin M (1995) Redox regulation of tyrosine phosphorylation in human spermatozoa and its role in the control of human sperm function *Journal of Cell Science* **108** 2017-2025

Aitken RJ, Buckingham DW and Irvine DS (1996a) The extragenomic action of progesterone on human spermatozoa: evidence for a ubiquitous response that is rapidly down regulated *Endocrinology* **137** 3999-4009

Aitken RJ, Buckingham DW, Harkiss D, Paterson M, Fisher H and Irvine DS (1996b) The extragenomic action of progesterone on human spermatozoa is influenced by redox regulated changes in tyrosine phosphorylation during capacitation *Molecular and Cellular Endocrinology* **117** 83-93

Aitken RJ, Harkiss D, Knox W, Paterson M and Irvine DS (1998a) A novel signal transduction cascade in capacitating human spermatozoa characterised by a redox-regulated, cAMP-mediated induction of tyrosine phosphorylation *Journal of Cell Science* **111** 645-656

Aitken RJ, Harkiss D, Knox W, Paterson M and Irvine DS (1998b) On the cellular mechanisms by which the bicarbonate ion mediates the extragenomic action of progesterone on human spermatozoa *Biology of Reproduction* **58** 186-196

Anderson L, Hoyland J, Mason WT and Eidne KA (1992) Characterization of the gonadotrophin-releasing hormone calcium response in single alpha T3-1 pituitary gonadotroph cells *Molecular and Cellular Endocrinology* **86** 167-175

Arnoult C, Cardullo RA, Lemos JR and Florman HM (1996) Activation of mouse sperm T-type Ca²⁺ channels by adhesion to the egg zona pellucida *Proceedings of the National Academy of Sciences USA* **93** 13004-13009

Austin CR (1951) Observations on the penetration of the sperm into the mammalian egg *Australian Journal of Scientific Research* **4** 581-596

Baldi E, Casano R, Falsetti C, Krausz C, Maggi M and Forti G (1991) Intracellular calcium accumulation and responsiveness to progesterone in capacitating human spermatozoa *Journal of Andrology* **12** 323-330

Biggers JD, Whitten WK and Whittingham DG (1971) The culture of mouse embryos *in vitro*. In *Methods in Mammalian Embryology* pp 86-116 Ed JC Daniels. Freeman, San Francisco

Blackmore PF and Eisoldt S (1999) The neoglycoprotein mannose-bovine serum albumin, but not progesterone, activates T-type calcium channels in human spermatozoa *Molecular Human Reproduction* **5** 498-506

Blackmore PF, Beebe SJ, Danforth DR and Alexander N (1990) Progesterone and 17α-hydroxy-progesterone. Novel stimulators of calcium influx in human spermatozoa *Journal of Biological Chemistry* **265** 1376-1380

Blackmore PF, Neulin J, Lattanzio F and Beebe SJ (1991) Cell surface binding sites for progesterone mediate calcium uptake in human sperm *Journal of Biological Chemistry* **266**, 18655-18659

Bonaccorsi L, Forti G and Baldi E (2001) Low-voltage-activated calcium channels are not involved in capacitation and biological response to progesterone in human sperm *International Journal of Andrology* **24** 341-351

Calogero AE, Burrello N, Ferrara E, Hall J, Fishel S and D'Agata R (1999) Gamma-aminobutyric acid (GABA) A and B receptors mediate the stimulatory effects of GABA on the human sperm acrosome reaction: interaction with progesterone *Fertility and Sterility* **71** 930-936

Chang MC (1951) Fertilizing capacity of spermatozoa deposited into the fallopian tubes *Nature* **168** 697-698

Dorval V, Dufour M and Leclerc P (2003) Role of protein tyrosine phosphorylation in the thapsigargin-induced intracellular Ca²⁺ store depletion during human sperm acrosome reaction *Molecular Human Reproduction* **9** 125-131

Fisher RA and Yates F (1963) *Statistical Tables for Biological, Agricultural and Medical Research*, 6th edition. Hafner Press, New York

Florman HM, Corron ME, Kim TDH and Babcock DF (1992) Activation of voltage sensitive calcium channels of mammalian sperm is required for zona pellucida-induced acrosomal exocytosis *Developmental Biology* 152 304-314

Foresta C, Rossato M and Di Virgilio F (1993) Ion fluxes through the progesterone activated channel of the sperm plasma membrane *Biochemical Journal* 294 279-283

Francavilla F, Romano R, Santucci R, Macerola B, Ruvolo G and Francavilla S (2002) Effect of human sperm exposure to progesterone on sperm-oocyte fusion and sperm-zona pellucida binding under various experimental conditions *International Journal of Andrology* 25 106-112

Fraser LR and McIntyre K (1989) Calcium channel antagonists modulate the acrosome reaction but not capacitation in mouse spermatozoa *Journal of Reproduction and Fertility* 86 223-233

Gu X, Olson EC and Spitzer NC (1994) Spontaneous neuronal calcium spikes and waves during early differentiation *Journal of Neuroscience* 14 6325-6335

Guzman-Grenfell AM and Gonzalez-Martinez MT (2004) Lack of voltage-dependent calcium channel opening during the calcium influx induced by progesterone in human sperm. Effect of calcium channel deactivation and inactivation *Journal of Andrology* 25 117-122

Harper CV and Publicover SJ (2005) Reassessing the role of progesterone in fertilization-compartmentalized calcium signalling in human spermatozoa? *Human Reproduction* 20 2675-2680

Harper CV, Kirkman-Brown JC, Barratt CL and Publicover SJ (2003) Encoding of progesterone stimulus intensity by intracellular $[Ca^{2+}]$ ($[Ca^{2+}]_i$) in human spermatozoa *Biochemical Journal* 372 407-417

Harper CV, Barratt CL and Publicover SJ (2004) Stimulation of human spermatozoa with progesterone gradients to simulate approach to the oocyte. Induction of $[Ca^{2+}]_i$ oscillations and cyclical transitions in flagellar beating *Journal of Biological Chemistry* 279 46315-46325

Hull MGR, Glazener CMA, Kelly NJ, Conway DI, Foster PA, Hunton RA, Coulson C, Lambert PA, Watt EM and Desai KM (1985) Population study of causes, treatment and outcome of infertility *British Medical Journal* 291 1693-1697

Jaiswal BS, Tur-Kaspa I, Dor J, Mashiach S and Eisenbach M (1999) Human sperm chemotaxis: is progesterone a chemoattractant? *Biology of Reproduction* 60 1314-1319

Jungnickel MK, Marrero H, Birnbaumer L, Lemos JR and Florman HM (2001) Trp2 regulates entry of $Ca^{2+}$ into mouse sperm triggered by egg ZP3 *Nature Cell Biology* 3 499-502

Kay VJ, Coutts JRT and Robertson L (1994) Effects of pentoxifylline and progesterone on human sperm capacitation and acrosome reaction *Human Reproduction* 9 2318-2323

Kirkman-Brown JC, Bray C, Stewart PM, Barratt CL and Publicover SJ (2000) Biphasic elevation of $[Ca^{2+}]_i$ in individual spermatozoa exposed to progesterone

*Developmental Biology* 222 326-335

Kirkman-Brown JC, Barratt CL and Publicover SJ (2004) Slow calcium oscillations in human spermatozoa *Biochemical Journal* 378 827-832

Kuroda Y, Kaneko S, Yoshimura Y, Nozawa S and Mikoshiba K (1999) Influence of progesterone and GABA A receptor on calcium mobilization during human sperm acrosome reaction *Archives of Andrology* 42 185-191

Losel R, Breiter S, Seyfert M, Wehling M and Falkenstein E (2005) Classic and non-classic progesterone receptors are both expressed in human spermatozoa *Hormone and Metabolic Research* 37 10-14

Lyttle CR and Kopf GS (2003) Status and future direction of male contraceptive development *Current Opinion in Pharmacology* 3 667-671

Morales P, Pizarro E, Kong M, Kerr B, Ceric F and Vigil P (2000) Gonadotropin-releasing hormone-stimulated sperm binding to the human zona is mediated by a calcium influx *Biology of Reproduction* 63 635-642

Osman RA, Andria ML, Jones AD and Meizel S (1989) Steroid induced exocytosis: the human sperm acrosome reaction *Biochemical and Biophysical Research Communications* 160 828-834

Patrat C, Serres C and Jouannet P (2002) Progesterone induces hyperpolarisation after a transient depolarisation phase in human spermatozoa *Biology of Reproduction* 66 1775-1780

Roldan ER, Murase T and Shi QX (1994) Exocytosis in spermatozoa in response to progesterone and zona pellucida *Science* 266 1578-1581

Rossato M, Di Virgilio F, Rizzuto R, Galeazzi C and Foresta C (2001) Intracellular calcium store depletion and acrosome reaction in human spermatozoa: role of calcium and plasma membrane potential *Molecular Human Reproduction* 7 119-128

Schuffner AA, Bastiaan HS, Duran HE, Lin ZY, Morshedi M, Franken DR and Oehninger S (2002) Zona pellucida-induced acrosome reaction in human sperm: dependency on activation of pertussis toxin-sensitive G(i) protein and extracellular calcium, and priming effect of progesterone and follicular fluid *Molecular Human Reproduction* 8 722-727

Senuma M, Yamano S, Nakagawa K, Irahara M, Kamada M and Aono T (2001) Progesterone accelerates the onset of capacitation in mouse sperm via T-type calcium channels *Archives of Andrology* 47 127-134

Shah C, Modi D, Sachdeva G, Gadkar S, D'Souza S and Puri C (2005) N-terminal region of progesterone receptor B isoform in human spermatozoa *International Journal of Andrology* 28 360-371

Shi Q-X and Roldan ERS (1995) Evidence that a GABAA-like receptor is involved in progesterone-induced acrosomal exocytosis in mouse spermatozoa *Biology of Reproduction* 52 373-381

Shimizu Y, Yorimitsu A, Maruyama Y, Kubota T, Aso T and Bronson RA (1998) Prostaglandins induce calcium influx in human spermatozoa *Molecular Human Reproduction* 4 555-561

Tesarik J, Carreras A and Mendoza C (1993) Differential sensitivity of progesterone- and zona pellucida- induced acrosome reactions to pertussis toxin *Molecular Reproduction and Development* **34** 183-189

Tesarik J, Carreras A and Mendoza C (1996) Single cell analysis of tyrosine kinase dependent and independent $Ca^{2+}$ fluxes in progesterone induced acrosome reaction *Molecular Human Reproduction* **2** 225-232

Thomas P, Tubbs C, Detweiler C, Das S, Ford L and Breckenridge-Miller D (2005) Binding characteristics, hormonal regulation and identity of the sperm membrane progestin receptor in Atlantic croaker *Steroids* **70** 427-433

Turner KO, Garcia MA and Meizel S (1994) Progesterone initiation of the human sperm acrosome reaction: the obligatory increase in intracellular calcium is independent of the chloride requirement *Molecular and Cellular Endocrinology* **101** 221-225

Urner F and Sakkas D (2003) Protein phosphorylation in mammalian spermatozoa *Reproduction* **125** 17-26

Visconti PE, Bailey JL, Moore GD, Pan D, Olds-Clarke P and Kopf GS (1995) Capacitation of mouse spermatozoa I. Correlation between the capacitation state and protein tyrosine phosphorylation *Development* **121** 1129-1137

Wang DD, Krueger DD and Bordey A (2003) GABA depolarizes neuronal progenitors of the postnatal subventricular zone via GABAA receptor activation *Journal of Physiology* **550** 785-800

White DR and Aitken RJ (1989) Relationship between calcium, cAMP, ATP and intracellular pH and the capacity to express hyperactivated motility by hamster spermatozoa *Gamete Research* **22** 163-178

Whyment AD, Wilson JM, Renaud LP and Spanswick D (2004) Activation and integration of bilateral GABA-mediated synaptic inputs in neonatal rat sympathetic preganglionic neurones *in vitro Journal of Physiology* **555** 189-203

World Health Organisation (1992) *WHO Laboratory Manual for the Examination of Human Semen and Semen-Cervical Mucus Interaction* Cambridge University Press, Cambridge

Yanagimachi R (1994) Mammalian fertilization. In *The Physiology of Reproduction 2nd ed* pp 189 – 317 Eds E Knobil and JD Neill. Raven Press, New York

# Mammalian sperm capacitation: role of phosphotyrosine proteins

S Shivaji, Vivek Kumar, Kasturi Mitra and Kula Nand Jha

*Centre for Cellular and Molecular Biology, Uppal Road, Hyderabad 500 007, India*

Capacitation was discovered independently by Austin and Chang in the early 1950s and was defined as the obligate period of residence of spermatozoa in the female reproductive tract, which confers on the spermatozoa the ability to fertilize an oocyte. The molecular basis of the phenomenon of capacitation is poorly understood despite the fact that it is an important event preceding fertilization. This review presents our current understanding of the signalling events involved in the process of capacitation.

Capacitation refers to the finite period of residence of spermatozoa in the female reproductive tract, prior to fertilization of the oocyte. During this period, the mature and motile spermatozoa acquire the ability to fertilize the oocyte. This phenomenon of capacitation was first observed independently by Chang (1951) and Austin (1951; 1952), who also emphasized that capacitation is a necessary event in the life cycle of the male gamete. The importance of capacitation also stems from the fact that a block in capacitation could cause male infertility. Thus, there is a need to understand the molecular basis of capacitation. Presently, capacitation is viewed as a culmination of molecular, cellular and physiological changes that occur in the spermatozoa in the female reproductive tract, to achieve the final competence to fertilize the oocyte. This review focuses on the key features of capacitation and also surveys the known factors which bring about these changes during capacitation.

## Hallmarks of capacitation

Hyperactivation and acrosome reactions are the two distinct hallmarks of sperm capacitation. Yanagimachi (1969; 1970) reported that the first observable physiological change during capacitation was a visible change in sperm trajectory from a linear to a non-linear track and a simultaneous increase in the flagellar activity thus bringing about an increase in velocity. Due to these distinct changes in their motility kinematics, hyperactivated spermatozoa can now be discriminated more objectively from the non-hyperactivated population during capacitation using a computer aided semen analyser (Shivaji et al., 1995; Jayaprakash et al., 1997: Uma Devi et al., 2000). Hyperactivation is considered to be critical to the success of fertilization as it enhances the ability of the spermatozoa to traverse viscoelastic zones in the female reproductive tract more effectively than non-hyperactivated ones and to penetrate mucous substances of the cumulus oophorous and zona pellucida of the oocyte (Suarez et al., 1991; Suarez and Dai, 1992). Hyperactivation has been studied extensively and needs to be dealt with separately. In brief, this process is an energy driven process and ATP required is derived from

E-mail: shivas@ccmb.res.in

glycolysis rather than mitochondrial ATP synthesis as observed in human (Williams and Ford, 2001) and rat (Bone et al., 2000) spermatozoa. Calcium is essential for hyperactivation of hamster (Kulanand and Shivaji, 2001) and mouse (Fraser et al., 2001) spermatozoa and Suarez et al. (1993) observed that calcium levels rise specifically in the flagella of hamster spermatozoa during hyperactivation. This increase in calcium during hyperactivation may be due to the entry of calcium from extracellular sources through voltage gated calcium channels or due to intracellular release of calcium (Mujica et al., 1994) indicating the presence of internal calcium store(s) in spermatozoa. The redundant Nuclear Envelope (RNE) has been proposed to be an intracellular store for calcium in bull spermatozoa and has also been implicated in hyperactivation (Ho and Suarez, 2003). Further, it has been proposed that in hamster spermatozoa cyclic adenosine monophosphate (cAMP) regulates optimum levels of calcium required for hyperactivation (Aoki et al., 1999). However, in bovine spermatozoa it has been shown that calcium, and not cAMP, is required for hyperactivation (Marquez and Suarez, 2004). In addition to calcium and cAMP, bicarbonate ions have also been implicated in the process of hyperactivation but the mechanism by which these three effectors ultimately convert a non-hyperactivated to a hyperactivated spermatozoon is still not clearly understood. Further, it is not clear, if all these molecules influence hyperactivation directly or indirectly through some other means like stimulating capacitation dependant tyrosine phosphorylation. Although hyperactivation occurs at some point during capacitation, both phenomena have been found to be uncoupled in some cases, like spermatozoa from tw32/+ mouse which show precocious hyperactivation while capacitation occurs on schedule (Olds-Clarke, 1989).

The acrosome reaction is another hallmark of capacitated spermatozoa and it is considered as the end point of capacitation. During the acrosome reaction the outer acrosomal membrane of the spermatozoon fuses with the overlying plasma membrane and releases the acrosomal contents which facilitate penetration of the oocyte (Curry and Watson, 1995). The physiological significance of this reaction is further highlighted by the fact that males with spermatozoa lacking the acrosome are infertile (Baccetti et al., 1991). An in vitro system for elucidating the fusion mechanisms during the acrosome reaction has been developed (Spungin et al., 1995). Recently, it has been shown that actin polymerizes during capacitation and the polymerized F-actin breaks down just before the acrosome reaction (Brener et al., 2003). The physiological trigger for the acrosome reaction is either the zona pellucida or progesterone (Patrat et al., 2000) and can also be induced non-physiologically by the calcium ionophore, A23187 (Talbot et al., 1976), thus implying an active role of the divalent cation in this event. Both extracellular and intracellular calcium sources are known to influence the acrosome reaction (Breitbart, 2002) and the effect may be mediated through its influence on calcium dependant molecules like protein kinase C, phospholipase C and calmodulin, which have been demonstrated to be involved in the acrosome reaction (Breitbart and Spungin, 1997). However, the interdependence of one molecule on the other is not yet clearly worked out in spermatozoa as in other cells.

### Molecular basis of capacitation

Crucial to our present understanding of the molecular basis of capacitation is the observation that the follicular fluid induced acrosome reaction that occurs in the female reproductive tract is successful only if the spermatozoal bound seminal plasma components are removed prior to exposure to the follicular fluid (Oliphant, 1976). Despite this finding, only with the advent of in vitro media, which could facilitate capacitation of spermatozoa, did our understanding of capacitation improved. The media allowed a single straight forward approach to identify components which are essential for capacitation in vitro and then work around these components so as to understand the process. Using this approach it has been demonstrated that sperm capacitation is influenced by cholesterol efflux, first messengers such as progesterone and bicarbonate ion, second messengers such as calcium and cAMP, and downstream events: such as, intracellular alkalinisation and hyperpolarization; membrane reorganization; the action of reactive oxygen species; nitric oxide; and protein tyrosine phosphorylation.

## Cholesterol efflux

Albumin was shown to trigger capacitation by reducing the cholesterol/phospholipids ratio of the sperm membrane (Davis, 1971; Davis *et al.*, 1980) thus providing evidence that the decapacitating capacity of seminal plasma of rat and rabbit semen is due to cholesterol bearing membrane vesicles present in the semen (Davis, 1971). Recently, Chiu *et al.* (2005) demonstrated that Glycodelin – S in human seminal plasma reduces cholesterol efflux and thus inhibits capacitation. It was demonstrated that the ability of uterine and follicular fluid to induce sperm capacitation is due to high density lipoproteins present in these fluids which act as cholesterol acceptors and thus induce capacitation *in vivo* (Manjunath and Therien, 2002). That cholesterol removal facilitates acrosome reaction was also substantiated in spermatozoa of many other species by using cholesterol depleting agents like ß-cyclodextrin (Choi and Toyoda, 1998; Cross, 1999; Osheroff *et al.*, 1999). In this context mention should be made here of the Bovine Seminal Plasma proteins (BSPs) which have been widely characterized in bovine species although similar proteins have been detected in several other species like stallion, boar, rat, mouse, hamster and human (Manjunath and Therien, 2002). BSPs have been found to be deposited on the surface of spermatozoa and bind to the cholesterol effluxing agents like heparin and High Density Lipoproteins (HDL). Thus it is hypothesized that their binding to spermatozoa increases the binding sites of these cholesterol effluxing agents in the female tract thus providing the trigger for capacitation. Cholesterol removal increases sperm membrane fluidity, and as a consequence, alterations in the bulk biophysical properties of the membrane would occur, which could obviously trigger the transmembrane signalling during capacitation. Besides, the increase in sperm membrane fluidity has also been correlated with lateral reorganization of the membrane proteins (Rochwerger and Cuasnicu, 1992) and increased permeability of the membrane to ions essential for capacitation (Visconti *et al.*, 1998).

## First messengers

### Progesterone

Attempts have been made to identify physiological inducers of capacitation in the female reproductive tract. Progesterone, which is present in high concentrations in the cumulus matrix surrounding the oocyte and in the follicular fluid, is capable of inducing the acrosome reaction in spermatozoa of almost all species studied (Thomas and Meizel, 1989; Libersky and Boatman, 1995) and is therefore a potential inducer. The effect of progesterone was not mediated through its cytoplasmic receptor thus indicating that its ability to induce the acrosome reaction is a non-genomic effect (Revelli *et al.*, 1998). The non-genomic effect of progesterone could be due to its ability to perturb membranes, induce fluidity changes, induce membrane fusion (Shivaji and Jagannadham, 1992) and also induce uptake of calcium (Baldi *et al.*, 1995) by interacting with surface receptors on the spermatozoa. In human spermatozoa, two surface receptors have been identified with different affinities for progesterone (Luconi *et al.*, 1998). Progesterone receptor mRNA has also been detected in human spermatozoa (Sachdeva *et al.*, 2000).

*Bicarbonate ions*

Bicarbonate is a major component of the uterine and oviductal fluid. Lee and Storey (1986) unequivocally established the bicarbonate ion as a first messenger in the process of capacitation based on the observation that replacement of the bicarbonate buffer in the *in vitro* fertilization medium with HEPES buffer effectively inhibited fertilization. It was eventually proved that bicarbonate is required for capacitation-associated processes like hyperactivation and tyrosine phosphorylation; the effect on hyperactivation being recently quantified (Wennemuth et al., 2003). It has been demonstrated that the absence of bicarbonate ions could be overcome by cAMP analogues during capacitation in mouse (Visconti et al., 1995a) and hamster (Visconti et al., 1999), suggesting strongly that cAMP is the second messenger for bicarbonate ions during capacitation. The recent cloning and characterization of the atypical soluble adenylate cyclase of spermatozoa (Chen et al., 2000) has conclusively proved that it is triggered by bicarbonate. Other effects of bicarbonate which support its involvement in capacitation are its ability to induce hyperpolarizing current during capacitation (Demarco et al., 2003); ability to increase the sensitivity of voltage dependant calcium channels (Wennemuth et al., 2000); ability to remodel the membrane lipid architecture during capacitation which is probably mediated through a bicarbonate phospholipid 'scramblase' (Gadella and Harrison, 2002); and its ability to increase the intracellular alkalinity, which is a very essential step early in capacitation (Zeng et al., 1996). The effects of bicarbonate could be blocked by anion transporter blockers, such as 4,4'-diisothiocyano-disulfonic stilbene (DIDS) and 4-acetamido-4'-isothiocyanostilbene-2,2'-disulfonic acid (SITS) (Visconti et al., 1999). A sodium dependant chloride/bicarbonate exchanger has been implicated in bicarbonate transport into the spermatozoa (Zeng et al., 1996) causing alkalinization during capacitation whereas the hyperpolarizing effect of bicarbonate might involve a sodium bicarbonate co-transporter (Demarco et al., 2003). Recently, the migration of sperm epididymal protein DE (CRISP-1) to the equatorial segment during capacitation of rat spermatozoa has been found to be bicarbonate ion dependant (Da Ros et al., 2004). The manifold importance of bicarbonate ions in sperm capacitation *in vivo* became evident in a very interesting report wherein the investigators postulate that the low female fertility in cystic fibrosis patients could be due to hampered uterine secretion of bicarbonate ions through defective cystic fibrosis transmembrane receptor (CFTR) in the patients thus, causing insufficient sperm capacitation (Wang et al., 2003).

*Others*

Heparin has been characterized as a first messenger in bovine species (Parrish et al., 1988) and is thought to cause cholesterol efflux from the spermatozoal plasma membrane and thus influence capacitation. An interesting class of agonists, the Fertilization Promoting Peptide (FPP) and adenosine, stimulate adenylyl cyclase activity in non-capacitated spermatozoa but inhibits its activity in capacitated spermatozoa (Fraser et al., 2003). FPP accomplishes this dual function by mediating its effect either through a stimulatory (A2) or an inhibitory (A1) receptor, depending on the status of capacitation (Fraser and Adeoya-Osiguwa, 1999). Although not well characterized, γ-amino butyric acid (GABA) (de las Heras et al., 1997; Ritta et al., 2004), neuraminidase (Srivastava et al., 1988), piperine (Piyachaturawat et al, 1991), interleukin-6 (Naz and Kaplan, 1994), platelet activating factor (Roudebush, 2001), atrial natriuretic peptide (Anderson et al., 1994) and epidermal growth factor (Furuya et al., 1993) have also been shown to induce capacitation in various species.

## Second messengers

*Calcium ions*

Calcium is known to influence various functions of the spermatozoa such as motility, capacitation and fertilizing ability. Elevation of calcium in the flagellum of spermatozoa drives hyperactivation in many species and this action of calcium could be at the level of the sperm flagella as revealed by experiments with demembranated rat (Lindemann and Goltz, 1988) and bull (Lindemann et al., 1991) spermatozoa. Calcium is also indispensable for the acrosome reaction. The acrosome itself is thought to be a calcium store, which is proposed to release calcium during the acrosome reaction by activating PKA and IP3 dependant calcium channels in the outer acrosomal membrane (Breitbart and Naor, 1999). Intracellular calcium surge in spermatozoa occurs at two stages, first from the internal stores and later from the extracellular medium (Spungin and Breitbart, 1996). Disruption of some voltage or cyclic nucleotide gated calcium channels, previously identified in sperm, has failed to render animals infertile (Quill et al., 2003). Knock out male mice of Catsper(s), which are sperm specific calcium channels localized to the sperm flagellum (Ren et al., 2001; Quill et al., 2003; Carlson et al., 2005) have been found to be defective in hyperactivation and are thus infertile. But, these knock out mice showed flagellar beating in the presence of bicarbonate (Carlson et al., 2005). The role of calcium in the capacitation dependant tyrosine phosphorylation event is quite controversial. In mouse (Visconti et al., 1995a,b) and human spermatozoa (Leclerc et al., 1998) increase in calcium concentration in the medium caused an increase in tyrosine phosphorylation. In human spermatozoa however, calcium negatively modulated tyrosine phosphorylation due to its inhibitory effect on kinases (Luconi et al., 1996) or by decreasing the availability of ATP by activating calcium dependant ATPases in the spermatozoa (Baker et al., 2004). In hamster, calcium has been shown to be required for hyperactivation and the acrosome reaction but not for tyrosine phosphorylation during capacitation (Kulanand and Shivaji, 2001).

*Cyclic adenosine monophosphate (cAMP)*

The intracellular concentrations of cAMP, which was first recognized as a probable player in capacitation by Toyoda and Chang (1974), have now been shown to increase in spermatozoa during capacitation (Visconti et al., 2002). This increase could probably be due to an increased activity of adenylyl cyclase and/or inhibition of the activity of cAMP phosphodiesterases. The increase in cAMP has also been shown to be dependant on the calcium/calmodulin system (Jaiswal and Conti, 2003) as well as bicarbonate ion (Visconti et al., 1995a, b). Sperm adenylyl cyclase is different from the somatic form in that it exists in many isoforms, is present both in the membrane fraction and in the soluble fraction (Baxendale and Fraser, 2003), is not coupled to any G protein (Rojas and Bruzzone, 1992) and is insensitive to hormonal regulation or fluoride (Hanoune et al., 1997). A truncated form of a soluble adenylyl cyclase from rat testis has been purified (Buck et al., 1999). It has been observed that the soluble enzyme was highly responsive to bicarbonate stimulation (and not to forskolin) and is substantially homologous to the cyanobacterial enzyme. The two well established molecular effects of cAMP during capacitation are: a) alteration of membrane phospholipid architecture (probably transduces the signal from the bicarbonate ions) (Gadella and Harrison, 2000) and b) protein tyrosine phosphorylation.

## Downstream events

*Intracellular alkalinization and hyperpolarization*

Increase in cytosolic pH is known to promote metabolic activity and swimming activity of the

the ejaculated spermatozoa (Babcock and Pfeiffer, 1987). In mouse spermatozoa, the alkalinization was shown to involve a sodium dependant chloride bicarbonate exchanger (Zeng et al., 1996)
while in bovine spermatozoa, glucose has been shown to cause a delay in capacitation by delaying
the intracellular alkalinization (Parrish et al., 1989). Alkalinization of the acrosome has also been
correlated to the acrosome reaction in mouse and hamster spermatozoa (Breitbart and Spungin,
1997). However, the exact stimuli bringing about the change in pH are still not known.

Hyperpolarization of the sperm plasma membrane has also been shown to accompany capacitation in mouse (through involvement of a bicarbonate sodium co-transporter) (Demarco et
al., 2003) and in bovine spermatozoa (by alteration of potassium permeability) (Zeng et al.,
1995) and has been suggested to be an indirect trigger for protein tyrosine phosphorylation
(Demarco et al., 2003). Hyperpolarization is also shown to recruit calcium channels from an
inactivated to an activated state which could be triggered later by the right stimuli, like zona
pellucida (Arnoult et al., 1996; Florman et al., 1998). Progesterone stimulation was also shown
to cause hyperpolarization in a certain population of human spermatozoa (Patrat et al., 2002).

### Membrane reorganization

Changes in sperm membrane fluidity and membrane reorganization triggered by cholesterol
efflux, are thought to render the male gamete "more fusible" with the female gamete. It has
been demonstrated that the sperm plasma membrane lipid architecture is controlled via a bicar-
bonate-cAMP-PKA dependant mechanism during capacitation in boar spermatozoa (Harrison et
al., 1996; Gadella and Harrison, 2000; Harrison and Miller, 2000). A caspase independant and
bicarbonate dependant externalization of phosphatidylserine and phosphatidylethanolamine in
human sperm membrane has also been demonstrated (de Vries et al., 2003). Further, a novel
aminophospholipid transporter in the acrosomal region of spermatozoa has been discovered
and shown to be critical for normal phospholipid distribution in the membrane bilayer as well
as for normal binding, penetration and signalling by the zona pellucida (Wang et al., 2004). In
addition to lipids, some membrane proteins like PH-20 also undergo reorganization during
capacitation. PH-20 migrates to the inner acrosomal membrane from the posterior head domain
after acrosome reaction (Cowan et al., 1986). Lipid rafts and their reorganization during capacitation have been observed in human spermatozoa (Travis et al., 2001a; Cross, 2004).

### Reactive oxygen species (ROS)

The ROS, superoxide anion and hydrogen peroxide, have been implicated in hyperactivation, the
acrosome reaction and protein tyrosine phosphorylation, which are associated with capacitation of
spermatozoa (de Lamirande et al., 1997; O'Flaherty et al., 2005). The action of ROS is probably
mediated by its ability to induce cAMP generation (Aitken et al., 1998) and by altering the sulphy-
dryl levels in sperm proteins (de Lamirande and Gagnon, 2003). The ROS generating machinery
in spermatozoa is uncharacterized but the NADPH oxidase system in spermatozoa has been hinted
to be responsible for generating ROS in rat (Vernet et al., 2001) and human (Aitken et al., 1997)
spermatozoa, although contradictory evidence to it also exists (Richer and Ford, 2001).

### Nitric oxide (NO)

Although high concentrations of the free radical, NO, have been reported to decrease viability
of spermatozoa by inhibiting respiration (Weinberg et al., 1995), lower concentrations of NO

seem to facilitate capacitation dependant events; such as, motility enhancement (Hellstorm *et al.*, 1994), stimulation of acrosome reaction in mouse spermatozoa (Herrero *et al.*, 1999) and enhancement of the zona pellucida binding ability in human spermatozoa (Sengoku *et al.*, 1998). NO inside a cell is mainly generated by the action of nitric oxide synthase (NOS) and inhibitors of the enzyme have been used to show that NO regulates capacitation dependant protein tyrosine phosphorylation in spermatozoa, and thus probably affects the capacitation dependant events (Herrero *et al.*, 1999; Kameshwari *et al.*, 2003).

*Protein tyrosine phosphorylation*

Concomitant with hyperactivation and the acrosome reaction, mammalian spermatozoa exhibit tyrosine phosphorylation in an array of sperm proteins (Visconti *et al.*, 1995a). However, the array of proteins that get phosphorylated varies from species to species (Visconti *et al.*, 1995a; Osheroff *et al.*, 1999; Si and Okuno, 1999; Kulanand and Shivaji, 2001; Bajpai *et al.*, 2003). Follicular fluid, an inducer of the acrosome reaction could induce protein tyrosine phosphorylation (Gye, 2003) while an epidymidal protein, CRISP-1, has been identified which when added to spermatozoa can inhibit capacitation dependant tyrosine phosphorylation (Roberts *et al.*, 2003). These findings prove the *in vivo* relevance of the protein tyrosine phosphorylation event.

The data from these initial studies also indicated that the increase in protein tyrosine phosphorylation associated with capacitation is species-dependent because constituents of the medium, such as $Ca^{2+}$, bovine serum albumin (BSA) and $HCO_3^-$, had different effects in different species. For instance, mouse spermatozoa required $Ca^{2+}$, BSA and $HCO_3^-$ for capacitation and the associated protein tyrosine phosphorylation (Visconti *et al.*, 1995a), whereas human spermatozoa required BSA and $HCO_3^-$ but not $Ca^{2+}$ (Carrera *et al.*, 1996; Luconi *et al.*, 1996). In golden hamster, in the absence of $Ca^{2+}$, protein tyrosine phosphorylation is not greatly affected if $HCO_3^-$ is present and the same holds true if $HCO_3^-$ is absent and $Ca^{2+}$ is present (Kulanand and Shivaji, 2001).

The increase in protein tyrosine phosphorylation during capacitation of mouse (Das Gupta *et al.*, 1993), bovine (Galantino-Homer *et al.*, 1997), human (Leclerc *et al.*, 1996; Osheroff *et al.*, 1999), boar (Kalab *et al.*, 1998), hamster (Visconti *et al.*, 1999; Kulanand and Shivaji, 2001), porcine and cynomologus monkey (Mahony and Gwathmey, 1999) spermatozoa has been shown to be regulated by a cAMP-dependent pathway involving protein kinase A (PKA) (Urner and Sakkas, 2003). This signalling pathway, involving protein tyrosine phosphorylation and cAMP, is unique to spermatozoa and has been confirmed by the fact that inhibitors of PKA are able to inhibit protein tyrosine phosphorylation observed during capacitation. On the other hand, a cell-permeant analogue of cAMP (dibutyryl cAMP) can also bring about the capacitation associated protein tyrosine phosphorylation. It should be emphasized here that PKA being a serine/ threonine kinase, cannot phosphorylate a protein at tyrosine residue(s) and this implicates the involvement of a protein tyrosine kinase or of a protein tyrosine phosphatase or both. It is possible that PKA influences the activity of the above enzymes either directly or indirectly. Attempts have been made to purify and characterize the protein tyrosine kinases from boar (Berruti and Martegani, 1989) and hamster (Uma Devi *et al.*, 2000) spermatozoa.

Metabolic components of the *in vitro* capacitation medium, like glucose, have also been shown to influence tyrosine phosphorylation in mouse and glycolysis has been shown to be the source of ATP for protein phosphorylation (Travis *et al.*, 2001b). In almost all species studied, immunolocalization studies have indicated that the capacitation dependant tyrosine phosphorylation goes up in the flagella of the spermatozoa (Mahony and Gwathmey, 1999; Si and Okuno, 1999; Jha and Shivaji, 2002a). Although the importance of a balance between protein tyrosine kinase and phosphatase activities has been demonstrated in causing a successful acrosome reaction (Tomes et al., 2004), the tyrosine phosphorylated proteins involved in the process are yet to be identified.

The involvement of calcium in capacitation hinted at a possible role for protein kinase C (PKC) in capacitation. This was confirmed by using phorbol ester, a potent stimulator (and other weaker stimulators) of PKC, to induce acrosome reaction which could be brought down by PKC inhibitors. Immunolabelling studies have revealed different locations of the kinase in different regions of the acrosome in different species (Breitbart and Naor, 1999). Moreover, PKC has also been implicated in the acrosome reaction stimulated by physiological agents like progesterone and zona pellucida (Breitbart and Spungin, 1997). The stimulation of PKC is proposed to be driven by diacyl glycerol produced by the action of phospholipases on the membrane phospholipids (Breitbart and Naor, 1999). Although protein tyrosine phosphorylation has been accepted as a hallmark of capacitation, direct involvement of any particular tyrosine kinase has not been shown in sperm capacitation. Epidermal Growth Factor (EGF) Receptor, a receptor tyrosine kinase, has been detected in the head of the bovine spermatozoa and the EGF induced acrosome reaction in the same species has been found to be completely inhibited by tyrosine kinase inhibitors (Lax et al., 1994). In hamster spermatozoa, components of the Ras pathway have been detected although their functional significance is still a matter of conjecture (NagDas et al., 2002). The non-receptor tyrosine kinase c-yes has been shown to be highly expressed in the spermatid acrosome among many mammalian tissues checked (Zhao et al., 1990) and its activity appears to be positively and negatively regulated by cAMP and calcium respectively during human sperm capacitation (Leclerc and Goupil, 2002). Partial purification of a tyrosine kinase has also been done from hamster spermatozoa which might be responsible for a capacitation dependant signalling event (Uma Devi et al., 2000). To increase the net tyrosine phosphorylation in a cell the normal balance between the kinases and the phosphatases has to be tilted; thus studies on the phosphatases are as important as that on kinases. An isoform of PPI (PPI gamma) or the inhibitor of the PKA regulatory subunit has been identified in spermatozoa (Smith et al., 1996) and has been found to be associated with AKAP220 in the spermatozoa. However, another abundant isoform of the same inhibitor, PPI beta, when knocked out, does not result in any detectable infertility in both male and female mice (Belyamani et al., 2001). The calcium/calmodulin dependant phosphatase, calcineurin (PP2B), has been detected in dog spermatozoa (Tash et al., 1988) and also appears to participate in the dephosphorylation of tyrosine phosphorylated substrates in human spermatozoa (Carrera et al., 1996). This area of research demands more focus and would eventually help in a better understanding of the regulation of the phenomenon of capacitation in spermatozoa.

*Protein serine/threonine phosphorylation during capacitation*

Although, serine/threonine kinases such as protein kinase A (PKA) and protein kinase C (PKC) have been implicated in sperm capacitation, comparatively little is known about serine/threonine phosphorylation during capacitation when compared to tyrosine phosphorylation. Nevertheless, using monoclonal antibodies against phosphoserine and phosphothreonine residues, a capacitation-associated increase in protein serine/threonine phosphorylation has been reported in human (Naz, 1999) and hamster (Jha and Shivaji, 2002b) spermatozoa. In addition to this, de Lamirande and Gagnon (2002) have demonstrated the presence of various components of the extracellular signal-regulated protein kinase (ERK1/2) pathway in human spermatozoa, which is a serine/threonine kinase. Furthermore, they have also shown the phosphorylation of putative substrates of ERK1/2 during capacitation by using antibodies against the phosphopeptide, Thr-Glu-Tyr. The above findings indicate that serine/threonine phosphorylation may have important physiological role in capacitation and warrants further investigation.

## The phosphorylated protein

Each tyrosine phosphorylated protein in spermatozoa is a tool to understand the cause and effect relationship between tyrosine phosphorylation and capacitation. However, till now, few of these proteins have been identified and to even fewer of them is attached any functional significance of the tyrosine phosphorylation. The most characterized protein in this aspect is the A Kinase Anchoring Protein 82 (AKAP82/AKAP4) which has been found to be the major structural protein in the fibrous sheath of spermatozoa (Johnson et al., 1997). This protein has been identified as a capacitation dependant phosphorylated protein in human (Carrera et al., 1996), mouse (Johnson et al., 1997), rat (Brito et al., 1989) and hamster (Jha and Shivaji, 2002a) spermatozoa, although the type of phosphorylation (tyrosine/serine) varied depending on the species; in human and hamster this protein has been shown to be tyrosine phosphorylated and in mouse it is serine phosphorylated. The state of phosphorylation of the AKAP could be a variant in the docking of the kinase (PKA) which would thus facilitate its localized action. However, this theory awaits proper evidence. Recently, the tyrosine phosphorylation of another AKAP, AKAP3, has been shown to result in the recruitment of protein kinase A to the sperm flagella causing an increase in motility (Luconi et al., 2004). It is worthwhile mentioning here that AKAP isoforms have also been localized in the acrosome of mammalian spermatozoa (Vijayaraghavan et al., 1999). Recently Bajpai et al. (2003) demonstrated that AKAP3 binds to phosphodiesterase EA4 in bovine sperm and thus may influence capacitation. However, the role of AKAP has been questioned since AKAP knock out mice showed no obvious fertility defects (Burton et al., 1997).

CABYR, a novel calcium-binding tyrosine phosphorylation-regulated protein, is another testis and spermatozoa-specific protein, which gets strongly phosphorylated at the tyrosine residue during capacitation. It has been characterized recently from human spermatozoa (Naaby-Hansen et al., 2002) and has the ability to bind calcium. These properties make CABYR a very interesting molecule as far as signalling in spermatozoa is concerned. One may very well speculate that CABYR is involved in the cross-talk between tyrosine phosphorylation and $Ca^{2+}$ during signal transduction. However, the exact role of CABYR remains to be elucidated.

Recently a candidate protein, dihydrolipoamide dehydrogenase, which is a post-pyruvate metabolic enzyme, exhibiting tyrosine phosphorylation during hamster spermatozoal capacitation was identified (Mitra and Shivaji, 2004). Downregulation of the activity of the enzyme blocked the acrosome reaction completely and hyperactivation partially, confirming the role of dihydrolipoamide dehydrogenase in hamster spermatozoal capacitation. Localization studies indicated that hamster spermatozoal dihydrolipoamide dehydrogenase along with its host complex, the pyruvate dehydrogenase complex, is localized in the acrosome and in the principal piece of the sperm flagellum (Mitra et al., 2005). The localization of dihydrolipoamide dehydrogenase, however, appears to be in the mitochondria in the spermatocytes but in spermatids it appears to show a juxta nuclear localization (like Golgi). The capacitation dependent time course of tyrosine phosphorylation of dihydrolipoamide dehydrogenase appears to be different in the principal piece of the flagellum and the acrosome in hamster spermatozoa. Activity assays of this bi-directional enzyme suggest a strong correlation between the tyrosine phosphorylation and the bi-directional enzyme activity (Mitra et al., 2005). This is the first report of a direct correlation of the localization, tyrosine phosphorylation and activity of the important metabolic enzyme, dihydrolipoamide dehydrogenase, implicating dual involvement and regulation of the enzyme during sperm capacitation (Mitra and Shivaji, 2004; Mitra et al., 2005).

A chaperone, HSP-90α (Ecroyd et al., 2003), and a mitochondrial sheath protein, phospholipid hydroperoxidase glutathione peroxidase (PHGPx) (NagDas et al., 2004), have also been identified as being tyrosine phosphorylated proteins in capacitated mouse and hamster spermatozoa, respectively. A global phosphoproteome analysis of human spermatozoa (Ficarro et al., 2003), led to the identification of a set of eighteen proteins which get tyrosine phosphorylated during sperm capacitation and includes ion channels, metabolic enzymes, structural proteins et cetera, as well as HSP-90α and PHGPx. In this report a valosin containing protein (VCP) was studied in more detail revealing relocalization of the protein from the neck in the non-capacitated to the anterior head region in the capacitated human spermatozoa.

## Metabolism during capacitation

The metabolism of spermatozoa is unique compared to other cells because of two primary reasons: a) presence of sperm specific isoforms of some metabolic enzymes (Goldberg et al., 1977; Blanco, 1980; Travis et al., 1998) and b) the metabolism in sperm is compartmentalized with the mitochondrial Electron Transport Chain (ETC) in the midpiece and the glycolytic apparatus in the principal piece (Travis et al., 2001a, b). In mouse and man, glucose has been shown to be the primary carbon source and glycolysis the primary metabolic pathway for capacitation of spermatozoa (Fraser and Quinn, 1981; Williams and Ford, 2001). A recent study also indicated the involvement of the pentose phosphate pathway and redox regulation by the fertilizing spermatozoon (Urner and Sakas, 2005). On the other hand, in guinea pig spermatozoa the role of glycolytic pathway is not clear (Rogers and Yanagimachi, 1975; Mujica et al., 1991). In human spermatozoa, it has been postulated that glucose could be linked to the increase in superoxide generation, which is an essential step in hyperactivation and protein phosphorylation during capacitation (Aitken et al., 1997). Glycolysis, and not the ETC, has also been indicated as a source of ATP for phosphorylation in mouse spermatozoa (Travis et al., 2001a). Oxygen consumption has also been shown to decline during capacitation (Fraser and Lane, 1987). Kknock-out mice of the sperm specific glyceraldehyde 3-phosphate dehydrogenase (GAPDH-S) are infertile however; the oxygen consumption in the sperm of the knock out animals was normal (Miki et al., 2004). In hamster, it was shown recently that all three carbon sources present in the in vitro capacitation medium namely glucose, pyruvate and lactate are required during capacitation; the former being required in the early hours and the latter two at the later hours of capacitation (Mitra and Shivaji, 2004). This confirms an earlier observation which indicated that inhibition of lactate dehydrogenase C4 blocks capacitation of mouse spermatozoa (Duan and Goldberg, 2003).

## Conclusion

A number of phosphotyrosine proteins have been identified and implicated during capacitation but the available information is not adequate to understand the exact mechanism by which these proteins influence capacitation. The situation at the moment is even more complex since several more proteins, which are upregulated in phosphorylation during capacitation are being added to the already complex scenario of capacitation. These partially characterized players of capacitation have been mentioned in a tabular form in Table 1. A schematic representation of

Table 1. Proteins* implicated in mammalian sperm capacitation

| Name of the candidate protein | Involvement | Species |
|---|---|---|
| Sperm c-myc proto oncogene product | Hyperactivation and penetration of zona-free hamster eggs | Human |
| | in-vitro fertilization | Mouse |
| | in-vivo fertilization | Rabbit |
| Soluble acrosome inducing factor in egg-cumulus complex | ZP induced acrosome reaction and spermatozoa penetration | Hamster |
| A guanine-nucleotide binding regulatory protein | ZP induced acrosome reaction | Human |
| Phospholipase A2 | Acrosome reaction and fertilizing ability | Hamster |
| Lipid transfer protein1 | Capacitation | Human |
| Glycine receptor/Chloride Channel | ZP induced acrosome reaction | Human, pig and mouse |
| Calmodulin | Capacitation and ZP induced acrosome reaction | Mouse |
| FSP95 | Capacitation | Human |
| Rab3a | ZP induced acrosome reaction | Mouse |
| Proacrosin binding protein, sp32 | Acrosome reaction | Pig |
| Gelsolin | Acrosome reaction | Guinea pig |
| HSP-90 | Capacitation | Human and rat |
| LDH C-4 | Capacitation | Bovine |
| gC1q-R/p33 | Capacitation | Human |
| Caveolin-1 | Capacitation | Guinea pig |
| Plasma membrane associated hyaluronidase | Acrosome reaction | Horse |
| Sp17 (calmodulin binding protein) | Acrosome reaction | Rabbit |
| T-type calcium channels | Progesterone induced capacitation | Mouse |
| PI3K | Negatively regulates hyperactivation | Human |
| VLA-6 integrin | Cumulus (laminin) induced acrosome reaction | Pig |
| Calcitonin | Capacitation and in-vitro fertilization | Mouse |
| Angiotensin II | Capacitation and in-vitro fertilization | Mouse |
| SVS VII phospholipid binding protein | Motility | Mouse |
| Vitronectin | Acrosome reaction | Human |
| Glycosyl transferses | Modification of spermatozoa surface glycoproteins | Hamster |

*Proteins not included in the text are only mentioned (Modified from Mitra and Shivaji, 2005).

the molecular basis of capacitation, which clearly highlights that the molecular basis of capacitation is ultimately dependent on the phosphotyrosine proteins which are the end products of the signalling cascade involving membrane reorganization, increase in cytosolic cAMP and

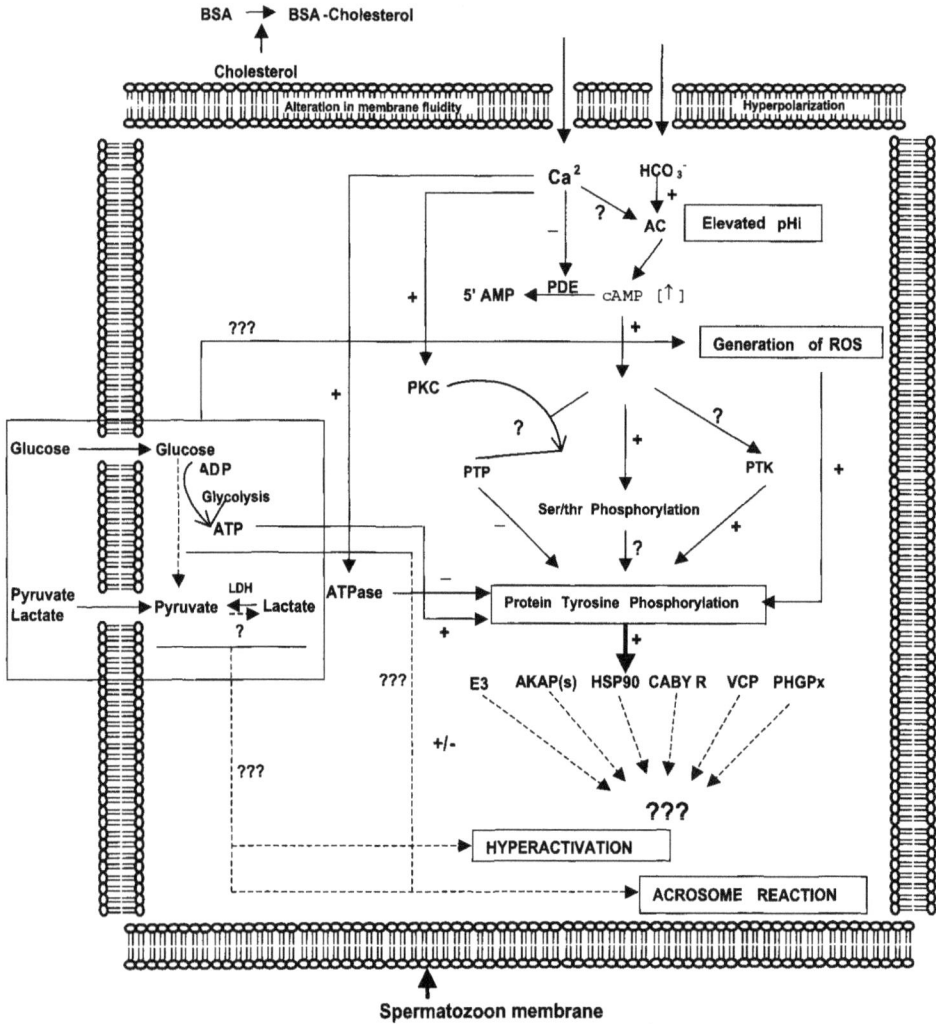

**Fig. 1** Schematic representation of the signalling events occurring during mammalian sperm capacitation. The cellular changes happening during capacitation are boxed. Hallmarks of capacitation are also boxed and are in capitals. Cholesterol effluxing agents like BSA, turn on capacitation and thereafter bicarbonate (HCO$_3^-$) induces adenylate cyclase (AC) to synthesize cyclic AMP (cAMP). Protein kinase A (PKA) is then trigerred by cAMP which drives protein tyrosine phosphorylation in an array of sperm proteins such as dihydrolipoamide dehydrogenase (E3), A kinase anchoring proteins (AKAPs), heat shock protein - 90α (HSP90), calcium-binding tyrosine phosphorylation-regulated protein (CABYR), valosin containing protein (VCP) and phospholipids hydroperoxide glutathione peroxidase (PHGPx). PDE, ROS, PKC, PTP, PTK and LDH refer to phosphodiesterase, reactive oxygen species, protein kinase C, protein tyrosine phosphatase, protein tyrosine kinase and lactate dehydrogenase, respectively. The metabolic components of capacitation and their links with the signalling components of capacitation are also mentioned. Bold arrow lines indicate pathways already established whereas discontinuous arrow lines are pathways that are probably operating. Modified from Jha et al. (2003) and Mitra et al. (2005).

ultimately protein tyrosine phosphorylation is presented in Fig. 1. The last event being regulated by a possible cross-talk between PKA, PTK and protein phosphatase. Thus identification of each and every tyrosine phosphorylated protein is essential. Further, the realization that metabolic components such as glucose, pyruvate, lactate et cetera. are essential for capacitation and the observation that glucose influences capacitation dependent tyrosine phosphorylation and that metabolic enzymes are tyrosine phosphorylated during capacitation implies that we need to understand the cross-talk between the metabolites, the metabolizing enzyme and the enzymes involved in signal transduction. Another untouched area is the identification of the PKA dependant tyrosine phosphatases and tyrosine kinases which are directly involved in protein tyrosine phosphorylation during capacitation. Thus understanding capacitation would help not only our understanding of the intricacies of sperm function but could also open up avenues for contraception oriented research.

## References

Aitken RJ, Fisher HM, Fulton N, Gomez E, Knox W, Lewis B and Irvine S (1997) Reactive oxygen species generation by human spermatozoa is induced by exogenous NADPH and inhibited by the flavoprotein inhibitors diphenylene iodonium and quinacrine *Molecular Reproduction and Development* **47** 468-482

Aitken RJ, Harkiss D, Knox W, Paterson M and Irvine DS (1998) A novel signal transduction cascade in capacitating human spermatozoa characterised by a redox-regulated, cAMP-mediated induction of tyrosine phosphorylation *Journal of Cell Science* **111** 645-656

Anderson RA, Feathergill KA, Drisdel RC, Rawlins RG, Mack SR and Zaneveld LJ (1994) Atrial natriuretic peptide (ANP) as a stimulus of the human acrosome reaction and a component of ovarian follicular fluid: correlation of follicular ANP content with *in vitro* fertilization outcome *Journal of Andrology* **15** 61-70

Aoki F, Sakai S and Kohmoto K (1999) Regulation of flagellar bending by cAMP and Ca²⁺ in hamster sperm *Molecular Reproduction and Development* **53** 77-83

Arnoult C, Zeng Y and Florman HM (1996) ZP3-dependent activation of sperm cation channels regulates acrosomal secretion during mammalian fertilization *Journal of Cell Biology* **134** 637-645

Austin CR (1951) Observations on the penetration of the sperm into mammalian egg *Australian Journal of Scientific Research* **4** 581–596

Austin CR (1952) The capacitation of mammalian spermatozoa *Nature* **170** 326

Babcock DF and Pfeiffer DR (1987) Independent elevation of cytosolic [Ca²⁺] and pH of mammalian sperm by voltage-dependent and pH-sensitive mechanisms *Journal of Biological Chemistry* **262** 15041-15047

Baccetti B, Burrini AG, Collodel G, Piomboni P and Renieri T (1991) A "miniacrosome" sperm defect causing infertility in two brothers *Journal of Andrology* **12** 104-111

Bajpai M, Asin S and Doncel GF (2003) Effect of tyrosine kinase inhibitors on tyrosine phosphorylation and motility parameters in human sperm *Archives of Andrology* **49** 229-246

Baker MA, Hetherington L, Ecroyd H, Roman SD and Aitken RJ (2004) Analysis of the mechanism by which calcium negatively regulates the tyrosine phosphorylation cascade associated with sperm capacitation *Journal of Cell Science* **117** 211-222

Baldi E, Krausz C, Luconi M, Bonaccorsi L, Maggi M and Forti G (1995) Actions of progesterone on human sperm: a model of non-genomic effects of steroids *Journal of Steroid Biochemistry and Molecular Biology* **53** 199-203

Baxendale RW and Fraser LR (2003) Evidence for multiple distinctly localized adenylyl cyclase isoforms in mammalian spermatozoa *Molecular Reproduction and Development* **66** 181-189

Belyamani M, Gangolli EA and Idzerda RL (2001) Reproductive function in protein kinase inhibitor-deficient mice *Molecular and Cellular Biology* **21** 3959-3963

Berruti G and Martegani E (1989) Identification of proteins cross-reactive to phosphotyrosine antibodies and of a tyrosine kinase activity in boar spermatozoa *Journal of Cell Science* **3** 667-674

Blanco A (1980) On the functional significance of LDH X *Johns Hopkins Medical Journal* **146** 231-235

Bone W, Jones NG, Kamp G, Yeung CH and Cooper TG (2000) Effect of ornidazole on fertility of male rats: inhibition of a glycolysis-related motility pattern and zona binding required for fertilization *in vitro Journal of Reproduction and Fertility* **118** 127-135

Breitbart H (2002) Role and regulation of intracellular calcium in acrosomal exocytosis *Journal of Reproductive Immunology* **53** 151-159

Breitbart H and Naor Z (1999) Protein kinases in mammalian sperm capacitation and the acrosome reaction *Reviews of Reproduction* **4** 151-159

Breitbart H and Spungin B (1997) The biochemistry of the acrosome reaction *Molecular Human Reproduction* **3** 195-202

Brener E, Rubinstein S, Cohen G, Shternall K, Rivlin J and Breitbart H (2003) Remodeling of the actin cytoskeleton during mammalian sperm capacitation and acrosome reaction *Biology of Reproduction* **68** 837-845

Brito M, Figueroa J, Maldonado EU, Vera JC and Burzio LO (1989) The major component of the rat sperm fibrous sheath is a phosphoprotein *Gamete Research* **22** 205-217

Buck J, Sinclair ML, Schapal L, Cann MJ and Levin LR (1999) Cytosolic adenylyl cyclase defines a unique signaling molecule in mammals *Proceedings of National Academy of Sciences USA* **96** 79-84

Burton KA, Johnson BD, Hausken ZE, Westenbrock RE, Idzerda RL, Scheuer T, Scott JD, Catterall WA and McKnight GS (1997) Type II regulatory subunits are not required for the anchoring depending modulation of Ca²⁺ channel activity by cAMP dependent protein kinase *Proceedings of National Academy of Sciences USA* **94** 11067-11072

Carlson AE, Quill TA, Westenbroek RE, Schuh SM, Hille B and Babcock DF (2005) Identical phenotypes of CatSper1 and CatSper2 null sperm *Journal of Biological Chemistry* **280** 32238-32244

Carrera A, Moos J, Ning XP, Gerton GL, Tesarik J, Kopf GS and Moss SB (1996) Regulation of protein tyrosine phosphorylation in human sperm by a calcium/calmodulin-dependent mechanism: identification of A kinase anchor proteins as major substrates for tyrosine phosphorylation *Developmental Biology* **180** 284-296

Chang MC (1951) Fertilizing capacity of spermatozoa deposited into the fallopian tubes *Nature* **168** 697-698

Chen Y, Cann MJ, Litvin TN, Iourgenko V, Sinclair ML, Levin LR and Buck J (2000) Soluble adenylyl cyclase as an evolutionarily conserved bicarbonate sensor *Science* **289** 625-628

Chiu PC, Chung MK, Tsang HY, Koistinen, R, Koistinen H, Seppala M, Lee KF and Yeung WS (2005) Glycodelin-S in human seminal plasma reduces cholesterol efflux and inhibits capacitation of spermatozoa *Journal of Biological Chemistry* **280** 25580-25589

Choi YH and Toyoda Y (1998) Cyclodextrin removes cholesterol from mouse sperm and induces capacitation in a protein-free medium *Biology of Reproduction* **59** 1328-1333

Cowan AE, Primakoff P and Myles DG (1986) Sperm exocytosis increases the amount of PH-20 antigen on the surface of guinea pig sperm *Journal of Cell Biology* **103** 1289-1297

Cross NL (1999) Effect of methyl-beta-cyclodextrin on the acrosomal responsiveness of human sperm *Molecular Reproduction and Development* **53** 92-98

Cross NL (2004) Reorganization of lipid rafts during capacitation of human sperm *Biology of Reproduction* **7** 1367-1373

Curry MR and Watson PF (1995) Sperm structure and function. In *Gametes – The Spermatozoon* pp 64-66 Eds JG Grudzinskas and JL Yovich. Syndicate Press, New York

Da Ros VG, Munuce MJ, Cohen DJ, Marin-Briggiler CI, Busso D, Visconti PE and Cuasnicu PS (2004) Bicarbonate is required for migration of sperm epididymal

protein DE (CRISP-1) to the equatorial segment and expression of rat sperm fusion ability *Biology of Reproduction* **70** 1325-1332

DasGupta S, Mills CL and Fraser LR (1993) Ca²⁺ -related changes in the capacitation state of human spermatozoa assessed by a chlortetracycline fluorescence assay *Journal of Reproduction and Fertility* **99** 135-143

Davis BK (1971) Macromolecular inhibitor of fertilization in rabbit seminal plasma *Proceedings of National Academy of Sciences USA* **68** 951-955

Davis BK, Byrne R and Bedigian K (1980) Studies on the mechanism of capacitation: albumin-mediated changes in plasma membrane lipids during *in vitro* incubation of rat sperm cells *Proceedings of National Academy of Sciences USA* **77** 1546-1550

de Lamirande E and Gagnon C (2002) The extracellular signal-regulated kinase (ERK) pathway is involved in human sperm function and modulated by the superoxide anion *Molecular Human Reproduction* **8** 124-135

de Lamirande E and Gagnon C (2003) Redox control of changes in protein sulfhydryl levels during human sperm capacitation *Free Radical Biology and Medicine* **35** 1271-1285

de Lamirande E, Jiang H, Zini A, Kodama H and Gagnon C (1997) Reactive oxygen species and sperm physiology *Reviews of Reproduction* **2** 48-54

de las Heras MA, Valcarcel A and Perez LJ (1997) *In vitro* capacitating effect of gamma-aminobutyric acid in ram spermatozoa *Biology of Reproduction* **56** 964-968

de Vries KJ, Wiedmer T, Sims PJ and Gadella BM (2003) Caspase-independent exposure of aminophospholipids and tyrosine phosphorylation in bicarbonate responsive human sperm cells *Biology of Reproduction* **68** 2122-2134

Demarco IA, Espinosa F, Edwards J, Sosnik J, De La Vega-Beltran JL, Hockensmith JW, Kopf GS, Darszon A and Visconti PE (2003) Involvement of a Na+/HCO-3 cotransporter in mouse sperm capacitation *Journal of Biological Chemistry* **278** 7001-7009

Duan C and Goldberg E (2003) Inhibition of lactate dehydrogenase C4 (LDH-C4) blocks capacitation of mouse sperm *in vitro* *Cytogenetic and Genome Research* **103** 352-359

Ecroyd H, Jones RC and Aitken RJ (2003) Tyrosine phosphorylation of HSP-90 during mammalian sperm capacitation *Biology of Reproduction* **69** 1801-1807

Ficarro S, Chertihin O, Westbrook VA, White F, Jayes F, Kalab P, Marto JA, Shabanowitz J, Herr JC, Hunt DF and Visconti PE (2003) Phosphoproteome analysis of capacitated human sperm. Evidence of tyrosine phosphorylation of a kinase-anchoring protein 3 and valosin-containing protein/p97 during capacitation *Journal of Biological Chemistry* **278** 11579-11589

Florman HM, Arnoult C, Kazam IG, Li C and O'Toole CM (1998) A perspective on the control of mammalian fertilization by egg-activated ion channels in sperm: a tale of two channels *Biology of Reproduction* **591** 12-16

Fraser LR and Adeoya-Osiguwa S (1999) Modulation of

adenylyl cyclase by FPP and adenosine involves stimulatory and inhibitory adenosine receptors and G proteins *Molecular Reproduction and Development* **53** 459-471

Fraser LR and Lane MR (1987) Capacitation- and fertilization-related alterations in mouse sperm oxygen consumption *Journal of Reproduction and Fertility* **81** 385-393

Fraser LR and Quinn PJ (1981) A glycolytic product is obligatory for initiation of the sperm acrosome reaction and whiplash motility required for fertilization in the mouse *Journal of Reproduction and Fertility* **61** 25-35

Fraser LR, Pondel MD and Vinson GP (2001) Calcitonin, angiotensin II and FPP significantly modulate mouse sperm function *Molecular Human Reproduction* **7** 245-253

Fraser LR, Adeoya-Osiguwa SA and Baxendale RW (2003) First messenger regulation of capacitation via G protein-coupled mechanisms: a tale of serendipity and discovery *Molecular Human Reproduction* **9** 739-748

Furuya S, Endo Y, Oba M, Suzuki S and Nozawa S (1993) Effect of epidermal growth factor on human sperm capacitation *Fertility and Sterility* **60** 905-910

Gadella BM and Harrison RA (2000) The capacitating agent bicarbonate induces protein kinase A-dependent changes in phospholipid transbilayer behavior in the sperm plasma membrane *Development* **127** 2407-2420

Gadella BM and Harrison RA (2002) Capacitation induces cyclic adenosine 3',5'-monophosphate-dependent, but apoptosis-unrelated, exposure of aminophospholipids at the apical head plasma membrane of boar sperm cells *Biology of Reproduction* **67** 340-350

Galantino-Homer H, Visconti PE and Kopf GS (1997) Regulation of protein tyrosine phosphorylation during bovine sperm capacitation by a cyclic adenosine 3',5'-monophosphate-dependent pathway *Biology of Reproduction* **56** 707-719

Goldberg E, Sberna D, Wheat TE, Urbanski GJ and Margoliash E (1977) Cytochrome c: immunofluorescent localization of the testis-specific form *Science* **196** 1010-1012

Gye MC (2003) Changes in sperm phosphotyrosine proteins by human follicular fluid in mice *Archives of Andrology* **49** 417-422

Hanoune J, Pouille Y, Tzavara E, Shen T, Lipskaya L, Miyamoto N, Suzuki Y and Defer N (1997) Adenylyl cyclases: structure, regulation and function in an enzyme superfamily *Molecular and Cellular Endocrinology* **128** 179-194

Harrison RA and Miller NG (2000) cAMP-dependent protein kinase control of plasma membrane lipid architecture in boar sperm *Molecular Reproduction and Development* **55** 220-228

Harrison RA, Ashworth PJ and Miller NG (1996) Bicarbonate/$CO_2$, an effector of capacitation, induces a rapid and reversible change in the lipid architecture of boar sperm plasma membranes *Molecular Reproduction and Development* **45** 378-391

Hellstrom WJ, Bell M, Wang R and Sikka SC (1994) Effect of sodium nitroprusside on sperm motility, viability, and lipid peroxidation *Fertility and Sterility* **61** 1117-1122

Herrero MB, de Lamirande E and Gagnon C (1999) Nitric oxide regulates human sperm capacitation and protein-tyrosine phosphorylation *in vitro Biology of Reproduction* **61** 575-581

Ho HC and Suarez SS (2003) Characterization of the intracellular calcium store at the base of the sperm flagellum that regulates hyperactivated motility *Biology of Reproduction* **68** 1590-1596

Jaiswal BS and Conti M (2003) Calcium regulation of the soluble adenylyl cyclase expressed in mammalian spermatozoa *Proceedings of National Academy of Sciences USA* **100** 10676-10681

Jayaprakash D, Kumar KS, Shivaji S and Seshagiri PB (1997) Pentoxifylline induces hyperactivation and acrosome reaction in spermatozoa of golden hamsters: changes in motility kinematics *Human Reproduction* **12** 2192-2199

Jha KN and Shivaji S (2002a) Identification of the major tyrosine phosphorylated protein of capacitated hamster spermatozoa as a homologue of mammalian sperm a kinase anchoring protein *Molecular Reproduction and Development* **61** 258-270

Jha KN and Shivaji S (2002b) Protein serine and threonine phosphorylation, hyperactivation and acrosome reaction in *in vitro* capacitated hamster spermatozoa *Molecular Reproduction and Development* **63** 119-130

Jha KN, Kameshwari DB and Shivaji S (2003) Role of signaling pathways in regulating the capacitation of mammalian spermatozoa *Cellular and Molecular Biology* **49** 329-340

Johnson LR, Foster JA, Haig-Ladewig L, VanScoy H, Rubin CS, Moss SB and Gerton, GL (1997) Assembly of AKAP82, a protein kinase A anchor protein, into the fibrous sheath of mouse sperm *Developmental Biology* **192** 340-350

Kalab P, Pernicova J, Geussova G and Moos J (1998) Regulation of protein tyrosine phosphorylation in boar sperm through a cAMP-dependent pathway *Molecular Reproduction and Development* **51** 304-314

Kameshwari DB, Siva AB and Shivaji S (2003) Inhibition of *in vitro* capacitation of hamster spermatozoa by nitric oxide synthase inhibitors *Cellular and Molecular Biology* **49** 421-428

Kulanand J and Shivaji S (2001) Capacitation-associated changes in protein tyrosine phosphorylation, hyperactivation and acrosome reaction in hamster spermatozoa *Andrologia* **33** 95-104

Lax Y, Rubinstein S and Breitbart H (1994) Epidermal growth factor induces acrosomal exocytosis in bovine sperm *FEBS Letters* **339** 234-238

Leclerc P and Goupil S (2002) Regulation of the human sperm tyrosine kinase c-yes. Activation by cyclic adenosine 3',5'-monophosphate and inhibition by $Ca^{2+}$ *Biology of Reproduction* **67** 301-307

Leclerc P, de Lamirande E and Gagnon C (1996) Cyclic adenosine 3',5'-monophosphate-dependent regulation of protein tyrosine phosphorylation in relation to human sperm capacitation and motility *Biology of Reproduction* **55** 684-692

Leclerc P, de Lamirande E and Gagnon C (1998) Interaction between Ca²⁺, cyclic 3',5' adenosine monophosphate, the superoxide anion, and tyrosine phosphorylation pathways in the regulation of human sperm capacitation *Journal of Andrology* **19** 434-443

Lee MA and Storey BT (1986) Bicarbonate is essential for fertilization of mouse eggs: mouse sperm require it to undergo the acrosome reaction *Biology of Reproduction* **34** 349-356

Libersky EA and Boatman DE (1995) Effects of progesterone on *in vitro* sperm capacitation and egg penetration in the golden hamster *Biology of Reproduction* **53** 483-487

Lindemann CB and Goltz JS (1988) Calcium regulation of flagellar curvature and swimming pattern in triton X-100—extracted rat sperm *Cell Motility and the Cytoskeleton* **10** 420-431

Lindemann CB, Kanous KS and Gardner TK (1991) The interrelationship of calcium and cAMP mediated effects on reactivated mammalian sperm models. In *Comparative Spermatology : 20 Years After* pp 491-496 Ed BE Bacetti. Raven Press Ltd., New York

Luconi M, Krausz C, Forti G and Baldi E (1996) Extracellular calcium negatively modulates tyrosine phosphorylation and tyrosine kinase activity during capacitation of human spermatozoa *Biology of Reproduction* **55** 207-216

Luconi M, Bonaccorsi L, Maggi M, Pecchioli P, Krausz C, Forti G and Baldi E (1998) Identification and characterization of functional nongenomic progesterone receptors on human sperm membrane *Journal of Clinical Endocrinology and Metabolism* **83** 877-885

Luconi M, Carloni V, Marra F, Ferruzzi P, Forti G and Baldi E (2004) Increased phosphorylation of AKAP by inhibition of phosphatidylinositol 3-kinase enhances human sperm motility through tail recruitment of protein kinase A *Journal of Cell Science* **117** 1235-1246

Mahony MC and Gwathmey T (1999) Protein tyrosine phosphorylation during hyperactivated motility of cynomolgus monkey (*Macaca fascicularis*) spermatozoa *Biology of Reproduction* **60** 1239-1243

Manjunath P and Therien I (2002) Role of seminal plasma phospholipid-binding proteins in sperm membrane lipid modification that occurs during capacitation *Journal of Reproductive Immunology* **53** 109-119

Marquez B and Suarez SS (2004) Different signaling pathways in bovine sperm regulate capacitation and hyperactivation *Biology of Reproduction* **70** 1626-1633

Miki K, Qu W, Goulding EH, Willis WD, Bunch DO, Strader LF, Perreault SD, Eddy EM and O'Brien D (2004) A glyceraldehyde 3-phosphate dehydrogenase-S, a sperm-specific glycolytic enzyme, is required for sperm motility and male fertility *Proceedings of National Academy of Sciences USA* **101** 16501-16506

Mitra K and Shivaji S (2004) Novel tyrosine-phosphorylated post-pyruvate metabolic enzyme,

dihydrolipoamide dehydrogenase, involved in capacitation of hamster spermatozoa *Biology of Reproduction* **70** 887-899

Mitra K and Shivaji S (2005) Proteins implicated in sperm capacitation *Indian Journal of Experimental Biology* **43** 1001-1015

Mitra K, Rangaraj N and Shivaji S (2005) Novelty of the pyruvate metabolic enzyme dihydrolipoamide dehydrogenase in spermatozoa : Correlation of its localization, tyrosine phosphorylation and activity during sperm capacitation *Journal of Biological Chemistry* **280** 25743-25753

Mujica A, Moreno-Rodriguez R, Naciff J, Neri L and Tash JS (1991) Glucose regulation of guinea-pig sperm motility *Journal of Reproduction and Fertility* **92** 75-87

Mujica A, Neri-Bazan L, Tash JS and Uribe S (1994) Mechanism for procaine-mediated hyperactivated motility in guinea pig spermatozoa *Molecular Reproduction and Development* **38** 285-292

Naaby-Hansen S, Mandal A, Wolkowicz MJ, Sen B, Westbrook VA, Shetty J, Coonrod SA, Klotz KL, Kim YH, Bush LA, Flickinger CJ and Herr JC (2002) CABYR, a novel calcium-binding tyrosine phosphorylation-regulated fibrous sheath protein involved in capacitation *Developmental Biology* **242** 236-254

NagDas SK, Winfrey VP and Olson GE (2002) Identification of ras and its downstream signaling elements and their potential role in hamster sperm motility *Biology of Reproduction* **67** 1058-1066

NagDas SK, Winfrey VP and Olson GE (2004) Tyrosine phosphorylation generates multiple isoforms of the mitochondrial capsule protein, phospholipid hydroperoxide glutathione peroxidase (PHGPx), during hamster sperm capacitation *Biology of Reproduction* **67** 1058-1066

Naz RK (1999) Involvement of protein serine and threonine phosphorylation in human sperm capacitation *Biology of Reproduction* **60** 1402-1409

Naz RK and Kaplan P (1994) Interleukin-6 enhances the fertilizing capacity of human sperm by increasing capacitation and acrosome reaction *Journal of Andrology* **15** 228-233

O'Flaherty C, Breininger E, Beorlegui N, Beconi MT (2005) Acrosome reaction in bovine spermatozoa : Role of reactive oxygen species and lactate dehydrogenase C4 *Biochimica et Biophysica Acta* **1726** : 96-101

Olds-Clarke P (1989) Sperm from tw32/+ mice: capacitation is normal, but hyperactivation is premature and nonhyperactivated sperm are slow *Developmental Biology* **131** 475-482

Oliphant G (1976) Removal of sperm-bound seminal plasma components as a prerequisite to induction of the rabbit acrosome reaction *Fertility and Sterility* **27** 28-38

Osheroff JE, Visconti PE, Valenzuela JP, Travis AJ, Alvarez J and Kopf GS (1999) Regulation of human sperm capacitation by a cholesterol efflux-stimulated signal transduction pathway leading to protein kinase A-mediated up-regulation of protein tyrosine phosphorylation *Molecular Human Reproduction* **5** 1017-1026

**Parrish JJ, Susko-Parrish J, Winer MA and First NL** (1988) Capacitation of bovine sperm by heparin *Biology of Reproduction* **38** 1171-1180

**Parrish JJ, Susko-Parrish JL and First NL** (1989) Capacitation of bovine sperm by heparin: inhibitory effect of glucose and role of intracellular pH *Biology of Reproduction* **41** 683-699

**Patrat C, Serres C and Jouannet P** (2000) The acrosome reaction in human spermatozoa *Biology of the Cell* **92** 255-266

**Patrat C, Serres C and Jouannet P** (2002) Progesterone induces hyperpolarization after a transient depolarization phase in human spermatozoa *Biology of Reproduction* **66** 1775-1780

**Piyachaturawat P, Sriwattana W, Damrongphol P and Pholpramool C** (1991) Effects of piperine on hamster sperm capacitation and fertilization *in vitro* *International Journal of Andrology* **14** 283-290

**Quill TA, Sugden SA, Rossi KL, Doolittle LK, Hammer RE and Garbers DL** (2003) Hyperactivated sperm motility driven by CatSper2 is required for fertilization *Proceedings of the National Academy of Sciences USA* **100** 14869-14874

**Ren D, Navarro B, Perez G, Jackson AC, Hsu S, Shi Q, Tilly JL and Clapham DE** (2001) A sperm ion channel required for sperm motility and male fertility *Nature* **413** 603-609

**Revelli A, Massobrio M and Tesarik J** (1998) Nongenomic actions of steroid hormones in reproductive tissues *Endocrinology Reviews* **19** 3-17

**Richer SC and Ford WC** (2001) A critical investigation of NADPH oxidase activity in human spermatozoa *Molecular and Human Reproduction* **7** 237-244

**Ritta MN, Bas DE and Tartaglione CM** (2004) *In vitro* effect of gamma-aminobutyric acid on bovine spermatozoa capacitation *Molecular Reproduction and Development* **67** 478-486

**Roberts KP, Wamstad JA, Ensrud KM and Hamilton DW** (2003) Inhibition of capacitation-associated tyrosine phosphorylation signaling in rat sperm by epididymal protein Crisp-1 *Biology of Reproduction* **69** 572-581

**Rochwerger L and Cuasnicu PS** (1992) Redistribution of a rat sperm epididymal glycoprotein after *in vitro* and *in vivo* capacitation *Molecular Reproduction and Development* **31** 34-41

**Rogers BJ and Yanagimachi R** (1975) Retardation of guinea pig sperm acrosome reaction by glucose: the possible importance of pyruvate and lactate metabolism in capacitation and the acrosome reaction *Biology of Reproduction* **13** 568-575

**Rojas FJ and Bruzzone ME** (1992) Regulation of cyclic adenosine monophosphate synthesis in human ejaculated spermatozoa. I. Experimental conditions to quantitate membrane-bound adenylyl cyclase activity *Human Reproduction* **7** 1126-1130

**Roudebush WE** (2001) Role of platelet-activating factor in reproduction: sperm function *Asian Journal of Andrology* **3** 81-85

**Sachdeva G, Shah CA, Kholkute SD and Puri CP** (2000) Detection of progesterone receptor transcript in human spermatozoa *Biology of Reproduction* **62** 1610-1614

**Sengoku K, Tamate K, Yoshida T, Takaoka Y, Miyamoto T and Ishikawa M** (1998) Effects of low concentrations of nitric oxide on the zona pellucida binding ability of human spermatozoa *Fertility and Sterility* **69** 522-527

**Shivaji S and Jagannadham MV** (1992) Steroid-induced perturbations of membranes and its relevance to sperm acrosome reaction *Biochimica et Biophysica Acta* **1108** 99-109

**Shivaji S, Peedicayil J and Girija Devi L** (1995) Analysis of the motility parameters of *in vitro* hyperactivated hamster spermatozoa *Molecular Reproduction and Development* **42** 233-247

**Si Y and Okuno M** (1999) Role of tyrosine phosphorylation of flagellar proteins in hamster sperm hyperactivation *Biology of Reproduction* **61** 240-246

**Smith GD, Wolf DP, Trautman KC, da Cruz e Silva EF, Greengard P and Vijayaraghavan S** (1996) Primate sperm contain protein phosphatase 1, a biochemical mediator of motility *Biology of Reproduction* **54** 719-727

**Spungin B and Breitbart H** (1996) Calcium mobilization and influx during sperm exocytosis *Journal of Cell Science* **109** 1947-1955

**Spungin B, Margalit I and Breitbart H** (1995) Sperm exocytosis reconstructed in a cell-free system: evidence for the involvement of phospholipase C and actin filaments in membrane fusion *Journal of Cell Science* **108** 2525-2535

**Srivastava PN, Kumar VM and Arbtan KD** (1988) Neuraminidase induces capacitation and acrosome reaction in mammalian spermatozoa *Journal of Experimental Zoology* **245** 106-110

**Suarez SS and Dai X** (1992) Hyperactivation enhances mouse sperm capacity for penetrating viscoelastic media *Biology of Reproduction* **46** 686-691

**Suarez SS, Katz DF, Owen DH, Andrew JB and Powell RL** (1991) Evidence for the function of hyperactivated motility in sperm *Biology of Reproduction* **44** 375-381

**Suarez SS, Varosi SM and Dai X** (1993) Intracellular calcium increases with hyperactivation in intact, moving hamster sperm and oscillates with the flagellar beat cycle *Proceedings of National Academy of Sciences USA* **90** 4660-4664

**Talbot P, Summers RG, Hylander BL, Keough EM and Franklin LE** (1976) The role of calcium in the acrosome reaction: an analysis using ionophore A23187 *Journal of Experimental Zoology* **198** 383-392

**Tash JS, Krinks M, Patel J, Means RL, Klee CB and Means AR** (1988) Identification, characterization, and functional correlation of calmodulin-dependent protein phosphatase in sperm *Journal of Cell Biology* **106** 1625-1633

**Thomas P and Meizel S** (1989) Phosphatidylinositol 4,5-bisphosphate hydrolysis in human sperm stimulated with follicular fluid or progesterone is dependent upon $Ca^{2+}$ influx *Biochemical Journal* **264** 539-546

**Tomes CN, Roggero CM, De Blas G, Saling PM and**

Mayorga LS (2004) Requirement of protein tyrosine kinase and phosphatase activities for human sperm exocytosis *Developmental Biology* **265** 399-415

Toyoda Y and Chang MC (1974) Capacitation of epididymal spermatozoa in a medium with high K-Na ratio and cyclic AMP for the fertilization of rat eggs *in vitro* *Journal of Reproduction and Fertility* **36** 125-134

Travis AJ, Foster JA, Rosenbaum NA, Visconti PE, Gerton GL, Kopf GS and Moss SB (1998) Targeting of a germ cell-specific type 1 hexokinase lacking a porin-binding domain to the mitochondria as well as to the head and fibrous sheath of murine spermatozoa *Molecular Biology of the Cell* **9** 263-276

Travis AJ, Merdiushev T, Vargas LA, Jones BH, Purdon MA, Nipper RW, Galatioto J, Moss SB, Hunnicutt GR and Kopf GS (2001a) Expression and localization of caveolin-1, and the presence of membrane rafts, in mouse and Guinea pig spermatozoa *Developmental Biology* **240** 599-610

Travis AJ, Jorgez CJ, Merdiushev T, Jones BH, Dess DM, Diaz-Cueto L, Storey BT, Kopf GS and Moss SB (2001b) Functional relationships between capacitation-dependent cell signaling and compartmentalized metabolic pathways in murine spermatozoa *Journal of Biological Chemistry* **276** 7630-7636

Uma Devi K, Jha K, Patil SB, Padma P and Shivaji S (2000) Inhibition of motility of hamster spermatozoa by protein tyrosine kinase inhibitors *Andrologia* **32** 95-106

Urner F and Sakkas D (2003) Protein phosphorylation in mammalian spermatozoa *Reproduction* **125** 17-26

Urner F and Sakkas D (2005) Involvement of the pentose phosphate pathway and redox regulation in fertilization in the mouse *Molecular Reproduction and Development* **70** 494-503

Vernet P, Fulton N, Wallace C and Aitken RJ (2001) Analysis of reactive oxygen species generating systems in rat epididymal spermatozoa *Biology of Reproduction* **65** 1102-1113

Vijayaraghavan S, Liberty GA, Mohan J, Winfrey VP, Olson GE and Carr DW (1999) Isolation and molecular characterization of AKAP110, a novel, sperm-specific protein kinase A-anchoring protein *Molecular Endocrinology* **13** 705-717

Visconti PE, Moore GD, Bailey JL, Leclerc P, Connors SA, Pan D, Olds-Clarke P and Kopf GS (1995a) Capacitation of mouse spermatozoa II Protein tyrosine phosphorylation and capacitation are regulated by a cAMP-dependent pathway *Development* **121** 1139-1150

Visconti PE, Bailey JL, Moore GD, Pan D, Olds-Clarke P and Kopf GS (1995b) Capacitation of mouse spermatozoa. I. Correlation between the capacitation state and protein tyrosine phosphorylation *Development* **121** 1129-1137

Visconti PE, Galantino-Homer H, Moore GD, Bailey JL, Ning X, Fornes M and Kopf GS (1998) The mo-lecular basis of sperm capacitation *Journal of Andrology* **19** 242-248

Visconti PE, Stewart-Savage J, Blasco A, Battaglia L, Miranda P, Kopf GS and Tezon JG (1999) Roles of bicarbonate, cAMP, and protein tyrosine phosphorylation on capacitation and the spontaneous acrosome reaction of hamster sperm *Biology of Reproduction* **61** 76-84

Visconti PE, Westbrook VA, Chertihin O, Demarco I, Sleight S and Diekman AB (2002) Novel signaling pathways involved in sperm acquisition of fertilizing capacity *Journal of Reproductive Immunology* **53** 133-150

Wang L, Beserra C and Garbers DL (2004) A novel aminophospholipid transporter exclusively expressed in spermatozoa is required for membrane lipid asymmetry and normal fertilization *Developmental Biology* **267** 203-215

Wang XF, Zhou CX, Shi QX, Yuan YY, Yu MK, Ajonuma LC, Ho LS, Lo PS, Tsang LL, Liu Y, Lam SY, Chan LN, Zhao WC, Chung YW and Chan HC (2003) Involvement of CFTR in uterine bicarbonate secretion and the fertilizing capacity of sperm *Nature Cell Biology* **5** 902-926

Weinberg JB, Doty E, Bonaventura J and Haney AF (1995) Nitric oxide inhibition of human sperm motility *Fertility and Sterility* **64** 408-413

Wennemuth G, Westenbroek RE, Xu T, Hille B and Babcock DF (2000) CaV2.2 and CaV2.3 (N- and R-type) Ca²⁺ channels in depolarization-evoked entry of Ca²⁺ into mouse sperm *Journal of Biological Chemistry* **275** 21210-21217

Wennemuth G, Carlson AE, Harper AJ and Babcock DF (2003) Bicarbonate actions on flagellar and Ca²⁺-channel responses: initial events in sperm activation *Development* **130** 1317-1326

Williams AC and Ford WC (2001) The role of glucose in supporting motility and capacitation in human spermatozoa *Journal of Andrology* **22** 680-695

Yanagimachi R (1969) *In vitro* capacitation of hamster spermatozoa by follicular fluid *Journal of Reproduction and Fertility* **18** 275-286

Yanagimachi R (1970) The movement of golden hamster spermatozoa before and after capacitation *Journal of Reproduction and Fertility* **23** 193-196

Zeng Y, Clark EN and Florman HM (1995) Sperm membrane potential: hyperpolarization during capacitation regulates zona pellucida-dependent acrosomal secretion *Developmental Biology* **171** 554-563

Zeng Y, Oberdorf JA and Florman HM (1996) pH regulation in mouse sperm: identification of Na+-, Cl--, and HCO₃⁻-dependent and arylaminobenzoate-dependent regulatory mechanisms and characterization of their roles in sperm capacitation *Developmental Biology* **173** 510-520

Zhao YH, Krueger JG and Sudol M (1990) Expression of cellular-yes protein in mammalian tissues *Oncogene* **5** 1629-1635

# Tyrosine phosphorylated proteins in mammalian spermatozoa: molecular and functional aspects

P B Seshagiri, D Mariappa and R H Aladakatti

*Department of Molecular Reproduction, Development and Genetics, Indian Institute of Science, Bangalore 560 012, India*

During mammalian fertilization, spermatozoa must undergo capacitation and the acrosome reaction. These processes of sperm function are critically associated with various molecular events and one such process is protein tyrosine phosphorylation (PYP). This event is downstream of increases in intracellular $Ca^{2+}$ and activities of $HCO_3^-$ activated adenylate cyclase, cAMP-dependent-protein kinase-A and reactive oxygen species. Though, PYP is known to be mediated by tyrosine kinases and phosphatases, only a few of them have been identified and characterized in spermatozoa. Since most identified tyrosine kinases are soluble proteins from somatic cells, it is believed that distinct mechanisms could exist in spermatozoa for PYP. Such sperm-specific protein tyrosine kinases/ phosphatases still remain to be thoroughly characterized in most species, including hamsters. Nevertheless, a few tyrosine phosphorylated sperm proteins have been identified in hamsters and in other mammals as well. There is very limited information available on our understanding of the molecular and ultrastructural localization, as well as the characteristics of tyrosine phosphorylated proteins. Functionally, how sperm motility is regulated by PYP is also poorly understood. Knowledge of tyrosine phoshorylated proteins and how they regulate sperm function is of immense significance in our understanding of male (in)fertility and clinical management of fertility; especially, in the light of studies that implicate the hypo-tyrosine phosphorylated state of sperm proteins with asthenozoospermic condition in humans. This article provides a comprehensive review on PYP and its regulation by kinases and phosphatases.

Mammalian spermatozoa, following ejaculation, must undergo a final maturational process called capacitation and the acrosome reaction. Physiologically, capacitation involves sperm acquiring hyperactivated motility and the capability to undergo the acrosome reaction (AR) (Yanagimachi, 1994), thereby rendering the spermatozoa competent to fertilize. Among the key molecular events associated with the acquisition of hyperactivated motility is protein tyrosine phosphorylation (PYP), the importance of which was demonstrated first in the mouse spermatozoa (Visconti et al., 1995). Though identification and characterization of tyrosine-phosphorylated proteins has been progressing with several critical proteins being sequenced, a

---

E-mail: polani@mrdg.iisc.ernet.in

significant number of them are yet to be thoroughly characterized. Moreover, only very limited information is available on protein tyrosine-kinases (PTKs) and phosphatases (PTPs), which apparently are involved in the modulation of tyrosine-phosphorylation status of a number of sperm proteins (Harayama *et al.*, 2004). These studies have implications in our understanding as to how flagellar bending of mammalian spermatozoa occurs and how this could regulate male fertility in the human. In this article, we provide a comprehensive review on the status of sperm PYP, PTKs and tyrosine-phosphorylated proteins, associated with sperm capacitation with a brief description of the events leading to sperm PYP, with particular relevance to the golden hamster spermatozoa.

### Sperm capacitation and signalling mechanisms associated with protein phosphorylation

A number of studies are being pursued to investigate the signalling events during sperm capacitation and the AR: particularly involving sperm protein phosphorylation. Signalling molecules, like $Ca^{2+}$ and cAMP, have been strongly implicated as key mediators of the molecular pathways regulating these events. Besides $Ca^{2+}$ (Dorval *et al.*, 2002), $HCO_3^-$ (Demarco *et al.*, 2003) and bovine serum albumin (BSA) (Visconti *et al.*, 1997) regulate adenylate cyclase activity and thereby the levels of cAMP. The primary downstream target of cAMP is protein kinase-A (PKA); targeted disruption of the sperm-specific catalytic subunit that is, $C\alpha_2$ of PKA, leads to the hypo-tyrosine phosphorylation of sperm proteins, accompanied by a lack of hyperactivation in mice (Nolan *et al.*, 2004). When we used the cAMP phosphodiesterase inhibitor, pentoxifylline (PF; used in the treatment of male factor infertility in humans), we observed an early onset of capacitation and an increased hyperactivation of hamster spermatozoa (Jayaprakash *et al.*, 1997). We showed this to occur via various cell-signalling molecules such as cAMP, $Ca^{2+}$ and protein kinases (Ain *et al.*, 1999). Sperm signalling pathways also involve changes in the levels of reactive oxygen species (ROS) and an optimal concentration of sperm-generated ROS appears to be required for various sperm functions, including PYP (Seshagiri *et al.*, 2003). PYP appears to be influenced as a consequence of sperm-generated ROS modulating adenylate cyclase activity (Baker and Aitken, 2004).

   The functional importance of sperm PYP during sperm capacitation has been demonstrated in the mouse (Visconti *et al.*, 1995), rat (Lewis and Aitken, 2001), hamster (Si and Okuno, 1999; Kulanand and Shivaji, 2001), monkey (Mahony and Gwathmey, 1999) and the human (Leclerc *et al.*, 1996). In the hamster, hyper-PYP was associated with PF-induced early onset of capacitation and, PKA activity, modulated by cAMP and $Ca^{2+}$, was shown to be upstream of this event (Seshagiri *et al.*, 2003). When the role of PYP in sperm capacitation was addressed using a specific PTK inhibitor, tyrphostin-A47 (TP-47), we observed that TP-47, apart from effectively inhibiting capacitation in a dose-dependent manner (Fig. 1), also induced circular motility (Mariappa *et al.*, 2006). Associated with such a circular motility pattern was a clear hypo-tyrosine phosphorylation of structural components of the hamster sperm flagellum (Fig. 2). Interestingly, a capacitation-associated increase in the flagellar localization of PYP is a feature observed across most mammalian species (Urner and Sakkas, 2003). A number of studies are being actively pursued to provide insights into the molecular identity and functional significance of PYP in mammalian spermatozoa. We are making our initial steps to identify some of the TP-47-inhibitable tyrosine phosphorylated flagellar proteins in the hamster. It is however clear that some of these proteins could be key modulators of sperm flagellar bending, depending on tyrosine-phosphorylated status. Molecular characterization of such sperm flagellar proteins, which are directly involved in sperm motility, is still in its infancy. Described in the following sections are the immediate upstream molecules that dictate the amount of PYP and the activities of PTKs and PTPs.

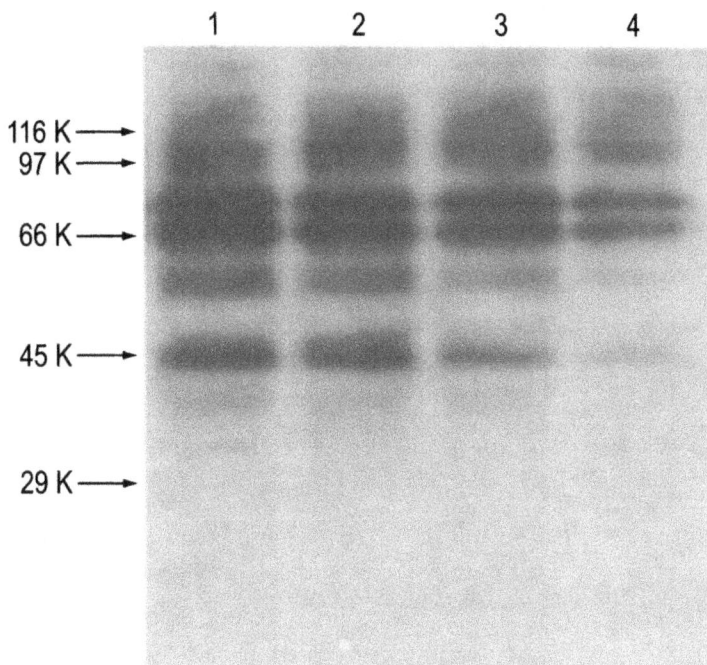

**Fig. 1** Effect of tyrphostin-A47 on hamster sperm protein tyrosine phosphorylation. Total sperm proteins were extracted in Laemmli buffer after 2 hours of culture in the presence of varying concentrations of tyrphostin-A47 and Western blot analysis was performed with anti-phosphotyrosine antibody. Samples loaded are DMSO-treated control (lane 1), 0.1 (lane 2), 0.25 (lane 3) and 0.5 mM TP-47-treated (lane 4) spermatozoa. (Reprinted from Mariappa *et al.*, Mol. Reprod. Dev. 73: 215-225 Copyright © [2006] Wiley-Liss Inc.)

### Capacitation-associated tyrosine phosphorylated proteins in the sperm flagellum

Recently, sequencing of a few of the identified tyrosine phosphorylated proteins of mammalian spermatozoa is being achieved primarily by the 2D-PAGE-MS/MS based approaches. A few examples of proteins with various functional attributes, among the sequenced tyrosine phosphorylated proteins are shown in Tables 1-2. Among these, A-kinase anchoring protein 4 (AKAP-4) was the first hamster sperm protein that was shown to be tyrosine phosphorylated. AKAPs are known to function as scaffold proteins that aid in spatially segregating PKA and its substrates, facilitating rapid compartmentalization and efficient signal transduction (Jha and Shivaji, 2002). Targeted disruption of the *Akap-4* gene in mice results in males being infertile, as a consequence of their spermatozoa lacking progressive motility (Miki *et al.*, 2002). Development of the fibrous sheath was compromised in *Akap4*$^{-/-}$ sperm, which indicates the importance of this protein during spermiogenesis as well (Miki *et al.*, 2002).

Among other proteins reported to be tyrosine phosphorylated in hamster spermatozoa is a mitochondrial/metabolic enzyme, namely, phospholipid hydroperoxide glutathione peroxidase (PHGPx). This is a 19 kDa mitochondrial capsule protein, identified as the only tyrosine-phosphorylated protein that is completely Triton X-100-dithiothreitol (DTT)-soluble (Nagdas *et al.*, 2005). Our observations with DT-solubilisation of hamster spermatozoa revealed complementary results. Intact spermatozoa showed hypo-tyrosine phosphorylation when treated with

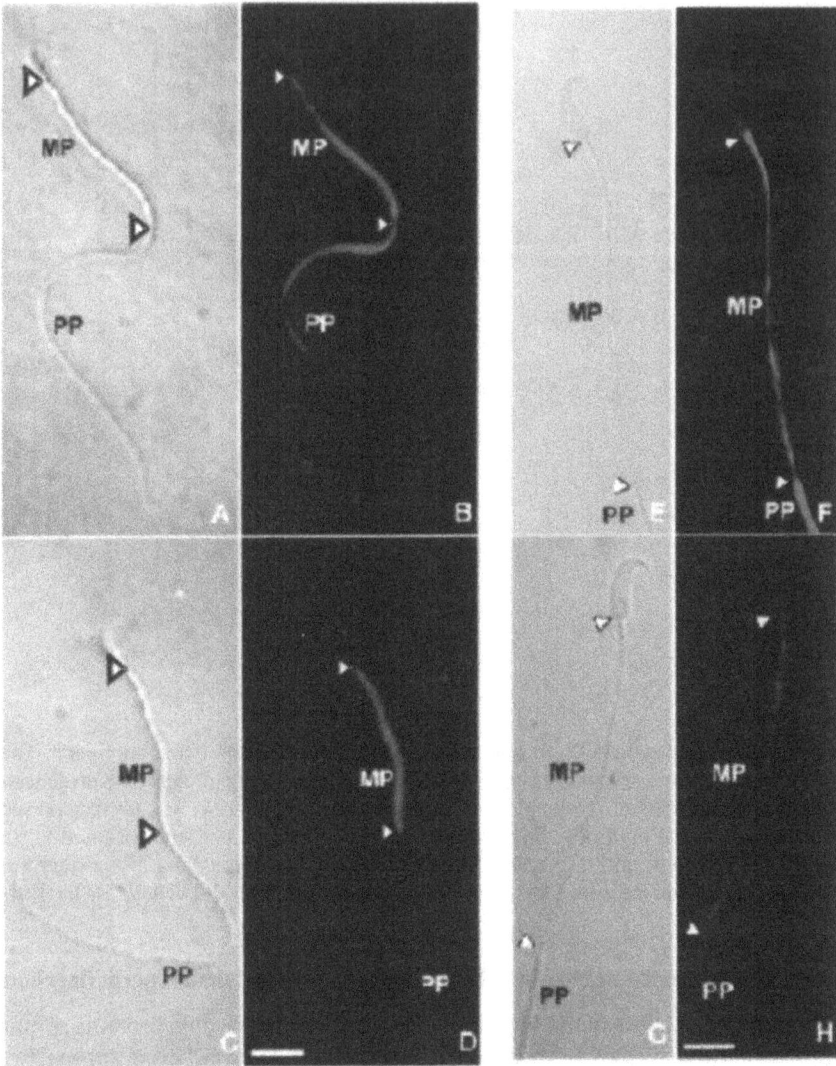

**Fig. 2** Localisation of tyrphostin-A47-inhibited protein tyrosine phosphorylation in hamster sperm flagellum. Immunocytochemistry was performed using antiphosphotyrosine antibody on control or 0.5 mM tyrphostin-A47-treated whole (A-D) or demembranated (E-H) spermatozoa cultured for 1 hour. Photomicrographs of fluorescence images of DMSO-treated (B, F) and tyrphostin-A47 (D, H) samples and their respective DIC images (A,E,C,G) are shown. MP: Mid Piece, PP: Principle Piece. Bar: 10 $\mu$m (A-D); 6 $\mu$m (E-H). Arrow heads indicate extremities of mid-piece. *(Reprinted from Mariappa et al., Mol. Reprod. Dev. 73: 215-225 Copyright © [2006] Wiley-Liss, Inc.)*

0.5 mM TP-47 specifically in the principal piece of the flagellum compared to the controls. There was no appreciable change in the amount of PYP in the mid-piece (Fig. 2). Interestingly, we observed that the solubilization-resistant structures (that is, the axoneme, outer dense fibres (ODFs) or fibrous sheath), after DT-solubilization had equally reduced phosphotyrosine levels in both the mid- and principal pieces of the sperm tail (Fig. 2). The phosphotyrosine levels of a ~21 kDa protein, possibly the reported PHGPx (Nagdas *et al.*, 2005), in the DT-soluble

**Table 1.** Tyrosine phosphorylated proteins identified in the hamster spermatozoa

| Protein | Localisation | Function | Reference |
|---------|--------------|----------|-----------|
| A kinase anchoring protein-4 (AKAP-4) | PP | scaffold protein | Jha and Shivaji, 2002 |
| Dihydrolipoamide dehydrogenase | Ac, MP, PP | post-pyruvate metabolic enzyme | Mitra and Shivaji, 2004 Mitra et al., 2005 |
| Phospholipid hydroperoxide glutathione peroxidase (PHGPx) | MP | component of mitochondrial capsule | Nagdas et al., 2005 |

Abbreviations: Ac: acrosome, MP: mid piece, PP: principal piece

**Table 2.** Tyrosine phosphorylated proteins identified in other mammalian spermatozoa

| Protein | Localisation | Function | Species | References |
|---------|--------------|----------|---------|------------|
| A kinase anchoring protein-4 (AKAP-4) | PP | scaffold protein | human mouse | Carrera et al., 1996 Ficarro et al., 2003 |
| A kinase anchoring protein-3 (AKAP-3) | PP | scaffold protein | human mouse | Mandal et al., 1999 Ficarro et al., 2003 |
| Calcium binding tyrosine phosphorylation-regulated (CABYR) protein | PP | signalling molecule | human | Naaby-Hansen et al., 2002 |
| Valosin containing Protein (VCP) | AH[a] Ac[b], PH[b], tail[b] | chaperone, membrane fusion | human boar | Ficarro et al., 2003 Geussova et al., 2002 |

Abbreviations: Ac: Acrosome, AH: anterior head, PH: posterior head, PP: principal piece.
[a] *localization changes from neck to anterior head on capacitation in human sperm*
[b] *in boar sperm*

fraction did not vary with TP-47 treatment (Fig. 3). It therefore appears from the foregoing that, in the hamster spermatozoa, detergent-resistant fractions contain most of the tyrosine-phospho-rylated proteins and the tyrosine-phosphorylation status of the detergent-soluble proteins might not be critical for flagellar bending. On obtaining hamster sperm 2D-PAGE proteome, followed by MS/MS based sequencing, we identified a few of the TP-47-inhibitable tyrosine phosphory-lated spots: a 51 kDa spot being tektin-2 (an axonemal protein) and a couple of 45 kDa spots being identified as ODF-2 protein (Mariappa et al., unpublished) which is a part of the ODFs of the sperm flagellum (Schalles et al., 1998). Interestingly, targeted disruption of tektin gene results in infertility due to compromised sperm motility in mice (Tanaka et al., 2004).

More recently dihydrolipoamide dehydrogenase, a post-pyruvate metabolic enzyme was identified to be tyrosine phosphorylated during capacitation in hamster spermatozoa. This pro-tein was localized to the principal piece and acrosome, apart from its canonical localization to the mitochondria (Mitra and Shivaji, 2004; Mitra et al., 2005). Activity of this enzyme was shown to be critical for sperm hyperactivation and the AR. In the light of available evidence that ATP, generated by extra mitochondrial glycolytic enzymes, could be utilized for sperm PYP in mouse (Travis et al., 2001), these results have great significance.

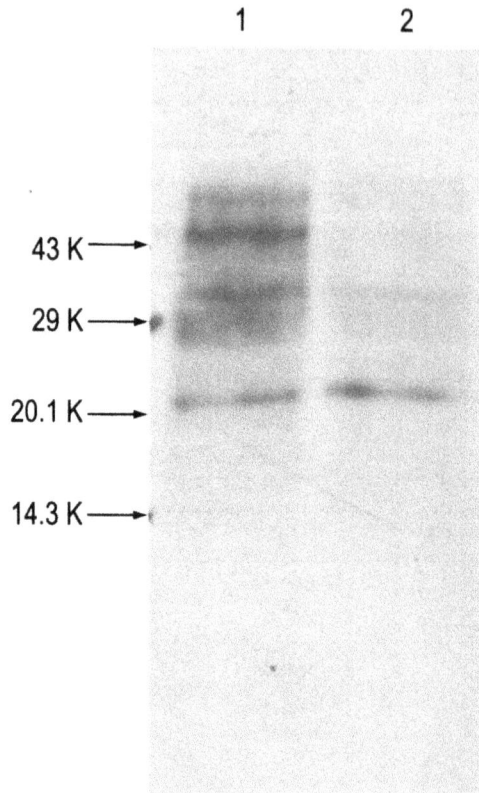

**Fig. 3** Effect of tyrphostin-A47 on tyrosine phosphorylation of DTT-Triton solubilised proteins of hamster spermatozoa. After one hour in culture spermatozoa were extracted with 2 mM DTT, 1% Triton for 15 min. DTT-Triton extracted DMSO- (lane 1) and tyrphostin-A47-treated (lane 2) samples were subjected to Western blot analysis with anti-phosphotyrosine antibody.

There are a few tyrosine-phosphorylated proteins in the hamster spermatozoa that are resistant to DT-solubilization but are solubilized on concomitant extraction by $\geq 2$ M urea. A sizable number of these proteins are probably components of the cytoskeleton. Except for PGHPx (Nagdas et al., 2005), all other tyrosine phosphorylated proteins in hamster spermatozoa are enriched in the urea-extracted fraction (Mariappa et al., unpublished data). Hence, there is compelling evidence to believe that several of the urea-enriched tyrosine phosphorylated proteins could be cytoskeletal components and their tyrosine phosphorylation status could bring about regulation of the flagellar bend in hamster spermatozoa. Of significance here is our immunogold labeling studies that show that most of the tyrosine phosphorylated proteins are localized to fibrous sheath, ODF and axoneme in the hamster (Mariappa et al., unpublished data).

In human spermatozoa, a $HCO_3^-$ dependent, soluble adenylate cyclase-mediated tyrosine phosphorylation of AKAP-3 was shown to be important for sperm motility (Luconi et al., 2005). These studies show that at least both AKAP-3 and AKAP-4 are tyrosine phosphorylated during sperm capacitation (Carrera et al., 1996; Mandal et al., 1999; Ficarro et al., 2003). It appears that AKAPs, components of the fibrous sheath, could be important signal transducing molecules, which are required for proper assembly of the sperm flagellum and maintenance of its structural integrity.

There is still a severe paucity of information with regard to the mechanism as to how cytoskeletal proteins of the mammalian sperm tail are tyrosine phosphorylated, if at all. Most proteins, other than AKAPs, which are reportedly tyrosine phosphorylated, are signaling molecules such as calcium-binding tyrosine phosphorylation-regulated protein (CABYR; Naaby-Hansen *et al.*, 2002) and glycogen synthase kinase-3α (GSK-3; Vijayaraghavan *et al.*, 2000), metabolic enzymes (dihydrolipoamide dehydrogenase and PHGPx) or chaperoning proteins such as valosin-containing protein (VCP; Geussova *et al.*, 2002; Ficarro *et al.*, 2003). Other than PYP, the evidence for serine/threonine phosphorylation of ODF-proteins, extracted from cauda epididymal spermatozoa, by a complex of Cdk5 and p35, is quite interesting (Rosales *et al.*, 2004). Because of the low specificities of the commercially available anti-phospho-serine/threonine antibodies, the total phosphoserine/phosphothreonine sperm protein phosphorylation changes during capacitation have not been well worked out. Nevertheless, the importance of PKA, a serine/threonine kinase, in sperm capacitation, as discussed earlier is well established (Nolan *et al.*, 2004). In the event of ODF phosphorylation being ascertained *in vitro*, efforts to identify similar post-translational modifications of cytoskeletal proteins would provide a better understanding of the regulation of sperm motility.

### Protein tyrosine phosphorylation in the sperm head

Localisation of tyrosine phosphorylated proteins in the sperm head has been reported in the mice (Leyton *et al.*, 1992), rat (Lewis and Aitken, 2001) and human (Ficarro *et al.*, 2003). The interaction of mouse spermatozoa with solubilised zona pellucida has been shown to increase tyrosine phosphorylation of a 95 kDa protein in spermatozoa and the protein localizes to the acrosome (Leyton *et al.*, 1992). In contrast, it was also observed that a high percentage of mouse spermatozoa bound to intact zona pellucida did not exhibit PYP in the acrosome (Urner *et al.*, 2001). However, rat caudal spermatozoa, on fresh recovery or post-culture in BWW medium alone showed immunolocalization of PYP in the posterior margin of the acrosome (Lewis and Aitken, 2001). Increased intracellular cAMP led to the change of this immunolocalisation to the sperm tail (Lewis and Aitken, 2001). Therefore, the lack of sustained or increasing PYP in the head of rat spermatozoa is not indicative of a role of this process in hyperactivation or more importantly in the AR. In human spermatozoa, PYP in the acrosomal region of the head has been demonstrated (Ficarro *et al.*, 2003). Significantly, the tyrosine-phosphorylated form of VCP has been shown to relocate to the anterior sperm head in capacitated human spermatozoa from the neck region in the non-capacitated state. Since tyrosine phosphorylation of VCP has been shown to regulate membrane fusion events, a similar mechanism could be operational for VCP in the acrosome during the AR (Ficarro *et al.*, 2003). Interestingly, using the streptolysin-O permeabilised human spermatozoa, it was shown that PTK inhibitors, such as TP-47, tyrphostin-A51 and genistein, inhibit the AR in association with reduced PYP in the sperm head (Tomes *et al.*, 2004).

When we looked at localisation of PYP to the head in capacitated hamster spermatozoa, we observed that the immunostaining was posterior to the acrosomal region, which was inhibitable by 0.5 mM TP-47 (Fig. 4). Interestingly, demembranated spermatozoa did not exhibit immunostaining in the head (Fig. 2 & 4), indicating that these tyrosine phosphorylated protein(s) in hamster sperm were probably membrane components. Our data suggests that, at least in the hamster spermatozoa, PYP in the head could play a role in the AR, which however, remains to be confirmed. Of relevance here is the fact that there is no unequivocal evidence yet for the need for head region PYP as a prerequisite for the AR in any mammalian spermatozoa.

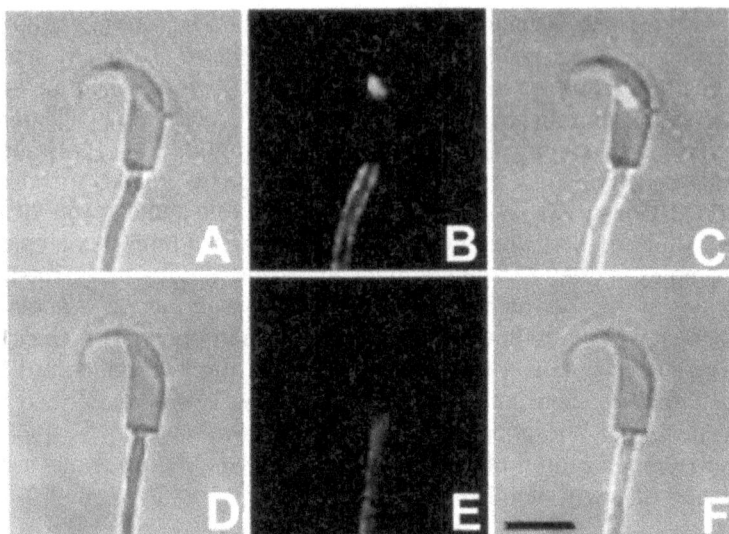

**Fig. 4** Localisation of tyrphostin-A47-inhibited protein tyrosine phosphorylation in hamster sperm head. Immunocytochemistry was performed using antiphosphotyrosine antibody on control (A-C) or 0.5 mM tyrphostin-A47-treated (D-F) spermatozoa cultured for 1 h. Photomicrographs of fluorescence images of DMSO-treated (B) and tyrphostin-A47-treated (E) samples, their respective DIC (A, D) and merged images (C, F) are shown. Bar: 5 $\mu$m.

## Protein tyrosine kinases and phosphatases involved in sperm capacitation

To define which PTK could be involved in sperm capacitation, a number of specific PTK inhibitors are being widely employed. However, there are virtually no studies describing identities of PTKs involved in mammalian sperm capacitation, and more importantly, their functional significance. All currently available reports have made identifications of PTKs based mostly on immunological techniques (Table 3). Though, over a decade ago, a 95 kDa sperm tyrosine kinase was reported in the mouse, the identity of the kinase was not clear (Leyton et al., 1992). It is assumed that PYP via an extracellular ligand-transmembrane receptor signalling mechanism is one of the potential means of initiating sperm capacitation. Even though, epidermal growth factor receptor-tyrosine kinase (EGFR-TK) was among the earliest receptor PTKs reported in bovine (Lax et al., 1994) and human (Damjanov et al., 1993) sperm, the functional significance of the presence of EGFR-TK in the sperm has not yet been clearly assessed. In this context, our data on tyrphostin-inhibitable PYP indicates that, at least in hamster spermatozoa, EGFR-TK does not appear to be involved in capacitation-associated PYP (Mariappa et al., 2006).

In a recent report, it was shown that a testis-specific knockdown of the fibroblast growth factor receptor-1 (FGFR1) gene does not lead to any increase in PYP during capacitation and that there is a possibility of a capacitation-inhibitory role for basic fibroblast growth factor-mediated signalling in wild-type spermatozoa in the mouse (Cotton et al., 2006). However, the PYP-phenotype observed in the FGFR1-knockdown mice could be as a consequence of more complex events occurring during spermatogenesis, which was also found to be impaired in these animals (Cotton et al., 2006). Therefore, there is a need to investigate into the role of receptor tyrosine kinases in mammalian spermatozoa which could identify more candidates and clarify further how already identified kinases contribute functionally to PYP and capacitation.

**Table 3.** Protein tyrosine kinases and phosphatases in mammalian spermatozoa

| Kinase/Phosphatase | Localization | Detected using | Species | References |
|---|---|---|---|---|
| **Kinase:** | | | | |
| EGFR | head apex, entire tail | IB, IF | bovine | Lax *et al.*, 1994 |
| sp42[a] | MP | IB, IF | boar | Berrutti and Borgonovo, 1996 |
| Glycogen synthase kinase-3α (GSK-3)[b] | PH, entire tail | IB, IF | bovine | Vijayaraghavan *et al.*, 2000 |
| c-yes | Ac | IB, IF | human | Leclerc and Goupil, 2002 |
| PYK-2[b] | Ac, entire tail | IB, IF | mouse | Chieffi *et al.*, 2003 |
| TK-32 | ND | In gel kinase assay | porcine | Tardif *et al.*, 2003 |
| Syk[b] | PP | IB, IF[c] | boar | Harayama *et al.*, 2004 |
| FGFR-1 | MP, PP | IB, IF, dominant Negative | mouse | Cotton *et al.*, 2006 |
| **Phosphatase:** | | | | |
| PTP1B | Equatorial segment, MP | IB, IF | human | Tomes *et al.*, 2004 |
| PTP1B | ND | IB | rat | Seligman *et al.*, 2004 |

Abbreviations: Ac: acrosome, IB: immunoblot, IF: immunofluorescence, MP: mid piece, ND: not determined, PH: posterior head, PP: principal piece.
[a] testis-specific
[b] these enzymes are also tyrosine phosphorylated
[c] only the tyrosine phosphorylated form has been detected

Identification of sperm-associated non-receptor PTKs, namely boar Syk (Harayama *et al.*, 2004), mouse PYK2 (Chieffi *et al.*, 2003) and human c-yes (Leclerc and Goupil, 2002) adds another dimension to signalling pathways regulating PYP during sperm capacitation. Interestingly, some of these sperm PTKs are responsive to regulation by increases in intracellular $Ca^{2+}$ (Chieffi *et al.*, 2003) and cAMP (Leclerc and Goupil, 2002; Harayama *et al.*, 2004) that are upstream of PYP. Therefore, it appears that there is a strong possibility for a sperm cell intrinsic activation of PYP, without the requirement for an extracellular ligand/signalling molecule, which otherwise would be a prerequisite for the receptor PTKs. Interestingly, immunoreactive PYK2 and Syk have been localized to sperm tails (Chieffi *et al.*, 2003; Harayama *et al.*, 2004). Moreover, Syk, which was shown to be activated by tyrosine phosphorylation, was present in the NP-40 insoluble fraction, expectedly, with most other tyrosine-phosphorylated proteins. Thus, indicating their localization to the cytoskeletal components of the sperm tail as shown in the porcine species (Harayama *et al.*, 2004).

Another non-receptor tyrosine kinase, TYK 2, a member of the JAK/STAT pathway was also reported to be present in the human sperm, localized to the principal piece of the sperm tail (D'Cruz *et al.*, 2001). Many members of the extracellular signal-regulated kinase (ERK) pathway (Shc, Grb2, RasP21, Raf and ERK 1/2) have been shown to be present in human spermatozoa and superoxide anion is able to activate the pathway (Lamirande and Gagnon, 2002). Additionally, most proteins identified in the ERK pathway were predominantly localized to the triton-insoluble fraction, suggesting a role for this pathway in phosphorylation of structural components of the sperm cytoskeleton (Lamirande and Gagnon, 2002). Such enriched pools of these non-receptor PTKs in various subcellular compartments indicate a probable location-specific function for these enzymes, likely to be aided by scaffold proteins like AKAPs.

While most of the above identifications of PTKs are based on immunological techniques, a 32 kDa protein was found to have tyrosine kinase activity in porcine spermatozoa on the basis of an in-gel kinase assay (Tardif *et al.*, 2003). Though PTKs have been identified across various mammalian species, information is scanty on their molecular and functional characterization, which is particularly so with the hamster spermatozoa. To date, there has not been any identification of sperm-specific PTK(s).

Among PTPs, the presence of PTP1B in spermatozoa has been established in the human (Tomes *et al.*, 2004) and the rat (Seligman *et al.*, 2004), by immunological methods. Crude human and mouse sperm lysates contain appreciable PTP activity which is capacitation state-dependent. Moreover, the localization of PTP1B to the equatorial segment lends its importance to a role in the AR, which appears to be PTP activity-dependent (Tomes *et al.*, 2004). There is a need to maintain a physiologically optimal status of PYP, regulated by sperm-derived PTKs and PTPs, and dysregulation of the PTP component could have a profound impact on normal sperm function. Therefore, there is a need to have detailed knowledge on sperm-associated PTKs and PTPs.

Knowledge on the molecular characteristics of PTK and PTPs is particularly important in terms of application to male contraception. For example, sperm-specific PTKs, like the boar sp42 (Berruti and Borgonovo, 1996), would serve as ideal drugable candidates, especially, if they have a capacitation-initiating/-sustaining/-promoting function(s). The quest for identities of PTK substrates, that is specific tyrosine phosphorylated proteins, on the other hand, is being pursued quite vigorously, with a few of them being molecularly characterized by sequencing.

## Protein tyrosine phosphorylation and sperm motility defects in spermatozoa

Evidence is accumulating to implicate PYP status with human sperm function and consequently human infertility. A known case of asthenozoospermia with sperm fibrous sheath dysplasia was linked to a genetic defect, wherein the partial deletion of AKAP-3/-4 was observed leading to defective assembly of the fibrous sheath and complete immotility of spermatozoa (Baccetti *et al.*, 2005). Understandably, such a defect in a scaffold protein could manifest into a structurally malformed sperm. Capacitation-associated PYP was compromised when spermatozoa from asthenozoospermic individuals were cultured *in vitro*, compared to samples from normozoospermic or fertile individuals (Buffone *et al.*, 2005). There was a partial rescue of the status of PYP on introduction of dibutyrl cAMP and PF, both of which bring about increased intracellular cAMP, indicating that inappropriate PYP could lead to asthenozoospermia in some cases. Therefore, the significance of PYP with the incidence of asthenospermia needs to be recognized, particularly because its prevalence in infertile men is estimated to be around 18% (Curi *et al.*, 2003). There are also a few other key signalling molecules, if not present during spermatogenesis, could lead to impaired formation of spermatozoa. It is observed that immotile spermatozoa from mice null for soluble adenylate cyclase (sAC; most critical among all of sperm adenylate cyclases for motility) could not be rescued as far as their motility is concerned by supplementation of dibutyrl cAMP (Hess *et al.*, 2005). Spermatozoa from these mice exhibited angulation of the sperm tail, suggesting a role for sAC during spermatogenesis.

In conclusion, there is a critical need for a thorough molecular and functional characterization of sperm-specific PTKs and PTPs and overall profile of sperm PYP. This will not only be important in our understanding of the biology of PYP-dependent sperm function, but also could provide crucial leads to develop targets for designing specific drugs for male contraception. Presently, there are only a few of these enzymes but most of them do not meet the sperm-specific criteria and are not drugable contraceptive targets. Therefore, it is important to understand molecular mechanisms of capacitation-associated tyrosine phosphorylation. There is great promise in the possibility that tyrosine phosphorylation of many of the detergent-insoluble proteins have a direct bearing on sperm motility apparatus. Instances of male infertility resulting from inadequate PYP would also find solutions by approaches of this kind.

## Acknowledgements

Financial support from the Council of Scientific and Industrial Research, New Delhi is gratefully acknowledged. In part, the study was supported by Indian Council of Medical Research, New Delhi. Our thanks are due to Ms. M. S. Padmavathi (IISc) for her help in the preparation of the manuscript.

## References

Ain R, Umadevi K, Shivaji S and Seshagiri PB (1999) Pentoxifylline-stimulated capacitation and acrosome reaction in hamster spermatozoa: involvement of intracellular signaling molecules *Molecular Human Reproduction* 5 618-626

Baccetti B, Collodel G, Estenoz M, Manca D, Moretti E and Piomboni P (2005) Gene deletions in an infertile man with sperm fibrous sheath dysplasia *Human Reproduction* 20 2790-2794

Baker MA and Aitken RJ (2004) The importance of redox regulated pathways in sperm cell biology *Molecular and Cellular Endocrinology* 216 47-54

Berruti G and Borgonovo B (1996) sp42, the boar sperm tyrosine kinase, is a male germ cell-specific product with a highly conserved tissue expression extending to other mammalian species *Journal of Cell Science* 109 851-858

Buffone MG, Calamera JC, Verstraeten SV and Doncel GF (2005) Capacitation-associated protein tyrosine phosphorylation and membrane fluidity changes are impaired in the spermatozoa of asthenozoospermic patients *Reproduction* 129 697-705

Carrera A, Moos J, Ning XP, Gerton GL, Tesarik J, Kopf GS and Moss SB (1996) Regulation of protein tyrosine phosphorylation in human sperm by a calcium/calmodulin dependent mechanism: Identification of A kinase anchor proteins as major substrates for tyrosine phosphorylation *Developmental Biology* 180 284–296

Chieffi P, Barchi M, Agostino SD, Rossi P, Tramontano D and Geremia, R (2003) Proline-rich tyrosine kinase 2 (PYK2) expression and localization in mouse testis *Molecular Reproduction and Development* 65 330-335

Cotton L, Gibbs GM, Sanchez-Partida LG, Morrison JR, de Krester DM and O'Bryan MK (2006) FGRF-1 signaling is involved in spermiogenesis and sperm capacitation *Journal of Cell Science* 119 75-84

Curi SM, Ariagno JI, Chenlo PH, Mendeluk GR, Pugliese MN, Sardi-Segovia LM, Repetto HE, Blanco AM (2003) Asthenozoospermia: analysis of a large population *Archives of Andrology* 49 343-349

Damjanov I, Solter D and Knowles BB (1993) Functional epidermal growth factor receptor localises to the postacrosomal region of human spermatozoa *Biochemical and Biophysics Research Communications* 190 901-906

D'Cruz OJ, Vassilev AO and Uckun FM (2001) Members of the Janus kinase/signal transducers and activators of transcription (JAK/STAT) pathway are present and active in human sperm *Fertility and Sterility* 76 258-266

Demarco IA, Espinosa F, Edwards J, Sosnik J, De La Vega-Beltran JL, Hockensmith JW, Kopf GS, Darszon A and Visconti PE (2003) Involvement of a Na$^+$/HCO$_3^-$ cotransporter in mouse sperm capacitation *Journal of Biological Chemistry* 278 7001-7009

Dorval V, Dufour M and Leclerc P (2002) Regulation of the phosphotyrosine content of human sperm proteins by intracellular Ca$^{2+}$: role of Ca$^{2+}$-adenosine triphosphatases *Biology of Reproduction* 67 1538-1545

Ficarro S, Chertihin O, Westbrook VA, White F, Jayes F, Kalab P, Marto JA, Shabanowitz J, Herr JC, Hunt DF and Visconti PE (2003) Phosphoproteome analysis of capacitated human sperm. Evidence of tyrosine phosphorylation of a kinase-anchoring protein 3 and valosin-containing protein/p97 during capacitation *Journal of Biological Chemistry* 278 11579-11589

Geussova G, Kalab P and Peknicova J (2002) Valosine

containing protein is a substrate of cAMP-activated boar sperm tyrosine kinase *Molecular Reproduction and Development* **78** 366-375

Harayama H, Muroga M and Miyake MA (2004) Cyclic adenosine 3', 5'-monophosphate-induced tyrosine phosphorylation of Syk protein tyrosine kinase in the flagella of boar spermatozoa *Molecular Reproduction and Development* **69** 436-447

Hess KC, Jones BH, Marquez B, Chen Y, Ord TS, Kamenetsky M, Miyamoto C, Zippin JH, Kopf GS, Suarez SS, Levin LR, Williams CJ, Buck J and Moss SB (2005) The "soluble" adenylyl cyclase in sperm mediates multiple signaling events required for fertilization *Developmental Cell* **9** 249-259

Jayaprakash D, Kumar KS, Shivaji S and Seshagiri PB (1997) Pentoxifylline induces hyperactivation and acrosome reaction in spermatozoa of golden hamster: changes in motility kinematics *Human Reproduction* **12** 2192-2199

Jha KN and Shivaji S (2002) Identification of the major tyrosine phosphorylated protein of capacitated hamster spermatozoa as a homologue of mammalian sperm a kinase anchoring protein *Molecular Reproduction and Development* **61** 258-270

Kulanand J and Shivaji S (2001) Capacitation-associated changes in protein tyrosine phosphorylation, hyperactivation and acrosome reaction in hamster spermatozoa *Andrologia* **33** 95-104

Lamirande E and Gagnon C (2002) The extracellular signal-regulated kinase (ERK) pathway is involved in human sperm function and modulated by the superoxide anion *Molecular Human Reproduction* **8** 124-135

Lax Y, Rubinstein S and Breitbart H (1994) Epidermal growth factor stimulates acrosomal exocytosis in bovine sperm *FEBS Letters* **339** 234-238

Leclerc P and Goupil S (2002) Regulation of the human sperm tyrosine kinase c-yes. Activation by cyclic adenosine 3', 5'-monophosphate and inhibition by $Ca^{2+}$ *Biology of Reproduction* **67** 301-307

Leclerc P, de Lamirande E and Gagnon C (1996) Cyclic adenosine 3', 5'-monophosphate dependent regulation of protein tyrosine phosphorylation in relation to human sperm capacitation and motility *Biology of Reproduction* **55** 684-692

Lewis B and Aitken RJ (2001) Impact of epididymal maturation on the tyrosine phosphorylation patterns exhibited by rat spermatozoa *Biology of Reproduction* **64** 1545-1556

Leyton L, LeGuen P, Bunch D and Saling PM (1992) Regulation of mouse gamete interaction by a sperm tyrosine kinase *Proceedings of the National Academy of Sciences USA* **89** 11692-11695

Luconi M, Porazzi I, Feruzzi P, Marchiani S, Forti G and Baldi E (2005) Tyrosine phosphorylation of the A kinase anchoring protein 3 (AKAP3) and soluble adenylate cyclase are involved in the increase of human sperm motility by bicarbonate *Biology of Reproduction* **72** 22-32

Mahony MC and Gwathmey TY (1999) Protein tyrosine phosphorylation during hyperactivated motility of cynomolgus monkey (*Macaca fascicularis*) spermatozoa *Biology of Reproduction* **60** 1239-1243

Mandal A, Naaby-Hansen S, Wolkowicz MJ, Klotz K, Shetty J, Retief JD, Coonrod SA, Kinter M, Sherman N, Cesar F, Flickinger CJ and Herr JC (1999) FSP95, a testis-specific 95-kilodalton fibrous sheath antigen that undergoes tyrosine phosphorylation in capacitated human spermatozoa *Biology of Reproduction* **61** 1184-1197

Mariappa D, Siva AB, Shivaji S and Seshagiri PB (2006) Tyrphostin-A47 inhibitable tyrosine phosphorylation of flagellar proteins is associated with distinct alteration of motility pattern in hamster spermatozoa *Molecular Reproduction and Development* **73** 215-225

Miki K, Willis WD, Brown PR, Goulding EH, Fulcher KD and Eddy EM (2002) Targeted disruption of the *Akap4* gene causes defects in sperm flagellum and motility *Developmental Biology* **248** 331-342

Mitra K and Shivaji S (2004) Novel tyrosine-phosphorylated post-pyruvate metabolic enzyme, dihydrolipoamide dehydrogenase, involved in capacitation of hamster spermatozoa *Biology of Reproduction* **70** 887-899

Mitra K, Rangaraj N and Shivaji S (2005) Novelty of the pyruvate metabolic enzyme dihydrolipoamide dehydrogenase in spermatozoa: correlation of its localization, tyrosine phosphorylation, and activity during sperm capacitation *Journal of Biological Chemistry* **280** 27543-27553

Naaby-Hansen S, Mandal A, Wolkowicz MJ, Sen B, Westbrook VA, Shetty J, Coonrod SA, Klotz KL, Kim Y, Bush LA, Flickinger CJ and Herr JC (2002) CABYR, a novel calcium-binding tyrosine phosphorylation-regulated fibrous sheath protein involved in capacitation *Developmental Biology* **242** 236-254

Nagdas SK, Winfrey VP and Olson GE (2005) Tyrosine phosphorylation generates multiple isoforms of the mitochondrial capsule protein, phospholipid hydroperoxide glutathione peroxidase (PHGPx), during hamster sperm capacitation *Biology of Reproduction* **72** 164-171

Nolan MA, Babcock DF, Wennemuth G, Brown W, Burton KA and McKnight GS (2004) Sperm-specific protein kinase A catalytic subunit $C\alpha 2$ orchestrates cAMP signaling for male fertility *Proceedings of the National Academy of Sciences USA* **101** 13483-13488

Rosales JL, Lee BC, Modarressi M, Sarker KP, Lee KY, Jeong YG, Oko R and Lee KY (2004) Outer dense fibers serve as a functional target for Cdk5.p35 in the developing sperm tail *Journal of Biological Chemistry* **279** 1224-1232

Schalles U, Shao X, van er Hoorn FA and Oko R (1998) Developmental expression of the 84 kDa ODF sperm protein: localization to both the cortex and medulla of outer dense fibers and to the connecting piece *Developmental Biology* **199** 250-260

Seligman J, Zipser Y and Kosower NS (2004) Tyrosine phosphorylation, thiol status, and protein tyrosine phosphatase in rat epididymal spermatozoa *Biology of Reproduction* **71** 1009-1015

**Seshagiri PB, Thomas M, Sreekumar A, Ray PS and Mariappa D** (2003) Pentoxifylline induced signaling events during capacitation of hamster spermatozoa: significance of protein tyrosine phosphorylation *Cellular and Molecular Biology (Noisy-le-grand)* **49** 371-380

**Si Y and Okuno M** (1999) Role of tyrosine phosphorylation of flagellar proteins in hamster sperm hyperactivation *Biology of Reproduction* **61** 240-246

**Tanaka H, Iguchi N, Toyama Y, Kitamura K, Takahashi T, Kaseda K, Maekawa M and Nishimune Y** (2004) Mice deficient in axonemal protein tektin-t exhibit male infertility and immotile-cilium syndrome due to impaired inner arm dynein function *Molecular and Cellular Biology* **24** 7958-7964

**Tardif S, Dube C and Bailey JL** (2003) Porcine sperm capacitation and tyrosine kinase activity are dependent on bicarbonate and calcium but protein tyrosine phosphorylation is only associated with calcium *Biology of Reproduction* **68** 207-213

**Tomes CN, Roggero CM, De Blas G, Saling PM and Mayorga LS** (2004) Requirement of protein tyrosine kinase and phosphatase activities for human sperm exocytosis *Developmental Biology* **265** 399-415

**Travis AJ, Jorgez CJ, Merdiushev T, Jones BH, Dess DM, Diaz-Cueto L, Storey BT, Kopf GS and Moss SB** (2001) Functional relationships between capacitation-dependent cell signaling and compartmentalised metabolic pathways in murine spermatozoa *Journal of Biological Chemistry* **276** 7630-7636

**Urner F and Sakkas D** (2003) Protein phosphorylation in mammalian spermatozoa *Reproduction* **125** 17-26

**Urner F, Leppens-Luisier G and Sakkas D** (2001) Protein tyrosine phosphorylation in sperm during gamete interaction in the mouse: the influence of glucose *Biology of Reproduction* **64** 1350–1357

**Vijayaraghavan S, Mohan J, Gray H, Khatra B and Carr DW** (2000) A role for phosphorylation of glycogen synthase kinase-3α in bovine sperm motility regulation *Biology of Reproduction* **62** 1647-1654

**Visconti PE, Moore GD, Bailey JL, Leclerc P, Connors SA, Pan D, Olds-Clarke P and Kopf GS** (1995) Capacitation in mouse spermatozoa II. Correlation between the capacitation state and protein tyrosine phosphorylation *Development* **121** 1129-1137

**Visconti PE, Johnson LR, Oyaski M, Fornes M, Moss SB, Gerton GL and Kopf GS** (1997) Regulation, localization and anchoring of protein kinase A subunits during mouse sperm capacitation *Developmental Biology* **192** 351-363

**Yanagimachi R** (1994) Mammalian fertilization. In *The Physiology of Reproduction* pp189-317 Eds E Knobil and JD Neill. Raven Press, New York

# Cellular and molecular events during oocyte maturation in mammals: molecules of cumulus-oocyte complex matrix and signalling pathways regulating meiotic progression

N Kimura[1], Y Hoshino[2], K Totsukawa[1] and E Sato[2]

[1]Laboratory of Animal Reproduction, Faculty of Agricultural Science, Yamagata University, 1-23 Wakaba-machi,Tsuruoka 997-8555, Japan (naonao@tds1.tr.yamagata-u.ac.jp); [2]Laboratory of Animal Reproduction, Graduate School of Agricultural Science, Tohoku University, 1-1 Tsutsumidori-amamiyamachi, Aoba-ku, Sendai 981-8555, Japan

Mammalian oocytes acquire their intrinsic ability in a stepwise manner through ovarian folliculogenesis, ultimately reaching the competence to undergo complete oocyte maturation at the final stage of Graafian follicle development. The fully-grown oocyte is tightly surrounded by compact layers of specialized granulosa cells (cumulus cells) to form a cumulus-oocyte complex (COC). After a preovulatory gonadotrophin surge, the COCs rapidly organize a special muco-elastic extracellular matrix (ECM) consisting of large amounts of hyaluronan (HA) and HA binding matrix glycoproteins. Simultaneously, the oocytes undergo meiotic resumption and cytoplasmic modification and attain the fertilizable metaphase II (MII) stage. These cellular events that immediately occur in COCs in the ovulatory phase are strictly regulated by pituitary hormones, steroids, growth factors and so on. Knowledge of the efficient mechanisms and the downstream cascades of the key molecules controlling oocyte maturation may gradually lead to improvement of the present oocyte/embryo culture systems and gamete biotechnology. Recent studies by our group imply that i) the interaction of HA-CD44 identified in the porcine COC matrix is likely to participate in gap junctional communication and meiotic progression, and that ii) phosphatidylinositol 3-kinase (PI3-K) and Akt contribute to the progress of follicle stimulating hormone (FSH)-induced meiosis in mice. Furthermore, this review focuses on the current understanding of biosynthetic regulation, the presumptive role of COC matrix molecules and the signalling pathways for meiotic modulators, such as the protein kinase A (PKA) pathway, the PI3-K/Akt pathway and the mitogen activated protein kinase (MAPK) pathway.

E-mail: eimei@bios.tohoku.ac.jp

## Introduction

During ovarian folliculogenesis, growth of the oocyte and its companion somatic cells takes place in a highly coordinated and mutually dependent manner via both paracrine and gap-junctional signalling under the specific regulation of pituitary gonadotrophins, steroids and various growth factors (Eppig, 2001; van den Hurk and Zhao, 2005). The cumulus-oocyte complex (COC), a structural unit of the later antral follicle, consists of several layers of cumulus cells (approximately 1000-3000 cells/COC in mice) around the oocyte (Salustri *et al.*, 1992). The oocyte and the inner layers of cumulus cells (termed corona radiata) maintain extensive communication through transzonal projections by which cumulus cells traverse the zona pellucida (ZP) and terminate on the oocyte plasma membrane to form both adhesive and gap-junctional contacts (Albertini *et al.*, 2001). Also, intercellular communication among cumulus cells is maintained by gap junctions composed of the connexin family of proteins (Kidder and Mhawi, 2002). The gap junctional communication between the oocyte and the innermost layer of cumulus cells, as well as between individual cumulus cells, manages the bidirectional transfer of the regulatory molecules concerned with meiotic arrest, organic ions, small nutrients and metabolites.

Following the endogenous luteinizing hormone (LH) surge just before ovulation, a Graafian follicle rapidly increases in volume by inflow and accumulation of follicular fluid. Concomitantly, the COC synthesizes a muco-elastic extracellular matrix (ECM) containing large amounts of hyaluronan (HA) and specific HA binding proteins. This brings about volumetric enlargement called cumulus expansion (Dekel and Beers, 1978; Eppig, 1979). The expanded COC becomes detached from the membrana granulosa and is finally released from the follicle into the fallopian tube. Almost simultaneously with cumulus expansion after hormonal signalling, the oocyte undergoes meiotic resumption and cytoplasmic modification and attains the mature metaphase II (MII) stage.

It is well established that the progress of oocyte maturation is closely linked to the formation of the COC matrix (Fig. 1). Certainly, these consecutive events to which COCs are subjected to in a short time are complicated and precisely regulated via multiple cascades under key stimulators. However, many questions about the mechanisms of action and the downstream pathway of the related molecules on oocyte maturation remain unclear. Here, we describe our views and recent findings gleaned from the literature on the cellular and molecular basis of oocyte maturation, as well as the signalling pathways for meiotic progression.

## Components of the COC matrix mass and factors regulating cumulus expansion

### HA, the major material of the COC matrix

In many species, the drastic enlargement of the COC matrix characterized by the intercellular deposition of HA derived from cumulus cells have been well established both *in vivo* and *in vitro* (Eppig, 1979; Salustri *et al.*, 1989). In mouse preovulatory Graafian follicles before gonadotrophin treatment, there is very little HA-enriched matrix in the compact COCs and mural granulosa cells. Five hours after the injection of human chorionic gonadotrophin (hCG), the COC matrix is partially synthesized. HA is found in the mural granulosa cells closest to the COC and in those cells adjacent to the antrum. Shortly before ovulation, the COC is fully expanded (20 to 40-fold of the initial volume) and occupies most of the antral cavity. HA is abundantly localized in the rim of the antrum in contact with the mural granulosa cells, although, it is not present in the mural granulosa cells closest to the basement membrane (Salustri *et al.*, 1992). A number of studies have been undertaken on the physiological

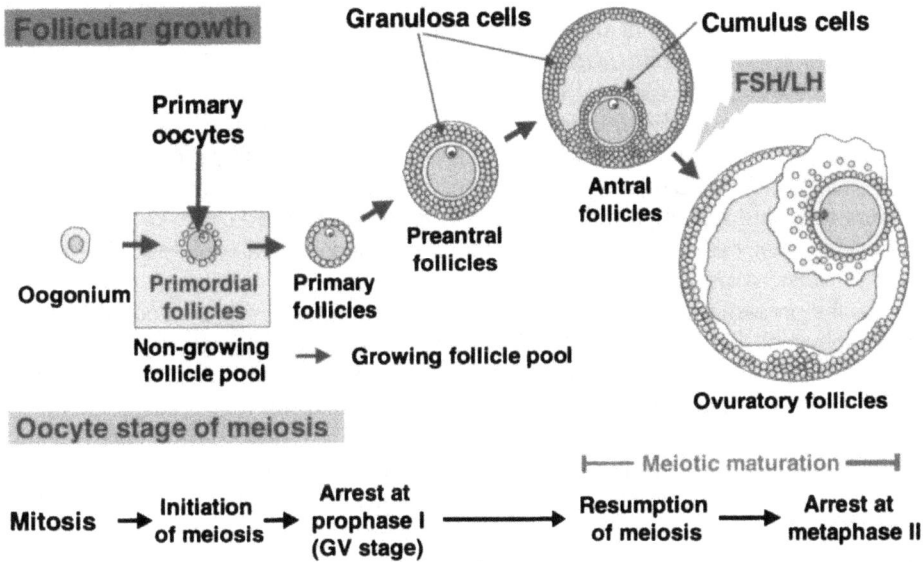

**Fig. 1** Follicular growth and oocyte meiotic maturation

importance of cumulus expansion leading to oocyte maturation, ovulation, fertilization and subsequent embryo development.

HA is widely distributed as the hydrophilic material of ECMs in nearly all vertebrate tissues. It is characterized by a linear macromolecule glycosaminoglycans (GAGs), ranging in molecular mass from $10^5$ to $10^7$ Da, composed of the repeating disaccharide units of D-glucuronic acid ß1→3 linked *N*-acetylglucosamine ß1→4, and is not sulfated nor assembled on core proteins for a proteoglycan (PG). Besides its structural role in maintaining the hydration and physical properties of tissues, HA influences cell shape and behaviour (for example, adhesion, migration, proliferation, differentiation, cell death and cell anchorage) via binding to specific cell surface receptor, such as CD44 and RHAMM, in many biological processes of tissue organization, tissue morphogenesis, cancer metastasis, wound healing, inflammation and angiogenesis (Knudson and Knudson, 1993; McDonald and Camenisch, 2002). HA is synthesized in the plasma membrane and directly extends into the ECM (Philipson and Schwartz, 1984). The ends of growing HA chains are elongated by the addition of sugar residues derived from UDP-glucuronate and UDP-*N*-acetylglucosamine. Several groups have succeeded in cloning and characterizing three separate genes for vertebrate HA synthase (HAS), the enzyme responsible for HA biosynthesis: *Has1*, *Has2*, *Has3* (Spicer and McDonald, 1998). The mammalian HAS proteins, predicted to be approximately 63 kDa in molecular mass, share between 55% and 71% amino acid identity and are encoded by distinct genes located on separate autosomes.

*Expression of Has genes in COCs*

Studies in mice, pigs and cattle have demonstrated that *Has2* mRNA is expressed in COCs with cumulus expansion during oocyte maturation *in vivo* and *in vitro* (Fulop *et al.*, 1997a; Kimura *et al.*, 2002; Schoenfelder and Einspanier, 2003). We have characterized two kinds of porcine cDNAs, *shas2* expressed in cumulus cells and *shas3* expressed in oocytes. The cDNA encoding the open reading frame (ORF) of *shas2* contained 1656 nucleotides (GenBank accession No.: AB050389) coding for 552 amino acids. The ORF of *shas3* contained 1659 nucleotides (GenBank accession No.: AB159675) coding for 553 amino acids. Each sequence of *shas2* and *shas3* cDNAs indicated exceedingly high similarity to the *Has* family of other mammals. Stock *et al.* (2002) have shown that HAS2 in equine mural granulosa cells are specifically expressed in the ovulatory process after hCG injection. Therefore, it is now generally accepted that HA secreted by cumulus cells and mural granulosa cells can be attributed to HAS2. On the other hand, HAS3 in porcine oocytes is expressed maternally from the germinal vesicle (GV) stage up to 8-cell stage (N. Kimura, unpublished).

Whereas the preovulatory LH surge triggers the cumulus expansion of preovulatory follicles *in vivo*, follicle-stimulating hormone (FSH) is a key factor *in vitro* (Eppig, 1979). FSH increases intracellular cyclic adenosine monophosphate (cAMP), activators of cAMP-dependent protein kinase (PKA), and promotes maturation of epidermal growth factor (EGF) receptors that activate the tyrosine kinase cascade (Dekel and Kraicer, 1978; Richards and Rolfes, 1980; Prochazka *et al.*, 2003). Therefore, it seems likely that their sequential response is mediated by cAMP and the tyrosine kinase cascade. In *in vitro* studies, the positive effects of EGF, transforming growth factor (TGF)-α and growth hormone on cumulus expansion in FSH-free culture medium have also been demonstrated in several species (Izadyar *et al.*, 1996). Similarly, expression of HAS2 in cumulus cells and granulosa cells is stimulated by pregnant mare's gonadotrophin (PMSG), FSH and porcine follicular fluid (pFF) *in vitro* (Kimura *et al.*, 2002; Schoenfelder and Einspanier, 2003) (Fig. 2). EGF-like growth factor family members induced by LH, such as amphiregulin, epiregulin and betacellulin, stimulate expression of *Has2* (Park *et al.*, 2004). In porcine COCs, oocytectomy slightly reduces the level of *Has2* mRNA in the presence of PMSG and pFF, suggesting that the oocyte up-regulates HAS2 expression during cumulus expansion (Kimura *et al.*, 2002). Mouse oocytectomized COCs synthesize very little HA and fail to expand in the presence of FSH or EGF (Buccione *et al.*, 1990; Tirone *et al.*, 1997). Incubation of mouse cumulus cells with isolated oocytes (co-culture) or cultured in denuded-oocyte conditioned medium stimulates FSH-dependent HA synthesis and cumulus expansion (Buccione *et al.*, 1990; Salustri *et al.*, 1990). These facts suggest that the secretion of a soluble factor from oocytes is involved in HA synthesis and cumulus expansion *in vitro*. On the other hand, an oocyte factor does not influence the production of other GAGs, dermatan sulfate and chondroitin sulfate in cumulus cells (Tirone *et al.*, 1993; Nakayama *et al.*, 1996). Porcine and bovine oocytes also secrete a cumulus expansion-enabling factor, though cumulus expansion progresses steadily without this factor (Vanderhyden, 1993). It is well known that growth differentiation factor-9 (GDF-9), bone morphogenetic protein 15 (BMP-15, also called GDF-9B) and BMP-6 are likely candidate molecules for oocyte-secreted factor (Gilchrist *et al.*, 2004). These growth factors are members of the TGF-ß superfamily that shows a high level of expression in mammalian oocytes. Elvin *et al.* (1999) reported that recombinant GDF-9 induces HAS2 expression and cumulus expansion *in vitro* in the presence of FSH, indicating that GDF-9 interacts with FSH and modulates the downstream target gene *Has2*. On the contrary, *shas3* expression in oocytes is not remarkably affected by gonadotrophins (Kimura *et al.*, 2002; Schoenfelder and Einspanier, 2003).

The expression of both FSH and LH receptors in cumulus cells has been established (Chen *et al.*, 1994a; Meduri *et al.*, 2002). While the expression of LH receptors is present at a very low level in cumulus cells isolated from preovulatory follicles, that of FSH receptors is abundantly present (Chen *et al.*, 1994a). Meduri *et al.* (2002) have documented the presence of FSH (but not LH) receptors in oocytes during follicular development from the primary stage to the

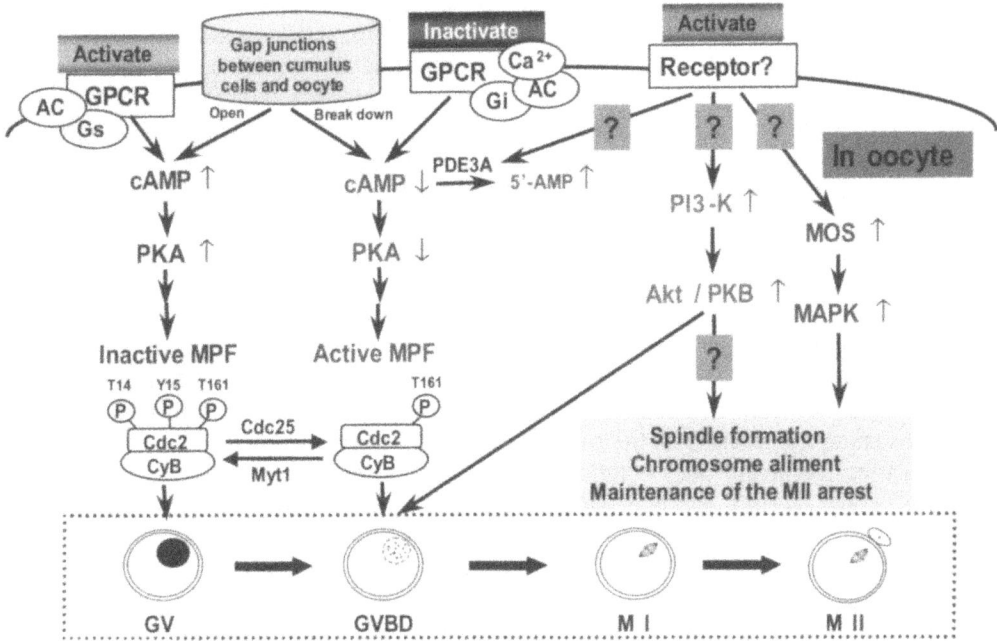

**Fig. 2** A schematic diagram of the signalling cascades regulating meiotic progression in oocytes, such as the PKA pathway, the PI3-K/Akt pathway and the MAPK pathway. The GPCRs (GPR3 and GPR12 in oocytes) interact with an unknown ligand and activates G protein Gs, which stimulates AC to keep enough levels of cAMP for the maintenance of meiotic arrest (Mehlmann *et al.*, 2004, Hinckley *et al.*, 2005). cAMP activates PKA, which finally inactivates maturation promoting factor (MPF) by phosphorylation of Cdc2 kinase. The potential stimulation inhibits the coupling of GPCR-Gs-AC and simultaneously activates PDE3A and reduces cAMP levels. An alternative cascade, PI3-K/Akt pathway, in which the upstream signalling is unknown, also activates MPF and leads to GVBD. The MAPK cascade is involved in normal transition from MI to MII stage.

preovulatory stage in human and porcine ovaries. Patsoula *et al.* (2001) found both FSH and LH receptor mRNAs in mice oocytes, zygotes and preimplantation embryos. The *shas3* expression is not considerably affected by gonadotrophins (N. Kimura, unpublished), regardless of the fact that oocytes express both FSH and LH receptors, although, the *shas2* expression in cumulus cells showing abundant FSH receptors is stimulated by FSH. This evidence implies that 1) there is a different type of signalling for transcriptional induction between HAS2 and HAS3, and that 2) HAS3 is independent of HA synthesis for cumulus expansion.

*Proteoglycans (PGs) and HA-associated proteins retaining COC matrix*

A volumetric COC matrix is organized not only by synthesis of HA and mucoid materials, but also their retention within COCs mass. A number of proteoglycans (PGs) and HA-binding proteins secreted by cumulus cells and granulosa cells contribute to stabilization of the matrix and may yield functional characteristics through cell-matrix and cell-cell interaction (Salustri *et al.*,

1999; Zhuo and Kimata, 2001). PGs are macromolecules composed of a core protein with covalently attached variable GAG chains (Salustri et al., 1999). While producing a large amount of HA, cumulus cells also synthesize moderate amounts of chondroitin sulfate, dermatan sulfate and heparan sulfate that are associated with a core protein to form proteoglycan. However, these GAGs, except for HA, do not appear to be directly involved in the synthesis of the COC matrix (Salustri et al., 1989; Nakayama et al., 1996).

Camaioni et al. (1996) have shown that a dermatan sulfate PG of large hydrodynamic size (> 1 million Da and a core protein of about 280 kDa) and a protein approximately 46 kDa in molecular mass are secreted by cumulus cells and accumulated in the COC matrix. The properties of this PG are similar to aggrecan and versican that interact specifically with HA. The 46 kDa protein is the same size as the cartilage link protein that is found in all hyaline cartilage and in the aortic intima, and interacts with HA to stabilize binding of PG monomers to HA for formation of a stable structure. Link protein markedly localizes in the COC matrix and mural granulosa cells in large preovulatory follicles after hCG injection (Kobayashi et al., 1999). In vitro studies have suggested that the synthesis of link protein in cumulus cells requires gonadotrophin stimulation and oocyte-derived factor. Fetal bovine serum is likely to improve the production of link protein and PGs and their deposition in the COC matrix (Camaioni et al., 1996; Kobayashi et al., 1999). Sun et al. (2002) have indicated that link protein enhances cumulus expansion by stabilizing heavy chains (HCs) of an inter-α-inhibitor (IαI) family-HA complex. Versican isoforms are predominantly expressed in the cumulus cells and mural granulosa cells during the ovulation (Russell et al., 2003).

Tumor necrosis factor-induced protein-6 (TNFIP-6, also known as tumor necrosis factor-stimulated gene-6) is a matrix protein that is essential for formation of the normal COC matrix assembly (Fulop et al., 2003). TNFIP-6, a glycoprotein with a mass of about 35 kDa is synthesized by cumulus cells and mural granulosa cells of antral follicles after the LH surge and is incorporated into the expanded COC matrix (Fulop et al., 1997b). There are two types of TNFIP-6: 1) a monomer binding HA through its link module, and 2) a covalent complex (of approximately 125 kDa) formed with the heavy chains of the IαI family (Mukhopadhyay et al., 2001; Carrette et al., 2001). The heavy chains of the IαI family are covalently transferred to HA by catalysis of TNFIP-6. HA oligosaccharides with eight or more monosaccharide units are potent acceptors in the transfer of heavy chains (Mukhopadhyay et al., 2004). Serum and follicular fluid (FF) also contain factors that stabilize the organization of the COC matrix. It is well known that serum-derived glycoprotein, the IαI family, is pivotal for the formation of the COC matrix both in vivo (Hess et al., 1999; Zhuo et al., 2001) and in vitro (Chen et al., 1992). After gonadotrophin stimulation, the IαI family flows into the predominant antral follicles from the plasma due to increased permeability of the blood-follicle barrier and is accumulated in the FF (Hess et al., 1998). The IαI family consists of two homologous heavy chains (HCs), HC1 and HC2 (approximately 65 kDa and 70 kDa, respectively), that are covalently linked to one light chain of about 30 kDa (termed bikunin) and to a chondroitin sulfate chain. Additionally, inter-α-like inhibitor (IαLI) and pre-α-inhibitor (PαI) are complexes of one HC (HC2 or HC3) (about 90 kDa) and a bikunin associated with a chondroitin sulfate chain (Bost et al., 1998). These major members of the IαI family are synthesized by hepatocytes and circulate in the blood (0.15 to 0.5 mg/ml plasma) (Mizon et al., 1996). One or two HCs are connected to the chondroitin sulfate chain of bikunin via an ester bond that forms between the C-terminal aspartic acid residues of HCs and the N-acetylglucosamine residues of the chondroitin sulfate chain. On the other hand, the covalent linkage between HCs of the IαI family and HA, which is referred to as serum-derived HA-associated protein (SHAP)-HA complex, has been found in the HA-rich matrix of cultured mice dermal fibroblasts supplemented with serum and in pathological synovial fluid from human arthritis patients (Yoneda et al., 1990; Zhao et al., 1995). These findings suggest that the transesterification model of covalent binding between HCs of the Iαl family and HA involves the exchange of the chondroitin sulfate chain with HA at the HCs/chondroitin sulfate junction followed by the release of chondroitin sulfate-bikunin. Bikunin with chondroitin sulfate moiety is indispensable for the arrangement of the SHAP-HA complex (Zhuo et al., 2001). In mice studies, it has been demonstrated that HCs of the Iαl family are covalently linked to HA and are bound to the COC matrix in ovulated COCs, while the Iαl family is incorporated into the HA-enriched matrix by a non-covalent mechanism in *in vitro* expanded COCs, even if serum is added to the medium (Chen et al., 1994b; 1996). Conversely, Nagyova et al. (2004) showed covalent transfer of HCs of the Iαl family to HA in both *in vivo* and *in vitro* expanded porcine COC matrices cultured with porcine serum and FF. A granulosa cell secreted factor and FF may contain the essential components that catalyze the covalent binding of HCs and HA (Chen et al., 1996; Odum et al., 2002). On the cumulus cell surface, the interaction among HA, PGs such as aggrecan and versican, extracellular matrix HA-binding proteins including link protein, TNFIP-6, the Iαl family and additional unknown factors appears to induce and strengthen COC expansion and lead to ovarian follicle maturation.

### Features of meiotic and cytoplasmic maturation

The two kinds of haploid germ cells, the spermatozoon and the oocyte, resulted from two consecutive divisions, leading to the generation of a new diploid progeny after successful fertilization. After meiosis, the four haploid gametes are unlikely to be genetically identical on account of homologous recombination at the diplotene stage of the first meiotic prophase (prophase I). Interestingly, only one haploid gamete is produced as a result of two asymmetric divisions in female mammals compared with four in males. The oocytes complete meiosis over an extended time of their life span including two meiotic arrests (Khan-Dawood, 2003) (Fig. 1). The first meiotic cell cycle starts either before birth (rodents, human, cow, sheep) or shortly after birth (hamster, ferret, dog). The mechanisms involved in the triggering of meiotic initiation are not understood; however, it has been proposed that a gradual diffusion or reduction of meiosis-initiating or preventing factor in the central-periphery area of the ovary activates meiosis (Byskov and Nielsen, 2003). The oocytes interrupt meiosis at the G2 phase of the cell cycle, the diplotene/dictyate stage of prophase I commonly known as the GV stage and remain transcriptionally and translationally inactive until puberty. After sexual maturity, fully-grown oocytes undergo the latter half of meiosis via LH signalling shortly before ovulation, which is again interrupted at the MII stage while awaiting fertilization, this is termed oocyte maturation. The MII oocytes have acquired fertilizability and complete developmental potential as a result of cellular events, including both nuclear and cytoplasmic alteration (Eppig, 1996). The traits of nuclear maturation, also termed meiotic maturation are as follows: 1) Following the surge of pituitary LH, meiotic resumption is morphologically characterized by GV breakdown (GVBD), chromosome condensation and spindle formation. 2) Organization of the bipolar spindle, separation of homologous chromosomes and exclusion of the polar body take place, described as metaphase I (MI), anaphase I and telophase I, respectively. 3) Transition to meiosis II occurs without an intermediate phase of DNA replication but is arrested at the MII stage until fertilization. On the other hand, cytoplasmic maturation involves metabolic and structural changes in the individual organelles, accompanied by the synthesis of biochemical molecules and their modification, to ensure precise meiosis, normal fertilization and subsequent development. Thus, unlike meiosis in the male, the process of oocyte maturation influences oocyte quality and the normal developmental potential of the embryo.

It is well known that maturation promoting factor (MPF), a complex formed by cyclin B and P34$^{cdc2}$ kinase, is a key regulator of M phase, resulting in meiotic resumption (Sun and Nagai, 2003; van den Hurk and Zhao, 2005). MPF is activated at GVBD and increases until it reaches a plateau at the end of the first meiotic M phase. A transient decline in MPF activity takes place during transition between MI and MII. MPF is reactivated rapidly to enter MII and is maintained at a high level during MII arrest. The equilibrium between pre-MPF and MPF is universally controlled at the post-transcriptional level by the balance between double specificity kinases that phosphorylates residues Thr14 and Tyr15 of Cdc2 and maintains cyclin B-Cdc2 complexes in the inactive pre-MPF form, and the dual specificity phosphatase Cdc25, which dephosphorylates these inhibitory residues and converts pre-MPF into active MPF.

## Preservation at arrest of the GV stage

Even if the fully-grown oocytes in the antrum or secondary follicles obtain complete maturation potential, they remain at the GV stage, awaiting the LH surge. In comparison with *in vivo* meiotic resumption in response to LH signalling, it is also stimulated by liberation of COC from the follicle into a suitable culture medium (Edwards, 1965). These facts imply that the follicular environment affects the oocyte in maintaining meiotic arrest at the GV stage with an inhibitory factor. It is considered that some factors such as oocyte-maturation inhibitor (OMI) are produced by granulosa cells, accumulate in follicular fluid and sustain meiotic arrest in preovulatory follicles (Tsafriri and Channing, 1975; Sato and Koide, 1984). However, the biochemical properties of this protein remain unknown. Another potential mechanism is that a meiotic inhibitor such as cAMP is transmitted from the follicular cells to the immature oocytes efficiently through the gap junctions. It has been demonstrated that a rise in intracellular cAMP using analogues of cAMP, folskolin, cAMP phosphodiesterase inhibitor, such as 3-isobutyl-1-methylxantine (IBMX) and hypoxanthine, blocks spontaneous *in vitro* maturation (Dekel and Kraicer, 1978). Also, purines such as hypoxanthine in follicular fluid, have been proposed as a candidate for this inhibitory activity (Eppig *et al.*, 1985). Thus, cAMP is now well established as the major inhibitory substance and its inhibitory action has been demonstrated (Aberdam *et al.*, 1987; Dekel *et al.*, 1988). Recently, an alternative hypothesis is suggested that the oocyte produces its own cAMP through G-protein coupled receptors (GPCRs) in oocyte plasma membrane. Mehlmann *et al.* (2004) have found that maintenance of GV arrest in fully grown oocytes requires the heterotrimetic G protein Gs and then have identified the oocyte G-protein coupled receptors GPR3 that activates Gs. Gs stimulates adenylyl cyclase (AC) in the oocyte to keep cAMP elevated (Horner *et al.*, 2003).

## Signal transduction for GVBD in oocyte meiotic maturation

The LH surge induced by increased levels of serum estradiol rapidly acts on each structural cell of Graafian follicles *in vivo*, including meiotic maturation of oocytes, expansion of cumulus cells, differentiation of mural granulosa cells with alternation of the expression of genes required for ovulation and luteinization. Although the signalling pathways for LH in oocyte and follicular cells affecting meiotic resumption remain unclear, it most likely involves a decrease in oocyte cAMP level, cessation of production of meiosis inhibitory molecules by follicular cells and stimulation of synthesis or activation of meiosis inducing/promoting molecules. Recently, the understanding of several signalling cascades that lead to GVBD has been improved as described below (Fig. 2).

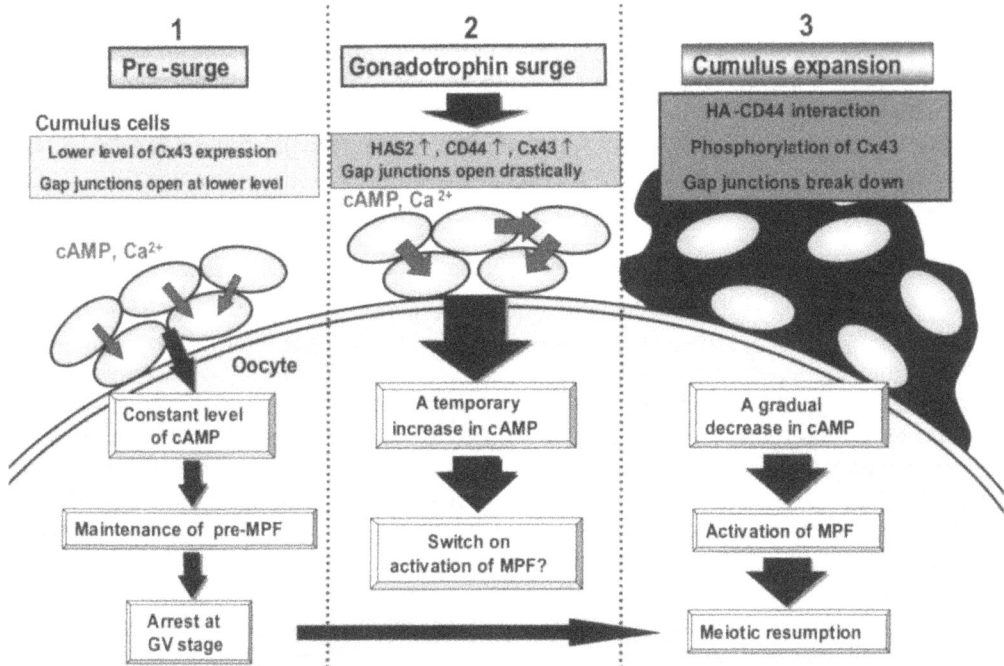

**Fig. 3** A schematic diagram of the changes in gap junctions consist of $C \times 43$ accompanied by cumulus expansion, and timing of GVBD.

*PKA signalling pathway*

The downstream cascade(s) by which an adequate cAMP concentration within the oocyte regulates meiotic arrest has been partially elucidated; ultimately, a decrease in oocyte cAMP concentrations precedes and leads to meiotic resumption accompanied by GVBD. The uncoupling of cumulus cells from the oocytes by interruption of gap junctions accompanied by cumulus expansion may block elevation of intraoocyte cAMP concentrations and permit reinitiation of meiotic maturation (Larsen *et al.*, 1986; Thomas *et al.*, 2004). However, it has been found that cell-cell communication in COCs remains until after GVBD (Eppig, 1982; Motlik *et al.*, 1986). Moreover, later studies have shown an increase rather than a decrease in cAMP facilitates induction of meiotic maturation (Yoshimura *et al.*, 1992). *In vitro* studies have indicated that FSH induces an increase in cAMP in the cumulus cells and results in an increase in cAMP in the oocyte via diffusion from the somatic cells to the oocyte through gap junctions (Webb *et al.*, 2002). Thus, it is hypothesized that a certain concentration of cAMP is maintained in GV oocytes, while a transient increase in cAMP induced by hormonal stimulation is likely to trigger GVBD (Dekel *et al.*, 1988). The drastic change in cAMP may be an important stimulus for reinitiation of meiosis, but not important in the absolute level of cAMP reached in the oocyte. The action of cAMP within oocytes is mediated by cAMP-dependent protein kinase (PKA).

Downs and Hunzicker-Dunn (1995) have demonstrated that two major isozymes of PKA are involved in opposing functions of meiotic regulation in COCs. They suggest that elevation of type I PKA within the oocyte is related to maintaining meiotic arrest while type II PKA mediates cAMP-stimulated cumulus expansion and meiotic resumption. Additionally, cyclic nucleotide phosphodiesterases (PDEs) that degrade and inactivate cAMP are important enzymes controlling PKA activity. Phosphodiesterase 4 is present in cumulus, mural granulosa cells (PDE4D) and theca cells (PDE4B) and PDE3A is present in oocytes (Conti *et al.*, 2002). *Pde3a*-deficient or PDE3 inhibitor treated oocytes do not induce meiotic maturation even after ovulation or after prolonged incubation *in vitro* (Masciarelli *et al.*, 2004). The accumulation of 5'-AMP that is the product of PDE activity stimulates AMP-activated protein kinase (AMPK), leading to meiotic resumption (Downs *et al.*, 2002).

*Phosphatidylinositol 3-kinase (PI3-K) - Akt signalling pathway in meiotic maturation*

PI3-K is known to play critical roles in signal transduction processes related to a variety of cellular activities such as cytoskeletal re-arrangement, cellular migration, differentiation, protection against apoptosis and mitogenesis (Franke *et al.*, 1997). Previously, we reported that PI3-K participates in FSH-induced cumulus expansion and meiotic maturation in mouse oocytes but not spontaneous maturation (Hoshino *et al.*, 2004). Akt also known as protein kinase B, has been identified as a serine/threonine kinase. The activation of Akt is thought to be a critical step in the PI3-K pathway that regulates cell growth and differentiation. In fully-grown mouse oocytes, a decrease in cAMP concentration precedes and is linked to Cdc2 activation. In mouse oocytes, Akt is involved in Cdc2 activation and resumption of meiosis (Kalous *et al.*, 2006). We examined the distribution of phosphorylated Akt during meiotic maturation. Thr308 phosphorylated Akt was localized in pericentiolar material at MI and MII. In contrast, the distribution of Ser473 phosphorylated Akt was similar to that of microtubules at prometaphase I (PMI) and was localized in spindles at MI and MII. The activity of Akt would be related to spindle formation in mouse oocytes. When COCs were treated with PI3-K inhibitor (LY294002) in FSH-induced meiotic maturation, the amount of Thr308 phosphorylated Akt was decrease to very low to undetectable level in PMI, MI and MII oocytes. The distribution of Ser473 phosphorylated Akt in LY294002-treated PMI oocytes was similar to that in normal PMI oocytes, whereas aberrant distribution and very low to undetectable level of expression were seen in LY294002-treated MI and MII oocytes, respectively. These results suggest that Akt activity participates in gonadotrophin-induced meiotic maturation as a downstream effector of the PI3-K pathway in mouse oocytes. Tomek and Smiljakovic (2005) have demonstrated that Akt is also involved in the MI to MII transition in bovine oocytes.

*Mitogen activated protein kinase (MAPK) pathway*

The MAPKs, also termed extracellular-regulated kinases (ERKs), are another family of serine/threonine protein kinases that are involved in meiosis (Yokoo and Sato, 2004; van den Hurk and Zhao, 2005). There are two isoforms, ERK1 (44 kDa) and ERK2 (42 kDa), which are more abundantly expressed in both the oocytes and cumulus cells at the later stage of meiosis. MAPKs are activated by an upstream kinase identified as a dual specific MAPK kinase (MAPKK) or MAPK/ERK kinase (MEK). The activation of MAPK and MEK during oocyte maturation has been reported in several species: marine invertebrate, *Xenopus* and mammals. The proto-oncogene MOS, which is a member of the serine/threonine protein kinase family, appears to be required for activation of MAPK. In *Xenopus* oocytes, it is well established that MOS/MAPKs is crucial for the activation of MPF and GVBD through a progesterone-induced cAMP-PKA decrease (Duckworth *et al.*, 2002). The oocytes from the *c-mos* knockout mouse had a normal level of MPF activity despite deletion of the MAPK activity and had simply undergone GVBD. However, they produced an abnormally large polar body and showed irregular spindle formation. Furthermore, when some oocytes from *c-mos* knockout mice that reached the MII stage were fertilized or stimulated by ethanol, they emitted a second polar body and progressed into a third meiotic metaphase. These results have demonstrated that the Mos/MAPK pathway is independent of MPF activity and is not essential for GVBD but plays a critical role in normal spindle and chromosome morphology in mice (Araki et al., 1996; Choi et al., 1996). It is now well accepted that the Mos/MAPK pathway is responsible for proper spindle formation at MI and MII, repression of DNA replication at the MI-MII transition and maintenance of the MII arrest.

Presumptive roles of cumulus expansion in oocyte meiotic maturation

In the process of cumulus expansion, the rearrangement of microfilaments in the cumulus cell cytoplasm occurs before preceding the acceleration of ECM synthesis, the redistribution of microtubules and intermediate filaments and the change of the gap junctional pathway. Finally, it results in the apparent withdrawal of microfilament-filled transzonal projections of cumulus cells (Allworth and Albertini, 1993; Sutovsky et al., 1993; Suzuki et al., 2000). Therefore, it is generally accepted that intercellular communication does not occur in expanded COCs. The loss of gap junctions between cumulus cells and the oocyte during cumulus expansion are closely related to oocyte meiotic progression (Eppig, 1982; Larsen et al., 1986).

Each gap junction channel comprises two symmetrical hemispheres, named connexons, a hexamer of protein subunits called connexins (Cxs), derived from two neighbouring cells (Goodenough et al., 1996). Various kinds of Cxs are expressed in ovarian follicular cells, the presence of Cx32, Cx37, Cx43 and Cx45 have been identified in mouse COCs (Wright et al., 2001; Kidder and Mhawi, 2002). Cx37 is detected in the gap junction-like structure on the surface of oocytes, beneath the ZP. Cx32, Cx43 and Cx45 are found in corona radiata cell projections close to the oocyte surface. Particularly, Cx43 contributes through transzonal projections from the cumulus cell couples with oocyte Cx37 to form heterotypic junctions. Cx43 is predominantly expressed in cumulus-cumulus gap junctions. Studies of knockout mice have demonstrated that Cx37 and Cx43 seem to be essential at each step for normal folliculogenesis (Simon et al., 1997; Carabatsos et al., 2000; Ackert et al., 2001). In Cx37 deficient mice, follicular growth fails at the preantral-antral transition and oocytes are unable to initiate meiotic maturation (Simon et al., 1997; Carabatsos et al., 2000). To investigate folliculogenesis in Cx43 deficient mice, neonatal ovaries were grafted into the kidney capsules of ovariecto-mized, immunocompromised adult mice and allowed to develop for up to 3 weeks (Ackert et al., 2001). However, most mutant follicles failed to become multilaminar, development being arrested in an early preantral stage. Correspondingly, the mutant oocytes failed to undergo meiotic maturation and could not be fertilized. Vozzi et al. (2001) have shown the physiological role of Cx43 in meiotic maturation by inhibition of Cx43 expression using a recombinant adenovirus expressing the antisense Cx43 cDNA (Ad-asCx43). The rate of GVBD was decreased by about 50% of COCs infected with Ad-asCx43. On the contrary, 89% of COCs infected with Ad-GFP (positive control) matured to MII. This evidence suggests that the completion of oocyte growth and the acquisition of meiotic competence are wholly supported by the gap junctional communications between the oocyte and its cumulus cells.

Recently, our studies have focused on the effect of HA-CD44 interaction on the modification of gap junctional channels during cumulus expansion (Yokoo and Sato, 2004). The ubiquitous HA receptor, CD44, is a member of the link module superfamily containing a common structural domain which interacts with HA to form complexes that stabilize the extracellular matrices (Day and Prestwich, 2002) and displays numerous isoforms because of alternative splicing of 12 variant exons in different combinations (Jackson et al., 1992; Screaton et al., 1992). HA-CD44 interaction has been reported to result in the activation of signalling cascades that contribute to cell adhesion, proliferation, migration and differentiation (Knudson and Knudson, 1993; Turley et al., 2002). The expression of CD44 in porcine COCs during in vitro maturation has been identified (Kimura et al., 2002; Yokoo et al., 2002; Yokoo and Sato, 2004). CD44 was distributed on the cytoplasm along the perimembrane of cumulus cells and at the connections between cumulus cells and oocytes as shown by immunostaining. Only a single band (about 85 to 90 kDa) of CD44 standard was detected in numerous isoforms of different molecular sizes from 80 to 250 kDa, using Western blotting analysis (Yokoo et al., 2002). CD44 was identified from the cumulus cell extracts, but not in oocyte extracts. The level of CD44 expression reached a peak at 24 h of culture although its expression was very weak at 0 h (Kimura et al., 2002), implying that the level of CD44 expression depends on the degree of COC expansion. Furthermore, Yokoo et al. (2003) have investigated the effect of HA-CD44 interaction on MPF and GVBD (Fig. 3). In this study, porcine COCs were cultured with 6-diazo-5-oxo-L-norleucine (DON) or anti-CD44 antibody for 24 h. After culture for 24 h, the MPF activity and the rate of GVBD were significantly decreased by the treatment of DON and anti-CD44 antibody, suggesting that HA-CD44 interaction might have promoted the meiotic resumption. Furthermore, HA-CD44 interaction might affect the expression of Cx43. Exposure of COCs to DON and anti-CD44 antibody had no effect on the expression level of Cx43; however, they notably inhibited the tyrosine phosphorylation of Cx43. When Cx43 is phosphorylated on tyrosine residues the intercellular junctional communication is inhibited. Also, HA-CD44 interaction is involved in the phosphorylation of MAPK in oocytes. The treatment of DON or anti-CD44 antibody down-regulated MAPK phosphorylation (N. Kimura, unpublished). LH, a stimulator of meiotic maturation in vivo, induces down-regulation of Cx43 expression and phosphorylation through PKA and MAPK cascades (Granot and Dekel, 1994; Kalma et al., 2004). This evidence indicates that HA-CD44 interactions are likely to be involved in the interruption of Cx43-derived gap junctional channels and may lead to inhibition of the transfer of cAMP from the cumulus cells to the oocytes.

## Concluding remarks

We have described molecules of cumulus-oocyte complex matrix and signalling pathways regulating meiotic progression. The degree of cumulus expansion seems to be a useful parameter to predict development of oocyte, maturation and fertilization in vitro (Vanderhyden and Armstrong, 1989; Qian et al., 2003), suggesting that the morphology of COC matrix mass is important for nuclear and cytoplasmic maturation. It is hoped that increased understanding gained from advances in cellular and molecular biology during oocyte maturation in mammals will lead to a system by which fertilization and the development of embryos in vivo can be sufficiently realized, such a system being applicable to gamete biotechnology.

## References

Aberdam E, Hanski E and Dekel N (1987) Maintenance of meiotic arrest in isolated rat oocytes by the invasive adenylate cyclase of Bordetella pertussis Biology of Reproduction 36 530-535

Ackert CL, Gittens JE, O'Brien MJ, Eppig JJ and Kidder GM (2001) Intercellular communication via connexin 43 gap junctions is required for ovarian folliculogenesis in the mouse Developmental Biology 233 258-270

Albertini DF, Combelles CM, Benecchi E and Carabatsos MJ (2001) Cellular basis for paracrine regulation of ovarian follicle development Reproduction 121 647-653

Allworth AE and Albertini DF (1993) Meiotic maturation in cultured bovine oocytes is accompanied by remodelling of the cumulus cell cytoskeleton Developmental Biology 158 101-112

Araki K, Naito K, Haraguchi S, Suzuki R, Yokoyama M, Inoue M, Aizawa S, Toyoda Y and Sato E (1996) Meiotic abnormalities of c-mos knockout mouse oocytes: activation after first meiosis or entrance into third meiotic metaphase *Biology of Reproduction* **55** 1315-1324

**Bost F, Diarra-Mehrpour M and Martin JP** (1998) Inter-alpha-trypsin inhibitor proteoglycan family-a group of proteins binding and stabilizing the extracellular matrix *European Journal of Biochemistry* **252** 339-346

**Buccione R, Vanderhyden BC, Caron PJ and Eppig JJ** (1990) FSH-induced expansion of the mouse cumulus oophorus *in vitro* is dependent upon a specific factor(s) secreted by the oocyte *Developmental Biology* **138** 16-25

**Byskov AG and Nielsen M** (2003) Ontogeny of the mammalian ovary. In *Biology and Pathology of the Oocyte*. pp13-28 Eds AO Trounson and RG Gosden. Cambridge, UK

**Camaioni A, Salustri A, Yanagishita M and Hascall VC** (1996) Proteoglycans and proteins in the extracellular matrix of mouse cumulus cell-oocyte complexes *Archives of Biochemistry and Biophysics* **325** 190-198

**Carabatsos MJ, Sellitto C, Goodenough DA and Albertini DF** (2000) Oocyte-granulosa cell heterologous gap junctions are required for the coordination of nuclear and cytoplasmic meiotic competence *Developmental Biology* **226** 167-179

**Carrette O, Nemade RV, Day AJ, Brickner A and Larsen WJ** (2001) TSG-6 is concentrated in the extracellular matrix of mouse cumulus oocyte complexes through hyaluronan and inter-alpha-inhibitor binding *Biology of Reproduction* **65** 301-308

**Chen L, Mao SJ and Larsen WJ** (1992) Identification of a factor in fetal bovine serum that stabilizes the cumulus extracellular matrix. A role for a member of the inter-alpha-trypsin inhibitor family *The Journal of Biological Chemistry* **267** 12380-12386

**Chen L, Russell PT and Larsen WJ** (1994a) Sequential effects of follicle-stimulating hormone and luteinizing hormone on mouse cumulus expansion *in vitro* Biology of Reproduction **51** 290-295

**Chen L, Mao SJ, McLean LR, Powers RW and Larsen WJ** (1994b) Proteins of the inter-alpha-trypsin inhibitor family stabilize the cumulus extracellular matrix through their direct binding with hyaluronic acid *The Journal of Biological Chemistry* **269** 28282-28287

**Chen L, Zhang H, Powers RW, Russell PT and Larsen WJ** (1996) Covalent linkage between proteins of the inter-α-trypsin inhibitor family and hyaluronic acid is mediated by a factor produced by granulosa cells *The Journal of Biological Chemistry* **271** 19409-19414

**Choi T, Fukasawa K, Zhou R, Tessarollo L, Borror K, Resau J and Vande Woude GF** (1996) The Mos/mitogen-activated protein kinase (MAPK) pathway regulates the size and degradation of the first polar body in maturing mouse oocytes *Proceedings of the National Academy of Sciences USA* **93** 7032-7035

**Conti M, Andersen CB, Richard F, Mehats C, Chun SY, Horner K, Jin C and Tsafriri A** (2002) Role of cyclic nucleotide signaling in oocyte maturation *Molecular and Cellular Endocrinology* **187** 153-159

**Day AJ and Prestwich GD** (2002) Hyaluronan-binding proteins: tying up the giant *The Journal of Biological Chemistry* **277** 4585-4588

**Dekel N and Kraicer PF** (1978) Induction *in vitro* of mucification of rat cumulus oophorus by gonadotrophins and adenosine 3', 5'-monophosphate *Endocrinology* **102** 1797-1802

**Dekel N and Beers WH** (1978) Rat oocyte maturation *in vitro*: relief of cyclic AMP inhibition by gonadotropins *Proceedings of the National Academy of Sciences USA* **75** 4369-4373

**Dekel N, Galiani D and Sherizly I** (1988) Dissociation between the inhibitory and the stimulatory action of cAMP on maturation of rat oocytes *Molecular and Cellular Endocrinology* **56** 115-121

**Downs SM and Hunzicker-Dunn M** (1995) Differential regulation of oocyte maturation and cumulus expansion in the mouse oocyte-cumulus cell complex by site-selective analogs of cyclic adenosine monophosphate *Developmental Biology* **172** 72-85

**Downs SM, Hudson ER and Hardie DG** (2002) A potential role for AMP-activated protein kinase in meiotic induction in mouse oocytes *Developmental Biology* **245** 200-212

**Duckworth BC, Weaver JS and Ruderman JV** (2002) G2 arrest in *Xenopus* oocytes depends on phosphorylation of cdc25 by protein kinase A *Proceedings of the National Academy of Sciences USA* **99** 16794-16799

**Edwards RG** (1965) Maturation *in vitro* of mouse, sheep, cow, pig, rhesus monkey and human ovarian oocytes *Nature* **208** 349-351

**Elvin JA, Clark AT, Wang P, Wolfman NM and Matzuk MM** (1999) Paracrine actions of growth differentiation factor-9 in the mammalian ovary *Molecular Endocrinology* **13** 1035-1048

**Eppig JJ** (1979) FSH stimulates hyaluronic acid synthesis by oocyte-cumulus cell complexes from mouse preovulatory follicles *Nature* **281** 483-484

**Eppig JJ** (1982) The relationship between cumulus cell-oocyte coupling, oocyte meiotic maturation and cumulus expansion *Developmental Biology* **89** 268-272

**Eppig JJ** (1996) Coordination of nuclear and cytoplasmic oocyte maturation in eutherian mammals *Reproduction Fertility and Development* **8** 485-489

**Eppig JJ** (2001) Oocyte control of ovarian follicular development and function in mammals *Reproduction* **122** 829-838

**Eppig JJ, Ward-Bailey PF and Coleman DL** (1985) Hypoxanthine and adenosine in murine ovarian follicular fluid: concentrations and activity in maintaining oocyte meiotic arrest *Biology of Reproduction* **33** 1041-1049

**Franke TF, Kaplan DR and Cantley LC** (1997) PI3K: downstream AKTion blocks apoptosis *Cell* **88** 435-437

**Fulop C, Salustri A and Hascall VC** (1997a) Coding sequence of a hyaluronan synthase homologue expressed during expansion of the mouse cumulus-oocyte complex *Archives of Biochemistry and Biophysics* **337** 261-266

**Fulop C, Kamath RV, Li Y, Otto JM, Salustri A, Olsen BR, Glant TT and Hascall VC** (1997b) Coding sequence, exon-intron structure and chromosomal localization of murine TNF-stimulated gene 6 that is specifically expressed by expanding cumulus cell-oocyte complexes *Gene* **202** 95-102

**Fulop C, Szanto S, Mukhopadhyay D, Bardos T, Kamath RV, Rugg MS, Day AJ, Salustri A, Hascall VC, Glant TT and Mikecz K** (2003) Impaired cumulus mucification and female sterility in tumor necrosis factor-induced protein-6 deficient mice *Development* **130** 2253-2261

**Gilchrist RB, Ritter LJ and Armstrong DT** (2004) Oocyte-somatic cell interactions during follicle development in mammals *Animal Reproduction Science* **82-83** 431-446

**Goodenough DA, Goliger JA and Paul DL** (1996) Connexins, connexons and intercellular communication *Annual Review of Biochemistry* **65** 475-502

**Granot I and Dekel N** (1994) Phosphorylation and expression of connexin-43 ovarian gap junction protein are regulated by luteinizing hormone *The Journal of Biological Chemistry* **269** 30502-30509

**Hess KA, Chen L and Larsen WJ** (1998) The ovarian blood follicle barrier is both charge- and size-selective in mice *Biology of Reproduction* **58** 705-711

**Hess KA, Chen L and Larsen WJ** (1999) Inter-alpha-inhibitor binding to hyaluronan in the cumulus extracellular matrix is required for optimal ovulation and development of mouse oocytes *Biology of Reproduction* **61** 436-443

**Hinckley M, Vaccari S, Horner K, Chen R and Conti M** (2005) The G-protein-coupled receptors GPR3 and GPR12 are involved in cAMP signaling and maintenance of meiotic arrest in rodent oocytes *Developmental Biology* **287** 249-261

**Horner K, Livera G, Hinckley M, Trinh K, Storm D and Conti M** (2003) Rodent oocytes express an active adenylyl cyclase required for meiotic arrest *Developmental Biology* **258** 385-396

**Hoshino Y, Yokoo M, Yoshida N, Sasada H, Matsumoto H and Sato E** (2004) Phosphatidylinositol 3-kinase and Akt participate in the FSH-induced meiotic maturation of mouse oocytes *Molecular Reproduction and Development* **69** 77-86

**Izadyar F, Colenbrander B and Bevers MM** (1996) *In vitro* maturation of bovine oocytes in the presence of growth hormone accelerates nuclear maturation and promotes subsequent embryonic development *Molecular Reproduction and Development* **45** 372-377

**Jackson DG, Buckley J and Bell JI** (1992) Multiple variants of the human lymphocyte homing receptor CD44 generated by insertions at a single site in the extracellular domain *The Journal of Biological Chemistry* **267** 4732-4739

**Kalma Y, Granot I, Galiani D, Barash A and Dekel N** (2004) Luteinizing hormone-induced connexin 43 down-regulation: inhibition of translation *Endocrinology* **145** 1617-1624

**Kalous J, Solc P, Baran V, Kubelka M, Schultz RM and Motlik J** (2006) PKB/AKT is involved in resumption of meiosis in mouse oocytes *Biology of the Cell* **98** 111-123

**Khan-Dawood FS** (2003) The ovarian cycle. In *Introduction to Mammalian Reproduction*. pp. 155-186 Ed DR Tulsiani. Kluwer Academic Publisher, Boston

**Kidder GM and Mhawi AA** (2002) Gap junctions and ovarian folliculogenesis *Reproduction* **123** 613-620

**Kimura N, Konno Y, Miyoshi K, Matsumoto H and Sato E** (2002) Expression of hyaluronan synthases and CD44 messenger RNAs in porcine cumulus-oocyte complexes during *in vitro* maturation *Biology of Reproduction* **66** 707-717

**Knudson CB and Knudson W** (1993) Hyaluronan-binding proteins in development, tissue homeostasis and disease *The FASEB Journal* **7** 1233-1241

**Kobayashi H, Sun GW, Hirashima Y and Terao T** (1999) Identification of link protein during follicle development and cumulus cell cultures in rats *Endocrinology* **140** 3835-3842

**Larsen WJ, Wert SE and Brunner GD** (1986) A dramatic loss of cumulus cell gap junctions is correlated with germinal vesicle breakdown in rat oocytes *Developmental Biology* **113** 517-521

**McDonald JA and Camenisch TD** (2002) Hyaluronan: genetic insights into the complex biology of a simple polysaccharide *Glycoconjugate Journal* **19** 331-339

**Meduri G, Charnaux N, Driancourt MA, Combettes L, Granet P, Vannier B, Loosfelt H and Milgrom E** (2002) Follicle-stimulating hormone receptors in oocytes? *Journal of Clinical Endocrinology and Metabolism* **87** 2266-2276

**Mehlmann LM, Saeki Y, Tanaka S, Brennan TJ, Evsikov AV, Pendola FL, Knowles BB, Eppig JJ and Jaffe LA** (2004) The Gs-linked receptor GPR3 maintains meiotic arrest in mammalian oocytes *Science* **306** 1947-1950

**Masciarelli S, Horner K, Liu C, Park SH, Hinckley M, Hockman S, Nedachi T, Jin C, Conti M and Manganiello V** (2004) Cyclic nucleotide phosphodiesterase 3A-deficient mice as a model of female infertility *The Journal of Clinical Investigation* **114** 196-205

**Mizon C, Balduyck M, Albani D, Michalski C, Burnouf T and Mizon J** (1996) Development of an enzyme-linked immunosorbent assay for human plasma inter-alpha-trypsin inhibitor (ITI) using specific antibodies against each of the H1 and H2 heavy chains *Journal of Immunological Methods* **190** 61-70

**Motlik J, Fulka J and Flechon JE** (1986) Changes in intercellular coupling between pig oocytes and cumulus cells during maturation *in vivo* and *in vitro* *Journal of Reproduction and Fertility* **76** 31-37

**Mukhopadhyay D, Hascall VC, Day AJ, Salustri A and Fulop C** (2001) Two distinct populations of tumor necrosis factor-stimulated gene-6 protein in the extracellular matrix of expanded mouse cumulus cell-oocyte complexes *Archives of Biochemistry and Biophysics* **394** 173-181

**Mukhopadhyay D, Asari A, Rugg MS, Day AJ and Fulop C** (2004) Specificity of the tumor necrosis factor-induced protein 6-mediated heavy chain transfer from inter-alpha-trypsin inhibitor to hyaluronan: implications for the assembly of the cumulus extracellular matrix *The Journal of Biological Chemistry* **279** 11119-11128

**Nagyova E, Camaioni A, Prochazka R and Salustri A** (2004) Covalent transfer of heavy chains of inter-α-trypsin inhibitor family proteins to hyaluronan in *in vivo* and *in vitro* expanded porcine oocyte-cumulus complexes *Biology of Reproduction* 71 1838-1843

Nakayama T, Inoue M and Sato E (1996) Effect of oocytectomy on glycosaminoglycan composition during cumulus expansion of porcine cumulus-oocyte complexes cultured *in vitro Biology of Reproduction* 55 1299-1304

Odum L, Andersen CY and Jessen TE (2002) Characterization of the coupling activity for the binding of inter-alpha-trypsin inhibitor to hyaluronan in human and bovine follicular fluid *Reproduction* 124 249-257

Park JY, Su YQ, Ariga M, Law E, Jin SL and Conti M (2004) EGF-like growth factors as mediators of LH action in the ovulatory follicle *Science* 303 682-684

Patsoula E, Loutradis D, Drakakis P, Kallianidis K, Bletsa R and Michalas S (2001) Expression of mRNA for the LH and FSH receptors in mouse oocytes and preimplantation embryos *Reproduction* 121 455-461

Philipson LH and Schwartz NB (1984) Subcellular localization of hyaluronate synthetase in oligodendroglioma cells *The Journal of Biological Chemistry* 259 5017-5023

Prochazka R, Kalab P and Nagyova E (2003) Epidermal growth factor-receptor tyrosine kinase activity regulates expansion of porcine oocyte-cumulus cell complexes *in vitro Biology of Reproduction* 68 797-803

Qian Y, Shi WQ, Ding JT, Sha JH and Fan BQ (2003) Predictive value of the area of expanded cumulus mass on development of porcine oocytes matured and fertilized *in vitro The Journal of Reproduction and Development* 49 167-174

Richards JS and Rolfes AI (1980) Hormonal regulation of cyclic AMP binding to specific receptor proteins in rat ovarian follicles. Characterization by photoaffinity labeling *The Journal of Biological Chemistry* 255 5481-5489

Russell DL, Ochsner SA, Hsieh M, Mulders S and Richards JS (2003) Hormone-regulated expression and localization of versican in the rodent ovary *Endocrinology* 144 1020-1031

Salustri A, Yanagishita M and Hascall VC (1989) Synthesis and accumulation of hyaluronic acid and proteoglycans in the mouse cumulus cell-oocyte complex during follicle-stimulating hormone-induced mucification *The Journal of Biological Chemistry* 264 13840-13847

Salustri A, Yanagishita M and Hascall VC (1990) Mouse oocytes regulate hyaluronic acid synthesis and mucification by FSH-stimulated cumulus cells *Developmental Biology* 138 26-32

Salustri A, Yanagishita M, Underhill CB, Laurent TC and Hascall VC (1992) Localization and synthesis of hyaluronic acid in the cumulus cells and mural granulosa cells of the preovulatory follicle *Developmental Biology* 151 541-551

Salustri A, Camaioni A, Giacomo MD, Fulop C and Hascall VC (1999) Hyaluronan and proteoglycans in ovarian follicles *Human Reproduction update* 5 293-301

Sato E and Koide SS (1984) A factor from bovine granulosa cells preventing oocyte maturation *Differentiation* 26 59-62

Schoenfelder M and Einspanier R (2003) Expression of hyaluronan synthases and corresponding hyaluronan receptors is differentially regulated during oocyte maturation in cattle *Biology of Reproduction* 69 269-277

Screaton GR, Bell MV, Jakson DG, Cornelis FB, Gerth U and Bell JI (1992) Genomic structure of DNA encoding the lymphocyte homing receptor CD44 reveals at least 12 alternatively spliced exons *Proceedings of the National Academy of Sciences USA* 89 12160-12164

Simon AM, Goodenough DA, Li E and Paul DL (1997) Female infertility in mice lacking connexin 37 *Nature* 385 525-529

Spicer AP and McDonald JA (1998) Characterization and molecular evolution of a vertebrate hyaluronan synthase gene family *The Journal of Biological Chemistry* 273 1923-1932

Stock AE, Bouchard N, Brown K, Spicer AP, Underhill CB, Dore M and Sirois J (2002) Induction of hyaluronan synthase 2 by human chorionic gonadotropin in mural granulosa cells of equine preovulatory follicles *Endocrinology* 143 4375-4384

Sun QY and Nagai T (2003) Molecular mechanisms underlying pig oocyte maturation and fertilization *The Journal of Reproduction and Development* 49 347-359

Sun GW, Kobayashi H, Suzuki M, Kanayama N and Terao T (2002) Link protein as an enhancer of cumulus cell-oocyte complex expansion *Molecular Reproduction and Development* 63 223-231

Sutovsky P, Flechon JE, Flechon B, Motlik J, Peynot N, Chesne P and Heyman Y (1993) Dynamic changes of gap junctions and cytoskeleton during *in vitro* culture of cattle oocyte cumulus complexes *Biology of Reproduction* 49 1277-1287

Suzuki H, Jeong BS and Yang X (2000) Dynamic changes of cumulus-oocyte cell communication during *in vitro* maturation of porcine oocytes *Biology of Reproduction* 63 723-729

Thomas RE, Armstrong DT and Gilchrist RB (2004) Bovine cumulus cell-oocyte gap junctional communication during *in vitro* maturation in response to manipulation of cell-specific cyclic adenosine 3',5'-monophosophate levels *Biology of Reproduction* 70 548-556

Tirone E, Siracusa G, Hascall VC, Frajese G and Salustri A (1993) Oocytes preserve the ability of mouse cumulus cells in culture to synthesize hyaluronic acid and dermatan sulfate *Developmental Biology* 160 405-412

Tirone E, D'Alessandris C, Hascall VC, Siracusa G and Salustri A (1997) Hyaluronan synthesis by mouse cumulus cells is regulated by interactions between follicle-stimulating hormone (or epidermal growth factor) and a soluble oocyte factor (or transforming growth factor beta1) *The Journal of Biological Chemistry* 272 4787-4794

Tomek W and Smiljakovic T (2005) Activation of Akt (protein kinase B) stimulates metaphase I to metaphase II transition in bovine oocytes *Repro-*

*duction* **130** 423-430

Tsafriri A and Channing CP (1975) An inhibitory influence of granulosa cells and follicular fluid upon porcine oocyte meiosis *in vitro Endocrinology* **96** 922-927

Turley EA, Noble PW and Bourguignon LY (2002) Signaling properties of hyaluronan receptors *The Journal of Biological Chemistry* **277** 4589-4592

van den Hurk R and Zhao J (2005) Formation of mammalian oocytes and their growth, differentiation and maturation within ovarian follicles *Theriogenology* **63** 1717-1151

Vanderhyden BC (1993) Species differences in the regulation of cumulus expansion by an oocyte-secreted factor(s) *Journal of Reproduction and Fertility* **98** 219-227

Vanderhyden BC and Armstrong DT (1989) Role of cumulus cells and serum on the *in vitro* maturation, fertilization and subsequent development of rat oocytes *Biology of Reproduction* **40** 720-728

Vozzi C, Formenton A, Chanson A, Senn A, Sahli R, Shaw P, Nicod P, Germond M and Haefliger JA (2001) Involvement of connexin 43 in meiotic maturation of bovine oocytes *Reproduction* **122** 619-628

Webb RJ, Marshall F, Swann K and Carroll J (2002) Follicle-stimulating hormone induces a gap junction-dependent dynamic change in [cAMP] and protein kinase A in mammalian oocytes *Developmental Biology* **246** 441-454

Wright CS, Becker DL, Lin JS, Warner AE and Hardy K (2001) Stage-specific and differential expression of gap junctions in the mouse ovary: connexin-specific roles in follicular regulation *Reproduction* **121** 77-88

Yokoo M and Sato E (2004) Cumulus-oocyte complex interactions during oocyte maturation *International Review of Cytology* **235** 251-291

Yokoo M, Miyahayashi Y, Naganuma T, Kimura N, Sasada H and Sato E (2002) Identification of hyaluronic acid-binding proteins and their expressions in porcine cumulus-oocyte complexes during *in vitro* maturation *Biology of Reproduction* **67** 1165-1171

Yokoo M, Kimura N, Shimizu T, Naganuma T, Matsumoto H, Sasada H and Sato E (2003) Induction of porcine oocyte maturation by hyaluronan-CD44 system *Journal of Fertilization and Implantation* **20** 33-36

Yoneda M, Suzuki S and Kimata K (1990) Hyaluronic acid associated with the surfaces of cultured fibroblasts is linked to a serum-derived 85-kDa protein *The Journal of Biological Chemistry* **265** 5247-5257

Yoshimura Y, Nakamura Y, Ando M, Jinno M, Oda T, Karube M, Koyama N and Nanno T (1992) Stimulatory role of cyclic adenosine monophosphate as a mediator of meiotic resumption in rabbit oocytes *Endocrinology* **131** 351-356

Zhao M, Yoneda M, Ohashi Y, Kurono S, Iwata H, Ohnuki Y and Kimata K (1995) Evidence for the covalent binding of SHAP, heavy chains of inter-alpha-trypsin inhibitor, to hyaluronan *The Journal of Biological Chemistry* **270** 26657-26663

Zhuo L and Kimata K (2001) Cumulus oophorus extracellular matrix: Its construction and regulation *Cell Structure and Function* **26** 189-196

Zhuo L, Yoneda M, Zhao M, Yingsung W, Yoshida N, Kitagawa Y, Kawamura K, Suzuki T and Kimata K (2001) Defect in SHAP-hyaluronan complex causes severe female infertility. A study by inactivation of the bikunin gene in mice *The Journal of Biological Chemistry* **276** 7693-7696

# Reversible phosphorylation and regulation of mammalian oocyte meiotic chromatin remodeling and segregation

JE Swain[1, 4] and GD Smith[1, 2, 3, 4, 5]

[1]Departments of Molecular and Integrative Physiology, [2]Obstetrics and Gynecology, [3]Urology, [4]Reproductive Sciences Program, University of Michigan, Ann Arbor, MI 48109-0617, USA

The mammalian oocyte is notorious for high rates of chromosomal abnormalities. This results in subsequent embryonic aneuploidy, resulting in infertility and congenital defects. Therefore, understanding regulatory mechanisms involved in chromatin remodeling and chromosome segregation during oocyte meiotic maturation is imperative to fully understand the complex process and establish potential therapies. This review will focus on major events occurring during oocyte meiosis, critical to ensure proper cellular ploidy. Mechanistic and cellular events such as chromosome condensation, meiotic spindle formation, as well as cohesion of homologues and sister chromatids will be discussed, focusing on the role of reversible phosphorylation in control of these processes.

## Introduction

It is estimated that 10-25% of all human conceptions are chromosomally abnormal due to an error in chromosome segregation during meiosis (Hunt, 1998; Hassold and Hunt, 2001). Furthermore, studies indicate greater than 95% of this human aneuploidy is attributable to defects within the meiotic machinery of the oocyte, rather than that of the sperm (Hassold and Hunt, 2001). Nevertheless, causative factors and intraoocyte biochemical signalling pathways causing these perturbations are unknown. Therefore, elucidating regulatory mechanisms involved in chromatin remodeling and chromosome segregation during oocyte meiotic maturation is imperative to fully understand this complex process and establish potential therapies to circumvent chromosomal non-disjunction, embryonic aneuploidy and resulting infertility and congenital defects.

When discussing fidelity of chromatin remodeling events to achieve proper separation and segregation of genetic material during meiosis, it is important to distinguish differences between the meiotic and mitotic cell cycles. Meiosis generates haploid sex cells, or gametes, through specialized cell divisions that includes one round of DNA replication and two rounds of division. The first round of division involves segregation of homologous chromosomes, while the second entails separation and segregation of sister chromatids. Thus, meiosis results in the production of haploid gametes with half the genetic material of somatic cells, whereas mitosis yields daughter cells with the normal diploid chromosomal complement.

E-mail: smithgd@med.umich.edu

[5]Correspondence: Gary D. Smith, 6428 Medical Sciences Bldg I, 1301 E Catherine St, Ann Arbor, MI, USA 48109-0617

Support: Grant support for research on this subject within the author's laboratory provided by HD35125-01A1 and HD046768-01A2 (GDS); HD044461 (Training Grant; JES).

Although the basic features of meiosis are conserved in males and females, there are important sex-specific differences, which are important when considering chromosome non-disjunction and aneuploidy. Specifically, during fetal development oocytes initiate meiosis, but at the time of birth, mammalian oocytes are arrested at prophase of meiosis I and remain arrested in this state until prior to ovulation. A surge of gonadotropins preceding ovulation releases the oocyte from its quiescent state and signals re-initiation of meiosis, indicated by dissolution of the nuclear envelope in a process known as germinal vesicle breakdown (GVBD). During completion of prophase of the first meiotic division, homologous chromosomes undergo a process of pairing and recombination. Homologues then condense and form metaphase chromosomes. A bipolar meiotic spindle forms and attaches to homologues at their centromeres. Subsequently, the physical contact between homologous pairs at chiasmata counteracts forces pulling the homologues apart, resulting in alignment of chromosomes along the metaphase plate, an event which signals completion of metaphase I (MI). The meiotic spindle then facilitates chromosome segregation as homologues are pulled towards opposite poles at the beginning of anaphase. The oocyte progresses through telophase, resulting in disproportionate cytokinesis and extrusion of half the genetic material in the first polar body. Finally, the oocyte proceeds directly to metaphase II (MII) of meiosis II, forgoing interphase and DNA duplication, where it remains until fertilization. Following fertilization, sister chromatid cohesion is released and chromatids separate and segregate, resulting in extrusion of the second polar body, signalling completion of meiosis II. This results in an oocyte with two polar bodies and a haploid female pronucleus capable of undergoing syngamy with the haploid male pronucleus, thus producing a diploid zygote (Fig. 1).

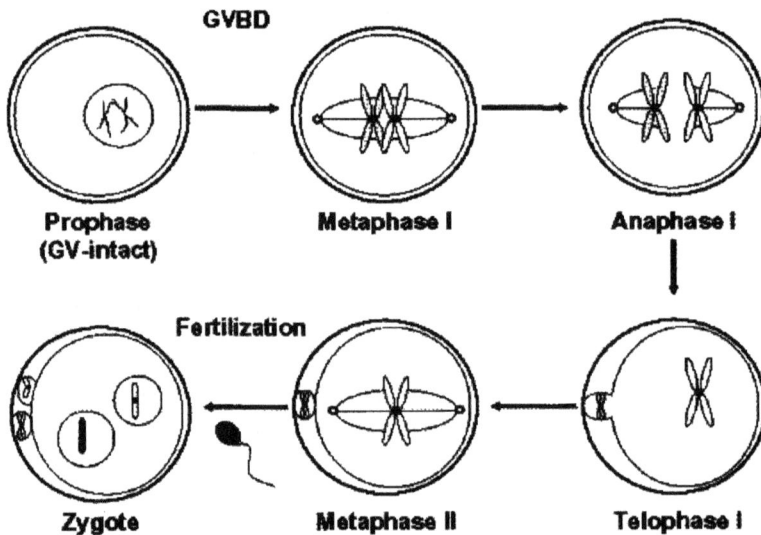

**Fig. 1** Prior to meiotic resumption, oocytes are arrested in prophase of meiosis I, also known as a GV-intact oocyte. In response to the preovulatory surge of gonadotropins, oocytes undergo dissolution of the nuclear envelope in a process known as germinal vesicle breakdown (GVBD). Oocytes then progress to metaphase I, where a meiotic spindle forms and condensed chromosomes align on the metaphase plate. Oocytes proceed to anaphase, where homologue cohesion is released and then to telophase where the oocyte undergoes disproportionate cytokinesis and extrusion of the first polar body. The oocyte re-arrests at metaphase II until fertilization occurs. Following fertilization, the oocyte undergoes completion of the second meiotic division, extruding the second polar body and undergoing syngamy of the male and female pronuclei to yield a diploid zygote.

Thus, female meiosis, or oocyte maturation, differs from meiosis in the male in the time of meiotic onset, length of the cell cycle, as well as the number of functional germ cells produced (Hassold and Hunt, 2001). These differences may account for the high rates of aneuploidy associated with the oocyte.

Fortunately, many meiotic mechanisms are evolutionarily conserved; so much of our understanding concerning regulation of mammalian oocyte meiotic events comes from studies focusing on mitosis or meiosis of lower organisms, such as yeast, *C. elegans* and *Xenopus*. These studies serve as a base to broaden our understanding of meiotic control in the mammalian oocyte. One such control mechanism is reversible phosphorylation. Reversible phosphorylation of proteins represents a highly conserved mechanism that regulates numerous cellular events. This process is influenced by the state of equilibrium of protein kinase and phosphatase (PP) activities, which phosphorylate and dephosphorylate phosphoproteins, respectively.

In this review, we will discuss various mechanistic events essential for proper cellular ploidy during oocyte meiosis. We will then outline molecular, cellular and biochemical processes, which govern the fidelity of these events, primarily focusing on reversible phosphorylation as a means of control.

## Chromosome condensation

Chromosome condensation is an essential prerequisite to ensure proper chromosome segregation. Condensation and compaction aid in the unencumbered movement of chromosomes during the metaphase-anaphase transition and reduces the likelihood of chromosome entrapment during cytokinesis (Swedlow and Hirano, 2003). As mentioned, during fetal development oocytes begin meiosis, entering the leptotene and zygotene substages of prophase I. During these stages, each set of homologues begins condensing and then pairs-up and aligns; steps crucial for subsequent recombination to occur in the pachytene stage. After completing the pachytene stage, homologues begin to undergo a period of decondensation as they enter the diplotene stage. It is at this point in the meiotic cell cycle that the oocyte arrests its development and remains quiescent until pubertal hormones signal meiotic resumption. Once this resumption begins, homologues undergo rapid condensation as oocytes complete the diakinesis stage of prophase I and form the typical cross-shape bivalent of metaphase chromosomes.

Chromatin condensation is the result of long strands of DNA coiling around an octamer of two each of four regulatory proteins known as histones: H2A, H2B, H3 and H4 (Alberts, 1994). These coiled structures of DNA and histones are referred to as nucleosomes. Additionally, a family of linker histones exist known as H1 histones, which function to fold DNA into a higher ordered structure as it is supercoiled. Phosphorylation of histone-H3 at Ser10 and Ser28 is tightly coupled to mitotic and meiotic chromosome condensation in yeast (Wei *et al.*, 1998; Goto *et al.*, 1999), as well as during meiotic condensation events in pig oocytes (Bui *et al.*, 2004). Phosphorylation of N-terminal tails of histone-H3 may act as a receptor or recruitment factor for condensation factors (Hirano, 2000). Alternatively, phosphorylation of the amino tail may reduce the affinity of histone-H3 for DNA and make the relatively compact chromatin fiber more readily accessible to condensation factors (Hirano, 2000), such as the condensin complex.

Condensins are multi-subunit protein complexes that play a central role in chromosome compaction and condensation. Recently, two condensin complexes have been identified (see review Hirano, 2005). Condensin II participates in early stages of chromosome condensation within the prophase nucleus, while condensin I regulates later condensation events once it gains access to chromosomes following nuclear envelope breakdown (Hirota *et al.*, 2004; Ono *et al.*, 2004). Condensins consist of a shared heterodimeric structural maintenance of chromosomes (SMC) core, but are distinguished by their unique set of three non-SMC subunits, essential for proper mitotic chromosome condensation from yeast to mammals. Core SMC proteins include SMC2 and SMC4 (Ono *et al.*, 2003). Non-SMC subunits belong to the chromosome-associated polypeptide family (CAP) and include CAP-D2, CAP-G and CAP-H in condensin I and CAP-D3, CAP-G2 and CAP-H2 in condensin II (Ono *et al.*, 2003). Condensins aid in chromosome condensation by inducing positive super coils in DNA (Kimura and Hirano, 1997). Additionally, it is suggested that condensins play a critical role in regulating chromosome rigidity or elasticity (see review Nasmyth and Haering, 2005). Interestingly, condensin co-localizes with phosphorylated histone-H3 (Schmiesing *et al.*, 2000). In fact, non-SMC condensin subunits directly bind to histone- H3 (Ball *et al.*, 2002), an event potentially regulated by the phosphorylated state of the histone or the non-SMC components. Although all condensin subunits have not yet been identified in mammalian oocytes, the complex is present and necessary for proper meiotic chromosome condensation in yeast and *C. elegans* (Yu and Koshland, 2003; Chan *et al.*, 2004). Thus, exploration of condensin's role in proper mammalian oocyte meiotic chromosome condensation remains an exciting area of future study.

### Aurora-B Kinase

The aurora kinases are evolutionarily conserved serine/threonine kinases that regulate multiple cellular processes. Originally discovered in *Drosophila* (Glover *et al.*, 1995), this kinase was soon found in a variety of other species. A single aurora kinase functions in yeast, two types of aurora kinase are present in *Xenopus*, *Drosophila* and *C. elegans*, while three are found in mammals (see review Carmena and Earnshaw, 2003). With several aurora kinases in a variety of species, a unified nomenclature system was soon adopted and aurora kinases were designated as either aurora-A, B or C, based primarily on sequence differences and sub-cellular localization (see review Nigg, 2001). Aurora-A localizes to centrosomes and spindle pole microtubules, aurora-B associates with chromosomes in mitosis and the spindle midzone in anaphase, while the less well studied aurora-C localizes to centrosomes.

Due to its localization to chromosomes, aurora-B is the aurora kinase of interest when discussing condensation (Fig. 2). In *Xenopus* mitotic extracts, aurora-B can phosphorylate histone-H3 at Ser10 (Murnion *et al.*, 2001) and Ser28 (Goto *et al.*, 2002). Interestingly, aurora-B may also be instrumental in condensin recruitment. Pharmacological inhibition of aurora kinases in *Xenopus* mitotic chromosome extracts results in abnormal chromosome condensation, manifested as the inability to maintain the condensed state (Gadea and Ruderman, 2005). Depletion of aurora-B by RNAi in cultured *Drosophila* cells results in partial condensation and defective loading of the condensin complex to condensing chromosomes (Giet and Glover, 2001). In *C. elegans* mitotic cells, condensin localization and activity is dependent on aurora-B kinase activity during metaphase (Hagstrom *et al.*, 2002). Additionally, distribution of both condensin I and condensin II along chromosome arms in HeLa cells is disrupted following aurora-B inhibition, which leads to defects in kinetochore structure and function (Ono *et al.*, 2004). However, conflicting data exists in *Xenopus* egg extracts, suggesting histone-H3 phosphorylation via aurora-B is not required for either condensation or condensin recruitment (Maccallum *et al.*, 2002), and an *in vitro* binding study suggests histone-H3 phosphorylation is not necessary for condensin binding (Kimura and Hirano, 2000).

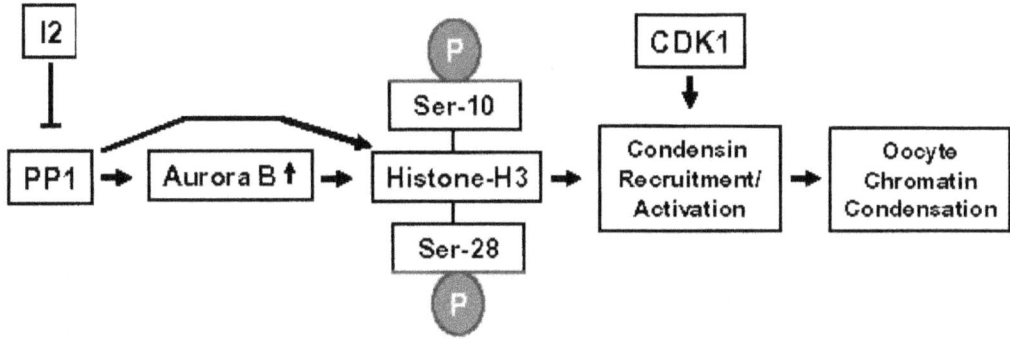

**Fig. 2** Proposed mechanism of how oocyte chromatin condensation is regulated via reversible phosphorylation. During times of oocyte chromatin condensation, chromatin associated I2 may inhibit PP1 activity. This decrease in chromatin associated PP1 activity results in increased histone-H3 phosphorylation through direct effects, or via possible activation of a histone-H3 kinase (aurora-B). This increased histone-H3 phosphorylation leads to chromatin condensation, possibly through recruitment and/or activation of condensation factors, such as the condensin complex. Additionally, direct actions of CDK1 on members of the condensin complex may affect its recruitment and/or activation, affecting chromosome condensation.

## CDK1 (MPF)

Another regulator of chromosome condensation in oocytes is maturation promoting factor (MPF). Discovered by Masui and Markert in frog oocytes (Masui and Markert, 1971), MPF is actually a heterodimer consisting of $p^{34}cdc2$ kinase (CDK1) and cyclin B. Association of CDK1 with cyclin-B is essential for MPF activity. Activation of MPF requires dephosphorylation of the kinase portion of the complex on tyrosine 15 and threonine 14 residues, as well as phosphorylation of cyclin B (Gautier and Maller, 1991). However, protease control of cyclin-B degradation has also been implicated in MPF regulation (Eppig, 1993). Histone-H1 is a known substrate for MPF and is often used for demonstration and quantification of MPF activity in mammalian oocytes. Histone-H1 kinase activity has been demonstrated in oocyte extracts and its activity is greatest during chromatin condensation at MI and MII, with an oscillatory pattern mimicking that of MPF (Jelinkova et al., 1994). Recall histone-H1 is instrumental in obtaining higher ordered condensation of chromosomes. Therefore, MPF may regulate chromatin condensation, in part, due to regulation of histone-H1 phosphorylation. It should be noted that oocyte chromatin condensation has been observed in the absence of CDK1 activity (Bui et al., 2004) and histone-H1 phosphorylation may not be required for meiotic chromatin condensation (Shen et al., 1995). However, normality of oocyte chromatin condensation under these circumstances may be compromised.

Additionally, CDK1 may regulate chromatin condensation through direct effects on components of the condensin complex (Fig. 2). Phosphorylation of condensin during mitosis stimulates the supercoiling activity of the complex and it is suggested that CDK1 is responsible for this activation through phosphorylation of CAP-D2 and CAP-H (Kimura et al., 1998). Furthermore, condensins are transported into the nucleus during mitosis in a CDK1-dependent manner (Sutani et al., 1999). One theory suggests cyclin-A/CDK1 phosphorylates and activates condensin II to initiate early condensation events within the nucleus. Following nuclear envelope dissolution, cyclin-B/CDK1 phosphorylates and activates condensin I and further acts upon condensin II (Kimura et al., 1998; Hirano, 2005).


*Protein Phosphatase-1 (PP1)*

Numerous kinases have been investigated and implicated in regulation of meiotic events such as chromosome condensation, while the role of specific serine/threonine and tyrosine protein phosphatases (PP) has been less well-studied. Classification of serine/threonine PPs is based on substrate specificity and sensitivity to defined inhibitors and results in four main types of PPs: PP1, PP2A, PP2B and PP2C (see review Cohen, 1989). In addition, four isoforms of the PP1 catalytic subunit have been identified, varying at their carboxyl termini: PP1α, PP1δ, PP1γ1 and PP1γ2 (Zhang et al., 1993; da Cruz e Silva et al., 1995; Tognarini and Villa-Moruzzi, 1998). Treatment of mammalian oocytes with PP1/PP2A inhibitors, such as okadaic acid (OA) or calyculin-A, induce rapid chromatin condensation (Picard et al., 1989; Gavin et al., 1991). These treatments also result in increased phosphorylation of pig oocyte histone-H3 at Ser10 (Bui et al., 2004). A PP1 isoform, PP1δ, localizes to condensed chromatin in mitotic cells (Andreassen et al., 1998). Studies in our laboratory demonstrate the localization of PP1δ and the endogenous PP1 regulator, inhibitor-2 (I2), to condensed chromatin in mouse oocytes (unpublished data). Thus PP1 may be the OA sensitive PP regulating phosphorylation of histone-H3 at serine residues important for condensin recruitment and chromatin condensation during oocyte meiosis (Fig. 2). Interestingly, in mammalian somatic cells, PP1 can directly dephosphorylate histone-H3, or regulate phosphorylation of the substrate through inhibition of aurora-B kinase activity (Murnion et al., 2001).


## Formation of the meiotic spindle apparatus

To ensure proper cellular ploidy, a dividing cell must form a functional spindle apparatus (Fig. 3). The spindle apparatus is a dynamic conglomerate of microtubules, comprised of α- and ß–tubulin and associated structural proteins, acting to coordinate cyto- and karyokinetic events essential for normal chromosome segregation. Microtubules persist in phases of elongation or shortening, where subunits are added or subtracted from tubule ends. As the cell enters the cell cycle, microtubules transition from radial arrays to an organized barrel-shaped bipolar structure containing a blend of dense material at either pole known as microtubule organizing centers (MTOCs: see review Schatten, 1994). Main structural proteins comprising oocyte MTOCs include γ-tubulin and pericentrin (Combelles and Albertini, 2001). It is hypothesized that the phosphorylated state of MTOC proteins regulates their function (Vandre et al., 1984; Centonze and Borisy, 1990). Indeed, studies indicate the phosphorylated state of oocyte MTOCs changes during oocyte meiotic maturation (Messinger and Albertini, 1991; Wickramasinghe and Albertini, 1992). Similarly, microtubule associated proteins (MAPs) are also regulated by cell cycle-dependent reversible phosphorylation (see review Cassimeris, 1999). These proteins can serve to both stabilize or destabilize the dynamic nature of microtubules. Furthermore, kinetochores are the major contact between spindle microtubules and chromosomes and play an important role in proper chromosome separation and segregation. Kinetochores are highly complex structures and consist of more than 50 different proteins, many of which have functions regulated by various protein kinases and phosphatases (see review Hauf and Watanabe, 2004).

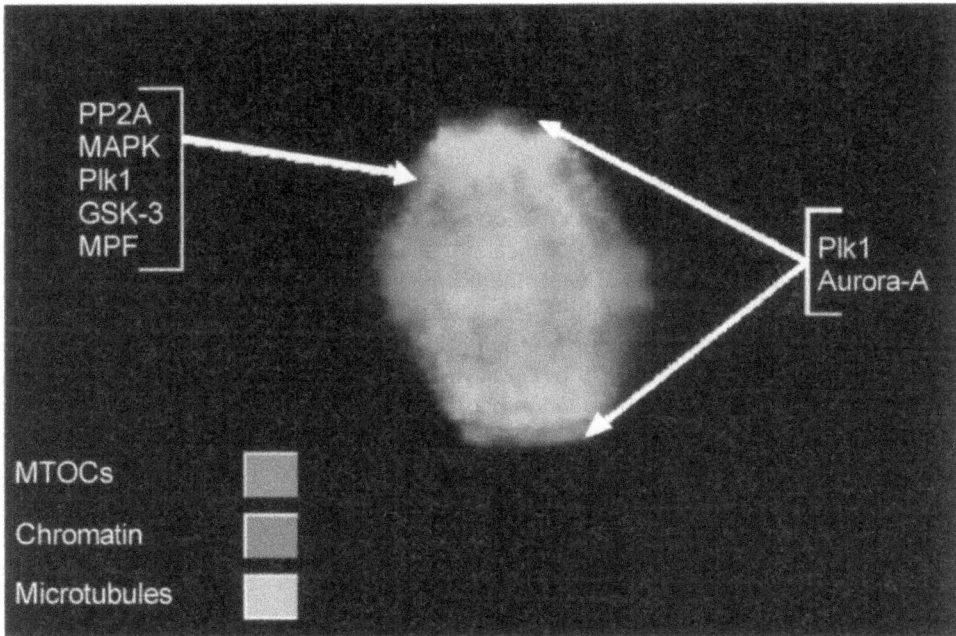

**Fig. 3** Micrograph of a mouse oocyte metaphase I meiotic spindle displaying microtubules (green/ ß-tubulin), condensed chromatin (blue/Hoescht) and MTOCs (red/ γ-tubulin). Potential regulators of components of the meiotic spindle are listed.

*Protein Phosphatase-2A (PP2A)*

OA inhibition of PP1/PP2A in mouse oocytes results in malformation of the meiotic spindle, dispersion of centrosomal staining from spindle poles and a disassociation of centrosomes from microtubules (De Pennart *et al.*, 1993). Thus, one or both of these PPs may regulate meiotic spindle formation and function. Immunocytochemical studies have shown PP2A localizes to polymerized microtubules in mouse oocytes (Lu *et al.*, 2002). Additionally, activity of PP2A increases at the time of first meiotic spindle formation (Lu *et al.*, 2002). In conjunction with various studies conducted in somatic cells (see review Sontag, 2001), PP2A appears to be the OA sensitive PP regulating meiotic spindle formation though regulation of the phosphorylated state of some unidentified microtubule associated protein. A likely candidate includes microtubule associated Tau protein (see review Sontag, 2001).

*MAP Kinase (MAPK)*

The family of MAPKs (p44MAPK1 and p42MAPK2), also known as ERK1 and ERK2, respectively, are serine/threonine kinases that require phosphorylation to be activated (see review Cobb *et al.*, 1991). Evidence suggests MAPKs are instrumental in regulation of microtubule dynamics and spindle formation in oocytes. Both ERK1 and ERK2 are activated after GVBD and remain activated throughout the MI/MII transition (Sobajima *et al.*, 1993), times when the meiotic spindle is formed and active. Additionally, active ERK1 and ERK2 are associated with mouse oocyte spindles and spindle poles (Lu *et al.*, 2002). Lastly, inhibition of MAPK activation results in compromised microtubule polymerization, no spindle formation and aberrant condensation of chromosomes (Lu *et al.*, 2002). Specific protein targets of MAPK within the meiotic spindle and/or oocyte MTOC remain to be elucidated.

## CDK1

The kinase portion of the MPF complex, CDK1, also plays a role in controlling microtubule dynamics essential for proper spindle formation and function. In mammalian somatic cells, the phosphorylated and active form of CDK1 localizes at MTOCs (Jackman *et al.*, 2003). Microtubule Associated Protein-4 (MAP4), a well-studied microtubule associated protein in mammalian somatic cells, promotes microtubule elongation and acts as a stabilizer. This stabilizing action is reduced after phosphorylation by CDK1 (Ookata *et al.*, 1995). Additionally, in *Xenopus* egg extracts, a microtubule destablizer, Op18, has been shown to alter its function based on its phosphorylated state. Phosphorylation of Op18 by CDK1 at two serine residues turns off its destabilizing action (Larsson *et al.*, 1997). Although neither of these proteins have been identified in mammalian oocytes, there is a high probability that MPF regulates the phosphorylation and function of oocyte microtubule associated proteins.

## Polo-like kinase-1 (Plk1)

Another serine/threonine kinase that appears to regulate meiotic spindle formation and function is Plk1. Much of the research focusing on Plk1 has focused on its role in regard to cancer. However, emerging data suggest Plk1 is also a regulator of oocyte meiosis. Levels of Plk1 protein increase following GVBD (Wianny *et al.*, 1998), a time when the meiotic spindle is forming. Additionally, Plk1 localizes to spindle poles in MI and MII oocytes and also to the middle of the spindle during the anaphase-telophase transition (Wianny *et al.*, 1998; Tong *et al.*, 2002). Neutralization of Plk1 activity in oocytes through antibody microinjection resulted in various MI spindle abnormalities and inability to progress to MII (Tong *et al.*, 2002; Fan *et al.*, 2003). In lower eukaryotes on mammalian mitotic cells, polo-like kinases have been found to target several microtubule associated proteins, as well as motor proteins essential for microtubule dynamics (see review Xie *et al.*, 2005).

## Aurora-A kinase

While aurora-B kinase appears to function during chromatin condensation, aurora-A functions to regulate spindle formation and microtubule dynamics. Aurora-A localizes to the spindle poles of MI and MII oocytes, and neutralization of kinase activity via antibody microinjection distorts MI spindle organization (Yao *et al.*, 2004). It is thought that the activity and regulatory function of aurora-A is dependent upon protein expression levels (Yao *et al.*, 2004). Specific protein targets for aurora-A within oocyte MTOCs remain to be elucidated. However, based on somatic cell studies, likely candidates include recruitment factors for γ-tubulin and other proteins essential for centrosome formation (see review Ducat and Zheng, 2004).

## Glycogen Synthase Kinase-3 (GSK-3)

A highly conserved serine/threonine protein kinase, GSK-3, was initially discovered as a kinase that phosphorylates and inactivates glycogen synthase (Embi *et al.*, 1980). The majority of research focusing on GSK-3 has been performed in neuronal cells, where it regulates phosphorylation of numerous microtubule associated proteins. This way, GSK-3 influences microtubule polymerization, stability, spindle formation and function (Ryves and Harwood, 2003; Doble and Woodgett, 2003; Jope and Johnson, 2004). Recently, GSK-3 isoforms have been identified in mammalian oocytes and inhibition of GSK-3 activity results in modified organization of microtubules and altered spindle function, resulting in compromised segregation of homologues (Wang *et al.*, 2003). Specific protein targets of GSK-3 in mammalian oocytes remain unknown.

## Homologue and sister chromatid cohesion

Regulation of separation of homologues and sister chromatids at meiosis I and II, respectively is as important as meiotic spindle formation in ensuring proper chromosome distribution in the oocyte. The majority of the understanding we have concerning mechanisms governing chromosome separation is derived from studies focusing on mitosis or non-mammalian meiosis (see reviews Hirano, 2000; Lee and Orr-Weaver, 2001). However, data is emerging on homologue and chromatid separation during mammalian oocyte meiosis.

It has long been observed that there is a physical association between replicated copies of chromosomes and an abrupt release of this association as chromosomes segregate during cell division. This cytologically visible attachment is referred to as cohesion (Lee and Orr-Weaver, 2001). Cohesion must be sustained until anaphase, because it is essential that each copy of the chromosome be able to resist forces of microtubules while aligned on the metaphase plate and remain in position to capture microtubules from opposite spindle poles, thus ensuring each cell retains only one copy of the genetic material. At anaphase, chromosome cohesion is released.

In mitosis, cohesion is released along chromatid arms in a single step. In contrast, meiosis utilizes a 2-step process. During prophase of meiosis I, homologous chromosomes undergo alignment and pairing or synapsis, dependent upon the synaptonemal complex. The synaptonemal complex (SC) is a proteinaceous structure consisting of two dense lateral elements (LE), flanking a less dense central element (CE). Lateral elements correspond to an axis along which sister chromatids of opposite homologues are organized by structures called axial elements (AE). Axial elements are eventually incorporated into the SC as part of the LEs. Transverse elements (TEs) cross the less dense region between LEs to create a striated, zipper like-appearance aligning homologues (Page and Hawley, 2004). Additionally, recombination events are occurring whereby a sister chromatid from one homologue attaches to a sister chromatid of the other homologue at points called chiasmata. Consequently, in order for homologues to segregate at anaphase I, cohesion along the chromosome arms must be released and chiasmata between homologues resolved. Release of cohesion along chromosome arms in meiosis coincides with the decondensation observed in the diplotene stage, during which the SC disassembles, except at chiasmata, which are resolved later during diakinesis as the chromosome condense prior to onset of metaphase I (Page and Hawley, 2004). Sister chromatid cohesion is maintained at centromeres, so each pair of sister chromatids moves to the same pole as a unit. Centromere cohesion is not released until anaphase of meiosis II, when sister chromatids are free to migrate to opposite poles.

Chromosome cohesion is regulated by a multi-subunit complex known as cohesin. The cohesion complex consists of at least four major subunits and is thought to form a ring-like structure, which wraps around sister chromatids, binding them together (Gruber *et al.*, 2003). Two subunits of the cohesin complex, Smc1p and Smc3p, belong to the SMC family; Scc1p/Rad21 belongs to the kleisin protein family; and the fourth component of cohesin is Scc3p (see reviews Hirano, 2000; Lee and Orr-Weaver, 2001; Nasmyth and Haering, 2005). In addition, meiosis-specific cohesin subunits exist (see review Marston and Amon, 2004). Mammalian oocytes contain meiosis specific cohesin subunits Rec8, STAG3 and Smc1 ß, which correspond to Scc1p/Rad21, Scc3p and Smc1p, respectively (Prieto *et al.*, 2004).

Regulation of cohesion via cohesin is extremely important considering the high incidence of chromosomal abnormalities in oocytes, the majority of which involve single chromatid aneuploidies (Hassold and Hunt, 2001). In mitotic cells, regulation of the spindle checkpoint plays a critical role regulating cohesion. In the absence of bipolar attachment, spindle-checkpoint proteins emit a global signal to inhibit onset of anaphase through formation of a Mad2, Bub and Cdc20 complex (see reviews by Musacchio and Hardwick, 2002; Dai and Moor, 2003). Attachment and tension sensors at kinetochores generate a stop signal causing phosphorylation of Mad2. While bound at the kinetochore, Mad2 binds and inhibits Cdc20 of the anaphase promoting complex (APC). Once proper attachment to kinetochores is complete, Mad2 is displaced and allowed to activate the APC. The APC then acts on and degrades securin. Once securin is degraded, separase becomes activated and induces anaphase by cleavage of the kleisin component of the cohesin complex; Scc1/Rad21 in mitotic cells or Rec8 in meiosis (Fig. 4).

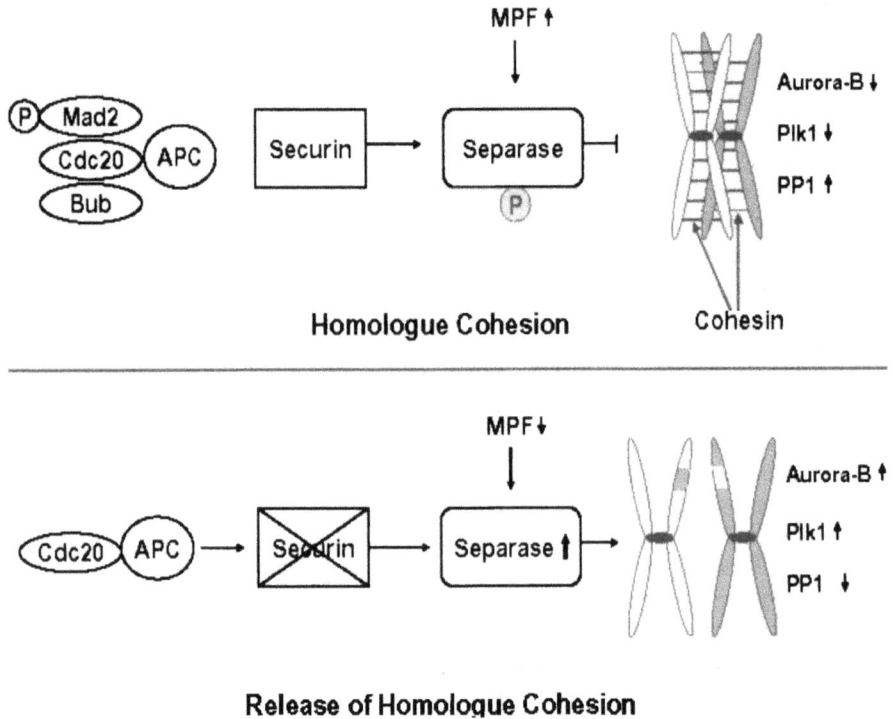

**Release of Homologue Cohesion**

**Fig. 4** Proposed mechanism of control of homologue cohesion during oocyte maturation. Formation of the Mad2/Bub/Cdc20 complex inhibits separation and the onset of anaphase. Phosphorylation of Mad2 allows it to bind and inhibit Cdc20 of the anaphase promoting complex (APC). In response to a start signal from sensors at kinetochores, Mad2 is displaced and the APC activated. The APC degrades securin which, in conjunction with decreased MPF activity via cyclin B degradation, leads to the activation of separase. Separase activity induces homologue separation via cleavage of the cohesin complex and resolution of chiasmata. Additionally, increases in the activity of aurora-B, polo-like kinase-1 and GSK-3, with a coincident decrease in PP1 activity, may contribute to separation by regulating the phosphorylated state of cohesin subunits.

It should be noted that due to the high aneuploidy rates observed in mammalian oocytes, the presence of a spindle checkpoint has been questioned. Mouse oocytes from XO females, with aberrant chromosome alignment, show no delay into anaphase entry as would be expected if a spindle checkpoint like that observed in mitotic cells existed (LeMaire-Adkins et al., 1997). However, oocyte fusion studies and a growing body of evidence identifying key checkpoint regulators in oocytes suggest a checkpoint may actually exist (see review by Dai and Moor, 2003). Indeed, securin degradation is required for homologue disjunction in mouse oocytes (Herbert et al., 2003) and Mad2 appears to regulate securin proteolysis (Homer et al., 2005). Depletion of Mad2 in mouse oocytes leads to premature securin degradation and increased incidence of aneuploidy (Homer et al., 2005). Both separase and the APC have been shown to function during mouse oocyte meiosis during the metaphase/anaphase transition (Herbert et al., 2003; Terret et al., 2003) and Mad2 and Cdc20 have been identified and regulate APC activity in oocytes (Dai and Moor, 2003). Importantly, it does appear as if key differences do exist between the sensory mechanisms which activate checkpoint control in somatic cells and mammalian oocytes. During meiosis, unattached kinetochores do not provide the same inhibitory signal that they do in mitosis, key checkpoint sensor proteins appear to have different localizations, and it is suggested that sensors may actually monitor defects in microtubule alignment within the spindle, rather than kinetechores (see review by Dai and Moor, 2003).

## Plk

As mentioned, Plks are evolutionarily conserved serine/threonine kinases. This family includes Cdc5 kinase in *Saccharomyces cerevisiae* and Plk1 in mouse (see review van Vugt and Medema, 2005). Dissociation of cohesin from chromosome arms in HeLa cells during early mitosis is dependent upon Plk1 phosphorylation of Scc3 (Hauf et al., 2005). This phosphorylation may lead to a conformational change in the cohesin ring structure, facilitating its removal from chromosomes, or act to recruit unloading factors which act upon cohesin components (Hauf et al., 2005). Additionally, Plk1/Cdc5 phosphorylation enhances the separase induced cleavability of cohesin in yeast and HeLa cells (Alexandru et al., 2001; Hauf et al., 2005). The fact that centromeric cohesion is maintained during meiosis I indicates cohesin subunits in this region are immune to the degrading effects of separase. One meiotic specific cohesin subunit that may be responsible for this immunity is Rec8. An attractive hypothesis is that Plk1 is prevented from phosphorylating centromeric cohesin proteins, like Rec8, which results in immunity to separase dependent cleavage (see review by Marston and Amon, 2004). This may be accomplished by the protein shugoshin (Sgo1p: Kitajima et al., 2004). Alternatively, it is suggested in mammals that Plk1 may play a role in cohesion through regulation of the APC (Descombes and Nigg, 1998). In mice, Plk1 has been identified in oocytes (Tong et al., 2002); however, effects on homologue and chromatid separation after its inhibition were not evaluated.

## Aurora-B

Aurora-B may also regulate chromatid cohesion along chromosome arms and at the centromere. Depletion of both Plk1 and aurora-B results in failure to load cohesin onto chromosomes (Losada et al., 2002). Additionally, in *C. elegans* aurora-B is located along chromosome arms in meiosis I and to centromeres in meiosis II, corresponding to cohesin location (Rogers et al., 2002). Furthermore, *in vitro*, aurora-B can directly phosphorylate Rec8, which may regulate its destruction by separase similar to Scc1/Rad21, and depletion of aurora-B by RNAi prevents cohesin disassociation, resulting in failure to enter anaphase (Rogers et al., 2002).

*PP1*

It is evident that PP1 and PP2A regulate multiple aspects of oocyte maturation. Indeed, PPs may also regulate cohesion and affect chromosome segregation. As mentioned, PP1 δ localizes to oocyte chromatin (unpublished data). Additionally, inhibition of PP1 and/or PP2A with OA during meiosis I results in MI oocytes with abnormal numbers of chromatids as well as MII oocytes with increased frequency of premature sister chromatid separation, single unpaired chromatids and hyperploidy (Mailhes *et al.*, 2003). During meiosis in *C. elegans*, depletion of PP1 results in cohesin disassociation and premature separation of sister chromatids, possibly through regulation of aurora-B (Rogers *et al.*, 2002). Thus PP1 may regulate the phosphorylated state and function of cohesin proteins on chromosomes, either directly, or through modification of kinases and/or cohesin recruitment factors.

Alternatively, inhibition of PP1 may induce aberrant chromosome segregation through regulation of the synaptonemal complex. Inhibition of PP1 and/or PP2A in rat spermatocytes with OA during pachytene results in premature homologue separation and dissolution of the synaptonemal complex (Tarsounas *et al.*, 1999). This separation and dissolution appear to be due to activation of a tyrosine kinase resulting in phosphorylation and subsequent relocation of SYN/SCP1 and/or COR1/SCP3, components of the transverse element and axial element of the synaptonemal complex, respectively (Tarsounas *et al.*, 1999). It should be noted that there appears to be sexual dimorphism in regulation of cohesion and the synaptonemal complex during mouse meiosis, with one difference being localization of SCP3 (Prieto *et al.*, 2004). Thus, effects of PP1 inhibition on oocyte synaptonemal complex components are unknown.

*GSK-3*

Recall that inhibition of oocyte GSK-3 results in abnormal segregation of homologues (Wang *et al.*, 2003). Recall also that PP1 δ and I2 localize to condensed oocyte chromatin (unpublished data). Residue Thr72 of I2 is a known substrate for GSK-3 phosphorylation (Sakashita *et al.*, 2003). Phosphorylation of Thr72 of I2, when complexed with PP1, results in PP activation. Therefore, inhibition of GSK-3 activity may result in reduced I2 phosphorylation and subsequent PP1 inactivation. Thus, errors in homologue segregation observed via inhibition of oocyte GSK-3 activity may not only be due to direct effects on microtubule associated proteins, but also due to premature release of cohesins through regulation of PP1 activity.

*MPF*

It appears as if MPF also regulates separation of chromosomes during oocyte maturation. Homologue disjunction in mouse oocytes requires proteolysis of cyclin B1, a component of MPF (Herbert *et al.*, 2003). Additionally, despite securin degradation, sister chromatids do not separate during meiosis II in oocytes (Madgwick *et al.*, 2004). It is hypothesized that MPF regulates phosphorylation and activity of separase to control chromosome segregation during oocyte maturation (Madgwick *et al.*, 2004) (Fig. 4).

## Conclusion

Determination of meiotic control of chromatin remodeling and chromosome separation and segregation in the mammalian oocyte is imperative, especially when one considers the high rates of aneuploidy associated with defects within the machinery of the oocyte. Although similarities exist between mitotic and meiotic cells in regard to control of chromatin dynamics, significant differences exist, such as spindle checkpoint. These differences illustrate the need for continued research into control of oocyte meiosis, with future investigations focusing on the human female gamete. With advances in assisted reproductive technology and with continued improvement of oocyte *in vitro* maturation systems, understanding these regulatory mechanisms will prove invaluable in circumventing meiotic errors.

# References

**Alberts B** (1994) *Molecular Biology of the Cell,*Garland Pub, New York

**Alexandru G, Uhlmann F, Mechtler K, Poupart M and Nasmyth K** (2001) Phosphorylation of the cohesin subunit scc1 by polo/cdc5 kinase regulated sister chromatid separation in yeast *Cell* 105 459-472

**Andreassen PR, Lacroix FB, Villa-Moruzzi E and Margolis RL** (1998) Differential subcellular localization of protein phosphatase-1 alpha, gamma1, and delta isoforms during both interphase and mitosis in mammalian cells *Journal of Cell Biology* 141 1207-1215

**Ball AR Jr, Schmiesing JA, Zhou C, Gregson HC, Okada Y, Doi T and Yokomori K** (2002) Identification of a chromosome-targeting domain in the human condensin subunit CNAP1/hCAP-D2/Eg7 *Molecular and Cellular Biology* 22 5769-5781

**Bui HT, Yamaoka E and Miyano T** (2004) Involvement of histone H3 (Ser10) phosphorylation in chromosome condensation without CDC2 kinase and mitogen-activated protein kinase activation in pig oocytes *Biology of Reproduction* 70 1843-1851

**Carmena M and Earnshaw WC** (2003) The cellular geography of aurora kinases *Nature Reviews Molecular Cell Biology* 4 842-854

**Cassimeris L** (1999) Accessory protein regulation of microtubule dynamics throughout the cell cycle *Current Opinion in Cell Biology* 11 134-141

**Centonze VE and Borisy GG** (1990) Nucleation of microtubules from mitotic centrosomes is modulated by a phosphorylated epitope *Journal of Cell Science* 95 405-411

**Chan RC, Severson AF and Meyer B J** (2004) Condensin restructures chromosomes in preparation for meiotic divisions *Journal of Cell Biology* 167 613-625

**Cobb M, TG Boulton and Robbins D** (1991) Extracellular signal-regulated kinases: ERKs in progress *Cell Regulation* 2 965-978

**Cohen P** (1989) The structure and regulation of protein phosphatases *Annual Review of Biochemistry* 58 453-508

**Combelles C and Albertini D** (2001) Microtubule patterning during meiotic maturation in mouse oocytes is determined by cell cycle specific sorting and redistribution of γ-tubulin *Developmental Biology* 239 281-294

**da Cruz e Silva EF, Fox CA, Ouimet CC, Gustafson E, Watson SJ and Greengard P** (1995) Differential expression of protein phosphatase 1 isoforms in mammalian brain *Journal of Neuroscience* 15 3375-3389

**Dai YFJ and Moor R** (2003) Checkpoint controls in mammalian oocytes. In *Biology and Pathology of the Oocyte* pp 120-133 Eds AO Trounson and RG Gosden. Cambridge, New York

**De Pennart H, Verlhac MH, Cibert C, Santa Maria A and Maro B** (1993) Okadaic acid induces spindle lengthening and disrupts the interaction of microtubules with the kinetochores in metaphase II-arrested mouse oocytes *Developmental Biology* 157 170-181

**Descombes P and Nigg EA** (1998) The polo-like kinase Plx1 is required for M phase exit and destruction of mitotic regulators in *Xenopus* egg extracts *EMBO Journal* 17 1328-1335

**Doble BW and Woodgett JR** (2003) GSK-3: tricks of the trade for a multi-tasking kinase *Journal of Cell Science* 116 1175-1186

**Ducat D and Zheng Y** (2004) Aurora kinases in spindle assembly and chromosome segregation *Experimental Cell Research* 301 60-67

**Embi N, Rylatt DB and Cohen P** (1980) Glycogen synthase kinase-3 from rabbit skeletal muscle. Separation from cyclic-AMP-dependent protein kinase and phosphorylase kinase *European Journal of Biochemistry* 107 519-527

**Eppig J** (1993) Regulation of mammalian oocyte maturation. In *The Ovary.* pp 185-208 Eds E Adashi and P Leung. Raven Press Ltd, New York

**Fan HY, Tong C, Teng CB, Lian L, Li S W, Yang ZM, Chen DY, Schatten H and Sun QY** (2003) Characterization of Polo-like kinase-1 in rat oocytes and early embryos implies its functional roles in the regulation of meiotic maturation, fertilization, and cleavage *Molecular Reproduction and Development* 65 318-329

**Gadea BB and Ruderman JV** (2005) Aurora kinase inhibitor ZM447439 blocks chromosome-induced spindle assembly, the completion of chromosome condensation, and the establishment of the spindle integrity checkpoint in *Xenopus* egg extracts *Molecular Biology of the Cell* 16 1305-1318

**Gautier J and Maller JL** (1991) Cyclin B in *Xenopus* oocytes: implications for the mechanism of pre-MPF activation *EMBO Journal* 10 177-182

**Gavin AC, Tsukitani Y and Schorderet-Slatkine S** (1991) Induction of M-phase entry of prophase-blocked mouse oocytes through microinjection of okadaic acid, a specific phosphatase inhibitor *Experimental Cell Research* 192 75-81

**Giet R and Glover DM** (2001) *Drosophila* aurora B kinase is required for histone H3 phosphorylation and condensin recruitment during chromosome condensation and to organize the central spindle during cytokinesis *Journal of Cell Biology* **152** 669-682

Glover DM, Leibowitz MH, Mclean DA and Parry H (1995) Mutations in aurora prevent centrosome separation leading to the formation of monopolar spindles *Cell* **81** 95-105

Goto H, Tomono Y, Ajiro K, Kosako H, Fujita M, Sakurai M, Okawa K, Iwamatsu A, Okigaki T, Takahashi T and Inagaki M (1999) Identification of a novel phosphorylation site on histone H3 coupled with mitotic chromosome condensation *The Journal of Biological Chemistry* **274** 25543-25549

Goto H, Yasui Y, Nigg EA and Inagaki M (2002) Aurora-B phosphorylates Histone H3 at serine28 with regard to the mitotic chromosome condensation *Genes to Cells* **7** 11-17

Gruber S, Haering CH and Nasmyth K (2003) Chromosomal cohesin forms a ring *Cell* **112** 765-777

Hagstrom KA, Holmes VF, Cozzarelli NR and Meyer BJ (2002) *C. elegans* condensin promotes mitotic chromosome architecture, centromere organization, and sister chromatid segregation during mitosis and meiosis *Genes and Development* **16** 729-742

Hassold T and Hunt P (2001) To err (meiotically) is human: the genesis of human aneuploidy *Nature Reviews. Genetics* **2** 280-291

Hauf S and Watanabe Y (2004) Kinetochore orientation in mitosis and meiosis *Cell* **119** 317-327

Hauf S, Roitinger E, Koch B, Dittrich CM, Mechtler K and Peters JM (2005) Dissociation of cohesin from chromosome arms and loss of arm cohesion during early mitosis depends on phosphorylation of SA2 *Public Library of Science Biology* **3** 419-430

Herbert M, Levasseur M, Homer H, Yallop K, Murdoch A and Mcdougall A (2003) Homologue disjunction in mouse oocytes requires proteolysis of securin and cyclin B1 *Nature Cell Biology* **5** 1023-1025

Hirano T (2000) Chromosome cohesion, condensation, and separation *Annual Review of Biochemistry* **69** 115-144

Hirano T (2005) Condensins: organizing and segregating the genome *Current Biology* **15** R265-275

Hirota T, Gerlich D, Koch B, Ellenberg J and Peters JM (2004) Distinct functions of condensin I and II in mitotic chromosome assembly *Journal of Cell Science* **117** 6435-6445

Homer HA, Mcdougall A, Levasseur M, Yallop K, Murdoch AP and Herbert M (2005) Mad2 prevents aneuploidy and premature proteolysis of cyclin B and securin during meiosis I in mouse oocytes *Genes and Development* **19** 202-207

Hunt PA (1998) The control of mammalian female meiosis: factors that influence chromosome segregation *Journal of Assisted Reproduction and Genetics* **15** 246-252

Jackman M, Lindon C, Nigg EA and Pines J (2003) Active cyclin B1-Cdk1 first appears on centrosomes in prophase *Nature Cell Biology* **5** 143-148

Jelinkova L, Kubelka M, Motlik J and Guerrier P (1994) Chromatin condensation and histone H1 kinase activity during growth and maturation of rabbit oocytes *Molecular Reproduction and Development* **37** 210-215

Jope RS and Johnson GV (2004) The glamour and gloom of glycogen synthase kinase-3 *Trends in Biochemical Sciences* **29** 95-102

Kimura K and Hirano T (1997) ATP-dependent positive supercoiling of DNA by 13S condensin: a biochemical implication for chromosome condensation *Cell* **90** 625-634

Kimura K and Hirano T (2000) Dual roles of the 11S regulatory subcomplex in condensin functions *Proceedings of National Academy of Sciences USA* **97** 11972-11977

Kimura K, Hirano M, Kobayashi R and Hirano T (1998) Phosphorylation and activation of 13S condensin by Cdc2 *in vitro* *Science* **282** 487-490

Kitajima TS, Kawashima SA and Watanabe Y (2004) The conserved kinetochore protein shugoshin protects centromeric cohesion during meiosis *Nature* **427** 510-517

Larsson N, Marklund U, Gradin HM, Brattsand G and Gullberg M (1997) Control of microtubule dynamics by oncoprotein 18: dissection of the regulatory role of multisite phosphorylation during mitosis *Molecular and Cellular Biology* **17** 5530-5539

Lee JY and Orr-Weaver TL (2001) The molecular basis of sister-chromatid cohesion *Annual Review of Cell and Developmental Biology* **17** 753-777

LeMaire-Adkins R, Radke K and Hunt PA (1997) Lack of checkpoint control at the metaphase/anaphase transition: a mechanism of meiotic nondisjunction in mammalian females *Journal of Cell Biology* **139** 1611-1619

Losada A, Hirano M and Hirano T (2002) Cohesin release is required for sister chromatid resolution, but not for condensin-mediated compaction, at the onset of mitosis *Genes and Development* **16** 3004-3016

Lu Q, Dunn RL, Angeles R and Smith GD (2002) Regulation of spindle formation by active mitogen-activated protein kinase and protein phosphatase 2A during mouse oocyte meiosis *Biology of Reproduction* **66** 29-37

Maccallum DE, Losada A, Kobayashi R and Hirano T (2002) ISWI remodeling complexes in *Xenopus* egg extracts: identification as major chromosomal components that are regulated by INCENP-aurora B *Molecular Biology of the Cell* **13** 25-39

Madgwick S, Nixon VL, Chang HY, Herbert M, Levasseur M and Jones KT (2004) Maintenance of sister chromatid attachment in mouse eggs through maturation-promoting factor activity *Developmental Biology* **275** 68-81

Mailhes JB, Hilliard C, Fuseler JW and London SN (2003) Okadaic acid, an inhibitor of protein phosphatase 1 and 2A, induces premature separation of sister chromatids during meiosis I and aneuploidy in mouse oocytes *in vitro* *Chromosome Research* **11** 619-631

Marston AL and Amon A (2004) Meiosis: cell-cycle controls shuffle and deal *Nature Reviews, Molecular Cell Biology* **5** 983-997

Masui Y and Markert CL (1971) Cytoplasmic control of

nuclear behavior during meiotic maturation of frog oocytes *Journal of Experimental Zoology* **177** 129-145

**Messinger SM and Albertini DF** (1991) Centrosome and microtubule dynamics during meiotic progression in the mouse oocyte *Journal of Cell Science* **100** 289-298

**Murnion M, Adams R, Callister D, Allis C, Earnshaw W and Swedlow J** (2001) Chromatin-associated protein phosphatase 1 regulates aurora-B and histone H3 phosphorylation *The Journal of Biological Chemistry* **276** 26656-26665

**Musacchio A and Hardwick KG** (2002) The spindle checkpoint: structural insights into dynamic signalling *Nature Reviews, Molecular Cell Biology* **3** 731-741

**Nasmyth K and Haering CH** (2005) The structure and function of SMC and kleisin complexes *Annual Review of Biochemistry* **74** 595-648

**Nigg E** (2001) Mitotic kinases as regulators of cell division and its checkpoints *Nature Reviews, Molecular Cell Biology* **2** 21-32

**Ono T, Losada A, Hirano M, Myers MP, Neuwald AF and Hirano T** (2003) Differential contributions of condensin I and condensin II to mitotic chromosome architecture in vertebrate cells *Cell* **115** 109-121

**Ono T, Fang Y, Spector DL and Hirano T** (2004) Spatial and temporal regulation of Condensins I and II in mitotic chromosome assembly in human cells *Molecular Biology of the Cell* **15** 3296-3308

**Ookata K, Hisanaga S, Bulinski J C, Murofushi H, Aizawa H, Itoh TJ, Hotani H, Okumura E, Tachibana K and Kishimoto T** (1995) Cyclin B interaction with microtubule-associated protein 4 (MAP4) targets p34cdc2 kinase to microtubules and is a potential regulator of M-phase microtubule dynamics *Journal of Cell Biology* **128** 849-862

**Page SL and Hawley RS** (2004) The genetics and molecular biology of the synaptonemal complex *Annual Review of Cell and Developmental Biology* **20** 525-558

**Picard A, Capony J, Brautigan D and Doree M** (1989) Involvement of protein phosphatases 1 and 2A in the control of M phase-promoting factor activity in starfish *Journal of Cell Biology* **109** 3347-3354

**Prieto I, Tease C, Pezzi N, Buesa JM, Ortega S, Kremer L, Martinez A, Martinez AC, Hulten MA and Barbero JL** (2004) Cohesin component dynamics during meiotic prophase I in mammalian oocytes *Chromosome Research* **12** 197-213

**Rogers E, Bishop JD, Waddle JA, Schumacher JM and Lin R** (2002) The aurora kinase AIR-2 functions in the release of chromosome cohesion in *Caenorhabditis elegans* meiosis *Journal of Cell Biology* **157** 219-229

**Ryves WJ and Harwood AJ** (2003) The interaction of glycogen synthase kinase-3 (GSK-3) with the cell cycle *Progress in Cell Cycle Research* **5** 489-495

**Sakashita G, Shima H, Komatsu M, Urano T, Kikuchi A and Kikuchi K** (2003) Regulation of type 1 protein phosphatase/inhibitor-2 complex by glycogen synthase kinase-3 ß in intact cells *Journal of Biochemistry* **133** 165-171

**Schatten G** (1994) The centrosome and its mode of inheritance: the reduction of the centrosome during gametogenesis and its restoration during fertilization *De-*

*velopmental Biology* **165** 299-335

**Schmiesing JA, Gregson HC, Zhou S and Yokomori K** (2000) A human condensin complex containing hCAP-C-hCAP-E and CNAP1, a homolog of *Xenopus* XCAP-D2, colocalizes with phosphorylated histone H3 during the early stage of mitotic chromosome condensation *Molecular and Cellular Biology* **20** 6996-7006

**Shen X, Yu L, Weir JW and Gorovsky MA** (1995) Linker histones are not essential and affect chromatin condensation *in vivo Cell* **82** 47-56

**Sobajima T, Aoki F and Kohmoto K** (1993) Activation of mitogen-activated protein kinase during meiotic maturation in mouse oocytes *Journal of Reproduction and Fertility* **97** 389-394

**Sontag E** (2001) Protein phosphatase 2A: the Trojan Horse of cellular signaling *Cell Signal* **13** 7-16

**Sutani T, Yuasa T, Tomonaga T, Dohmae N, Takio K and Yanagida M** (1999) Fission yeast condensin complex: essential roles of non-SMC subunits for condensation and Cdc2 phosphorylation of Cut3/SMC4 *Genes and Development* **13** 2271-2283

**Swedlow JR and Hirano T** (2003) The making of the mitotic chromosome: modern insight into classical questions *Molecular Cell* **11** 357-369

**Tarsounas M, Pearlman RE and Moens PB** (1999) Meiotic activation of rat pachytene spermatocytes with okadaic acid: the behavior of synaptonemal complex components SYN1/SCP1 and COR1/SCP3 *Journal of Cell Science* **112** 423-434

**Terret ME, Wassmann K, Waizenegger I, Maro B, Peters JM and Verlhac MH** (2003) The meiosis I-to-meiosis II transition in mouse oocytes requires separase activity *Current Biology* **13** 1797-1802

**Tognarini M and Villa-Moruzzi E** (1998) Analysis of the isoforms of protein phosphatase 1 (PP1) with polyclonal peptide antibodies *Methods in Molecular Biology* **93** 169-183

**Tong C, Fan HY, Lian L, Li SW, Chen DY, Schatten H and Sun QY** (2002) Polo-like kinase-1 is a pivotal regulator of microtubule assembly during mouse oocyte meiotic maturation, fertilization, and early embryonic mitosis *Biology of Reproduction* **67** 546-554

**van Vugt MA and Medema RH** (2005) Getting in and out of mitosis with Polo-like kinase-1 *Oncogene* **24** 2844-2859

**Vandre D, Davis F, Rao P and Borisy G** (1984) Phosphoproteins are components of mitotic microtubule organizing centers *Proceedings of the National Academy of Sciences USA* **81** 4439-4443

**Wang X, Liu X, Dunn R, Ohl D and Smith GD** (2003) Glycogen synthase kinase-3 regulates mouse oocyte homologue segregation *Molecular Reproduction and Development* **64** 96-105

**Wei Y, Mizzen CA, Cook R G, Gorovsky MA and Allis CD** (1998) Phosphorylation of histone H3 at serine 10 is correlated with chromosome condensation during mitosis and meiosis in Tetrahymena *Proceedings of the National Academy of Sciences USA* **95** 7480-7484

**Wianny F, Tavares A, Evans MJ, Glover DM and Zernicka-Goetz M** (1998) Mouse polo-like kinase 1 associates with the acentriolar spindle poles, meiotic chromosomes and spindle midzone during oocyte maturation *Chromosoma* **107** 430-439

**Wickramasinghe D and Albertini D** (1992) Centrosome phosphorylation and the developmental expression of meiotic competence in mouse oocytes *Developmental Biology* **152** 62-74

**Xie S, Xie B, Lee MY and Dai W** (2005) Regulation of cell cycle checkpoints by polo-like kinases *Oncogene* **24** 277-286

**Yao LJ, Zhong ZS, Zhang LS, Chen DY, Schatten H and Sun QY** (2004) Aurora-A is a critical regulator of microtubule assembly and nuclear activity in mouse oocytes, fertilized eggs, and early embryos *Biology of Reproduction* **70** 1392-1399

**Yu HG and Koshland DE** (2003) Meiotic condensin is required for proper chromosome compaction, SC assembly, and resolution of recombination-dependent chromosome linkages *Journal of Cell Biology* **163** 937-947

**Zhang Z, Bai G, Shima M, Zhao S, Nagao M and Lee EY** (1993) Expression and characterization of rat protein phosphatases-1 alpha, -1 gamma 1, -1 gamma 2, and -1 delta *Archives of Biochemistry and Biophysics* **303** 402-406

# The enigma of sperm-egg recognition in mice

Jurrien Dean

*Laboratory of Cellular and Developmental Biology, NIDDK, National Institutes of Health, Bethesda, Maryland 20892 USA*

Mouse fertilization occurs in the oviduct where spermatozoa bind to the zona pellucida, penetrate the extracellular matrix and fuse with the egg's plasma membrane. Following fertilization, the zona pellucida, is modified to prevent sperm binding and penetration. The molecular basis of these interactions have long intrigued investigators, but remain incompletely understood. The recent use of mouse genetics has prompted review of current knowledge and suggests new models for the critical events for fertilization and early embryo development.

## Introduction

Although millions of sperm are deposited in the lower female reproductive tract, relatively few encounter ovulated eggs in the ampulla of the oviduct where fertilization takes place. Mouse eggs are ovulated in a matrix of hyaluronan in which cumulus cells are interspersed (Yanagimachi, 1994). Although Sperm Adhesion Molecule 1(SPAM1/PH-20), a GPI-anchored hyaluronidase, has been implicated in the passage of acrosome-intact spermatozoa through the cumulus mass, male mice lacking the enzyme remain fertile (Baba et al., 2002). More recently, a second hyaluronidase encoded by *Hyal5,* which is abundant in epididymal spermatozoa has been implicated in cumulus penetration (Kim et al., 2005), although the effect of genetic ablation on fertility has yet to be reported.

Only acrosome-intact sperm penetrate through the cumulus mass and bind to the zona pellucida (Storey et al., 1984), an acellular glycocalyx that surrounds ovulated eggs and preimplantation embryos (Wassarman, 1988). This initial interaction triggers the exocytosis of the acrosome, a Golgi derived vesicle that underlies the plasma membrane of the anterior sperm head. The released contents are thought to modify the zona matrix, the surface of the spermatozoa or both so as to facilitate passage through the matrix (Abou-Haila and Tulsiani, 2000). It is also most probable that the forward motility of the sperm plays a prominent role in penetration of the zona pellucida (Stauss et al., 1995; Bedford, 1998) and the subsequent fusion of the gametes in the perivitelline space initiates fertilization (Yanagimachi, 1994).

Following fertilization, prevention of polyspermy entails blocks to: 1) gamete fusion at the egg's plasma membrane; 2) sperm penetration of the zona matrix; and 3) sperm binding at the surface of the zona pellucida. Little is known about mechanisms that block fusion and, although a number of biochemical changes in the zona matrix have been inferred, only cleavage of ZP2 has been experimentally observed in mice and humans (Bleil et al., 1981; Bauskin et al., 1999). The block to zona penetration occurs rapidly and is independent of ZP2 cleavage, whereas the block to sperm binding occurs over several hours and, in the absence of ZP2 cleavage, spermatozoa continue to bind even after fertilization (Rankin et al., 2003). Thus, the post-fertilization

E-mail: jurrien@helix.nih.gov

block to sperm binding is mediated by cleavage of ZP2, but the mechanisms that block zona penetration and fusion with the egg plasma membrane remain to be determined.

To elucidate the molecular basis of sperm-egg recognition, investigators have sought to take advantage of two dichotomous stages: 1) mouse sperm bind to eggs but not to two-cell embryos; and 2) human sperm bind to human eggs, but not to mouse eggs.

### Simple models of sperm-egg recognition

*Single zona proteins* - The presence of individual zona glycoproteins was first reported in the mouse (Bleil and Wassarman, 1980a) and subsequently confirmed in rat (Araki *et al.*, 1992) and human (Shabanowitz and O'Rand, 1988) among other species. There has been considerable debate as to the functional role of the different glycoproteins, particularly in terms of sperm-egg recognition and the post-fertilization block to polyspermy. Initially, investigators, using different vertebrate models, suggested that sperm binding was dependent on a single protein (Bleil and Wassarman, 1980b; Prasad *et al.*, 1996; Tian *et al.*, 1997). However, *Zp1* null mice are fertile, albeit with decreased fecundity (Rankin *et al.*, 1999) and the replacement of endogenous mouse ZP2, ZP3 or both with human homologues that are 60-70% conserved does not alter the specificity of sperm binding (Rankin *et al.*, 1998; 2003). These observations suggest that a single zona protein is not sufficient to support sperm binding to the zona pellucida.

*Glycan release models* - Investigators have also proposed that sperm-egg recognition is based on N- or O-glycans attached to specific zona proteins. Glycosidases released from cortical granule following fertilization are invoked in 'glycan release' models to account for the inability of sperm to bind to two-cell embryos. *Zp1* null mice are fertile (Rankin *et al.*, 1999) which indicates that if mouse sperm adhesion is mediated by carbohydrate side chains, they must reside on either ZP2 or ZP3. N-glycans occupy 11 of 12 potential sites on mouse ZP2 and ZP3 (Boja *et al.*, 2003) and the majority are high mannose or biantennary complex (Tulsiani *et al.*, 1992; Nagdas *et al.*, 1994; Easton *et al.*, 2000). Conditional null mice lacking N-acetylglucosaminyltransferase I (*MgatI* null) required to initiate hybrid and complex N-glycan synthesis in oocytes remain fertile, indicating that these sugars are not required for sperm-egg recognition (Shi *et al.*, 2004). Few, if any, O-glycans are present on ZP2 (Nagdas *et al.*, 1994; Boja *et al.*, 2003; Dell *et al.*, 2003) and only two clusters have been identified on ZP3 (Boja *et al.*, 2003), the majority of which have Core-2 structures (Easton *et al.*, 2000). However, mice lacking Core 2 ß1-6 N-acetylglucosaminyltransferase (*C2 GlcNAcT I* null) involved in formation of mucin Core 2 O-glycans remain fertile (Ellies *et al.*, 1998). Taken together, these data suggest that, if sperm adhesion is glycan mediated, it must be via high mannose N-glycans on ZP2 or ZP3, or O-glycans on ZP3 other than mucin Core 2 structures.

Specific sugar residues also have been proposed as mediators of sperm-egg recognition in mice (Bleil and Wassarman, 1988; Miller *et al.*, 1992), but biochemical and genetic data have not supported their candidacies (Thall *et al.*, 1995; Lu and Shur, 1997; Boja *et al.*, 2003). More generally, the ability of mouse spermatozoa to bind to embryos derived from hu*ZP2* rescue mice is seemingly inconsistent with 'glycan release' models of sperm-egg recognition in mice (Rankin *et al.*, 2003). In particular, it is difficult to envision a carbohydrate side chain that would remain accessible for sperm binding after fertilization and yet be inaccessible for cleavage by a glycosidase(s) released by cortical granule exocytosis. Taken together, these genetic data do not support an O- or N-glycan on the zona pellucida acting as a sperm ligand with its release by cortical granule glycosidase(s) accounting for the inability of sperm to bind to two-cell embryos.

## Higher-order zona structures

*Zona scaffold model*

The inability to ascribe sperm binding to a single protein or carbohydrate determinant in the zona pellucida has focused attention on the role that supramolecular structures might play in sperm-egg recognition. ZP3 is essential for formation of the zona pellucida matrix (Liu *et al.*, 1996; Rankin *et al.*, 1996) and can partner with either ZP1 (*Zp2* null), ZP2 (*Zp1* null) or both (normal), although the width and stability of the ZP1/ZP3 matrix are considerably thinner than the ZP2/ZP3 matrix due to limiting amounts of ZP1 (Rankin *et al.*, 1999; 2001). Genetically altered mice in which human ZP2 replaces endogenous mouse ZP2 (huZP2 rescue mice) form a zona pellucida and are fertile. Unexpectedly, sperm continue to bind to two-cell embryos after *in vitro* fertilization of eggs derived from huZP2 rescue mice (Fig. 1) and this observation correlates with intact huZP2 which is fortuitously not cleaved in the chimeric mouse-human zona pellucida (Rankin *et al.*, 2003). These observations form the basis of the 'zona scaffold' model of sperm binding in which the cleavage status of ZP2 regulates the three-dimensional structure of the zona matrix, rendering it permissive (intact ZP2) or non-permissive (cleaved ZP2) for sperm binding (Dean, 2004). The three-dimensional zona structure(s) to which sperm bind (and may involve protein, carbohydrate or both) have not been determined and will require further investigation.

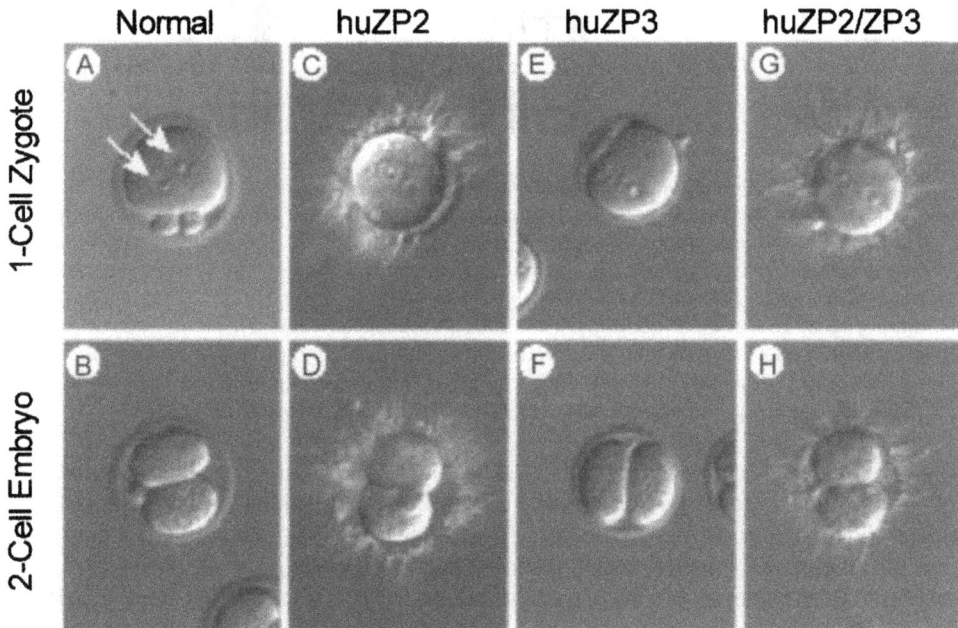

**Fig. 1** Persistent sperm binding to transgenic mouse embryos with intact human ZP2 Embryos from (A,B) normal, (C,D) huZP2 rescue, (E,F) huZP3 rescue, and (G,H) huZP2/ZP3 double rescue mice were examined morphologically 8 (A,C,E,G) and 24 (B,D,F,H) hours after insemination with capacitated, epididymal mouse sperm. Two pronuclei (arrows) were detected in each of the four genotypes in 1-cell embryos and cleavage into 2-cell embryos occurred by 24 hours. Persistence of sperm binding to the zona pellucida was observed in huZP2 and huZP2/huZP3 double rescue mice, but not in normal or huZP3 rescue mice. The sperm binding to the 2-cell embryos correlates with intact huZP2, even after fertilization and cortical granule exocytosis (Rankin *et al.*, 2003).

*Three versus four zona proteins*

Analyses of the mouse, rat and human genomes identified four loci encoding zona pellucida proteins located on syntenic chromosomes (Fig. 2). Microscale mass spectrometry of native zonae pellucidae from each species detected four zona proteins in rat (Hoodbhoy *et al.*, 2005) and human (Lefièvre *et al.*, 2004), but only three in mice (Boja *et al.*, 2003). Although aberrant transcripts from the fourth mouse locus are reported in EST databases, multiple stop codons in the three potential open reading frames preclude protein expression. The further observation that *Zp1* null mice are fertile and form a zona pellucida to which sperm will bind (Rankin *et al.*, 1999) indicates that a functional zona pellucida matrix can be formed with two (*Zp1* null mice), three (normal mice) or four (rat and human) zona proteins.

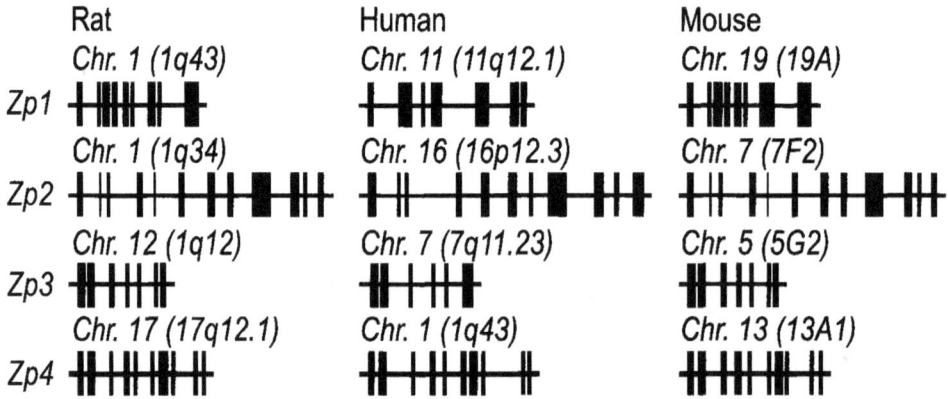

**Fig. 2** Syntenic loci of zona pellucida genes

Comparison of the mouse, rat and human genomes (Gibbs *et al.*, 2004; International Human Genome Sequencing Consortium 2004; Waterston *et al.*, 2002) identified genetic loci potentially encoding four zona pellucida proteins. The loci contain conserved exon maps of *Zp1*, *Zp2*, *Zp3* and *Zp4/ZPB* located on somatic chromosomes in regions syntenic among the three species (Hoodbhoy *et al.*, 2005). Vertical bars represent exons. Microscale mass spectrometry detects four proteins in rat (Hoodbhoy *et al.*, 2005), human (Lefièvre *et al.*, 2004), but only three in mouse (Boja *et al.*, 2003) due to multiple stop codons in each of the three reading frames encoded at the *Zp4/ZPB* locus (Lefièvre *et al.*, 2004).

Normally human spermatozoa will bind to human but not to mouse eggs whereas mouse spermatozoa will bind to both. To investigate if the number of proteins was sufficient to dictate order-specific recognition among gametes, *Zp1* null eggs (ZP2-3), normal mouse eggs (ZP1-3) and normal rat eggs (ZP1-4) were used in binding assays with human and mouse sperm. Mouse sperm bound to all the eggs, but human sperm bound to none (Fig. 3). Therefore, the presence of four zona glycoproteins is not sufficient to support human sperm binding even though the primary protein structures of the homologous rat and human zona proteins are 62-70% identical (Hoodbhoy *et al.*, 2005).

### Other potential determinants

Two variables beyond the composition of the zona pellucida may affect sperm-egg recognition, the geometry of the zona matrix and the sperm surface. One of the most evolutionarily dynamic aspects of sperm-egg recognition among mammals is morphology of the sperm head

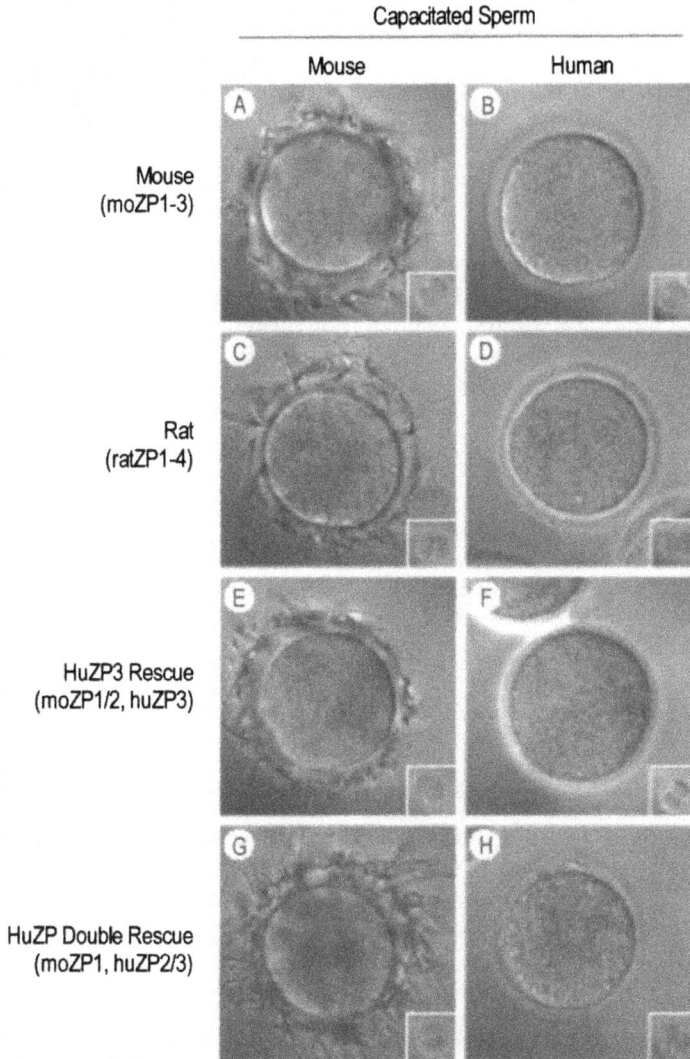

**Fig. 3** Heterologous sperm binding to mouse and rat eggs
To determine if human sperm binding is predicated on the number of glycoproteins in the zona pellucida, capacitated mouse or human sperm were incubated with ovulated rodent eggs using 2-cell embryos (insets) as wash controls. Mouse sperm bound to: (A) mouse (moZP1-3); (C) rat (ratZP1-4); (E) hu*ZP3* rescue (moZP1/ZP2, huZP3); and (G) hu*ZP2/3* double rescue (moZP1, huZP2/3) eggs. Human sperm did not bind to: (B) mouse; (D) rat; (F) hu*ZP3* rescue; or (H) hu*ZP2/3* double rescue eggs. Therefore, the presence of the four rat homologous zona pellucida proteins is not sufficient to support human sperm binding (Hoodbhoy *et al.*, 2005).

(Bedford, 1998). Mouse sperm (125 $\mu$m long) have a curved anterior head that resembles a hook and the length of the head (8 $\mu$m) is comparable to the thickness of the zona pellucidae (7-8 $\mu$m) that surrounds the 80 $\mu$m diameter egg. Human sperm (60 $\mu$m) are more compact and morphologically distinct from mice with an anterior head resembling a flattened spade the length of which (4 $\mu$m) is significantly less than the width of the human zona pellucidae (10-15 $\mu$m) surrounding the larger, 120 $\mu$m diameter human egg. Whether these distinct morphological differences between sperm head surfaces and zona matrices affect three-dimensional molecular structures important for taxon-specific sperm-egg recognition remains to be determined.

## Conclusion

Thus, despite 25 years of intense investigation, the molecular basis of sperm-recognition leading to successful fertilization remains an enigma. Great progress has been made in understanding the molecular biology of the zona pellucida and genetics continues to provide an important platform to test models that seek to clarify the molecular basis of sperm binding to the zona pellucida, a critical step in successful fertilization.

## References

Abou-Haila A and Tulsiani DR (2000) Mammalian sperm acrosome: formation, contents, and function *Archives of Biochemistry and Biophysics* 379 173-182

Araki Y, Orgebin-Crist MC and Tulsiani DR (1992) Qualitative characterization of oligosaccharide chains present on the rat zona pellucida glycoconjugates *Biology of Reproduction* 46 912-919

Baba D, Kashiwabara S, Honda A, Yamagata K, Wu Q, Ikawa M, Okabe M and Baba T (2002) Mouse sperm lacking cell surface hyaluronidase PH-20 can pass through the layer of cumulus cells and fertilize the egg *Journal of Biological Chemistry* 277 30310-30314

Bauskin A R, Franken DR, Eberspaecher U and Donner P (1999) Characterization of human zona pellucida glycoproteins *Molecular Reproduction and Development* 5 534-540

Bedford JM (1998) Mammalian fertilization misread? Sperm penetration of the eutherian zona pellucida is unlikely to be a lytic event *Biology of Reproduction* 59 1275-1287

Bleil JD and Wassarman PM (1980a) Structure and function of the zona pellucida: Identification and characterization of the proteins of the mouse oocyte's zona pellucida *Developmental Biology* 76 185-202

Bleil JD and Wassarman PM (1980b) Mammalian sperm-egg interaction: Identification of a glycoprotein in mouse egg zonae pellucidae possessing receptor activity for sperm *Cell* 20 873-882

Bleil JD and Wassarman PM (1988) Galactose at the nonreducing terminus of O-linked oligosaccharides of mouse egg zona pellucida glycoprotein ZP3 is essential for the glycoprotein's sperm receptor activity *Proceedings of the National Academy of Sciences USA* 85 6778-6782

Bleil JD, Beall CF and Wassarman PM (1981) Mammalian sperm-egg interaction: Fertilization of mouse eggs triggers modification of the major zona pellucida glycoprotein, ZP2 *Developmental Biology* 86 189-197

Boja ES, Hoodbhoy T, Fales HM and Dean J (2003) Structural characterization of native mouse zona pellucida proteins using mass spectrometry *Journal of Biological Chemistry* 278 34189-34202

Dean J (2004) Reassessing the molecular biology of sperm-egg recognition with mouse genetics *Bioessays* 26 29-38

Dell A, Chalabi S, Easton RL, Haslam SM, Sutton-Smith M, Patankar MS, Lattanzio F, Panico M, Morris HR and Clark GF (2003) Murine and human zona pellucida 3 derived from mouse eggs express identical O-glycans *Proceedings of the National Academy of Sciences USA* 100 15631-15636

Easton RL, Patankar MS, Lattanzio FA, Leaven TH, Morris HR, Clark GF and Dell A (2000) Structural analysis of murine zona pellucida glycans. Evidence for the expression of core 2-type O-glycans and the Sd(a) antigen *Journal of Biological Chemistry* 275 7731-7742

Ellies LG, Tsuboi S, Petryniak B, Lowe JB, Fukuda M and Marth JD (1998) Core 2 oligosaccharide biosynthesis distinguishes between selectin ligands essential for leukocyte homing and inflammation *Immunity* 9 881-890

Gibbs RA, Weinstock GM, Metzker ML, Muzny DM, Sodergren EJ, Scherer S, Scott G, Steffen D, Worley KC, Burch PE et al. (2004) Genome sequence of the Brown Norway rat yields insights into mammalian evolution *Nature* 428 493-521

Hoodbhoy T, Joshi S, Boja ES, Williams SA, Stanley P and Dean J (2005) Human sperm do not bind to rat zonae pellucidae despite the presence of four homologous glycoproteins *Journal of Biological Chemistry* 280 12721-12731

International Human Genome Sequencing Consortium (2004) Finishing the euchromatic sequence of the human genome *Nature* 431 931-945

Kim E, Baba D, Kimura M, Yamashita M, Kashiwabara SI and Baba T (2005) Identification of a hyaluronidase, Hyal5, involved in penetration of mouse sperm through cumulus mass *Proceedings of the National Academy of Sciences USA* 180 20828-20833

Lefièvre L, Conner SJ, Salpekar A, Olufowobi O, Ashton P, Pavlovic B, Lenton W, Afnan M, Brewis IA, Monk M, Hughes DC and Barratt CL (2004) Four zona pellucida glycoproteins are expressed in the human *Human Reproduction* **19** 1580-1586

Liu C, Litscher ES, Mortillo S, Sakai Y, Kinloch RA, Stewart CL and Wassarman PM (1996) Targeted disruption of the *mZP3* gene results in production of eggs lacking a zona pellucida and infertility in female mice *Proceedings of the National Academy of Sciences USA* **93** 5431-5436

Lu Q and Shur BD (1997) Sperm from ß1,4-galactosyltransferase-null mice are refractory to ZP3-induced acrosome reactions and penetrate the zona pellucida poorly *Development* **124** 4121-4131

Miller DJ, Macek MB and Shur BD (1992) Complementarity between sperm surface beta-1,4-galactosyltransferase and egg-coat ZP3 mediates sperm-egg binding *Nature* **357** 589-593

Nagdas SK, Araki Y, Chayko CA, Orgebin-Crist M-C and Tulsiani DRP (1994) O-linked trisaccharide and N-linked poly-N-acetyllactosaminyl glycans are present on mouse ZP2 and ZP3 *Biology of Reproduction* **51** 262-272

Prasad SV, Wilkins B, Skinner SM and Dunbar BS (1996) Evaluating zona-pellucida structure and function using antibodies to rabbit 55 kDa ZP protein expressed in baculovirus expression system *Molecular Reproduction and Development* **43** 519-529

Rankin T, Familari M, Lee E, Ginsberg AM, Dwyer N, Blanchette-Mackie J, Drago J, Westphal H and Dean J (1996) Mice homozygous for an insertional mutation in the *Zp3* gene lack a zona pellucida and are infertile *Development* **122** 2903-2910

Rankin TL, Tong Z-B, Castle PE, Lee E, Gore-Langton R, Nelson LM and Dean J (1998) Human *ZP3* restores fertility in *Zp3* null mice without affecting order-specific sperm binding *Development* **125** 2415-2424

Rankin T, Talbot P, Lee E and Dean J (1999) Abnormal zonae pellucidae in mice lacking ZP1 result in early embryonic loss *Development* **126** 3847-3855

Rankin TL, O'Brien M, Lee E, Wigglesworth K, Eppig JJ and Dean J (2001) Defective zonae pellucidae in *Zp2* null mice disrupt folliculogenesis, fertility and development *Development* **128** 1119-1126

Rankin TL, Coleman JS, Epifano O, Hoodbhoy T, Turner SG, Castle PE, Lee E, Gore-Langton R and Dean J (2003) Fertility and taxon-specific sperm binding persist after replacement of mouse 'sperm receptors' with human homologues *Developmental Cell* **5** 33-43

Shabanowitz RB and O'Rand MG (1988) Characterization of the human zona pellucida from fertilized and unfertilized eggs *Journal of Reproduction and Fertility* **82** 151-161

Shi S, Williams SA, Seppo A, Kurniawan H, Chen W, Zhengyi Y, Marth JD and Stanley P (2004) Inactivation of the *Mgat1* gene in oocytes impairs oogenesis, but embryos lacking complex and hybrid N-glycans develop and implant *Molecular and Cellular Biology* **24** 9920-9929

Stauss CR, Votta TJ and Suarez SS (1995) Sperm motility hyperactivation facilitates penetration of the hamster zona pellucida *Biology of Reproduction* **53** 1280-1285

Storey BT, Lee MA, Muller C, Ward CR and Wirtshafter DG (1984) Binding of mouse spermatozoa to the zonae pellucidae of mouse eggs in cumulus: Evidence that the acrosomes remain substantially intact *Biology of Reproduction* **31** 1119-1128

Thall AD, Maly P and Lowe JB (1995) Oocyte Galα 1,3Gal epitopes implicated in sperm adhesion to the zona pellucida glycoprotein ZP3 are not required for fertilization in the mouse *Journal of Biological Chemistry* **270** 21437-21440

Tian J, Gong H, Thomsen GH and Lennarz WJ (1997) Gamete interactions in *Xenopus laevis*: Identification of sperm binding glycoproteins in the egg vitelline envelope *Journal of Cell Biology* **136** 1099-1108

Tulsiani DR, Nagdas SK, Cornwall GA and Orgebin-Crist MC (1992) Evidence for the presence of high-mannose/hybrid oligosaccharide chain(s) on the mouse ZP2 and ZP3 *Biology of Reproduction* **46** 93-100

Wassarman PM (1988) Zona pellucida glycoproteins *Annual Review of Biochemistry* **57** 415-442

Waterston RH, Lindblad-Toh K, Birney E, Rogers J, Abril JF, Agarwal P, Agarwala R, Ainscough R, Alexandersson M, An P et al. (2002) Initial sequencing and comparative analysis of the mouse genome *Nature* **420** 520-562

Yanagimachi R (1994) Mammalian fertilization. In *The Physiology of Reproduction*, edn 2, pp 189-317 Eds E Knobil and J Neil. New York Raven Press

# Novel gamete receptors that facilitate sperm adhesion to the egg coat[1]

Michael A Ensslin, Robert Lyng, Adam Raymond, Susannah Copland
and Barry D Shur*

*Department of Cell Biology, Emory University School of Medicine, Atlanta, GA 30322, USA*

Mammalian fertilization is initiated by species-specific binding of the sperm to the zona pellucida, or egg coat. Previous studies suggested that sperm adhesion to the egg coat is facilitated, at least in part, through the binding of sperm surface ß1,4-galactosyltransferase I (GalT) to glycoside chains on the egg coat glycoprotein, ZP3. Binding of multiple ZP3 oligosaccharides induces aggregation of GalT within the sperm membrane, triggering, directly or indirectly, a pertussis toxin sensitive G-protein cascade leading to induction of the acrosome reaction. Consistent with this, spermatozoa bearing targeted deletions in GalT are unable to bind ZP3 or undergo ZP3-dependent acrosomal exocytosis; however, unexpectedly, GalT-null sperm are still able to bind to the egg coat. This indicates that sperm-egg binding requires at least two independent binding mechanisms; a GalT-ZP3-independent event that mediates initial adhesion, followed by a GalT-ZP3 interaction that facilitates acrosomal exocytosis. Our recent efforts have focused on the identification and characterization of these novel gamete receptors. One recently identified sperm protein that is required for sperm adhesion to the egg coat is SED1. SED1 is a bimotif protein composed of two Notch-like EGF repeats and two discoidin/complement F5/8 domains. SED1 is secreted by the epididymal epithelium and coats spermatozoa as they progress through the epididymis. Spermatozoa null for SED1 fail to bind the egg coat, illustrating its requirement for gamete adhesion. Interestingly, SED1 is also expressed by a variety of other epithelial tissues, where it appears to be required for epithelial morphogenesis and/or maintenance. A second novel gamete receptor has recently been identified on the coat of ovulated oocytes. This ZP3-independent, egg coat component is a high molecular weight, wheat germ agglutinin (WGA)-reactive glycoprotein that is derived from oviduct secretions and appears to participate in initial sperm adhesion. The amino acid sequence of this oviduct-derived ligand is currently being determined for the generation of peptide-specific antibodies and for the creation of knock out mice. The identification of

*Corresponding author
E-mail: barry@cellbio.emory.edu
[1]Adapted from Shur *et al.*, 2006 (*Molecular and Cellular Endocrinology*)

novel gamete receptors that are required for sperm-egg binding opens up new avenues for the development of specific contraceptive strategies.

## Introduction

Mammalian spermatozoa must undergo a series of morphological and physiological maturations before they can successfully fertilize the egg (Yanagimachi, 1994). During spermatogenesis and epididymal maturation, the sperm surface is customized to enable it to recognize and bind the egg coat, or zona pellucida (ZP). In the female reproductive tract, the spermatozoa undergo a second series of transformations during a process called "capacitation". Sperm capacitation is associated with changes in the fluidity and permeability of the sperm surface, as well as changes in its intermediary metabolism. Capacitation results in sperm that are capable of completing the acrosome reaction and successfully fertilizing eggs. A limited number of sperm eventually reach the ovulated eggs in the oviduct, and must then traverse the surrounding cumulus cells before binding to the egg ZP. Sperm binding to the ZP triggers the completion of the acrosome reaction, which releases hydrolytic enzymes that enable the spermatozoa to penetrate the ZP, bind to and fuse with the egg plasma membrane, and activate development.

The species-specificity of fertilization is thought to be determined, in large part, at the level of sperm binding to the egg ZP. The work of many investigators suggest that species-specific gamete recognition occurs between defined carbohydrate structures of the ZP and their corresponding receptors on the sperm plasma membrane (Yanagimachi, 1994; Tulsiani et al., 1997; Wassarman et al., 2001). In mouse, the best studied mammalian system, evidence suggests that initial gamete adhesion is mediated by an unusual sperm surface receptor, ß1,4-galactosyltransferase-I (GalT) binding to specific oligosaccharide chains on the zona pellucida glycoprotein ZP3 (Fig. 1) (Wassarman et al., 2001; Rodeheffer and Shur, 2002). Binding of ZP3 oligosaccharide chains induces aggregation of GalT, thus activating, directly or indirectly, acrosomal exocytosis (Macek et al., 1991; Gong et al., 1995). However, more recent studies have questioned whether sperm-egg binding can be solely accounted for by sperm GalT binding to ZP3 oligosaccharides and suggest that sperm-egg binding is likely mediated by receptors in addition to GalT and ZP3. This article will first review earlier studies that indicated a role for GalT and ZP3 oligosaccharides in sperm-egg binding, and then present newer studies that have identified novel gamete receptors that appear to contribute to initial gamete adhesion.

## Previous model for sperm-egg binding in mouse

*Identification of ZP3 as a sperm-binding ligand in the zona pellucida:* The mammalian ZP is synthesized and secreted by growing oocytes and forms an extracellular glycoprotein matrix with a variety of functions. In the mouse, the ZP is composed of three families of glycoproteins, referred to as ZP1, ZP2 and ZP3, with mean molecular weights of 200, 120, and 83 kDa, respectively (Bleil and Wassarman, 1980a; b). ZP3 isolated from unfertilized eggs is able to competitively inhibit sperm binding to the ZP of ovulated oocytes, whereas ZP1 and ZP2 cannot, suggesting that ZP3 possesses the sperm-binding activity of the egg coat. Furthermore, ZP3 isolated from fertilized eggs, which no longer support sperm binding, fails to competitively inhibit sperm binding to unfertilized eggs.

*ZP3 oligosaccharides are likely responsible for sperm binding activity:* The sperm-binding activity of ZP3 appears to reside in its oligosaccharide chains rather than the polypeptide backbone (Florman and Wassarman, 1985). Consistent with this, release of O-linked chains destroys ZP3 binding activity and the released O-linked oligosaccharides bind to spermatozoa and competi-

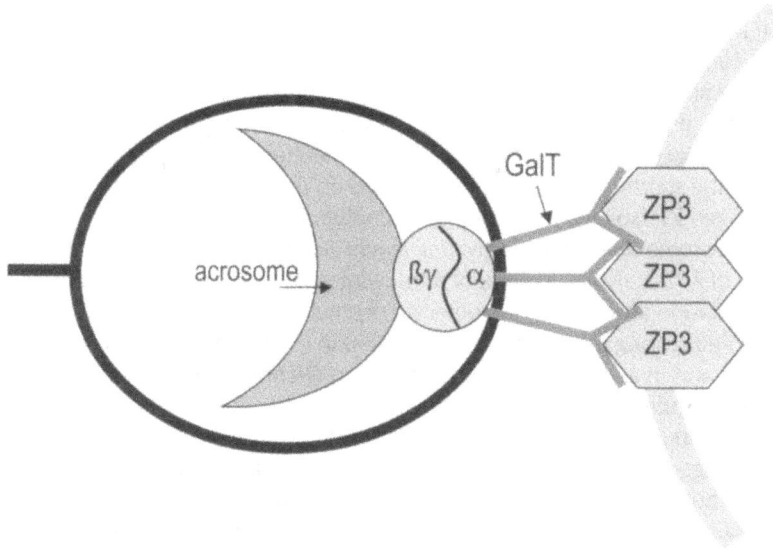

**Figure 1.** Previous model for mouse sperm binding to the egg zona pellucida. Initial gamete binding is proposed to be mediated by sperm surface ß1,4-galactosyltransferase (GalT) binding to specific oligosaccharide residues on the ZP3 glycoprotein. ZP3 oligosaccharides induce aggregation of GalT, thus activating, directly or indirectly, a heterotrimeric G-protein (α, βγ subunits illustrated) cascade leading to acrosomal exocytosis.

tively inhibit sperm-zona binding. In contrast, removal of *N*-linked oligosaccharides by endo-H glycosidase digestion does not affect ZP3 sperm-binding activity.

Although these studies suggest that *O*-linked oligosaccharide chains on ZP3 act as one of the primary binding epitopes recognized by spermatozoa, there remains contention surrounding the molecular composition of the sperm-binding oligosaccharide. In this regard, at least four different monosaccharide residues have been implicated as being critical for initial sperm-egg binding in mouse: α-galactose (Bleil and Wassarman, 1988), GlcNAc (Miller *et al.*, 1992), fucose (Kerr *et al.*, 2004) and mannose (Tulsiani *et al.*, 1992). Further complicating this issue, is the suggestion that *N*-linked oligosaccharides are involved in mouse sperm binding, in addition to *O*-linked oligosaccharides, since treatment of the intact ZP, as opposed to solubilized ZP3, with *N*-glycanase inhibits subsequent sperm binding (Yamagata, 1985). Similarly, *N*-linked oligosaccharides have been implicated in the binding of porcine and bovine spermatozoa to their egg ZP (Yonezawa *et al.*, 1995; 2005).

Collectively, these studies show that soluble glycosides with a variety of specific terminal residues are able to interfere with sperm-zona binding; however, the interpretation of these competitive inhibition assays is brought into question by two findings. First, immunocytochemical analyses show that the sugar composition of the ZP is heterogeneous (Avilés *et al.*, 2000). For example, α-galactosyl and *N*-acetylgalactosaminyl residues are confined to the inner portions of the mature ZP, both ovarian and ovulated, whereas other sugars, such as GlcNAc, are dispersed uniformly throughout the zona. The spatial heterogeneity of zona glycosides makes it impossible to know whether soluble glycosides added as competitors of sperm-zona binding necessarily mimic oligosaccharides available to the spermatozoa at initial binding, or rather, during later aspects of zona penetration.

A second recent observation that impacts on our identification of sperm-binding structures on the ZP, is that it now appears there are at least two distinct binding events during the early stages of sperm-egg interaction, a ZP3-dependent and a ZP3-independent event (see below). Thus, it is unknown which, if any, binding event is being perturbed by the addition of soluble competitive glycosides. Similarly, it is noteworthy that the ZP is modified after ovulation by the adsorption of oviduct-secreted glycoproteins (Verhage et al., 1998; Rodeheffer and Shur, 2004). This highlights the possibility that the sperm-binding glycans exposed on the zona surface are distinct from those assayed in solubilized ZP isolated form ovarian homogenates.

*ZP3 induces the acrosome reaction by activating sperm heterotrimeric G-proteins and cation channels:* The binding of ZP3 activates a range of intracellular signal cascades that culminate in fusion of the plasma membrane and underlying outer acrosomal membrane; that is, the acrosome reaction. The best studied among these are i) a pertussis toxin (PTx)-sensitive heterotrimeric G-protein (Ward et al., 1992) and ii) a voltage-independent cation channel that regulates membrane potential, which in turn, regulates a voltage-dependent cation channel (Florman et al., 1998). These cascades, among others, result in elevation in $[Ca^{2+}]_i$ and pHi, which are required for acrosomal exocytosis (Florman et al., 1998). Zona binding also activates a sperm glycine receptor/Cl- channel that may contribute to the control of voltage-sensitive calcium influx (Meizel, 1997).

As spermatozoa undergo the acrosome reaction, they must remain transiently attached to the ZP prior to initiation of zona penetration. The binding of acrosome-reacted spermatozoa to the zona may depend on ZP2, since acrosome-reacted spermatozoa lose affinity for ZP3 and gain affinity for ZP2 (Bleil et al., 1988; Mortillo and Wassarman, 1991). Recent developments suggest that the acrosome reaction may sequentially release and/or expose specific components of the acrosomal matrix, which may stabilize sperm adhesion to the zona matrix (Kim et al., 2001a). Candidates include sp56 (Kim et al., 2001b) and zonadhesin, among other zona-binding proteins associated with the acrosomal matrix.

*Acrosome-reacted spermatozoa penetrate the ZP, bind and fuse with the egg plasma membrane:* Eventually, the acrosome-reacted spermatozoa penetrates the ZP, presumably relying upon hydrolytic enzymes released from the acrosome to digest a penetration slit through the zona matrix (Yanagimachi, 1994). Most notable among these, is the trypsin-like protease acrosin. However, complete elimination of acrosin by homologous recombination still enables sperm penetration through the zona, suggesting that other acrosomal enzymes are likely important for penetration (Baba et al., 1994). The sperm arrives in the perivitelline space where it binds to and fuses with the egg plasma membrane. Recently, there has been considerable advancement in the identification of sperm surface receptors (for example, Izumo) and oocyte membrane tetraspanin proteins (for example, CD9), that appear critical for the binding and subsequent fusion of the spermatozoa with the egg plasma membrane (Le Naour et al., 2000; Miyado et al., 2000; Nishimura et al., 2001; Inoue et al., 2005).

*ZP3 sperm receptor activity is destroyed after egg activation:* Following egg activation, cortical granules are released to elicit the zona block to polyspermy. The contents of the cortical granules inactivate ZP3, among other effects on the ZP, which leads to a loss of ZP3's sperm binding activity; however, ZP3 is indistinguishable by SDS-PAGE before and after fertilization (Bleil and Wassarman, 1980b). Therefore, the loss of ZP3's sperm binding activity must result from subtle, but critical, modifications in its structure and/or sperm-binding glycosides.

## Sperm receptors for ZP3 oligosaccharides

*GalT functions as a ZP3 receptor on spermatozoa:* The sperm receptor for ZP3 oligosaccharides has been more difficult to identify. Thus far, among the sperm surface proteins thought to bind the egg coat, and ZP3 oligosaccharides in particular, sperm surface GalT satisfies virtually all of the criteria expected of a ZP3 receptor (Fig. 1). Although glycosyltransferases are generally considered to be intracellular biosynthetic enzymes, a few glycosyltransferases are also present on the cell surface where they function as cell adhesion molecules by binding their specific oligosaccharide substrates, or ligands, on adjacent cell surfaces or in the extracellular matrix (Shur, 1992). Since sugar donor substrates are normally not present in the extracellular fluids, it is presumed that the surface-localized glycosyltransferase-glycoside complex forms a stable adhesive bond. Among those glycosyltransferases reported to be present on the cell surface, the best studied is GalT. Most cells synthesize two GalT isoforms that differ in the length of their cytoplasmic domains. Results suggest that the GalT isoform possessing the shorter cytoplasmic domain is normally confined to the Golgi complex, where it serves a purely biosynthetic function. In contrast, the long GalT isoform can function both biosynthetically in the Golgi, and due to its additional cytoplasmic sequence, can also function as a signal transducing receptor when present on the cell surface (for review, Evans *et al.*, 1995; Rodeheffer and Shur, 2002).

Sperm are unusual in that they express exclusively the long GalT isoform (Shaper *et al.*, 1990; Pratt and Shur, 1993), and all of their GalT protein is confined to the dorsal, anterior aspect of the plasma membrane, where it behaves as an integral membrane protein (Shur and Neely, 1988). In the epididymis, GalT is masked by epididymal glycosides that are subsequently shed from the sperm surface during capacitation, thus unmasking GalT and enabling it to bind its GlcNAc-terminating ligand on ZP3 (Shur and Hall, 1982a; b; Lopez *et al.*, 1985). GalT specifically binds to the same class of ZP3 oligosaccharides that possess sperm-binding activity, and removing or masking the GalT I binding site on these oligosaccharides removes their sperm-binding activity (Miller *et al.*, 1992).

*ZP3-induced aggregation of sperm GalT contributes to acrosomal exocytosis:* Aggregation of GalT by multivalent ZP3 oligosaccharides causes activation of a heterotrimeric G-protein cascade, leading to the acrosome reaction (Lopez and Shur, 1987). The cytoplasmic domain of GalT I binds, directly or indirectly, to heterotrimeric G proteins that are activated following ZP3-induced aggregation of GalT I (Gong *et al.*, 1995). In support of this, ectopic expression of GalT I on *Xenopus* oocytes results in ZP3-specific binding and G-protein activation, and mutagenesis of a putative G-protein binding motif within the GalT I cytoplasmic domain prevents ZP3-dependent G-protein activation (Shi *et al.*, 2001).

*Altering the expression of sperm GalT impacts the efficacy of sperm-egg binding:* The function of sperm GalT in mediating binding to ZP3 oligosaccharides has been tested through the ability to selectively increase or decrease GalT expression on the sperm surface (Fig. 2). Transgenic sperm that overexpress GalT bind more ZP3 than do normal sperm, have accelerated G-protein activation, and undergo precocious acrosome reactions (Youakim *et al.*, 1994). In contrast, sperm in which surface GalT has been eliminated, but which maintain normal intracellular galactosylation, no longer bind ZP3 *in vitro* or undergo zona-induced acrosomal exocytosis (Lu and Shur, 1997).

## Successful sperm-egg adhesion requires receptors in addition to GalT and ZP3

Despite these and many other studies that support a role for sperm GalT and ZP3 oligosaccharides in initial sperm-egg binding, recent observations have forced a re-examination of their roles in mediating gamete adhesion. For example, even though GalT-null sperm are unable to bind ZP3 in solution or undergo zona-induced acrosome reactions, they still retain the ability to bind to the ovulated egg coat (Lu and Shur, 1997). This binding enables GalT-null spermatozoa to penetrate the zona pellucida matrix, presumably via spontaneous acrosome reactions, and fertilize eggs, although at only 7% the efficiency of wild-type spermatozoa, when assayed *in vitro*.

| **Wild-type** | **GalT over-expresser** | **GalT-null** |
|---|---|---|
| a) Binding of ZP3 | elevated over +/+ | near background levels |
| b) Zona-induced AR | accelerated AR | fails to undergo zona-induced AR |
| c) Penetration of the ZP | transient binding limited penetration | normal binding limited penetration |

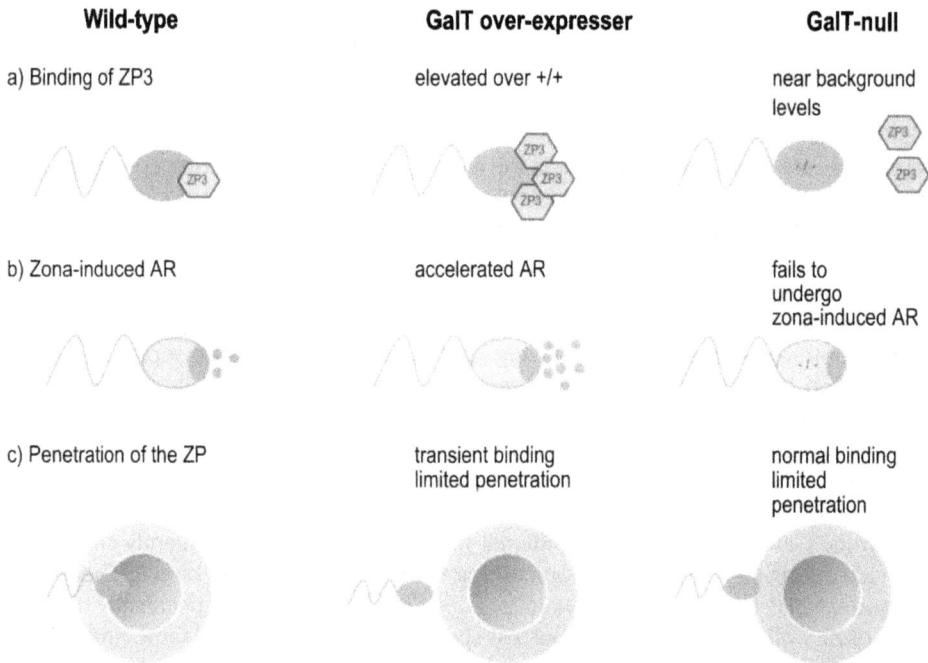

**Figure 2.** Altering GalT expression on the sperm surface leads to predictable consequences in binding ZP3 and acrosomal exocytosis. Diagrams are presented to summarize: a) the relative binding of radiolabeled ZP3 to normal spermatozoa (wild-type), increased ZP3 binding to spermatozoa that overexpress GalT (over-expresser) and reduced ZP3 binding to GalT-null spermatozoa; b) the binding of ZP3 induces acrosomal exocytosis in normal spermatozoa, but the increased levels of GalT on over-expresser spermatozoa lead to a hyperactivation of acrosomal exocytosis, whereas deletion of GalT makes spermatozoa refractory to ZP3-induced acrosomal exocytosis; c) acrosome-reacted sperm are now able to penetrate the ZP but the precocious acrosome reaction in GalT-over-expressers makes them bind to the ZP with low affinity, whereas GalT-null spermatozoa show near background levels of zona-induced acrosome reactions. However, despite the inability of GalT-null spermatozoa to bind ZP3, they still bind to the egg coat, although this does not lead to acrosomal exocytosis. Surprisingly, GalT-null males are fertile, suggesting that the reduced efficiency of zona penetration is sufficient for fertility in the absence of competing wild-type spermatozoa. In any event, the ability of GalT-null spermatozoa to bind to the egg coat indicates the presence of additional gamete receptors independent of GalT and ZP3.

Similarly, as discussed above, oligosaccharides that are not predicted to be substrates for GalT still competitively inhibit sperm-egg binding (Johnston et al., 1998; Loeser and Tulsiani, 1999; Amari et al., 2001; Kerr et al., 2004). Even though some of these inhibitory oligosaccharides do not appear to be present within the zona pellucida (Avilés et al., 2000), their ability to inhibit binding speaks against a role for GalT in initial gamete adhesion. Finally, the observation that mouse spermatozoa still bind to eggs in which the mouse ZP3 polypeptide has been replaced by human ZP3 argues against ZP3 as the only sperm-binding ligand in the egg coat (Rankin et al., 1998; 2003). Some of this confusion likely results from the fact that the biological activity of putative gamete receptors is usually assayed by competitive inhibition of sperm binding to the coats of ovulated eggs, whereas the source of competitive ligands for these experiments is frequently the ovarian zona pellucida. At any rate, all of these observations suggest that sperm binding to the zona pellucida may involve multiple receptor-ligand interactions of which ZP3 binding to GalT may be one component.

## Identifying novel gamete adhesins on the spermatozoon and the egg coat

Among the many sperm proteins implicated as receptors for the egg coat, as opposed to receptors specifically for ZP3 glycosides, two deserve particular attention: zonadhesin and SED1 (previously called p47, lactadherin, MFG-E8). Originally identified in the pig, zonadhesin is a sperm protein that binds to the ZP in a species-specific manner (Hardy and Garbers, 1995; Gao and Garbers, 1998). The cDNA for murine zonadhesin demonstrates that it encodes a novel precursor protein comprising a single transmembrane domain separating a short intracellular carboxy terminus from a relatively large extracellular region. The extracellular sequence specifies a mosaic protein comprising several MAM (meprin, A5 protein, receptor protein-tyrosine phosphatase mu), mucin, D- and EGF-like domains. Homologous domains have been identified in a number of membrane proteins and have been implicated in protein-protein interactions. As discussed above, recent studies suggest that zonadhesin is localized in the acrosomal matrix, rather than on the plasma membrane of acrosome-intact spermatozoa. Thus, zonadhesin may mediate the binding of spermatozoa to the egg coat during early stages of acrosomal exocytosis, rather than during initial sperm-egg binding.

*Identification of SED1 as a gamete receptor:* SED1 was first identified as a putative gamete receptor by affinity chromatography of solubilized boar sperm membranes applied to columns of immobilized zona pellucida glycoproteins (Ensslin et al., 1998). Elution and sequencing of the principal bound proteins identified p47, which proved to be identical to a milk fat globule membrane protein previously isolated from a variety of species (see Shur et al., 2004 for review). The mouse protein was cloned and named SED1 to bring uniformity to the various names assigned to the mammary gland proteins; SED denotes a secreted protein containing a cleavable signal sequence, N-terminal Notch-like type II EGF repeats and C-terminal Discoidin/F5/8 Complement domains (Fig. 3).

*SED1 expression, secretion and deposition on the sperm surface:* Within the sexually mature male reproductive tract, SED1 is found associated with two distinct secretory pathways. The first occurs within the Golgi complex of spermatogenic cells, and presumably results in SED1 secretion onto the maturing sperm surface; the second is within the initial segment of the epididymis where sperm are exposed to high levels of secreted SED1 as they traverse through the epididymal lumen (Ensslin and Shur, 2003). Spermatozoa retain SED1 as they progress through the epididymis, and excess soluble SED1 appears to be removed by the adsorptive cells of the cauda epididymis before spermatozoa enter the vas deferens. At this point, sperm are now able to fertilize an egg (following capacitation) and SED1 expression is confined to the sperm plasma membrane overlying the acrosome, the site of sperm binding to the egg coat.

*In vitro studies support SED1 function during sperm-egg binding:* Recombinant SED1 binds selectively to the sperm plasma membrane overlying the acrosome and competitively inhibits sperm-egg binding. Similarly, anti-SED1 antibodies localize endogenous SED1 to this same plasma membrane domain and block sperm binding to the egg coat. In an analogous manner, recombinant SED1 binds to the zona pellucida of unfertilized eggs, but not to eggs following fertilization, thus mimicking the inability of spermatozoa to bind the zona pellucida after fertilization. Studies with truncated SED1 proteins illustrate that the biological activity of SED1 requires the discoidin/F5/8 C domains, which are responsible for its binding to both the sperm membrane and to the zona pellucida matrix (Ensslin and Shur, 2003).

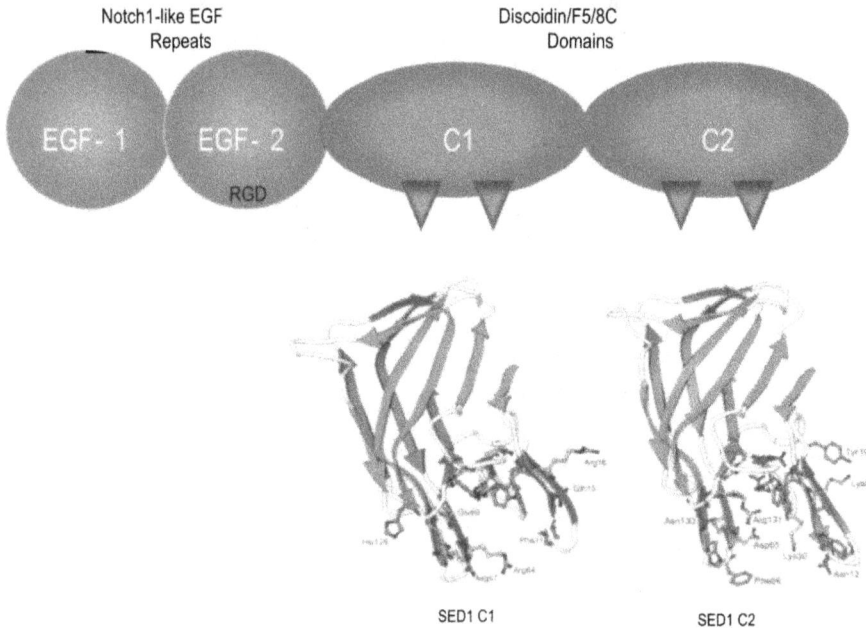

Figure 3. Structural motifs of SED1. SED1 contains a cleavable signal sequence (not shown), two Notch1 like-EGF repeats, the second of which possesses an RGD integrin-binding motif, as well as two discoidin/F5/8C domains that are able to bind phospholipid bilayers and/or extracellular glycosides via 2-3 hairpin loops projecting from the central barrel core. Molecular models of SED1 C1 and C2 domains based upon the crystallographic analysis of the C2 domain of Factor V and VIII are also shown. The discoidin/F5/8C domain core is composed of an eight-stranded ß sheet, five of which appear in the front (green arrows) and three in the rear of the barrel (red arrows). Each C domain projects two large hairpin loops (or spikes), as well as a smaller, irregular intermediate loop. The C1 and C2 domains of SED1 are more similar to the Complement C2 domain, than to the Complement C1 domain, in that they both possess the ß-hairpin loops required for association with lipid bilayers. However, the homologous loops in C1 and C2 display distinctly different amino acid side chains that may dictate unique binding specificities. The aromatic residues are shown in green, charged residues in white and uncharged polar residues in magenta. Sequence alignment of the SED1 C1 and C2 domains with Factor V and VIII C2 domains are pictured, with identical residues in red and conserved residues in green. The relative positions of the hairpin loops, or spikes, are indicated as determined by crystallographic analysis (adapted from Shur et al., 2004).

Overlay assays suggest that recombinant SED1 recognizes two of the three glycoprotein families, ZP2 and ZP3, that comprise the mouse zona pellucida. Given the ability of discoidin/F5/8 C domains to bind carbohydrate matrices (Reitherman et al., 1975; Fuentes-Prior et al., 2002), it is most likely that SED1 is binding to the carbohydrate residues of ZP2 and ZP3, although this awaits direct testing.

*In vivo analysis of SED1-null male fertility:* Independent of its zona binding specificity, the function of SED1 during fertilization was directly tested by analyzing the fertility of SED1-null mice. All SED1-null males produced smaller litters than controls, although the average litter size per male varied considerably, ranging from no pups born (apparent sterility) to litter sizes approaching the lower limit of normal. This phenotypic variability, that is incomplete penetrance, is most likely due to genetic factors segregating on the mosaic 129Sv/C57Bl6 background, as was shown to be the case for at least two other targeted mutations in spermatozoa (Pearse *et al.*, 1997; Nayernia *et al.*, 2002). In any event, despite the variable fertility of SED1-null males *in vivo*, sperm from all SED1-null males tested show near background levels of binding to the zona pellucida *in vitro* (Fig. 4). The reduced binding of SED1-null spermatozoa is not due to secondary effects on sperm morphology, number, acrosomal status, or motility, thus directly supports a role for SED1 in sperm-egg binding (Ensslin and Shur, 2003).

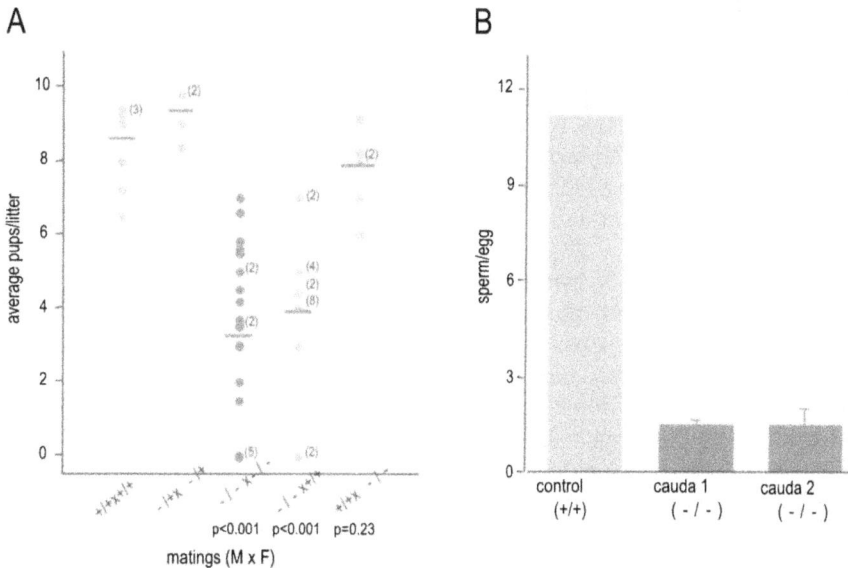

**Figure 4.** SED1-null males show reduced fertility *in vivo* and their spermatozoa fail to bind the zona pellucida *in vitro*. A) The frequency distribution of litter sizes resulting from each breeding pair is presented. Each dot represents the average litter size resulting from a single breeding pair (0-20 litters/pair). In instances where breeding pairs produced identical average litter sizes, the number of pairs is given in parentheses. All control breeding pairs produced an average of 8.5-9.3 pups/litter, whereas matings between SED1-null mice produced an average of 3.3 pups/litter. This average reflects wide variability in male fertility, ranging from apparent sterility in five males (that is, no pups) to litter sizes approaching the lower limit of normal. The reduced litter size in homozygous null matings is due to reduced fertility of the male ($p < 0.001$ relative to control matings) rather than the female, which is near normal ($p = 0.23$ relative to controls). B) Sperm isolated from the cauda epididymis of SED1 males (-/-) bind to the zona pellucida at very low levels ($1.45 \pm 0.56$ spermatozoa/egg), as compared to wild-type littermates (+/+) ($11.2 \pm 0.85$ spermatozoa/egg). Error bars represent Mean $\pm$ SEM (reprinted from Ensslin and Shur, 2003 with permission from Elsevier)

*SED1 functional domains and mechanism of action:* Discoidin/F5/8 C domains are remarkable structural motifs that mediate the binding of a wide variety of proteins to cell membranes and to extracellular matrix components (Fuentes-Prior *et al.*, 2002). Recent structural analysis indicates that discoidin/F5/8 C domains comprise a central barrel of eight antiparallel ß-strands from which two (or three) hairpin spikes project that present specific polypeptide sequences for interacting with various binding surfaces (Fig. 3). For example, the spikes found in the C2 domain of Complement possess hydrophobic residues that intercalate into the lipid bilayer and are surrounded by a ring of charged residues that dock with phosphatidylserine headgroups (Macedo-Ribeiro *et al.*,

1999; Pratt et al., 1999).

The discoidin/F5/8 C domains are also characteristic of a number of proteins that bind to galactose and other monosaccharide determinants, including the prototypical animal lectin "discoidin" from *Dictyostelium*. These sugar-binding motifs rely upon a similar eight-stranded antiparallel ß-barrel core, but they extend truncated hairpin loops that coordinate binding to glycoside residues rather than to membrane phospholipids (Reitherman et al., 1975; Lonhienne et al., 2001; Fuentes-Prior et al., 2002). In all instances, hypervariable loops projecting from the ß barrel core are thought to determine the specificity of binding to membranes or to the extracellular matrix. It is, therefore, not surprising that the discoidin/F5/8 C domains are thought to be required for SED1 biological activity by binding directly to the ZP and the sperm plasma membrane.

In contrast to the discoidin/F5/8 C domains, there is no evidence that SED1 function during sperm-egg binding requires the EGF repeats, or more specifically, RGD recognition by integrins, as has been reported for SED1; that is, lactadherin-mediated adhesion of mammary epithelial cells (Andersen et al., 2000). Nevertheless, one intriguing possibility is that RGD motifs within SED1 may participate in sperm interactions with the uterine and/or oviductal epithelium, which are known to express αvß3 integrins and to which sperm bind during regulated transit up the oviduct (Chegini et al., 2001).

*Models for SED1 function during gamete adhesion:* Given that the discoidin/F5/8 C domains can account for SED1 binding to both the sperm membrane and the zona pellucida, the simplest model proposes that SED1 functions as a monomer, whereby the two discoidin/F5/8 C domains of a single SED1 polypeptide are sufficient to bind spermatozoa to the zona pellucida. Since constructs containing both discoidin/F5/8 C domains (C1C2) show increased binding to the zona, relative to constructs with only the C1 domain (Ensslin and Shur, 2003), it is possible that the C2 domain has greater affinity for the zona, whereas the C1 domain mediates attachment to the sperm membrane. Alternatively, SED1 may facilitate sperm-egg binding as a dimer, or oligomer, in a manner analogous to that reported for the epithelial cell adhesion molecule, Ep-CAM (Balzar et al., 2001). Importantly, the two EGF repeats in both SED1 and Ep-CAM contain classic Notch-like consensus sequences and only two Notch-like EGF repeats are required for protein-protein binding (Lawrence et al., 2000; Balzar et al., 2001). The potential oligomerization of SED1 may be further regulated by glycosylation, since the first EGF repeat contains the Notch O-glycosylation consensus sequence that is known to influence Notch-dependent protein-protein binding (Haltiwanger, 2002).

In either scenario, one of the discoidin/F5/8 C domains is postulated to mediate SED1 attachment to spermatozoa, whereas the second discoidin/F5/8 C domain, either as part of a SED1 monomer or oligomer, is exposed for association with the ZP. Crystallographic data suggest that the exposed, or unoccupied, discoidin/F5/8 C domain remains in a cryptic, or "closed" conformation until presentation to its binding surface (that is, ZP), which stabilizes the active, or "open" binding conformation (Fuentes-Prior et al., 2002).

### Identification of an oviduct-derived, ZP3-independent sperm-binding ligand

The ability of GalT-null spermatozoa to bind to the coat of ovulated eggs, even though they fail to bind to ZP3 at normal levels, suggests that egg coats must contain sperm-binding ligands in addition to ZP3. Insight into the identification of ZP3-independent sperm-binding ligands came from the realization that spermatozoa bind to ovulated oocytes, rather than to the coat of ovarian oocytes which have served as the source for much of our understanding of ZP3 function. Consequently, the sperm-binding activity of egg coats isolated from ovulated and ovarian oocytes was compared.

*The zona pellucida of ovulated eggs contains two distinct sperm-binding ligands:* As expected, the ZP isolated from ovarian eggs inhibits the binding of normal spermatozoa to ovulated eggs, presumably due to the presence of ZP3, but does not inhibit the binding of GalT-null spermatozoa to ovulated eggs, consistent with the inability of GalT-null spermatozoa to bind ZP3. In contrast, the egg coats isolated from ovulated eggs competitively inhibit the binding of both normal and GalT-null spermatozoa to ovulated oocytes, indicating the presence of a ZP3-independent ligand on the ovulated egg that is recognized by both normal and GalT-null spermatozoa (Fig. 5) (Rodeheffer and Shur, 2004).

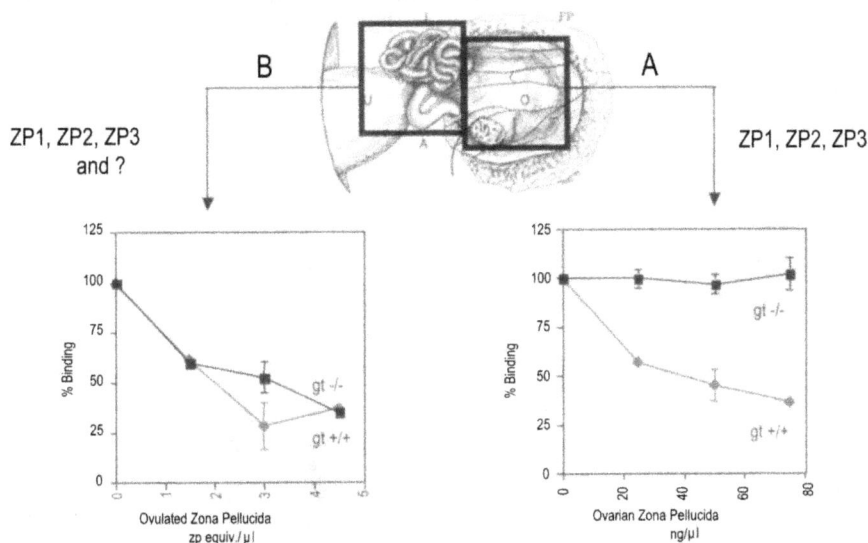

**Figure 5.** The ovulated egg coat contains at least two distinct sperm-binding ligands. Ovulated, but not ovarian, ZP glycoproteins contain a ligand that competitively inhibits GalT-null spermatozoa binding to ovulated eggs. A) Inhibition of wild-type (+/+) and GalT-null (-/-) sperm binding to ovulated eggs with 0-75 ng/ml solubilized ovarian ZP. B) Inhibition of wild-type (+/+) and GalT-null (-/-) spermatozoa binding to ovulated eggs with 0-5 zonae equivalents solubilized ovulated ZP. Each point represents the mean $\pm$ SEM; $n = 3$ experiments. In a single experiment, the value for each data point represents the average of three determinations (for a total of 9 assays per data point) (from Rodeheffer and Shur, 2004).

Since the ZP3-independent ligand is unique to the ovulated egg coat, and is presumably adsorbed onto the ZP as the oocyte enters the oviduct, it can be isolated away from the insoluble ZP matrix by extensive washing. Thus, the two distinct sperm-binding activities can be attributed to two distinct sperm-binding ligands present in distinct compartments of the ovulated egg coat: the ZP3 ligand in the insoluble zona matrix and a peripherally-associated, oviduct-derived, ZP3-independent ligand that can be removed by washing. Furthermore, since wild-type sperm are sensitive to both the peripheral and matrix fractions, this indicates that the ZP3-independent ligand is physiologically relevant to wild-type sperm-egg binding, and not a peculiarity of the GalT I-null sperm phenotype. In fact, oviductal glycoprotein secretions are known to permeate the zona pellucida, and, at least in hamster, there is evidence to suggest that the ovulated zona pellucida has biological activities that are distinctly different from those in the ovarian zona pellucida (Robitaille *et al.*, 1988; Kan *et al.*, 1990; St-Jacques *et al.*, 1992; Boatman and Magnoni, 1995).

*The ZP3-independent ligand is a high molecular weight, basic, WGA-reactive glycoprotein:* The novel egg coat ligand activity does not result from contamination by egg cortical granules or by

ZP3. In addition, this ligand is distinct from the mouse ortholog of Oviduct-specific Glycoprotein (OGP), which has been implicated in sperm-egg adhesion in hamster (Boatman and Magnoni, 1995). Lectin depletion studies and lectin-blotting of two-dimensional polyacrylamide gels indicate that the ZP3-independent ligand in ovulated egg coats is a relatively basic, high molecular weight, wheat germ agglutinin (WGA)-reactive glycoprotein that is not present in the ovarian zona pellucida. This protein, when eluted from isoelectric focusing gels, possesses sperm-binding activity for both wild-type and GalT-null spermatozoa, whereas ZP3 possesses sperm-binding activity for only wild-type spermatozoa (Fig. 6). These results suggest that the WGA-reactive, basic glycoprotein within the ovulated egg coat functions as a ZP3- and OGP-independent ligand that facilitates gamete adhesion. Current studies are aimed at purification of sufficient material for amino acid sequencing and identification of the full-length polypeptide.

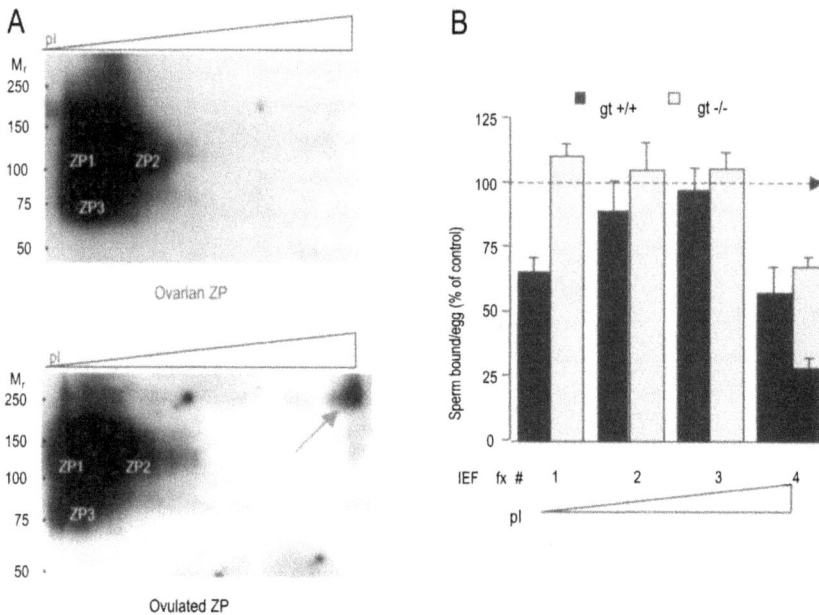

**Figure 6.** Ovulated, but not ovarian, zona pellucida glycoproteins contain a basic, WGA-reactive, 250 kDa glycoprotein that possesses sperm-binding activity. (A) 2.5 mg of ovarian and 500 zonae equivalents of ovulated zona proteins were separated by two-dimensional polyacrylamide gel electrophoresis, transferred to PVDF and visualized by staining with biotinylated-WGA. The arrow indicates a 250 kDa, basic, WGA-reactive protein that is present in ovulated ZP but not ovarian ZP. The pH gradient, the theoretical location of the matrix proteins according to Bleil and Wassarman (1980b), and molecular weight markers are shown. (B) Proteins isolated from the basic region of an IEF gel containing ovulated zona pellucida glycoproteins inhibit wild-type and GalT-null spermatozoa binding to ovulated eggs. Proteins were isolated from acidic (fraction #1), neutral (fractions #2 and #3) and basic (fraction #4) regions of an IEF gel containing 500 ovulated egg ZP and tested for biological activity in the sperm-egg binding assay. The acidic fraction (containing ZP3) inhibits wild-type (+/+), but not GalT-null (-/-), spermatozoa from binding to ovulated eggs. In contrast, the basic fraction (containing the basic ligand) prevents both wild-type and GalT-null spermatozoa from binding to ovulated eggs. Doubling the amount of starting material (2X) lead to 69% inhibition of sperm-egg binding by fraction #4. Each bar represents the mean ± SEM; n = 4 experiments. In a single experiment, the value for each bar represents the average of three determinations (for a total of 12 assays per bar). The dashed line indicates the number of spermatozoa bound in the control; that is, material obtained from a blank IEF gel treated identically to gels containing zona glycoproteins (from Rodeheffer and Shur, 2004).

## Revised model for mouse sperm-egg binding

The molecular mechanisms underlying gamete recognition are not yet fully understood. Until recently, evidence suggested that gamete recognition in the mouse is mediated by a single egg coat glycoprotein (ZP3) that is recognized by a specific sperm receptor, with most evidence implicating GalT as, at least one of, the ZP3 receptors. In this article, data are presented suggesting that gamete recognition is more complex than a single receptor-ligand interaction and can be resolved into at least two distinct binding events: a ZP3- and GalT-independent interaction responsible for gamete adhesion, and a ZP3- and GalT-dependent interaction that facilitates acrosomal exocytosis (Fig. 7). Since sperm are able to bind to ovarian eggs, ZP3 may support some degree of sperm adhesion as well, possibly via GalT. However, sperm normally fertilize ovulated rather than ovarian eggs, and therefore must encounter the ZP3-independent binding activity under normal physiological conditions. The presence of an oviduct-derived sperm-binding ligand can therefore account for the observation that mouse spermatozoa still bind to ovulated eggs in which the constituent mouse ZP glycoproteins are replaced by human homologues (Rankin *et al.*, 2003), since the fertilizing spermatozoa would bind the oviduct-derived ligand irrelevant of the species composition of the insoluble ZP matrix.

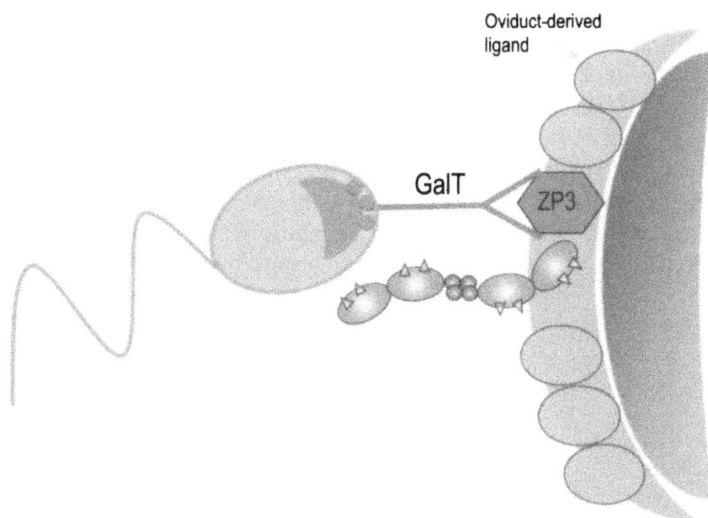

**Figure 7.** Revised model for mouse sperm-egg binding. Studies discussed in this article illustrate the requirement for at least two distinct mechanisms mediating sperm-egg binding in mouse: a GalT-ZP3-independent mechanism that mediates initial sperm-egg adhesion followed by a GalT-ZP3-dependent interaction that contributes to acrosomal exocytosis. At least two GalT-ZP3-independent receptors have been identified thus far. Sperm SED1 appears to be required for initial sperm adhesion to the egg coat and an oviduct-derived egg coat ligand that has sperm-binding activity, the necessity of which for successful fertilization is presently unknown. It is likely that further analysis of mice bearing targeted deletions in these and other sperm and egg components will reveal the presence of additional gamete receptors that orchestrate the multiple steps that contribute to successful fertilization.

The sperm receptor for the oviduct-derived, ZP3-independent ligand is of great interest. It is possible that SED1 described above, which contributes to initial sperm-egg adhesion, serves in this capacity. Alternatively, SED1 and the ZP3-independent ligand may interact with unique gamete components that have yet to be identified. In any event, the studies presented here and elsewhere (Rankin *et al.*, 2003) necessitate a revision of our earlier models for mouse gamete interactions. The most attractive models at present suggest that SED1 is required for the initial adhesion between spermatozoa and egg, and in so doing, brings ZP3 oligosaccharides into close enough proximity to bind and aggregate its receptor on the sperm membrane (that is, GalT among other candidate receptors) (Fig. 7). ZP3 binding then elicits activation of heterotrimeric G-proteins and ion channels that culminate in acrosomal exocytosis, zona penetration and oocyte activation.

## Potential utility of SED1 in human reproduction

The identification of SED1 as a requisite sperm receptor for binding to the ZP raises some interesting possibilities for contraceptive development that distinguish it from other potential sperm surface targets. Most importantly, SED1 is added to the morphologically mature spermatozoa as they traverse through the epididymis, rather than being incorporated into the sperm membrane solely during spermatogenesis. Thus, SED1 deposition on spermatozoa could be potentially eliminated by inhibiting epididymal secretion. In an analogous fashion, identification of the residues that mediate SED1 attachment to the sperm surface or to the ZP enables the screening of low molecular weight mimetics that may displace SED1 from spermatozoa and/or directly interfere with SED1-dependent binding to the egg coat. Since the binding specificity of the discoidin domains is encoded by relatively few amino acid residues that form the hairpin loops, it should be relatively straightforward to screen for molecular mimetics of these residues within the context of the crystallographic data. Although most of these approaches are amenable to systemic application, one significant advantage is that antagonists of sperm-egg interactions, such as SED1 mimetics and antibodies can be applied topically at the time of intercourse, thus offering a greater degree of personal control regarding the appropriate use of contraceptive agents. Finally, it should not be overlooked that since SED1 is added to morphologically mature sperm, it may be possible to increase the reproductive efficacy of infertile men with subnormal levels of SED1 by topical application of SED1 or the relevant functional domains.

## Acknowledgements

Original work conducted in the authors' laboratory was supported by grant RO1 HD23479 from the National Institutes of Health.

## References

Amari S, Yonezawa N, Mitsui S, Katsumata T, Hamano S, Kuwayama M, Hashimoto Y, Suzuki A, Takeda Y and Nakano M (2001) Essential role of the nonreducing terminal α-mannosyl residues of the N-linked carbohydrate chain of bovine zona pellucida glycoproteins in sperm-egg binding *Molecular Reproduction and Development* **59** 221-226

Andersen MH, Graversen H, Fedosov SN, Petersen TE and Rasmussen JT (2000) Functional analyses of two cellular binding domains of bovine Lactadherin *Biochemistry* **39** 6200-6206

Avilés M, Okinaga T, Shur BD and Ballesta J (2000) Differential expression of glycoside residues in the mammalian zona pellucida *Molecular Reproduction and Development* **57** 296-308

Baba T, Azuma S, Kashiwabara S and Toyoda Y (1994) Sperm from mice carrying a targeted mutation of the acrosin gene can penetrate the oocyte zona pellucida and effect fertilization *Journal of Biological Chemistry* **269** 31845-31849

Balzar M, Briaire-de Bruijn IH, Rees-Bakker HA, Prins FA, Helfrich W, de Leij L, Riethmuller G, Alberti S, Warnaar SO, Fleuren GJ and Litvinov SV (2001) Epidermal growth factor-like repeats mediate lateral and reciprocal interactions of Ep-CAM molecules in homophilic adhesions *Molecular and Cellular Biology* **21** 2570-2580

Bleil JD and Wassarman PM (1980a) Structure and function of the zona pellucida: identification and characterization of the proteins of the mouse oocytes zona pellucida *Developmental Biology* **76** 185-202

**Bleil JD and Wassarman PM** (1980b) Mammalian sperm-egg interaction: identification of a glycoprotein in mouse egg zonae pellucidae possessing receptor activity for sperm *Cell* **20** 873-882

**Bleil JD and Wassarman PM** (1988) Galactose at the non-reducing terminus of O-linked oligosaccharides of mouse egg zona pellucida glycoprotein ZP3 is essential for the glycoprotein's sperm receptor activity *Proceedings of the National Academy of Sciences USA* **85** 6778-6782

**Bleil JD, Greve JM and Wassarman PM** (1988) Identification of a secondary sperm receptor in the mouse egg zona pellucida: role in maintenance of binding of acrosome-reacted sperm to eggs *Developmental Biology* **128** 376-385

**Boatman DE and Magnoni GE** (1995) Identification of a sperm penetration factor in the oviduct of the golden hamster *Biology of Reproduction* **52** 199-207

**Chegini N, Kotseos K, Ma C, Williams RS, Diamond MP, Holmdahl L and Skinner K** (2001) Differential expression of integrin alpha v and beta 3 in serosal tissue of human intraperitoneal organs and adhesion *Fertility and Sterility* **75** 791-796

**Ensslin MA and Shur BD** (2003) Identification of mouse sperm SED1, a bi-motif EGF repeat and discoidin domain protein involved in sperm-egg binding *Cell* **114** 405-417

**Ensslin MA, Vogel T, Calvete JJ, Thole HH, Schidtke J, Matsuda T and Toepfer-Petersen E** (1998) Molecular cloning and characterization of P47, a novel boar sperm-associated zona pellucida-binding protein homologous to a family of mammalian secretory proteins *Biology of Reproduction* **58** 1057-1064

**Evans SC, Youakim A and Shur BD** (1995) Biological consequences of targeting ß1,4 galactosyltransferase to two different subcellular compartments *BioEssays* **17** 261-268

**Florman HM and Wassarman PM** (1985) O-linked oligosaccharides of mouse egg ZP3 account for its sperm receptor activity *Cell* **41** 313-324

**Florman HM, Arnoult C, Kazam IG, Li C and O'Toole CM** (1998) A perspective on the control of mammalian fertilization by egg-activated ion channels in sperm: a tale of two channels *Biology of Reproduction* **59** 12-16

**Fuentes-Prior P, Fujikawa K and Pratt KP** (2002) New insights into binding interfaces of coagulation factors V and VIII and their homologues – lessons from high resolution crystal structures *Current Protein and Peptide Science* **3** 313-339

**Gao Z and Garbers DL** (1998) Species diversity in the structure of zonadhesin, a sperm-specific membrane protein containing multiple cell adhesion molecule-like domains *Journal of Biological Chemistry* **273** 3415-3421

**Gong X, Dubois DH, Miller DJ and Shur BD** (1995) Activation of a G-protein complex by aggregation of ß1,4-galactosyltransferase on the surface of sperm *Science* **269** 1718-1721

**Haltiwanger RS** (2002) Regulation of signal transduction

pathways in development by glycosylation *Current Opinion in Structural Biology* **12** 593-598

**Hardy DM and Garbers DL** (1995) A sperm membrane protein that binds in a species-specific manner to the egg extracellular matrix is homologous to von Willebrand factor *Journal of Biological Chemistry* **270** 26025-26028

**Inoue N, Ikawa M, Isotani A and Okabe M** (2005) The immunoglobulin superfamily protein Izumo is required for sperm to fuse with eggs *Nature* **434** 234-238

**Johnston DS, Wright WW, Shaper JH, Hokke CH, van den Eijnden DH and Joziasse DH** (1998) Murine sperm-zona binding, a fucosyl residue is required for a high affinity sperm-binding ligand. A second site on sperm binds a nonfucosylated, ß-galactosyl-capped oligosaccharide *Journal of Biological Chemistry* **273** 1888-1895

**Kan FW, Roux E, St-Jacques S and Bleau G** (1990) Demonstration by lectin-gold cytochemistry of transfer of glycoconjugates of oviductal origin to the zona pellucida of oocytes after ovulation in hamsters *The Anatomical Record* **226** 37-47

**Kerr CL, Hanna WF, Shaper JH and Wright WW** (2004) Lewis X-containing glycans are specific and potent competitive inhibitors of the binding of ZP3 to complementary sites on capacitated, acrosome-intact mouse sperm *Biology of Reproduction* **71** 770-777

**Kim K-S, Foster JA and Gerton GL** (2001a) Differential release of guinea pig acrosomal components during exocytosis *Biology of Reproduction* **64** 148-156

**Kim K-S, Cha MC and Gerton GL** (2001b) Mouse sperm protein sp56 is a component of the acrosomal matrix *Biology of Reproduction* **64** 36-43

**Lawrence N, Klein T, Brennan K and Martinez Arias A** (2000) Structural requirements for notch signalling with Delta and Serrate during the development and patterning of the wing disc of Drosophila *Development* **127** 3185-3195

**Le Naour F, Rubinstein E, Jasmin C, Prenant M and Boucheix C** (2000) Severely reduced female fertility in CD9-deficient mice *Science* **287** 319-321

**Loeser CR and Tulsiani DRP** (1999) The role of carbohydrates in the induction of the acrosome reaction in mouse spermatozoa *Biology of Reproduction* **60** 94-101

**Lonhienne T, Zoidakis J, Vorgias CE, Feller G, Gerday C and Bouriotis V** (2001) Modular structure, local flexibility and cold-activity of a novel chitobiase from a psychrophilic Antarctic bacterium *Journal of Molecular Biology* **310** 291-297

**Lopez LC and Shur BD** (1987) Redistribution of mouse sperm galactosyltransferase after the acrosome reaction *Journal of Cell Biology* **105** 1663-1670

**Lopez LC, Bayna EM, Litoff D, Shaper NL, Shaper JH and Shur BD** (1985) Receptor function of mouse sperm surface galactosyltransferase during fertilization *Journal of Cell Biology* **101** 1501-1510

**Lu Q and Shur BD** (1997) Sperm from ß1,4-galactosyltransferase-null mice are refractory to ZP3-induced acrosome reactions and penetrate the zona pellucida poorly *Development* **124** 4121-4131

Macedo-Ribeiro S, Bode W, Huber R, Quinn-Allen MA, Kim SW, Ortel TL, Bourenkov GP, Bartunik HD, Stubbs MT, Kane WH and Fuentes-Prior P (1999) Crystal structures of the membrane-binding C2 domain of human coagulation factor V *Nature* **402** 434-439

Macek MB, Lopez LC and Shur BD (1991) Aggregation of ß-1,4-galactosyltransferase on mouse sperm induces the acrosome reaction *Developmental Biology* **147** 440-444

Meizel S (1997) Amino acid neurotransmitter receptor/chloride channels of mammalian sperm and the acrosome reaction *Biology of Reproduction* **56** 569-574

Miller DJ, Macek MB and Shur BD (1992) Complementarity between sperm surface ß1,4-galactosyltransferase and egg-coat ZP3 mediates sperm-egg binding *Nature* **357** 589-593

Miyado K, Yamada G, Yamada S, Hasuwa H, Nakamura Y, Ryu F, Suzuki K, Kosai K, Inoue K, Ogura A, Okabe M and Mekada E (2000) Requirement of CD9 on the egg plasma membrane for fertilization *Science* **287** 321-324

Mortillo S and Wassarman PM (1991) Differential binding of gold-labeled zona pellucida glycoproteins mZP2 and mZP3 to mouse sperm membrane compartments *Development* **113** 141-149

Nayernia K, Adham IM, Burkhardt-Gottges E, Neesen J, Rieche M, Wolf S, Sancken U, Kleene K and Engel W (2002) Asthenozoospermia in mice with targeted deletion of the sperm mitochondrion-associated cysteine-rich protein (Smcp) gene *Molecular and Cellular Biology* **22** 3046-3052

Nishimura H, Cho C, Branciforte DR, Myles DG and Primakoff P (2001) Analysis of loss of adhesive function in sperm lacking cyritestin or fertilin beta *Developmental Biology* **233** 204-213

Pearse RV 2nd, Drolet DW, Kalla KA, Hooshmand F, Bermingham JR Jr and Rosenfeld MG (1997) Reduced fertility in mice deficient for the POU protein sperm-1 *Proceedings of the National Academy of Sciences USA* **94** 7555-7560

Pratt SA and Shur BD (1993) ß1,4-galactosyltransferase expression during spermatogenesis: stage-specific regulation by t alleles and uniform distribution in +- spermatids and t-spermatids *Developmental Biology* **156** 80-93

Pratt KP, Shen BW, Takeshima K, Davie EW, Fujikawa K and Stoddard BL (1999) Structure of the C2 domain of human factor VIII at 1.5 Å resolution *Nature* **402** 439-442

Rankin TL, Tong ZB, Castle PE, Lee E, Gore-Langton R, Nelson LM and Dean J (1998) Human ZP3 restores fertility in Zp3 null mice without affecting order-specific sperm binding *Development* **125** 2415-2424

Rankin TL, Coleman JS, Epifano O, Hoodbhoy T, Turner SG, Castle PE, Lee E, Gore-Langton R and Dean J (2003) Fertility and taxon-specific sperm binding persist after replacement of mouse sperm receptors with human homologs *Developmental Cell* **5** 33-43

Reitherman RW, Rosen SD, Frasier WA and Barondes

SH (1975) Cell surface species-specific high affinity receptors for discoidin: developmental regulation in *Dictyostelium discoideum Proceedings of the National Academy of Sciences USA* **72** 3541-3545

Robitaille G, St-Jacques S, Potier M and Bleau G (1988) Characterization of an oviductal glycoprotein associated with the ovulated hamster oocyte *Biology of Reproduction* **38** 687-694

Rodeheffer C and Shur BD (2002) Targeted mutations in ß1,4 galactosyltransferase I reveal its multiple cellular functions *Biochimica Biophysica Acta Reviews* **1573** 258-270

Rodeheffer C and Shur BD (2004) Characterization of a novel ZP3-independent sperm-binding ligand that facilitates sperm adhesion to the egg coat *Development* **131** 503-512

Shaper NL, Wright WW and Shaper JH (1990) Murine ß1,4-galactosyltransferase: both the amounts and structure of the mRNA are regulated during spermatogenesis *Proceedings of the National Academy of Sciences USA* **87** 791-795

Shi X, Amindari S, Paruchuru K, Skalla D, Shur BD and Miller DJ (2001) Cell surface ß1,4-galactosyltransferase-I activates G-protein-dependent exocytotic signaling *Development* **128** 645-654

Shur BD (1992) Glycosyltransferases as cell adhesion molecules *Current Opinion in Cell Biology* **5** 854-863

Shur BD and Hall NG (1982a) Sperm surface galactosyltransferase activities during *in vitro* capacitation *Journal of Cell Biology* **95** 567-573

Shur BD and Hall NG (1982b) A role for mouse sperm surface galactosyltransferase in sperm binding to the egg zona pellucida *Journal of Cell Biology* **95** 574-579

Shur BD and Neely CA (1988) Plasma membrane association, purification and characterization of mouse sperm ß1,4 galactosyltransferase *Journal of Biological Chemistry* **263** 17706-17714

Shur BD, Ensslin MA and Rodeheffer C (2004) SED1 function during mammalian sperm–egg adhesion *Current Opinion in Cell Biology* **16** 477-485

Shur BD, Rodeheffer C, Ensslin MA, Lyng R and Raymond A (2006) Identification of novel gamete receptors that mediate sperm adhesion to the egg coat *Molecular and Cellular Endocrinology* **250** 137-148

St-Jacques S, Malette B, Chevalier S, Roberts KD and Bleau G (1992) The zona pellucida binds the mature form of an oviductal glycoprotein (oviductin) *Journal of Experimental Zoology* **262** 97-104

Tulsiani DRP, Nagdas SK, Cornwall GA and Orgebin-Crist MC (1992) Evidence for the presence of high mannose/hybrid oligosaccharide chain(s) on the mouse ZP2 and ZP3 *Biology of Reproduction* **46** 93-100

Tulsiani DRP, Yoshida-Komiya H and Araki Y (1997) Mammalian fertilization: a carbohydrate mediated event *Biology of Reproduction* **57** 487-494

Verhage HG, Mavrogianis PA, O'Day-Bowman MB, Schmidt A, Arias EB, Donnelly KM, Boomsma RA, Thibodeaux JK, Fazleabas AT and Jaffe RC (1998) Characteristics of an oviductal glycoprotein and its potential role in the fertilization process *Biology of Reproduction* **58** 1098-1101

**Ward CR, Storey BT and Kopf GS** (1992) Activation of a G$_i$ protein in mouse sperm membranes by solubilized proteins of the zona pellucida, the egg's extracellular matrix *Journal of Biological Chemistry* **267** 14061-14067

**Wassarman PM, Jovine L and Litscher ES** (2001) A profile of fertilization in mammals *Nature Cell Biology* **3** 59-64

**Yamagata T** (1985) The role of saccharides in fertilization of the mouse *Development, Growth and Differentiation* **27** 176-177

**Yanagimachi R** (1994) Mammalian fertilization. In *The Physiology of Reproduction*, edn 2, pp 189-317 Eds. E Knobil and J Neill. Raven Press, New York

**Yonezawa N, Aoki H, Hatanaka Y and Nakano M** (1995) Involvement of N-linked carbohydrate chains of pig zona pellucida in sperm-egg binding *European Journal of Biochemistry* **233** 35-41

**Yonezawa N, Amari S, Takahashi K, Ikeda K, Imai FL, Kanai S, Kikuchi K and Nakano M** (2005) Participation of the nonreducing terminal beta-galactosyl residues of the neutral N-linked carbohydrate chains of porcine zona pellucida glycoproteins in sperm-egg binding *Molecular Reproduction and Development* **70** 222-227

**Youakim A, Hathaway HJ, Miller DJ, Gong X and Shur BD** (1994) Overexpressing sperm surface ß1,4 galactosyltransferase in transgenic mice affects multiple aspects of sperm-egg interactions *Journal of Cell Biology* **126** 1573-1584

# Mechanism of sperm-zona pellucida penetration during mammalian fertilization: 26S proteasome as a candidate egg coat lysin

Young-Joo Yi[1], Gaurishankar Manandhar[1], Richard J Oko[2], William G Breed[3] and Peter Sutovsky[1,*]

[1]Division of Animal Sciences and Departments of Obstetrics & Gynecology, University of Missouri-Columbia, Columbia, MO 65211, USA; [2]Department of Anatomy and Cell Biology, Queen's University, Kingston, Ontario K7L 3N6, Canada; [3]Department of Anatomical Sciences, The University of Adelaide, Adelaide, South Australia 5005; Australia

Despite years of research work, biologists remain divided over the issue of zona pellucida function during fertilization and the mode of sperm-ZP penetration. The present review examines the emerging evidence for the participation of ubiquitin-proteasome pathway in the process of sperm-ZP penetration generated in the last five years in species of mammals, ascidians and invertebrates. The 26S proteasome, a multi-subunit protease, selectively recognizes, and degrades, egg coat substrate proteins tagged by a covalent ligation of a small, multimeric protein, ubiquitin. Our in vitro work with pig gametes indicates that the sperm-borne 26S proteasomes selectively degrade an ubiquitinated ZP (glyco)protein during fertilization. We suggest that one or more of the ZP proteins are ubiquitinated, and proteasomes associated with the inner acrosomal membrane, are exposed as a result of acrosomal exocytosis. Sperm-ZP penetration may involve the ZP-deubiquitination, with several proteasomal subunits becoming phosphorylated. Polyubiquitin chain recognition activities associated with the sperm acrosomal proteasome could also contribute to anti-polyspermy control after sperm-egg fusion. Here, we bring together the relevant recent data on the mechanism of sperm-ZP penetration in mammals. Such observations could possibly lead to the development of novel non-hormonal contraceptives, improvement of infertility diagnostics and optimization of assisted reproduction.

*If the ovum should be impregnated, several important changes take place which are as follows: The zona pellucida, or outer membrane of the egg, having thrown off its outer cell covering, and the spermatozoa have no difficulty in penetrating the soft albuminous membrane that encloses the yelk. When the spermatozoa penetrate the zona, the yelk contracts. This fact was first reported by Newport, who called the space "the respiratory chamber". This interspace is filled with a transparent fluid. After the*

---

*To whom correspondence should be addressed
E-mail: SutovskyP@missouri.edu

*contraction takes place, another remarkable change occurs, which is revolving of the yelk. This rotation is effected by the aid of cilia, which line the inner surface of the yelk. About this time a small body, or there may be several bodies, is seen in the "respiratory" space between the yelk and zona which is supposed to have some connection to the cleavage of the yelk, which is about commencing. An additional change observed taking place in the tubes is a deposit of albumen around zona pellucida, which takes place when the ovum is passing the middle and lower third of the tube. These occurrences are so uniform that the different offices for different portions of the Fallopian tube may be readily determined. The first or upper third is appropriated to the reception of the ovum, and for removing the adventitious covering of cells, while it also prepares the ovum for the operation of the spermatozoa. In the middle third, the respiratory chamber is formed, and here the rotation of the yelk commences. In the lower third the cleverage takes place, as also the deposit of albumen.*

Frederick Wilson Pitcairn and Elizabeth J. Williard (1906) *Woman's Guide to Health, Beauty and Happiness. What Every Woman Should Know* pp 96-97. Horace C. Fray Publisher, Washington DC.

### Introduction

While we often think that all the important discoveries concerning the process of fertilization were made in the last four decades, the citation introducing this review shows that our great grandmothers already had access to some knowledge of human procreation at the cellular level. One hundred years later, biologists remain divided over the functions of the egg coat, the zona pellucida (ZP), and how sperm-ZP penetration occurs. A recent review by Olds-Clarke (2003) identifies the lack of general agreement on the means of sperm adhesion to, and penetration of, the zona pellucida, as well as "the need for new approaches to this problem". Two schools of thought regarding sperm-zona interactions have evolved over last two decades. One school suggests that mechanical thrust of the sperm, the flagellar motility, provides the driving force necessary for sperm-ZP penetration (Bedford, 1998). By contrast others consider that sperm head protease actively digests ZP glycoprotein (Yanagimachi, 1994). Although the proteolytic hypothesis has gained some support in the last decade, the lack of a known protease that could be inhibited, or ablated genetically, to induce ZP penetration block (Yanagimachi, 1994) has taken the wind out of the sails of this viewpoint and spurred a comeback of the mechanical hypothesis (Bedford, 1998). Recently, several groups have identified a 26S proteasome as a candidate egg coat lysin, a protease that is present in the sperm head acrosome and can be inhibited by specific pharmacological inhibitors to prevent sperm ZP-binding and penetration. The purpose of the present review is to reevaluate our knowledge of sperm-ZP penetration in light of these recent findings.

### Mechanism of ZP assembly and primary sperm-ZP binding

The mammalian zona pellucida comprises a family of related, yet structurally diverse, glyco-proteins. Most of our current knowledge is derived from the laboratory mouse in which the

ZP2-ZP3 (=ZPA-ZPC) heterodimers are cross-linked by ZP1 (ZPB), with ZP3 alone serving as a sperm receptor for primary sperm-ZP binding and the induction of acrosomal exocytosis (Wassarman, 1990). Nevertheless, substantial differences exist between the laboratory mouse and many other species with investigators using species in other mammalian orders concluding that both the glycoproteins that make up the ZP and the ZP-sperm interactions differ from that of the mouse. While sharing the conserved ZP domains, the amino acid sequences of ZP proteins differ, even in some cases, between closely related species (Swann *et al.*, 2002; Shibata, 2005). In both the human and laboratory rat, four ZP proteins have been identified, whereas in the pig a complex of ZPC (~ZP3) and ZPB (~ZP1) proteins appears to function as the sperm receptor (Yurewicz *et al.*, 1998). Also, whereas in the mouse, ZP proteins are thought to be expressed exclusively by the oocyte (Lira *et al.*, 1990), *in situ* hybridization analysis of the major ZP proteins in rabbit (Lee and Dunbar, 1993), dog (Blackmore *et al.*, 2004) and cynomolgus monkey (Martinez *et al.*, 1996) suggest expression by surrounding cumulus/granulosa cells. In the pig, the localization of ZP3α (~ZPB) mRNA shifts progressively from oocyte to cumulus cells as maturation from the primordial to preovulatory follicle occurs (Kolle *et al.*, 1996). In other vertebrates, such as several species of fish, extra-ovarian (for example, liver) egg coat glycoprotein synthesis occurs (Shibata, 2005). While species differences may explain the various differences in fertilization mechanisms, it is possible that the interspecies variability of the egg coat composition, sperm morphology and signalling pathways determining sperm-zona interactions, may be one of the factors that can drive speciation. For instance, if sperm-zona adhesion is species specific, it may be determined by the amino acid sequence of ZP glycoproteins, by their posttranslational modification and/or by the 3D conformation of the matrix. In particular, species differences of ZP-protein glycosylation may explain why mouse, but not human, spermatozoa bind to the ZP surrounding oocytes of transgenic mice lacking murine ZP3 but expressing human ZP3 (Rankin *et al.*, 1996). Consequently, it has been suggested that the supramolecular structure of the ZP, rather than the primary amino acid sequence of the sperm receptor, may facilitate mammalian sperm-ZP binding (Dean, 2002; 2004).

## Importance of secondary sperm-ZP binding and the inner acrosomal membrane for sperm ZP-penetration

The proximal region of the sperm head leads in zona penetration. This region of the sperm head is covered by a cap-like vesicle, the acrosome. The acrosomal matrix (AM) is enveloped by the outer acrosomal membrane (OAM) that lies just beneath the cell membrane and an inner acrosomal membrane (IAM) that lies close to the outer nuclear envelope tightly bound to the subacrosomal perinuclear theca (PT). Upon sperm-ZP adhesion, often termed primary sperm-ZP binding, acrosomal exocytosis (AE) results, which is thought to be induced by the binding to a sperm receptor on the ZP (for example, ZP3 in mouse: Wassarman, 1990) although the receptor molecule on the sperm OAM and/or plasma membrane has yet to be identified (reviewed by Primakoff and Myles, 2002). The AE causes vesiculation of the OAM, followed by a stepwise dispersion of the AM, perhaps aided by the mobilization, and release, of the sperm protease, acrosin (Baba *et al.*, 1994). Subsequently, part of the OAM and AM complex is shed as an acrosomal shroud (Olds-Clarke, 2003), that typically remains on, or near, the outer matrix of the ZP (Fig. 1). This process results in the exposure of the IAM and subsequent secondary sperm-ZP binding thought necessary for sustaining sperm-ZP adhesion after AE (Bleil *et al.*, 1988). Recently, we have identified several molecules of the IAM that may be involved in this secondary binding of the sperm IAM to the ZP. Affinity purified antibodies against these IAM proteins reduced sperm-ZP adhesion when added to the IVF medium (Yu *et al.*, 2006).

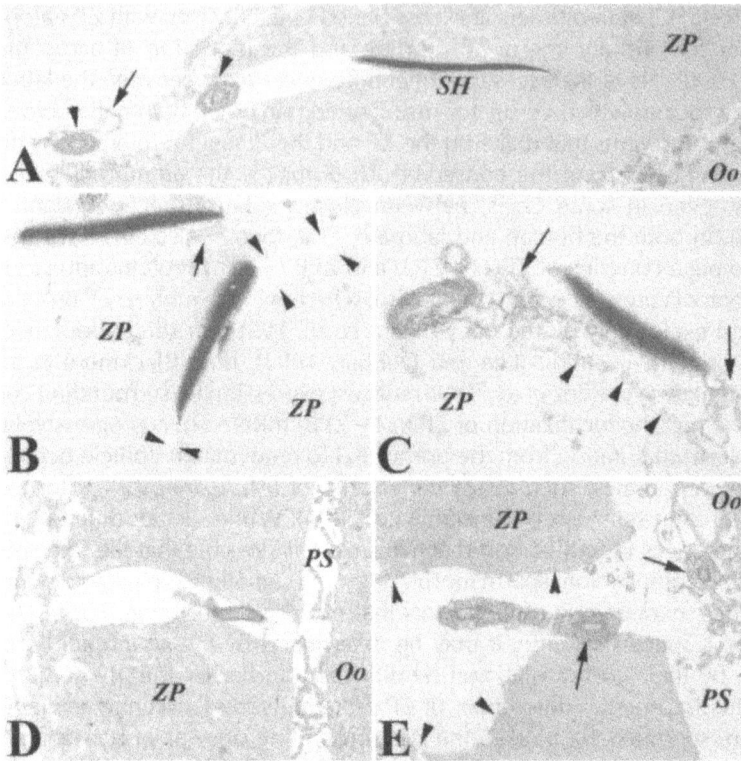

**Figure 1.** Transmission electron microscopy of sperm-zona pellucida interactions during porcine (A-C) and bovine (D-E) fertilization *in vitro*. (A) Narrow fertilization slit inside porcine zona pellucida *(ZP)* contains a longitudinal section of the sperm head *(SH)* and cross sections of the undulated sperm tail (arrowheads), suggestive of persistent flagellar motility at this stage of porcine fertilization. A remnant of acrosomal ghost (arrow) is seen on ZP-surface. Ooplasm *(Oo)* is indicated. (B) Fertilization slit (arrowheads) extending beyond the sperm head could be a result of active proteolysis due to undulating sperm head movement during ZP-penetration. Arrow indicates the remnants of acrosomal ghost. (C) Vesiculated outer acrosomal membrane (arrowheads) and the acrosomal ghost (arrows) are visible near a zona-bound boar sperm head. (D, E) In contrast to porcine, the diameter of bovine fertilization slit (arrowheads in panel E) appears much larger than the cross-sectional diameter of the sperm head or tail (arrows in panel E). Matrix shrinking during fixation and dehydration for transmission electron microscopy samples may not provide a sufficient explanation of this phenomenon, as it would also be expected to occur in other species. *ZP* = zona pellucida; *PS* = perivitelline space; *Oo* = oocyte cytoplasm/ooplasm; *SH* = sperm head.

The suggestion that secondary sperm zona binding is essential for subsequent sperm-zona penetration originated from studies carried out in mice by Wassarman and co-investigators (Wassserman *et al.*, 2004). In an initial investigation, Bleil and Wassarman (1986) observed that radio-iodinated ZP3 preferentially bound to the acrosome cap region of acrosome-intact spermatozoa, whilst ZP2 bound preferentially to the surface of the IAM of acrosome-reacted spermatozoa. Subsequently, these authors showed that antibodies directed against ZP2 inhibited the maintenance of sperm binding that had undergone the acrosome reaction on the ZP (Bleil *et al.*, 1988). These findings stimulated a search for the secondary binding receptor on the IAM (reviewed by McLeskey et al., 1998). Several candidate molecules have been considered including PH20, proacrosin, Sp38 and Sp17. However, none of these molecules fully met the requirements of a secondary binding receptor with the two most important requirements being the capability of sperm binding to the ZP and the localization of the molecules involved being to the IAM after the acrosome reaction. In our studies, we considered that the questions concerning the existence of a receptor on the IAM with zona secondary binding ability and zona penetrating ability (lytic capability) could be determined, if direct information was obtained of the peripheral and integral composition of the IAM.

To investigate this, we developed a rat sperm head fractionation procedure which allowed direct protein analysis of the IAM. For this, the apical tips of isolated rat falciform sperm heads were broken off by ultra-sonication and isolated on a sucrose density gradient (Fig. 2 insets; see Yu et al., 2006, for details). These sperm tips consist solely of the IAM attached to the detergent insoluble perforatorium (perinuclear theca) with the surface of the IAM displaying a peripheral layer of electron dense material that we refer to as the inner acrosomal membrane coat (IAMC) (Fig. 2A). High salt extraction removed this coat (Fig. 2B) and, coincident with the recovery, a prominent 38 kDa protein occurred in the supernatant (Fig. 2C). High salt extracts of isolated, and sonicated, sperm heads of bull, boar and rat (in which the tips were not broken) gave us the same SDS-PAGE protein profile as shown in Fig. 2C, lane 2. Antibodies were then raised, and affinity purified, against the isolated 38 kDa protein for the purpose of localizing this protein *in situ* and probing a testicular cDNA expression library for its clone. To our surprise, the deduced sequence of our clone coded for Sp38, an intra-acrosomal protein with ZP (ZP2) binding ability (Mori et al., 1995), whose precise localization in spermatozoa had previously been unknown. Although Sp38 was found to be intra-acrosomal by immunofluorescence, the immunoreactivity was short lived after induction of the acrosome reaction with ionophore, which did not lend support to its role of binding the IAM surface of the sperm to the zona pellucida during fertilization. Contrary to this finding, we found by LM and EM immunocytochemistry, that Sp38 was indeed an IAMC protein (Fig. 2 D, E), which, during IVF, was retained on the IAM after the acrosome reaction and zona pellucida penetration (Fig. 3). Thus we renamed the protein IAM38 to better reflect its location within the sperm head (Yu et al., 2006).

Polyclonal antibodies raised, and affinity purified, against a peptide segment of IAM38 (anti-C38) labeled only the acrosome reacted spermatozoa *in situ* (Fig. 3) and blocked *in vitro* fertilization by preventing secondary sperm binding to the zona pellucida and consequently prevented zona penetration (Yu et al., 2006). Control antibodies raised against other isolated IAM proteins on the other hand were ineffective in blocking fertilization. We were also able to extract, with non-ionic detergents, a prominent integral IAM protein that has been cloned and identified as a heat shock/chaperone protein (Richard J. Oko; unpublished data). This protein, termed IAM32, although not directly involved in binding and penetration of the zona, appears to be involved in anchoring the IAM to the underlying perinuclear theca and in organizing, and attaching, a number of extracellular proteins to the IAM (Richard J. Oko; unpublished data). Some of these IAM-coat proteins have serine protease activities, whilst others are associated with the 26S proteasome (Fig. 3). This sperm head fractionation scheme therefore allows us to obtain direct information on the peripheral and integral protein composition of the IAM for identifying candidate proteins responsible for sperm-zona interactions. Our IVF antibody blocking study suggests that IAM38 is a strong candidate for secondary binding and supports the hypothesis that secondary binding is a prerequisite for zona penetration.

The part of the exposed IAM which forms the leading edge of the penetrating sperm head through the ZP is located at the apical, or upper, convex region of the rat sperm head where the underlying perinuclear theca, or perforatorium, comes to a sharp point in a murid rodent spermatozoon. The shape of the apical region of the sperm head in itself imparts an advantage for

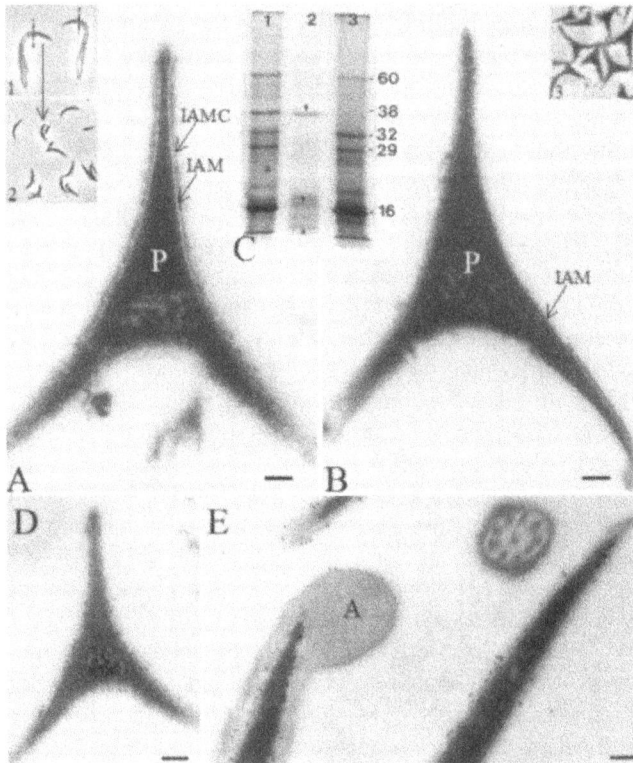

**Figure 2.** Ultrastructural and biochemical analysis of the inner acrosomal membrane. (A-C) Consequence of 1M KCl extraction of the laboratory rat sperm apical tips. Cross-section of an apical tip before (A) and after (B) salt extraction. The dashed lines in the phase contrast micrograph of isolated heads (inset 1) indicate the region of the head where tips were broken by sonication, while insets 2 and 3 indicate phase contrast and electron microscope fraction of isolated tips. SDS-PAGE (C), under reducing conditions, shows a decrease in the intensity of a 38 kDa protein band in the tips after KCL extraction (compare lanes 1 and 3) coinciding with both the appearance of a 38 kDa band (upper arrowhead) in the extract (lane 2) and the removal of most of the inner acrosomal membrane-coat (IAMC) from the inner acrosomal membrane (IAM) (compare panels A and B). Because the membranes are not osmicated in our tissue preparation, they appear as noticeable white lines only if they are contrasted by electron-dense cellular material on either side. Therefore once the IAMC is salt extracted the IAM becomes much less visible. Lane 1, protein profile of apical tips before extraction; Lane 2, protein profile of 1M KCl extract; Lane 3, protein profile of apical tips after 1M KCl extraction; P, perforatorium; IAM, inner acrosomal membrane; IAMC, inner acrosomal membrane coat. The low molecular weight salt extracted proteins in lane 2 (between the lower arrowheads) have been identified as core somatic histones that are assembled as part of the perinuclear theca (PT) (Tovich and Oko, 2003) – so far these histones have been the only proteins identified as salt extractable components of the PT. (D, E) Immunogold localization of IAM38 protein in the rat (D) and bull (E) spermatozoa. In cross-sections of isolated tips (D), the labeling is mainly associated with the IAM and IAMC. Likewise, in intact and acrosome disrupted bull spermatozoa (E) immunogold-labeling is found only over the IAM and IAMC. A = Acrosome. Bars = 0.2$\mu$m.

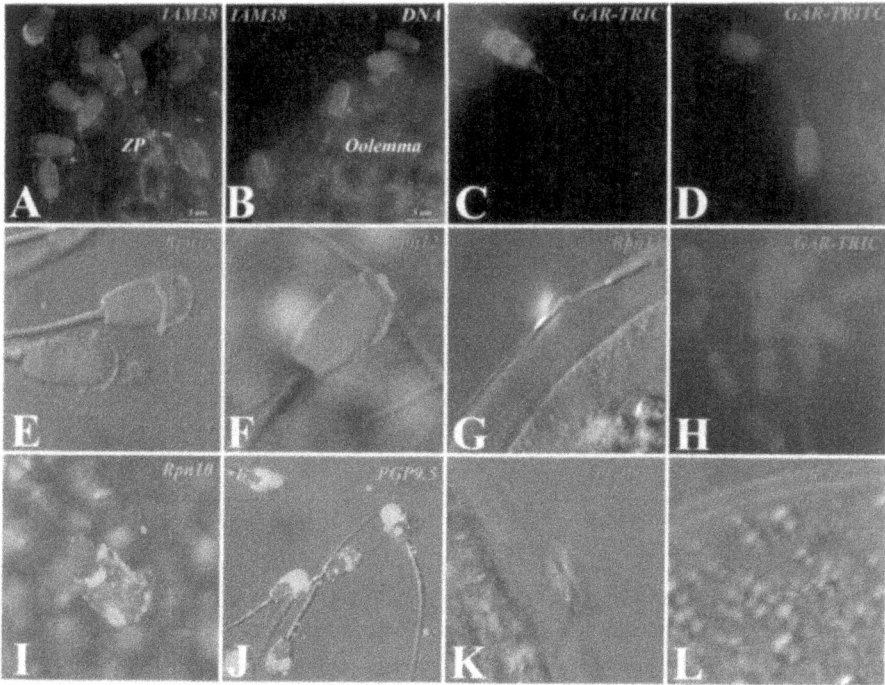

**Figure 3.** Co-localization of the components of ubiquitin-proteasome pathway with the inner acrosomal membrane. (A-D) Immunofluorescence labeling of IAM38 protein in boar spermatozoa during zona pellucida-induced acrosome reaction (A) and after zona penetration (B) of pig oocytes. Rabbit antibody C38, recognizing the inner acrosomal membrane protein, IAM38, was added directly to medium during *in vitro* fertilization, and antibody binding to IAM was detected after fertilization by a rabbit IgG-specific goat antibody conjugated to red-fluorescent tetramethyl-rhodamine-isothiocyanate (GAR-TRITC). In panel A, the acrosome reacted spermatozoa are adjacent to the zona pellucida, whilst in B the sperm have already penetrated the zona and are attached to oolemma. Figures C and D show the specific binding of anti-C38 antibody to sperm IAM during IVF-antibody blocking studies. The binding of anti-C38 (C) and a control, unrelated rabbit antibody (D) was detected by the incubation of inseminated, fixed ova with GAR-TRIC (red). DNA in all figures was stained with DAPI (blue). ZP = zona pellucida. (E-H) Detection of 19S proteasomal regulatory complex subunit Rpn12 (red) in the non-permeabilized, formaldehyde fixed, boar spermatozoon (E), in a Triton-permeabilized spermatozoon (F), in spermatozoa bound to, or penetrating, pig zona pellucida (G), and in negative control spermatozoa labeled with a non-immune rabbit serum in place of anti-Rpn12 antibody (H). (I) Detection of 19S regulatory complex subunit Rpn10 (green), responsible for the recognition and binding of polyubiquitin chains during proteasome-ubiquitinated substrate recognition. This antibody when added to the IVF medium inhibits *in vitro* fertilization in pigs (J-L) Detection of a deubiquitinating enzyme, the ubiquitin C-terminal hydrolase PGP9.5 in the boar sperm acrosome before (J) and during (K) acrosomal exocytosis and fertilization, and in a negative control sample (L). Blocking of deubiquitinating enzyme activities increases the rate of proteasomal proteolysis. During pig fertilization, the blocking of PGP9.5 activity increases the rate of sperm-zona penetration and significantly augments the rate of polyspermy. Bars = 5 μm.

penetration but the force generated by motility of the axoneme does not appear to be strong enough for completion of the ZP penetration (Green, 1987). If a lytic agent is indeed involved in mammalian sperm-egg coat penetration, as is the case of the sperm 26S proteasome in an ascidian species (Sawada *et al.*, 2002a; b) or as a hydrophobic, non-enzymatic, egg-lysin, protein in a sea urchin (Fridberger *et al.*, 1985), it would most likely be sequestered onto, or reside on the surface of, the IAM and be attached by either covalent or non-covalent bonds as is the case of IAM38. If the 26S proteasome is indeed the mammalian egg coat lysin, then there is already some evidence for its association with the inner and outer acrosomal membranes. In mouse, Berruti and Mantegani (2005) described the association between the 26S proteasome and the DnaJ protein of the acrosomal membrane. In our studies, we regularly observe the association of various proteasomal subunits with IAM of intact, as well as acrosome reacted, spermatozoa (Sutovsky *et al.*, 2004). The isolated IAM protein fraction is highly enriched in 19S and 20S proteasomal subunits (Peter Sutovksy and Richard J. Oko; unpublished), although this does not rule out the presence of proteasomes on the OAM as well.

### Is the hand driving a knife or pushing a drill?

The mechanism by which the fertilizing spermatozoon penetrates the ZP is an outstanding unresolved issue in developmental and reproductive biology (Olds-Clarke, 2003). The pendulum of opinion periodically appears to swing between a mechanical and a proteolytic mechanism of sperm-ZP penetration although, as noted by some, both activities may be involved (Yanagimachi, 1994). In EM studies of both Australian murine rodents and marsupials, individual electron micrographs of spermatozoa penetrating the ZP and the penetration hole left after passage of the spermatozoon, have suggested both some physical force as well as some enzymatic digestion of the zona matrix exist (Breed and Leigh, 1990; Breed, 1994) (see Fig. 4, 5). Egg coat penetration assisted by 26S proteasomal-proteolysis has been described in both invertebrate (Matsumura and Aketa, 1991) and ascidian (Sawada *et al.*, 2002a; b) gametes and protease-assisted egg coat lysis has been regarded for a long time to be a plausible mechanism in mammals (Yamagata *et al.*, 1985; Yanagimachi, 1994). This view has been supported by the identification of several proteolytic enzymes in the sperm acrosome (Tulsiani *et al.*, 1998). However, targeted mutations and knockouts of several candidate acrosomal enzymes and sperm surface receptors have not eliminated the ability of mouse spermatozoa to bind to, and penetrate, the ZP (reviewed by Bi *et al.*, 2002; Talbot *et al.*, 2003). The ablation of acrosin (Baba *et al.*, 1994), beta 1,4-galactosyl-tranferase (Lu and Shur, 1997) and SED1 (Ensslin and Shur, 2003) genes has not completely prohibited fertility in mutant mice. Thus an additional mechanism must be involved in sperm-ZP binding and penetration, thereby supporting the revival of the mechanical hypothesis (Bedford, 1998; Olds-Clarke, 2003). According to this school of thought the enhanced mechanical thrust exerted as a result of hyperactivity of the sperm tail results in a hole occurring in the zona pellucida (Bedford, 1998). Videomicroscopy has shown sidewise oscillation of the sperm head reminiscent of a cutting motion of a knife slicing through paper (Drobnis *et al.*, 1988; Bedford, 1998). Sperm tail beating persists during ZP penetration and stops only at time of sperm-oolemma fusion (Yanagimachi, 1994). Some investigators have found that the penetration slit in the ZP left behind by the fertilizing spermatozoon (Fig. 1, 4), has sharp edges and a diameter similar to that of the sperm head whereas, as in some marsupials at least, it may be considerably larger than this (Fig. 5). However, the mechanical hypothesis is seemingly not valid because the calculated motile force of the hyperactivated sperm flagellum may not be sufficient to push the sperm head through the zona (Green, 1987). Bedford (1998) has argued against the enzymatic hypothesis on the basis that, if protease is released from the

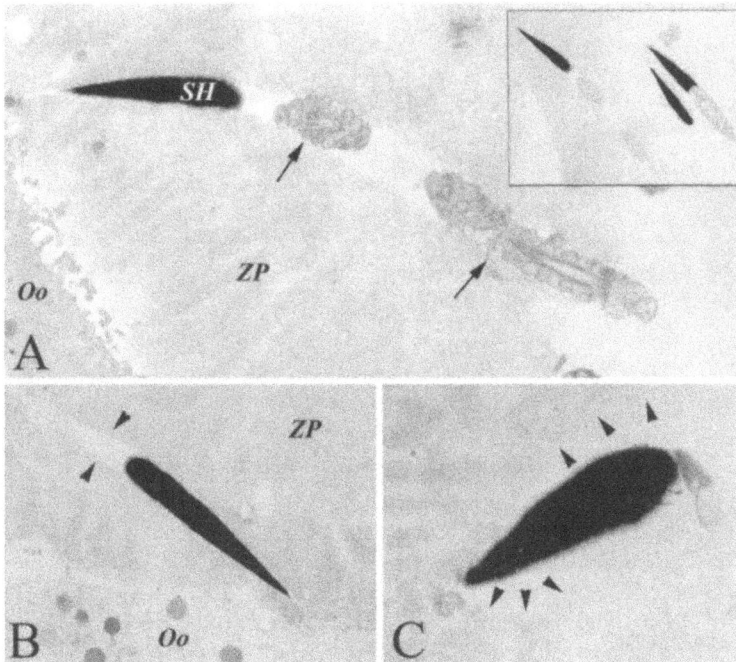

**Figure 4.** Transmission electron microscopy of *in vitro* fertilization in the Rhesus monkey, *Macaca mulatta*. Fertilization slit (arrowheads) in panels (A-C) appears similar in the diameter to that of the sperm head *(SH)* and tail (arrows in panel A), yet a gap between sperm head and zona matrix *(ZP)* is visible at an early stage of penetration near the outer zona surface (arrowheads in panel C). *Oo* = ooplasm.

sperm acrosome on the external matrix of the ZP, a crater formed there would be somewhat larger than the sperm head diameter. However, this may not necessarily be the case if a protease is sequestered onto the IAM surface and remains associated with the IAM throughout the process of sperm-ZP penetration. Several scanning electron microscopy studies (reviewed by Olds-Clarke, 2003), as well as our recent study using DIC microscopy (Figure 6B in Sutovsky et al., 2004) indicate a shallow depression underneath the ZP-bound sperm head suggestive of proteolytic digestion. This depression may guide the sperm head through the outer matrix as penetration is initiated (Olds-Clarke, 2003). While a narrow penetration slit is observed after porcine (Fig. 1) and rhesus monkey (Fig. 4) fertilization, the penetration hole left behind by the fertilizing spermatozoon in some of our micrographs of both bovine (Fig. 1) as well as native Australian rodent fertilization appears, like that in marsupials, to be larger than the sperm head cross-sectional diameter (Fig. 5, 6). However, electron micrographs of partly zona penetrated sperm of the plains rat suggest otherwise (Fig. 5). Shrinkage and/or stretching of the penetration slit during fixation and dehydration cannot be ruled out in some of these EM samples. It also seems possible that, if sperm-ZP penetration is solely due to a mechanical force, the elastic ZP fibers might snap back and close the slit after sperm penetration. In the course of our bovine fertilization studies, we made an incidental observation of a spermatozoon lodged in the perivitelline space with the acrosomal part of its head buried in the ZP, seemingly on its way out of perivitelline space (Fig. 7). This observation appears to support the view that a proteolytic activity of an egg coat lysin may be associated with IAM and not completely quenched

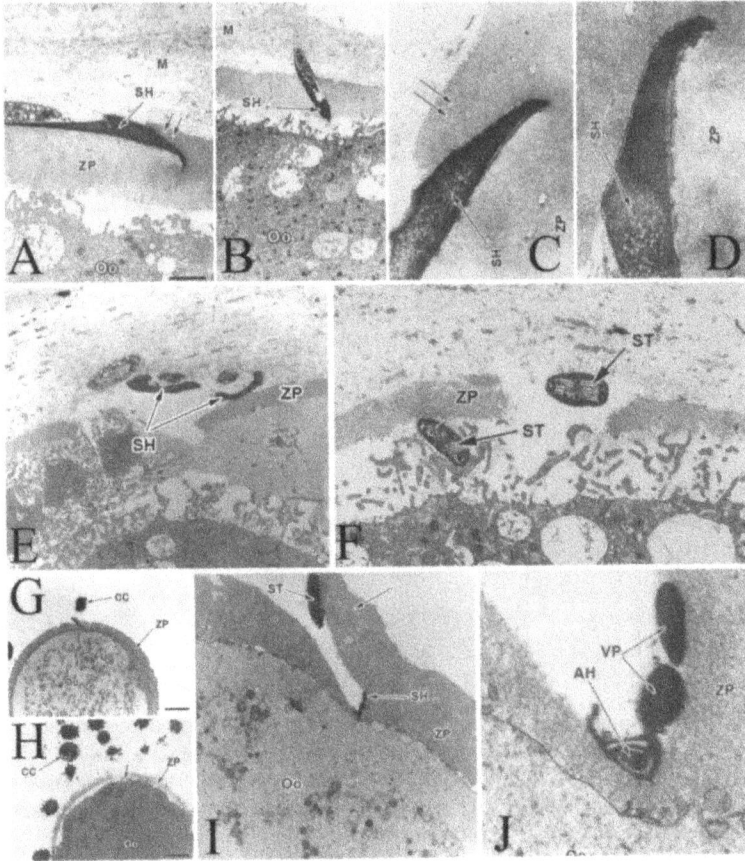

**Figure 5** (A-F) *In vivo* sperm-zona interactions in a dasyurid marsupial, the fat-tailed dunnart, *Sminthopsis crassicaudata*. (A-D) Sperm heads *(SH)* partly incorporated into the zona pellucida *(ZP)* matrix of recently fertilized oocytes *(Oo)* with some mucoid *(M)* material deposited on the outer surface of the zona. Note in (A) and (C) some zona matrix elevated around the partly incorporated sperm head. In (E) and (F) a large gap is present in the zona with transverse sections of sperm heads *(SH)* and sperm tail *(ST)* lying close by. The large hole in the zona suggests some enzymatic digestion of the matrix in spite of the other micrographs suggesting some physical displacement of zona matrix by the sperm head. (G-J) *In vivo* sperm-zona interactions in the Australian rodent, the plains rat, *Pseudomys australis*. The sperm head *(SH)* of this species has an apical hook *(AH)* together with two further processes, the 'ventral processes (VP)', that extend from its upper concave surface. Panel (G) shows sperm head partly incorporated in the ZP with zona matrix elevated around its convex surface; after the sperm head has penetrated the zona (arrow) a large hole in the zona matrix is evident (H). Panels (I) and (J) show a sperm head partly incorporated within the ZP with the matrix packed tightly around the leading edge of the spermatozoon; note also filaments close to the sperm head that appear to be somewhat stretched (J). Cumulus cells *(CC)* and oocyte *(Oo)*.

after acrosomal exocytosis. Instead of being a hand driving the knife, the motile force of the sperm flagellum may be more akin to a hand driving a spinning drill. In this analogy, the spinning drill bite would be the lytic protease that remains on IAM after acrosomal exocytosis.

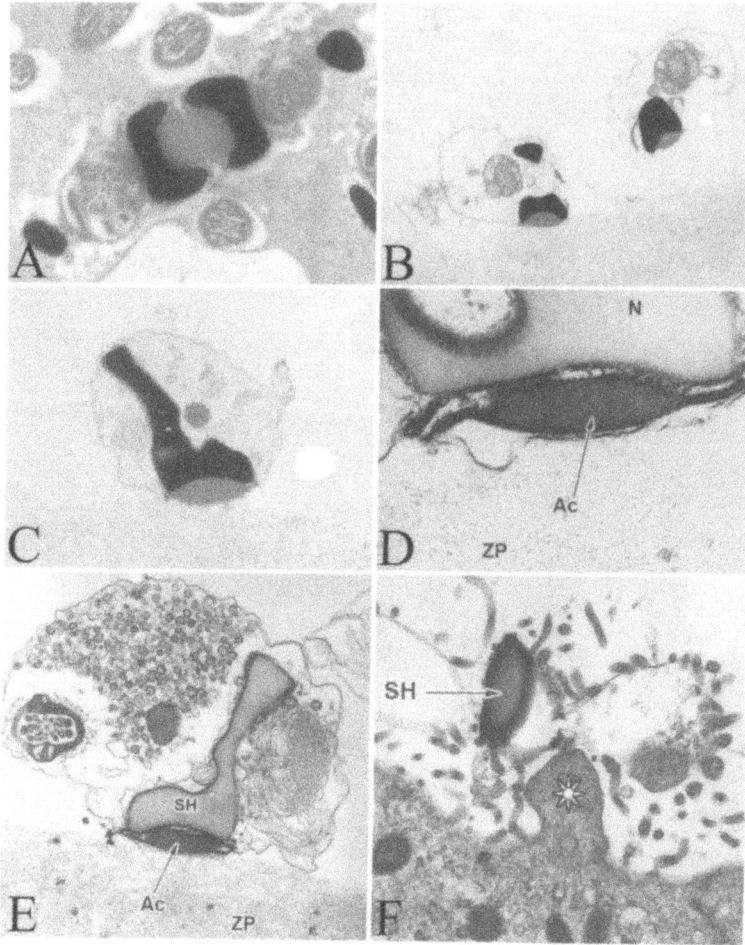

**Figure 6.** *In vitro* sperm-zona interactions in the didelphid marsupial, *Monodelphis domestica*. (A) Spermatozoa become joined by their plasma membranes over the acrosomes during epididymal maturation (A) and uncouple just prior to binding to zona pellucida at fertilization (B). (C-E) Acrosome *(Ac)* lying over nucleus of sperm head *(SH)* is still intact; spermatozoon head lies along outer surface of zona pellucida *(ZP)*. In (F) a projection of oocyte cytoplasm is elevated *(star)* in close proximity to the sperm head lodged in the perivitelline space after ZP-penetration. N = nucleus

## Is the 26S proteasome the mammalian egg coat lysin?

Conventional protease inhibitors decelerate, but do not prevent, sperm-ZP penetration (Saling, 1981; Yanagimachi, 1994; Olds-Clarke, 2003), whereas trypsin inhibitors block fertilization by preventing acrosomal exocytosis rather than interfering with the actual ZP-digestion and penetration (Llanos *et al.*, 1993). Delayed sperm-ZP penetration has also been observed in the acrosin-mutant mouse, which suggests that acrosin facilitates the dispersal of the AM rather than serving as an actual egg coat lysin (Baba *et al.*, 1994). Thus a yet to be identified protease, other than the conventional sperm acrosomal proteases, may be involved in ZP digestion during mammalian fertilization (Yanagimachi, 1994). In the last five years, evidence has been

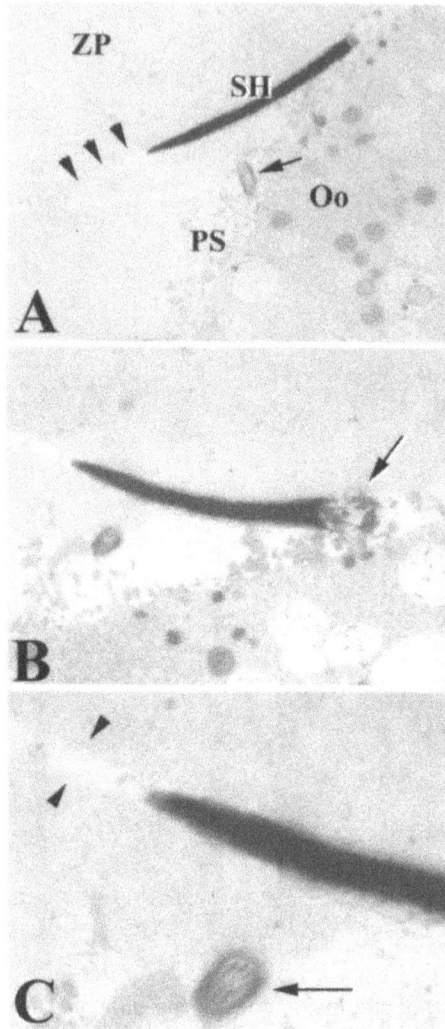

**Figure 7.** Serial sections of a bull spermatozoon *(SH* = sperm head), partially lodged in the perivitelline space *(PS)* of a fertilized ovum that started reverse penetration through zona pellucida *(ZP)*. While the sperm tail midpiece (arrow in panel B) is clearly lodged in the perivitelline space, the sperm head is forming an inside-out fertilization hole, extending beyond the tip of sperm head (arrowheads in panel C). A possible explanation for this observation is that this was a secondary fertilizing spermatozoon that failed to fuse with oolemma after reaching perivitelline space shortly after activation of anti-polyspermy defense by the primary fertilizing spermatozoon. This is supported by the lack of cortical granules in the oocyte cortex, suggesting the occurrence of cortical granule-exocytosis induced by sperm-induced oocyte activation after sperm-oolemma fusion.

accumulated that a sperm borne 26S proteasome participates in sperm-zona penetration (reviewed by Baska and Sutovsky, 2005). This is a multi-subunit, substrate-specific protease with action typically limited to proteinaceous substrates marked for degradation by covalent tandem ligation of a small chaperone ubiquitin protein. There are at least three lines of evidence favoring the 26S proteasome as a mammalian, as well as non-mammalian, egg coat lysin: 1) in ascidians, ubiquitination and proteasomal degradation of the sperm receptor Hrvp70, a homologue of mouse sperm receptor ZP3, occurs during sperm penetration of vitelline envelope (Sawada *et al.*, 2002a; b; Sakai *et al.*, 2003; 2004); 2) The proteasomal activity of the human and rodent sperm acrosome occurs at the time of sperm-ZP interactions (Morales *et al.*, 2004; Pizzaro *et al.*, 2004; Pasten *et al.*, 2005); and 3) in pigs, proteasomal proteolysis of the ZP occurs at the time of ZP penetration (Sutovsky *et al.*, 2003; 2004). Additional studies supporting this have been carried out in murine (Pasten *et al.*, 2005), human (Wojcik *et al.*, 2000; Bialy *et al.*, 2001) and sea urchin (Matsumura and Aketa, 1991) gametes.

## Ubiquitin system

Last year's Nobel Prize in chemistry was awarded to Aaron Ciechanover, Avram Hershko and Irwin Rose for their discovery of the ATP-dependent, proteolytic ubiquitin system. The fundamental event of ubiquitin-proteasome pathway is polyubiquitin tagging of defective (modified) proteins through a cascade of ATP-dependent coupling reactions mediated by a set of ubiquitin activating and conjugating enzymes. This is followed by recognition of the ubiquitin tagged proteins and the cleaving of them into short peptide fragments by proteasomes (Kisselev *et al.*, 1999). The 26S proteasome catalyses rapid degradation of cytosolic proteins such as rate-limiting enzymes (Hochstrasser, 1995), transcriptional regulators (Glickman and Ciechanover, 2002), short-lived regulatory proteins (Ciechanover, 1994) and mis-folded or abnormal proteins. Due to its unique combination of substrate specificity and high evolutionary conservation, the ubiquitin system is likely to play a role in many normal body functions and pathologies. Unlike the lysosomal system, the ubiquitin-proteasome pathway is not simply a terminal mechanism of protein degradation in many cells: rather it plays a central role in several cellular functions such as antigen presentation (Rock *et al.* 1994), cellular differentiation, signal transduction, cell cycle regulation, gene expression, nuclear remodeling *et cetera*. Alterations in the ubiquitin system accompany various pathologies such as Alzheimer disease, malignancies, liver cirrhosis and infertility (reviewed by Glickman and Ciechanover, 2002). Tetra- and poly-ubiquitination, that is the formation of a tetrameric or polymeric ubiquitin-isopeptide on a substrate protein, is the consensus signal for the docking of ubiquitinated proteins to 26S proteasome, resulting in proteolytic degradation of the substrate and return of the intact ubiquitin molecules to the cytosolic pool (Pickart, 1998). Liberated poly-ubiquitin chains are disassembled by ubiquitin C-terminal hydrolases to regenerate monoubiquitin and the substrate is hydrolyzed into small peptides of 3-23 amino acids that can be further broken down by cytosolic endopeptidases (reviewed by Wilkinson and Hochstrasser, 1998).

## The 26S proteasome

The canonical 26S proteasome is a holoenzyme composed of a barrel-shaped 20S core capped with two 19S regulatory complexes, on each side of the 20S barrel. The 19S regulatory complex is composed of at least 17 subunits of a total mass of ~1 MDa. The function of the 19S complex is to recognize the polyubiquitin tail on the ubiquitinated substrate protein. This is thought to be fulfilled by the 19S subunit Rpn10/S5a (Young *et al.*, 1998). Other 19S subunits are thought to have a role in the binding and removal of polyubiquitin chain from the ubiquitinated substrate before it is translocated to the 20S core (Glickman and Ciechanover, 2002). The 20S proteasomal core is composed of four concentric rings containing seven proteasomal subunits of the α-type (two outer rings) and seven subunits of the ß-type (two inner rings). The α-subunits form the accessory structures while the ß -subunits display the protease activities responsible for the break down of the linearized polypeptide chains by trypsin-like, chymotrypsin-like and postglutamyl peptidyl hydrolysis activities (Cardozo, 1993; Tanaka, 1998). The proteolytic activities of the ß-subunits can be targeted by specific inhibitors like lactacystin or MG132 (Fenteany et al., 1995). While most α-type and ß-type subunits are constitutively expressed in all tissues, three ß-type subunits, including ß1, ß2 and ß5, can be replaced by their inducible analogues, ß1i /LMP2, ß5i /LMP7 and ß2i /LMP10. These inducible ß-type subunits are preferentially expressed in eye lens and in the immunoproteasomes responsible for antigen presentation in leukocytes (Glickman and Ciechanover, 2002); their expression is also induced in other tissues by heat shock or interferon signal when there is an increased need for ubiquitin-dependent proteolysis. After the removal of polyubiquitin chain and denaturation of the substrate protein in the 19S complex, the substrate protein is transported to the lumen of the 20S core where it is cleaved into small oligopeptides of 3-23 amino acids. Both the ubiquitination and the proteasomal degradation are ATP-dependent. The role of ubiquitin-proteasome pathway in reproduction was initially recognized due to its involvement in spermatogenesis (Chen et al., 1998; Baarends et al., 1999) and the degradation of paternal mitochondria after fertilization in mammals (Sutovsky et al., 1999, 2003). Later its direct role in egg coat penetration was discovered during ascidian (Sawada et al., 2002a; b) and porcine fertilization (Sutovsky et al., 2004).

## Ubiquitin-proteasome pathway during sperm-ZP penetration

We first observed the ability of proteasomal inhibitors, lactacystin and MG132, to block zona penetration in mammals during our attempts to reversibly block the degradation of the paternal mitochondria inside the cytoplasm of the zygote (Sutovsky et al., 2003). For this to occur, proteasomal inhibitors had to be added to IVF medium only after the spermatozoa passed through the zona and entered oocyte cytoplasm, 6 hours after the addition of sperm. This was an unexpected observation because proteasomal inhibitors do not block the activity of conventional serine proteases (Fenteany et al., 1995; Goldberg et al., 1995) that could be present in the sperm acrosome. However, our experiments corroborated the findings of Sawada et al., (2002a; b), who showed that the ascidian sperm exudates contain unconjugated mono-ubiquitin, ubiquitin conjugating enzymes and proteasomes. These components of the ascidian sperm ubiquitin system are deployed on the vitelline envelope (VE) during fertilization to first ubiquitinate, and then degrade, the ascidian analogue of ZP3, a protein called HrVC70. Additional studies indicate that egg coat penetration by the spermatozoon can be blocked both by proteasomal inhibitors (Wang et al., 2002; Sutovsky et al., 2003; 2004) and anti-proteasome antibodies (Sutovsky et al., 2004) in both mammals and invertebrates (Matsumura and Aketa, 1991; Sawada et al., 2002b). Specific proteasomal inhibitors and anti-proteasomal antibodies block ZP penetration completely, without affecting sperm motility, sperm ZP-binding and AE (Sutovsky et al., 2004). Conventional protease inhibitors used in previous studies (for example, Llanos et al., 1993) may not inhibit 26S proteasome.

Consistent with the proposed function of 26S proteasome during fertilization, various proteasomal subunits can be detected in the acrosome of spermatozoa from humans (Wojcik et al., 2000; Bialy et al., 2001; Morales et al., 2004), boars, bulls and mice (Sutovsky et al., 2004; Baska and Sutovsky, 2005). Ubiquitin conjugating enzyme E2 and the deubiquitinating enzyme mUBPy (Berruti and Martegani, 2002; 2005) have been identified in the acrosome of mouse spermatozoa. In Drosophila, multiple, unique, testis-specific proteasomal subunits are expressed that may play a role in spermatogenesis and sperm function (Belote *et al.*, 1998). Anti-ubiquitin antibody-immunoreactive proteins that could be targeted by sperm proteasomes were detected on the outer face of the pig (Sutovsky *et al.*, 2004), rabbit and primate ZP (Peter Sutovsky; unpublished data). Sequential ubiquitination and proteasomal degradation of the putative sperm receptor protein HrVC70, the *ascidian* homologue of mammalian ZP3 sperm receptor, has been shown on the ascidian vitelline envelope (Sawada *et al.*, 2002a; b). An interesting idea of ZP aging has been put forward by Olds-Clarke (2003). She suggested that, in the mouse, since the ZP glycoprotein molecules are secreted by the oocyte over a relatively long period of time, the "old" outer layers may contain "decayed" glycoproteins whose sugar residues differ from those of the "young" proteins on the inner face of the ZP. This may account for differences in lectin binding patterns between the inner and outer layers of the ZP. Consequently, it may be that these "old" ZP proteins are recognized as being defective and hence ubiquitinated by ubiquitin system present in the follicular fluid. The finding of ubiquitin in bovine follicular fluid (Einspanier *et al.*, 1993) is consistent with such an ubiquitination of ZP-surface during mammalian oogenesis although it is more likely that the proteins deposited onto ZP surface are already ubiquitinated during post-translational modification in the endoplasmic reticulum (ER), as there is no evidence of ubiquitin activating and conjugating enzymes being present in follicular fluid. The resident endoplasmic reticulum-associated protein degradation (ERAD) protein quality control system of the ER is based on branched sugar trimming of glycoproteins by alpha mannosidase, followed by ubiquitination facilitated by a saccharide-specific ubiquitin ligase recognizing the trimmed, mannose-rich, sugars of such glycoproteins . This could account for the high affinity of porcine ZP to both anti-ubiquitin antibodies (Sutovsky *et al.*, 2004) and the alpha-mannosyl binding lectin LCA (Maymon *et al.*, 1994). In contrast to the mouse, porcine ZP proteins may be deposited by both oocyte and cumulus cells with the localization of porcine ZP3α mRNA shifting progressively from the oocyte to cumulus cells as follicular maturation progresses (Kolle *et al.*, 1996). It is possible that the ubiquitination machinery associated with the secretory pathway could efficiently ubiquitinate the nascent ZP proteins in the cumulus cells before they are secreted and deposited onto the ZP with such proteins being synthesized by cumulus/granulosa cells in rabbits (Lee and Dunbar, 1993), dogs (Blackmore *et al.*, 2004), cynomolgus monkeys (Martinez *et al.*, 1996) and pigs (Kolle *et al.*, 1996).

## Model for the study of sperm-ZP interaction in the pig

Although the laboratory mouse has frequently been used as the "model" for studying sperm-ZP interaction in mammals, recent evidence has suggested that, unlike in many eutherian mammals, the mouse ZP is composed of only three, not four, ZP glycoproteins. This may limit the validity of extrapolating mouse studies to other mammals, including primates. A non-rodent, eutherian mammal is thus needed for comparative studies if the mechanisms of sperm-ZP interactions are to be extrapolated to humans. Due to prohibitive cost of non-human primates, the pig has become a model animal for these studies and important advances have been made in studies of the biology of egg coats in this species (examples are: Hedrick, 1996; Yurewicz *et al.*, 1987). At the present time, the pig is increasingly becoming the biomedical model of choice and, with the upcoming opening of the NIH-funded National Swine Resource and Research Center at the University of Missouri-Columbia, we will have the opportunity to further develop our studies of porcine 26S proteasome and other porcine sperm proteins.

Sperm 26S proteasome has been studied most intensively in ascidian fertilization where the proteasome-containing exudate degrades the sperm receptor molecule on the vitelline envelope (Sawada *et al.*, 2002a; b). In the pig, we have preliminary evidence that the sperm receptor protein ZPC is ubiquitinated which would make it recognizable to 26S proteasomes in the acrosome (Sutovsky *et al.*, 2005). Compared to the mammals, the major difference appears to be that the ubiquitination of the ascidian sperm receptor occurs during fertilization, immediately prior to its degradation. In the pig, the sperm receptor already may be ubiquitinated during ZP deposition inside the ovarian follicle. Beside the pig, we have detected ZP ubiquitination in the rabbit and primate ovary, but have failed to find it in the mouse after immunohistochemical staining of ovarian sections (Peter Sutovsky, unpublished observations). Likewise, ubiquitination was not reported in LC-MS analysis of deglycosylated mouse ZP proteins (Boja *et al.*, 2003), although such a posttranslational modification may not have been accounted for in the analysis of MS spectra. There is evidence that the sperm proteasomal activity is required for murine fertilization (Wang *et al.*, 2002; Pasten *et al.*, 2005) and we have confirmed the presence of proteasomes in the murine sperm acrosome (Baska and Sutovsky, 2005; and unpublished data). Specific proteasomal inhibitors have no effect on sperm-ZP binding and acrosomal exocytosis in human and porcine ova (Sutovsky *et al.*, 2004), but ZP penetration is inhibited. In mouse, the ZP penetration is only partially reduced but sperm-ZP binding is greatly affected by proteasomal inhibitors (Pasten *et al.*, 2005). Further investigations are, however necessary to determine whether, or not, sperm penetration through a murine ZP is dependent on an ubiquitin system.

## Anti-polyspermy defense and 26S proteasome

After spermatozoa incorporation into the oocyte, fertilization normally involves the formation of one male and one female pronucleus. When two or more spermatozoa enter the ooplasm, polyspermy occurs, which results in aberrant development of the early embryo. In *in vivo* porcine oocytes, the polyspermy rate is substantially higher than other species commonly investigated (Polge, 1978) and in the *in vitro* system it often exceeds 50% (Wang *et al.*,1994). The hardening of ZP associated with exocytosis of the cortical granules (CG: cortical reaction) and cleavage of ZP2 protein is thought to be the major protective mechanism against polyspermy in mouse oocytes (McLeskey *et al.*, 1998; Wassarman *et al.*, 2004). The hardened murine ZP becomes refractory to additional sperm binding. In contrast to the mouse, this mechanism of ZP-controlled anti-polyspermy defense does not appear to be effective in the rabbit (Gould *et al.*, 1971), whereas in porcine oocytes, the block of polyspermy occurs at both the level of ZP and at the surface of the vitelline membrane, the oolemma (Hunter, 1990). The high incidence of polyspermy in the pig may result from incomplete cortical granule exocytosis in oocytes matured *in vitro* (Wang *et al.*, 1997). Anti-polyspermy control at the level of the ZP may in part be due to other factors such as the deposition of oviductal glycoproteins in the pig (McCauley *et al.*, 2003). In this species, we have found that ZP2 cleavage occurs and can be prevented by proteasomal inhibitors (Fig. 8). However, the parthenogenetically activated porcine ova can be fertilized despite CG exocytosis (Young-Joo Yi and Peter Sutovsky; unpublished data). We have new data showing that a deubiquitinating enzyme, PGP9.5, is present in the boar sperm acrosome, possibly associated with the sperm proteasome. In addition, PGP9.5 is present in the oocyte cortex and anti-PGP9.5 antibodies significantly increase polyspermy rate during porcine IVF. Thus, it appears that PGP9.5, and possibly other sperm deubiquitinating enzymes (DUBs), have a rate-limiting effect on the sperm proteasome during fertilization. This could contribute to reducing the chances of polyspermy. The protein gene product 9.5 (PGP9.5 or L1) belongs to a family of ubiquitin C-terminal hydrolases, which regenerate monoubiquitin from ubiquitin-protein complexes or from polyubiquitin chains by cleaving the amide linkage next to the C-terminal glycine of ubiquitin. The PGP9.5 is expressed in the spermatogonia of Japanese monkeys, and it is present in the cytoplasm of the primate spermatogonia when their proliferation

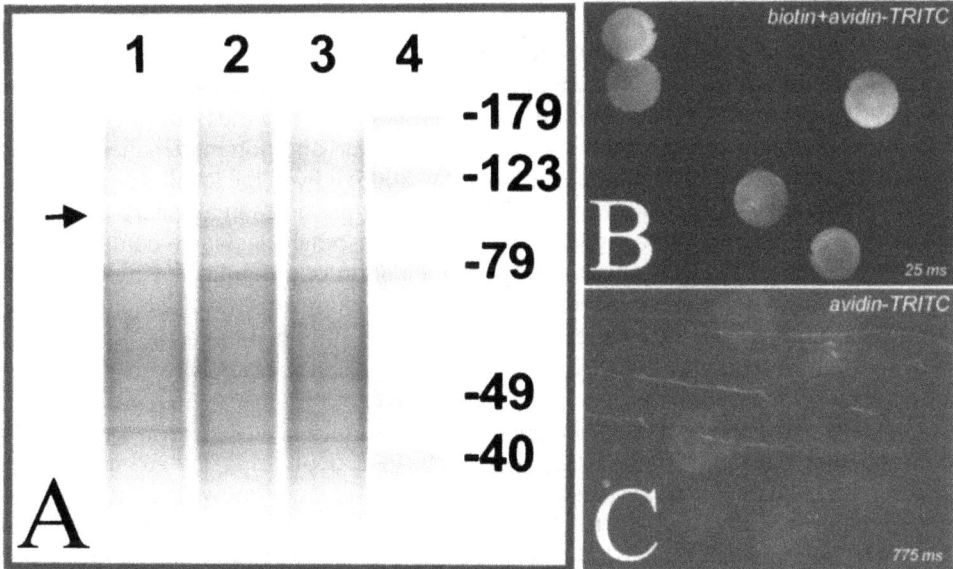

**Figure 8.** Fertilization induced cleavage of porcine zona pellucida protein visualized by streptavidin detection of biotinylated ZP protein extracts after SDS-PAGE and Western blotting. (A) The band indicative of ZP2-protein cleavage after fertilization and cortical reaction is seen in lane 2 (100 fertilized ova), but not in lanes 1 (100 unfertilized, metaphase-II ova) or in lane 3 (100 ova exposed to spermatozoa in presence of a proteasomal inhibitor MG132, which block sperm penetration through ZP, thus preventing oocyte activation and ZP-protein cleavage). Positive (B) and negative (C) control samples of biotinylated ova, detected with fluorescently labeled streptavidin are show. Even though ZP-protein cleavage occurs during porcine fertilization *in vitro,* it does not prevent polyspermy. Anti-polyspermy defense in this species likely involves additional modification of ZP-surface during oocyte descent down the oviduct. Regulation of proteasomal activity during acrosomal exocytosis and sperm-ZP penetration may also contribute to anti-polyspermy defense.

activity is suppressed. The PGP9.5 is absent from the spermatogonial cytoplasm when the proliferation activity is at the normal level and reversible quiescence of type Ap spermatogonia is thought to be mediated by PGP9.5, which regulates cell cycle activities by deconjugating ubiquitin from the substrate proteins (Tokunaga *et al.*, 1999). In general, the role of DUBs is to disassemble polyubiquitin chains removed from substrates at the 19S proteasomal regulatory complex, thus regenerating the pool of unconjugated monoubiquitin for further protein ubiquitination. Some of the subunits within the 19S regulatory complex of the proteasome have a DUB activity; cytosolic DUBs can associate with the 19S regulatory complex (Stone *et al.*, 2004).

In crustaceans, the sperm proteasomes take part in acrosomal exocytosis triggered by the egg coat glycoproteins (Mykles, 1998). In the solitary ascidian (*Halocynthia roretzi*) it also facilitates sperm binding and penetration through the vitelline coat (Yokosawa *et al.*, 1987; Sawada *et al.*, 2002a; b). In human spermatozoa, proteasome is involved in ZP and progesterone-induced acrosomal exocytosis and in the sustained phase of the $Ca^{2+}$ influx provoked by progesterone (Morales *et al.*, 2003), whereas in *H. roretzi*, Sawada *et al.* (2002a) have reported that the sperm proteasome is localized to the plasma membrane surface over its head region. Sutovsky *et al.* (2004) inhibited sperm penetration through porcine ZP by adding reversible proteasomal inhibitors and antibodies specific to 20S proteasomal core subunits to the culture medium. They detected proteasomes both within the acrosomal matrix of intact boar spermatozoa and on the inner acrosomal membrane after ZP-induced acrosomal exocytosis (Figure 3). Thus, in the recent past, evidence has accumulated that shows that the 26S proteasome is a protease involved in the digestion of egg coat during both mammalian and non-mammalian fertilization (reviewed by Sakai *et al.*, 2004; Baska and Sutovsky, 2005). Given the localization of both 26S proteasome (Sutovsky *et al.*, 2004) and PGP9.5 on the inner acrosomal membrane, it is possible that PGP9.5 interacts directly with the 19S proteasomal regulatory complex subunits responsible for the initial recognition and deubiquitination of ubiquitinated proteins. Similar association has been described between yeast deubiquitinating enzyme Ubp6 and 19S regulatory complex subunit Rpn12 (Stone *et al.*, 2004). It is possible that the association with non-proteasomal DUBs regulates the deubiquitinating activity of various types of proteasomes, including the sperm proteasome.

The subunits of the 19S proteasomal regulatory complex play diverse roles in the recognition, binding and deubiquitination of the tetra/polyubiquitinated substrates by the 26S proteasome. In particular, proteasomal subunit Rpn10 is the only proteasomal subunit known to bind polyubiquitin chains. Consequently, we were able to block porcine IVF by the addition of a specific anti-Rpn10 antibody to an IVF medium (Young-Joo Yi and Peter Sutovsky; unpublished data). Consistent with our previous immuno-inhibition studies (Sutovsky *et al.*, 2004), this treatment blocked neither sperm-ZP binding nor acrosomal exocytosis. Contrary to the action of Rpn10, the 19S complex-subunit Rpn12 may be involved in the deubiquitination of the ubiquitinated substrates upon their binding to 19S complex. Accordingly, antibodies against Rpn12 increase the rate of polyspermic fertilization in a fashion similar to anti-PGP9.5 antibody.

In *in vivo* pig zygotes, the polyspermy rate is less than 5% (Hunter, 1990) which compares to >50% in *in vitro* zygotes. In *in vitro* fertilized porcine oocytes there is incomplete CG exocytosis, whereas in *in vivo* fertilization in the oviduct, complete dispersion of CG contents in the perivitelline space occurs after exocytosis (Cran and Cheng, 1986). Spermatozoa co-cultured with oviductal epithelial cells resulted in reduced polyspermy when used for *in vitro* fertilization (Anderson and Killian, 1994; Dubuc and Sirard, 1995), whereas co-culture of oocytes with oviductal epithelial cells (Kano *et al.*, 1994) or preincubation of oocytes with oviductal fluid (Kim et al., 1996) significantly reduces polyspermy. This may be due to the oviductal proteins in preincubation, or fertilization media competing with sperm receptors for the binding of zona pellucida ligands, that might stimulate the rate of sperm acrosome reaction and reduce the number of capacitated spermatozoa attaching to the surface of pig oocytes (Funahashi and Day, 1997). We are currently examining the presence of deubiquitinating enzymes in porcine oviductal fluid as they could play a role in controlling sperm penetration and the prevention of polyspermic fertilization. An alternative possibility is that the CG matrix could contain PGP9.5 or other DUBs that would deubiquitinate the ZP after penetration of the first spermatozoon. Thus far, our data do not indicate the involvement of PGP9.5, but there could be other DUBs present in the CG exudates. Alternatively, CGs could contain protein phosphatases that could dephosphorylate the sperm-proteasomal core subunits, thus rendering them inactive. Morales *et al.*, (2005) recently reported that fibronectin, an extracellular matrix protein present in the cumulus extracellular matrix (ECM), increases the phosphorylation of sperm proteins and enhances proteolytic activity of the human sperm proteasome. Protein phosphatases or other proteins could be present in the CG exudates to down-regulate sperm proteasomal activity. Horiguchi and Tokumoto (2005) reviewed the role of phosphorylation in the modulation of proteasomal activity in the meiotic oocyte extracts and somatic cells. Serine, threonine and tyrosine phosphorylation occurs in five of the seven alpha type subunits of the 20S proteasomal core. The phosphorylation of proteasomal subunits modulates the capability of the 20S core to degrade specific substrate proteins. Casein-kinase-II is responsible for the phosphorylation of multiple 20S alpha-subunits (Horiguchi and Tokumoto, 2005) and has also been implicated in phosphorylation of other sperm proteins (Chaudhry *et al.*, 1991). Phosphorylated 26S proteasomes isolated from prophase-I ooplasm of immature oocytes can degrade cyclin B but the dephosphorylated proteasomes from the ooplasm of mature metaphase-II oocytes cannot (Tokumoto *et al.*, 2000). The loss of proteasomal proteolytic activity is dependent upon dephosphorylation of the 20S proteasomal core subunit $\alpha$–4 (Tokumoto *et al.*, 1999).

## Proposed proteasomal mechanism corroborates the established models of ZP penetration

Is it possible to reconcile the proteasomal concept of sperm-ZP binding and penetration with the current models? The enzymatic hypothesis of ZP penetration by the spermatozoon although originally an attractive proposition was somewhat diminished in its attractiveness by the acrosin mutation experiments (Baba *et al.*, 1994). Nevertheless, the acrosin gene ablation results do not rule out the possible involvement of 26S proteasomes in zona penetration. Yanagimachi (1994) favors an enzymatic over mechanical hypothesis for sperm-ZP penetration although he points out that some conventional protease inhibitors do not prevent ZP penetration. Our recent work has shown that proteasome-specific inhibitors do indeed prevent fertilization. These proteasomal inhibitors (such as MG132, lactacystin and clasto-lactacystin-beta-lactone/CLBL) do not inhibit conventional, non-proteasomal proteases (Fenteany *et al.*, 1995): if zona penetration is dependent on the sperm proteasomal activity, conventional protease inhibitors are not likely to affect fertilization, although some may interfere with acrosomal function or ZP hardening.

There is some support for sperm forward thrust facilitating ZP penetration (Bedford, 1998). However, the mechanical force generated by sperm motility alone may not be sufficient for sperm zona penetration (Green, 1987) although it possibly assists in moving the sperm head forward during the digestion of the slit in the ZP. Continued or even an enhanced motility during sperm-ZP penetration (Drobnis *et al.*, 1988) would assure that the sperm head apex with proteasomes on the IAM would move forward through the ZP during penetration, thus assuring progressive digestion of the fertilization slit. Flagellar motility as a sole force for sperm penetration does not explain our proteasomal inhibitor data. In the presence of proteasomal inhibitors and anti-proteasomal antibodies, the sperm-ZP binding and acrosomal exocytosis occurred at the usual rate, even though ZP penetration was inhibited. We also observed a depression in the ZP beneath the bound sperm heads (Sutovsky *et al.*, 2004) suggesting a proteolytic event.

## Concluding remarks and future directions

Our findings at this stage provide an outline of an alternative mechanism of sperm-zona pellucida penetration during mammalian fertilization. These findings also provide a basis for developing new contraceptive agents based on targeting sperm proteasomal activity. In order to further validate the sperm proteasome as the egg coat lysin in mammals, several conditions must be met. First, ZP ubiquitination has to occur during oogenesis or during fertilization to render the ZP recognizable to sperm proteasomes. We are working on proteomic analysis of porcine ZP fractions that have been affinity purified by using recombinant, polyubiquitin chain-binding proteins immobilized on agarose matrix. Mass spectroscopy followed by protein database search for Gly-Gly-modified, ubiquitinated proteins (Peng *et al.*, 2003) is being employed in this work. Concomitantly, it should be considered that partially assembled proteasomes, composed solely of a 20S core can degrade proteins that are not ubiquitinated (Sheaff *et al.*,

2000). Second, proteasomes should remain on the inner acrosomal membrane after acrosomal exocytosis to provide protease activity for the digestion of the fertilization slit at all stages of ZP penetration. Towards this end, we are examining the presence of proteasomes in the isolated IAM fractions, as well as in the whole spermatozoa, during capacitation, AE, ZP-penetration and sperm-oolemma binding. A 26S proteasome with one 19S complex on each side of the 20S core could be anchored to the IAM matrix via the interaction of 19S subunits with IAM proteins. Such an interaction could be revealed by co-immunoprecipitation and mass spectroscopic amino acid sequence analysis of the isolated IAM protein fractions. The 19S complex on the other side of the 20S barrel would be exposed after AE and free to interact with polyubiquitin chains on the ZP. Finally, a rate limiting activity should be associated with the sperm proteasome to contribute to anti-polyspermy defense at the ZP level. The preliminary work by our group and others suggests that deubiquitinating enzymes and regulated phosphorylation of 19S and 20S proteasomal subunits could contribute to anti-polyspermy defense. A variety of approaches will be employed to further examine these pathways during fertilization and in a cell free system. In the near future, this work will lead likely to a genetic model for the functional analysis of the sperm proteasome. At present, this is a difficult task because the proteasomal subunits are highly conserved and a conditional deletion or mutation of one of them in the testis would likely derail spermatogenesis. Thus we are focusing our effort on identifying and genetically altering proteins responsible for anchoring the proteasomes to the inner acrosomal membrane.

## Acknowledgements

Our sincere gratitude belongs to our collaborators, Drs. Randall S. Prather, Billy N. Day, Gustavo Doncel and Mary Zelinski-Wooten for their support of our research on the mechanism of mammalian sperm-zona pellucida penetration. We thank Kathryn Craighead, Nicole Leitman and Miriam Sutovsky for administrative and technical assistance. We gratefully acknowledge the support by National Research Initiative Competitive Grant #2002-35203-12237 from the USDA Cooperative State Research, Education and Extension Service to PS, and Canadian CIHR and NSERC grants to RO. Additional funding to PS was provided by the Food for the 21st Century Program.

## References

**Anderson SH and Killian GJ** (1994) Effect of macromolecules from oviductal conditioned medium on bovine sperm motion and capacitation *Biology of Reproduction* **51** 795-799

**Baarends WM, Roest HP and Grootegoed JA** (1999) The ubiquitin system in gametogenesis *Molecular and Cellular Endocrinology* **151** 5-16

**Baba T, Azuma S, Kashiwabara S and Toyoda Y** (1994) Sperm from mice carrying a targeted mutation of the acrosin gene can penetrate the oocyte zona pellucida and effect fertilization *Journal of Biological Chemistry* **269** 31845-31849

**Baska KM and Sutovsky P** (2005) Protein modification by ubiquitination and its consequences for spermatogenesis, sperm maturation, fertilization and pre-implantation embryonic development. In: *New Impact on Protein Modifications in the Regulation of Repro-*

*ductive System* pp 83-114 Ed. *T* Tokumoto. Research Signpost, Kerala

**Bedford JM** (1998) Mammalian fertilization misread? Sperm penetration of the eutherian zona pellucida is unlikely to be a lytic event *Biology of Reproduction* **59** 1275-1287

**Belote JM, Miller M and Smyth KA** (1998) Evolutionary conservation of a testes-specific proteasome subunit gene in *Drosophila Gene* **215** 93-100

**Berruti G and Martegani E** (2002) mUBPy and MSJ-1, a deubiquitinating enzyme and a molecular chaperone specifically expressed in testis, associate with the acrosome and centrosome in mouse germ cells *Annals of the New York Academy of Science* **973** 5-7

**Berruti G and Martegani E** (2005) The deubiquitinating enzyme mUBPy interacts with the sperm-specific molecular chaperone MSJ-1: the relation with the proteasome, acrosome, and centrosome in mouse male germ cells *Biology of Reproduction* **72** 14-21

**Bi M, Wassler MJ and Hardy DM** (2002) Sperm adhesion to the extracellular matrix of the egg. In: *Fertilization* pp 153-180 Ed. DM Hardy. Academic Press, San Diego

**Bialy LP, Ziemba HT, Marianowski P, Fracki S, Bury M and Wojcik C** (2001) Localization of a proteasomal antigen in human spermatozoa: immunohistochemical electron microscopic study *Folia Histochemica et Cytobiologica* **39** 129-130

**Blackmore DG, Baillie LR, Holt JE, Dierkx L, Aitken RJ and McLaughlin EA** (2004) Biosynthesis of the canine zona pellucida requires the integrated participation of both oocytes and granulosa cells *Biology of Reproduction* **71** 661-668.

**Bleil JD and Wassarman PM** (1986) Autoradiographic visualization of the mouse egg's sperm receptor bound to sperm *Journal of Cell Biology* **102** 1363-1371

**Bleil JD, Greve JM and Wassarman PM** (1988) Identification of a secondary sperm receptor in the mouse egg zona pellucida: role in maintenance of binding of acrosome-reacted sperm to eggs *Developmental Biology* **128** 376-385

**Boja ES, Hoodbhoy T, Fales HM and Dean J** (2003) Structural characterization of native mouse zona pellucida proteins using mass spectrometry *Journal of Biological Chemistry* **278** 34189-34202

**Breed WG** (1994) How does sperm meet egg?-in a marsupial *Reproduction Fertility and Development* **6** 485-506

**Breed WG and Leigh CM** (1990) Morphological changes in the oocyte and its surrounding vestments during *in vivo* fertilization in the dasyurid marsupial *Sminthopsis rassicaudata Journal of Morphology* **204** 177-196

**Cardozo C** (1993) Catalytic components of the bovine pituitary multicatalytic proteinase complex (proteasome) *Enzyme and Protein* **47** 296-305

**Chaudhry PS, Newcomer PA and Casillas ER** (1991) Casein kinase I in bovine sperm: purification and characterization *Biochemical and Biophysiological Research Communications* **179** 592-598

**Chen HY, Sun JM, Zhang Y, Davie JR and Meistrich ML** (1998) Ubiquitination of histone H3 in elongating spermatids of rat testes *Journal of Biological Chemistry* **273** 13165-13169

**Ciechanover A** (1994) The ubiquitin-proteasome proteolytic pathway *Cell* **79** 13-21

**Cran DG and Cheng WTK** (1986) The cortical reaction in pig oocytes during *in vivo* and *in vitro* fertilization *Gamete Research* **13** 241-251

**Dean J** (2002) Oocyte-specific genes regulate follicle formation, fertility and early mouse development *Journal of Reproductive Immunology* **53** 171-180

**Dean J** (2004) Reassessing the molecular biology of sperm-egg recognition with mouse genetics *Bioessays* **26** 29-38

**Drobnis EZ, Yudin AI, Cherr GN and Katz DF** (1988) Hamster sperm penetration of the zona pellucida: kinematic analysis and mechanical implications *Developmental Biology* **130** 311-323

**Dubuc A and Sirard MA** (1995) Effect of coculturing spermatozoa with oviductal cells on the incidence of polyspermy in pig *in vitro* fertilization *Molecular Reproduction and Development* **41** 360-367

**Einspanier R, Schuster H and Schams D** (1993) A comparison of hormone levels in follicle-lutein-cyst and in normal bovine ovarian follicles *Theriogenology* **40** 181-188

**Ensslin MA and Shur BD** (2003) Identification of mouse sperm SED1, a bimotif EGF repeat and discoidin-domain protein involved in sperm-egg binding *Cell* **114** 405-417

**Fenteany G, Standaert RF, Lane WS, Choi S, Corey EJ and Schreiber SL** (1995) Inhibition of proteasome activities and subunit-specific amino-terminal threonine modification by lactacystin *Science* **268** 726-731

**Fridberger A, Sundelin J, Vacquier VD and Peterson PA** (1985) Amino acid sequence of an egg-lysin protein from abalone spermatozoa that solubilizes the vitelline layer *Journal of Biological Chemistry* **260** 9092-9099

**Funahashi H and Day BN** (1997) Advances in *in vitro* production of pig embryos *Journal of Reproduction and Fertility Supplement* **52** 271-283

**Glickman MH and Ciechanover A** (2002) The ubiquitin-proteasome proteolytic pathway: destruction for the sake of construction *Physiological Reviews* **82** 373-428

**Goldberg AL, Stein R and Adams J** (1995) New insights into proteasome function: from archaebacteria to drug development *Chemistry and Biology* **2** 503-508

**Gould K, Zaneveld LJ, Srivastava PN and Williams WL** (1971) Biochemical changes in the zona pellucida of rabbit ova induced by fertilization and sperm enzymes *Proceedings of the Society for Experimental Biology and Medicine* **136** 6-10

**Green DP** (1987) Mammalian sperm cannot penetrate the zona pellucida solely by force *Experimental Cell Research* **169** 31-38

**Hedrick JL** (1996) Comparative structural and antigenic properties of zona pellucida glycoproteins *Journal of Reproduction and Fertility Supplement* **50** 9-17

**Hochstrasser M** (1995) Ubiquitin, proteasomes, and the regulation of intracellular protein degradation *Current Opinion in Cell Biology* **7** 215-223

**Horiguchi R and Tokumoto T** (2005) Modifications to proteasomal subunits during meiotic cell cycle-implications in the regulation of fertilization throughout proteasome activity. In: *New Impact on Protein Modifications in the Regulation of Reproductive System* pp37-60 Ed. T Tokumoto, Research Signpost, Kerala

**Hunter RHF** (1990) Fertilization of pig eggs *in vivo* and *in vitro Journal of Reproduction and Fertility Supplement* **40** 211-226

**Kano K, Miyano T and Kato S** (1994) Effect of oviductal epithelial cells on fertilization of pig oocytes *in vitro Theriogenology* **42** 1061-1068

**Kim NH, Funahashi H, Abeydeera LR, Moon SJ, Prather RS and Day BN** (1996) Effects of oviductal fluid on sperm penetration and cortical granule exocytosis during fertilization of pig oocytes *in vitro Journal of*

*Reproduction and Fertility* **107** 79-86

Kisselev AF, Akopian TN, Woo KM and Goldberg AL (1999) The sizes of peptides generated from protein by mammalian 26 and 20 S proteasomes. Implications for understanding the degradative mechanism and antigen presentation *Journal of Biological Chemistry* **274** 3363-3371

Kolle S, Sinowatz F, Boie G, Totzauer I, Amselgruber W and Plendl J (1996) Localization of the mRNA encoding the zona protein ZP3 alpha in the porcine ovary, oocyte and embryo by non-radioactive in situ hybridization *The Histochemical Journal* **28** 441-447

Lee VH and Dunbar BS (1993) Developmental expression of the rabbit 55-kDa zona pellucida protein and messenger RNA in ovarian follicles *Developmental Biology* **155** 371-382

Lira SA, Kinloch RA, Mortillo S and Wassarman PM (1990) An upstream region of the mouse ZP3 gene directs expression of firefly luciferase specifically to growing oocytes in transgenic mice *Proceedings of the National Academy of Sciences USA* **87** 7215-7219

Llanos M, Vigil P, Salgado AM and Morales P (1993) Inhibition of the acrosome reaction by trypsin inhibitors and prevention of penetration of spermatozoa through the human zona pellucida *Journal of Reproduction and Fertility* **97** 173-178

Lu Q and Shur BD (1997) Sperm from beta 1,4-galactosyltransferase-null mice are refractory to ZP3-induced acrosome reactions and penetrate the zona pellucida poorly *Development* **124** 4121-4131

Martinez ML, Fontenot GK and Harris JD (1996) The expression and localization of zona pellucida glycoproteins and mRNA in cynomolgus monkeys (*Macaca fascicularis*) *Journal of Reproduction and Fertility Supplement* **50** 35-41

Matsumura K and Aketa K (1991) Proteasome (multicatalytic proteinase) of sea urchin sperm and its possible participation in the acrosome reaction *Molecular Reproduction and Development* **29** 189-199

Maymon BB, Maymon R, Ben-Nun I, Ghetler Y, Shalgi R and Skutelsky E (1994) Distribution of carbohydrates in the zona pellucida of human oocytes *Journal of Reproduction and Fertility* **102** 81-86

McCauley TC, Buhi WC, Wu GM, Mao J, Caamano JN, Didion BA and Day BN (2003) Oviduct-specific glycoprotein modulates sperm-zona binding and improves efficiency of porcine fertilization *in vitro Biology of Reproduction* **69** 828-834

McLeskey SB, Dowds C, Carballada R, White RR and Saling PM (1998) Molecules involved in mammalian sperm-egg interaction *International Review of Cytology* **177** 57-113

Morales P, Kong M, Pizarro E and Pasten C (2003) Participation of the sperm proteasome in human fertilization *Human Reproduction* **18** 1010-1017

Morales P, Pizarro E, Kong M and Jara M (2004) Extracellular localization of proteasomes in human sperm *Molecular Reproduction and Development* **68** 115-124

Morales P, Diaz S, Kong M and Perez B (2005) Effect of fibronectin (FN) on proteasomal activity, acrosome reaction (AR) and tyrosine phosphorylation in human sperm *Biology of Reproduction* Special Issue, pp. 150

Mori E, Kashiwabara S, Baba T, Inagaki Y and Mori T (1995) Amino acid sequences of porcine Sp38 and proacrosin required for binding to the zona pellucida *Developmental Biology* **168** 575-583

Mykles DL (1998) Intracellular proteinases of invertebrates: calcium-dependent and proteasome/ubiquitin-dependent systems *International Review of Cytology* **184** 157-289

Olds-Clarke P (2003) Unresolved issues in mammalian fertilization *International Review of Cytology* **232** 129-184

Pasten C, Morales P and Kong M (2005) Role of the sperm proteasome during fertilization and gamete interaction in the mouse *Molecular Reproduction and Development* **71** 209-219

Peng J, Schwartz D, Elias JE, Thoreen CC, Cheng D, Marsischky G, Roelofs J, Finley D and Gygi SP (2003) A proteomics approach to understanding protein ubiquitination *Nature Biotechnology* **21** 921-926

Polge C (1978) Fertilization in the pig and the horse *Journal of Reproduction and Fertility* **54** 461-470

Pickart CM (1998) Polyubiquitin chains. In: *Ubiquitin and the Biology of the Cell* pp 19-63 Eds J-M Peters, JR Harris and D Finley. Plenum Press, New York

Pizarro E, Pasten C, Kong M and Morales P (2004) Proteasomal activity in mammalian spermatozoa *Molecular Reproduction and Development* **69** 87-93

Primakoff P and Myles DG (2002) Penetration, adhesion, and fusion in mammalian sperm-egg interaction *Science* **296** 2183-2185

Rankin T, Familari M, Lee E, Ginsberg A, Dwyer N, Blanchette-Mackie J, Drago J, Westphal H and Dean J (1996) Mice homozygous for an insertional mutation in the *Zp3* gene lack a zona pellucida and are infertile *Development* **122** 2903-2910

Rock KL, Gramm C, Rothstein L, Clark K, Stein R, Dick L, Hwang D and Goldberg AL (1994) Inhibitors of the proteasome block the degradation of most cell proteins and the generation of peptides presented on MHC class I molecules *Cell* **78** 761-771

Sakai N, Sawada H and Yokosawa H (2003) Extracellular ubiquitin system implicated in fertilization of the ascidian, *Halocynthia roretzi*: isolation and characterization *Developmental Biology* **264** 299-307

Sakai N, Sawada MT and Sawada H (2004) Non-traditional roles of ubiquitin-proteasome system in fertilization and gametogenesis *International Journal of Biochemistry and Cell Biology* **36** 776-784

Saling PM (1981) Involvement of trypsin-like activity in binding of mouse spermatozoa to zonae pellucidae *Proceedings of the National Academy of Sciences USA* **78** 6231-6235

Sawada H, Sakai N, Abe Y, Tanaka E, Takahashi Y, Fujino J, Kodama E, Takizawa S and Yokosawa H (2002a) Extracellular ubiquitination and proteasome-mediated degradation of the ascidian sperm receptor *Proceedings of the National Academy of Sciences USA* **99** 1223-1228

Sawada H, Takahashi Y, Fujino J, Flores SY and Yokosawa H (2002b) Localization and roles in fertilization of sperm proteasomes in the ascidian *Halocynthia roretzi Molecular Reproduction and Development* 62 271-276

Sheaff RJ, Singer JD, Swanger J, Smitherman M, Roberts JM and Clurman BE (2000) Proteasomal turnover of p21Cip1 does not require p21Cip1 ubiquitination *Molecular Cell* 5 403-410

Shibata Y (2005) Modification of egg envelope proteins during fertilization in teleost fish. In: *New Impact on Protein Modifications in the Regulation of Reproductive System* pp 115-134 Ed. T Tokumoto. Research Signpost, Kerala

Stone M, Hartmann-Petersen R, Seeger M, Bech-Otschir D, Wallace M and Gordon C (2004) Uch2/Uch37 is the major deubiquitinating enzyme associated with the 26S proteasome in fission yeast *Journal of Molecular Biology* 344 697-706

Sutovsky P, Moreno RD, Ramalho-Santos J, Dominko T, Simerly C and Schatten G (1999) Ubiquitin tag for sperm mitochondria *Nature* 402 371-372

Sutovsky P, McCauley TC, Sutovsky M and Day BN (2003) Early degradation of paternal mitochondria in domestic pig *(Sus scrofa)* is prevented by selective proteasomal inhibitors lactacystin and MG 132 *Biology of Reproduction* 68 1793-1800

Sutovsky P, Manandhar G, McCauley TC, Caamaño JN, Sutovsky M, Thompson WE and Day BN (2004) Proteasomal interference prevents zona pellucida penetration and fertilization in mammals *Biology of Reproduction* 71 1625–1637

Sutovsky P, Manandhar G, Miller D and Sutovsky M (2005) ZPB/ZPC is the ubiquitinated sperm receptor on porcine zona pellucida *Biology of Reproduction*: Special Issue: 89

Swann CA, Hope RM and Breed WG (2002) cDNA nucleotide sequence encoding the ZPC protein of Australian hydromyine rodents: a novel sequence of the putative sperm-combining site within the family Muridae *Zygote* 10 291-299

Talbot P, Shur BD and Myles DG (2003) Cell adhesion and fertilization: Steps in oocyte transport, sperm-zona pellucida interactions, and sperm-egg fusion *Biology of Reproduction* 68:1-9

Tanaka K (1998) Molecular biology of the proteasome *Biochemical and Biophysical Research Communications* 247 537-541

Tokumoto M, Horiguchi R, Nagahama Y and Tokumoto T (1999) Identification of the *Xenopus* 20S proteasome alpha4 subunit which is modified in the meiotic cell cycle *Gene* 239 301-308

Tokumoto M, Horiguchi R, Nagahama Y, Ishikawa K and Tokumoto T (2000) Two proteins, a goldfish 20S proteasome subunit and the protein interacting with 26S proteasome, change in the meiotic cell cycle *European Journal of Biochemistry* 267 97-103

Tokunaga Y, Imai S, Torii R and Maeda T (1999) Cytoplasmic liberation of protein gene product 9.5 during the seasonal regulation of spermatogenesis in the mon-

key *(Macaca fuscata) Endocrinology* 140 1875-1883

Tovich PR and Oko RJ (2003) Somatic histones are components to the perinuclear theca in bovine spermatozoa *Journal of Biological Chemistry* 278 32431-32438

Tulsiani DR, Abou-Haila A, Loeser CR and Pereira BM (1998) The biological and functional significance of the sperm acrosome and acrosomal enzymes in mammalian fertilization *Experimental Cell Research* 240 151-164

Wang WH, Abeydeera LR, Okuda K and Niwa K (1994) Penetration of porcine oocytes during maturation *in vitro* by cryopreserved, ejaculated spermatozoa *Biology of Reproduction* 50 510-515

Wang WH, Hosoe M and Shioya Y (1997) Induction of cortical granule exocytosis of pig oocytes by spermatozoa during meiotic maturation *Journal of Reproduction and Fertility* 109 247-255

Wang HM, Song CC, Duan CW, Shi WX, Li CX, Chen DY and Wang YC (2002) Effects of ubiquitin-proteasome pathway on mouse sperm capacitation, acrosome reaction and *in vitro* fertilization *Chinese Science Bulletin* 47 127-132

Wassarman PM (1990) Profile of a mammalian sperm receptor *Development* 108 1-17

Wassarman PM, Jovine L and Litscher ES (2004) Mouse zona pellucida genes and glycoproteins *Cytogenetic and Genome Research* 105 228-234

Wilkinson KD and Hochstrasser M (1998) The deubiquitinating enzymes. In: *Ubiquitin and the Biology of the Cell* pp 99-125 Eds J-M Peters, JR Harris and D Finley. Plenum Press, New York

Wojcik C, Benchaib M, Lornage J, Czyba JC and Guerin JF (2000) Proteasomes in human spermatozoa *International Journal of Andrology* 23 169-177

Yamagata T, Ito M and Takahashi N (1985) Involvement of an asparagine-linked oligosaccharide located in the zona pellucida in mouse fertilization *in vitro Zoological Science* 1 933 (abst)

Yanagimachi R (1994) Mammalian fertilization. In *The Physiology of Reproduction* pp 189-317 Eds E Knobil and JD Neil, second edition. Raven Press, New York

Yokosawa H, Numakunai T, Murao S and Ishii S (1987) Sperm chymotrypsin-like enzymes of different inhibitor-susceptibility as lysine in ascidians *Experientia* 43 925-927

Young P, Deveraux Q, Beal RE, Pickart CM and Rechsteiner M (1998) Characterization of two polyubiquitin binding sites in the 26S protease subunit 5a *Journal of Biological Chemistry* 273 5461-5467

Yu Y, Xu W, Kazemie M, Yi YJ, Sutovsky P and Oko RJ (2006) The extracellular protein coat of the inner acrosomal membrane is involved in zona pellucida binding and penetration during fertilization: Characterization of its most prominent polypeptide (IAM38) *Developmental Biology* 290 32-43

Yurewicz EC, Sacco AG and Subramanian MG (1987) Structural characterization of the Mr = 55,000 antigen (ZP3) of porcine oocyte zona pellucida. Purification and characterization of alpha- and beta-glycoproteins following digestion of lactosaminoglycan with endo-beta-galactosidase *Journal of Biological Chemistry* **262** 564-571

**Yurewicz EC, Sacco AG, Gupta SK, Xu N and Gage DA** (1998) Pig zona pellucida pZPB-pZPC heterocomplexes, but not subunit glycoproteins, binds with high affinity to boar sperm membrane vesicles *Journal of Biological Chemistry* **273** 7488-7494

# A comparative analysis of molecular mechanisms for blocking polyspermy: identification of a lectin-ligand binding reaction in mammalian eggs

Jerry L Hedrick

Department of Animal Science, University of California, Davis, CA 95616, USA

Fertilization is a critically important event to the creation of a new individual organism and to the propagation of a species. Evolutionarily conserved cellular and molecular mechanisms exist to modify the glycoproteins composing the egg extracellular matrix at fertilization. These matrix modifications regulate the cellular interactions of sperm and egg, maintain the diploid state of the nucleus after successful union of the two gametes (block to polyspermy) and control the environment for the developing embryo. Only recently have mammals been studied regarding extracellular matrix block to polyspermy mechanisms compared to the long term investigations of the same in sea urchins, fish and amphibians - knowledge of evolutionary conserved mechanisms in these animal groups can be used to predict the existence of mechanisms in mammals. Experimental evidence exists for the conservation of proteolytic, glycolytic, cross linking, conformational and binding mechanisms for establishing extracellular matrix blocks to polyspermy at fertilization. Analogous to a binding mechanism in anurans, a lectin-ligand binding mechanism for establishing an extracellular matrix block to polyspermy in mammalian eggs has been discovered. This binding mechanism involves the exocytotic release of a cortical granule lectin in the sperm-induced egg cortical reaction, diffusion and binding of the lectin to its ligand associated with the zona pellucida, and prevention of sperm-zona pellucida binding by the lectin-ligand reaction, thereby resulting in a block to polyspermy at fertilization. The glycoproteins involved in the lectin-ligand polyspermy block can potentially be used as targets for contraception.

### Evolutionary perspective on a primary biological process

*"Analogy would lead me one step further, namely to the belief that all animals and plants have descended from some one prototype...and living things have much in common in their chemical composition, their germinal vesicles, their cellular structures, and their laws of growth and reproduction".* Charles Darwin "Origin of the Species" (1859)

E-mail: jlhedrick@ucdavis.edu

Charles Darwin, proponent of natural selection and the founder of the modern theory of evolution, recognized almost 150 years ago that all living organisms share certain biological properties. Biological processes that are fundamental and essential to animal and plant life are evolutionary conserved. As stated above, this conservation includes molecular and cellular structures and fundamental physiological or biological processes, including reproduction. Some 60 years later, Frank Lillie (1919) emphasized, within the domain of reproduction, the fertilization event is important to natural selection and evolutionary processes. *"Fertilization is essentially the phenomenon of the union of two cells. Considered in this broad sense, it is practically a universal phenomenon among animals and plants. There is perhaps no phenomenon in the field of biology that touches so many fundamental questions as the union of the germ cells in the act of fertilization; in this supreme event all the strands of the webs of two lives are gathered in one knot, from which they diverge again and are rewoven in a new individual life history. It is the central decisive event in the genesis of all sexually produced animals and plants. Thus, from one point of view it envisages the entire problem of sex; from another point of view it constitutes the basis of all development and inheritance"*. The recent history of biology has amply demonstrated the perspective of Theodosius Dobzhansky (1973) regarding the relation of evolution to biological events: *"Nothing in biology makes sense except in the light of evolution"*. However, in spite of the fundamental biological importance of the fertilization event and the perspective of Dobzhansky, mammalian fertilization has been insufficiently studied from an evolutionary perspective.

Some researchers in gamete biology have incorporated an evolutionary perspective into their experimental strategies. For instance, Albert Tyler (1967), enunciated and justified 'The Comparative Approach' to biological research. *"There are several reasons why the investigation of many diverse kinds of organisms is important to the understanding of biological processes. One is that some organisms may exhibit better than others one or another of the special features of the biological process under investigation. A further and perhaps the most cogent reason for adopting a comparative approach is that it permits broad generalizations. It is clear that the most important features of any biological process are those which are common to diverse organisms. Thus, this must hold for the various biological processes under consideration here, namely those of gametogenesis, those of maintenance of the functional state of gametes, those of fertilization and the ensuing processes. This statement does not imply that special adaptive features that have evolved in various species are not important areas of investigation. However, it does imply that these features must be properly assessed and not permitted to obscure the basic features of gametogenesis and fertilization"*. A reminder to scholars and researchers with primary interests in mammalian reproduction to understand and acknowledge the reproductive biology of non-mammalian gametes was issued by Ryuzo Yanagimachi (1990), one of the most productive contemporary gamete biologists, *"We who engage in the study of mammalian gametes and fertilization tend to limit our attention to mammals. I recommend that you read both classic and modern textbooks, from time to time, which cover reproductive and developmental biology of a wide variety of non-mammalian vertebrates and invertebrates. What we discover from the study of "lower" animals may not be directly applicable to mammals, but it can nevertheless be very informative. The reproductive and developmental capacities that are present in lower animals may also be present but dormant in mammals. We may be able to reactivate these capacities"*.

In contrast, some gamete biologists consider aspects of mammalian gametes and their fertilization process to be unique or unrelated to those used in other animal systems. For instance Bedford (1982) stated that a variety of elements make fertilization in mammals a more complex matter than in non-mammalian species. Parenthetically, he stated *"A variety of elements make fertilization in mammals a more complex matter. Until recently it was believed that mammalian fertilization closely resembled simple creatures. We now know that this is not so. The physiological relationships of mammalian gametes, their environment, mode of fertilization, differ in important details from those established for the sea urchin and other invertebrates in the classical studies of Frank Lillie, Jacques Loeb and others".* He cautioned that "a *fact for one species may not be so for others".* This non-comparative or anti-evolutionary perspective, which is contrary to that of Tyler, was specifically applied by others to the block to polyspermy (the fertilization step of primary interest in this paper) and the extracellular matrix surrounding eggs (variously known as the zona pellucida in mammals, the vitelline envelope in frogs, the vitelline layer in sea urchins and the chorion in fish). *"Since the morphology and the molecular properties of the mammalian zona pellucida are remarkable different from those of lower vertebrates and invertebrates, it is difficult to propose a universal mechanism for the block to polyspermy"* (O'Rand and Dunbar, 1991); and *"The mammalian ZP is the unique extracellular glycoprotein matrix which surrounds the mammalian egg...."* (Schwoebel et al., 1991). The perspective that fertilization in non-mammalian systems is unrelated to mammalian systems was also echoed in the US peer review system used for evaluating research grant applications, *"It seems probable that Xenopus laevis fertilization uses different mechanisms than mammals"* (personal communication, 1994). In addition, the "apparent evolutionary uniqueness" of genes encoding the zona pellucida glycoproteins was asserted *"...the ZP3 gene appears to be found exclusively in mammalian genomes"* (Ringuette, 1988). However, when gene sequences corresponding to egg envelope glycoproteins from different organisms were determined, aligned and phylogenetically analyzed, the ZP genes were unequivocally evolutionarily conserved (Harris et al., 1994; Spargo and Hope, 2003). In addition, sequence comparisons of other proteins and glycoproteins involved in fertilization and expressed in gametes, refute the assertion that proteins and molecular processes in mammalian fertilization are unrelated to those in non-mammalian fertilization (for example, hyaluronidase, Gmachl and Kreil, 1993; acrosin, Kodama et al., 2002; zonadhesin, Hunt et al., 2005).

Contemporary evidence that molecular mechanisms and gamete macromolecules involved in animal fertilization are conserved verifies the correctness of Darwin's belief that animals have descended from some one prototype and have much in common in their *"...laws of reproduction".* In addition, contemporary evidence validates the comparative approach to fertilization and gamete biology research as useful and relevant. This approach can uncover the fundamental or evolutionarily conserved mechanisms of fertilization (*...those which are common to diverse organisms..,* Tyler) that will lead to broad generalizations and discovery of Darwin's laws of reproduction.

Embracing the conservation of molecules and mechanisms in the fertilization process and the comparative approach to research, a search was made to identify conserved molecular mechanisms for establishing an extracellular matrix block to polyspermy. Five different molecular mechanisms for establishing a block to polyspermy were compared in four different groups of organisms (Table 1). A particular mechanism for example, proteolysis, may involve different proteases in different groups of organisms. A + indicates the molecular mechanism has been experimentally verified and direct evidence for the molecules involved is available for example; biochemical or molecular biology evidence such as isolation and characterization of an enzyme and its biologically relevant substrate as well as cell biology evidence exists that the molecules are associated with gametes and that the proposed molecular changes occur *in situ* e.g., proteolysis of glycoproteins in the egg envelope. A + does not necessarily indicate that the molecules involved have been evolutionarily conserved but only that the mechanism has been conserved. Also, a + does not require that all species in a given animal group possess a particular molecular mechanism. In most animal groups, only a few species have been investigated as "models"; for example, amongst frogs (anurans), *Xenopus laevis*, *Bufo japonicus* and *Bufo arenarum* are the most commonly used animals for gamete biology and fertilization research although more than 3900 amphibian species have been identified. A ?(+) indicates that experimental evidence is incomplete for a particular molecular mechanisms; for example, glycosidase hydrolysis of a specific glycosidic bond in an egg extracellular matrix glycoprotein has not been demonstrated, but the evidence is sufficient from the example above that cortical granules are associated with glycosidase activity, to speculate that the mechanism is likely involved. A – indicates experimental evidence is consistent with the absence of such a mechanism (that is, predicted experimental observations were not observed) while a ? represents an unknown situation due to a lack of published results.

**Table 1** Molecular mechanisms for extracellular matrix blocks to polyspermy

|            | Proteases | Glycosidases | X-linking | Δ Conformation | Binding |
|------------|-----------|--------------|-----------|----------------|---------|
| Sea urchin | +         | +            | +         | +              | +       |
| Fish       | +         | ?(+)         | +         | +              | ?       |
| Frog       | +         | +            | –         | +              | +       |
| Mammal     | ?(+)      | +            | –         | +              | ?       |

References included are to original observations and to more subsequent publications. They are not inclusive of all relevant publications due to space limitations. For additional references see reviews by: Schmell *et al.*, 1983; Hedrick and Nishihara, 1991; Yamagami *et al.*, 1992; Hoodboy and Talbot, 1994; Wessel *et al.*, 2001; Talbot and Dandekar, 2003.

Sea urchins:
    Proteases – Hagstrom, 1956; Vacquier *et al.*, 1973; Haley and Wessel, 1999
    Glycosidases – Epel *et al.*, 1969; Bachman and McClay, 1996
    X linking – Foerder and Shapiro, 1977; Hall, 1978; Wessel *et al.*, 2000
    Δ conformation – Motomura, 1941; Nomura and Suzuki, 1995
    Binding – Wiedman *et al.*, 1985; Somers and Shapiro, 1991
Fish:
    Proteases – Inoue and Inoue, 1986; Hyllner and Haux, 1992; Sugiyama *et al.*, 1999; Darie *et al.*, 2005
    Glycosidases – Ishii *et al.*, 1989; Seko *et al.*, 1999
    X linking – Hagenmaier *et al.*, 1976; Oppen-bernsten *et al.*, 1990; Ha and Iuchi, 1998
    Δ conformation – Zotin, 1958; Iwamatsu, 1969
Frogs:
    Proteases – Miceli *et al.*, 1978; Gerton and Hedrick, 1986a; b; Lindsay and Hedrick, 1989; Lindsay *et al.*, 1999a; b; Lindsay and Hedrick, 2004
    Glycosidases – Prody *et al.*, 1985; Vo *et al.*, 2003
    Δ conformation – Bakos *et al.*, 1990a; b; Hardy and Hedrick, 1992; Lindsay and Hedrick, 2004
    Binding – Wyrick *et al.*, 1974; Quill and Hedrick, 1996; Tseng *et al.*, 2001; Chang *et al.*, 2004
Mammals:
    Proteases – Repin and Akimova, 1976; Hartman and Gwatkin, 1971; Cherr *et al.*, 1988
    Glycosidases – Miller *et al.*, 1993
    Δ conformation – Cholewa-Steward and Massaro, 1972; Inoue and Wolf, 1974; Drobnis *et al.*, 1988

Three conclusions are readily apparent from the data in Table 1. 1) Proteolytic, glycolytic, and conformational changes are molecular mechanisms for altering the egg extracellular matrix that have been evolutionarily conserved amongst these groups of animals. 2) A cross linking mechanism, used by more evolutionarily ancient organisms (sea urchins and fish), is not used in more recently evolved organisms (frogs and mammals). It seems likely that the genes coding for some of the molecules involved in egg extracellular matrix cross linking mechanisms are present in frogs and mammals since these cross linking mechanisms are used in other biological processes; for example, production of isopeptide bonds by transglutaminase in blood clotting, keratin formation and copulatory plug formation (for references see Ha and Iuchi, 1998), and hydrogen peroxide-peroxidase mechanisms used in phagocytosis as an antimicrobial system (Babior, 1999). 3) Binding mechanisms for modifying the egg extracellular matrix have not been sufficiently investigated in fish and mammals. However, the use of binding mechanisms for modification of the egg extracellular matix in sea urchins and frogs has been well documented. Since frogs are evolutionarily more closely related to mammals than sea urchins, we applied the experimental strategy used for frogs (*Xenopus laevis*) to mammals for investigating the presence of binding mechanisms for effecting a block to polyspermy in mammalian eggs.

### A lectin-ligand binding mechanism as a mammalian egg polyspermy block

In the case of frogs, an extracellular matrix block to polyspermy involves a lectin-ligand mechanism (Fig. 1). A cortical granule lectin (CGL) is released into the perivitelline space by the fertilizing-sperm induced cortical reaction.

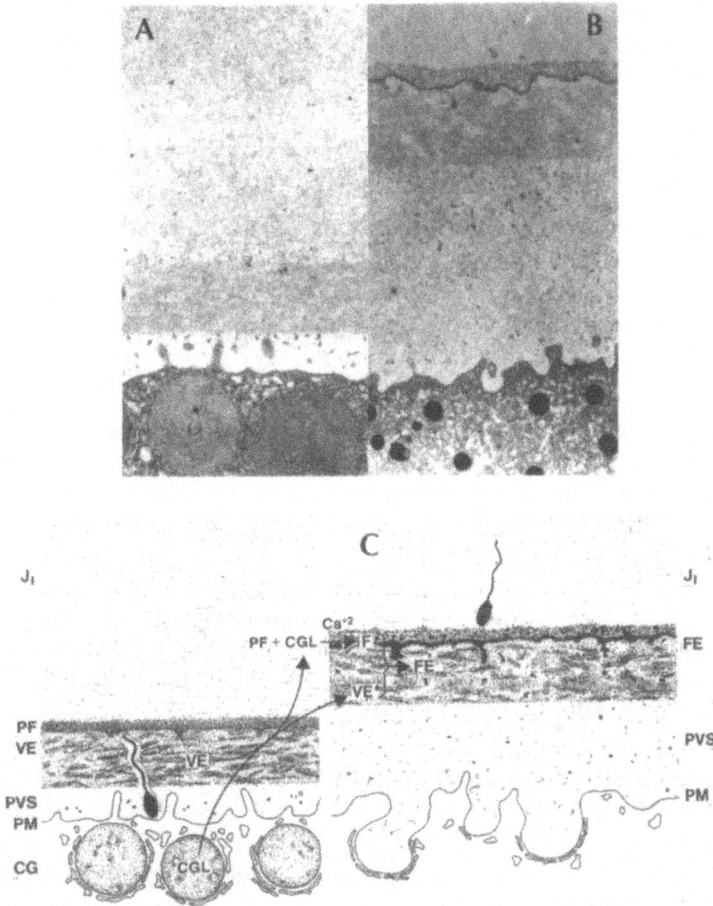

**Fig. 1** The extracelluar matrix structure in *Xenopus laevis* eggs and the lectin-ligand mechanism for a block to polyspermy. Abbreviations used are : F, fertilization layer of the FE; FE, fertilization envelope; CG, cortical granule; CGL, cortical granule lectin; $J_1$, innermost jelly coat layer; PF, prefertilization layer; PM, plasma membrane; PVS, perivitelline space; VE, vitelline envelope: VE*, vitelline envelope component of the FE. a) the unfertilized egg, b) the fertilized or activated egg, c) the lectin-ligand hypothesis. Figure adapted from Grey *et al.*, 1974 and Greve and Hedrick, 1978.

The CGL diffuses through the vitelline envelope (VE) and binds to its ligand (PF). PF is located in the innermost aspect of the jelly coat layer, $J_1$, adjacent to the VE. Noncovalent binding of CGL to its ligand (Gal specific and $Ca^{2+}$ requiring) produces a heteropolymer readily seen in electron microscopy and termed the fertilization (F) layer of the fertilization envelope (FE). The F layer is impenetrable by spermatozoa thereby providing a block to polyspermy. Experimental evidence in support of this hypothesis is the following:

1)    A cortical granule lectin was isolated from the cortical granule exudate (Wyrick et al., 1974; Nishihara et al., 1986). The CGL was cloned and shown to be a charter member of a recently discovered class of lectins, the eglectins (Chang et al., 2004).

2)    CGL was present in the cortical granules prior to the cortical reaction, and in the perivitelline space and F layer after the cortical reaction (Greve and Hedrick, 1978).

3)    The isolated FE contained CGL and the CGL ligand (Gerton and Hedrick, 1986a; b). The F layer could be dissociated with Gal and EDTA, consistent with the specificity and reaction conditions of CGL-ligand binding (Grey et al., 1974; Hedrick and Nishihara, 1991).

4)    Addition of isolated CGL to unfertilized eggs produced an F layer and prevented fertilization of eggs (Hedrick and Nishihara, 1991).

5)    The CGL ligand was isolated from solubilized egg jelly, an oviductal secretory product (Quill and Hedrick, 1996). The ligand glycoprotein contained O-linked oligosaccharides. The O-linked glycan structures responsible for ligand activity contain terminal galactosyl residues associated with fucosyl and sulfate residues (Hedrick et al., 1993; Tseng et al., 2001).

The presence of a CGL homologue was investigated in the eggs (oocytes) of mice (*Mus muscularis*), pigs (*Sus scrofa*) and rhesus macaque monkeys (*Macaca mulatta*), and in human ovaries (*Homo sapiens*) using molecular biology, biochemical and cell biological methods (Chang et al., 2004; Peavy and Hedrick, 2006). Using mouse and human ovarian cDNA libraries, oligonucleotide primers to conserved amino acid sequences and PCR methods, full length cDNAs were cloned. The human and mouse translated cDNA sequences exhibited 63% identity to the *Xenopus laevis* CGL sequence. Using the CGL sequence, some 21 human and 23 mouse cDNA sequences in EST databases were identified. Thus, genes homologous to that of *Xenopus laevis* CGL exist in mouse and human genomes and the CGL gene is expressed in the ovary.

Using lysed pig eggs, a CGL lectin homologue was isolated using *Xenopus laevis* egg jelly affinity chromatographic methods. The isolated pig egg CGL had the same subunit molecular weight (SDS-PAGE), the same sugar specificity (Gal) and required $Ca^{2+}$ for ligand binding. In addition, CGL lectin homologues were isolated from the lysates of pig and mouse eggs using immunoprecipitation methods (antibody to *Xenopus laevis* CGL). The precipitated CGLs comigrated with *Xenopus laevis* CGL on SDS-PAGE. Thus, mouse and pig eggs contained a lectin homologous with the *Xenopus laevis* CGL.

Using confocal microscopy, the cytochemical localization of a CGL homologue in ovulated mouse, and *in vitro* matured pig and rhesus macaque eggs, and employing antibodies to deglycosylated *Xenopus laevis* CGL was determined. Identification of CGs in the egg cortex used fluorescence conjugated *Lens culinarus* agglutinin for mouse eggs and fluorescence conjugated peanut lectin agglutinin for pig eggs. In all three species, CGL was present in egg cortical granules. Thus, a CGL homologue is localized in the CGs of mouse, pig and macaque eggs. Additionally, in mouse and pig eggs, three populations of cortical granules were detected: CGs that reacted with antibodies to *Xenopus laevis* CGL only, CGs that reacted with fluorescent conjugated plant lectins only and CGs that reacted with both reagents. Observations that CGs are heterogeneous with regard to their macromolecular compositions and staining properties has been previously observed in *Xenopus laevis*, mice and lobster eggs (Grey et al., 1974; Nicosia et al., 1977; Talbot and Goudeau, 1988; Liu et al., 2003). The functional significance to fertilization or development of a heterogeneous population of CGs is unknown.

When a *Xenopus laevis* egg is activated or fertilized, CGL is released from the CGs and appears in the extracellular matrix (Mozingo and Hedrick, 1996). Therefore, using the cytochemical methods above, CGL localization in activated or fertilized mouse and pig eggs was determined. In the case of pig eggs, after egg activation, CGL was localized in the perivitelline space and in the zona pellucida. In the 2-cell mouse embryo, CGL was localized in the perivitelline space and in the zona pellucida. In both pig and mouse eggs, addition of melibiose to activated eggs or embryos eliminated the visualization of CGL. Melibiose reversed the lectin-ligand reaction and the dissociated CGL washed away in the staining procedures. These results in mouse and pig eggs are in keeping with those in *Xenopus laevis* eggs and support a CG lectin-extracellular matrix ligand binding mechanism. A notable difference however, is the extracellular location of the CGL ligand. In *Xenopus laevis*, the CGL ligand is located on the outer aspect of the vitelline envelope (the zona pellucida equivalent) and the ligand is a jelly coat glycoprotein (secreted by the oviduct). In mouse and pig eggs, the CGL ligand is associated with the zona pellucida itself. It seems likely that a zona pellucida glycoprotein (biosynthesized in the oocyte in most mammals) is a CGL ligand. However, other cellular sources of a CGL ligand are also possible including the cumulus cells (follicular fluid) and epithelial cells of the oviduct. Both of these cell types are highly active in secreting glycoproteins during the "fertilization phase" of reproduction (Phillips and Dekel, 1982; Zhuo and Kimata, 2001).

Addition of exogenous CGL to an unfertilized egg prevents fertilization in a *Xenopus laevis* egg (Hedrick and Nishihara, 1991). To test the functional properties of CGL to block egg fertilization, *Xenopus laevis* CGL was added to mouse eggs prior to the addition of spermatozoa. CGL prevented fertilization and the CGL effect was concentration dependent. Addition of galactose to eggs prevented CGL inhibition of fertilization, whereas mannose, a monosaccharide not bound by CGL, did not prevent the CGL inhibition of fertilization.

From the above molecular biology, biochemical and cell biology observations, it can be concluded that a CGL is present in the CGs of mammalian eggs (mouse, pig, monkey) and it functions as an extracellular block to polyspermy at fertilization. The binding mechanism of the mammalian polyspermy block is equivalent to that found in *Xenopus laevis* with the exception of the CGL ligand location. The ligand oligosaccharide structure in mammalian eggs is likely to be the same as in *Xenopus laevis* eggs since the binding properties of CGL are apparently the same. However, isolation and structural determination of the CGL ligand in *Xenopus laevis* and mammalian eggs (pig) is in progress, but has not yet been accomplished. Identification of the CGL ligand and determination of its structure is critically important to the understanding of the molecular mechanism for establishing an extracellar matrix block to polyspermy.

### Potential targets for contraception

Identification of target molecules in gametes potentially useful for contraception continues to be a primary area of research (see articles in this issue by G.S. Kopf, S. Choudhury and C.M.

Hardy; Nass and Strauss, 2004). Since a block to polyspermy is a normal process associated with most animal fertilization, understanding the mechanisms of this process and the molecules involved should provide information with high potential for application to contraceptive research. With the identification of a lectin-ligand binding mechanism in mammals, including humans, applied research can focus on the use of these molecules and related biological molecules (such as antibodies) for controlling conception. For instance, once the structure of the glycan functioning as a ligand for CGL is known, it may be possible to produce an antibody to such a glycan. Such an antibody could function in an analogous binding manner to CGL binding the egg extracellular matrix such as the zona pellucida, thereby compromising sperm-egg interaction; that is, contraception. This possibility has been demonstrated by the use of isolated pig zona pellucida (Dunbar et al., 1980) for immunization and the control of reproduction in wild and zoo animals (Kirkpatrick et al., 2002; Frank et al., 2005; Liu et al., 2005). The pig zona pellucida glycoprotein has proven very effective, being used for immunocontraception in 112 mammalian species. However, when deglycosylated pig zona pellucida was used for immunization, its immunogenicity and biological effects were altered. Immunogenicity of the zona pellucida was shown to be directly related to its carbohydrate content and carbohydrate has an important role in maintaining the functional integrity of zona pellucida glycoproteins (Sacco et al., 1986; 1989).

A possible explanation for these observations is that the glycan moiety, as well as the polypeptide moiety, of the pig zona pellucida is an important epitope for the immunocontraceptive action of anti zona pellucida antibodies. If the glycan moiety of the zona pellucida were the same as, or related structurally to the glycan ligand for CGL, then antibodies to such a glycan may mimic the action of CGL; that is, antibody binding to CGL ligand glycans prevent fertilization (sperm binding or penetration of the zona pellucida) thereby effecting contraception. Knowledge of the oligosaccharide structure of the CGL ligand would permit this suggestion to be experimentally tested. Such experiments should take advantage of the many different mammalian species used for zona pellucida induced immunocontraception.

## Acknowledgements

The author is indebted to an exceptional group of students, postdoctoral scholars and fellow scientists for the experimental work summarized here. Primary amongst these are Ron Wyrick, Tatsuro Nishihara, Heather Fabry Harris, the late Nate Wardrip, Y. Betty Chang and Tom Peavy.

## References

Babior BM (1999) NADPH oxidase: an update *Blood* **93** 1464-1476

Bachman ES and McClay DR (1996) Molecular cloning of the first metazoan beta-1-3 glucanase from eggs of the sea urchin *Strongylocentrotus purpuratus Proceedings of the National Academy of Sciences USA* **93** 6808-6813

Bakos M, Kurosky A and Hedrick JL (1990a) Enzymatic and envelope-converting activities of pars recta oviductal fluid from *Xenopus laevis Developmental Biology* **138** 169-176

Bakos M, Kurosky A and Hedrick JL (1990b) Physicochemical characterization of progressive changes in the *Xenopus laevis* egg envelope following oviductal transport and fertilization *Biochemistry* **29** 609-615

Bedford J (1982) Fertilization. In *Reproduction in Mammals: 1 Germ Cells and Fertilization* p 130 Eds CR Austin and RV Short. Cambridge University Press, Cambridge

Chang YB, Peavy TR, Wardrip NJ and Hedrick JL (2004) The *Xenopus laevis* cortical granule lectin: cDNA cloning, developmental expression, and a human homolog *Comparative Biochemistry and Physiology* **137** 115-129

Cherr GN, Drobnis EZ and Katz DF (1988) Localization of cortical granule constituents before and after exocytosis in the hamster egg *Journal of Experimental Zoology* **246** 81-93

Cholewa-Stewart J and Massaro EJ (1972) Thermally induced dissolution of the murine zona pellucida *Biology of Reproduction* **7** 166-169

Darwin C (1859) *Origins of the Species by Means of*

Natural Selection (6th edition 1972) Murray, London

Darie CC, Biniosske ML, Gawinowicz MA, Milgrom Y, Thumfart JO, Jovine L, Litscher ES and Wassarman PM (2005) Mass spectrophotometric evidence that proteolytic processing of rainbow trout egg vitelline envelope proteins takes place on the egg Journal of Biological Chemistry 280 37585-37598

Dobzhansky T (1973) Nothing in biology makes sense except in the light of evolution American Biology Teacher 35 125-129

Drobnis EZ, Andrew JB and Katz DF (1988) Biophysical properties of the zona pellucida measured by capillary suction: is zona hardening a mechanical phenomenon? Journal of Experimental Zoology 245 206-219

Dunbar BS, Wardrip NJ and Hedrick JL (1980) Isolation, physicochemical properties, and macromolecular composition of zona pellucida from porcine oocytes Biochemistry 19 356-365

Epel D, Weaver AM, Muchmore AV and Schimke RT (1969) Beta 1-3 glucanase of sea urchin eggs: release from particles at fertilization Science 163 294-296

Foerder CA and Shapiro BM (1977) Release of ovoperoxidase from sea urchin eggs hardens the fertilization membrane with tyrosine cross-links Proceedings of the National Academy of Sciences USA 74 4214-4218

Frank KM, Lyda RO and Kirkpatrick JF (2005) Immunocontraception of captive exotic speicies IV. Species differences in response to the porcine zona pellucida vaccine, timing of booster inoculations, and procedural failures Zoo Biology 24 349-358

Gerton GL and Hedrick JL (1986a) The vitelline envelope to fertilization envelope conversion in eggs of Xenopus laevis Developmental Biology 116 1-7

Gerton GL and Hedrick JL (1986b) The coelomic envelope to vitelline envelope conversion in eggs of Xenopus laevis Journal of Cellular Biochemistry 30 341-350

Gmachl M and Kreil G (1993) Bee venom hyaluronidase is homologous to a membrane protein of mammalian sperm Proceedings of the National Academy of Sciences USA 90 3569-3573

Greve LC and Hedrick JL (1978) An immunocytochemical localization of the cortical granule lectin in fertilized and unfertilized eggs of Xenopus laevis Gamete Research 1 13-18

Grey RD, Wolf DP and Hedrick JL (1974) Formation and structure of the fertilization envelope in Xenopus laevis Developmental Biology 36 44-61

Ha C and Iuchi I (1998) Enzyme responsible for egg envelope (chorion) hardening in fish: purification and partial characterization of two transglutaminases associated with their substrate, unfertilized egg chorion, of the rainbow trout, Onchoryncus mykiss Journal of Biochemistry 124 917-926

Hagenmaier HE, Smitz I and Fohles J (1976) Zum vorkommen von isopeptidbindungen in der eihulle der regenbogenforelle (Salmo gairdneri Rich.) Hoppe-Seyler's Zeitschrift fur Physiologica Chemie 357 1435-1438

Hagstrom BE (1956) Studies on polyspermy in sea urchins Arkiv. For. Zool. 10 307-315

Haley SA and Wessel GM (1999) The cortical granule serine protease CGSPI of the sea urchin Strongylocentrotus purpuratus, is autocatalytic and contains a low-density lipoprotein receptor-like domain Developmental Biology 211 1-10

Hall HG (1978) Hardening of the sea urchin fertilization envelope by peroxidase-catalyzed phenolic coupling of tyrosine Cell 15 343-355

Hardy DM and Hedrick JL (1992) Oviductin: Purification and properties of the oviductal protease that processes the molecular weight 43,000 glycoprotein of the Xenopus laevis egg protein Biochemistry 31 4466-4472

Harris JD, Hibler DW, Fontenot GK, Hsu KT, Yurewicz EC and Sacco AG (1994) Cloning and characterization of zona pellucida genes and cDNAs from a variety of mammalian species: The ZPA, ZPB, and ZPC gene families DNA Sequence Journal of Sequence Map 4 361-393

Hartman JF and Gwatkin RBL (1971) Alteration of sites on the mammalian sperm surface following capacitation Nature (London) 234 479-481

Hedrick JL and Nishihara T (1991) Structure and function of the extracellular matrix of anuran eggs Journal of Electron Microscopy Technique 17 319-335

Hedrick J, Chang B, Wardrip N and Quill T (1993) The Xenopus laevis cortical granule lectin and its ligand Journal of Reproduction and Development 39 (Suppl) 81-82

Hoodboy T and Talbot P (1994) Mammalian cortical granules: contents, fate, and function Molecular Reproduction and Development 39 439-448

Hunt PND, Wilson MD, von Schalburg KR, Davidson WS and Koop BF (2005) Expression and genomic organization of zonadhesin-like genes in three species of fish give insight into the evolutionary history of a mosaic protein BMC Genomics 6 Article Number 165 http://www.biomedcentral.com/1471-2164/6/165

Hyllner SJ and Haux C (1992) Immunochemical detection of the major vitelline envelope proteins in the plasma and oocytes of the maturing female rainbow trout (Onchorynchus mykiss) Journal of Endocrinology 135 303-309

Inoue S and Inoue Y (1986) Fertilization (activation)-induced 200-9-kDa depolymerization of polysialoglycoprotein, a distinct component of cortical alveoli of rainbow trout eggs Journal of Biological Chemistry 261 5256-5261

Inoue M and Wolf DP (1974) Comparative solubility properties of the zonae pellucidae of unfertilized and fertilized mouse ova Biology of Reproduction 11 558-565

Ishii K, Iwasaki M, Inoue S, Kenny PTM, Komura H and Inoue Y (1989) Free sialooligo-saccharides found in the unfertilized eggs of a freshwater trout, Plecoglossu altivelis. A large storage pool of complete-type bi-, tri-, and tetraantennary sialooligosaccharides Journal of Biological Chemistry 264 1623-1630

Iwamatsu T (1969) Changes of the chorion upon fertilization in medaka, Oryzias latipes Bulletin of Aichi

*University Education* **18** 43-64

Kirkpatrick JF, Lasley BL, Allen WR, Doberska C Eds (2002) Fertility control in wildlife – Proceedings of the fifth international symposium on fertilility control in wildlife – Skukuza, the Kruger National Park, South Africa *Reproduction V – V Suppl* **60**

Kodama E, Baba T, Kohno N, Satoh S, Yokosawa H and Sawada H (2002) Spermosin, a trypsin-like protease from ascidian sperm: cDNA cloning, protein structures and functional analysis *European Journal of Biochemistry* **269** 657-663

Lillie FR (1919) *Fertilization* p31 The University of Chicago Press, Chicago

Lindsay LL and Hedrick JL (1989) Proteases released from *Xenopus laevis* eggs at activation and their role in envelope conversion *Developmental Biology* **135** 202-211

Lindsay LL and Hedrick JL (2004) Proteolysis of ZPA triggers egg envelope hardening in *Xenopus laevis* *Biochemical and Biophysical Research Communications* **324** 648-654

Lindsay LL, Wieduwelt MJ and Hedrick JL (1999a) Oviductin, the *Xenopus laevis* oviductal protease that processes egg envelope glycoprotein gp43, increases sperm binding to envelopes, and is translated as part of an unusual mosaic protein composed of two protease and several CUB domains *Biology of Reproduction* **60** 989-995

Lindsay LL, Yang JC and Hedrick JL (1999b) Ovochymase, an Xenopus laevis egg extracellular protease, is translated as part of a polyprotein *Proceedings of the National Academy of Sciences USA* **96** 11253-11258

Liu M, Sims D, Calarco P and Talbot P (2003) Biochemical heterogeneity, migration, and prefertilization release of mouse oocyte cortical granules *Reproductive Biology and Endocrinology* **1** 77 http://www.rbej.com/content/3/1/42

Liu IKM, Turner JW, VanLeeuwen EMG, Flanagan DR, Hedrick JL, Murata K, Kirkpatrick JF, Lane VM and Morales-Levy MP (2005) Persistence of anti-zona pellucidae antibodies following a single inoculation of porcine zona pellucidae in the domestic equine *Reproduction* **129** 181-190

Miceli DC, Fernandez SN, Raisman JS and Barbieri FD (1978) A trypsin-like oviducal proteinase involved in *Bufo arenarum* fertilization *Journal of Embryology and Experimental Morphology* **48** 79-91

Miller DJ, Gong X, Decker G and Shur BD (1993) Egg cortical granule N-acetylglucsoaminidase is required for the mouse zona block to polyspermy *Journal of Cell Biology* **123** 1431-1440

Motomura I (1941) Materials of the fertilization membrane in the eggs of echinoderms *Sci. Rep. Tohoku Imperial University* **4** 345-363

Mozingo NM and Hedrick JL (1996) Localization of the cortical granule lectin ligand in *Xenopus laevis* eggs *Development Growth & Differentiation* **38** 647-652

Nass SJ and Strauss JF Eds (2004) *New frontiers in contraceptive research* The National Academies Press, Washington, D.C. pp 27-77

Nicosia SV, Wolf DP and Inoue M (1977) Cortical granule distribution and cell surface characteristics in mouse eggs *Developmental Biology* **57** 56-74

Nishihara T, Wyrick RE, Working PK, Chen YK and Hedrick JL (1986) Isolation and characterization of a lectin from the cortical granules of *Xenopus laevis* eggs *Biochemistry* **25** 6013-6020

Nomura K and Suzuki N (1995) Sea urchin ovoperoxidase: solubilization and isolation from the fertilization envelope, some structural and functional properties, and degradation by hatching enzyme *Archives of Biochemistry and Biophysics* **319** 525-534

Oppen-bernsten DO, Helvik JV and Walther BT (1990) The major structural proteins of cod *(Gadus morhua)* eggshells and protein crosslinking during teleost egg hardening *Developmental Biology* **137** 258-265

O'Rand MG and Dunbar BS (1991) *A Comparative View of Mammalian Fertilization* Plenum Press

Peavy TR and Hedrick JL (2006) A homologue of the *Xenopus laevis* cortical granule lectin is in mammalian egg cortical granules and participates in the block to polyspermy *Biology of Reproduction* in press

Phillips DM and Dekel N (1982) Effect of gonadotropins and prostaglandin on cumulus mucification in cultures of intact follicles *Journal of Experimental Zoology* **221** 275-28

Prody GA, Greve LC and Hedrick JL (1985) Purification and characterization of an N-acetyl-α-D-glucosaminidase from cortical granules of *Xenopus laevis* eggs *The Journal of Experimental Zoology* **235** 335-340

Quill TA and Hedrick JL (1996) The fertilization layer mediated block to polyspermy in *Xenopus laevis*: isolation of the cortical granule lectin ligand *Archives of Biochemistry and Biophysics* **333** 326-332

Repin VS and Akimova IM (1976) The microelectrophoretic analysis of protein patterns of mammalian oocyte and zygote zonae pellucidae *Biokhimiia (USSR)* **41** 5057

Ringuette MJ, Chamberlin ME, Baur AW, Sobieski DA and Dean J (1988) Molecular analysis of cDNA coding for ZP3, a sperm binding protein of the mouse zona pellucida *Developmental Biology* **127** 287-295

Sacco AG, Yurewicz EC and Subramanian MG (1986) Carbohydrate influences the immunogenic and antigenic characteristics of the ZP3 macromolecule (Mr = 55000) of the pig zona pellucida *Journal of Reproduction and Fertility* **76** 575-586

Sacco AG, Yurewicz EC, Subramanian MG and Matzat PD (1989) Porcine zona pellucida: association of sperm receptor activity with the α-glycoprotein component (ZP3α) of the Mr = 55000 family (ZP3) *Biology of Reproduction* **41** 523-532

Schmell ED, Gulyas BJ and Hedrick JL (1983) Egg surface changes during fertilization and the molecular mechanism of the block to polyspermy. In *Mechanism and Control of Animal Fertilization* pp 365-413 Ed JF Hartmann. Academic Press, New York

Schwoebel E, Prasad S, Timmons TM, Cook R, Kimura H, Niu EM, Cheung P, Skinner S, Avery SE and

**Wilkins B** (1991) Isolation and characterization of a full-length cDNA encoding the 55- kDa rabbit zona pellucida protein *Journal of Biological Chemistry* **266** 7214-7219

**Seko A, Kitajima K, Iwamatsu T, Inoue Y and Inoue S** (1999) Identification of two discrete peptide: N-glycanases in *Oryzias latipes* during embryogeneis *Glycobiology* **9** 887-895

**Somers CE and Shapiro BM** (1991) Functional domains of proteoliasin, the adhesive protein that orchestrates fertilization envelope assembly *Journal of Biological Chemistry* **226** 16870-16875

**Spargo SC and Hope RM** (2003) Evolution and nomenclature of the zona pellucida gene family *Biology of Reproduction* **68** 358–362

**Sugiyama H, Murata K, Iuchi, I, Nomura K and Yamagami K** (1999) Formation of mature egg envelope subunit proteins from their precursors (choriogenins) in the fish, *Oryzias latipes*: loss of partial C-terminal sequences of the choriogenins *Journal of Biochemistry* **125** 469-475

**Talbot P and Goudeau M** (1988) A complex cortical reaction leads to formation of the fertilization envelope in the lobster, Homarus *Gamete Research* **19** 1-18

**Talbot P and Dandekar P** (2003) Perivitelline space: does it play a role in blocking polyspermy in mammals? *Microscopy Research and Technique* **61** 349-357

**Tseng K, Wang H, Lebrilla CB, Bonnell B and Hedrick JL** (2001) Identification and structural elucidation of lectin-binding oligosaccharides by bioaffinity matrix-assisted laser desorption/ionization fourier transform mass spectrometry *Analytical Chemistry* **73** 3556-3561

**Tyler A** (1967) Introduction: problems and procedures of comparative gametology and syngamy. In *Fertilization: Comparative Morphology, Biochemistry and Immunology* Eds CB Metz and A Monroy Academic Press New York

**Vacquier VD, Tegner MJ and Epel D** (1973) Protease released from sea urchin eggs at fertilization alters the vitelline layer and aid in preventing polyspermy *Experimental Cell Research* **80** 111-119

**Vo LH, Yen T-Y, Macher BA and Hedrick JL** (2003) Identification of the ZPC oligosaccharide ligand involved in sperm binding and the glycan structures of *Xenopus laevis* vitelline envelope glycoproteins *Biology of Reproduction* **69** 1822–1830

**Weidman PJ, Kay ES and Shapiro BM** (1985) Assemby of the sea urchin fertilization membrane: isolation of proteoliasin, a calcium-dependent ovoperoxidase binding protein *Journal of Cell Biology* **100** 938-946

**Wessel GM, Conner S, Laidlaw M, Harrison J and LaFleur GJ** (2000) SFE1, a constituent of the fertilization envelope in the sea urchin is made by oocytes and contains low-density lipoprotein-receptor-like repeats *Biology of Reproduction* **63** 1706-1712

**Wessel GM, Brooks JM, Green E, Haley S, Voronina E, Wong J, Zaydfudim V and Conner S** (2001) The biology of cortical granules *International Review of Cytology* **2 09** 117-206

**Wyrick RE, Nishihara T and Hedrick JL** (1974) Agglutination of jelly coat and cortical granule components and the block to polyspermy in the amphibian *Xenopus laevis* *Proceedings of the National Academy of Sciences USA* **71** 2067-2071

**Yamagami K, Hamazaki TS, Yasumasu S, Masuda K and Iuchi I** (1992) Molecular and cellular baiss of formation, hardening, and breakdown of the egg envelope in fish *International Journal of Cytology* **136** 51-92

**Yanagimachi R** (1990) Research on mammalian gametes and fertilization: my personal view. In *Fertilization in Mammals* p410 Eds BD Bavister, J Cummins and ERS Roldan Serano Symposia, Norwell

**Zhuo LS and Kimata K** (2001) Cumulus oophorus extracellular matrix: its construction and regulation *Cell Structure & Function* **26** 189-196

**Zotin AI** (1958) The mechanism of hardening of the salmonid egg membrane after fertilization or spontaneous activation *Journal of Embryology and Experimental Morphology* **6** 546-568

# Contraceptive development:
# targets, approaches and challenges

GS Kopf

*Women's Health and Musculoskeletal Biology, Women's Health Research Institute,
Wyeth Research, Collegeville, Pennsylvania 19426, USA*

The control of fertility constitutes a global health issue, since overpopulation and unintended pregnancy have both major personal and societal impact. Although some regions of the world are seeing neutral or negative population growth, many developing countries are seeing explosive growth of their populations and these population changes will affect the entire globe. It is estimated that in a decade, the largest cohort of young women worldwide in human history will reach adolescence thus necessitating the need for a wide range of contraceptive options that can be used by both females and males. The contraceptive revolution that occurred in the 1960s with the development of the hormonal-based oral contraceptive for women has subsequently made a significant impact on societal dynamics in several cultures, yet there has been virtually no innovation in this field since that time. This lack of innovation contrasts dramatically with the vast enhancement of our knowledge base of the basic processes of reproduction. The genomic and proteomic revolutions have provided new tools and new targets for contraceptive development, and the results of such approaches have identified gene products that play critical roles in female and male reproduction, thus expanding the array of targets for novel and innovative female- and male-based contraceptives. This normally would herald a renaissance in contraceptive development, yet the commitment of industry to this endeavor is limited to a few firms due to the economics of contraceptive development. This chapter will consider the types of targets being considered in the development of new generations of contraceptives and will also focus on the challenges that industry has in meeting these goals.

## Introduction

It is well recognized that the control of fertility is a significant and essential component of reproductive health. Over-population in under developed countries poses significant societal, geopolitical and economic impact, and issues of fertility control are not just confined to these countries. For example, it is estimated that up to 50% of all pregnancies in the U.S. are unintended and half of these occur in a majority of sexually active couples who state that they are using some form of contraception (Harrison and Rosenfield, 1996; Henshaw, 1998; Nass and Strauss, 2004). Worldwide, this number averages about 25%. Moreover, unplanned preg-

E-mail: kopfg@wyeth.com

nancies result in a significant mortality rate (for example between 1995 and 2000, approximately 700,000 women worldwide died as a result of unplanned pregnancies), and a higher risk of depression and physical abuse (Nass and Strauss, 2004). The consequences of unplanned pregnancies, therefore, are far reaching. Although it can be argued that compliance, cost and availability of contraceptives play a major role in this problem, the types and choices of contraceptives currently available are limited. Significant advances in contraceptive development have been hampered by liability issues, politics, religion and public/private investment. Most of the development that has occurred has focused on different steroidal formulations and/or combinations, lower doses of steroid hormones, and different delivery systems based on the initial steroidal platform for female contraception. However, the fact remains that there has not been a fundamental shift in the paradigm for contraception for the past 45 years. Given the fact that the next decade will see the single largest number of young women in human history reaching adolescence (~ 600 million), the availability of effective contraceptives that have different modes of action, are easy to obtain and use, and are acceptable by many individuals and cultures is of paramount importance.

As stated above, most advances made in the development of contraceptives have concentrated on female-based contraception, since women bear the responsibility of child bearing and issues of trust of the male partner are consistently articulated. A conundrum exists, however, since one-third of the world's population rely on male-based methods of contraception (that is withdrawal, condoms and vasectomy) (UNFPA, 1998). Given this number, efforts in both academia and industry have been initiated to identify targets with the goal to develop male-based contraceptives. Given the fact that women bear the responsibility of child bearing, there has been serious debate over the value of developing male contraceptives. Some recent studies have demonstrated that men in many societies are willing to share the responsibility of contraception with their partner (Martin et al., 2000) and have engendered the trust of their partner to utilize male methods in a responsible manner (Glasier et al., 2000). Moreover, with genotyping methods now existing for the establishment of paternity, monetary liability issues for the male have become a reality. The need for, and interest in, the development of new male-based contraceptive methods was clearly articulated in the 1996 and 2004 Institute of Medicine Committee reports (Harrison and Rosenfield, 1996; Rosenfield, 1999; Nass and Strauss, 2004). Work is currently ongoing with the development of male-based hormonal contraceptives to capture this unmet need (Anderson and Baird, 2002; Lyttle and Kopf, 2003) and these approaches will be discussed in greater detail below.

The explosion of information dealing with reproduction now available as a result of the genomic and proteomic revolutions has opened up the possibility that new targets for fertility control can be exploited for the development of novel contraceptives that can shift the paradigm for contraception away from the steroidal modulation of the hypothalamic-pituitary-gonadal axis (Matzuk and Lamb, 2002). Academia and industry are now poised to enter such a new arena with new tools and targets. However, only a few companies have positioned themselves to exploit this new knowledge, due to the economics of the global contraceptive market with respect to the costs of current contraceptives, over-the-counter availability, issues related to drug re-importation and liability issues. Although academic initiatives to identify and develop new contraceptive modalities are ongoing around the world, it is highly unlikely that a viable and marketable product will evolve without the partnership with industry. The 2004 report of the Institute of Medicine of the National Academy of Sciences entitled, **"New Frontiers in Contraceptive Research: A Blueprint for Action"**, delineates a course of action as a blueprint for the development of new generations of contraceptives (Nass and Strauss, 2004). As an initial strategy for identifying new targets, a recommendation was made to generate complete reproductive transcriptomes, proteomes, lipidomes and glycomes, followed by integration of these data with genetic and protein networks. Second, initiatives to enhance contraceptive drug discovery, development and clinical testing should be actively pursued. This would include high throughput screening, translational research, examination of different drug delivery systems, pursuit of additional non-contraceptive health benefits, and early integration of behavioral research in target identification and development. Since this clearly would require a partnership between academia and industry, the final recommendation was to expand public-private partnerships, to increase training and career options in contraception development, to increase the participation of developing countries in contraceptive development, and to establish alliances for contraceptive development that would require academic-government-industry integration. Such approaches will be needed in order to reach this goal.

This review will consider in general terms the types of targets that could/should be considered in the development of contraceptives with new modes of action, the approaches that one could take in identifying and validating such targets, and the scientific and economic challenges that must be addressed if new and novel products are to be brought to the marketplace.

## Targets

Several factors need to be considered in the identification of a viable target for contraception (see Table 1). From a pharmaceutical perspective, historically one considers the "drug ability" of the target; that is, is the target amenable for modulation by a therapeutic entity, be it a small molecule or a biologic (for example: antibody, recombinant protein, peptide). Most therapeutics in the market today are targeted towards receptors, ion channels, exchangers and enzymes. Ideally such targets are expressed on the membrane of the target cell or in the extracellular space, although there are several examples of therapeutic agents directed towards intracellular targets. However, it is becoming clear from work that is in the pipelines of several companies that different types of targets (for example, transcription factors, protein-protein interactions) are also being pursued the success of these types of targets remain to be established.

In addition to drug ability, one also needs to consider the type of population and the therapeutic indication when considering the characteristics of any target. In contrast to therapeutics for oncology, where drugs are being administered to sick individuals for finite periods of time for a disease that is life threatening, any target to be considered for contraception needs to take into account the fact that the therapeutic entity would be administered to healthy individuals for very long periods of time. The safety and side effect profile of a contraceptive needs to be extremely high and low, respectively, and this dictates the type of target that would be considered. When considering new and novel contraceptive targets this would mean that the expression profile of the target be very highly tissue-selective or completely tissue-specific. This is a rather high hurdle to achieve.

When considering a new target one must also consider the efficacy of any therapeutic entity towards that target. Although the hormonal based birth control pill does not have an ideal side effect profile, its efficacy is very high (~98% effectiveness against pregnancy with normal compliance). This high efficacy hurdle would be the starting point for any new contraceptive, whether it be for females or males.

As shown in Table 1 below, there are many different criteria that need to be considered: not all of these will be considered in this review for the sake of brevity. However, it is important to note especially with contraceptives that religious, ethical and social issues need to be considered. Although these issues are important in the selection of a target for any therapeutic indication, it is especially important in contraceptive development since both the science and medicine can get intertwined with the politics of religion and ethics: this is reality.

**Table 1.** Criteria to consider when identifying new contraceptive targets

Tissue specific expression
Knockout shows desired phenotype
Human orthologue
Pathway information
Drug able properties, degree of difficulty
Reference compound availability
Intellectual property issues
Closely related family members to test specificity (number of members in the family)
Reversibility/onset time
Non-contraceptive health benefits
Religious/ethical/social issues
Biomarkers for activity
Cross-therapeutic applications

*Female-based targets*

There are several points of intervention that one can consider when mining for novel female-based contraceptive targets. The present gold standard is the steroidal combined estrogen/progestin or progestin only therapies that function at the level of the hypothalamic-pituitary-ovarian (HPO) axis to inhibit ovulation. The primary mechanisms of action of the progestin-only regimens are at the level of the endometrium and the cervical mucus. Nonsteroidal compounds that function either as progesterone agonists or antagonists are also being pursued.

If one were to target pathways other than the HPO axis, one could target genes playing key roles in ovarian function, ovulation, oviductal function, sperm-egg interaction, cervical mucus function or implantation. With respect to ovarian function, regulation of later follicular development represents an attractive intervention point since it is likely to interfere with proper oocyte development and ovulation. Intervention at this level, however, must not affect normal estradiol production. Likewise, modulation at the level of follicle-oocyte communication represents an attractive target. It is only now being appreciated that two way crosstalk between the granulosa cells and the oocyte occurs, and that these regulatory networks regulate several aspects of oocyte and follicle function (Buccione et al., 1990; Rankin et al., 2001; Eppig et al., 2002; Matzuk et al., 2002; Eppig et al., 2005). Genes involved in the ovulation process itself might also serve as attractive targets. Since follicular development is associated with the development of oocytes competent to undergo meiotic maturation in response to the proper environment/regulatory factors, oocyte-specific genes that are involved in conferring meiotic competence might also be an attractive target for intervention. To date, very little is known about these genes.

Since the oviduct (fallopian tube) provides an environment for gamete maturation and transport, fertilization and early pre-implantation development, a thorough understanding of oviductal function and genes involved in setting up an environment conducive to sperm-egg interaction and fertilization may ultimately yield novel targets that could be exploited (Suarez, 1998; Buhi et al., 2000; Buhi, 2002).

The fertilization process itself involves a finely orchestrated series of cell-matrix (sperm-zona pellucida), exocytotic (zona pellucida-mediated acrosomal exocytosis) and cell-cell (sperm-egg) interactions, and these interactions are mediated by cell surface proteins on both gametes. This process is still not completely understood and it appears as if multiple proteins mediate this process. Targets based on inhibiting the interaction of sperm with the zona pellucida are more plausible contraceptive targets than those based on inhibiting sperm-egg fusion, since the later event represents a target very late in the stages of fertilization.

Finally, targeting implantation is a viable approach that clearly produces a contraceptive effect and there are current therapeutics that act at this level (for example, progestins). From a commercial standpoint, many companies are very uncomfortable with developing new contraceptives that act at the level of implantation due mostly to political fallout.

### Male-based targets

Male-based contraceptive targets, while gaining interest from a commercial value, bring with them a unique set of challenges. Unlike female contraception, where the goal of hormone-based therapies is to suppress the ovulation of one or several oocytes, successful development of hormonal or non-hormonal male contraceptives involves either the suppression of spermatogenesis or the functional inactivation of up to $250 \times 10^6$ sperm in the ejaculate. Although a lower limit of suppression/inactivation is likely to be functionally acceptable for contraceptive efficacy, the criteria that must be met for successful product development is an efficacy at least equal to the contraceptive efficacy of the current oral contraceptives for women. This criterion, therefore, sets a high bar for their successful development. Furthermore, whereas the timing for the onset of contraceptive efficacy in women taking oral contraceptives is one menstrual cycle, the time to efficacy in the male could be up to several months depending on the mode of action (see below) since spermatogenesis and epididymal sperm transport in the human can take up to 11-12 weeks (Clermont, 1963; Clermont et al., 1993; Robaire and Hinton, 2002): this timing also impacts on the time to reversibility.

As delineated in both the 1996 and 2004 Institute of Medicine reports on Contraceptive Research and Development (Harrison and Rosenfield, 1996; Nass and Strauss, 2004), new male contraceptives will be based on either hormonal or non-hormonal regimens. Hormonal suppression of spermatogenesis (to oligospermic or azoospermic levels) by administration of exogenous hormones to interfere with the hypothalamic-pituitary-testicular axis has been the primary hormonal approach, and several regimens are currently being examined with varied degrees of success. These include androgen/progestin combinations, androgen monotherapy, GnRH antagonist/androgen combinations and selective androgen and progestin receptor modulators (Anderson and Baird, 2002; Brady and Anderson, 2002; Jensen, 2002; Wang and Swerdloff, 2002) . Several of these approaches, while effective, have additional side effects due primarily to the suppression of testicular steroidogenesis. Such effects include decreased muscle mass, decreased bone mass and effects on hematopoiesis: it is for this reason that androgen supplementation is essential. Delivery of these is through injection, depot formulations or through various transdermal delivery systems. There are several challenges to these hormonal approaches, including the mode of drug delivery that will undoubtedly impact on user acceptability, differential responses in different ethnic male populations, issues related to side effect profiles and health risks that currently accompany hormonal contraceptive therapies. Nevertheless, progress has been made, with several of these regimens at various stages of clinical trials both in the public and private sector, and it is possible that a product will be in the market within the next five to seven years.

Immunological approaches to male contraception, based on immunization with sperm antigens or immunoneutralization of trophic hormones, is also being considered. This approach evolved from similar approaches taken to develop female-based immunocontraception, and has suffered from many of the same concerns and pitfalls. It is known that circulating antisperm antibodies have been shown to contribute to sub-fertility in humans and in several experimental animal models. In this case, selection of the sperm antigen for use as an immunogen is critical. The strategy for selection would include the localization of the antigen on the spermatozoa (intracellular or cell surface), as well as its function in the fertilization process (such as, epididymal sperm maturation, sperm motility, sperm acrosomal exocytosis, sperm-zona pellucida binding, sperm-egg plasma membrane binding). Although several studies in animal models have been published, the overall efficacy of this approach has been extremely variable and has, in some cases, lead to undesirable side effects (Harrison and Rosenfield, 1996; Primakoff et al., 1997; Tung et al., 1997; Anderson and Baird, 2002; O'Rand et al., 2004). Passive immunization with antisera to FSH and GnRH have been proposed as a way to reduce/ablate spermatogenesis without altering testosterone production, but, to date, such approaches have not met with success (Anderson and Baird, 2002): a similar lack of success has been obtained with active immunization with heterologous FSH (Moudgal et al., 1997). As with female-based immunocontraception, variability in the response and reversibility, as well as concerns regarding autoimmune responses, have dampened enthusiasm for this approach.

Non-hormonal approaches represent a more innovative strategy for the development of male-based contraceptives. These approaches take advantage of those cellular and physiological processes that are unique to the reproductive organs, as well as those genes and gene products that are either unique to the reproductive tracts or are highly selective for reproductive tract tissues. The rationale for this approach is further strengthened by the recent report that up to 4% of the mouse genome is dedicated to expression in post-meiotic germ cells of the testis and that a significant number of these genes appear to be expressed only in the male germ cells (Schultz et al., 2003). The idea is to target a specific process or gene product that interferes with or abrogates key regulatory processes associated with testicular spermatogenesis, epididymal sperm maturation and sperm function. The goal is to produce a therapeutic agent that intervenes in a highly specific manner, the effect of which is to reduce/abolish sperm production or inhibit sperm functional maturation or sperm function without affecting hormonal balance. It is likely that these approaches will provide the novelty, innovation and selectivity necessary to shift the paradigm for contraception.

As previously stated, there are three major targets for non-hormonal intervention in the male (that is, spermatogenesis, epididymal sperm maturation and sperm function), and each target has its unique advantages and disadvantages. From the standpoint of reducing/abolishing the production of sperm, targeting events associated with pre- and post-meiotic spermatogenesis is the most effective strategy and, in theory, should not affect hormonal status. In order to ensure reversibility, such strategies should also not interfere with the function of the spermatogonial stem cell population in the testis. There are increasing numbers of meiotically and post-meiotically expressed germ cell-specific gene products involved in several different aspects of spermatogenesis (Cooke and Saunders, 2002) that, theoretically, could serve as targets for intervention. Since meiosis only occurs in the male and female gonads, targeting processes such as cell cycle regulators and meiotic checkpoints offers potential. One, however, must be cautious of any potential effects that negatively affect chromosomal dynamics. Germ cell-specific or selective gene products involved in the post-meiotic phase of spermatogenesis (that is, spermiogenesis) might also serve as attractive targets; such targets might include genes involved in acrosome biogenesis, histone-protamine exchange, tail assembly, glycolipid biosynthesis and membrane organization (Roest et al., 1996; Martianov et al., 2001; 2002; van der Spoel et al., 2002). Advantages of these targets over meiotic targets is the lower likelihood of genomic effects and the fact that reversibility is likely to be more rapid. A significant hurdle to targeting differentiating germ cells is the delivery of the contraceptive to its site of action due to the presence of a blood-testis barrier. This hurdle, however, is not insurmountable as the alkylated imino sugar N-butyldeoxynojirimycin, which can affect testicular glycolipid biosynthesis, is likely to traverse this barrier to exert its effects on spermiogenesis (van der Spoel et al., 2002). Any contraceptive that successfully functions at the level of the germ cells will, by design, have a latency period for both the initiation of the contraceptive effect and the return to fertility following cessation of the drug due to the extended period of time over which spermatogenesis takes place.

The epididymis can also be considered as a target for male contraception (Cooper and Yeung, 1999; Cooper, 2002). This organ is responsible for the storage and maturation of the spermatozoa following spermiogenesis and before their entrance into the vas deferens and the ejaculatory ducts. The post-testicular maturation events that occur to spermatozoa traversing the epididymis are critical to the subsequent fertilizing capacity of these cells since epididymal transit is associated with the acquisition of sperm motility and the capacity to fertilize eggs (Kopf et al., 1999; Robaire and Hinton, 2002). The intraluminal ionic, organic solute, substrate and protein environment of the epididymis is known to be quite unique and is responsible for these aforementioned changes, but the mechanism by which this occurs is not completely understood. Several novel epididymal proteins that are secreted into the lumen at discrete regions of this organ are thought to interact with the sperm surface and mediate processes that are important in ultimately conferring fertilization capacity to the spermatozoa (Cooper and Yeung, 1999; Cohen et al., 2000): these proteins represent potential candidates for the development of contraceptives. Alternative approaches such as altering the peritubular muscular activity of the epididymis in order to affect normal sperm transport through this organ and interfering with the secretory activity of the epididymal epithelium in order to create a hostile extracellular luminal environment for the sperm are also potential targets (Breton et al., 1996; Cooper, 2002). There are several advantages to targeting epididymal function for male contraception. Intervention at this level could be very effective since it would be possible to generate immotile and nonfunctional sperm. Second, intervention would not impact on spermatogenesis or any other testicular function. Third, endocrine function is likely not to be affected. Finally, both the initiation and reversibility of the contraceptive effect would be significantly shortened given the fact that epididymal transport is a much more rapid event than spermatogenesis. As with testicular targets, the presence of a blood-epididymal barrier could pose a hurdle with respect to the delivery of agents to this organ.

The mature spermatozoa, itself, also represents a target for intervention. In this case, the rationale for a particular target might be based on the uniqueness of a protein or a physiological process that the spermatozoa must undergo following deposition into the female reproductive tract and prior to fertilization. Such targets for intervention might include unique cell surface proteins that function as acceptors or receptors involved in motility, chemosensory signalling, sperm-female reproductive tract interactions, sperm-zona pellucida interaction and sperm-egg plasma membrane interaction (Parmentier et al., 1992; Vanderhaeghen et al., 1993; 1997; Eisenbach, 1999; Evans and Florman, 2002; Primakoff and Myles, 2002; Bahat et al., 2003; Spehr et al., 2003). Novel components of sperm signal transduction cascades regulating capacitation, motility and acrosomal exocytosis also represent potential targets for intervention (Buck et al., 1999; Fukami et al., 2001; 2003; Quill et al., 2001; Ren et al., 2001; Wang et al., 2003; Hess et al., 2005). Finally, targeting scaffolding components essential for cellular compartmentalization (Moss and Gerton, 2001) and novel components of the flagellar apparatus (Sapiro et al., 2002) could result in the generation of non-functional sperm. Advantages to targeting the mature sperm cell include rapid and complete reversibility. Issues of contraceptive onset and efficacy will be important considerations when intervening at this level, given the numbers of spermatozoa in the human ejaculate. An additional consideration is the point of intervention with regard to the sequence of events leading to fertilization. For example, intervening at the level of sperm-egg plasma membrane binding/fusion might be considered too late to be considered safe as a contraceptive.

## Approaches

As stated above, the ideal target profile for non-hormonal contraceptive development would be a gene product that: 1) is highly tissue-selective or tissue-specific; 2) plays a key role in reproductive function such that its modulation would reversibly affect fertility; 3) has properties that would make it amenable for modulation by pharmaceutical intervention (that is, has drug able properties); and 4) for the purposes of a pharmaceutical company, is novel so that intellectual property around the target can be obtained. Several approaches are taken to identify targets with these ideal profiles.

Trancriptional profiling of the reproductive tissues is a very powerful way to identify genes (both novel and known) that are uniquely expressed in that tissue under various physiological conditions (Jervis and Robaire, 2001; Chauvin and Griswold, 2004; Johnston et al., 2005). Unknown genes would then be cloned and experiments designed to determine their function. For example, transcriptional profiling of the mouse epididymis has revealed that this tissue expresses over 17,000 genes, many of them in a segment-dependent manner (Johnston et al., 2005). Binning of these genes into different classes has enabled us to identify novel epididymal-specific and -selective transcripts that have very distinct expression patterns. These transcripts are currently being examined in greater detail to determine whether they meet many of the aforementioned criteria. The results of these studies have also provided a wealth of information to the general scientific community (see data from Johnston et al., 2005 that is posted on the Mammalian Reproductive Genetics database: http://mrg.genetics.washington.edu).

Although transcriptional profiling yields information regarding gene expression profiles and the potential protein that the gene encodes, it does not necessarily tell anything about the protein itself. Therefore, proteomic analyses are carried out in parallel with transcriptional profiling and the data generated from both approaches utilized together in identifying targets. This has proven to be extremely valuable with respect to new target identification. For example, we have performed an in-depth proteomic analysis of human sperm by liquid chromatography and tandem mass spectrometry (Johnston et al., unpublished). Proteomic analysis of bands cut from one dimensional SDS-PAGE yielded more than 1,760 high confidence proteins with 1,350 proteins, 719 proteins and 309 proteins identified in the soluble fraction, insoluble fraction and both fractions, respectively. These proteins were then classified based on their known functional properties in other cell types, and it is interesting to note that at least 17% of the identified proteins were novel, consistent with the data reported by Schultz and co-workers (Schultz et al., 2003). These proteins, as well as others that have been identified, are currently being further characterized to see whether they meet the criteria outlined above. It is likely that the proteomic analysis of human sperm, as well as other regions of both the male and female reproductive tracts, will provide a wealth of targets, a small percentage of which might ultimately be targets for contraceptive development.

Once targets meet the aforementioned profile with respect to tissue-specificity, tissue-selectivity, novelty and potential drug ability, they then need to be validated to ensure that they play a key role in fertility. The ideal target would be one which, when modulated (either inhibited or activated) would result in complete, but reversible, infertility. Target validation is usually carried out in animal models that are amenable to genetic manipulation. Targeted gene disruption ("knockouts") is the most common approach, although transgenic RNAi and lentiviral RNAi approaches clearly have been used. The disadvantage to several of these approaches is the cost and the time involved. Clearly, faster and more economic approaches are needed. Perhaps the recent advances in the culture, maintenance and genetic manipulation of spermatogonial stem cells will provide us with better tools (Hamra et al., 2005; Ryu et al., 2005).

## Challenges

As discussed above, the challenges facing the development of new generations of contraceptives are numerous and, in all aspects, daunting. Identifying new targets that are amenable to therapeutic intervention is a time consuming, difficult and expensive proposition. Once a target is identified and validated, the ability to identify and generate lead compounds with the proper pharmacological, pharmacokinetic and pharmacodynamic properties requires a large and coordinated chemistry effort. Such compounds need to cross the blood-ovarian, blood-testis or blood epididymal barrier and once there must modulate their target with the appropriate efficacy and selectivity. The safety hurdles for a new therapeutic entity are extremely high and many compounds fail at this stage. It is for these reasons that research and development in the pre-clinical phases for most products can take any where from 2-10 years and that the time to registration of a product can take up to a total of 16 years. The risk is very high and the costs are spiraling upward. The average costs of discovering and developing a new drug has now approached $1.7 billion dollars.

One cannot consider the drug development process in isolation when considering the development of new generations of contraceptives. Equally, if not more, important is the economics that surround contraception. No one doubts the need for contraceptives and most recognize the global health issues that can/will arise as a consequence of over population. With costs of research and development spiraling upwards at a rapid rate, the economics of contraceptive development need to be considered. How are the costs of research and development recouped? From an industry perspective the current contraceptive markets are highly segmented with low cost products, some of which are sold over-the-counter. As with the development of any new therapeutic product, issues of consumer market, market segmentation, product liability and pricing must be weighed against innovation, cost of development and profitability. These problems, in addition to the research and development involved in developing new contraceptive products, are issues that industry, academia and governments need to face together, if we are ultimately to be successful.

## References

Anderson RA and Baird DT (2002) Male Contraception *Endocrine Review* 23 735-762

Bahat A, Tur-Kaspa I, Gakamsky A, Giojalas LC, Breitbart H and Eisenbach M (2003) Thermotaxis of mammalian sperm cells: A potential navigation mechanism in the female genital tract *Nature Medicine* 9 149-150

Brady BM and Anderson RA (2002) Advances in male contraception *Expert Opinion on Investigational Drugs* 11 333-344

Breton S, Smith PJ, Lui B and Brown D (1996) Acidification of the male reproductive tract by a proton pumping (H+) - ATPase *Nature Medicine* 2 470-472

Buccione R, Vanderhyden BC, Caron PJ and Eppig JJ (1990) FSH-induced expansion of the mouse cumulus oophorus *in vitro* is dependent upon a specific factor(s) secreted by the oocyte *Developmental Biology* 138 16-25

Buck J, Sinclair ML, Schapal L, Cann MJ and Levin LR (1999) Cytosolic adenylyl cyclase defines a unique signaling molecule in mammals *Proceedings of the National Academy of Sciences USA* 96 79-84

Buhi WC (2002) Characterization and biological roles of oviduct-specific, oestrogen-dependent glycoprotein *Reproduction* 123 355-362

Buhi WC, Alvarez IM and Kouba AJ (2000) Secreted proteins of the oviduct *Cells Tissues Organs* 166 165-179

Chauvin TR and Griswold MD (2004) Androgen-regulated genes in the murine epididymis *Biology of Reproduction* 71 560-569

Clermont Y (1963) The cycle of the seminiferous epithlium in man *American Journal of Anatomy* 112 35-51

Clermont Y, Oko R and Hermo L (1993) Cell biology of mammalian spermatogenesis. In "*Cell and Molecular Biology of the Testis*" pp 332-376 Eds C Desjardins and L Ewing. Oxford University Press, New York

Cohen DJ, Rochwerger L, Ellerman DA, Morgenfeld MM, Busso D and Cuasnicu PS (2000) Relationship between the association of rat epididymal protein "DE" with spermatozoa and the behavior and function of the protein *Molecular Reproduction and Development* 56 180-188

Cooke HJ and Saunders PT (2002) Mouse models of male infertility *Nature Reviews. Genetics* **3** 790-801

Cooper TG (2002) The epididymis as a target for male contraception. In "*The Epididymis: From Molecules to Clinical Practice*" pp 483-502 Eds B Robaire and B T Hinton. Kluwer Academic/Plenum Publsihers, New York

Cooper TG and Yeung CH (1999) Recent biochemical approaches to post-testicular, epididymal contraception *Human Reproduction Update* **5** 141-152

Eisenbach M (1999) Sperm chemotaxis *Reviews of Reproduction* **4** 56-66

Eppig JJ, Wigglesworth K and Pendola FL (2002) The mammalian oocyte orchestrates the rate of ovarian follicular development *Proceedings of the National Academy of Sciences USA* **99** 2890-2894

Eppig JJ, Pendola FL, Wigglesworth K and Pendola JK (2005) Mouse oocytes regulate metabolic cooperativity between granulosa cells and oocyes: amino acid transport *Biology of Reproduction* **73** 351-357

Evans JP and Florman HM (2002) The state of the union: the cell biology of fertilization *Nature Cell Biology* **Supplement** s57-s63

Fukami K, Nakao K, Inoue T, Kataoka Y, Kurokawa M, Fissore RA, Nakamura K, Katsuki M, Mikoshiba K, Yoshida N and Takenawa T (2001) Requirement of phospholipase Cδ4 for the zona pellucida-induced acrosome reaction *Science* **292** 920-3

Fukami K, Yoshida M, Inoue T, Kurokawa M, Fissore RA, Yoshida N, Mikoshiba K and Takenawa T (2003) Phospholipase Cδ4 is required for Ca$^{2+}$ mobilization essential for acrosome reaction in sperm *Journal of Cell Biology* **161** 79-88

Glasier AF, Anakwe R, Everington D, Martin CW, van der Spuy Z, Cheng L, Ho PC and Anderson RA (2000) Would women trust their partners to use a male pill? *Human Reproduction* **15** 646-649

Hamra FK, Chapman KM, Nguyen DM, Williams-Stephens AA, Hammer RE and Garbers DL (2005) Self renewal, expansion, and transfection of rat spermatogonial stem cells in culture *Proceedings of the National Academy of Sciences USA* **102** 17430-17435

Harrison PF and Rosenfield A (1996) *Contraceptive research and development: looking to the future*, National Academy Press, Washington DC

Henshaw SK (1998) Unintended pregnancy in the United States *Family Planning Perspectives* **30** 24-29

Hess KC, Jones BH, Marquez B, Chen Y, Ord TS, Kamenetsky M, Miyamoto C, Zippin JH, Kopf GS, Suarez SS, Levin LR, Williams CJ, Buck J and Moss SB (2005) The "soluble" adenylyl cyclase in sperm mediates multiple signaling events required for fertilization *Developmental Cell* **9** 249-259

Jensen JT (2002) Male contraception *Current Women's Health Reports* **2** 338-345

Jervis KM and Robaire B (2001) Dynamic changes in gene expression along the rat epididymis *Biology of Reproduction* **65** 696-703

Johnston DS, Jelinsky SA, Bang HJ, DiCandeloro P, Wilson E, Kopf GS and Turner TT (2005) The mouse epididymal transcriptome: transcriptional profiling of segmental gene expression in the epididymis *Biology of Reproduction* **73** 404-413

Kopf GS, Visconti PE and Galantino-Homer H (1999) Capacitation of the mammalian spermatozoon *Advances in Developmental Biochemistry* **5** 81-105

Lyttle CR and Kopf GS (2003) Status and future direction of male contraceptive development *Current Opinion in Pharmacology* **3** 667-671

Martianov I, Fimia GM, Dierich A, Parvinen M, Sassone-Corsi P and Davidson I (2001) Late arrest of spermiogenesis and germ cell apoptosis in mice lacking the TBP-like TLF/TRF2 gene *Molecular Cell* **7** 509-515

Martianov I, Brancorsini S, Gansmuller A, Parvinen M, Davidson I and Sassone-Corsi P (2002) Distinct functions of TBP and TLF/TRF2 during spermatogenesis: requirement of TLF for heterochromatic chromocenter formation in haploid round spermatids *Development* **129** 945-955

Martin CW, Anderson RA, Cheng L, Ho PC, van der Spuy Z, Smith KB, Glasier AF, Everington D and Baird DT (2000) Potential impact of hormonal male contraception: cross-cultural implications for development of novel preparations *Human Reproduction* **15** 637-645

Matzuk MM and Lamb DJ (2002) Genetic dissection of mammalian fertility pathways *Nature Cell Biology* **4** s41-49

Matzuk MM, Burns KH, Viveiros MM and Eppig JJ (2002) Intercellular communication in the mammalian ovary: oocytes carry the conversation *Science* **296** 2178-2180

Moss SB and Gerton GL (2001) A-kinase anchor proteins in endocrine systems and reproduction *Trends in Endocrinology and Metabolism* **12** 434-440

Moudgal NR, Murthy GS, Prasanna Kumar KM, Martin F, Suresh R, Medhamurthy R, Patil S, Sehgal and Saxena BN (1997) Responsiveness of human male volunteers to immunization with ovine follicle stimulating hormone vaccine: results of a pilot study *Human Reproduction* **12** 457-463

Nass SJ and Strauss JF III (2004) "*New Frontiers in Contraceptive Research: A Blueprint for Action*" National Academy Press, Washington, DC

O'Rand MG, Widgren EE, Sivashanmugam P, Richardson RT, Hall SH, French FS, VandeVoort CA, Ramachandra SG, Ramesh V and Jagannadha Rao A (2004) Reversible immunocontraception in male monkeys immunized with eppin *Science* **306** 1189-1190

Parmentier M, Libert F, Schurmans S, Schiffmann S, Lefort A, Eggerickx D, Ledent C, Mollereau C, Gerard C, Perret J, Grootegoed A and Vassart G (1992) Expression of members of the putative olfactory receptor gene family in mammalian germ cells *Nature* **355** 453-455

Primakoff P, Woolman-Gamer L, Tung KS and Myles DG (1997) Reversible contraceptive effect of PH-20 immunization in male guinea pigs *Biology of Reproduction* **56** 1142-1146

Primakoff P and Myles DG (2002) Penetration, adhesion and fusion in mammalian sperm-egg interaction *Science* **296** 2183-2185

Quill TA, Ren D, Clapham DE and Garbers DL (2001) A voltage-gated ion channel expressed specifically in spermatozoa *Proceedings of the National Academy of Sciences USA* **98** 12527-12531

Rankin TL, O'Brien M, Lee E, Wigglesworth K, Eppig J and Dean J (2001) Defective zonae pellucidae in ZP2-null mice disrupt folliculogenesis, fertility and development *Development* **128** 1119-1126

Ren D, Navarro B, Perez G, Jackson AC, Hsu S, Shi Q, Tilly JL and Clapham DE (2001) A sperm ion channel required for sperm motility and male fertility *Nature* **413** 603-609

Robaire B and Hinton BT (2002) *The epididymis: from molecules to clinical practice*, pp 575 Kluwer Academic/Plenum Publishers, New York

Roest HP, van Klaveren J, de Wit J, van Gurp CG, Koken MH, Vermey M, van Roijen JH, Hoogerbrugge JW, Vreeburg JT, Baarends WM, Bootsma D, Grootegoed JA and Hoeijmakers JH (1996) Inactivation of the HR6B ubiquitin-conjugating DNA repair enzyme in mice causes male sterility associated with chromatin modification *Cell* **86** 799-810

Rosenfield A (1999) Pushing the frontiers of science: reflections on an Institute of Medicine study *International Journal of Gynaecology and Obstetrics* **67** Suppl 2 S93-S99

Ryu B-Y, Kubota H, Avarbock MR and Brinster RL (2005) Conservation of spermatogonial stem cell self-renewal signaling between mouse and rat *Proceedings of the National Academy of Sciences USA* **102** 14302-14307

Sapiro R, Kostetskii I, Olds-Clarke P, Gerton GL, Radice GL and Strauss IJ (2002) Male infertility, impaired sperm motility, and hydrocephalus in mice deficient in sperm-associated antigen 6 *Molecular and Cellular Biology* **22** 6298-6305

Schultz N, Hamra FK and Garbers DL (2003) A multitude of genes expressed solely in meiotic or postmeiotic spermatogenic cells offers a myriad of contraceptive targets *Proceedings of the National Academy of Sciences USA* **100** 12201-2206

Spehr M, Gisselmann G, Poplawski A, Riffell JA, Wetzel CH, Zimmer RK and Hatt H (2003) Identification of a testicular odorant receptor mediating human sperm chemotaxis *Science* **299** 2054-2058

Suarez SS (1998) The oviductal sperm reservoir in mammals: mechanisms of formation *Biology of Reproduction* **58** 1105-1107

Tung KS, Primakoff P, Woolman-Gamer L and Myles DG (1997) Mechanism of infertility in male guinea pigs immunized with sperm PH-20 *Biology of Reproduction* **56** 1133-1141

UNFPA (1998) *Levels and Trends of Contraceptive Use as Assessed in 1998* UNFPA, New York

van der Spoel AC, Jeyakumar M, Butters TD, Charlton HM, Moore HD, Dwek RA and Platt FM (2002) Reversible infertility in male mice after oral administration of alkylated imino sugars: A nonhormonal approach to male contraception *Proceedings of the National Academy of Sciences USA* **99** 17173-17178

Vanderhaeghen P, Schurmans S, Vassart G and Parmentier M (1993) Olfactory receptors are displayed on dog mature sperm cells *Journal of Cell Biology* **123** 1441-1452

Vanderhaeghen P, Schurmans S, Vassart G and Parmentier M (1997) Specific repertoire of olfactory receptor genes in the male germ cells of several mammalian species *Genomics* **39** 239-246

Wang C and Swerdloff RS (2002) Male contraception best practice and research *Clinical Obstetrics and Gynaecology* **16** 193-203

Wang D, King SM, Quill TA, Doolittle LK and Garbers DL (2003) A new sperm-specific $Na^+/H^+$ exchanger required for sperm motility and fertility *Nature Cell Biology* **5** 1117-1122

# Family of sperm associated antigens: relevance in sperm-egg interaction and immunocontraception

A Suri

Genes and Proteins Laboratory, National Institute of Immunology, Aruna Asaf Ali Marg,
New Delhi-110 067, India

Overpopulation is a global problem of significant magnitude, with grave implications for the future. Development of new contraceptives is necessary, since current forms of birth control are unavailable, impractical and/or too expensive to many individuals due to sociological, financial, or educational limitations. A novel contraceptive strategy that is receiving considerable attention is that of immunocontraception. The targeting of antibodies to gamete-specific antigens implicated in sperm function, sperm-egg binding and fertilization offers an attractive approach to the growing global problem of over population. The sermatozoon has proteins that are unique, cell specific, immunogenic and accessible to antibodies. Immunological interaction with such molecules can cause block of sperm binding to the oocyte and thus fertilization. Modern biotechnologies (such as sperm proteomics, the determination of molecular and structural details of sperm proteins, and the modelling of protein-ligand interaction using X-ray and/or NMR structures to name a few) are trying to make intervention into the domain of human reproduction possible through the development of a variety of new methods and products to control fertility. The present article highlights the various sperm associated antigens involved in various aspects of sperm-egg interaction.

## Introduction

Around the world, human population is experiencing unprecedented demographic changes. The most obvious and best known example is the expansion in human numbers. World population today stands at 6.4 billion and another 3 billion will likely be added by 2050 (United Nations Population Information Network; _http://www.un.org/popin_). This growth in human numbers has been a principal cause of rising demand for food, water and other natural resources in the past, and this will continue to be the case for the foreseeable future. Modern methods of contraception and birth control used are impractical and/or too expensive for widespread use in the different parts of the world. Therefore, it is necessary to develop new, safer, effective and more economical methods of contraception. Contraceptive vaccines have been proposed as one of the possible strategies for controlling fertility (Talwar et al., 1994).

Immunocontraception as a means to control fertility has gathered a growing number of advocates in recent decades. The pursuit of this objective has involved the selection of appropriate

E-mail: anil@nii.res.in

targets within the reproductive process that are amenable to interference with antibodies. Cases of naturally occurring sperm-agglutinating antibodies produced by infertile couples indicate that such strategies should be effective (van Voorhis and Stovall, 1997) and the lack of any additional effects other than infertility due to the presence of these antibodies is reassuring. In the normal physiological condition, the immune system does not respond to spermatozoa, egg or the fetus. Identification of sperm proteins capable of eliciting functionally relevant sperm antibodies is a first step toward a more complete understanding of the mechanism(s) underlying immunologic infertility. Thus, efforts to develop an immunocontraceptive for human application will ideally target sperm molecule(s).

The development of contraceptive vaccines depends on various factors such as the sperm protein(s) as antigen(s), adjuvant in humans for an anti-fertility vaccine, maintenance of high antibody titer, termination of contraceptive effect and with no side-effects (Suri, 2004; 2005a). A large number of candidate molecules for sperm based vaccine have been identified using several approaches, including using antibodies obtained from infertile couples to screen testis cDNA expression libraries (Shankar et al., 1998) or high-resolution two-dimensional electrophoresis gels (Shetty et al., 1999). Although antifertility effects have been obtained in animal models with several spermatozoal antigens, sperm based vaccines have yet to enter clinical trials. The present review highlights the various studies on the sperm based antigens intended to develop immuno-contraceptive vaccine.

## Candidate sperm vaccinogens

Although whole spermatozoa can produce an antibody response that is capable of inducing infertility in humans, they *per se* cannot be employed for the development of a vaccine. Besides the presence of numerous antigens common to somatic and sperm cells, there are several proteins on the sperm surface that are likely to be shared with various somatic cell plasma membranes. Thus, only those antigens that have been carefully analyzed for sperm specificity should be employed for the development of an antisperm contraceptive vaccine (Suri, 2004; 2005a, b). Various sperm-specific antigens that are relevant to fertility and have potential application in immuno-contraception are discussed in the following section.

## Sperm Associated Antigen 9 (SPAG9)

Recently, we characterized a novel highly conserved testis specific gene designated as sperm associated antigen 9 (SPAG9), a new member of the JNK-interacting protein (JIP) family involved in molecular interactions during sperm-egg fusion and MAPK signalling pathway (Shankar et al., 1998; Jagadish et al., 2005a, b, c). SPAG9 is an acrosomal molecule, which is not only restricted to a specific region (domain) of the acrosome but also undergoes relocation in a stage-specific manner during the acrosome reaction (Jagadish et al., 2005a). Recently, MAPKs interaction studies revealed that SPAG9 interacts with higher binding affinity to JNK3 and JNK2 compared with JNK1 (Jagadish et al., 2005a). Sperm specific SPAG9 protein is classified as JIP4 (Morrison and Davis, 2003), which is structurally related to JIP3, and the involvement of SPAG9 mediated signal transduction pathways in reproductive processes is speculated. SPAG9 is therefore a potential candidate for development of a contraceptive vaccine attributed to its sperm specificity and role in sperm binding to zona pellucida. An antibody generated against recombinant human SPAG9 showed cross reactivity with spermatozoa from other animal species, further resulting in inhibition of sperm binding to oocytes in *in vitro* rodent and human systems (Jagadish et al., 2005a; 2006). In a separate study, mice elaborating antibody response to immunization with the pcDNA-hSPAG9 plasmid and recombinant hSPAG9 protein (Jagadish *et al.*, 2006) also revealed inhibition of sperm adherence or penetration to zona free hamster eggs.

*SPAG9* gene encoding SPAG9 protein has also been cloned and sequenced from non-human primates (baboon and macaque), which have shown that SPAG9 protein exhibited conserved amino acid sequences. The presence of high level of amino acid homology and recognition of macaque SPAG9 and baboon SPAG9 by hSPAG9 antibody implies their common function and common origin in the biological past (Shankar *et al.*, 2004, Jagadish *et al.*, 2005d). Since the h*SPAG9* was cloned from human testis and its contraceptive effects cannot be examined in humans at the present time, a suitable animal model is needed to investigate its immuno-contraceptive effect. The high homology in SPAG9 protein from human and non-human primates provide important information for the rational design of contraceptive vaccine formulations and suggest the appropriateness of fertility trials in these primates using hSPAG9 as a contraceptive vaccinogen. Currently, we are in the process of investigating the immuno-contraceptive potential of SPAG9 protein in a macaque model.

## PH-20

A widely conserved sperm antigen, PH-20, is a glycosylphosphatidyl inositol-linked protein with multiple roles in mammalian fertilization. It has been shown to be dually expressed in testis and epididymis and is conserved in mouse, rat, fox, rabbit, macaque and human (Lathrop *et al.*, 1990; Ten Have *et al.*, 1998; Evans *et al.*, 2003; Zhang *et al.*, 2004). Immunization trials in both male and female guinea pigs were reported to lead to 100% effective contraception in all immunized animals (Primakoff *et al.*, 1988). However, assessment of contraceptive vaccines in mice based on recombinant mouse sperm protein PH20 failed to show a significant reduction in fertility indicating that recombinant PH20 is not a useful antigen for inclusion in immuno-contraceptive vaccines in mice (Hardy *et al.*, 2004). Lack of immuno-contraceptive effect for PH20 has also been reported for bacterially produced rabbit PH20 in rabbits (Pomering *et al.*, 2002). The strong immuno-contraceptive effect reported for guinea pig PH20 in guinea pigs (Primakoff *et al.*, 1997), but not seen for mice and rabbit, could reflect fundamental differences in the biological role of PH20 among the various species and raise the questions against the creditability of employing rodents for assessing the contraceptive effect of candidate contraceptive vaccinogens. These studies reflect the need of amendment in the present scientific approach of employing rodents or animal species other than primates for carrying out fertility trials. The better approach would be to exploit the potential candidate sperm antigens in non-human primates before undertaking the human trials.

## Sperm Protein-10 (SP-10)

SP-10 is a sperm-specific acrosomal protein that was first identified in the human using a monoclonal antibody (Herr *et al.*, 1990) and subsequently cloned and sequenced from human (Wright *et al.*, 1993), baboon and macaque (Freemerman *et al.*, 1993). Active immunization trials have been performed in female baboons using the human recombinant SP-10. These baboons developed antibodies that were reactive with the cognate antigen. However, in spite of the presence of high titers of anti-SP-10 antibodies, there was only a partial reduction in fertility in a few animals. Further studies demonstrated that recombinant sperm specific antigen SP-10 induces IgG and IgA antibodies in primate oviductal fluids after systemic immunization and that these antibodies recognise the endogenous SP-10 molecule on both human and macaque spermatozoa (Kurth *et al.*, 1997). An important application of the human intra-acrosomal protein SP-10 has been to insert the SP-10 gene into avirulent *Salmonella typhimurium* for use in oral administration. Since *S. typhimurium* naturally invades and persists in gut-associated lymphoid tissue (GALT), oral immunization with attenuated *Salmonella* expressing foreign antigens stimulated antigen-specific secretory, humoral and cellular immune responses. These results are the first indication that a gene encoding a human sperm antigen can be delivered by an oral immunogen vector and induce a secretory immune response against sperm-specific antibodies in the reproductive tract (Srinivasan et al., 1995). This discovery could lead to the development of a simple, safe, efficient and easy to use immuno-contraceptive and opens the way for development of additional vectors that induce secretory immunity in the female reproductive tract.

## Sperm Acrosome Membrane Proteins (SAMPs)

Human SAMP32, specifically expressed in testis, was shown to be associated with the inner acrosomal membrane of principal and equatorial segments of the sperm acrosome (Hao et al., 2002). Antibodies generated in rat in response to the recombinant SAMP32 protein significantly suppressed the binding and the fusion of capacitated human spermatozoa with zona-free hamster eggs. These results suggested that SAMP32 might have a role in one or more events of primary and secondary binding and in fusion of the spermatozoon with the oolemma or in sperm internalisation. Serum from an ASA-positive infertile man strongly reacted with the recombinant SAMP32 antigen, suggesting that it might be one antigen related to immune infertility (Hao et al., 2002). It is appropriate to add here that another acrosomal membrane protein SAMP14, represents a glycosylphosphatidylinositol (GPI) -anchored putative receptor in the Ly-6/uPAR family that is exposed on the inner acrosomal membrane after the acrosome reaction and has also been shown to be involved in sperm-egg interaction (Shetty et al., 2003). The definition of the molecular constituents of the acrosomal matrix and acrosomal membranes is clearly of importance to understanding the molecular mechanisms that mediate induction of the acrosome reaction and sperm-egg interactions. Therefore, molecules identified as key players in these events are justified as candidates for targeting by rational drug design or for inclusion in a sperm based contraceptive vaccine.

## Fertilization Antigen-1 (FA-1) and Testis Specific Antigen-1 (TSA-1)

Fertilization Antigen-1 (FA-1) is a sperm specific glycoprotein localized on the post-acrosomal region of human spermatozoa (Naz and Zhu, 2002). Studies have reported that sera from infertile women and not from fertile women react strongly with FA- 1, indicating its involvement in infertility in humans (Naz, 1996). Another study demonstrated the immuno-adsorption of autoantibodies by FA-1 from the surface of sperm of immunoinfertile men permitting an increased acrosome reaction (AR) and thereby fertilization capacity of now antibody-free sperm (Menge et al., 1999). Further, antibodies to recombinant FA-1 antigen inhibited *in vitro* fertilization by interfering with sperm-zona interaction (Naz et al., 1992) and immunization of female rabbits and mice with FA-1 does appear to reduce fertility *in vivo* (Naz and Zhu, 1998). Recently, testis specific antigen (TSA-1) expressed in murine spermatozoa (Trivedi and Naz, 2002) and human spermatozoa (Santhanam and Naz, 2001) has been cloned and characterized. In functional bioassays, recombinant TSA-1 antibodies inhibited the acrosome reaction in a concentration dependent manner (Santhanam and Naz, 2001) and sperm-egg binding in *in vitro* assays (Trivedi and Naz, 2002).

## Proteins from a disintegrin and metalloprotease domain (ADAM) family

A group of sperm specific molecules involved in cell-cell adhesion mainly belong to the family of ADAM proteins. Sperm-egg plasma membrane fusion is preceded by sperm adhesion to the egg plasma membrane. Cell-cell adhesion frequently involves multiple adhesion molecules on the adhering cells. One such sperm surface protein with a role in sperm-egg plasma membrane adhesion is fertilin, a transmembrane heterodimer (alpha and beta subunits). Fertilin alpha and beta are the first identified members of a new family of membrane proteins that each has the following domains: pro-, metalloprotease, disintegrin, cysteine-rich, EGF-like, transmembrane and cytoplasmic domain. This protein family has been named ADAM because all members contain a disintegrin and metalloprotease domain. Intriguingly, many of the antigens that are sequestered into specific surface domains in the testis and subsequently become targets for post testicular processing are those found to be important for gamete interaction during fertilization. Particularly instructive in this respect is the behaviour of PH20 and PH30 antigens on guinea-pig spermatozoa. PH30 antigen renamed 'fertilin' also localizes to the whole head of testicular spermatozoa, undergoes site-specific cleavage in the epididymis and as a result relocates to the postacrosomal domain on cauda spermatozoa (Phelps et al., 1990). Fertilin has a putative role in sperm-oolema binding and fusion via integrin/disintegrin-like interactions (Evans, 2001). However, in vivo trials revealed that the induction of very high serum IgG antibody titers to the sperm auto-antigen fertilin is insufficient to cause infertility in the rabbit (Hardy et al., 1997). Further experiments are needed to measure the relationship between the levels of fertilin in sperm, antibody titers and the nature of the immune response in the various regions of the reproductive tract (Hardy et al., 1997) for the eventual development of candidate molecule for contraception.

## A-kinase anchoring protein (AKAP)

Infertility due to sperm antibodies may be attributed to a number of mechanisms, which essentially fall into two categories. The first is antibody disrupting pivotal events of the fertilisation process such as capacitation, acrosome reaction or sperm-egg fusion (Marshburn and Kutteh, 1994). The second is impairment of sperm motility, preventing their normal progress through the female reproductive tract. Thus proteins that play a role in sperm motility could be an enticing target for contraceptive that could be taken by either men or women to block the process of fertilization. The mammalian sperm tail, which is mainly responsible for sperm motility, is characterized by its complex cytoskeletal structure (Oko and Clermont, 1990). The nine outer dense fibers that surround the microtubular axonemes are encompassed by a mito-chondrial sheath in the midpiece and the fibrous sheath in the principal piece. The fibrous sheath is believed to influence the degree of flexibility, plane of flagellar motion and the shape of the flagellar beat. Various proteins associated with the fibrous sheath identified in recent studies indicate that it also has an active role in sperm motility (Oko and Clermont, 1990; Vijayaraghavan et al., 1997; Harrison et al., 2000; Miki et al., 2002; Brown et al., 2003; Luconi et al., 2005).

Sperm motility is regulated by the cAMP-dependent protein kinase (protein kinase-A)-mediated phosphorylation of a group of flagellar proteins. This phosphorylation is facilitated by a group of proteins known as A-kinase anchoring proteins (AKAPs). AKAPs tether cyclic AMP-dependent protein kinases and thereby localize phosphorylation of target proteins and initiation of signal-transduction processes triggered by cyclic AMP. AKAPs can also be scaffolds for kinases and phosphatases and form macromolecular complexes with other proteins involved in signal transduction (Brown et al., 2003). Nearly half of the protein in fibrous sheaths isolated from mouse sperm is AKAP4. AKAP4 is transcribed only in the post-meiotic phase of spermatogenesis and encodes the most abundant protein in the fibrous sheath. Similarly, a human testis specific gene (EMBL nomenclature AKAP4), cloned from a testis cDNA expression library, encodes a protein having regional homology to the domain of AKAP and may act as a regulatory protein in the flagellum for sperm motility (Mohapatra et al., 1998). Gene targeting was used to test the hypothesis that AKAP4 is a scaffold for protein complexes involved in regulating flagellar function (Miki et al., 2002). Sperm numbers were not reduced in male mice lacking AKAP4, but sperm failed to show progressive motility and male mice were infertile. The fibrous sheath anlagen formed, but the definitive fibrous sheath did not develop, the flagellum was shortened, and proteins usually associated with the fibrous sheath were absent or substantially reduced in amount. However, the other cytoskeletal components of the flagellum were present and appeared fully developed concluding that AKAP4 is a scaffold protein required for the organization and integrity of the fibrous sheath. Effective sperm motility is lost in the absence of AKAP4 due to failure of signal transduction and glycolytic enzymes to become associated with the fibrous sheath (Miki et al., 2002) suggesting the possible use of AKAP4 as a target for a male contraceptive. Despite several attempts to elucidate the role of different components of the PKA holoenzyme in regulation of sperm motility, it is not yet clear which regulatory subunit is important for sperm motility. Cyclic AMP-dependent protein kinase (PKA) is anchored at specific subcellular sites through the interaction of the regulatory subunit (R) with protein kinase A-anchoring proteins (AKAP3) via an amphipathic helix-binding motif. Synthetic peptides containing this amphipathic helix domain competitively disrupt PKA binding to AKAPs, cause a loss of PKA modulation of cellular responses and thus inhibit sperm motility (Vijayaraghavan et al., 1997). Phosphatidylinositol 3-kinase (PI3-kinase) has been recently suggested to negatively regulate sperm motility by interfering with AKAP3-PKA binding (Luconi et al., 2005). Studies infer that interaction of the regulatory subunit of PKA with sperm AKAP3 is a key regulator of sperm motility and that disruption of this interaction using cell-permeable anchoring inhibitor peptides may form the basis of a sperm-targeted contraceptive (Vijayaraghavan et al., 1997; Harrison et al., 2000).

## Equatorial Segment Protein (ESP)

The equatorial segment of the acrosome is of considerable functional importance to fertilization as it remains intact following the acrosomal reaction, underlies the domain of the plasma membrane involved in fusion with the egg membrane and is the site where breakdown of the sperm nuclear envelope is initiated after fertilization (Yanagimachi and Noda, 1970). ESP, a testis-specific protein, is unique to the equatorial segment and can serve as a marker for early specification of the equatorial segment (Wolkowicz et al., 2003). Several research groups have previously identified immunoreagents that react with equatorial segment proteins similar in size to ESP (Toshimori et al., 1992; Noor and Moore 1999; Auer et al., 2000). The identification of a molecule such as ESP will be useful in understanding the molecular mechanisms underlying key aspects of equatorial segment biology: including 1) membrane trafficking during acrosome development: 2) the basis for stability and retention of the equatorial segment during the acrosome reaction: 3) molecular mediators of sperm-egg binding and fusion, and 4) the fate of the equatorial segment after fertilization, including its role as the initiation site for breakdown of the sperm nuclear envelope (Wolkowicz et al., 2003).

## Epididymal proteins

Sperm are produced in the testis and gain the ability to fertilize an egg as they transit the epididymis. During spermatogenesis many germ cell proteins are produced that are necessary for the function of mature spermatozoa. Likewise, as sperm transit the epididymis they acquire proteins necessary for normal function from the epididymal fluid, which is a secretory product of the epididymis. These epididymal proteins may act directly in the process of gamete recognition and interaction or play vital roles in sperm-egg fusion. Additionally, there is evidence suggesting an indirect participation of epididymal proteins upon fertilization (for example, the alteration of the sperm surface by glycosylation, hydrolysis, lipid removal *et cetera*).

### i) Eppin

Eppin (Epididymal protease inhibitor) represents the first member of a family of protease inhibitors on human chromosome 20 characterized by dual inhibitor consensus sequences (Richardson *et al.*, 2001). There are three splice variants of Eppin that are expressed differently; Eppin-1 is expressed in the testis and epididymis, Eppin-2 is expressed in the epididymis and Eppin-3 in the testis. A study has shown that Eppin on the surface of spermatozoa and in semen is bound to semenogelin (Wang *et al.*, 2005). Robert and Gagnon (1996) have demonstrated that semenogelin is bound to the sperm surface and that the sperm motility inhibitory factor in semen is the N-terminal (amino acids 45-136) of semenogelin-I. Liquefaction removes semenogelin-I from the sperm surface, allowing spermatozoa to migrate along the female reproductive tract. However, failure to remove semenogelin from the sperm surface results in the inhibition of capacitation (de Lamirande *et al.*, 2001). Recently, a study demonstrated that effective and reversible male immunocontraception with Eppin, a testis/epididymis-specific protein in primates is an attainable goal (O'Rand *et al.*, 2004). Seven out of nine males (78%) developed sustained high antibody titers (> 1:1000) to Eppin, though it required boosters approximately every 3 weeks. All these high-titered monkeys were infertile. Further, five out of seven (71%) high–anti-Eppin titer males recovered fertility when immunization was stopped. This may be explained by the fact that the antibodies to Eppin interfere in the interaction of normal Eppin with the sperm surface and with semenogelin. Although this method has shown promise in male monkeys, its immunocontraceptive effect in humans needs to be extended.

### ii) Epididymal protein (DE)

The epididymal sperm protein, DE, originally identified in rat (Rochwerger and Cuasnicu, 1992), is localized on the dorsal region of the acrosome and migrates to the equatorial segment concomitantly with the occurrence of the acrosome reaction (Cohen *et al.*, 2000). Relocation of DE to the equatorial segment, the region through which the sperm fuses with the egg (Cohen *et al.*, 2000), together with the results of experiments showing that the polyclonal anti-DE antibody significantly inhibited the percentage of penetrated zona-free rat eggs, supported a role for this protein in sperm-egg fusion. Immunization of rats with protein DE produced a significant and reversible reduction in male and female fertility by a specific inhibition of sperm fertilizing ability (Martinez *et al.*, 1995; Ellerman *et al.*, 1998). Given the potential use of DE for fertility regulation, the availability of the recombinant DE protein will further allow research on the structure-function relationship of DE, and also provide an important tool to continue exploring the use of this protein for contraceptive development.

### iii) Epididymis-specific secretory protein (E-3)

Epididymis-specific secretory protein (E-3) is a novel and epididymis-specific secretory protein localized to the rat sperm flagellum. It is an isoantigen resembling ß-defensins, a lectin, or both. Results suggested that the E-3 gene is predominantly expressed in the corpus and cauda of the epididymis, and that the secreted E-3 protein is associated with spermatozoa. However, the nature of the association between E-3 and the maturing spermatozoa is unknown. Based on its predicted secondary structure, E-3 is similar to ß-defensins, having an alpha helical structure followed by three beta sheets. The presence of these defensin-like proteins in the male and female reproductive tracts in humans and rodents may protect sperm and the epididymis from bacterial infections (Quayle et al., 1998; Li et al., 2001). Confirmation of this possibility for E-3 awaits tests of antibacterial activity against a variety of organisms. Recently, it has been shown that ß-defensins also play multifunctional roles apart from antibacterial activity (Lehrer and Ganz, 1999); including interaction with plasma membranes of Xenopus oocytes and as a chemo-attractant to macrophages (Garcia et al., 2001) indicating a possible role in sperm-egg interaction and fertilization. Thus, it is possible that E-3 (and perhaps other similar family members) may have assumed different roles in spermatozoa. Although the functional properties of E-3 are yet to be defined, its molecular characteristics and pattern of expression suggest that it might act in sperm maturation, sperm-egg binding, as a decapacitation factor or as a defensin to protect spermatozoa from bacterial infection.

## Conclusion and future direction

Several candidate sperm antigens have been tested for immunogenicity and anti-fertility effects, and many have shown promise in experimental animals. However, no single antigen has yet shown the levels of efficacy demanded of an anti-fertility vaccine. Therefore, an effective immunocontraceptive vaccine would probably consist of several sperm-specific antigenic epitopes in a single formulation. Pre-formed monospecific antibodies to sperm antigens may also be combined in intravaginal sperm-specific spermicides for immunocontraceptive purposes. A recombinant mini-antibody has been engineered to the tissue-specific carbohydrate epitope located on the sperm glycoform of the CD52 antigen (Domagala and Kurpisz, 2001): its efficacy in agglutinating human sperm in a tangled pattern might make it a useful candidate for such applications (Norton et al., 2001). By a judicious use of delivery vehicles, carriers and adjuvants, it may become possible to challenge the gut associated lymphatic tissues (GALT) for the elicitation of local immunity in the reproductive tract. In conjunction with a systemic vaccine, this strategy could ensure the persistence of antibodies at all sites of spermatozoa transit.

Recombinant DNA and bioprocess technologies have made available recombinant proteins for immunogenicity and fertility studies. There now exists experimental data in animals (including non-human primates) that firmly establishes the basis for contraceptive vaccine development based upon recombinant proteins or synthetic peptides. The technology underpinning vaccine development is constantly being developed and the introduction of DNA/RNA vaccines are certain to impact upon the field of immunocontraception. Current and emerging strategies and methodologies are expected to provide the experimental foundation for the design of small molecule having anti-fertility effects. Moreover, gene knockout and RNA interference (RNAi) strategies will no doubt contribute towards a better understanding of the molecular biology of sperm function and sperm-egg interaction. Such data would not only aid in the identification of novel contraceptive targets, but would also be critical in any future clinical advances for treatment of infertility.

## Acknowledgements

The author wishes to acknowledge Dr Ritu Rana for assisting in the preparation of this manuscript. This work was supported by grants from the Department of Biotechnology, Government of India, Indo-US Program on Contraceptive and Reproductive Health Research (CRHR), Mellon Foundation and CONRAD, USA.

## References

Auer J, Senechal H, Desvaux FX, Albert M and De Almeida M (2000) Isolation and characterization of two sperm membrane proteins recognized by sperm-associated antibodies in infertile men *Molecular Reproduction and Development* 57 393-405

Brown PR, Miki K, Harper DB and Eddy EM (2003) A-Kinase Anchoring Protein 4 binding proteins in the fibrous sheath of the sperm flagellum *Biology of Reproduction* 68 2241-2248

Cohen DJ, Ellerman DA, Morgenfeld MM, Busso D and Cuasnicu PS (2000) Relationship between the association of rat epididymal protein DE with spermatozoa and the behavior and function of the protein *Molecular Reproduction and Development* 56 180-188

de Lamirande E, Yoshida K, Yoshiike TM, Iwamoto T and Gagnon C (2001) Semenogelin, the main protein of semen coagulum, inhibits human sperm capacitation by interfering with the superoxide anion generated during this process *Journal of Andrology* 22 672-679

Domagala A and Kurpisz M (2001) CD52 antigen – a review *Medical Science Monitor* 7 325–331

Ellerman DA, Brantua VS, Martinez SP, Cohen DJ, Conesa D and Cuasnicu PS (1998) Potential contraceptive use of epididymal proteins: immunization of male rats with epididymal protein DE inhibits sperm fusion ability *Biology of Reproduction* 59 1029-1036

Evans JP (2001) Fertilin beta and other ADAMs as integrin ligands: insights into cell adhesion and fertilization *Bioessays* 23 628-639

Evans EA, Zhang H and Martin-Deleon PA (2003) SPAM1 (PH-20) protein and mRNA expression in the epididymides of humans and macaques: utilizing laser microdissection/RT-PCR *Reproductive Biology and Endocrinology* 1 54-65

Freemerman AJ, Wright RM, Flickinger CJ and Herr JC (1993) Cloning and sequencing of baboon and cynomolgus monkey intra-acrosomal protein SP-10: homology with human SP-10 and a mouse sperm antigen (MSA-63) *Molecular Reproduction and Development* 34 140-148

Garcia JR, Jaumann F, Schulz S, Krause A, Rodriguez-Jimenez J, Forssmann U, Adermann K, Kluver E, Vogelmeier C, Becker D, Hedrich R, Forssmann WG and Bals R (2001) Identification of a novel, multifunctional beta-defensin (human beta-defensin 3) with specific antimicrobial activity. Its interaction with plasma membranes of *Xenopus* oocytes and the induction of macrophage chemoattraction *Cell and Tissue Research* 306 257-264

Hao Z, Wolkowicz MJ, Shetty J, Klotz K, Bolling L, Sen B, Westbrook VA, Coonrod S, Flickinger CJ and Herr JC (2002) SAMP32, a testis-specific, isoantigenic sperm acrosomal membrane-associated protein *Biology of Reproduction* 66 735-744

Hardy CM, Clarke HG, Nixon B, Grigg A, Hinds A and Holland MK (1997) Examination of the immunocontraceptive potential of recombinant rabbit fertilin subunits in rabbit *Biology of Reproduction* 57 879-886

Hardy CM, Clydesdale G, Mobbs KJ, Pekin J, Lloyd ML, Sweet C, Shellam GR and Lawson MA (2004) Assessment of contraceptive vaccines based on recombinant mouse sperm protein PH20 *Reproduction* 127 325-334

Harrison DA, Carr DW and Meizel S (2000) Involvement of protein kinase A and A kinase anchoring protein in the progesterone-initiated human sperm acrosome reaction *Biology of Reproduction* 62 811-820

Herr JC, Flickinger CJ, Homyk M, Klotz K and John E (1990) Biochemical and morphological characterization of intra-acrosomal antigen SP-10 from human sperm *Biology of Reproduction* 42 181-189

Jagadish N, Rana R, Selvi R, Mishra D, Garg M, Yadav S, Herr JC, Okumura K, Hasegawa A, Koyama K and Suri A (2005a) Characterization of a novel human sperm-associated antigen 9 (SPAG9) having structural homology with c-Jun N-terminal kinase-interacting protein *Biochemical Journal* 389 73-82

Jagadish N, Rana R, Mishra D, Kumar M and Suri A (2005b) Sperm associated antigen 9 (SPAG9): a new member of c-Jun NH2 -terminal kinase (JNK) interacting protein exclusively expressed in testis *Keio Jounal of Medicine* 54 66-71

Jagadish N, Rana R, Mishra D, Garg M, Chaurasiya D, Hasegawa A, Koyama K and Suri A (2005c) Immunogenicity and contraceptive potential of recombinant human sperm associated antigen (SPAG9) *Journal of Reproductive Immunology* 67 69-76

Jagadish N, Rana R, Selvi R, Mishra D, Shankar S, Mohapatra B and Suri A (2005d) Molecular cloning and characterization of the macaque sperm associated antigen 9 (SPAG9): an orthologue of human SPAG9 gene *Molecular Reproduction and Development* 71 58-66

Jagadish N, Rana R, Mishra D, Garg M, Selvi R and Suri A (2006) Characterization of immune response in mice to plasmid DNA encoding human sperm associated antigen 9 (SPAG9) *Vaccine* 24 3695-3703

Kurth BE, Weston C, Reddi PP, Bryant D, Bhattacharya R, Flickinger CJ and Herr JC (1997) Oviductal antibody response to a defined recombinant sperm antigen in macaques *Biology of Reproduction* **57** 981-989

Lathrop WF, Carmichael EP, Myles DG and Primakoff P (1990) cDNA cloning reveals the molecular structure of a sperm surface protein, PH-20, involved in sperm–egg adhesion and the wide distribution of its gene among mammals *Journal of Cell Biology* **111** 2939–2949

Lehrer RI and Ganz T (1999) Antimicrobial peptides in mammalian and insect host defence *Current Opinion in Immunology* **11** 23-27

Li P, Chan HC, He B, So SC, Chung YW, Shang Q, Zhang YD and Zhang YL (2001) An antimicrobial peptide gene found in the male reproductive system of rats *Science* **291** 1783-1785

Luconi M, Porazzi I, Ferruzzi P, Marchiani S, Forti G and Baldi E (2005) Tyrosine phosphorylation of the A Kinase Anchoring Protein 3 (AKAP3) and soluble adenylate cyclase are involved in the increase of human sperm motility by bicarbonate *Biology of Reproduction* **72** 22-32

Marshburn PB and Kutteh WH (1994) The role of antisperm antibodies in infertility *Fertility and Sterility* **61** 799-811

Martinez SP, Conesa D and Cuasnicu PS (1995) Potential contraceptive use of epididymal proteins: evidence for the participation of specific antibodies against rat epididymal protein DE in male and female fertility inhibition *Journal of Reproductive Immunology* **29** 31-45

Menge AC, Christman GM, Ohl DA and Naz RK (1999) Fertilization antigen-1 removes antisperm autoantibodies from spermatozoa of infertile men and results in increased rates of acrosome reaction *Fertility and Sterility* **71** 256-260

Miki K, Willis WD, Brown PR, Goulding EH, Fulcher KD and Eddy EM (2002) Targeted disruption of the *Akap4* gene causes defects in sperm flagellum and motility *Developmental Biology* **248** 331-342

Mohapatra B, Verma S, Shankar S and Suri A (1998) Molecular cloning of human testis mRNA specifically expressed in haploid germ cells, having structural homology with the A-kinase anchoring proteins *Biochemical Biophysical Research Communications* **244** 540-545

Morrison DK and Davis RJ (2003) Regulation of MAP Kinase signaling modules by scaffold proteins in mammals *Annual Review of Cell and Developmental Biology* **19** 91-118

Naz RK (1996) Application of sperm antigens in immunocontraception *Frontiers in Bioscience* **1** 87-95

Naz R and Zhu X (1998) Recombinant fertilization antigen-1 causes a contraceptive effect in actively immunized Mice *Biology of Reproduction* **59** 1095-1100

Naz RK and Zhu X (2002) Molecular cloning and sequencing of cDNA encoding for human FA-1 antigen *Molecular Reproduction and Development* **63** 256-268

Naz RK, Brazil C and Overstreet JW (1992) Effects of antibodies to sperm surface fertilization antigen-1 on human sperm-zona pellucida interaction *Fertility and Sterility* **57** 1304-1310

Noor MM and Moore HD (1999) Monoclonal antibody that recognizes an epitope of the sperm equatorial region and specifically inhibits sperm-oolemma fusion but not binding *Journal of Reproduction and Fertility* **115** 215-224

Norton EJ, Diekman AB, Westbrook VA, Flickinger CK and Herr JC (2001) RASA, a recombinant single-chain variable fragment (scFV) antibody directed against the human sperm surface: implications for the novel contraceptives *Human Reproduction* **16** 1854-1860

Oko R and Clermont Y (1990) Mammalian spermatozoa: structure and assembly of the tail. In: *Controls of Sperm Motility: Biological and Clinical Aspects* pp 4-21 Ed C Gagnon. CRC Press, Florida

O'Rand MG, Widgren EE, Sivashanmugam P, Richardson RT, Hall SH, French FS, VandeVoort CA, Ramachandra SG, Ramesh V and Jagannadha Rao A (2004) Reversible immunocontraception in male monkeys immunized with Eppin *Science* **306** 1189-1190

Phelps BM, Koppel DE, Primakoff P and Myles DG (1990) Evidence that proteolysis of the surface is an initial step in the mechanism of formation of sperm cell surface domains *Journal of Cell Biology* **111** 1839–1847

Pomering M, Jones RC, Holland MK, Blake AE and Beagley KW (2002) Restricted entry of IgG into male and female rabbit reproductive ducts following immunization with recombinant rabbit PH-20 *American Journal of Reproductive Immunology* **47** 174–182

Primakoff P, Woolman-Gamer L, Tung KS and Myles DG (1997) Reversible contraceptive effect of PH-20 immunization in male guinea pigs *Biology of Reproduction* **56** 1142–1146

Primakoff P, Lathrop W, Woolman L, Cowan A and Myles D (1988) Fully effective contraception in male and female guinea pigs immunized with the sperm protein PH-20 *Nature* **335** 543-547

Quayle AJ, Porter EM, Nussbaum AA, Wang YM, Brabec C, Yip KP and Mok SC (1998) Gene expression, immunolocalization, and secretion of human defensin-5 in human female reproductive tract *American Journal of Pathology* **152** 1247-1258

Richardson RT, Sivashanmugam P, Hall SH, Hamil KG, Moore PA, Ruben SM, French FS and O'Rand MG (2001) Cloning and sequencing of human Eppin: a novel family of protease inhibitors expressed in the epididymis and testis *Gene* **270** 93-102

Robert M and Gagnon C (1996) Purification and characterization of the active precursor of a human sperm motility inhibitor secreted by the seminal vesicles: identity with semenogelin *Biology of Reproduction* **55** 813-821

Rochwerger L and Cuasnicu PS (1992) Redistribution of a rat sperm epididymal glycoprotein after *in vivo* and *in vitro* capacitation *Molecular Reproduction and Development* **31** 34-41

**Santhanam R and Naz RK** (2001) Novel human testis-specific cDNA: molecular cloning, expression and immunological effects of the recombinant protein *Molecular Reproduction and Development* **60** 1-12

**Shankar S, Mohapatra B and Suri A** (1998) Cloning of a novel human testis mRNA specifically expressed in testicular haploid germ cells, having unique palindromic sequences and encoding a leucine zipper dimerization motif *Biochemical Biophysical Research Communications* **243** 561-565

**Shankar S, Mohapatra B, Verma S, Selvi R, Jagadish N and Suri A** (2004) Isolation and characterization of a haploid germ cell specific sperm associated antigen 9 (SPAG9) from the baboon *Molecular Reproduction and Development* **69** 186-193

**Shetty J, Naaby-Hansen S, Shibahara H, Bronson R, Flickinger CJ and Herr JC** (1999) Human sperm proteome: immunodominant sperm surface antigens identified with sera from infertile men and women *Biology of Reproduction* **61** 61–69

**Shetty J, Wolkowicz MJ, Digilio LC, Klotz KL, Jayes FL, Diekman AB, Westbrook VA, Farris EM, Hao Z, Coonrod SA, Flickinger CJ and Herr JC** (2003) SAMP14, a novel, acrosomal membrane-associated, glycosylphosphatidylinositol-anchored member of the Ly-6/urokinase-type plasminogen activator receptor superfamily with a role in sperm-egg interaction *Journal of Biological Chemistry* **278** 30506-30515

**Srinivasan J, Tinge S, Wright R, Herr JC and Curtiss R** 3rd (1995) Oral immunization with attenuated *Salmonella* expressing human sperm antigen induces antibodies in serum and the reproductive tract *Biology of Reproduction* **53** 462-471

**Suri A** (2004) Sperm specific proteins-potential candidate molecules for fertility control *Reproductive Biology and Endocrinology* **2** 10-15

**Suri A** (2005a) Sperm-based contraceptive vaccines: current status, merits and development *Expert Reviews in Molecular Medicine* **7** 1-16

**Suri A** (2005b) Contraceptive vaccines targeting sperm *Expert Opinion on Biolological Therapy* **5** 381-392

**Talwar GP, Singh O, Pal R, Chatterjee N, Sahai P, Dhall K, Kaur J, Das SK, Suri S, Buckshee K, Saraya L and Saxena BN** (1994) Vaccine that prevents pregnancy in women. *Proceedings National Academy of Sciences USA* **91** 8532-8536

**Ten Have J, Beaton S and Bradley MP** (1998) Cloning and characterization of the cDNA encoding the PH20 protein in the European red fox *Vulpes vulpes Reproduction Fertility and Development* **10** 165-172

**Toshimori K, Tanii I, Araki S and Oura C** (1992) Characterization of the antigen recognized by a monoclonal antibody MN9: unique transport pathway to the equatorial segment of the sperm head during spermiogenesis *Cell and Tissue Research* **270** 459-468

**Trivedi RN and Naz RK** (2002) Testis-specific antigen (TSA-1) is expressed in murine sperm and its antibodies inhibit fertilization *American Journal of Reproductive Immunology* **47** 38-45

**United Nations Population Information Network**; *http://www.un.org/popin*

**van Voorhis BJ and Stovall DW** (1997) Autoantibodies and infertility: a review of the literature *Journal of Reproductive Immunology* **33** 239–256

**Vijayaraghavan S, Goueli SA, Davey MP and Carr DW** (1997) Protein Kinase A-anchoring inhibitor peptides arrest mammalian sperm motility *Journal of Biological Chemistry* **272** 4747-4752

**Wang Z, Widgren EE, Sivashanmugam P, O'Rand MG and Richardson RT** (2005) Association of eppin with semenogelin-I on human spermatozoa *Biology of Reproduction* **72** 1064-1070

**Wolkowicz MJ, Shetty J, Westbrook A, Klotz K, Jayes F, Mandal A, Flickinger CJ and Herr JC** (2003) Equatorial segment protein defines a discrete acrosomal subcompartment persisting throughout acrosomal biogenesis *Biology of Reproduction* **69** 735–745

**Wright RM, Suri AK, Kornreich B, Flickinger CJ and Herr JC** (1993) Cloning and characterization of the gene coding for the human acrosomal protein SP-10 *Biology of Reproduction* **49** 316-325

**Yanagimachi R and Noda YD** (1970) Ultrastructural changes in the hamster sperm head during fertilization *Journal of Ultrastructure Research* **31** 465-485

**Zhang H, Morales CR, Badran H, El-Alfy M and Martin-DeLeon PA** (2004) Spam1 (PH-20) expression in the extratesticular duct and accessory organs of the mouse: A possible role in sperm fluid reabsorption *Biology of Reproduction* **71** 1101-1107

# Eppin: an epididymal protease inhibitor and a target for male contraception

MG O'Rand*, EE Widgren, Zengjun Wang[1] and RT Richardson

*Department of Cell & Developmental Biology, University of North Carolina at Chapel Hill, Chapel Hill, North Carolina-27599, USA*

Eppin (epididymal protease inhibitor) is one of several serine protease (or serine protease-like) inhibitors that are encoded by genes on human chromosome 20 and on mouse chromosome 2. Here we review our current knowledge of human and mouse Eppin genes and the Eppin protein in the context of protease inhibitors. Antibodies to Eppin in immunized male monkeys provide an effective and reversible contraceptive and these antibodies may be effective by interfering with Eppin's interaction with semenogelin during ejaculation. We review Eppin-semenogelin interaction and present a working model in the context of the hydrolysis of semenogelin by prostate specific antigen.

Eppin (epididymal protease inhibitor) is one of several serine protease (or serine protease-like) inhibitors that are encoded by genes on human chromosome 20 and the conserved synteny regions on mouse chromosome 2. Our interest in these inhibitors began with our discovery of a new epididymal serine protease inhibitor, which we called Eppin, that is characterized by both whey acidic protein (WAP)-type and Kunitz-type consensus sequences. The mouse-human genome homology map described by Gregory et al. (2002) indicates that there is alignment between mouse chromosome 2 and human chromosome 20 in the region of the Eppin gene cluster. The official gene symbol for human Eppin is *SPINLW1* and *Spinlw1* for mouse Eppin and their localization on their respective chromosomes may be seen on the web through the NCBI website: http://www.ncbi.nih.gov/entrez/query.fcgi?db = gene&cmd = search&term = SPINLW1. The Eppin gene is a member of the WAP-type four-disulfide core (*WFDC*) gene family. The *WFDC* genes are on chromosome 20q12-q13 in two clusters, one centromeric and one telomeric. They are numbered *WFDC1-WFDC1-n* and Eppin is *WFDC7* in the telomeric cluster. *WFDC* genes are expressed in numerous tissues, however, *WFDC6, 7* and *8* are predominantly expressed in the epididymis and testis with the caveat that *WFDC7* transcripts have been reported in the trachea (Clauss et al., 2002). Actual secreted proteins from these tracheal transcripts have not been reported. The mouse homologues of *WFDC6, 7* and *8* are expressed only in the epididymis and/or testis.

---

*Corresponding author
E-mail: morand@unc.edu
[1]ZW was supported by a Fogarty Postdoctoral Fellowship; permanent address The First Affiliated Hospital of Nanjing Medical University, China, Department of Urology, Nanjing Medical University, China
Grant support: Supported by grant CIG-96-06 from the CICCR Program of CONRAD (Contraceptive Research and Development Program), the Andrew W. Mellon Foundation, and by D43TW-HD00627, Program for International Training and Research in Population and Health from the Fogarty International Center and the National Institute of Child Health and Human Development.

## Human Eppin

*Eppin* is a gene on human chromosome 20 characterized by three mRNAs encoding two isoforms of a cysteine-rich protein containing both Kunitz-type and WAP-type four disulfide core protease inhibitor consensus sequences (Richardson *et al.*, 2001). Analysis of Eppin's genomic sequence from chromosome 20q12-13.2 predicts the existence of all three splice variants of *Eppin*. The three splice variants are expressed differently; Eppin-1 is expressed in the testis and epididymis, Eppin-2 is expressed in the epididymis and Eppin-3 in the testis. These variants represent the first members of a family of protease inhibitors characterized by dual inhibitor consensus sequences on human chromosome 20. *Eppin* is a single copy gene and TATA box transcription initiation sites are present for both of the different *Eppin* 5' UTRs. The differential tissue expression of the various *Eppin* isoforms is most likely controlled by regulatory regions upstream of the 5' UTRs. Examination of the promoter region 1800 bp upstream of the start codon revealed a number of putative transcription enhancer binding sites relevant to the expression of epididymal and testis genes.

Northern blot and tissue specific PCR data indicate *Eppin* is differentially expressed in the testis and epididymis (Richardson *et al.*, 2001). The proteins encoded by *Eppin-1* and *-3* are identical; however the protein encoded by *Eppin-2* lacks a secretory signal sequence. Hence, in the testis all Eppin would be secreted either in a constitutive or regulated (vesicle) pathway, whereas in the epididymis, Eppin could be both secreted and intracellular. On the basis of protein expression data from Western blots and from immunohistochemical localization, a significant quantity of Eppin appears to be bound to spermatozoa. Western blots of extracts of human caput and corpus epididymal tissue demonstrate that Eppin is predominantly a dimer ($\sim$ 36-46 kDa), while in extracts from ejaculated spermatozoa and seminal plasma multiples of the monomer form (18-23 kDa) are present (Wang *et al.*, 2005). Not surprisingly, the 14-cysteine residues (10.5% cysteine content) account for the ease with which Eppin can form multimers.

In tissue sections of human epididymis, Eppin is located in the ciliated cells of the epithelium of the ductuli efferentes, to a lesser extent in the non-ciliated cells, and along most of the apical surfaces of the epithelium lining the lumen (Richardson *et al.*, 2001). Additionally, strong staining of spermatozoa within the lumen was observed (see figure 7A, B in Richardson *et al.*, 2001). Washed ejaculated human spermatozoa stained with affinity purified anti-Eppin antibodies showed intense staining on the acrosome, post-acrosome, midpiece and tail, while antibody preabsorbed with recombinant Eppin showed no staining (see Figure 7D, F in Richardson *et al.*, 2001; Figure 2A, B in Wang *et al.*, 2005).

## Mouse Eppin

In addition to human *Eppin*, monkey (*Macaca mulatta*, Accession # AF346414), rabbit (*Oryctolagus cuniculus*, Accession # AF346415) and mouse (Accession # AF 346413) *Eppin* has been cloned, sequenced and characterized (Sivashanmugam *et al.*, 2003).

As described in Sivashanmugam *et al.* (2003), the *Eppin* gene on mouse chromosome 2 has an open reading frame that encodes a 134 amino acid protein with a calculated molecular mass of 15.4 kDa and an isoelectric point of 8.28. The deduced protein sequences for rabbit, monkey and mouse Eppin have consensus sequences for both WAP-type disulfide core domains and Kunitz-type domains (Fig. 1). Comparative sequence analysis of the Eppin protein family members reveals that mouse Eppin is 60% identical to monkey, 62% identical to human and 51% identical to rabbit.

M——SGLLS-LVLF-LL——VQGPGL-D-LFPRRCP-IREECE—ERD

-C̲T̲R̲-R̲-C̲P̲D̲—̲K̲C̲C̲VF-CGKKCLDL-QDVC-MPKETGPCLA-F-R

*WWYDK—TC——* F-YGGCQGNNNNFQS-A-CL̲N-C——K——

**Figure 1.** The consensus sequence of Eppin from mouse, human, rabbit, baboon, and rhesus monkey is shown, the underlined sequence **CTR-R-CPD—KCC** is the signature sequence for WAP-type disulfide core domain and the boxed sequence **F-YGGCQGNNNNFQS-A-C** is the signature sequence for the Kunitz-type domain. Modified from Sivashanmugam *et al.* (2003)

Northern blots indicate that 1.0 and 1.7 kb mouse *Eppin* transcripts are detected only in the testis and epididymis. RT-PCR analysis for *Eppin* tissue specificity also indicates that Eppin is expressed specifically in the testis and epididymis. To determine the relative size of the expressed protein, mouse Eppin was expressed *in vitro* and the translated, $^{35}$S-methionine labeled Eppin separated by SDS-PAGE and detected by autoradiography. The apparent molecular weight of mouse Eppin under reducing conditions is 18-23 kDa indicating that its high cysteine content influences its apparent molecular weight increase over its deduced molecular weight of 15.4 kDa.

In order to establish the localization of mouse Eppin in the testis, tissue sections were probed with affinity purified anti-Eppin antibodies. Eppin was present in the acrosomal region of round spermatids, the cytoplasm of Sertoli cells and the cytoplasm of Leydig cells. In Sertoli cells, Eppin was seen throughout the cytoplasm, extending from the basal aspect of the cell to the luminal surface. In the epididymis, the caudal epididymal principle cells showed intense staining within the apical cytoplasm and luminal staining was observed on the long microvilli/stereocilia. No staining was seen with antiserum absorbed with recombinant Eppin (rEppin). Caput and corpus epididymis did not show any staining. On *in vitro* capacitated cauda epididymal mouse spermatozoa, Eppin was localized on the postacrosomal region of the head and the tail.

## Protease inhibitors

Protease inhibitors play key roles in many physiological processes by specifically regulating proteases and thereby maintaining homeostasis (Potempa *et al.*, 1994). Serine proteases in particular are a diverse family that includes components of the blood coagulation cascade and the complement system and whose activities require the appropriate inhibitors to ensure their regulation. The presence of proteases in the epididymis and ejaculate requires the presence of their inhibitors as an important regulatory mechanism for regional maturation and processing of spermatozoa.

Spermatozoa undergo a variety of post-testicular changes in the epididymis to acquire both progressive motility and fertilizing ability (Bedford, 1967; Orgebin-Crist, 1967). Absorption and secretion by epididymal epithelia lead to a changing fluid environment (Hinton and Palladino, 1995) that surrounds spermatozoa during their epididymal transit, changing the spermatozoon membrane composition (Scott *et al.*, 1967), including the modification, acquisition and removal of various surface glycoproteins (Cooper, 1995; Legare *et al.*, 1999; Robaire *et al.*, 2000). Glycosyltransferases and glycosidases found in epididymal fluid are responsible for many of the sperm surface modifications (Eddy and O'Brien, 1994), while proteases, such as procathepsin L (Okamura *et al.*, 1995), ACE and serine proteases also play roles in both sperm surface modifications and sperm maturation (Phelps *et al.*, 1990; He *et al.*, 1995; Kirchhoff *et al.*, 1997; Dacheux *et al.*, 1998). Concomitant with the expression of proteases is the expression of their inhibitors: α2-macroglobulin, cystatin C, cystatin-like proteins of the *Cres* gene family (Cornwall *et al.*,1992; 1999; Dacheux *et al.*, 1998), HE4 (Kirchhoff *et al.*, 1991) and Eppin (Richardson *et al.*, 2001). We now know that Eppin is a specific inhibitor of prostate specific antigen (PSA).

## Prostate specific antigen (PSA)

PSA is a prostate specific kallikrein, which is a member of the serine protease gene family. Kallikreins are involved in a number of physiological processes such as blood clotting; the human glandular kallikreins, of which there are three members, process precursor proteins to produce, for example, kinin from kininogen, and EGF and NGF from their precursors. In seminal plasma PSA exists in a soluble form that is inhibited by $Zn^{++}$, spermine and spermidine (Carvalho *et al.*, 2002). The structure of PSA predicts a chymotrypsin-like specificity because of a serine residue at position S1 of the specificity pocket (Carvalho *et al.*, 2002). This chymotrypsin-like specificity is supported by experimental evidence and in particular PSA hydrolysis of its natural substrate, semenogelin, occurs C-terminally to leucine or tyrosine residues (Robert *et al.*, 1997). Occasionally, histidine and glutamine residues may also be cleaved (Lovgren *et al.*, 1999). Our interest in PSA binding to Eppin arose from two observations: 1) A report in the literature by Wu *et al.* (2000) that PSA specific binding peptides can modulate PSA activity and that one identified peptide contained a critical amino acid sequence that was also present in Eppin (CVF), and 2) the PSA hydrolysis of semenogelin that was bound to Eppin.

## Semenogelin

During ejaculation spermatozoa encounter a change in extracellular environmental conditions, particularly osmotic changes that require cell volume adjustments when mixing with the secretions from the seminal vesicles and prostate. Bound surface molecules undergo a variety of modifications, even removal; as it has been known for many years that peripheral membrane proteins (decapacitation factors) are removed as spermatozoa begin capacitation in the female reproductive tract (see for example O'Rand, 1979; 1980). In rodents, transglutaminases cross-link proteins to form the copulatory plug, while in primates, including humans, non-covalent interactions form the semen coagulum. In humans, caudal epididymal spermatozoa are in an immotile state and immediately following ejaculation, sperm continue to appear immotile in the semen coagulum until liquefaction occurs and progressive motility begins (Amelar, 1962). In spite of these differences between rodent and primate ejaculates, the two major proteins, semenoclotin in mice and semenogelin I and II in humans, are in the same gene family, perform similar coagulation functions, are under androgen control and are rapidly degraded by human prostate-specific antigen (Lundwall *et al.*, 1997). The human genes for semenogelin-I and II are located on chromosome 20 approximately 145 kb on the centromeric side of Eppin, within another cluster of serine protease inhibitor genes (Clauss *et al.*, 2002). Indeed it has been suggested that human semenogelin-I and mouse semenoclotin arose by gene duplication at the same time as *Eppin* and *Eppin*-like genes were undergoing gene duplication during the time of the separation of murine and primate lineages (Lundwall, 1997; Clauss *et al.*, 2002).

The role of semenogelin (Sg) during ejaculation has been reviewed by Robert and Gagnon (1999). Briefly, it is currently thought that Sg I and II are initially protected from proteolysis by protein C inhibitor (PCI) in the seminal vesicles and that PSA (a serine protease), in the prostatic secretions is inactive, inhibited by high (7 mM) concentrations of zinc. PCI, a serine protease inhibitor, binds to both PSA and Sg II (Christensson and Lilja, 1994; Kise *et al.*, 1996). During ejaculation the mixing of Sg with prostatic secretions chelates most of the free zinc, which triggers release of PCI, aggregation of Sg and fibronectin, and activates PSA. PSA cleaves the coagulum proteins, resulting in the release of Sg proteolytic fragments. Extensive work by Gagnon's laboratory (Iwamoto and Gagnon, 1988a, b; Luterman *et al.*, 1991; Robert and Gagnon, 1994; 1995; 1996) has demonstrated that Sg is bound to the sperm surface and that the sperm motility inhibitory factor in semen is the N-terminal (amino acids 45-136) of Sg I. Cleavage of Sg I by PSA during liquefaction removes Sg I from the sperm surface. Failure to remove Sg I from the sperm surface results in loss of motility and inhibition of capacitation (de Lamirande *et al.*, 2001).

## Interaction of Eppin and semenogelin

Experiments by Wang *et al.*, (2005) demonstrate that Eppin is bound to semenogelin I in seminal plasma and on human spermatozoa following ejaculation. Six different experimental approaches demonstrate that Eppin and Sg bind to each other: i) immunoprecipitation from spermatozoa and seminal plasma with anti-Eppin; ii) co-localization in semen and spermatozoa; iii) incubation of rEppin and recombinant Sg (rSg) and immunoprecipitation with either anti-Eppin or anti-Sg; iv) far-Western blotting of Eppin and Sg; v) saturation binding of $^{125}$I-Sg to Eppin, which is competed by unlabeled Sg; and vi) direct binding of $^{125}$I-Sg to Eppin on a blot, which is competed with unlabeled Sg.

In the human ductuli efferentes spermatozoa encounter epididymal Eppin and Eppin continues to coat the spermatozoa (Richardson *et al.*, 2001). As spermatozoa enter the ejaculatory duct, the copious secretions formed in the seminal vesicles mix with the spermatozoa and shortly thereafter the spermatozoa and seminal fluid mix with prostatic secretions. In the ejaculatory ducts the spermatozoon surface coating of Eppin should become saturated with Sg, the major protein constituent of seminal fluid (de Lamirande *et al.*, 2001). Significantly, the binding of Sg to Eppin (Wang *et al.*, 2005) is the major protective event that occurs to ejaculated spermatozoa in primates, providing antimicrobial activity to spermatozoa in a coagulum. The cleavage of Sg on spermatozoa and in the coagulum is accomplished by PSA (Robert *et al.*, 1997). Immediately following ejaculation when spermatozoa are part of the ejaculate coagulum, Eppin is an integral part of that coagulum. On the surface of fertile, live, human spermatozoa there is a saturable binding site (receptor) for Eppin. This has been demonstrated by the saturation kinetics of $^{125}$I-recombinant human Eppin binding to live human "swim-up" spermatozoa (O'Rand *et al.*, 2006).

The Eppin-semenogelin complex bound on the surface of ejaculate spermatozoa provides anti-microbial activity for spermatozoa, which has been reported for both Eppin (Yenugu *et al.*, 2004) and semenogelin derived peptides (Bourgeon *et al.*, 2004), and for the preparation of spermatozoa for fertility in the female reproductive tract (Robert *et al.*, 1997). Additionally, Eppin may protect spermatozoa from proteolytic attack by modulating the cleavage of Sg bound to Eppin. Semenogelin that is co-immunoprecipitated with Eppin from spermatozoa and seminal plasma has an apparent molecular weight of < 27 kDa, smaller than the full-length Sg (52 kDa). The recovered Sg fragments bound to Eppin have probably been cleaved by PSA (Robert and Gagnon, 1996), which usually cleaves Sg adjacent to tyrosine and leucine residues (Robert *et al.*, 1997). Leu246 and Leu261 are known PSA cleavage sites on Sg and either would provide an N-terminal semenogelin peptide fragment of approximately 27 kDa. Our finding (Wang *et al.*, 2005) that the Eppin binding site on Sg is within amino acids 164-283 that would allow PSA to cleave Sg and the 27 kDa fragment would still remain attached to Eppin on the sperm surface. However, continued PSA activity would most likely remove Sg from its Eppin binding site. Similarly, the sperm motility inhibitory factor in semen (Robert and Gagnon, 1996; Sg amino acids 85-136 in Robert et al., 1997) is located within the Sg fragment bound to Eppin and most likely would be removed by PSA activity since its presence inhibits capacitation (de Lamirande et al., 2001).

As reported previously (Richardson et al., 2001; Wang et al., 2005), native Eppin occurs as multimers, which are thought to form by the intermolecular interaction of the 14-cysteine residues. Mass spectroscopy studies on reduced and carboxymethylated recombinant forms of Eppin determined that the actual mass of the dimer is 33 kDa (O'Rand, Widgren and Richardson, unpublished). The Sg fragment (Sg164-283) that contains the only cysteine in human Sg I (Cys239) is necessary for binding the Eppin75-133 C-terminal fragment. If a disulfide linkage occurs between Sg and Eppin it might allow several Sg molecules (or fragments) to bind Eppin simultaneously, multiplying Eppin's effectiveness as a binding site.

Our conclusions from studying Eppin-Sg interaction (Wang et al., 2005) are that the Eppin-Sg complex found on human ejaculate spermatozoa is part of a network of protein complexes on the sperm surface that provides a protective shield prior to capacitation in the female reproductive tract. Such sperm coating proteins function in innate immunity, anti-microbial activity and inhibition of proteases that may directly attack the sperm plasma membrane (Yenugu et al., 2004). Co-localization of Eppin and Sg in the post-ejaculation coagulum (see Figure 2C, D in Wang et al., 2005) demonstrates the need for spermatozoa to escape this coagulum in order to migrate along the female reproductive tract and become capacitated.

## Eppin and male fertility

O'Rand et al., (2004) have demonstrated that effective and reversible male immunocontraception in primates is an obtainable goal. We find that a high serum titer (> 1:1000) and sustaining that titer over several months achieves an effective level of contraception. Seven out of nine males (78%) developed high titers to Eppin, and all these high titer monkeys were infertile. Five out of 7 (71%) high anti-Eppin titer males recovered fertility when immunization was stopped.

Following the immunization of two adult Macaca males at UC-Davis, observations were made (Dr. C.A. VandeVoort) on the sperm motility in their ejaculates. Monkey # 28309 showed 82-95% progressively motile sperm after three injections of Eppin, which was not different from the pre-immunization rate. However, semen collected eleven days after two additional boosters contained sperm whose progressive motility had dropped to 74% and three months after the third booster the progressive motility dropped to 65%. The first observations of reduced motility coincided with the appearance of an anti-Eppin titer in the semen (see Figure S3: monkey #28309 in Supporting Online Material, O'Rand et al., 2004). Spermatozoa with reduced motility were observed to have curled tails and significantly, semen collected after the third booster had no coagulum. Monkey # 24059, which had a considerably lower semen titer than # 28309, did not show significant changes in the percent of progressively motile sperm, but did exhibit a loss of a semen coagulum three months after the third booster. Spermatozoa from both monkeys appeared to exhibit normal binding to zona pellucida in vitro.

Anti-Eppin antibodies may disrupt the Eppin-Sg complex and inhibit the proper removal of Sg by PSA, forming the basis for the inhibition of fertility by anti-Eppin antibodies (O'Rand et al., 2004).

## Eppin-semenogelin protein complex on the ejaculated sperm surface

It is our hypothesis that Eppin, found in a complex bound to the sperm surface, is a surface receptor for semenogelin and the presence of Eppin modulates the activity of PSA on semenogelin. Sg on the sperm surface is bound to Eppin and therefore the cleavage of Sg by PSA must occur while Sg is bound to Eppin. *In vitro*, when recombinant Sg is digested with PSA many low molecular weight fragments are produced. However, when Eppin is bound to Sg before digestion, digestion by PSA is modulated; producing incomplete digestion and an undigested 15-16 kDa fragment (O'Rand *et al.*, 2006). *In vivo*, during semen liquefaction, activated PSA cleaves semenogelin (Sg) bound to the sperm surface, releasing the sperm motility inhibitory factor (Robert and Gagnon, 1996; Robert *et al.*, 1997). The activation of sperm motility during liquefaction is a direct result of the loss or inactivation of Eppin-semenogelin binding due to the degradation of semenogelin by PSA.

Our working model (Fig. 2) depicts the interactions between Eppin, semenogelin and PSA. The model shown depicts observations supported by current data: [box #1] Eppin binding to the sperm surface (Richardson *et al.*, 2001; Wang *et al.*, 2005); [box #2] Eppin75-133 C-terminal binding to semenogelin (cys239; Wang *et al.*, 2005); [box #3] PSA degrades semenogelin by hydrolysis at leucine and tyrosine residues (Robert *et al.*, 1997); [box #4] Eppin's modulation of PSA activity (O'Rand *et al.*, 2006). In the presence of semenogelin (sperm motility inhibitory factor) sperm motility is blocked. When PSA degrades semenogelin, motility is initiated and capacitation can occur (de Lamirande *et al.*, 2001). The model depicts the relationship among proteins in the complex. Identification of the receptor(s) for Eppin is currently under investigation.

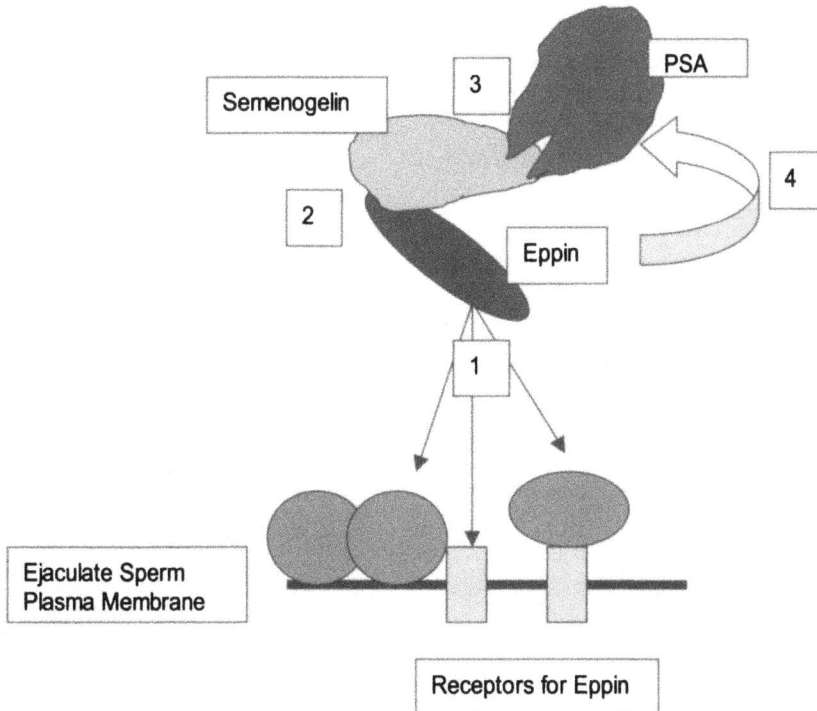

**Figure 2.** Our working model depicts the interactions between Eppin, semenogelin and prostate specific antigen (PSA). See text for details

# References

Amelar RD (1962) Coagulation, liquefaction and viscosity of human semen *Journal of Urology* **87** 187-190

Bedford JM (1967) Effects of duct ligation on the fertilizing ability of spermatozoa from different regions of the rabbit epididymis *Journal of Experimental Zoology* **166** 271-282

Bourgeon F, Evrard B, Brillard-Bourdet M, Colleu D, Jegou B and Pineau C (2004) Involvement of semenogelin-derived peptides in the antibacterial activity of human seminal plasma *Biology of Reproduction* **70** 768-774

Carvalho AL, Sanz L, Barettino D, Romero A, Calvete JJ and Romao MJ (2002) Crystal structure of a prostate kallikrein isolated from stallion seminal plasma: a homologue of human PSA *Journal of Molecular Biology* **322** 325-337

Christensson A and Lilja H (1994) Complex formation between protein C inhibitor and prostate-specific antigen *in vitro* and in human semen *European Journal of Biochemistry* **220** 45-53

Clauss A, Lilja H and Lundwall A (2002) A locus on human chromosome 20 contains several genes expressing protease inhibitor domains with homology to whey acidic protein *Biochemistry Journal* **368** 233-242

Cooper TG (1995) Role of the epididymis in mediating changes in the male gamete during maturation *Advances in Experimental Medicine and Biology* **377** 87-101

Cornwall GA, Orgebin-Crist M-C and Hann SR (1992) The CRES gene: a unique testis-regulated gene related to the cystatin family is highly restricted in its expression to the proximal region of the mouse epididymis *Molecular Endocrinology* **6** 1653-1664

Cornwall GA, Hsia N and Sutton HG (1999) Structure, alternative splicing and chromosomal localization of the cystatin-related epididymal spermatogenic gene *Biochemistry Journal* **340** 85-93

Dacheux J-L, Druart X, Fouchecourt S, Syntin P, Gatti J-L, Okamura N and Dacheux F (1998) Role of epididymal secretory proteins in sperm maturation with particular reference to the boar *Journal of Reproduction and Fertility (Supplement)* **53** 99-107

de Lamirande E, Yoshida K, Yoshiike TM, Iwamoto T and Gagnon C (2001) Semenogelin, the main protein of semen coagulum, inhibits human sperm capacitation by interfering with the superoxide anion generated during this process *Journal of Andrology* **22** 672-679

Eddy EM and O'Brien D (1994) The Spermatozoon. In *The Physiology of Reproduction* pp 29-77 Eds E Knobil et al., Raven Press, New York

Gregory SG et al. (86 authors), (2002) A physical map of the mouse genome *Nature* **418** 743-750

He X, Shen L, Bjartell A, Malm J, Lilja H and Dahlback B (1995) The gene encoding vitamin K-dependent anticoagulant protein C is expressed in human male reproductive tissues *Journal of Histochemistry and Cytochemistry* **43** 563-570

Hinton BT and Palladino MA (1995) Epididymal epithelium: its contribution to the formation of a luminal fluid microenvironment *Microscope Research Technology* **30** 67-81

Iwamoto T and Gagnon C (1988a) Purification and characterization of a sperm motility inhibitor in human seminal plasma *Journal of Andrology* **9** 377-383

Iwamoto T and Gagnon C (1988b) A human seminal plasma protein blocks the motility of human spermatozoa *Journal of Urology* **140** 1045-1048

Kirchhoff C, Habben I, Ivell R and Krull N (1991) A major human epididymis-specific cDNA encodes a protein with sequence homology to extracellular proteinase inhibitors *Biology of Reproduction* **45** 350-357

Kirchhoff C, Pera I, Derr P, Yeung CH and Cooper T (1997) The molecular biology of the sperm surface. Post-testicular membrane remodeling *Advances in Experimental Medicine and Biology* **424** 221-232

Kise H, Nishioka J, Kawamura J and Suzuki K (1996) Characterization of semenogelin II and its molecular interaction with prostate-specific antigen and protein C inhibitor *European Journal of Biochemistry* **238** 88-96

Legare C, Gaudreautl C, St-Jacques S and Sullivan R (1999) P34H sperm protein is preferentially expressed by the human corpus epididymis *Endocrinology* **140** 3318-3327

Lovgren J, Airas K and Lilja H (1999) Enzymatic action of human glandular kallikrein 2 (hK2) *European Journal of Biochemistry* **262** 781-789

Lundwall A (1997) The cloning of a rapidly evolving seminal-vesicle-transcribed gene encoding the major clot-forming protein of mouse semen *European Journal of Biochemistry* **235** 424-430

Lundwall A, Peter A, Lovgren J, Lilja H and Malm J (1997) Chemical characterization of the predominant proteins secreted by mouse seminal vesicles *European Journal of Biochemistry* **249** 39-44

Luterman M, Iwamoto T and Gagnon C (1991) Origin of the human seminal plasma motility inhibitor within the reproductive tract *International Journal of Andrology* **14** 91-98

Okamura N, Tamba M, Uchiyama Y, Sugita Y, Dacheux F, Syntin P and Dacheux J (1995) Direct evidence for the elevated synthesis and secretion of procathepsin L in the distal caput epididymis of boar *Biochemistry Biophysics Acta* **1245** 221-226

O'Rand MG (1979) Changes in sperm surface properties correlated with capacitation. In *The Spermatozoon: Maturation, Motility and Surface Properties* pp 412-428 Eds DW Fawcett and JM Bedford. Urban and Schwarzenberg, Baltimore

O'Rand MG (1980) Antigens of spermatozoa and their environment. In *Immunological Aspects of Infertility and Fertility Regulation* pp 155-171 Eds DS Dhindsa and GFB Schumacher. Elsevier/North Holland, New York

O'Rand MG, Widgren EE, Sivashanmugam P, Richardson RT, Hall SH, French FS, VandeVoort CA, Ramachandra SG, Ramesh V and Rao AJ (2004) Reversible immunocontraception in male monkeys immunized with Eppin *Science* **306** 1189-1190

O'Rand MG, Widgren EE, Wang Z and Richardson RT (2006) Eppin: An effective target for male contraception *Molecular and Cellular Endocrinology* **250** 157-162

Orgebin-Crist M-C (1967) Sperm maturation in rabbit epididymis *Nature* **216** 816-818

Phelps BM, Koppel DE, Primakoff P and Myles DG (1990) Evidence that proteolysis of the surface is an initial step in the mechanism of formation of sperm cell surface domains *Journal of Cell Biology* **111** 1839-1847

Potempa J, Korzus E and Travis J (1994) The serpin superfamily of proteinase inhibitors: structure, function, and regulation *Journal of Biological Chemistry* **269** 15957-15960

Richardson RT, Sivashanmugam P, Hall SH, Hamil KG, Moore PA, Ruben SM, French FS and O'Rand MG (2001) Cloning and sequencing of human *Eppin*: A novel family of protease inhibitors expressed in the epididymis and testis *Gene* **270** 93-102

Robaire B, Syntin P and Jarvis K (2000) The coming of age of the epididymis. In *Testis, Epididymis and Technologies in the Year 2000* pp 229-262 Eds B Jegou, C Pineau and J Saez. Springer, Berlin

Robert M and Gagnon C (1994) Sperm motility inhibitor from human seminal plasma: presence of a precursor molecule in seminal vesicle fluid and its molecular processing after ejaculation *International Journal of Andrology* **17** 232-240

Robert M and Gagnon C (1995) Sperm motility inhibitor from human seminal plasma: association with semen coagulum *Human Reproduction* **10** 2192-2197

Robert M and Gagnon C (1996) Purification and charac-terization of the active precursor of a human sperm motility inhibitor secreted by the seminal vesicles: identity with semenogelin *Biology of Reproduction* **55** 813-821

Robert M and Gagnon C (1999) Semenogelin I: a coagulum forming, multifunctional seminal vesicle protein *Cellular and Molecular Life Science* **55** 944-960

Robert M, Gibbs BF, Jacobson E and Gagnon C (1997) Characterization of prostate-specific antigen proteolytic activity on its major physiological substrate, the sperm motility inhibitor precursor/semenogelin I *Biochemistry* **36** 3811-3819

Scott TW, Vogylmeyr JK and Setchell BP (1967) Lipid composition and metabolism in testicular and ejaculated ram spermatozoa *Biochemistry Journal* **102** 456-461

Sivashanmugam P, Hall SH, Hamil KG, French FS, O'Rand MG and Richardson RT (2003) Characterization of mouse Eppin and a gene cluster of similar protease inhibitors on mouse chromosome 2 *Gene* **312** 125-134

Wang Z, Widgren EE, Sivashanmugam P, O'Rand MG and Richardson RT (2005) Association of Eppin with Semenogelin on human spermatozoa *Biology of Reproduction* **72** 1064-1070

Wu P, Leinonen J, Koivunen E, Lankinen H and Stenman U-H (2000) Identification of novel prostate-specific antigen-binding peptides modulating its enzyme activity *European Journal of Biochemistry* **267** 6212-6220

Yenugu S, Richardson RT, Sivashanmugam P, Wang Z, O'Rand MG, French FS and Hall SH (2004) Antimicrobial activity of human EPPIN, an androgen regulated sperm bound protein with a whey acidic protein motif *Biology of Reproduction* **71** 1484-1490

# Effect of immunization with six sperm peptide vaccines on fertility of female mice

R Naz and A Aleem

*Reproductive Immunology and Molecular Biology Laboratory, Department of Obstetrics and Gynecology, The West Virginia University, School of Medicine, Morgantown, West Virginia 26505, USA*

Spermatozoon is an exciting target for contraceptive vaccine development. Several sperm antigens (native or recombinant) and sperm peptides cause various degrees of contraceptive effect in female mice. No single antigen/ peptide has shown to cause a complete block in fertility in the mouse model. To enhance the efficacy of the vaccine, six sperm peptides were selected for the present study namely $mFA-1_{2-19}$, $mFA-1_{117-136}$, $YLP_{12}$, P10G, A9D and SP56. These have been shown to cause > 50% to >80% reduction in fertility when used individually for immunization. The present study was undertaken to test the hypothesis that the vaccination with all the six peptides together will enhance the contraceptive efficacy by an additive effect resulting in a complete block of fertility in the mouse model. Six vaccines were prepared by conjugating the six synthetic peptides with the recombinant binding subunit of cholera toxin (rCTB). Female CD-1 mice were immunized intramuscularly with all the six peptide vaccines. Each animal received a total of five injections at 2- to 3- week intervals of all of the six vaccines and each vaccine was injected at a separate site. Approximately four weeks after the last injection, the animals were mated. Immunization of each mouse with all six peptides resulted in a dose-dependent inhibition of fertility. At 150 μg dose, there was an overall 45% reduction compared to controls. Several mice produced antibodies ($\geq$ 2SD units) against these peptides in the serum and the genital tract but the titers were low, and many animals did not respond to several peptides. No animal produced antibodies to all six peptides in serum or the genital tract. When the antibody titers against all six peptides disappeared after > 10 months from circulation and the genital tract, all the animals regained fertility. These findings indicate that the immunization with the six sperm peptide vaccines induce antibodies in serum and the genital tract that cause a reversible long-term contraceptive effect in female mice. The inhibition in fertility was up to 45% rather than a complete block that seems to be due to low antibody titers, especially in the genital tract. It was interesting to note that even with such low titers there was a significant reduction in fertility after immunization with multipeptide vaccine. Multipeptide vaccination is an exciting approach and the present preliminary data warrant further studies.

E-mail: Rnaz@hsc.wvu.edu

*R. Naz and A. Aleem*

## Introduction

A vaccine targeting spermatozoa represents a promising approach to contraception (Naz, 2005; Naz *et al.*, 2005). Deliberate immunization of various species of animals and humans with spermatozoa or their extracts raise antisperm antibodies that cause infertility (Baskin, 1932; Edwards, 1964; Menge, 1970). However, the whole spermatozoon cannot be used for the vaccine development, since sperm have several antigens that are shared with several somatic cells. Only sperm-specific antigens can be used. Several sperm antigens have been characterized; antibodies to which affect fertilization *in vitro*, and some of them also cause infertility *in vivo* after active immunization (Goldberg, 1986; Primakoff *et al.*, 1988; Naz and Zhu, 1998). To obtain Food and Drug Administration approval and to conduct appropriate multicenter fertility trials in a quality-controlled manner, recombinant or synthetic peptide molecules are required.

Several studies have examined the effect of synthetic peptides based upon the amino acid sequences of various sperm antigens in several species of animals including rats, baboons and mice. The sperm peptides YAL-198 (SEEIPPFHPFHPFPSL) and RSMP-230 (MRISVSEGGSSGLFFSRAFSGVLNVEEY) caused a 59% and 83%, respectively, reduction in fertility of female rats (Vanage *et al.*, 1992; 1994). Sperm peptides $YLP_{12}$ peptide (YLPVGGLRRIGG) (Naz and Chauhan, 2002), P10G peptide (PGGGTLPPSG) (O'Rand *et al.*, 1993), A9D peptide (AEWGAKVED) (Lea *et al.*, 1998) and SP56 peptide (VLFGHEENSTEHAMKG) (Hardy and Mobbs, 1999) have shown to cause 70%, >80%, >50% and >80%, respectively, reduction in fertility of various strains of female mice. A contraceptive vaccine to be utilized for humans has to have almost 100% efficacy in all the recipients for it to be acceptable. No single antigen, recombinant or peptide, has induced 100% reduction in fertility in the mouse model. Various strategies such as increasing the immunogenicity to obtain high antibody titers, especially in the genital tract, and combining various sperm antigens/peptides in a single vaccine formulation are envisaged to enhance the efficacy of contraceptive vaccines. There is no study at the present time that has examined the contraceptive effect of more than one sperm antigen/peptide vaccine. The aim of the present study was to investigate the effect of multiple sperm vaccines on the antibody response and fertility using the female mouse model. The objective was to test the hypothesis: whether or not immunization with more than one sperm peptide will enhance the contraceptive efficacy by an additive effect in the mouse model. For this purpose, four peptides namely $YLP_{12}$, P10G, A9D and SP56, and two peptides based upon the cloned murine FA-1 sequence, from amino acids 2-19 and 117-136, respectively, (Zhu and Naz, 1997) were used for immunization. $YLP_{12}$, P10G, A9D, SP56 peptides and the recombinant murine FA-1 antigens have been shown to reduce (> 50% to < 80%) fertility of female mice (Table 1). The long term objective was to generate a vaccine that can cause 100% contraceptive effect in the mouse model.

**Table 1.** Immunobiological parameters of six peptides

| Peptide | Amino Acids (n) | Sequence | Carrier used | Mouse strain tested | Fertility effect | Reference |
|---------|-----------------|----------|--------------|---------------------|------------------|-----------|
| 1. $mFA-1_{2-19}$ | 18 | TEADVNPKPIPSQMPTSP | — | — | — | Zhu and Naz, 1997 |
| 2. $mFA-1_{117-136}$ | 20 | QSIQQSIERLWCRLWPLPFP | — | — | — | Zhu and Naz, 1997 |
| 3. $YLP_{12}$ | 12 | YLPVGGLRRIGG | rCTB | CD-1 | Up to 70% | Naz and Chauhan, 2002 |
| 4. P10G | 10 | PGGGTLPPSG | KLH | BALB/c | >80% | O'Rand *et al.*, 1993 |
| 5. A9D | 9 | AEWGAKVED | RNAse | CD-1/ BALB/c | >50% | Lea *et al.*, 1998 |
| 6. SP56 | 16 | YLFGHEENSTEHAMKG | KLH | BALB/c | >80% | Hardy and Mobbs, 1999 |

## Materials and methods

*Vaccine preparation*

All the six peptides (mFA-1$_{2-19}$, mFA-1$_{117-136}$, YLP$_{12}$, P10G, A9D, and SP56) were synthesized by solid-phase synthesis using Fmoc chemistry at the Biosynthesis Inc. (Lewisville, TX). Deprotection was achieved by 20% piperidine in dimethylformamide and the peptide was cleaved from the resin by 85% triflouroacetic acid (TFA). The peptide was then precipitated in methyl tert-butyl ether and purified by using reverse-phase high performance liquid chromatography. The fractions eluted with 0.5% TFA in acetonitrile were dried in a speed vacuum, redissolved in water and lyophilized. All the peptides were water-soluble and had >95% purity level. Each peptide was individually coupled to the recombinant binding subunit of cholera toxin (rCTB) (SBL Vaccin, Stockholm, Sweden) by the two-step glutaraldehyde procedure (McKenzie and Halsey, 1984). Briefly, 10 mg of each peptide was dissolved in 200 µl of phosphate buffer (0.1 M pH 6.8) and 10 µl of 25% glutaraldehyde was slowly added. The mixture was incubated overnight at room temperature in the absence of light. Then, 2 mg of rCTB was added and the mixture was incubated for another 24 h at 4°C.

*Active immunization and fertility trial*

Virgin CD-1, 10- to 12- week old female mice were immunized (designated as day zero) intramuscularly against all the six peptide vaccines. Each animal received all the six vaccines and each vaccine was injected at a separate site. Each animal received a total of five injections at 2- to 3- week intervals. Each injection consisted of 75-100 µl of PBS containing 100 µg or 150 µg of the vaccine. The animals injected with rCTB served as controls. Starting three weeks after the last injection, the animals were bled biweekly by retro-orbital puncture to collect the serum, and the vaginas were rinsed with 50 µl of PBS to collect the vaginal washings in order to examine the antibody titer. When the antibody titers peaked, the animals were mated overnight with male animals of proven fertility (two females with one male in each cage). The next morning, the mating was confirmed by the presence of a vaginal plug and the mated animals were separated until they delivered pups. The number of pups delivered by each mated animal was counted and pups were killed by $CO_2$ asphyxiation.

Fertility was defined as the mean number of pups born by the peptide vaccinated group divided by the mean number of pups born by the rCTB-injected group, multiplied by 100. Some of the animals were kept up to one year to examine the reversibility of the contraceptive effect.

*Analysis of antibodies*

The presence and titers of antibodies (IgG, IgA and IgM) were analyzed in sera and vaginal washings using an enzyme linked immunosorbent assay (ELISA). ELISA was performed as described earlier (Naz and Chauhan, 2002). Each well was coated overnight (at 4°C) with peptide (4 µg/well in 200 µl) diluted in carbonate buffer (0.1 M, pH 9.6). The wells were washed three times (for 5 minutes each) with PBS containing 0.05% Tween-20 (PBS-T). To block non-specific binding sites, the wells were incubated with PBS-T containing 1% BSA at 37°C for 45 min and washed three times (for 5 min each) with PBS-T. The wells were incubated (at 37°C for 3 hr) with serum (1:50 dilution) or vaginal washing (1:10 dilution) (200 µl/well), diluted in PBS-T containing 0.5% BSA. The wells were washed (five times) with PBS-T and then incubated (at 37°C for 1.5 h) with alkaline phosphatase conjugated anti-mouse IgG, IgA and IgM (γ-, α-, or μ-chain specific) immunoglobins (Sigma Chemical Co., St. Louis, MO, USA) diluted (1:1000) in PBS-T containing 0.5% BSA (200 μl/well). The wells were washed as before, and then incubated (37°C for 30 min) with the substrate solution (1 mg/ml disodium p-nitrophenyl phosphate diluted in 0.05 M carbonate buffer, pH 9.8). The reaction product was read at 405 nm. The absorbance readings were converted to standard deviation units by using the following formula: SD units = absorbance (test) – mean absorbance (control group)/SD of control group. The test samples with = 2 SD units were considered as having a positive reaction with a peptide. The titers of antibodies against the two FA-1 peptides were analyzed by coating both peptides together in the same well.

*Statistical analysis*

The significance of difference between the mean pups born in the vaccinated and control groups was analyzed by using unpaired Student's t-test. Correlation between the antibody titer (SD units) and fertility (number of pups born) was analyzed by linear regression. A P value of < 0.05 was considered significant.

## Results

The female mice injected with six vaccines produced antibodies to various peptides. The antibodies peaked at 3- to 4- weeks after the fourth booster injection. Taking $\geq$ 2SD units as a cutoff for positivity, in the sera of Group 1 animals that were injected with 100 μg of each vaccine: a) for IgG class, 55% of the vaccinated animals were positive for FA-1 peptides, 55% were for $YLP_{12}$ peptide, 45% were for P10G, 36% were for A9D and 36% were for SP56 peptide; b) for IgA class, 18% of the vaccinated animals were positive for FA-1 peptides, 36% were for $YLP_{12}$ peptide, 18% were for P10G, 0% were for A9D and 9% were for SP56 peptide; and c) for IgM class, 55% of the vaccinated animals were positive for FA-1 peptides, 27% were for $YLP_{12}$ peptide, 0% were for P10G, 0% were for A9D and 18% were for SP56 peptide (unpublished data). None of the sera from control animals reacted positively with any of the peptides (Figure 1). There was more IgG antibody response compared to IgA in the serum.

In the sera of the Group 2 animals injected with 150 μg of each vaccine : a) for IgG class, 75% of the vaccinated animals were positive for FA-1 peptides, 75% were for $YLP_{12}$ peptide, 67% were for P10G, 33% were for A9D and 58% were for SP56 peptide; b) for IgA class, 0% of the vaccinated animals were positive for FA-1 peptides, 42% were for $YLP_{12}$ peptide, 25% were for P10G, 0% were for A9D and 17% were for SP56 peptide; and c) for IgM class, 50% of the vaccinated animals were positive for FA-1 peptides, 17% were for $YLP_{12}$ peptide, 0% were for P10G, 0% were for A9D and 17% were for SP56 peptide (unpublished data). None of the sera from control animals reacted positively with any of the peptides (Figure 2). There was more IgG antibody response compared to IgA in the serum.

The animals in the Group 2 produced a significantly higher antibody response compared to those in Group 1. Sera from more animals were positive ($\geq$ 2SD units) for each of the six peptides in the Group 2.

In the vaginal washings of Group 1 animals injected with 100 μg of each vaccine: a) for IgG class, 0% of the vaccinated animals were positive for FA-1 peptides, 36% were for $YLP_{12}$ peptide, 27% were for P10G, 0% were for A9D and 55% were for SP56 peptide; and b) for IgA class, 9% of the vaccinated animals were positive for FA-1 peptides, 36% were for $YLP_{12}$ peptide, 0% were for P10G, 18% were for A9D and 55% were for SP56 peptide. None of the vaginal washings from control animals reacted positively with any of the peptides (Figure 3). There was more IgA antibody response compared to IgG in the vaginal washings.

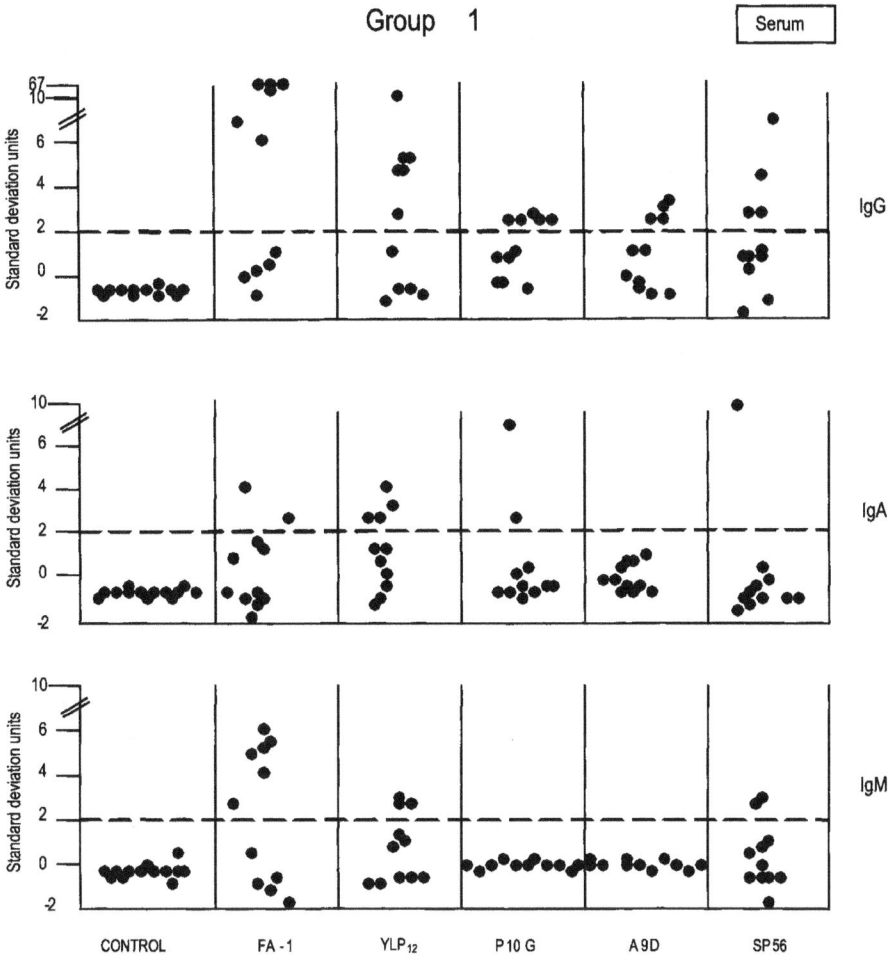

**Fig. 1** Antibody titers against FA-1, YLP$_{12}$, P10G, A9D and SP56 peptides in sera of Group 1 mice immunized with 100 μg of each peptide. Controls were immunized with rCTB alone. The horizontal dotted line represents two SD units. The samples with standard deviation (SD) units ≥ +2 were considered positive.

In the vaginal washings Group 2 animals injected with 150 μg of each vaccine: a) for IgG class, 17% of the vaccinated animals were positive for FA-1 peptides, 42% were for YLP$_{12}$ peptide, 8% were for P10G, 17% were for A9D and 42% were for SP56 peptide; and b) for IgA class, 33% of the vaccinated animals were positive for FA-1 peptides, 33% were for YLP$_{12}$ peptide, 8% were for P10G, 0% were for A9D and 58% were for SP56 peptide (unpublihsed data). None of the vaginal washings from control animals reacted positively with any of the peptides (Figure 4). There was more IgA antibody response compared to IgG in the vaginal washings.

There were no significant differences in the antibody responses (IgG and IgA) in vaginal washings between Group 1 and Group 2 mice for any of the six peptides. There was no significant linear correlations between the antibody (IgG and IgA) titers in sera with those in the

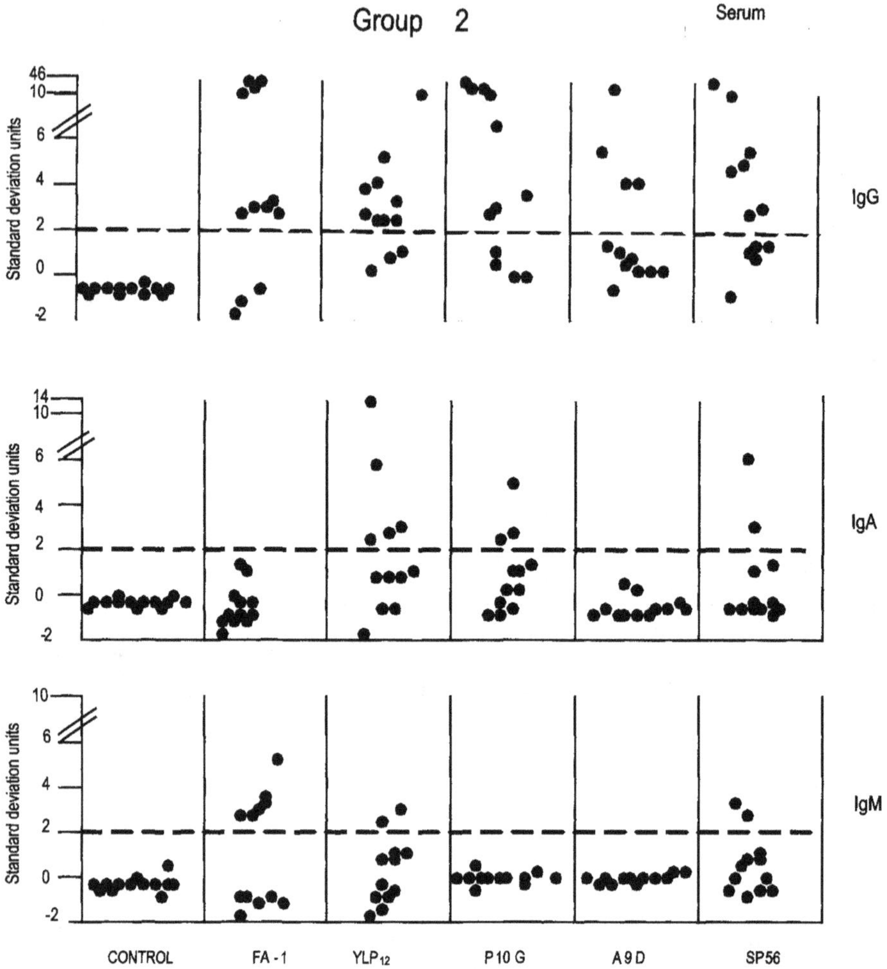

**Fig. 2** Antibody titers against FA-1, YLP$_{12}$, P10G, A9D and SP56 peptides in sera of Group 2 mice immunized with 150 μg of each peptide. Controls were immunized with rCTB alone. The horizontal dotted line represents two SD units. The samples with standard deviation (SD) units ≥ +2 were considered positive.

vaginal washings for any of the six peptides. Although some of the animals having high titers in sera also showed high titers in vaginal washings for some of the peptides, most of the animals showed no correlation. The peptides P10G and A9D were the least immunogenic among the six peptides.

When the animals were mated at 110-120 days after the first injection, they were in a contraceptive state and delivered significantly less pups than the control animals injected with rCTB alone (Table 2). In Group 1, there was a 25.8% reduction in fertility and in Group 2, there was a 45.4% reduction compared to the control. There were no significant linear correlations between the number of pups born and the antibody titers (IgG, IgA, or IgM) in sera or vaginal washings against any of the six peptides. The antibody titers against all the six peptides disappeared from sera and vaginal washing after >10 months. When the animals were mated at

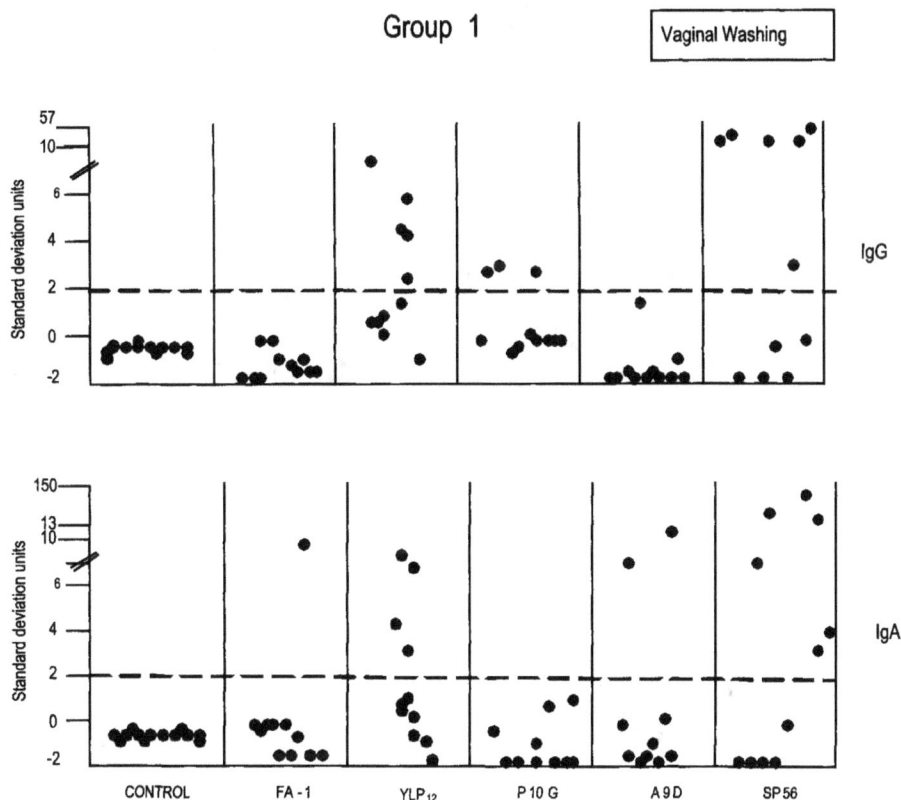

**Fig. 3** Antibody titers against FA-1, YLP$_{12}$, P10G, A9D and SP56 peptides in vaginal washings of Group 1 mice immunized with 100 μg of each peptide. Controls were immunized with rCTB alone. The horizontal dotted line represents two SD units. The samples with standard deviation (SD) units ≥ +2 were considered positive.

that time, the contraceptive effect was gone and the vaccinated animals in both Group 1 and 2 delivered the same number of pups as the control group.

## Discussion

Vaccines based upon four of the six peptides (YLP$_{12}$, P10G, A9D and SP56) conjugated to various carriers (rCTB/KLH/RNAse) have been shown to cause > 50% to >80% reduction in fertility of mice (CD-1 or BALB/c strains) (O'Rand et al., 1993; Lea et al., 1998; Hardy and Mobbs, 1999; Naz and Chauhan, 2002). In all these studies (O'Rand et al., 1993; Lea et al., 1998; Hardy and Mobbs, 1999), except YLP$_{12}$ peptide vaccine (Naz and Chauhan, 2002), Freund's adjuvant was used for immunization even after conjugation of the peptide with the carrier. The remaining two peptides (mFA-1$_{2-19}$ and mFA-1$_{117-136}$) were based upon the sequence of mFA-1 (Zhu and Naz, 1997). The whole recombinant mFA-1 molecule causes a reduction (>70%) in the fertility of mice (B6D2F1/J, the strain tested) using Freund's adjuvant (Naz and Zhu, 1998). The two FA-1 peptides have not been tested for their immunocontraceptive potentials before.

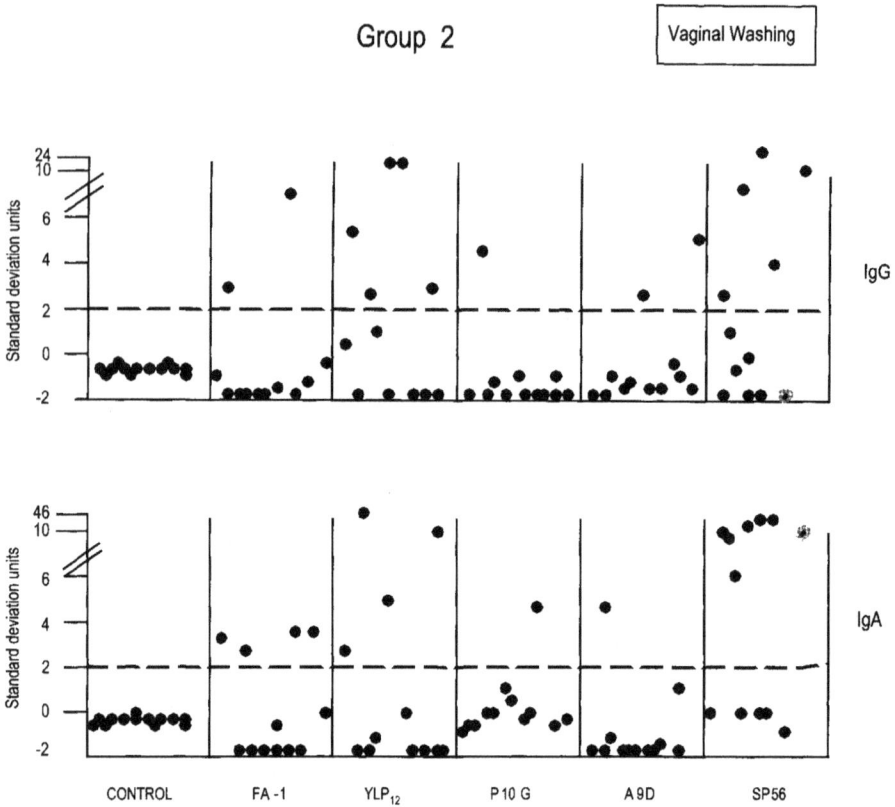

**Fig. 4** Antibody titers against FA-1, YLP$_{12}$, P10G, A9D and SP56 peptides in vaginal washing of Group 2 mice immunized with 150 µg of each peptide. Controls were immunized with rCTB alone. The horizontal dotted line represents two SD units. The samples with standard deviation (SD) units $\geq$ +2 were considered positive.

**Table 2.** Fertility of female mice after immunizations with six peptide vaccines

| Group | Immunization dose of each peptide vaccine | At 110-120 days | | | After > 10 months | | |
|---|---|---|---|---|---|---|---|
| | | Mice (n) | Pups born (n) | Pups born/animal (mean ± SD) | Mice (n) | Pups born (n) | Pups born/animal (mean ± SD) |
| Group I | 100 µg | 11 | 133 | 12.1 ± 3.3** | 9 | 67 | 7.4 ± 3.0** |
| Group II | 150 µg | 12 | 107 | 8.9 ± 4.7* | 7 | 57 | 8.1 ± 1.9** |
| Control | rCTB alone | 13 | 212 | 16.3 ± 2.3 | 12 | 97 | 8.1 ± 3.05 |

*P < 0.001; **P > 0.05, significantly different from respective control group

CTB acts as a carrier as well as an adjuvant. It has been successfully used to enhance mucosal IgA and systemic IgG without the use of an exogenous adjuvant (McKenzie and Halsey, 1984; Russel et al., 1996; Wu and Russel, 1998). Previously, our laboratory prepared a vaccine by conjugating YLP$_{12}$ peptide with rCTB and tested it in CD-1 female mice (Naz and Chauhan,

2002). The $YLP_{12}$-rCTB vaccine produced very high antibody titers against dodecamer $YLP_{12}$, both in serum as well as in the genital tract, that caused up to 70% reduction in fertility. No additional adjuvant was administered with the vaccine. It was rationalized that if we conjugate the other peptides with rCTB, similar results would be obtained as related to their immunogenicity. The contraceptive efficacy would be enhanced by an additive effect causing almost a complete block in the vaccinated animals.

The findings of the present study indicate that immunization with the multiple peptides conjugated to rCTB without any additional adjuvant did not cause a complete block in fertility in the vaccinated animals. There was a dose-dependent inhibition in fertility. At the 150 µg dose, there was a 45% reduction in the number of pups born compared to controls. The reason for only 45% reduction, instead of a complete block, may be due to one or all of the following reasons:

1) It is possible that in this study, we used a different strain of mice than that has been previously used for these peptides. For the $YLP_{12}$ and A9D peptides, the CD1 strain of mice that has been used previously is the same as used in the present study. But for the PG10G and SP56 peptides, the BALB/c strain has been used, and for the mFA-1 recombinant protein, the B6D2F1/J strain of mice has been used previously. Both of these are different strains than what was used in the present study. Strain-dependent variations in the immunobiological effect of a sperm vaccine have been reported before (Lea *et al.*, 1998).

2) It is possible that the antibody titers were insufficient, especially in the genital tract, against all the peptides after vaccination to completely block fertility. This may be due to the facts that: a) we did not use any adjuvant, such as Freund's adjuvant, for immunization as has been previously used for many of these peptides to enhance antibody titers; and/or b) there may be peptide-induced immunodominance and/or carrier-induced immunosuppression. Antibody responses in the genital tract were very poor as one can see in Figures 3 and 4. No animal produced antibodies ($\geq$ 2SD units) in serum or the genital tract against all the six peptides. If one compares the antibody response against $YLP_{12}$ peptide in serum/genital tract between the present study and the previous report (Naz and Chauhan, 2002), one can see that the antibody titers in the present study are very low compared to what were obtained in the previous study. The variation in the antibody titers was not due to the type of carrier used for conjugation, the dose of the vaccine injected, or strain of the mice used for immunization. In both these studies, $YLP_{12}$ was conjugated to rCTB, the same dose was used for injections and the same strain of mice, CD1, was employed for immunization. These findings strongly indicate that the low titers obtained after immunization with the six peptides may be due to peptide-induced immunodominance and/or carrier-induced immunosuppression. Carrier-induced immunosuppression has been reported for many vaccines including birth control vaccine based upon human chorionic gonadotropin (ß-hCG) (Gaur *et al.*, 1990; Naz *et al.*, 2005).

3) It may be due to the inherent nature of the mouse model in which it is difficult to achieve a complete block of fertility.

The antibody titers against all the peptides disappeared both from the circulation as well as from the genital tract in > 10 months post-immunization. The contraceptive effect disappeared and the fertility was regained. These findings also indirectly indicate that none of the six peptides cause any irreversible immunopathological damage to the ovaries of the vaccinated animals.

In conclusion, both the immunization with the six sperm peptide vaccines induced antibodies in serum and genital tract that caused a reversible long-term contraceptive effect in female mice. The inhibition in fertility was up to 45% rather than a complete block which seems to be due to low antibody titers, especially in the genital tract. Presently, we are investigating different strategies to enhance the immunogenicity and efficacy of the multiepitope vaccines. However, it was interesting to observe that even with these low titers, there was a 45% reduction after immunization with the six peptide vaccines. These findings warrant further research on the multiepitope vaccines.

## Acknowledgement

We thank Ashleigh Pegg, Krista Robrecht and Maegan Cook for help in typing the manuscript. This work was supported by NIH grant HD24425 to RKN.

## References

**Baskin MJ** (1932) Temporary sterilization by injection of human spermatozoa: a preliminary report *American Journal of Obstetrics and Gynecology* 24 892-897

**Edwards RG** (1964) Immunological control of fertility in female mice *Nature* 203 50-53

**Gaur A, Arunan K, Singh O and Talwar GP** (1990) Bypass by an alternate 'carrier' of acquired unresponsiveness to hCG upon repeated immunization with tetanus conjugated vaccine *International Immunology* 2 151-155

**Goldberg E** (1986) Sperm specific lactate dehydrogenase and development of contraceptive vaccine. In *Reproductive Immunology*, pp 137-142 Eds DA Clark and BA Croy. Elsevier Science Publishers, New York

**Hardy CM and Mobbs KJ** (1999) Expression of recombinant mouse sperm protein sp56 and assessment of its potential for use as an antigen in an immunocontraceptive vaccine *Molecular Reproduction and Development* 52 216-224

**Lea IA, van Lierop MJC, Widgren EE, Grootenhuis A, Wen Y, van Duin M and O'Rand MG** (1998) A chimeric sperm peptide induces antibodies and strain-specific reversible infertility in mice *Biology of Reproduction* 59 527-536

**Mckenzie SJ and Halsey JF** (1984) Cholera toxin B subunit as a carrier protein to stimulate a mucosal immune response *Journal of Immunology* 133 1818-1824

**Menge AC** (1970) Immune reactions and infertility *Journal of Reproduction and Fertility Supplement* 42 171-182

**Naz RK** (2005) Antisperm vaccine for contraception *American Journal of Reproductive Immunology* 54 378-383

**Naz RK and Chauhan SC** (2002) Human sperm-specific peptide vaccine that causes long-term reversible contraception *Biology of Reproduction* 67 674-680

**Naz RK and Zhu X** (1998) Recombinant fertilization antigen-1 causes a contraceptive effect in actively immunized mice *Biology of Reproduction* 59 1095-1100

**Naz RK, Gupta SK, Gupta JC, Vyas HK and Talwar GP** (2005) Recent advances in contraceptive vaccine development: a mini-review *Human Reproduction* 20 3271-3283

**O'Rand MG, Beavers J, Widgren E and Tung K** (1993) Inhibition of fertility in female mice by immunization with a B-cell epitope, the synthetic sperm peptide, P10G *Journal of Reproductive Immunology* 25 89-102

**Primakoff P, Lathrop W, Wollman L, Cowan A and Myles D** (1988) Fully effective contraception in male and female guinea pigs immunized with the sperm protein PH-20 *Nature* 335 543-547

**Russel MW, Moldoveanu Z, White PL, Sibert GJ, Mestecky J and Michalek SM** (1996) Salivary, nasal, genital, and systemic antibody response in monkeys immunized intranasally with a bacterial protein antigen and the cholera toxin B subunit *Infection and Immunology* 64 1272-1282

**Vanage G, Lu YA, Tam JP and Koide SS** (1992) Infertility induced in rats by immunization with synthetic peptide segments of a sperm protein *Biochemical and Biophysical Research Communications* 183 538-543

**Vanage G, Jaiswal YK, Lu YA, Tam JP, Wang LF and Koide SS** (1994) Immunization with synthetic peptide segments of a sperm protein impair fertility in rats *Research Communications in Chemical Pathology and Pharmacology* 84 3-15

**Wu HY and Russel MW** (1998) Induction of mucosal and systemic immune responses by intranasal immunization using recombinant cholera toxin B subunit as an adjuvant *Journal of Clinical Investigations* 16 286-292

**Zhu X and Naz RK** (1997) Fertilization antigen-1: cDNA cloning, testis specific expression, and immunocontraceptive effects *Proceedings of the National Academy of Sciences USA* 94 4704-4709

# Expression of a recombinant human sperm-agglutinating mini-antibody in tobacco (*Nicotiana tabacum* L.)

Bingfang Xu[1,2], Michael Copolla[2], John C Herr[2] and Michael P Timko[1]

[1]Departments of Biology and [2]Cell Biology, University of Virginia, Charlottesville, Virginia, 22904, USA

The murine monoclonal antibody (mAB) S19 recognizes an N-linked carbohydrate antigen designated sperm agglutination antigen-1 (SAGA-1) located on the membrane protein CD52. This antigen is added to the sperm surface during epididymal maturation. Binding of the S19 mAB to SAGA-1 causes the rapid agglutination of sperm and blocks pre-fertilization events. Previous studies indicated that the S19 mAB may be a potential specific spermicidal agent (termed a spermistatic) capable of replacing current spermicidal products that contain harsh detergents with harmful side effects. The nucleotide sequences encoding the heavy (H) and light (L) chains of the S19 antibody were cloned. A chimeric gene was constructed using the nucleotide sequences encoding the variable regions of both the H and L chains, and this gene (scFv19) was expressed in transgenic tobacco (*Nicotiana tabacum* L.) to produce a recombinant anti-sperm antibody (RASA). Highest levels of RASA expression were observed in BY-2 plant cell suspension cultures and regenerated *N. tabacum* cv. Xanthi plants transformant in which the RASA coding sequences were expressed under the control of the Cauliflower Mosaic Virus 35S promoter containing a double-enhancer sequence (2X CaMV 35S). Subsequent modifications of the transgene including the addition of a 5′-untranslated sequence from the tobacco etch virus (TEV leader sequence), N-terminal fusion of the coding region with an endoplasmic reticulum targeting signal of patatin (pat) and C-terminal fusion with the endoplasmic reticulum retention signal peptide KDEL showed further enhancement of RASA expression. The plant-expressed RASA formed intrachain disulfide bonds and was primarily soluble in the cytoplasmic fraction of the cells. Introduction of a poly-histidine (6xHIS) tag in the recombinant RASA protein allowed for rapid purification of the recombinant protein using Ni-NTA chromatography. Optimization of scale-up production and purification of this plant-derived recombinant protein should provide large quantities of an inexpensive spermistatic plantibody.

Corresponding author: Michael P. Timko
E-mail: mpt9g@virginia.edu

## Introduction

All spermicides currently marketed in the United States are formulations whose active ingredients are non-specific, non-ionic detergents. In recent years these products have come under increased scrutiny, as the potential harmful side effects of their use became known. In particular, use of one of the most common ingredients, nonoxynol-9 (N-9), has been correlated with an increased incidence of urogenital infections, cervicovaginal inflammation and epithelial changes in women repeatedly using this substance for birth control (Gupta *et al.*, 2005). Additionally, the use of detergent-based spermicides may increase a woman's risk of HIV transmission. As an alternative to detergent-based spermicidal products, the use of sperm-reactive monoclonal antibodies in topical spermicides has been suggested as an attractive and safe alternative (Cone and Whaley, 1994; Castle *et al.*, 1997). In fact, the contraceptive potential of sperm-agglutinating monoclonal antibodies has been demonstrated in a rabbit model (Castle *et al.*, 1997).

S19 is a murine IgG1 monoclonal antibody (mAb) that recognizes a male reproductive tract-specific carbohydrate epitope on sperm agglutination antigen-1 (SAGA-1), a highly acidic glycoprotein localized on the entire surface of ejaculated human spermatozoa (Diekman *et al.*, 1997; 1999). The binding of the S19 antibody to SAGA-1 causes the rapid agglutination of sperm, blocks pre-fertilization events including cervical mucus penetration and sperm-egg interaction, and has a complement-dependent cytotoxic effect on spermatozoa (Diekman *et al.*, 1997; 1999). In an attempt to produce S19 antibody economically, a recombinant single-chain antibody (scFv) was constructed in which the heavy (VH) and light (VL) chain variable domains of S19 were tethered by a flexible polypeptide linker (Norton *et al.*, 2001). Functional expression of the scFv S19 in bacteria demonstrated that the recombinant anti-sperm antibody (dubbed RASA) retains the binding affinity of the S19 mAb (Norton *et al.*, 2001). Levels of expression and recovery of functional RASA were low; therefore, alternative platforms for its production were explored.

The initial reports describing the successful expression of antibodies in transgenic plants occurred in the late 1980s and early 1990s (Hiatt *et al.*, 1989; During *et al.*, 1990). Since that time, the combination of two rapidly advancing technologies (immunology and plant genetic engineering) has resulted in the expression of a diverse range of antibodies (often referred to as plantibodies) in numerous vascular plant and green algal species (Fischer *et al.*, 1999a; 1999b; 2003; Vaquero *et al.*, 1999; 2002; Fischer and Emans, 2000; Giddings *et al.*, 2000; Smith and Glick, 2000; Stoger *et al.* 2000; 2002; Daniell *et al.*, 2001; Larrick and Thomas, 2001; Larrick *et al.*, 2001; Peeters *et al.*, 2001; Teli and Timko, 2004). Based on published findings, transgenic plants and/or plant cell cultures offer a number of advantages for the production of antibodies over extraction from human or animal fluids/tissues, use of recombinant microbes, transfected animal cell lines or transgenic animals. These advantages include (i) low cost of production; (ii) ease of scaling up or down to meet market demand; and (iii) freedom from possible contamination with associated blood-borne pathogens (Smith, 1996; Smith and Glick, 2000; Sharp and Doran, 2001a, b). In addition, transgenic plant systems allow for the expression of heavy and light-chains and assembly of functional dimeric antibodies *in vivo* or *in vitro*, similar to mammalian secretory antibodies (Hiatt *et al.*, 1989; During *et al.*, 1990; Hein *et al.*, 1991; De Neve *et al.*, 1993; Ma *et al.*, 1995; Smith and Glick, 2000). Antigen-binding fragments (Fab), single-chain binding fragments (scFv) and functional full size antibodies can be expressed in leaves and seeds of plants without loss of binding specificity or affinity (Hiatt *et al.*, 1989; De Neve *et al.*, 1993; Ma *et al.*, 1995; 1998; Baum *et al.*, 1996; Fischer *et al.*, 1999a, b; Smith and Glick, 2000; Xu *et al.*, 2002).

Several research groups have successfully expressed active, full-length antibodies in plants by targeting the antibodies to the apoplastic space (that is, the space between adjacent plant cells) or the endoplasmic reticulum (ER) (Conrad and Fiedler, 1998; Smith and Glick, 2000; Vaquero et al., 2002). It appears that the apoplast and ER provides an environment suitable for the accurate assembly of these complex molecules.

In this study, we demonstrate expression of the human sperm agglutinating RASA in transgenic tobacco cell suspensions and regenerated whole plants. The studies presented here are a first step to the development of plant-based expression platforms for the low-cost production of antibody-based topical spermistatic contraceptives.

## Materials and methods

### Bacterial strains

The *Escherichia coli* strains DH5α and BL21DE3 were grown at 37°C in LB medium, and *Agrobacterium tumefaciens* strain EH105 was grown at 28°C in YEB medium.

### Construction of RASA containing plant expression plasmids

The RASA gene coding region (Norton et al., 2001) was used to construct a series of plant expression vectors based on the backbone of the plant binary expression vector pCAMBEL1300A (Figure 1). In these constructs, the RASA coding region was placed under the control of either the wild-type Cauliflower Mosaic Virus 35S (CaMV 35S) promoter or a modified version of the CaMV 35S promoter containing a dual enhancer region (2X CaMV 35S). The 5'-untranslated region from the tobacco etch virus (TEV leader sequence) was inserted between the end of the 2X CaMV 35S promoter and 5'-end of the coding region of the RASA gene, and the *vspB* gene terminator sequences and polyadenylation site were placed at the 3'-end of the RASA gene. Additional modifications to this basic chimeric gene construct included the C-terminal addition of the endoplasmic reticulum retention signal peptide (KDEL) or a E-epitope tag (E-tag), and N-terminal additions of the endoplasmic reticulum targeting signal of the potato tuber storage patatin gene (pat), or a signal sequence from the mouse light chain cDNA (LP), or a plant-optimized version of the signal sequence from the mouse light chain cDNA (oLP), and/or a poly-histidine (6xHis) tag.

The fidelity of each construct was confirmed by nucleotide sequencing using double-stranded plasmid DNA templates prepared by utilizing the Qiaprep Spin Plasmid Kit (Qiagen USA, Valencia, CA). Sequence analyses were performed manually using the Sequenase Version 2.0 protocols as described by the manufacturer (United States Biochemical, Cleveland, OH) or on an ABI 310 Genetic Analyzer (PE Applied Biosystems, Foster City, CA).

### Plant cell growth and transformation

Tobacco (*N. tabacum* cv. Xanthi) plants used for transformation were grown in a soil:vermiculite mixture in the greenhouse under natural lighting conditions. Cell suspension cultures of the tobacco cell line BY-2 (*N. tabacum* cv. Bright Yellow) were grown in Murashigee-Skoog (MS) medium containing 3% sucrose and 0.2 mg/L 2,4-dichlorophenoxyacetic acid, pH 5.8, at 23°C. Cultures were maintained on an orbital shaker at 130 rpm. Wild-type and transgenic plant cell suspensions were subcultured into fresh media (plus or minus antibiotics) every 7 days.

Selectable marker for cell transformation                    RASA encoding transgene
               (antibiotic resistance)

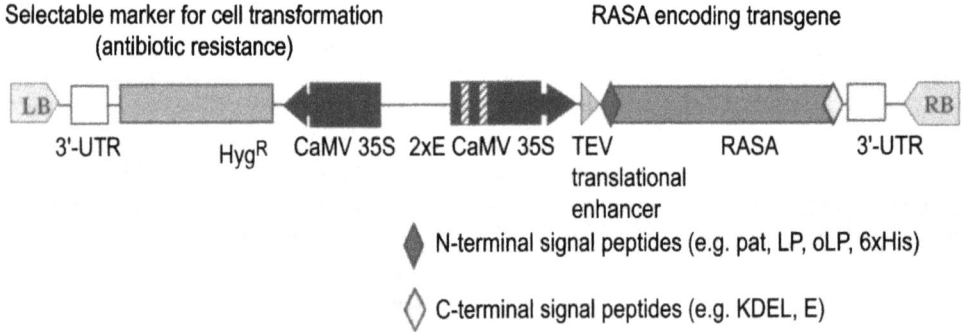

3'-UTR          Hyg$^R$   CaMV 35S  2xE CaMV 35S  TEV              RASA          3'-UTR
                                                  translational
                                                  enhancer

◆ N-terminal signal peptides (e.g. pat, LP, oLP, 6xHis)

◇ C-terminal signal peptides (e.g. KDEL, E)

**Fig 1.** The basic structure of plant cell expression vectors carrying RASA transgenes. The various plasmids used in these experiments were formed from the backbone of the plant binary expression vector pCAMBEL1300A that contains the Cauliflower Mosaic Virus 35S (CaMV 35S) promoter fused to the tobacco etch virus (TEV) translational enhancer sequence (providing the gene a 5'-untranslated region, 5'-UTR), a multicloning site, and the vegetative storage protein ß subunit (*vspB*) gene transcription terminator [providing the gene a 3'-untranslated region (3'-UTR) and polyadenylation signal]. A hygromycin resistance gene under the control of the CaMV 35S promoter is present as a selectable marker for plant cell transformation. The transgenes are bounded by the left (LB) and right (RB) border sequences of the T-DNA. The various constructs of RASA transgene are described in the text. Abbreviations are as follows: RASA, RASA coding region; 2XE CaMV 35S, CaMV 35S promoter containing a dual enhancer region; pat, leader sequence from the potato tuber storage patatin gene; LP, signal sequence from the mouse light chain cDNA; KDEL, endoplasmic reticulum (ER) retention signal; 6xHis, hexa-histidine tag; E, E-epitope antibody tag.

Plasmids encoding the various RASA transgenes were introduced into *Agrobacterium tumefaciens* using standard low-temperature transformation procedures. Leaf-disc transformation was performed as described previously by Horsch et al., (1985). Transformed cells were selected on the solid MS medium containing 50 mg/L hygromycin and 500 mg/L cefotaxime, and transgenic plantlets (T0) were regenerated as described by Horsch et al. (1985). Subsequently, T0 plantlets were transplanted and moved to the greenhouse. Young leaves were collected from transgenic plants and frozen into liquid nitrogen prior to use.

Agrobacterium-mediated transformation of BY2 cells was carried out as described by Gynheung (1985). Transformants were selected on the solid MS medium containing 50 mg/L hygromycin and 500 mg/L cefotaxime. More than one hundred individual transformed calli were picked for each transgene construct analyzed and transgenic cells were selected through at least another two rounds of replanting prior to analysis. To evaluate expression, 50 individual BY-2 microcalli expressing the same transgene construct were pooled and introduced into the liquid MS medium containing 50 mg/L hygromycin and 500 mg/L cefotaxime. The cell suspensions were grown to the logarithmic phase, harvested by vacuum filtration and the cell mass quickly frozen by immersion in liquid nitrogen.

*RNA gel blot analysis*

Total RNA was extracted from tobacco leaves and BY-2 cells using a Qiagen Plant RNeasy kit according to the manufacturer's protocol. For RNA gel blot analysis, aliquots (10 μg) of total extracted RNA was fractionated by electrophoresis through a 1.2% agarose-formaldehyde gel and blotted onto Nytran nylon membranes (Schleicher & Schuell, Keene, NH) using 10 X SSC. The transferred RNA was UV cross-linked to the membrane using a UV Stratalinker (Stratagene, La Jolla, CA) and the membranes were prehybridized in Church-Gilbert solution for 2 hours at 65°C. Hybridization was carried out in the same buffer in the presence of $^{32}$P-labeled probes for 16 hrs at 65°C. The [$^{32}$P]-dCTP -labeled probe was prepared from 25-50 ng of RASA DNA fragment by random primer labeling (Random Primed Labeling Kit, Boehringer Mannheim, Indianapolis, IN). The membranes were washed under high stringency conditions and subjected to autoradiography at - 80°C for approximately 48 hrs.

### Preparation of anti-RASA polyclonal antibodies

The general procedures for the construction of recombinant RASA in *E. coli*, purification of the recombinant protein and generation of anti-RASA polyclonal antiserum in chicken were performed as follows. Briefly, the RASA coding region (Norton *et al.*, 2001) was cloned into the expression vector pET28b (Novagen, Inc.). The resulting plasmid containing the complete RASA coding sequence and C-terminal histidine tag (His6x) was transformed into *E. coli* BL21DE3, and the recombinant protein was expressed and purified according to the manufacturer's protocol Novagen pET Expression System (Novogen, Inc.). Following purification by affinity chromatography on Ni-NTA agarose (Qiagen, Germany), the purified RASA protein was used to immunize chickens (Alpha Diagnostic International, Inc.). Sera and egg yolks from immunized chickens were collected. IgY antibodies from egg yolks were purified by Alpha Diagnostic International, Inc. according to the protocol described by Verdoliva *et al.* (2000).

### Immunoblot analysis

Frozen BY-2 cells and transgenic tobacco leaf tissues were pulverized in a mortar and pestle, and the frozen powder was extracted in phosphate buffered saline (PBS) solution. Following centrifugation, the supernatant, which contained soluble proteins, was recovered and used for Western analysis. Alternatively, ground powder was dissolved in 1 X Laemmli sample buffer (Bio-Rad, Inc.) and boiled for 5 min to extract total proteins. Protein concentrations were determined with the BCA reagent assay kit (Pierce Chemical Company, Rockford, IL) and equivalent amounts of proteins were separated by SDS-PAGE on 10 - 15% polyacrylamide gels. Following electrophoresis, the separated proteins were transferred to nitrocellulose membranes. Western hybridization was carried out using polyclonal antisera against the RASA protein generated in chickens and a secondary anti-chicken antibody. Finally, an enhanced chemiluminescence detection kit (Amersham, UK) or a DAB reagent was used to detect the signals.

### Immunofluorescence microscopy

Human spermatozoa were harvested from semen by the swim-up method. After counting the sperm concentration, appropriate numbers of sperm were spotted on slides, air-dried and fixed in 4% paraformaldehyde/PBS. Non-specific protein binding sites were blocked by incubating slides in 10% normal goat serum (NGB)/PBS. Subsequently, the positive control slides were incubated with S19 mAb (1:100 dilution in 5% NGB/PBS), followed by FITC-conjugated anti-mouse secondary antibody (Jackson ImmunoResaerch Laboratories). The test slides were incubated with plant extract (1:1 dilution in 5% NGB/PBS) either from transgenic or nontransgenic plants. The specimens were then incubated with anti-RASA antibody and finally were exposed to FITC-conjugated anti-chicken secondary antibodies (Jackson ImmunoResaerch Laboratories). For negative controls, the first step of plant protein incubation was omitted, and the slides were incubated with anti-RASA antibody and secondary antibody. All slides were mounted with Slow Fade (Molecular Probes, Eugene, OR). The preparations were visualized with a Zeiss Axioplan microscope.

## Results

*Construction design for RASA expression and expression analysis of RASA miniantibody in transgenic tobacco*

A recombinant anti-sperm antibody (RASA) was expressed in transgenic tobacco cells under the control of either the wild-type Cauliflower Mosaic Virus 35S promoter (CaMV 35S) or a variant of this promoter containing a dual enhancer sequence (2XE CaMV 35S) (Fig. 1). In addition, the effects of various N- and C-terminal modifications to the RASA coding regions were evaluated, including the C-terminal addition of the endoplasmic reticulum retention signal peptiode (KDEL) (Schouten et al., 1996; Fischer et al., 1999a, b; Stoger et al., 2000) or a E-epitope tag (E-tag), and N- terminal additions of the endoplasmic reticulum targeting signal of the potato tuber storage patatin gene (pat) (Iturriaga et al., 1989), a signal sequence from the mouse light chain cDNA (LP) (Zimmermann et al., 1998; Fischer et al. 1999a, b; Stoger et al., 2000), a plant-optimized version of the signal sequence from the mouse light chain cDNA (oLP), and/or a poly-histidine (6xHis) tag.

*Agrobacterium* strains containing the various RASA transgene expression constructs were used to transform *N. tabacum* (tobacco) cells. In the first series of experiments, BY-2 cells were transformed with the RASA transgenes, and antibiotic-resistant (hygromycin resistant) microcalli expressing the various transgenes were selected and used to establish transgenic cell suspension cultures. Following growth of the cell suspension cultures to the logarithmic phase, the BY-2 cells were harvested and the presence of RASA mRNA and protein were analyzed. In the second series of experiments, leaf-disc transformation was carried out using explants prepared from *N. tabacum* cv. Xanthi plants. For each construct 10 or more independently derived transgenic plantlets (T0 ) expressing each of the different RASA constructs were selected, grown to maturity, and RASA mRNA and protein levels were analyzed in leaf tissues. In both series of experiments, the presence of the RASA transgene in transformed tobacco calli and regenerated plants was confirmed by genomic DNA gel blot analysis and/or PCR-based amplification using a RASA coding region-specific probe or gene specific oligonucleotide primers, respectively (data not shown).

The mRNA encoding RASA can be easily detected in total RNA extracts prepared from transformed BY-2 microcalli (Fig. 2) and the leaves of transgenic tobacco plants (data not shown) by RNA gel blot analysis using a RASA coding region-specific probe. The highest levels of mRNA expression were observed in transgenic BY-2 microcalli in which the RASA transgene was expressed under the control of the 2X CaMV 35S promoter and in which the RASA coding region was fused at the N-terminus with the patatin ER targeting sequence (pat) and the C-terminus with the KDEL ER retention signal (Fig. 2, lane 2PK). The combination of CaMV 35S promoter with leader peptide (LP) at the N-terminus of RASA, and/or KDEL ER retention signal at the C terminus of RASA also provided relative high RNA signals (Fig. 2, lanes LK and L). Transgenic cell suspensions expressing RASA constructs with only the E-epitope tag or 6xHis tag (Fig. 2, lanes E and HK) or KDEL ER retention signal alone (Fig. 2, lane K) showed significantly lower levels of RASA mRNA accumulation. Similar results were observed by analysis of RASA mRNA expression in leaves of transformed tobacco plants (data not shown). In contrast to

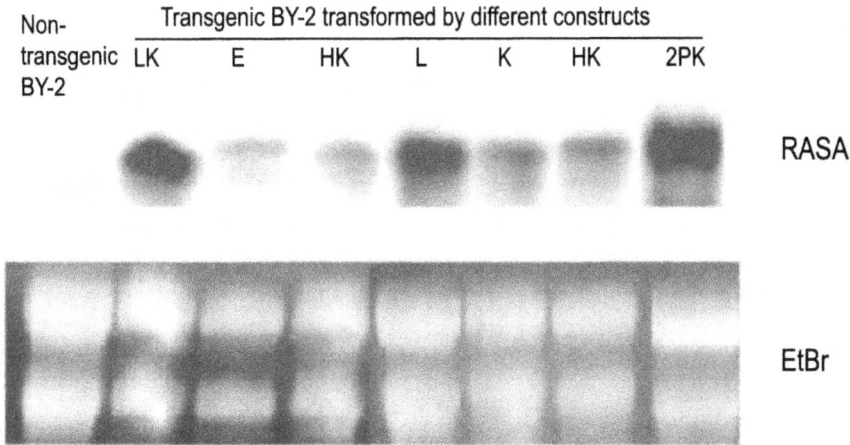

**Fig 2.** RNA gel blot analysis of RASA mRNA accumulation in transgenic tobacco cells. RNA gel blot analysis of RASA mRNA levels in transgenic BY-2 cells was performed as described in the Materials and Methods using a $^{32}$P-labeled RASA coding region fragment as probe. Abbreviations of transgene constructs: LK (CaMV 35S-LP-RASA-KDEL); E (CaMV 35S-RASA-E); HK (CaMV 35S-6xHis-RASA-KDEL); L (CaMV 35S-LP-RASA); K (CaMV 35S-RASA-KDEL); 2PK (2XE CaMV 35S-Pat-RASA-KDEL).

the ease by which RASA mRNA could be detected in transgenic BY-2 cells and the leaves of transformed tobacco plants, detection of RASA protein was highly dependent on the nature of the transgene construct. To determine whether the RASA protein also accumulated in the various transgenic tobacco cell lines and plants, polyclonal IgY antiserum capable of recognizing the purified RASA protein was prepared from immunized chickens, and immunoblot assays were carried out. As shown in Fig. 3A, the anti-RASA polyclonal IgY antiserum was capable of easily recognizing recombinant RASA generated in *E. coli* at dilutions as high as 1:20,000 (unpublished data). As shown in Fig. 3B and 3C, respectively, a 29 kDa RASA protein was observed in extracts of transgenic BY-2 cells and the leaves of transgenic tobacco plants. The highest levels of RASA were observed in transgenic BY-2 cells and the leaves of transgenic tobacco plants with constructs containing the 2X CaMV 35S promoter and in which the coding region of the RASA protein was fused with the pat signal sequence and KDEL ER retention signal.

*RASA is present in the cytosolic fraction of the plant cell*

To examine the intracellular localization of the RASA protein in plant cells, young leaves of transgenic tobacco plants expressing the pat-RASA-KDEL or pat-6xHis-RASA-KDEL constructs under the control of the 2X CaMV 35S promoter were harvested, frozen in liquid nitrogen, and pulverized in PBS buffer either in the presence or absence of 1% (v/v) Triton X-100. Immunoblot analysis was then carried out using anti-RASA polyclonal IgY antiserum to determine whether the RASA protein fractionated with the soluble or insoluble fraction. As shown in Fig. 4A, greater than 50% of the immunodetectable RASA protein was recovered in the soluble cell fraction (unpublished data). Recovery of the RASA protein in the soluble (supernatant) fraction was not dependent upon inclusion of detergent in the extraction buffer (data not shown), indicating that the majority of the expressed protein was likely localized to be within the cytosol.

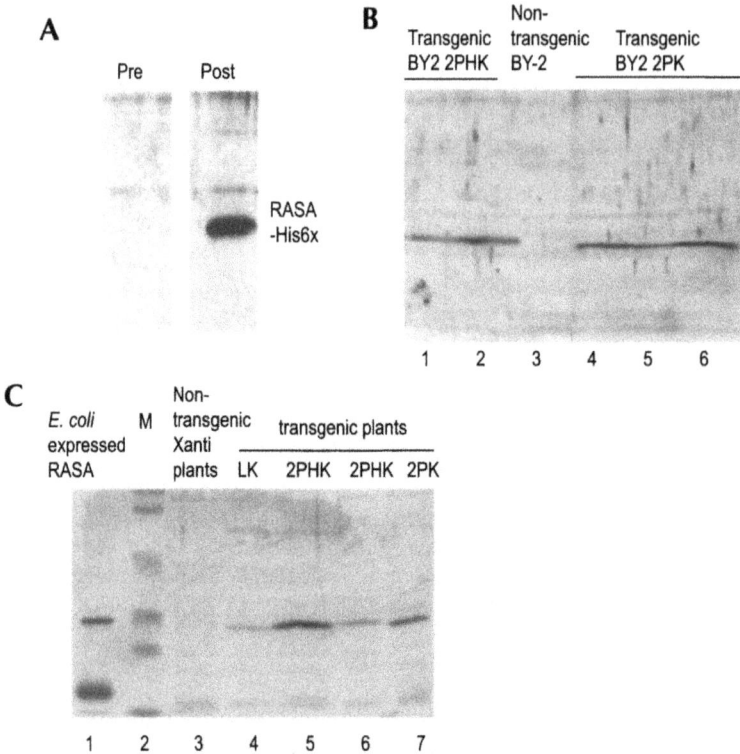

Fig 3. Immunoblot analyis of RASA protein levels in transgenic BY-2 cells and leaf tissues of transgenic *N. tabacum* cv. Xanthii plants. (A) Immunoblot analysis demonstrating the specificity of the anti-RASA antibody to 6xHis-tagged RASA generated in *E. coli*. Working dilution was 1:10,000. (B) BY-2 cell suspensions expressing various RASA transgene constructs; construct 2PHK (2XE CaMV 35S-Pat-6xHis-RASA-KDEL, lane 1 and 2), or 2PK (2XE CaMV 35S-Pat-RASA-KDEL, lane 4, 5 and 6). (C) Transgenic *N. tabacum* cv. Xanthi plants expressing various RASA transgene constructs; construct LK (CaMV 35S-LP-RASA-KDEL, lane 4), 2PHK (2XE CaMV 35S-Pat-6xHis-RASA-KDEL, lane 5 and 6), 2PK (2XE CaMV 35S-Pat-RASA-KDEL, lane 7).

In order to demonstrate that the overexpressed RASA protein could be easily purified from leaf tissue, young leaves of transgenic tobacco plants expressing the pat-6xHis-RASA-KDEL construct were harvested, frozen in liquid nitrogen, and pulverized in PBS buffer either in the presence or absence of 1% (v/v) Triton X-100. The soluble extracts were passed over a Ni-NTA column and the histidine-tagged RASA protein was recovered (Fig. 4B).

*Functional properties of plant-expressed RASA*

We previously reported that recombinant RASA expressed in *E. coli* was capable of forming monomers, dimers and trimers despite the fact that the RASA protein was engineered to contain only a single antigen-binding site (Norton *et al.*, 2001). ScFv fragments normally contain two intrachain disulphide bridges, one in the $V_H$ domain and the other in the $V_L$ domain, and the presence of these disulfide bridges can be readily ascertained by changes in protein mobility during electrophoresis in the presence and absence of sulfhydryl reagents (Tavladoraki *et al.*, 1999). To determine whether plant-expressed RASA formed intra- and intermolecular disul-

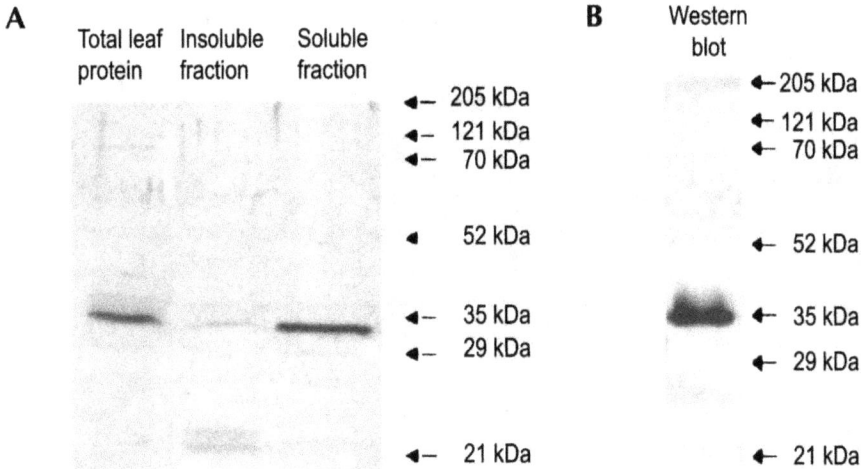

**Fig 4.** Cellular localization of plant-expressed RASA and purification of His-tagged RASA. (A) Cellular localization of plant -expressed RASA. Leaf tissue taken from transgenic tobacco plants expressing 2PK (2XE CaMV 35S-Pat-RASA-KDEL) transgene was frozen in liquid nitrogen, ground to a fine powder and extracted with PBS buffer. The extract was clarified by centrifugation, and the supernatant fraction (soluble fraction) and pellet (insoluble fraction) were dissolved in SDS-PAGE sample buffer. Alternatively, ground powder was dissolved in SDS-PAGE sample buffer directly for total protein extraction (total leaf protein). Immunoblot analysis was carried out using anti-RASA serum as described in the Materials and Methods. (B) Purification of His-tagged RASA. Leaf soluble proteins were extracted from transgenic tobacco plants (2PHK, 2XE CaMV 35S-Pat-6xHis-RASA-KDEL) expressing the RASA protein fused to an N-terminal 6xHis tag. The clarified soluble leaf extracts were passed over a Ni-NTA column and the bound RASA protein eluted. The eluted protein was examined by immunoblotting with the anti-RASA antibody.

fide bridges, RASA protein was extracted from the leaves of transgenic tobacco plants, which were expressing transgene pat-RASA-KDEL, using PBS buffers devoid of any sulfhydryl reducing agents. The plant-expressed RASA existed primarily as a monomer, with no indication of any higher molecular weight multimers (data not shown). The electrophoretic mobility of the plant-expressed RASA was slightly faster in the absence of added reducing agent, indicating that the protein contained intrachain disulphide bonds.

To determine whether the plant-expressed RASA possessed the same specificity as the bacteria-expressed RASA and native S19 mAb, His-tagged RASA was extracted from the leaves of transgenic tobacco plants and analyzed for its ability to specifically bind human spermatozoa using an indirect immunofluorescence assay (Norton et al., 2001). As shown in Fig. 5, native S19 mAb bound the entire surface of the human sperm in a punctate (uneven) fluorescence pattern (unpublished observations). RASA-containing plant extracts produced a similar fluorescence pattern, although the intensity of the signal was not as high as that observed with the S19 mAb. Extracts from untransformed controls showed no specific signal.

## Discussion

In the studies described above, we demonstrate that it is possible to express recombinant human sperm-agglutinating mini-antibody (RASA) in transgenic tobacco cell suspensions and

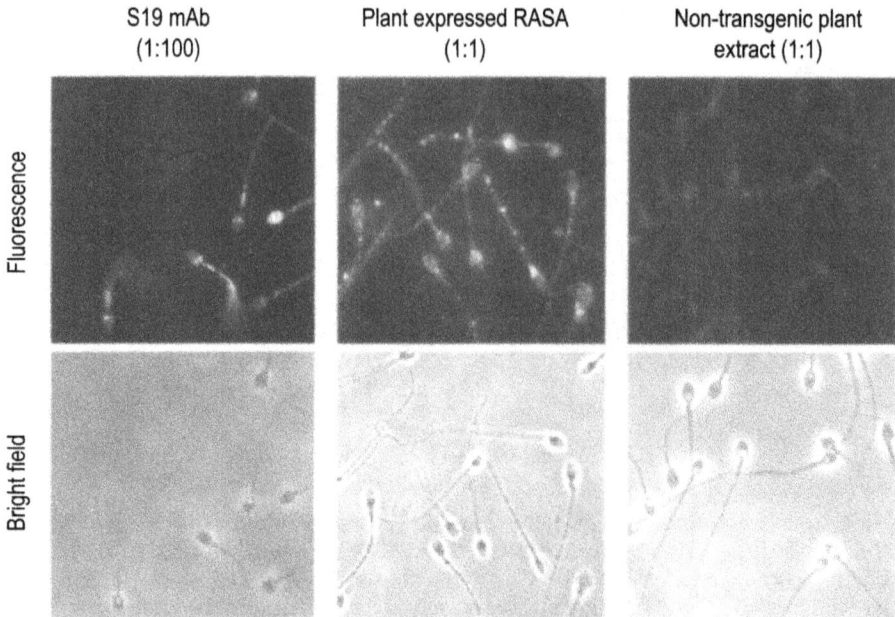

**Fig 5.** Indirect immunofluorescent staining of human spermatozoa with the S19 mAb and plant expressed RASA. Paraformaldehyde-fixed, air dried spermatozoa were incubated with (left panels) the S19 mAb (1:100), (middle pannels) RASA plantibody (1:1) or (right panels) nontransgenic plant extract (1:1). The upper panels are immunofluorescence images; the lower panels represent phase contrast images of the same fields.

regenerate whole plants and recover active antibody that retains its recognition specificity. These results are consistent with our previous work demonstrating the successful expression and recovery of recombinant RASA from bacteria (Norton et al., 2001) and with previous studies demonstrating the utility of plant-based expression platforms for the generation of useful quantities of recombinant antibodies (Fischer et al., 1999a, b; Fischer and Emans, 2000; Smith and Glick, 2000; Stoger et al., 2000; 2002; Daniell et al., 2001; Larrrick and Thomas, 2001; Larrick et al., 2001; Peeters et al., 2001; Teli and Timko, 2004). Similar to previous studies, we found that both the nature of the transcriptional regulatory sequences used to express the transgene, as well as the nature of the transgene construct influenced antibody production.

In our studies, RASA expression was most efficient when the 2XE CaMV 35S promoter was used to drive transgene expression. Inclusion of an ER targeting signal from the patatin gene at the N-terminus of the RASA protein and the tetrapeptide "KDEL" ER retention signal at the C-terminus gave the highest levels of immunoreactive protein in both BY-2 cells and transgenic tobacco leaves. There are several studies showing that the highest transgenic protein expression was established when the tetrapeptide "KDEL" ER retention signal was added at the C-terminus of the protein, and an apoplastic targeting signal, such as the leader peptide from mouse light chain (LP), was added at the N terminus (Conrad and Fiedler, 1998; Fisher et al., 1999b; Smith and Glick, 2000; Vaquero et al., 2002). In our case, the combination of 2XE CaMV 35S promoter, LP peptide and KDEL gave little detectable RASA protein expression (data not shown). In contrast, the successful new combination of 2XE CaMV 35S promoter, pat leader peptide and KDEL may offer opportunities for wider application with other antibodies.

Potential differences in activity and specificity might exist between native antibodies and recombinant antibodies produced in plant-based systems (Ko *et al.*, 2003). Such differences could arise because of subtle differences in antibody structure caused by small changes in secondary modifications to the protein. The plant-expressed RASA protein is soluble and capable of forming intrachain disulfide bonds, similar to the native S19 antibody and recombinant RASA expressed in *E. coli* (Norton *et al.*, 2001).

Previous studies have noted that subtle changes in glycosylation occur when recombinant proteins are expressed in plant versus animal cells. Cabanes-Macheteau *et al.* (1999) analyzed the glycosylation patterns of a Guy's 13 monoclonal antibody produced in transgenic tobacco plants and IgG1 of murine origin. The number of Guy's 13 glycoforms found in the transgenic plants was higher than that found in the mammalian cells. In addition to high-mannose-type N-glycans, 60% of the oligosaccharides N-linked to the plantibody have beta (1,2)-xylose and alpha (1,3)-fucose residues linked to the core Man-3-GlcNAc-2. Since these linkages are not found in mammalian N-linked glycans, they are potentially immunogenic. Thus, these plantibodies have potential toxicity in humans. Bakker *et al.* (2001) reported the stable expression of human ß-1,4-galactosyl transferase in tobacco plants. This enzyme is responsible for the conversion of typical plant N-glycans into mammalian-like N-glycans. Crossing a tobacco plant expressing human ß-1,4-galactosyl transferase with a plant expressing the heavy and light-chains of a mouse antibody resulted in the expression of a plantibody that exhibits partially galactosylated N-glycan (50%). This level of carbohydrate incorporation is approximately that obtained when the same antibody is produced by hybridoma cells.

In plant cells, high mannose-type glycosylation may be favored by the addition of a C-terminal KDEL sequence and a N-terminal pat leader, which subsequently target the plantibodies to the proximal endoplasmic reticulum. The KDEL sequence is not cleaved during processing and the potential of a KDEL tag to alter immunogenicity, pharmacokinetics *et cetera* is not known (Larrick *et al.*, 2001). In our work, no evidence was found indicating that the specificity of RASA was altered due to the presence of the KDEL sequence, and the plant-expressed RASA was shown to bind the entire human sperm surface in immunoflurorescence assays, similar to that observed for the native S19 mAb.

There is currently little information available about the effects of development stage and physiological changes on the yield and quality of antibodies produced in plants. Stevens *et al.* (2000) showed that the levels of a mouse IgG1 antibody heterologously expressed in tobacco paralleled total soluble protein levels and that the ratio of the heterologous IgG1 and total soluble leaf protein was constant throughout the development of the leaf. Although proteolytic degradation can be a serious obstacle for the production of antibodies in plant by negatively affecting product homogeneity (Stevens *et al.*, 2000), we found no evidence for proteolytic breakdown products of the RASA protein in our experiments.

## References

**Bakker H, Bardor M, Molthoff JW, Gomard VV, Elbers II, Stevens LH, Jordi W, Lommen A, Faye L, Lerouge P and Bosch D** (2001) Galactose-extended glycans of antibodies produced by transgenic plants *Proceedings of the National Academy of Sciences USA* **98** 2899-2904

**Baum TJ, Hiatt A, Parrott WA, Pratt LH and Hussey RS** (1996) Expression in tobacco of a functional monoclonal antibody specific to stylet of the root-knot nematode *Molecular Plant-Microbe Interactions* **9** 382-387

**Cabanes-Macheteau M, Fitchette-Laine AC, Loutelier-Bourhis C, Lange C, Vine ND, Ma JKC, Lerouge P and Faye L** (1999) N-glycosylation of a mouse IgG expressed in transgenic tobacco plants *Glycobiology* **9** 365-372

**Castle PE, Whaley KJ, Hoen TE, Moench TR and Cone RA** (1997) Contraceptive effect of sperm-agglutinating monoclonal antibodies in rabbits *Biology of Reproduction* **56** 153-159

**Cone RA and Whaley KJ** (1994) Monoclonal antibodies for reproductive health: Preventing sexual transmission of disease and pregnancy with topically applied antibodies *American Journal of Reproductive Immu-*

nology **32** 114-131

**Conrad U and Fiedler U** (1998) Compartment-specific accumulation of recombinant immunoglobulins in plant cells: an essential tool for antibody production and immunomodulation of physiological functions and pathogen activity *Plant Molecular Biology* **38** 101-109

**Daniell H, Streatfield J and Wycoff K** (2001) Medical molecular farming: production of antibodies, biopharmaceuticals and edible vaccines in plants *Trends in Plant Science* **6** 219-226

**De Neve M, De Loose M, Jacobs A, Van Houdt H, Kaulza B, Weidle U, Van Montagu M and Depicker A** (1993) Assembly of an antibody and its derived antibody fragment in *Nicotiana* and *Arabidopsis Transgenic Research* **2** 227-237

**Diekman AB, Westbrook-Case VA, Naaby-Hansen S, Klotz KL, Flickinger CJ and Herr JC** (1997) Biochemical characterization of sperm agglutination antigen-1, a human sperm surface antigen implicated in gamete interactions *Biology of Reproduction* **57** 1136-1144

**Diekman AB, Norton EJ, Klotz KL, Westbrook VA, Shibahara H, Naaby-Hansen S, Flickinger CJ and Herr JC** (1999) N-linked glycan of a sperm CD52 glycoform associated with human infertility *The FASEB Journal* **13** 1303-1313

**During K, Hippe S, Kreutzaler F and Schell J** (1990) Synthesis and self-assembly of a functional monoclonal antibody in transgenic *Nicotiana tabacum Plant Molecular Biology* **15** 281-293

**Fischer R and Emans N** (2000) Molecular farming of pharmaceutical proteins *Transgenic Research* **9** 279-299

**Fischer R, Drossard J, Emans N, Commandeur U and Hellwig S** (1999a) Towards molecular farming in the future: pichia pastoris-based production of single-chain antibody fragments *Biotechnology and Applied Biochemistry* **2** 117-120

**Fischer R, Schumann D, Zimmermann S, Drossard J, Sack M and Schillberg S** (1999b) Expression and characterization of bispecific single-chain Fv fragments produced in transgenic plants *European Journal of Biochemistry* **262** 810-816

**Fischer R, Twyman RM and Schillberg S** (2003) Production of antibodies in plants and their use for global health *Vaccine* **21** 820-825

**Giddings G, Allison G, Brooks D and Carter C** (2000) Transgenic plants as factories for biopharmaceuticals *Nature Biotechnology* **18** 1151-1155

**Gupta G, Jain RK, Maikhuri JP, Shukla PK, Kumar M, Roy AK, Patra A, Singh V and Batra S** (2005) Discovery of substituted isoxazolecarbaldehydes as potent spermicides, acrosin inhibitors and mild anti-fungal agents *Human Reproduction* **20** 2301-2308

**Gynheung A** (1985) High Efficiency transformation of cultured tobacco cells *Plant physiology* **79** 568-570

**Hein MB, Tang Y, McLeod DA, Janda KD and Hiatt A** (1991) Evaluation of immunoglobulins from plant cells *Biotechnology Progress* **7** 455-461

**Hiatt AC, Cafferkey R and Bowdish K** (1989) Production of antibodies in plants *Nature* **342** 76-78

**Horsch RB, Fry JE, Hoffman NL, Rogers SG and Fraley RT** (1985) A simple and general method for transferring genes into plants *Science* **227** 568-570

**Iturriaga G, Jefferson RA and Bevan MW** (1989) Endoplasmic reticulum targeting and glycosylation of hybrid proteins in transgenic tobacco *Plant Cell* **1** 381-393

**Ko K, Tekoah Y, Rudd PM, Harvey DJ, Dwek RA, Spitsin S, Hanlon CA, Rupprecht C, Dietzschold B, Golovkin M and Koprowski H** (2003) Function and glycosylation of plant-derived antiviral monoclonal antibody *Proceedings of the National Academy of Sciences USA* **100** 8013-8018

**Larrick JW and Thomas DW** (2001) Producing proteins in transgenic plants and animals *Current Opinion in Biotechnology* **12** 411-418

**Larrick JW, Yu L, Naftzger C, Jaiswal S and Wycoff K** (2001) Production of secretory IgA antibodies in plants *Biomolecular Engineering* **18** 87-94

**Ma JKC, Hiatt A, Hein M, Vine MD, Wang F, Stabila P, van Dolleweerd C, Mostov K and Lehner T** (1995) Generation and assembly of secretory antibodies in plants *Science* **268** 716-719

**Ma JKC, Hikmat B, Wycoff K, Vine M, Chargelegue D, Yu L, Hein M and Lehner T** (1998) Characterization of a recombinant plant monoclonal secretary antibody and preventive immunotherapy in humans *Nature Medicine* **4** 601-606

**Norton EJ, Diekman AB, Westbrook VA, Flickinger CJ and Herr JC** (2001) RASA, a recombinant single-chain variable fragment (scFv) antibody directed against the human sperm surface: implications for novel contraceptives *Human Reproduction* **16** 1854-1860

**Peeters K, De Wilde C, De Jaeger G, Angenon G and Depicker A** (2001) Production of antibodies and antibody fragments in plants *Vaccine* **19** 2756-2761

**Schouten A, Roosien J, van Engelen FA, de Jong GA, Borst-Vrenssen AW, Zilverentant JF, Bosch D, Stiekema WJ, Gommers FJ, Schots A and Bakker J** (1996) The C-terminal KDEL sequence increases the expression level of a single-chain antibody designed to be targeted to both the cytosol and the secretory pathway in transgenic tobacco *Plant Molecular Biology* **30** 781-793

**Sharp JM and Doran PM** (2001a) Characterization of monoclonal antibody fragments produced by plant cells *Biotechnology and Bioengineering* **73** 338-346

**Sharp JM and Doran PM** (2001b) Strategies for enhancing monoclonal antibody accumulation in plant cell and organ cultures *Biotechnology Progress* **17** 979-992

**Smith MD** (1996) Antibody production in plants *Biotechnology Advances* **14** 267-281

**Smith MD and Glick BR** (2000) The production of antibodies in plants: An idea whose time has come *Biotechnology Advances* **18** 85-89

**Stevens LH, Stoopen GM, Elbers IJW, Molthoff JW, Bakker HAC, Lommen A, Bosch D and Jordi W** (2000) Effect of climate conditions and plant developmental stage on the stability of antibodies expressed in transgenic tobacco *Plant Physiology* **124** 173-182

**Stoger E, Vaquero C, Torres E, Sack M, Nicholson L,**

Drossard J, Williams S, Keen D, Perrin Y, Christou P and Fischer R (2000) Cereal crops as viable production and storage systems for pharmaceutical scFv antibodies *Plant Molecular Biology* **42** 583-590

Stoger E, Sack M, Fischer R and Christou P (2002) Plantibodies: applications, advantages and bottlenecks *Current Opinion in Biotechnology* **13** 161-166

Tavladoraki P, Girotti A, Donini M, Arias FJ, Mancini C, Morea V, Chiaraluce R, Consalvi V and Benvenuto E (1999) A single-chain antibody fragment is functionally expressed in the cytoplasm of both *Escherichia coli* and transgenic plants *European Journal of Biochemistry* **262** 617-624

Teli NP and Timko MP (2004) Recent developments in the use of transgenic plants for the production of human therapeutics and biopharmaceuticals *Plant Cell Tissue and Organ Culture* **79** 125-145

Verdoliva A, Basile G and Fassina G (2000) Affinity purification of immunoglobulins from chicken egg yolk using a new synthetic ligand *Journal of Chromatography. B, Biomedical Sciences and Applications* **749** 233-242

Vaquero C, Sack M, Chandler J, Drossard J, Schuster F, Monecke M, Schillberg S and Fischer R (1999) Transient expression of a tumor-specific single chain fragment and a chimeric antibody in tobacco leaves *Proceedings of the National Academy of Sciences USA* **96** 11128-11130

Vaquero C, Sack M, Schuster F, Finnern R, Drossard J, Schumann D, Reimann A and Fischer R (2002) A carcinoembryonic antigen-specific diabody produced in tobacco *The FASEB Journal* **16** 161-182

Xu H, Montoya FU, Wang ZP, Lee JM, Reeves R, Linthicum DS and Magnuson NS (2002) Combined use of regulatory elements within the cDNA to increase the production of a soluble mouse single-chain antibody, scFv, from tobacco cell suspension cultures *Protein Expression and Purification* **24** 384-394

Zimmermann S, Schillberg S, Liao Y and Fisher R (1998) Intracellular expression of TMV-specific single-chain Fv fragments leads to improved virus resistance in *Nicotiana tabacum* *Molecular Breeding: new strategies in plant improvement* **4** 369-379

# Feasibility and challenges in the development of immunocontraceptive vaccine based on zona pellucida glycoproteins

Sangeeta Choudhury[1], Neelu Srivastava[1], PS Narwal[2,a], Archana Rath[1], Sonika Jaiswal[1] and Satish K Gupta[1,b]

[1]Gamete Antigen Laboratory, National Institute of Immunology, Aruna Asaf Ali Marg, New Delhi 110 067, INDIA; [2]Central Military Veterinary Laboratory, Meerut Cantt 250 001, UP, India; [a]Present Address: APO 42, Military Veterinary Hospital, Chandimandir, Haryana, India

The zona pellucida (ZP) glycoproteins play a crucial role during fertilization and thus are considered as important target antigens for the development of immunocontraceptive vaccines aiming to inhibit fertility at a pre-fertilization stage. In order to evaluate the immunocontraceptive potential of ZP glycoproteins, bonnet monkey (*Macaca radiata*) ZP2, ZP3 and ZP4 have been cloned and expressed using either *E. coli* or baculovirus expression systems. Active immunization studies with the recombinant ZP glycoproteins in female baboons (*Papio anubis*) and bonnet monkeys revealed curtailment of fertility. In order to minimize the ovarian pathology, synthetic peptides corresponding to B cell epitopes that are devoid of 'oophoritogenic' T cell epitopes were designed and their *in vitro* immunocontraceptive potential explored. There are several issues that need to be addressed before ZP glycoproteins based immunocontraceptive vaccines become feasible for use in humans. Nonetheless, the utility of such a vaccine is imminent for controlling wild life population. In this direction, active immunization of female non-descript dogs with recombinant canine ZP3 conjugated to diphtheria toxoid led to curtailment of fertility. Further, canine ZP3 has also been expressed in insect cells as a fusion protein with rabies virus glycoprotein G (RV-G), an antigen that is involved in providing protection against rabies. The immunogenicity of such a recombinant protein and its potential to curtail fertility was explored both in female mice and dogs. Simultaneously, DNA vaccine encoding canine ZP3 and RV-G have been made and evaluated for their immunogenicity. The results obtained so far, current shortcomings and the possible ways to circumvent these have been discussed in the present manuscript.

[b]Corresponding author
E-mail: skgupta@nii.res.in

## Introduction

Immunocontraceptive vaccines based on zona pellucida (ZP) glycoproteins, aiming to curtail fertility by acting at pre-fertilization stage, have been proposed (for review see Gupta et al., 1997a; 2004). This approach is based on the fact that the ZP glycoproteins play a critical role during fertilization (Wassarman, 1999). In mouse, the ZP is primarily composed of three bio-chemically distinct glycoproteins designated as ZP1, ZP2 and ZP3 based on their mobility in SDS-PAGE. Due to variations in the extent of glycosylation of ZP proteins from various species, varied migration profiles have been observed for the same ZP glycoprotein from different species. To avoid this ambiguity, an alternate nomenclature of ZPA (ZP2), ZPB (ZP1) and ZPC (ZP3), based on their mRNA transcript has also been proposed (Harris et al., 1994). The classi-fication of ZP glycoproteins has been further complicated by the recent documentation of the presence of a fourth protein in human eggs (Lefievre et al., 2004). The new designation for ZP glycoproteins as ZP1, ZP2, ZP3 and ZP4 (previously designated as ZP1/ZPB in non-human primates as well as human) is used in the present manuscript. Comparison of the deduced amino acid (aa) sequence of the ZP glycoproteins from various species revealed a variable degree of sequence identity (Table 1). Due to this sequence homology, antibodies generated against ZP glycoproteins of a given species recognize ZP glycoproteins of the other species and it has made heterologous immunization a feasible approach. In this regard, porcine ZP glyco-proteins have been extensively studied as candidate antigens due to the observations that antibodies against porcine ZP glycoproteins show a high degree of immunological cross-reac-tivity with the human ZP (Sacco et al., 1981). Moreover, the procedure to purify ZP glycopro-teins from native source has been standardized (Yurewicz et al., 1987) and porcine ovaries from abattoirs are also easily accessible.

**Table 1.** Sequence identity at the deduced amino acid level of the four ZP glycoproteins from various species with their respective human homologues

| *Species* | *% identity with human ZP glycoproteins* | | | |
| | *ZP1* | *ZP2* | *ZP3* | *ZP4* |
| --- | --- | --- | --- | --- |
| Mouse | 64.00 | 57.00 | 67.00 | NA |
| Rat | 66.00 | 57.00 | 68.00 | 64.00 |
| Hamster | NA | NA | 67.00 | NA |
| Dog | NA | 67.00 | 70.00 | NA |
| Fox | NA | 65.00 | 71.00 | NA |
| Furo | NA | NA | 74.00 | NA |
| Cat | NA | 67.00 | 72.00 | 63.00 |
| Cow | NA | 67.00 | 74.00 | 69.00 |
| Pig | NA | 64.00 | 74.00 | 68.00 |
| Macaque | NA | 94.20 | 93.90 | 92.00 |
| Marmoset | NA | 85.00 | 91.00 | NA |
| Rabbit | NA | 72.00 | 69.00 | 71.00 |
| Chicken | 55.00 | NA | 54.00 | 53.00 |
| Quail | 52.00 | NA | 54.00 | NA |
| Possum | NA | 53.00 | 48.00 | 55.00 |
| *Xenopus* | NA | NA | 45.00 | 44.00 |

NA, denotes respective sequences are Not Available in the protein sequence database of National Centre of Biotechnology Information (NCBI)

Active immunization studies in female rabbits, dogs and non-human primates with heat-solubi-lized porcine ZP demonstrated the efficacy of this approach to curtail fertility (Wood et al., 1981; Mahi-Brown et al., 1982; Gulyas et al., 1983). However, infertility observed in these studies is not due to inhibition at the sperm-oocyte interaction level, but due to ovarian dysfunction itself. To rule out the possibility that observed side-effects are not due to the contamination of the other ovarian associated proteins that may be present as contaminant, subsequent studies were performed with highly purified forms of porcine zona proteins. Such investigations carried out in female dogs and non-human primates further confirmed the contraceptive potential of ZP glycoproteins based immunocontraception (Sacco et al., 1987; Mahi-Brown et al., 1985; 1988; Bagavant et al., 1994). In addition, these studies demonstrated that the use of the purified ZP glycoproteins as immunogens decreased the side effects on ovarian functions but were not eliminated altogether (Sacco et al., 1987; Mahi-Brown et al., 1985; 1988). The apprehension that immunization with zona proteins may lead to disturbances in the follicular development or autoimmune oophoritis hampered further progress on their use as candidate antigens with respect to the development of immunocontraceptive vaccines meant for human use. In order to assess further the contraceptive potential of zona proteins and in depth analysis of the autoimmune side-effects, our group has developed a homologous non-human primate model (Govind et al., 2002).

### Non-human primate model to assess immunocontraceptive potential and safety of zona proteins based contraceptive vaccine

The cDNA clones encoding bonnet monkey (*Macaca radiata*) zona pellucida glycoprotein-2 (ZP2, 745 aa), zona pellucida glycoprotein-3 (ZP3, 424 aa), zona pellucida glycoprotein-4 (ZP4, 540 aa; previously designated as ZP1/ZPB) were isolated. The bonnet monkey recombinant ZP2, ZP3 and ZP4 excluding the N-terminal signal sequence and the C-terminal domain after furin cleavage site were expressed as polyhistidine-tag fusion proteins in *Escherichia coli* (Kaul et al., 1997; Gupta et al., 1997b; Jethanandani et al., 1998). SDS-PAGE and Western blot analysis revealed major bands of 68 kDa for ZP2, 50 kDa for ZP3 and 51 & 40 kDa for ZP4. The purified recombinant proteins were conjugated to diphtheria toxoid (DT). Female bonnet monkeys were immunized with ZP2-DT and ZP4-DT conjugates using Arlacel-A and Squalene (Sigma Chemical Co., St Louis, MO, USA) as adjuvants (Govind et al., 2002). A sodium phthlyl derivative of lipopolysaccharide (SPLPS) was also used as an additional adjuvant but only in the first injection. All the immunized animals generated good antibody response against respective recombinant protein as well as the DT used as a carrier. Immunized monkeys failed to conceive when mated with males of proven fertility. During the first year of active immunization studies, immunized animals showed normal ovulatory progesterone hormonal peak. No disturbances in the duration of the menstrual cycles were observed except during summer amenorrhea. To examine if the contraception thus achieved was reversible or not, further boosting of the immunized animals were stopped. However, immunized animals failed to conceive in spite of undetectable circulating antibodies against the respective zona proteins. Evaluation of the ovarian histopathology revealed that immunization with both ZP3-DT and ZP4-DT led to disturbances in follicular growth. The follicles were atretic with degenerated oocytes (Govind et al., 2002).

Additional studies have been carried out by various investigators in non-human primates (VandeVoort et al., 1995; Paterson et al., 1998; Govind and Gupta, 2000; Martinez and Harris, 2000). Immunization of cynomolgous monkeys (*Macaca fasicularis*) with recombinant rabbit rec55 (~ZP1) and a partial rec75 (~ZP2) conjugated to either protein-A or keyhole limpet haemocyanin (KLH) generated adequate antibodies (VandeVoort et al., 1995). Normal ovarian follicular development and hormonal profile were observed in rec55 immunized animals whereas immunization with rec75 led to disturbances in folliculogenesis. Antibodies against rec55 also inhibited *in vitro* monkey sperm-egg interaction. Immunization of female marmoset (*Callithrix jacchus*) with mammalian-expressed human recombinant ZP3 also resulted in long-term infertility associated with ovarian pathology characterized by depletion of primordial follicles (Paterson *et al.*, 1998). A reversible block in fertility has been documented in female baboons (*Papio anubis*) immunized with recombinant bonnet monkey ZP4 (described as ZPB) conjugated to DT (Govind and Gupta, 2000). In another study, female cynomolgous monkeys (*Macaca fascicularis*) and baboons (*Papio cynocephalus*) were immunized with purified human ZP2, ZP3 and ZP4 (described as ZPB) expressed in CHO cells. These studies revealed enhanced contraceptive efficacy in animals immunized with ZP4 as compared to those animals immunized with ZP2 and ZP3. The animals immunized with recombinant human ZP4 remained infertile for 9-35 months and during the time of high antibody titers, some animals experienced disruption of the menstrual cycle, which eventually returned to normal (Martinez and Harris, 2000).

## Synthetic peptides or recombinant proteins based on B-cell epitope of zona proteins as immunogens to circumvent ovarian dysfunction

It is imperative that the immunocontraceptive vaccines based on ZP glycoproteins should be safe and devoid of any ovarian dysfunction. In a series of elegant experiments, it was demonstrated that the observed changes in hormonal profiles and ovarian dysfunction associated with zona proteins immunization is in fact mediated through the 'oophoritogenic' T cell epitopes (Luo *et al.*, 1993). If synthetic or recombinant proteins are designed corresponding to critically mapped B-cell epitopes and devoid of 'oophoritogenic' T cell epitopes, it may be feasible to develop an immunocontraceptive vaccine based on zona proteins without any considerable side effects. The evidence for this principle was first demonstrated by employing a chimeric peptide comprising of a 'promiscuous' T cell epitope of bovine RNase (NCAYKTTQNK), co-linearly synthesized with the minimal B-cell epitope of mouse ZP3 corresponding to aa residues 335-342, in which phenylalanine was substituted with alanine (QAQIHGPR). Immunization of mice of eight different haplotypes with this chimeric peptide resulted in infertility without any concomitant side effects (Lou *et al.*, 1995). The observed infertility correlated with antibody titers.

Disruption of ovarian function was also not observed in female marmosets immunized with synthetic peptides corresponding to either marmoset or human ZP3 (Paterson *et al.*, 1998; 1999). The antibodies against marmoset ZP3 peptide (aa residues 301-320) also inhibited *in vitro* the binding of human sperm to human oocytes (Paterson *et al.*, 1999). However, the *in vivo* studies did not show consistent reduction in fertility (Paterson *et al.*, 1999). Immunization of female bonnet monkey with a synthetic peptide corresponding to bonnet monkey ZP3 (aa residues 324-347) conjugated with DT, however resulted in curtailment of fertility (Kaul *et al.*, 2001). The immunized animals exhibited normal ovulatory cycles without any disturbances in the menstrual cycles. Further, no ovarian pathology was observed in this group of immunized animals.

Synthetic peptide immunogens corresponding to ZP2 and ZP4 (previously described as ZP1/ZPB in bonnet monkey and human) families of zona proteins have also been identified (Hinsch *et al.*, 1998; Sun *et al.*, 1999; Govind *et al.*, 2000; Hasegawa *et al.*, 2002; Sivapurapu *et al.*, 2002). Antibodies against synthetic peptides corresponding to human ZP2 (aa residues 50-67 and 541-555) significantly inhibited *in vitro* binding of human sperm to human ZP (Hinsch *et al.*, 1998; Hasegawa *et al.*, 2002). In another study, mice immunized with a mouse ZP2 peptide (aa residues 121-140) showed no adverse effect on the ovaries (Sun *et al.*, 1999). Synthetic peptide corresponding to bonnet monkey ZP4 (aa residues 251-273) has also been identified which generates antibodies with significant *in vitro* contraceptive potential (Sivapurapu *et al.*, 2002).

Generally, synthetic peptide based immunogens have low immunogenicity. A combination of synthetic peptides can be used for immunization to enhance the immunogenicity and thereby the contraceptive efficacy of antibodies thus generated (Afzalpurkar *et al.*, 1997). The observed higher efficacy of antibodies to inhibit sperm-oocyte binding may be due to cooperative effect among the antibodies pertaining to different domains. Immunogenicity can also be enhanced by designing a chimeric synthetic peptide or recombinant antigen incorporating multiple epitopes (Sivapurapu *et al.*, 2003; 2005). To evaluate this concept, a chimeric recombinant protein encompassing the epitopes of bonnet monkey ZP2 (aa residues 86-113), ZP3 (aa residues 324-347) and ZP4 (aa residues 132-147) was designed and expressed in *E. coli* (Sivapurapu *et al.*, 2003). In another study, chimeric peptides encompassing the epitopes of bonnet monkey ZP3 (aa residues 324-347) and ZP4 (aa residues 251-273) were also synthesized (Sivapurapu *et al.*, 2005). Antibodies generated against the above chimeric recombinant protein/peptide recognized independently their constituent epitopes. Antibodies also showed a very good *in vitro* contraceptive efficacy (Sivapurapu *et al.*, 2003; 2005). It will be of interest to evaluate the *in vivo* contraceptive efficacy of the above chimeric peptide/recombinant protein.

### Logistic hurdles in using zona proteins based immunocontraceptive vaccine for application to human

Traditionally, vaccines have been used in 'herd' immunization approaches to combat infectious diseases. Immunization with any given vaccine may not result in generating immune response in 100% of the recipients. It may be acceptable with regard to immunoprophylactic vaccines meant for infectious diseases. Concerning contraceptive vaccines however, it is imperative that all recipients should respond to vaccination. Further, it should generate an adequate protective antibody response as an inadequate immune response may result in conception. It has been observed in Phase-II clinical trials with an immunocontraceptive vaccine based on ß subunit of human chorionic gonadotrophin (ß-hCG) that even the duration of protective antibody titers varies among immunized women (Talwar *et al.*, 1994). It further poses a challenge to determine the timing of the booster injections. Finally, the potential risk of ovarian dysfunction associated with ZP-based immunocontraceptive vaccines must be resolved beyond any doubt before their use for controlling fertility in human can be proposed.

### Potential of ZP glycoproteins based immunocontraceptive vaccines for controlling wild life population

Nonetheless, immunocontraceptive vaccines based on zona proteins have great potential for controlling wild life population. In this case, it is not imperative that the contraceptive vaccine elicits protective antibody response in 100% of the recipients. Further, irreversible block in fertility as observed in some studies subsequent to immunization with zona proteins (Govind *et al.*, 2002) may be desirable with respect to controlling populations of certain animal species. Our group has explored the feasibility of using a ZP glycoprotein-based immunocontraceptive vaccine for controlling a street dog population. The rationale for developing a contraceptive vaccine for controlling street dog populations is to reduce the burden of rabies which is a major zoonosis of significant public health concern in many parts of the world and is endemic among dogs (Meslin *et al.*, 1994).

*(i)  Cloning and expression of canine zona proteins*

The cDNA encoding dog ZP2 (dZP2) and ZP3 (dZP3) were cloned in prokaryotic expression vector and the respective recombinant proteins were expressed as histidine-tag fusion proteins in *E. coli* (Santhanam et al., 1998; Srivastava et al., 2002). SDS-PAGE and Western blot analysis revealed that the purified recombinant dZP2 appear as ~ 70 kDa band and dZP3 as a major band of ~ 42 kDa along with another band of ~ 32 kDa. Female dogs were immunized with *E. coli*–expressed purified recombinant dZP2 and dZP3 conjugated with DT. Immunization with the respective recombinant protein–DT conjugates resulted in the generation of high antibody titers against their respective zona proteins as well as the carrier (Srivastava et al., 2002). However, curtailment of fertility was observed only in the dZP3-DT immunized group of dogs. Female dogs immunized with either DT or dZP2-DT conjugate conceived when mated during the period of 'heat' with males of proven fertility. Ovarian histopathology revealed that the block in fertility in the group immunized with recombinant dZP3-DT, is probably manifested by inhibition in the development of follicles and is due to atretic changes in the ZP.

Keeping in view the contraceptive potential of dZP3 expressed in *E. coli*, a baculovirus expression system was used to express dZP3 with the premise that the glycosylated form of the protein may be more immunogenic and thus confer improved contraceptive potential. The cDNA encoding dZP3 excluding the N-terminal signal sequence and the C-terminal domain following furin cleavage site was cloned in-frame and downstream of the polyhedrin promoter in pAcHLTA-His baculovirus transfer vector (Fig. 1). The pAcHLTA-dZP3 plasmid DNA and BaculoGold™ viral DNA were co-transfected into *Sf* 21 cells. Further, a single viral clone was obtained by performing plaque assay. The plaque-picks were used to propogate a single recombinant virus to obtain recombinant glycoprotein. The Western blot analysis of the Ni-NTA purified recombinant dZP3 with monoclonal antibody MA-451 (generated against porcine ZP3ß which cross-reacted with dZP3; Santhanam et al., 1998) revealed the presence of ~ 50 kDa band (Fig. 1).

As an alternate to chemical conjugation of the recombinant dZP3 with DT for active immunization studies, attempts have been made to clone and express a fusion protein comprising of dZP3 and rabies glycoprotein–G (RV-G). RV-G was selected as a carrier protein with the aim of developing a contraceptive vaccine with dual efficacy by providing additional protection against rabies infection in dogs immunized with this fusion recombinant protein for controlling street dog population. It has been demonstrated that immunization with RV-G indeed leads to the induction of rabies virus neutralizing antibodies (Wiktor et al., 1973; Prehaud et al., 1989). The cDNA encoding dZP3 (978 bp; excluding the N-terminal signal sequence and the C-terminal domain following furin cleavage site) and RV-G (1383 bp; excluding the N-terminal signal sequence but including the C-terminal transmembrane domain without cytoplasmic tail) were PCR amplified separately using appropriate primers from the parent clones (pQE30-dZP3 and pKB3-JE-13 respectively). The general strategy to assemble the cDNA encoding dZP3-RV-G by PCR is schematically shown in Fig. 2. Two rounds of PCR were carried out to assemble the hybrid cDNA. In the first round, cDNA corresponding to dZP3 encompassing part of the N-terminal segment of RV-G and RV-G cDNA encompassing part of C-terminal segment of dZP3 were PCR amplified using the respective parent clones as templates. In the second round of PCR, the amplified fragments of dZP3 cDNA containing a part of N-terminal end of RV-G at its 3' end and RV-G cDNA containing a part of C-terminal end of dZP3 at its 5' end were used as templates to finally assemble the hybrid cDNA encoding dZP3-RV-G (2360 bp). This hybrid cDNA, having *Bam* HI – *Eco* RI restriction sites were finally cloned in pAcHLTA baculovirus transfer vector. A single recombinant viral clone was propogated that expressed the recombinant protein in *Sf* 21 insect cells (Fig. 2). The purified recombinant dZP3-RV-G migrated as a ~ 100 kDa band when analyzed on SDS-PAGE (data not shown). The presence of dZP3 and RV-

**A**

**pAcHLTA-dZP3**

**B**

Western blot

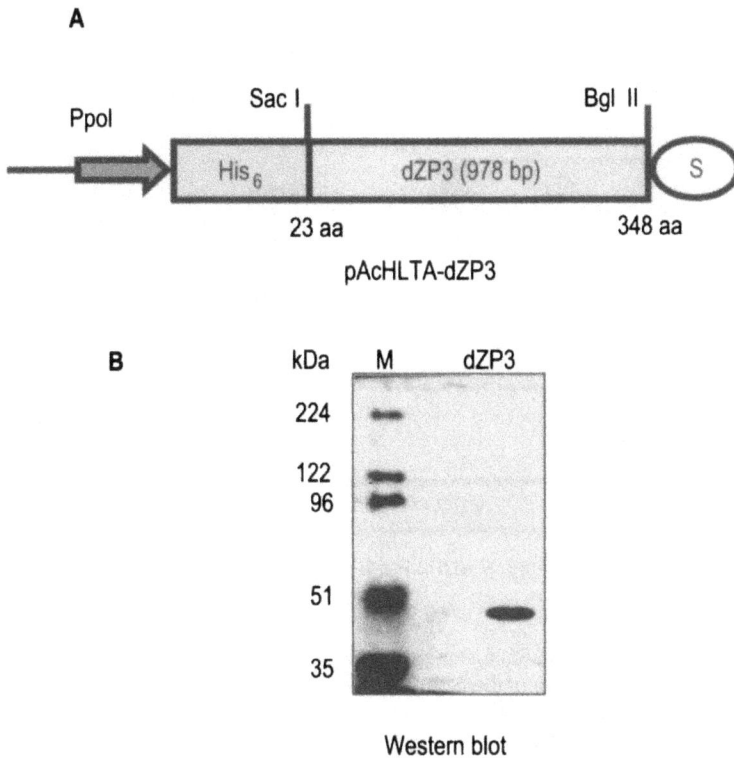

**Fig 1.** Cloning and expression of dog zona pellucida glycoprotein-3 (dZP3) in baculovirus. An internal fragment of dZP3 (79-1056 nt; 978 bp; nt numbering based on Harris *et al.*, 1994), excluding the signal sequence and transmembrane domain, was PCR amplified from pQE30-dZP3 clone (Santhanam *et al.*, 1998) and subsequently cloned in pAcHLTA vector (pAcHLTA-dZP3). (A) Schematic representation of pAcHLTA-dZP3 construct. The dZP3 cDNA was cloned downstream of a polyhedrin promoter (Ppol). (B) Western blot profile of the purified recombinant dZP3 protein expressed in baculovirus expression system. The recombinant protein was purified by Ni-NTA affinity chromatography, resolved by SDS-PAGE, transferred to nitrocellulose membrane and detected by using monoclonal antibody (MA 451) reactive with dZP3. M = molecular weight marker

G in the fusion protein was confirmed by performing Western blot analysis employing a monoclonal antibody reactive with dZP3 (Fig. 2) and rabbit polyclonal antibodies generated against RV-G (data not shown).

*ii) Immunogenicity studies of dZP3-DT conjugate and dZP3-RV-G fusion protein in mice*

Female mice (BALB/cJ; 5 animals/group) were immunized with dZP3-DT conjugate and a denatured form of dZP3-RV-G fusion protein. The mice were immunized subcutaneously with the above recombinant proteins (equivalent to 50 µg protein/mice/injection) employing complete Freund's adjuvant. In addition, to determine the efficacy of adjuvant to generate an antibody response, a group of mice were also immunized with denatured recombinant dZP3-RV-G fusion protein employing TiterMax (Sigma Chemical Co., St Louis, MO, USA) (1:1 ratio) or

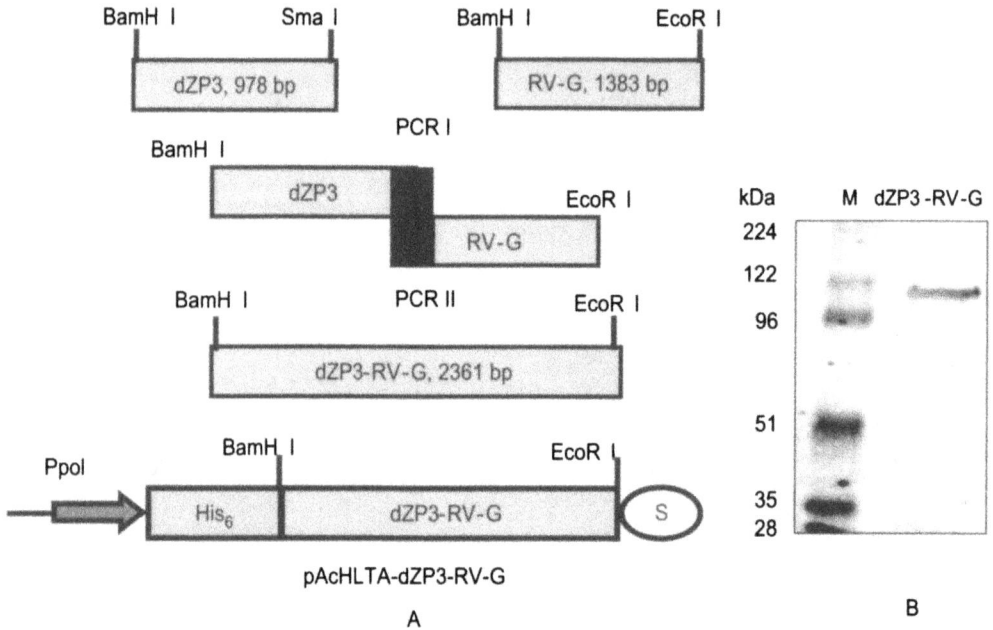

**Fig 2.** Expression of dZP3-RV-G fusion protein in a baculovirus expression system. Panel A is a schematic representation of the assembly of cDNA encoding dZP3-RV-G fusion protein in pAcHLTA vector. Initially, dZP3 cDNA encompassing a part of the N-terminal segment of RV-G cDNA and RV-G cDNA comprising a part of the C-terminal end of dZP3 cDNA were amplified from pQE30-dZP3 (Santhanam et al., 1998) and pQE30-RV-G plasmid (unpublished observation) respectively. The cDNA corresponding to dZP3-RV-G was assembled by second round of PCR using the above PCR-amplified fragments corresponding to dZP3 and RV-G as templates. Panel B represents Western blot profile of the Ni-NTA affinity column purified recombinant dZP3-RV-G fusion protein expressed in Sf 21 cells and resolved by 0.1% SDS – 10% PAGE under reducing conditions and analyzed by a monoclonal antibody MA 451 reactive with dZP3. M = molecular weight marker.

Squalene:Arlacel-A (4:1 ratio with 10 µg SPLPS/mice only in the first injection as additional adjuvants). Two booster injections of 50 µg protein/mice/injection with appropriate adjuvants were administered intraperitoneally at 30 day intervals. Mice were bled from the retro-orbital venous plexus before and at day 15 after the second and third injections for determination of antibody titers in ELISA. The results are shown in Table 2 and Fig 3. Immunization of mice with the denatured recombinant dZP3-RV-G fusion protein resulted in generating antibodies against both recombinant dZP3 and RV-G (Table 2). A higher antibody response was observed in the group of mice immunized using Squalene:Arlacel-A as an adjuvant whereas the lowest antibody response was observed in TiterMax group of immunized mice (Table 2). Mice immunized with dZP3-DT conjugate also generated antibodies against recombinant dZP3 as well as DT (Fig. 3). However, antibody titers generated against DT were higher than the antibody titers generated against recombinant dZP3.

*iii) Characterization of antibodies generated by dZP3-DT conjugate and recombinant dZP3-RV-G fusion protein*

It is imperative that the antibodies generated by immunization with recombinant proteins

**Table 2.** Comparison of antibody response in female mice immunized with denatured recombinant dZP3-RV-G using three different adjuvant in ELISA

| Group | Experiment* | Absorbance at 492 nm** | |
| | | dZP3 | RV-G |
| --- | --- | --- | --- |
| dZP3-RV-G in CFA/IFA | Preimmune (1 : 100) | 0.59 | 0.35 |
| | 1st bleed (1 : 3200) | 1.62 | 0.94 |
| | 2nd bleed (1 : 3200) | 2.04 | 0.95 |
| dZP3-RV-G in TiterMax | Preimmune (1 : 100) | 0.53 | 0.35 |
| | 1st bleed (1 : 3200) | 0.54 | 0.13 |
| | 2nd bleed (1 : 3200) | 1.01 | 0.19 |
| dZP3-RV-G in Squalene and Arlacel-A | Preimmune (1 : 100) | 0.32 | 0.22 |
| | 1st bleed (1 : 3200) | 1.71 | 1.23 |
| | 2nd bleed (1 : 3200) | 2.14 | 1.58 |

Figures in parenthesis indicates the dilution of serum used in ELISA
* *Mice received three injections (~ 50 μg recombinant protein/injection/mice) on days 0, 30, 60 respectively. Preimmune serum represents the bleed before initiating the immunization schedule. 1st and 2nd bleed represent the serum samples obtained from bleeds 15 days after the 2nd and 3rd injections respectively*
** Absorbance at 492 nm is the geometric mean obtained from 5 mice/group

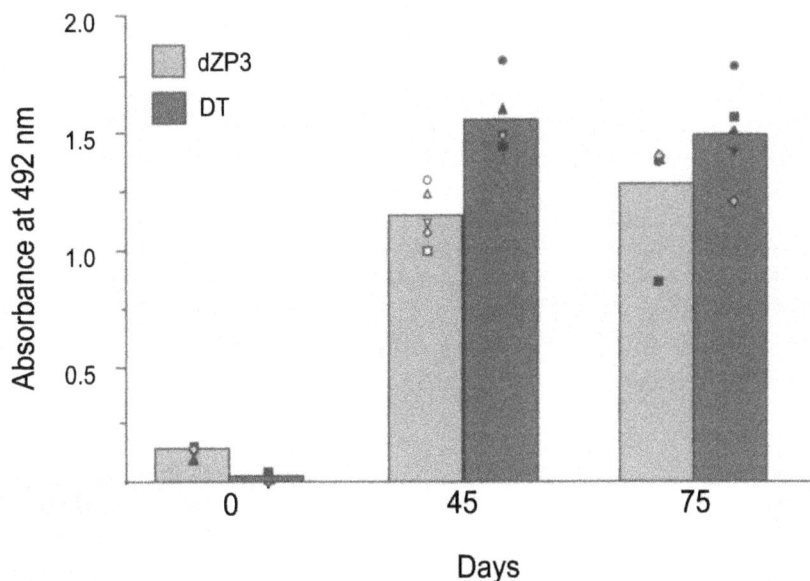

**Fig 3.** Humoral immune response in female mice immunized with dZP3-DT conjugate. Groups (5 animals/group) of female mice (BALB/cJ) were immunized with the recombinant dZP3-DT (equivalent to 50 μg protein/mice/injection) using complete and incomplete Freund's adjuvants. On day 30 and 60, mice were boosted with an equivalent amount of dZP3-DT conjugate as employed in the first injection. Mice were bled retro-orbitally on days 0, 45 and 75 for analysis of antibodies reactive with dZP3 and DT in ELISA. The serum samples were tested in ELISA at 1:400 dilution.

should react with the native ZP. Analysis by indirect immunofluorescence revealed that the antibodies generated by recombinant dZP3-RV-G fusion protein recognize canine ZP (Fig. 4). Antibodies failed to show any reactivity with the granulosa cells and other ovarian associated

Preimmune            Immune

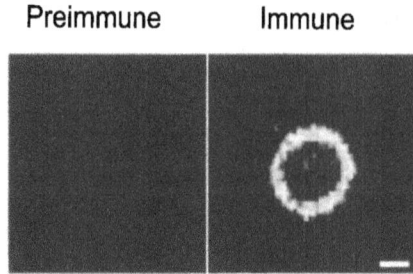

**Fig 4.** Reactivity of antibodies against dZP3-RV-G fusion protein with native ZP. Reactivity of the serum samples from female mice immunized with recombinant dZP3-RV-G fusion protein were tested with dog ovarian cryosections by indirect immunofluorescence. Results with the serum samples of one of the immunized animal are presented. The left panel represents the reactivity with the preimmune serum samples at 1:50 dilution and the right panel at 1:50 dilution with immune serum sample. Scale bar represents 50 μm.

cells. No fluorescence was observed when preimmune serum from the same immunized animal was used. The specificity of the antibody reactivity with the dog zona was further confirmed by inhibition of the fluorescence in the presence of E. coli expressed recombinant dZP3 (data not shown). Immune serum samples from mice immunized with dZP3-DT conjugate also recognized native ZP when evaluated by indirect immunofluoresence assay on cryosections of dog ovaries (data not shown).

The ability of the immune sera from mice immunized with denatured recombinant dZP3-RV-G fusion protein (employing various adjuvant) to neutralize the rabies virus in vitro was tested by Rapid Fluorescent Focus Inhibition Test (RFFIT) assay (Smith et al., 1996). The highest RVNA titers were observed with the recombinant protein in Squalene:Arlacel-A group whereas, the mice sera immunized with the protein in TiterMax generated the lowest RVNA titers (Table 3).

**Table 3.** Rabies virus neutralizing antibody (RVNA) titers in mice immunized with denatured recombinant dZP3-RV-G fusion protein using three different adjuvant

| Experimental Group | *RVNA titers | |
|---|---|---|
| | Day 45 | Day 75 |
| dZP3-RV-G in CFA/IFA | 70 | 75 |
| dZP3-RV-G in TiterMax | 13 | 19 |
| dZP3-RV-G in Squalene and Arlacel A | 180 | 440 |

*RVNA titers are indicated as geometric mean of 5 mice/group

*iv) Immunization studies in female dogs*

In light of the results obtained from the mice experiments, active immunization studies in female non-descript dogs were initiated. The female dogs (aged 3-5 years) used for immunization were reared at the Central Military Veterinary Laboratory, Meerut Cantt, UP, India. The animals were immunized intramuscularly in the hind limbs at two sites with dZP3-RV-G (n = 6; 250 μg protein/animal), physical mixture of dZP3-DT conjugate and baculovirus–expressed recombinant RV-G (unpublished observation) (n = 4; 250 μg dZP3 and 250 μg RV-G/animal) and DT (n = 3; 250 µg DT/animal) emulsified in Squalene and Arlacel-A in 4:1 ratio. SPLPS (1 mg/injection/animal) was also included in the first injection as an additional adjuvant. Primary immunization of dogs comprised three injections of the respective proteins at monthly intervals. Blood samples were collected before and at day 30 after first and second injections and on day 15 after third injection from the ante-cubital vein for determination of progesterone concentrations and antibody titers.

Immunization of female dogs with the recombinant dZP3-RV-G resulted in generation of antibodies against both dZP3 and RV-G as determined by ELISA. Typical data from one of the immunized female dog is shown in Fig. 5. Similarly, antibodies against dZP3, RV-G and DT were observed in animals immunized with the physical mixture of dZP3-DT and RV-G. Representative data from one of the immunized female dog from this group is shown in Fig. 6. An increase in the antibody response was observed in all the animals after the booster injections. However, no significant RVNA titers were observed in any of the immunized groups of female dogs. The circulating progesterone levels in the immunized female dogs are also represented in the Figs. 5, 6 and in Table 4. Once the immunized female dogs came into 'heat', they were mated with male dogs of proven fertility. The results are summarized in Table 4. In the group immunized with dZP3-DT conjugate, two animals came 'on heat' and both failed to conceive on mating. In the group immunized with recombinant dZP3-RV-G fusion protein, 4 animals came 'on heat', of which two failed to conceive and the other two animals became pregnant. In this group, the contraceptive effect was not related to antibody titers.

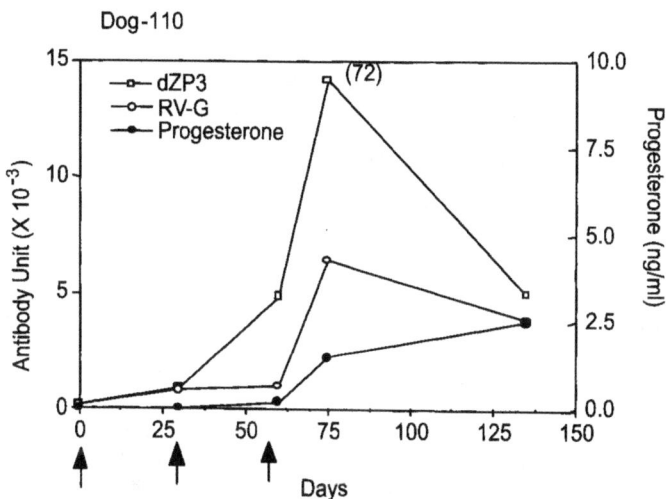

**Fig 5.** Antibody response and progesterone concentrations in a female dog immunized with dZP3-RV-G fusion protein. Female dog (Dog # 110) was immunized intramuscularly in the hind limbs at two sites with dZP3-RV-G (250 µg dZP3-RV-G/ animal), emulsified in Squalene:Arlacel in 4:1 ratio. SPLPS (1 mg/animal) was also included in the first injection as an additional adjuvant. Injections were given at monthly intervals. Blood samples were collected before and at day 30 after first and second injection and on day 15 after third injection from the ante-cubital vein for determination of progesterone concentrations and antibody titers. Serum progesteron concentrations were measured by radioimmunoassay (Sufi et al., 1999). For each serum sample tested in ELISA for determining antibody titers against dZP3 and RV-G, a reciprocal of the dilution giving an absorbance of 1.0 was calculated by regression analysis and has been represented as antibody units (AU). Values in the figure represented as AU X $10^{-3}$.

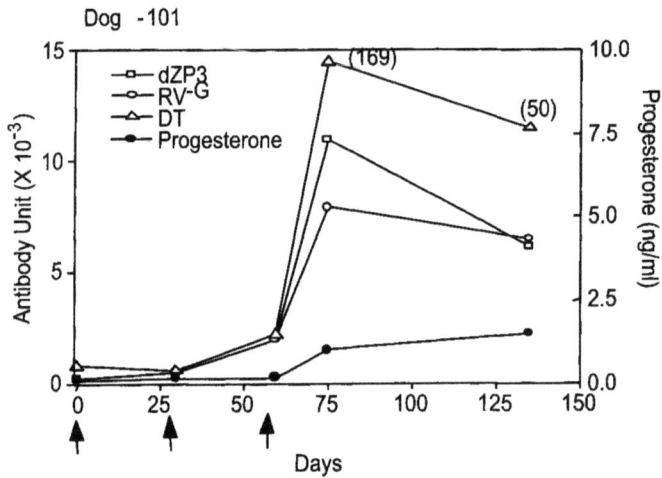

**Fig 6.** Antibody response and progesterone concentrations in a female dog immunized with a physical mixture of dZP3-DT conjugate and RV-G protein. The female dog (Dog # 101) was immunized intramuscularly in the hind limbs at two sites with physical mixture of dZP3-DT and RV-G (250 µg dZP3 and 250 µg RV-G per animal). Other details are same as Fig. 5.

**Table 4.** Summary of active immunization studies in non-descript female dogs

| Experiment/ dog identification # | Dog in heat | Number of mating | Antibody titers of 3rd bleed by ELISA (10⁻³) | | | Progesterone value after mating (ng/ml) | Mating results |
|---|---|---|---|---|---|---|---|
| | | | dZP3 | RV-G | DT | | |
| Immunized with dZP3-RV-G | | | | | | | |
| 103 | Yes | 3 | 13.7 | 22.6 | NA | 2.00 | NP |
| 104 | Yes | 3 | 14.7 | 5.0 | NA | 5.60 | P |
| 105 | - | - | 4.0 | 4.0 | NA | 0.00 | - |
| 110 | Yes | 3 | 72.0 | 6.4 | NA | 2.50 | NP |
| 122 | Yes | 3 | 11.5 | 20.0 | NA | 8.80 | P |
| 125 | - | - | 8.0 | 6.6 | NA | 0.04 | - |
| Immunized with dZP3-DT + RV-G | | | | | | | |
| 101 | Yes | 3 | 10.9 | 7.9 | 169.5 | 1.50 | NP |
| 106 | Yes | 1 | 6.0 | 4.6 | 8.0 | 2.50 | NP |
| 124 | - | - | 16.6 | 11.7 | 8.2 | 1.09 | - |
| 126 | - | - | 4.0 | 4.0 | 10.5 | 1.48 | - |
| Immunized with DT | | | | | | | |
| 100 | Yes | 3 | NA | NA | 81.7 | 7.80 | Aborted |
| 115 | Yes | 3 | NA | NA | 67.7 | 6.50 | Delivered (4) |
| 116 | Yes | 1 | NA | NA | 92.2 | 6.00 | P |

The number of figures in parenthesis represents number of pups; P, denotes for pregnant and NP, denotes non-pregnant; NA- Not Applicable

*v) Alternate strategies for immunization with ZP - based contraceptive vaccines*

An alternate to conventional immunization with protein/peptide is to use plasmid DNA encoding zona proteins. The cDNA encoding bonnet monkey ZP4 and dog ZP3 has been cloned in VR1020 – mammalian expression vectors, downstream of a tissue plasminogen activator (TPA) signal sequence under cytomegalovirus (CMV) promoter (Rath *et al.*, 2002; 2003). Immunization of mice with plasmid DNA encoding bonnet monkey ZP4 (previously described as ZPB) in saline resulted in antibodies with *in vitro* contraceptive potential (Rath *et al.*, 2002). Immunization of mice with plasmid DNA encoding dZP3 resulted in antibodies that reacted with native canine ZP in cryostat ovarian sections by indirect immunofluorescence (Rath *et al.*, 2003). The *in vivo* contraceptive potential of both of the above DNA vaccines is awaited. However, the ability of a DNA vaccine encoding partial sequence of rabbit ZP3 (aa residues 263-415) to inhibit fertility in mice, without any interference in follicular development, has been demonstrated (Xiang *et al.*, 2003). The use of live recombinant viruses as vectors for delivery of ZP glycoprotein to control the population of wild life populations such as mice have been extensively studied as another approach (Jackson *et al.*, 1998; Smith *et al.*, 2005). The advantage of this approach is the natural dissemination of the recombinant viruses from one animal to the other. However, the issue of establishing infection by recombinant virus by competing with the wild virus and certain environmental concerns have to be addressed before it is permitted for large scale application for controlling of pest animals.

## Concluding comments

Immunization with native or recombinant zona proteins lead to curtailment of fertility. The use of a zona–based contraceptive vaccine for fertility regulation in the human has to wait until the concerns of ovarian dysfunction and disturbances in hormonal concentrations are resolved beyond doubt. Immunization with ZP3 has shown promise in curtailing fertility in dogs. In future, it is pertinent to carry out additional active immunization studies in female dogs using various doses of the recombinant proteins in conjunction with various adjuvants. The ability of baculovirus expressed recombinant dZP3-RV-G to generate RVNA titers has to be established. It may require additional inputs to produce appropriate immunogens capable of generating good RVNA titers and curtailment of pregnancy. It may be interesting to develop a new attenuated rabies virus variant by a reverse genetics approach, in which attempts can be made to express dZP3. Finally, the success of this approach in controlling a wild life population depends on the successful development of strategies for vaccine delivery with minimal injections, preferably a single inoculum. Alternate vaccine delivery routes, such as oral, nasopharyngeal *et cetera* are worth considering.

## Acknowledgement

Financial support for these studies has been provided by the Department of Biotechnology, Government of India either independently or under the Indo-US co-operative agreement on 'Vaccine Action Program (VAP)' and 'Contraceptive and Reproductive Health Research'. The views expressed by the authors do not necessarily reflect the views of the funding agencies.

## References

Afzalpurkar A, Shibahara H, Hasegawa A, Koyama K and Gupta SK (1997) Immunoreactivity and *in-vitro* effect on human sperm-egg binding of antibodies against peptides corresponding to bonnet monkey zona pellucida-3 glycoprotein *Human Reproduction* 12 2664-2670

Bagavant H, Thillai-Koothan P, Sharma MG, Talwar GP and Gupta SK (1994) Antifertility effects of porcine zona pellucida-3 immunization using permissible adjuvant in female bonnet monkeys (*Macaca radiata*): reversibility, effect on follicular development and hormonal profiles *Journal of Reproduction and Fertility* **102** 17-25

Govind CK and Gupta SK (2000) Failure of female baboons (*Papio anubis*) to conceive following immunization with recombinant non-human primate zona pellucida glycoprotein-B expressed in *Escherichia coli* *Vaccine* **18** 2970-2978

Govind CK, Hasegawa A, Koyama K and Gupta SK (2000) Delineation of a conserved B cell epitope on bonnet monkey (*Macaca radiata*) and human zona pellucida glycoprotein-B by monoclonal antibodies demonstrating inhibition of sperm-egg binding *Biology of Reproduction* **62** 67-75

Govind CK, Srivastava N and Gupta SK (2002) Evaluation of the immunocontraceptive potential of *Escherichia coli* expressed recombinant non-human primate zona pellucida glycoproteins in homologous animal model *Vaccine* **21** 78-88

Gulyas BJ, Yuan LC, Gwatkin RB and Schmell ED (1983) Response of monkeys to porcine zona pellucida as detected by solid-phase radioimmunoassay *Journal of Medical Primatology* **12** 331-342

Gupta SK, Jethanandani P, Afzalpurkar A, Kaul R and Santhanam R (1997a) Prospects of zona pellucida glycoproteins as immunogens for contraceptive vaccine *Human Reproduction Update* **3** 311-324

Gupta SK, Sharma M, Behera AK, Bisht R and Kaul R (1997b) Sequence of complementary deoxyribonucleic acid encoding bonnet monkey (*Macaca radiata*) zona pellucida glycoprotein-ZP1 and its high-level expression in *Escherichia coli* *Biology of Reproduction* **57** 532-538

Gupta SK, Srivastava N, Choudhury S, Rath A, Sivapurapu N, Gahlay GK and Batra D (2004) Update on zona pellucida glycoproteins based contraceptive vaccine *Journal of Reproductive Immunology* **62** 79-89

Harris JD, Hibler DW, Fontenot GK, Hsu KT, Yurewicz EC and Sacco AG (1994) Cloning and characterization of zona pellucida genes and cDNAs from a variety of mammalian species: the ZPA, ZPB and ZPC gene families *DNA Sequence* **4** 361-393

Hasegawa A, Hamada Y, Shigeta M and Koyama K (2002) Contraceptive potential of synthetic peptides of zona pellucida protein (ZPA) *Journal of Reproductive Immunology* **53** 91-98

Hinsch E, Oehninger S, Schill WB and Hinsch KD (1998) Evaluation of ZP2 domains of functional importance with antisera against synthetic ZP2 peptides *Journal of Reproduction and Fertility* **114** 245-251

Jackson RJ, Maguire DJ, Hinds LA and Ramshaw IA (1998) Infertility in mice induced by a recombinant ectromelia virus expressing mouse zona pellucida glycoprotein-3 *Biology of Reproduction* **58** 152-159

Jethanandani P, Santhanam R and Gupta SK (1998) Molecular cloning and expression in *Escherichia coli* of cDNA encoding bonnet monkey (*Macaca radiata*) zona pellucida glycoprotein-ZP2 *Molecular Reproduction and Development* **50** 229-239

Kaul R, Afzalpurkar A and Gupta SK (1997) Expression of bonnet monkey (*Macaca radiata*) zona pellucida-3 (ZP3) in a prokaryotic system and its immunogenicity

Kaul R, Sivapurapu N, Afzalpurkar A, Srikant V, Govind CK and Gupta SK (2001) Immunocontraceptive potential of recombinant bonnet monkey (*Macaca radiata*) zona pellucida-C expressed in *Escherichia coli* and its corresponding synthetic peptide *Reproductive BioMedicine Online* **2** 33-39

Lefievre L, Conner SJ, Salpekar A, Olufowobi O, Ashton P, Pavlovic B, Lenton W, Afnan M, Brewis IA, Monk M, Hughes DC and Barratt CL (2004) Four zona pellucida glycoproteins are expressed in the human *Human Reproduction* **19** 1580-1586

Lou Y, Ang J, Thai H, McElveen F and Tung KS (1995) A zona pellucida-3 peptide vaccine induces antibodies and reversible infertility without ovarian pathology *Journal of Immunology* **155** 2715-2720

Luo AM, Garza KM, Hunt D and Tung KS (1993) Antigen mimicry in autoimmune disease: sharing of amino acid residues critical for pathogenic T cell activation *Journal of Clinical Investigation* **92** 2117-2123

Mahi-Brown CA, Huang TTF Jr and Yanagimachi R (1982) Infertility in bitches induced by active immunization with porcine zonae pellucidae *Journal of Experimental Zoology* **222** 89-95

Mahi-Brown CA, Yanagimachi R, Hoffman JC and Huang TT Jr (1985) Fertility control in the bitch by active immunization with porcine zonae pellucidae: use of different adjuvants and patterns of estradiol and progesterone levels in estrous cycles *Biology of Reproduction* **32** 761-772

Mahi-Brown CA, Yanagimachi R, Nelson ML, Yanagimachi H and Palumbo N (1988) Ovarian histopathology of bitches immunized with porcine zonae pellucidae *Journal of Reproductive Immunology* **13** 85-95

Martinez ML and Harris JD (2000) Efectiveness of zona pellucida protein ZPB as an immunocontraceptive antigen *Journal of Reproduction and Fertility* **120** 19-32

Meslin FX, Fishbein D and Matter HC (1994) Rationale and prospects for rabies elimination in developing countries In: *Lyssaviruses* pp 1-26 Eds CE Rupprecht, B Dietzschold and H Koprowski. Springer, Berlin

Paterson M, Wilson MR, Morris KD, van Duin M and Aitken RJ (1998) Evaluation of contraceptive potential of recombinant human ZP3 and human ZP3 peptides in primate model: their safety and efficacy *American Journal of Reproductive Immunology* **40** 198-209

Paterson M, Wilson MR, Jennings ZA, van Duin M and Aitken RJ (1999) Design and evaluation of a ZP3 peptide vaccine in a homologous primate model *Molecular Human Reproduction* **5** 342-352

Prehaud C, Takehara K, Flamand A and Bishop DH (1989) Immunogenic and protective properties of rabies virus glycoprotein expressed by baculovirus vectors *Virology* **173** 390-399

Rath A, Choudhury S, Hasegawa A, Koyama K and Gupta SK (2002) Antibodies generated in response to plasmid DNA encoding zona pellucida glycoprotein-B inhibit *in vitro* human sperm-egg binding *Molecular Reproduction and Development* **62** 525-533

Rath A, Batra D, Kaur R, Vrati S and Gupta SK (2003) Characterization of immune response in mice to plasmid DNA encoding dog zona pellucida glycoprotein-3 *Vaccine* **21** 1913-1923

Sacco AG, Yurewicz EC, Subraminian MG and DeMayo FJ (1981) Zona pellucida composition: species cross reactivity and contraceptive potential of antiserum to a purified pig zona antigen (PPZA) *Biology of Reproduction* **25** 997-1008

Sacco AG, Pierce DL, Subraminian MG, Yurewicz EC and Dukelow WR (1987) Ovaries remain functional in squirrel monkeys (*Saimiri sciureus*) immunized with porcine zona pellucida 55,000 macromolecule *Biology of Reproduction* **36** 481-490

Santhanam R, Panda AK, Kumar VS and Gupta SK (1998) Dog zona pellucida glycoprotein-3 (ZP3): expression in *Escherichia coli* and immunological characterization *Protein Expression and Purification* **12** 331-339

Sivapurapu N, Upadhyay A, Hasegawa A, Koyama K and Gupta SK (2002) Native zona pellucida reactivity and *in-vitro* effect on human sperm-egg binding with antisera against bonnet monkey ZP1 and ZP3 synthetic peptides *Journal of Reproductive Immunology* **56** 77-91

Sivapurapu N, Upadhyay A, Hasegawa A, Koyama K and Gupta SK (2003) Efficacy of antibodies against *Escherichia coli* expressed chimeric recombinant protein encompassing multiple epitopes of zona pellucida glycoproteins to inhibit *in vitro* human sperm-egg binding *Molecular Reproduction and Development* **65** 309-317

Sivapurapu N, Hasegawa A, Gahlay GK, Koyama K and Gupta SK (2005) Efficacy of antibodies against a chimeric synthetic peptide encompassing epitopes of bonnet monkey (*Macaca radiata*) zona pellucida-1 and zona pellucida-3 glycoproteins to inhibit *in vitro* human sperm-egg binding *Molecular Reproduction and Development* **70** 247-254

Smith JS, Yager PA and Baer GM (1996) A rapid fluorescent focus inhibition test (RFFIT) for determining rabies virus- neutralizing antibodies. In *Laboratory Techniques in Rabies* pp 181-191 Eds FX Meslin, MM Kaplan and H Koprowski. WHO, Geneva, Switzerland

Smith LM, Lloyd ML, Harvey NL, Redwood AJ, Lawson MA, and Shellam GR (2005) Species-specificity of a murine immunocontraceptive utilising murine cy-

tomegalovirus as a gene delivery vector *Vaccine* **23** 2959-2969

Srivastava N, Santhanam R, Sheela P, Mukund S, Thakral SS, Malik BS and Gupta SK (2002) Evaluation of the immunocontraceptive potential of *Escherichia coli*-expressed recombinant dog ZP2 and ZP3 in a homologous animal model *Reproduction* **123** 847-857

Sufi SB, Donaldson A and Jeffcoate SL (1999) *WHO matched reagent programme: steroid radioimmunoassay method manual* pp 36-46, WHO, Geneva

Sun W, Lou YH, Dean J and Tung KS (1999) A contraceptive peptide vaccine targeting sulfated glycoprotein ZP2 of the mouse zona pellucida *Biology of Reproduction* **60** 900-907

Talwar GP, Singh O, Pal R, Chatterjee N, Sahai P, Dhall K, Kaur J, Das SK, Suri S, Buckshee K, Saraya L and Saxena BN (1994) A vaccine that prevents pregnancy in women *Proceedings of the National Academy of Sciences USA* **91** 8532-8536

VandeVoort CA, Schwoebel ED and Dunbar BS (1995) Immunization of monkeys with recombinant complimentary deoxyribonucleic acid expressed zona pellucida proteins *Fertility and Sterility* **64** 838-847

Wassarman PM (1999) Mammalian fertilization: molecular aspects of gamete adhesion, exocytosis, and fusion *Cell* **96** 175-183

Wiktor TJ, Gyorgy E, Schlumberger D, Sokol F and Koprowski H (1973) Antigenic properties of rabies virus components *Journal of Immunology* **110** 269-276

Wood DM, Liu C and Dunbar BS (1981) Effect of alloimmunization and heteroimmunization with zonae pellucidae on fertility in rabbits *Biology of Reproduction* **25** 439-450

Xiang RL, Zhou F, Yang Y and Peng JP (2003) Construction of the plasmid pCMV4-rZPC DNA vaccine and analysis of its contraceptive potential *Biology of Reproduction* **68** 1518-1524

Yurewicz EC, Sacco AG and Subramanian MG (1987) Structural characterization of the Mr = 55,000 antigen (ZP3) of porcine oocyte zona pellucida. Purification and characterization of alpha- and beta-glycoproteins following digestion of lactosaminoglycan with endo-beta-galactosidase *The Journal of Biological Chemistry* **262** 564-571

# Current status of virally vectored immunocontraception for biological control of mice

Christopher M Hardy

Invasive Animals Cooperative Research Centre, CSIRO Entomology, GPO Box 1700, Canberra, ACT 2601, Australia

Over the past fifteen years, considerable progress has been made in developing biological agents to control wild pest animals that limit fertility rather than increase mortality. The approach, termed virally vectored immunocontraception (VVIC), involves genetically engineering viruses that stimulate the immune system of an infected animal to attack its own reproductive cells, thus rendering it sterile. Our program has focused on the development of mouse-specific viruses that cause infertility by triggering an autoimmune response against the zona pellucida proteins that surround the developing oocyte. The immunocontraceptive vaccine is intended to be transmissible (self-disseminating), and in conjunction with other management practices, will be used to help prevent mouse plagues in Australia. Results from laboratory and field studies so far support the feasibility of applying a recombinant self-disseminating murine cytomegalovirus expressing mouse zona pellucida subunit 3 for mouse control.

## Introduction

Virally vectored contraceptive vaccines are being developed as a means of controlling wild pest animals, particularly in Australia where a wide variety of pests, including rabbits, foxes, cats and mice, cause enormous environmental and economic damage in situations where conventional control is difficult (Tyndale-Biscoe, 1991; 1994; Robinson and Holland, 1995; Jackson et al., 1998; Seamark, 2001; Lloyd et al., 2003; Reubel et al., 2005; Hardy et al., 2006; Mackenzie et al., 2006; Strive et al., 2006). These are an attractive and potentially more humane alternative to conventional methods including poisoning, trapping, and lethal biological controls. For mice, the current annual average loss to the Australian grain industry due to mouse plagues is estimated at 10-15 million AUS$ (McLeod, 2004). Controlling the frequency and severity of eruptions of mouse populations would not only lead to economic benefits but would also reduce the broad scale application of chemical rodenticides and alleviate the social stress experienced by rural communities during mouse plagues.

Many issues need to be addressed in the design of contraceptive vaccines that target wild populations of pest animals (Cowan, 1996; Nettles, 1997; Williams, 1997; Cooper and Herbert, 2001). The vaccine must be highly effective and reduce fertility for long periods of time, target

E-mail: chris.hardy@csiro.au

large numbers of animals over wide areas and not require repeated administration. Additional desirable features include causing minimal side effects, such as those that affect behaviour and low likelihood of leading to genetic resistance (Magiafoglou *et al.*, 2003). Reversibility of the contraceptive effect may not be required. If a live or disseminating virus vaccine is proposed, then consideration must be given to ensure safety and identify any risks associated with the use of genetically modified organisms and quantify their consequences (Hinds *et al.*, 2003). International concerns have also been raised about the potential ecological and economic consequences of using genetically modified immunocontraceptive viruses inappropriately (Williams, 1997; Angulo and Cooke, 2002). Most importantly, the vaccine must be species-specific and unable to affect non-target animals. A lead product contraceptive virus vaccine has been identified for mice based on recombinant murine cytomegalovirus (MCMV) engineered to express mouse zona pellucida subunit 3 (mZP3). This vaccine appears to meet most of the criteria for acceptability listed above (Shellam, 1994).

## Candidate antigens for wildlife contraception

Studies in many species have shown that immunization with antigens derived from the ovary, spermatozoa, and embryo as well as peptide hormones leads to infertility. There are three commercially available immunocontraceptive vaccines currently registered for use in domestic animals: one (SpayVac) based on porcine zona pellucida (PZP) and two (Improvac and Vaxstrate) based on gonadotrophin releasing hormone (GnRH) (Delves, 2004; Ferro and Mordini, 2004). They prevent fertility in a wide variety of species, but require direct injection and are only appropriate for domestic animals or limited wildlife applications. They often do not provide 100% infertility even after repeated doses. None are available for use in humans for these reasons or due to unacceptable side-effects.

Solubilised PZP (SPZP) is probably the most widely used antigen derived from ovaries to reduce fertility in captive animals (Frank *et al.*, 2005) and is widely supported for its humaneness compared with surgical techniques (Grandy and Rutberg, 2002). It requires direct injection of antigen prepared from pig ovaries and therefore its application in wild animals is limited to small, isolated populations (Kirkpatrick *et al.*, 1997; Miller and Killian, 2002; Turner *et al.*, 2002; Rutberg *et al.*, 2004). Additionally, SPZP antigen is not effective in some species (Harrenstien *et al.*, 2004; Levy *et al.*, 2005) and there are conflicting reports on its effectiveness in mice (Sacco *et al.*, 1981; Li *et al.*, 2002). However, complete infertility has been reported in female mice immunized with mouse ZP2 (O'Rand and Lea, 2002) and mouse ZP3 (Chambers *et al.*, 1999a; Redwood *et al.*, 2005).

Sperm antigens would appear to be attractive targets for immunocontraception due to the low likelihood of side-effects and the potential for use in both males and females. Contraceptive potential has frequently been demonstrated *in vitro* using anti-sperm antibodies and it could be expected that these effects are reproducible *in vivo*. Indeed the use of sperm antigens with 100% contraceptive efficacy has been reported in both male and female guinea pigs (Primakoff *et al.*, 1988; Ramarao *et al.*, 1996; Primakoff *et al.*, 1997), rats (Bandivdekar *et al.*, 2001) and female mice (Gupta and Syal, 1997). However, despite these promising results, no recombinant sperm-derived antigen or synthetic peptide has been shown to be sufficiently efficacious *in vivo* to justify their use in any species (Suri, 2004; Naz, 2005). Lack of efficacy has been attributed to poor immune responses or antibody penetration in the reproductive tracts, reduced antigenicity compared to native proteins or to functional differences between species (Frayne and Hall, 1999; Pomering *et al.*, 2002; Hardy *et al.*, 2004a).

Embryo antigens have received less attention due to ethical concerns associated with termination of pregnancy. Nevertheless, one antigen, chicken riboflavin carrier protein (RCP) has strong contraceptive effects in a range of different species including mice (Subramanian et al., 2000; 2003) and recombinant human placental immunomodulatory ferritin has been reported to cause infertility in female mice (Nahum et al., 2004).

An alternative to targeting reproductive tract antigens is to target peptide hormones or their receptors. Key hormonal antigenic targets such as GnRH (LHRH), follicle stimulating hormone (FSH) and luteinizing hormone (LH) regulate gametogenesis in males and females and several synthetic peptides have been reported to cause infertility in male and female mice (Remy et al., 1996; Ghosh and Jackson, 1999; Zeng et al., 2002). However, there are a number of problems associated with the use of hormone antigens. First, they are required for steroidogenesis and secondly, they have significant homology to proteins unrelated to fertility with wider regulatory roles in mammals. Interfering with their function therefore generally leads to undesirable side effects, particularly in males (Awoniyi, 1994; Delves, 2004; Ferro and Mordini, 2004). Finally, there is a high degree of sequence conservation between peptide hormones across species and this presents potentially unacceptable risks to non target animals, particularly if delivered using virus vectors.

Targeting males for most wildlife immunocontraception programs poses the problem that even a few fertile males that escape contraception might breed with numerous females and so compensate for infertility in relatively large numbers of males. Fertilization and oogenesis in females on the other hand are targets that promise high efficacy, comparatively lower risks of causing deleterious side-effects and the potential to be species-specific. Our focus has therefore been on immunocontraception targeting female gametes for pest animal control.

## A virally vectored immunocontraceptive vaccine for mice

Various peptides and purified recombinant antigens, immunization regimes and virus vectors have been tested to determine the most appropriate delivery mechanism for immunocontraception in mice. Experiments conducted in Australia within the Invasive Animals (formerly the Pest Animal Control) Cooperative Research Centre have now firmly established mZP3 as the preferred contraceptive antigen for use in the European house mouse and MCMV as the most effective virus vector (Table 1). The mZP3 antigen has been shown to induce infertility in female mice using a variety of immunization regimes, including direct inoculation with purified protein (Hardy et al., 2003; Clydesdale et al., 2004) and infection with recombinant viruses (Jackson et al., 1998; Lloyd et al., 2003; Redwood et al., 2005). The main physiological effect in response to immunization with mZP3 is the elimination of developing follicles from the ovary (Figure 1). Recombinant mouse-specific viruses such as MCMV and ectromelia virus (ECTV), genetically engineered to express mZP3, are able to induce long-term infertility (> 250 days) in mice following a single dose with virus (Jackson et al., 1998; Lloyd et al., 2003; Redwood et al., 2005). Live recombinant mammalian viruses expressing mZP3 are so far the only vaccines to cause high levels of long-term infertility without adjuvants.

A potential impediment to the use of MCMV as a vaccine vector is the existence of genetic resistance in mice to this virus. Genetic resistance in mice to MCMV is determined by several gene loci, including the *Cmv1*, *Cmv2* and others mapping to the MHC region (Scalzo et al., 2005). The *Cmv1* resistance locus controls splenic replication of MCMV and confers natural killer (NK) cell-mediated resistance to otherwise lethal infection (Scalzo et al., 1995). This resistance gene is present in C57/BL6J but not BALB/c mice. C57/BL6J mice are considerably

**Table 1.** Fertility of mice immunized with reproductive antigens.

| Antigen target | Antigen[a] | Expression system | Carrier protein | Adjuvants | Doses[b] | Mouse strain | Total mice | % fertile | Infertility duration | Reference |
|---|---|---|---|---|---|---|---|---|---|---|
| Egg | ZP3 | ECTV | | Live virus | 1 | BALB/c | 4 | 0 | L | Jackson et al., 1998 |
| | ZP3 | MCMV | | Live virus | 1 | BALB/c | 9 | 0 | L | Chambers et al., 1999a |
| | ZP3 | MCMV | | Live virus | 1 | BALB/c | 6 | 0 | L | Lloyd et al., 2003 |
| | ZP3 | MCMV | | Live virus | 1 | BALB/c | 6 | 0 | L | Redwood et al., 2005 |
| | ZP3 | MCMV (del:07-12) | | Live virus | 1 | BALB/c | 6 | 0 | L | Redwood et al., 2005 |
| | ZP3 | MCMV | | Live virus | 1 | A/J | 9 | NP | L | Chambers et al., 1999a |
| | ZP3 | MCMV | | Live virus | 1 | ARC | 9 | NP | L | Chambers et al., 1999a |
| | ZP3 | MCMV | | Live virus | 1 | C57/BL6 | 9 | NP | L | Chambers et al., 1999a |
| | ZP3 | MCMV | | Live virus | 1 | BALB/c | 10 | 10 | S | Lloyd et al., 2003 |
| | ZP3 | Vaccinia | 6XHis | FCA+FIA | 1+3 | BALB/c | 10 | 20 | S | Hardy et al., 2003 |
| | ZP3 | Synthetic peptide | KLH | FCA+FIA | 1+4 | Wild | 30 | 23 | S | Hardy et al., 2002a |
| | ZP3 | ECTV | | Live virus | 1 | BALB/c | 13 | 31 | S | Jackson et al., 1998 |
| | ZP3 | Vaccinia | 6XHis | FCA+FIA | 1+3 | CBA | 15 | 40 | S | Clydesdale et al., 2004 |
| | ZP3 | Vaccinia | 6XHis | FCA+FIA | 1+3 | BALB/c | 18 | 56 | S | Clydesdale et al., 2004 |
| | ZPC (porcine) | Vaccinia | | FCA+FIA | 1+3 | BALB/c | 14 | 64 | S | Clydesdale et al., 2004 |
| | ZPC (porcine) | Vaccinia | | FCA+FIA | 1+3 | CBA | 7 | 71 | None | Clydesdale et al., 2004 |
| | ZP3 | Baculovirus | 6XHis | FCA+FIA | 1+3 | BALB/c | 20 | 75 | None | Hardy et al., 2003 |
| | ZP3 | Synthetic peptide | KLH | FCA+FIA | 1+4 | BALB/c | 30 | 77 | None | Hardy et al., 2002a |
| | AgZ | E. coli | MBP | FCA+FIA | 1+3 | BALB/c | 9 | 78 | S[c] | Hardy et al., 2004b |
| | ZP1 | E. coli | MBP | FCA+FIA | 1+3 | BALB/c | 10 | 90 | None | Hardy et al., 2004b |
| | ZP3 | Baculovirus | 6XHis | FCA+FIA | 1+3 | Wild | 9 | 100 | None | Hardy et al., 2003 |
| Sperm | SP56 | Synthetic peptide | KLH | FCA+FIA | 1+3 | BALB/c | 5 | 40 | S | Hardy et al., 2004b |
| | SP56 | E. coli | Flag | FCA+FIA | 1+5 | BALB/c | 10 | 40 | S | Hardy and Mobbs, 1999 |
| | SP56 | E. coli | Flag | FCA+FIA | 1+3 | BALB/c | 5 | 60 | None | Hardy and Mobbs, 1999 |
| | PH20 | E. coli | MBP | FCA+FIA | 1+3 | BALB/c | 10 | 80 | None | Hardy et al., 2004a |
| | PH20 | E. coli | MBP | FCA+FIA | 1+3 | BALB/c | 5 | 100 | None | Hardy et al., 2004a |
| | PH20 | MCMV | | Live virus | 1 | BALB/c | 6 | 100 | None | Hardy et al., 2004a |
| | PH20 | MCMV | | Live virus | 1 | BALB/c | 6 | 100 | None | Hardy et al., 2004a |
| Embryo | GMCSF | Synthetic peptide | KLH | FCA+FIA | 1+3 | BALB/c | 5 | 40 | S | Hardy et al., 2004b |
| | PRL | Synthetic peptide | KLH | FCA+FIA | 1+3 | BALB/c | 5 | 40 | None | Hardy et al., 2004b |
| | LIFR | Synthetic peptide | KLH | FCA+FIA | 1+3 | BALB/c | 5 | 60 | None | Hardy et al., 2004b |
| | PLF | Synthetic peptide | KLH | FCA+FIA | 1+3 | BALB/c | 5 | 60 | None | Hardy et al., 2004b |
| | HSP27 | Synthetic peptide | KLH | FCA+FIA | 1+3 | BALB/c | 5 | 80 | None | Hardy et al., 2004b |
| | OGP | Synthetic peptide | KLH | FCA+FIA | 1+4 | BALB/c | 5 | 100 | None | Hardy et al., 2004b |

**Table 1.** Contd.

| Antigen target | Antigen[a] | Expression system | Carrier protein | Adjuvants | Doses[b] | Mouse strain | Total mice | % fertile | Infertility duration | Reference |
|---|---|---|---|---|---|---|---|---|---|---|
| Multiple | AgB | *E. coli* | MBP | FCA+FIA | 1 + 3 | BALB/c | 10 | 40 | S | Hardy et al., 2002b |
| | Several[d] | Synthetic peptide | KLH | FCA+FIA | 1 + 3 | BALB/c | 6 | 50 | None | Hardy et al., 2004b |
| | AgA | *E. coli* | MBP | FCA+FIA | 1 + 3 | BALB/c | 10 | 60 | None | Hardy et al., 2002b |
| | AgA | *E. coli* | 6XHis | FCA+FIA | 1 + 3 | BALB/c | 10 | 100 | None | Hardy et al., 2002b |

[a]Antigens are murine in origin except where stated. [b]Immunizations were by intraperitoneal inoculation, except ECTV-ZP3 (footpad injection). [c]Reduced litter size only. [d]Combination of SP45, GMCSF and PRL peptides. MCMV, murine cytomegalovirus; ECTV, ectromelia virus; KLH, keyhole limpet hemocyanin; MBP, maltose binding protein; FCA, Freund's complete adjuvant; FIA, Freund's incomplete adjuvant; NP, data not provided; S, short-term; L, long-term (experiments run for greater than 100 days).

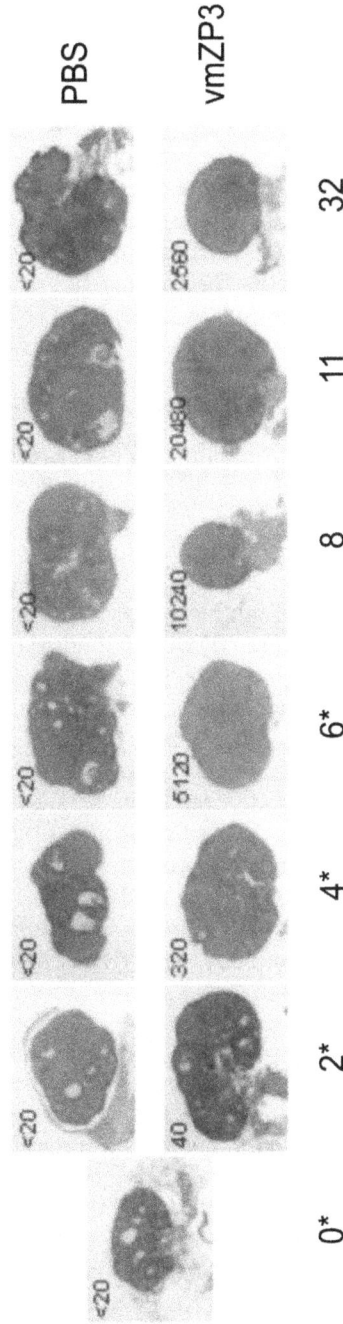

**Fig 1.** Hematoxylin and eosin stained sections of ovaries (X40 magnification) showing progression of ovarian pathology in mice immunized with PBS (top panels) or recombinant mouse ZP3 (bottom panels). BALB/c mice (6 weeks old) were immunized by intraperitoneal injection and boosted at 2 week intervals with 20 µg of recombinant mouse ZP3 (vmZP3) produced in a vaccinia virus expression system or PBS using Freund's adjuvants as previously described (Hardy et al., 2003). Injection points are indicated by asterisks. Mice immunized with vmZP3 (n=8) were mated between weeks 8 and 32 and were infertile. End-point serum IgG titres to mouse ZP3 are shown for each mouse at autopsy.

less susceptible to VVIC when challenged with recombinant virus expressing mZP3 derived from the laboratory strain K181 of MCMV (Chambers et al., 1999a). Several wild isolates of MCMV are insensitive to the Cmv1 resistance mechanism present in C57/BL6J mice (Voigt et al., 2003). One of these, strain G4 (Booth et al., 1993), has been selected as the preferred virus vector for delivery of mZP3 to mice in the field.

The first intended product recombinant virus has now been constructed based on the G4 isolate of MCMV, using an early tissue culture passage to avoid attenuation in vitro. This virus (MCMV-eG4-mZP3) expresses the mZP3 under the control of the constitutive human cytomegalovirus immediate early 1 gene promoter. The MCMV-eG4-mZP3 is able to induce infertility in wild mice as well as laboratory strains of mice resistant to earlier recombinant viruses based on the K181 strain of MCMV (Pest Animal Control CRC, 2004; M.L. Lloyd, unpublished). Thus, the MCMV-eG4-mZP3 virus appears able to overcome at least one potential resistance mechanism, although MCMV resistance genes do not appear prevalent in populations of wild mice from Australia (Scalzo et al., 2005).

## Mechanism of mZP3-induced infertility

The mechanisms whereby immunization with mZP3 leads to infertility are still unclear. Several experiments have demonstrated that anti-mZP3 serum antibodies are primarily responsible for the contraceptive effect, although the precise mechanism appears to differ depending on the vaccine. Sera taken from mice immunized with mZP3 peptides, mZP3 protein and ECTV and MCMV expressing mZP3 have all been shown to bind to the zona in sections of mouse ovaries using indirect immunofluorescence (Jackson et al., 1998; Hardy et al., 2002a, b; Lloyd et al., 2003). Likewise, immunoglobulin binds directly to the ZP in the ovaries of infected but not control mice immunized with ECTV-mZP3 and MCMV-mZP3 (Jackson et al., 1998; Lloyd et al., 2003). Passive immunization of mice using serum antibodies from mice immunized with recombinant mZP3 leads to a delay in breeding and mice return to fertility, presumably as antibody titres fall below a critical level (Lloyd et al., 2003).

In MCMV-mZP3 infected mice, fragmentation of the ZP has been observed and the most likely cause of infertility is destruction of maturing oocytes (Lloyd et al., 2003). In ECTV-mZP3 infected mice, however, at least two different contraceptive processes appear to operate. Loss of developing oocytes comparable to that seen in ovaries of MCMV-mZP3 infected mice is present in 50% of BALB/c mice. In these mice, disruption of folliculogenesis has been proposed to occur through antibody-dependent lysis of the oocyte by complement or T-cells or to disruption of signalling between the oocyte and surrounding granulosa cells (Jackson et al., 1998).

However, in mice where ovaries appear normal, prevention of sperm-binding and fertilization or failure of embryos to implant appears responsible for infertility. Comparable but antigen-dependent effects on the ovary have also been observed in mice immunized with purified recombinant mZP3 or porcine ZPC (pZPC) proteins (Clydesdale et al., 2004). Ovaries of infertile mice immunized with mZP3 lack mature follicles whereas the ovaries from infertile pZPC immunized mice appear normal, indicating that immune responses other than antibodies contribute to infertility. High titres of anti-ZP3 antibodies are not always associated with infertility in mice (Hardy et al., 2003; Clydesdale et al., 2004) and infertility following immunization with mZP3 proteins has been proposed to require the development of predominantly anti-inflammatory (Th2) responses (Clydesdale et al., 2004).

No inflammatory (Th1) responses have been reported in the ovaries of mice immunized with purified mZP3 proteins or recombinant viruses expressing mZP3 (Jackson et al., 1998; Lloyd et al., 2003; Clydesdale et al., 2004). This contrasts with reports that inflammatory T-cell mediated damage to the ovary (oophoritis) occurs rapidly in some strains of mice immunized with ZP3 peptides (Lou et al., 1995; 2000; Lou and Borillo, 2003). However, mice with oophoritis induced by passive transfer of T-cells remain fertile in the absence of mZP3 antibodies (Bagavant et al., 1999), reinforcing the relationship between mZP3-specific antibodies and contraception. The difference between these two systems is likely to reflect differences between the antigens used, mouse strains and immunization protocols. Nevertheless, results obtained using live recombinant viruses to deliver mZP3 have shown that contraceptive responses are primarily antibody mediated and do not cause significant oophoritis.

## Vaccine transmission

One of the key requirements for an immunocontraceptive virus to be successful at controlling mice in the field is the ability to transmit and infect large numbers of animals. The growth characteristics of recombinant MCMVs expressing mZP3 are indistinguishable to wild type viruses *in vitro*, but all are significantly attenuated *in vivo*, particularly for their ability to replicate in the salivary glands of mice (Lloyd et al., 2003; Redwood et al., 2005). Recombinant MCMVs are therefore deficient in the ability to transmit between mice compared to parental viruses as the salivary gland is considered the main source of infectious MCMV (Baker, 1998). Experiments are currently being conducted to determine whether recombinant MCMVs expressing mZP3 are nevertheless able to transmit between mice and cause infertility by natural transmission using both laboratory and wild strains of mice.

Naturally infected wild-caught mice in the Murrumbidgee Irrigation Area (MIA) in southeastern Australia transmit MCMV to naive mice more rapidly than mice infected with laboratory or previous field virus isolates. Several isolates of these MIA field viruses have been plaque purified and one isolate in particular has retained the ability to transmit readily within social groups of mice (L.A. Hinds, unpublished results). This isolate is currently being used to construct a new recombinant MIA MCMV expressing mZP3 and may transmit more readily than previous recombinant viruses.

The ability of insects to transmit MCMV is being assessed. This is because insect vectors potentially could enhance transmission of MCMV by either supporting viral replication or by direct mechanical transfer. Experiments where uninfected mice have been exposed to fleas that have fed on mice infected with wild MCMV have produced no evidence of transmission (L.A. Hinds, personal communication). Data from these experiments will be used to inform regulatory authorities on the risks of testing recombinant MCMVs in field pens that may not be insect-proof.

## Species-specificity and safety

Species-specificity is a prerequisite for release of any biological control agent. Demonstration that recombinant MCMV-mZP3 is strictly mouse-specific will be essential before it can be considered for release and initial testing in non-target species has been conducted in rats (Smith et al., 2005). Direct inoculation of laboratory rats using large doses of recombinant MCMV-mZP3 leads to long-lived antibody responses to both MCMV and mZP3, but there was no evidence for replicating virus. The anti-mZP3 antibodies were specific for mouse ZP3, did not cross-react with rat ZP3 and had no effect on the fertility of the rats. These results are encouraging as they support claims that MCMV is strictly mouse-specific (Shellam, 1994) and show that additional specificity can be conferred through the use of the mZP3 antigen. Further testing will be required on likely non target species both in Australia and overseas.

## Ecology of mice and modelling release of immunocontraceptive viruses

The factors that determine mouse abundance and occurrence of mouse plagues in Australia have been the subject of long-term experimental and modelling studies (Singleton et al., 2005). Many viruses spread during mouse plagues and MCMV in particular has epidemiological properties that make it an ideal vector for delivery of immunocontraceptive antigens to wild mice (Singleton et al., 1993; 2000; 2002). MCMV seroprevalence is variable in wild mice but is greatest during times of plague (Singleton et al., 2000; 2002). Rapid seroconversion of wild mice to MCMV in Australia is reported to occur once mouse densities reach 40-100 mice/ha, indicating naturally high transmission rates for this virus (Jacob and Sutherland, 2004; Sutherland et al., 2005). Transmission of wild strains of MCMV has also been shown to occur readily amongst populations of mice held in enclosures (Farroway et al., 2005) and MCMV itself does not appear to affect fertility or survival (Farroway et al., 2002).

Several models have been advanced to determine the levels of infertility required to control mice with estimates ranging from as low as 30% to 90% in some cases (Hone, 1999; Davis et al., 2003; Arthur et al., 2005). In particular, results from the latest models (Arthur et al., 2005) combined with experimental data (Chambers et al., 1999b; Singleton et al., 2002) indicate that approximately 70% of mice must be made infertile to keep the population below a level where significant damage would be caused. This estimate relies on there being no competitive disadvantage to the introduced immunocontraceptive MCMV compared to wild-type viruses. If prior infection with MCMV prevents infertility on subsequent infection with an immunocontraceptive MCMV, there is a significant reduction in the effectiveness of VVIC and close to 100% of susceptible mice must be made infertile. Trials are underway to determine the extent to which prior infection of wild mice with wild-type MCMV affects infertility and transmission of MCMV-eG4-mZP3. Simulations also indicate the need to retain high transmission rates of the immunocontraceptive MCMV.

A mathematical model of mouse dynamics and epidemiology of immunocontraceptive MCMV and field strains has also been constructed and used to evaluate the risk of inadvertent export of a sterilising MCMV from Australia to related mouse species overseas (C.K. Williams, personal communication). This analysis will provide key documentation determining some safety aspects of an application to conduct field trials.

## Conclusions

Virally vectored immunocontraception remains to be proven as a viable approach for the biological control of pest animals. Nevertheless, at least one contraceptive vaccine, a recombinant murine cytomegalovirus (MCMV) engineered to express mouse zona pellucida subunit 3 (mZP3) fulfils many of the requirements of an acceptable biological control agent for mice. If this vaccine can be confirmed as mouse-specific and able to spread to large numbers of mice under experimental field conditions, then commercial application will become an option. Nevertheless, many technical and social challenges remain before this goal can be achieved. Staged transmission, fertility and species-specificity trials must show efficacy and safety and it will be several years before any commercial vaccine can be produced. The benefits of deploying a genetically modified biological control agent must be clearly demonstrated and potential risks understood and accepted by the public, both in Australia and overseas, if VVIC is to proceed.

## Acknowledgements

This work was supported by funds provided by the Australian Grains Research and Development Corporation (CSV16) and the Australian Government's Cooperative Research Centres Program. I am also grateful for critical comments on the manuscript provided by Tony Robinson and Lyn Hinds.

## References

**Angulo E and Cooke B** (2002) First synthesize new viruses then regulate their release? The case of the wild rabbit *Molecular Ecology* **11** 2703-2709

**Arthur AD, Pech RP and Singleton GR** (2005) Predicting the effect of virally vectored recMCMV immunocontraception on house mice (*Mus musculus domesticus*) in mallee wheatlands *Wildlife Research* **32** 631-637

**Awoniyi CA** (1994) GnRH immunization and male infertility: immunocontraception potential *Advances in Contraceptive Delivery Systems:CDS* **10** 279-290

**Bagavant H, Adams S, Terranova P, Chang A, Kraemer FW, Lou Y, Kasai K, Luo AM and Tung KS** (1999) Autoimmune ovarian inflammation triggered by proinflammatory (Th1) T cells is compatible with normal ovarian function in mice *Biology of Reproduction* **61** 635-642

**Bandivdekar AH, Vernekar VJ, Moodbidri SB and Koide SS** (2001) Characterization of 80 kDa human sperm antigen responsible for immunoinfertility *American Journal of Reproductive Immunology* **45** 28-34

**Baker DG** (1998) Natural pathogens of laboratory mice, rats, and rabbits and their effects on research *Clinical Microbiology Reviews* **11** 231-266

**Booth TW, Scalzo AA, Carrello C, Lyons PA, Farrell HE, Singleton GR and Shellam GR** (1993) Molecular and biological characterization of new strains of murine cytomegalovirus isolated from wild mice *Archives of Virology* **132** 209-220

**Chambers LK, Lawson MA and Hinds LA** (1999a) Biological control of rodents-the case for fertility control using immunocontraception. In *Ecologically-based Rodent Management* pp 215-242 Eds G Singleton, L Hinds, H Leirs and Z Zhang. Australian Centre for International Agricultural Research, Canberra

**Chambers LK, Singleton GR and Hinds LA** (1999b) Fertility control of wild mouse populations: the effects of hormonal competence and an imposed level of sterility *Wildlife Research* **26** 579-591

**Clydesdale G, Pekin J, Beaton S, Jackson RJ, Vignarajan S and Hardy CM** (2004) Contraception in mice immunized with recombinant zona pellucida subunit 3 proteins correlates with Th2 responses and the levels of interleukin 4 expressed by CD4+ cells *Reproduction* **128** 737-745

**Cooper DW and Herbert CA** (2001) Genetics, biotechnology and population management of over- abundant mammalian wildlife in Australasia *Reproduction, Fertility and Development* **13** 451-458

**Cowan PE** (1996) Possum biocontrol: prospects for fertility regulation *Reproduction, Fertility and Development* **8** 655-660

**Davis SA, Pech RP and Singleton GR** (2003) Simulation of fertility control in an eruptive house mouse (*Mus domesticus*). In *Rats, mice and people: rodent biology and management* pp 320-324 Eds GR Singleton, LA Hinds, CJ Krebs and DM Spratt. Australian Centre for International Agricultural Research, Canberra

**Delves PJ** (2004) How far from a hormone-based contraceptive vaccine? *Journal of Reproductive Immunology* **62** 69-78

**Farroway LN, Singleton GR, Lawson MA and Jones DA** (2002) The impact of murine cytomegalovirus (MCMV) on enclosure populations of house mice (*Mus domesticus*) *Wildlife Research* **29** 11-17

**Farroway LN, Gorman S, Lawson MA, Harvey NL, Jones DA, Shellam GR and Singleton GR** (2005) Transmission of two Australian strains of murine cytomegalovirus (MCMV) in enclosure populations of house mice (*Mus domesticus*) *Epidemiology and Infection* **133** 701-710

**Ferro VA and Mordini E** (2004) Peptide vaccines in immunocontraception *Current Opinion in Molecular Therapeutics* **6** 83-89

**Frank KM, Lyda RO and Kirkpatrick JF** (2005) Immunocontraception of captive exotic species - IV. Species differences in response to the porcine zona pellucida vaccine, timing of booster inoculations, and procedural failures *Zoo Biology* **24** 349-358

**Frayne J and Hall L** (1999) The potential use of sperm antigens as targets for immunocontraception: past, present and future *Journal of Reproductive Immunology* **43** 1-33

**Ghosh S and Jackson DC** (1999) Antigenic and immunogenic properties of totally synthetic peptide-based antifertility vaccines *International Immunology* **11** 1103-1110

**Grandy JW and Rutberg AT** (2002) An animal welfare view of wildlife contraception *Reproduction Supplement* **60** 1-7

**Gupta GS and Syal N** (1997) Immune responses of chemically modified homologous LDH-C4 and their effect on fertility regulation in mice *American Journal of Reproductive Immunology* **37** 206-211

**Hardy CM and Mobbs KJ** (1999) Expression of recombinant mouse sperm protein sp56 and assessment of its potential for use as an antigen in an immuno-contraceptive vaccine *Molecular Reproduction and Devel-*

*opment* **52** 216-224

Hardy CM, ten Have JF, Mobbs KJ and Hinds LA (2002a) Assessment of the immunocontraceptive effect of a zona pellucida 3 peptide antigen in wild mice *Reproduction, Fertility and Development* **14** 151-155

Hardy CM, Pekin J and ten Have J (2002b) Mouse-specific immunocontraceptive polyepitope vaccines *Reproduction Supplement* **60** 19-30

Hardy CM, ten Have JF, Pekin J, Beaton S, Jackson RJ and Clydesdale G (2003) Contraceptive responses of mice immunized with purified recombinant mouse zona pellucida subunit 3 (mZP3) proteins *Reproduction* **126** 49-59

Hardy CM, Clydesdale G, Mobbs KJ, Pekin J, Lloyd ML, Sweet C, Shellam GR and Lawson MA (2004a) Assessment of contraceptive vaccines based on recombinant mouse sperm protein PH20 *Reproduction* **127** 325-334

Hardy CM, Clydesdale G and Mobbs KJ (2004b) Development of mouse-specific contraceptive vaccines: infertility in mice immunized with peptide and polyepitope antigens *Reproduction* **128** 395-407

Hardy CM, Hinds LA, Kerr PJ, Lloyd ML, Redwood AJ, Shellam GR and Strive T (2006) Biological control of vertebrate pests using virally vectored immunocontraception *Journal of Reproductive Immunology* **71** 102-111

Harrenstien LA, Munson L, Chassy LM, Liu IK and Kirkpatrick JF (2004) Effects of porcine zona pellucida immunocontraceptives in zoo felids *Journal of Zoo and Wildlife Medicine* **35** 271-279

Hinds LA, Hardy CM, Lawson MA and Singleton GS (2003) Developments in fertility control for pest animal management. In *Rats, mice and people: Rodent biology and management* pp 31-36 Eds GR Singleton, LA Hinds, CJ Krebs and DM Spratt. Australian Centre for International Agricultural Research, Canberra

Hone J (1999) On the rate of increase: patterns of variation in Australian mammals and the implications for wildlife management *Journal of Applied Ecology* **36** 709-718

Jackson RJ, Maguire DJ, Hinds LA and Ramshaw IA (1998) Infertility in mice induced by a recombinant ectromelia virus expressing mouse zona pellucida glycoprotein 3 *Biology of Reproduction* **58** 152-159

Jacob J and Sutherland DR (2004) Murine cytomegalovirus (MCMV) infections in house mice: a matter of age or sex? *Wildlife Research* **31** 369-373

Kirkpatrick JF, Turner JW Jr, Liu IK, Fayrer-Hosken R and Rutberg AT (1997) Case studies in wildlife immunocontraception: wild and feral equids and white-tailed deer *Reproduction, Fertility and Development* **9** 105-110

Levy JK, Mansour M, Crawford PC, Pohajdak B and Brown RG (2005) Survey of zona pellucida antigens for immunocontraception of cats *Theriogenology* **63** 1334-1341

Li D, Sun X, Li C, Cai L and Meng Y (2002) Effects on fertility of immunizing mice with anti-idiotypic antibodies to porcine zona pellucida antigen *Journal of*

*Reproductive Immunology* **54** 81-92

Lloyd ML, Shellam GR, Papadimitriou JM and Lawson MA (2003) Immunocontraception is induced in BALB/c mice inoculated with murine cytomegalovirus expressing mouse zona pellucida 3 *Biology of Reproduction* **68** 2024-2032

Lou YH and Borillo J (2003) Migration of T cells from nearby inflammatory foci into antibody bound tissue: a relay of T cell and antibody actions in targeting native autoantigen *Journal of Autoimmunity* **21** 27-35

Lou YH, McElveen F, Adams S and Tung KS (1995) Altered target organ. A mechanism of post-recovery resistance to murine autoimmune oophoritis *Journal of Immunology* **155** 3667-3673

Lou YH, Park KK, Agersborg S, Alard P and Tung KS (2000) Retargeting T cell-mediated inflammation: a new perspective on autoantibody action *Journal of Immunology* **164** 5251-5257

Mackenzie SM, McLaughlin EA, Perkins HD, French N, Sutherland T, Jackson RJ, Inglis B, Muller WJ, van Leeuwen BH, Robinson AJ and Kerr PJ (2006) Immunocontraceptive effects on female rabbits infected with recombinant myxoma virus expressing rabbit ZP2 or ZP3 *Biology of Reproduction* **74** 511-521

Magiafoglou A, Schiffer M, Hoffmann AA and McKechnie SW (2003) Immunocontraception for population control: will resistance evolve? *Immunology and Cell Biology* **81** 152-159

McLeod R (2004) *Counting the Cost: Impact of Invasive Animals in Australia 2004* Cooperative Research Centre for Pest Animal Control, Canberra http://www.invasiveanimals.com/images/pdfs/Mcleod.pdf (accessed 5 December 2005)

Miller LA and Killian GJ (2002) In search of the active PZP epitope in white-tailed deer immunocontraception *Vaccine* **20** 2735-2742

Nahum R, Brenner O, Zahalka MA, Traub L, Quintana F and Moroz C (2004) Blocking of the placental immune-modulatory ferritin activates Th1 type cytokines and affects placenta development, fetal growth and the pregnancy outcome *Human Reproduction* **19** 715-722

Naz RK (2005) Contraceptive vaccines *Drugs* **65** 593-603

Nettles VF (1997) Potential consequences and problems with wildlife contraceptives *Reproduction, Fertility and Development* **9** 137-143

O'Rand MG and Lea IA (2002) Update on contraceptive vaccines: understanding the immune response *Infertility and Reproductive Medicine Clinics of North America* **13** 221-232

Pest Animal Control CRC (2004) Innovative anti-fertility products and strategies developed for mouse control. In *Annual Report 2003-2004: Pest Animal Control CRC* pp 8-11 Eds. Cooperative Research Centre for Pest Animal Control, Canberra http://www.invasiveanimals.com/images/pdfs/FINALlr.pdf (accessed 5 December 2005)

Pomering M, Jones RC, Holland MK, Blake AE and Beagley KW (2002) Restricted entry of IgG into male and female rabbit reproductive ducts following immunization with recombinant rabbit PH-20 *American Journal of Reproductive Immunology* **47** 174-182

Primakoff P, Lathrop W, Woolman L, Cowan A and Myles D (1988) Fully effective contraception in male and female guinea pigs immunized with the sperm protein PH-20 *Nature* **335** 543-546

Primakoff P, Woolman-Gamer L, Tung KS and Myles DG (1997) Reversible contraceptive effect of PH-20 immunization in male guinea pigs *Biology of Reproduction* **56** 1142-1146

Ramarao CS, Myles DG, White JM and Primakoff P (1996) Initial evaluation of fertilin as an immunocontraceptive antigen and molecular cloning of the cynomolgus monkey fertilin beta subunit *Molecular Reproduction and Development* **43** 70-75

Redwood AJ, Messerle M, Harvey NL, Hardy CM, Koszinowski UH, Lawson MA and Shellam GR (2005) Use of a murine cytomegalovirus K181-derived bacterial artificial chromosome as a vaccine vector for immunocontraception *Journal of Virology* **79** 2998-3008

Reubel GH, Beaton S, Venables D, Pekin J, Wright J, French N and Hardy CM (2005) Experimental inoculation of European red foxes with recombinant vaccinia virus expressing zona pellucida C proteins *Vaccine* **23** 4417-4426

Remy JJ, Couture L, Rabesona H, Haertle T and Salesse R (1996) Immunization against exon 1 decapeptides from the lutropin/choriogonadotropin receptor or the follitropin receptor as potential male contraceptive *Journal of Reproductive Immunology* **32** 37-54

Robinson AJ and Holland MK (1995) Testing the concept of virally vectored immunosterilisation for the control of wild rabbit and fox populations in Australia *Australian Veterinary Journal* **72** 65-68

Rutberg AT, Naugle RE, Thiele LA and Liu IKM (2004) Effects of immunocontraception on a suburban population of white-tailed deer *Odocoileus virginianus Biological Conservation* **116** 243-250

Sacco AG, Subramanian MG and Yurewicz EC (1981) Active immunization of mice with porcine zonae pellucidae: immune response and effect on fertility *Journal of Experimental Zoology* **218** 405-418

Scalzo AA, Lyons PA, Fitzgerald NA, Forbes CA, Yokoyama WM and Shellam GR (1995) Genetic mapping of Cmv1 in the region of mouse chromosome 6 encoding the NK gene complex-associated loci Ly49 and musNKR-P1 *Genomics* **27** 435-441

Scalzo AA, Manzur M, Forbes CA, Brown MG and Shellam GR (2005) NK gene complex haplotype variability and host resistance alleles to murine cytomegalovirus in wild mouse populations *Immunology and Cell Biology* **83** 144-149

Seamark RF (2001) Biotech prospects for the control of introduced mammals in Australia *Reproduction, Fertility and Development* **13** 705-711

Shellam GR (1994) The potential of murine cytomegalovirus as a viral vector for immunocontraception *Reproduction, Fertility and Development* **6** 401-409

Singleton GR, Smith AL, Shellam GR, Fitzgerald N and Muller WJ (1993) Prevalence of viral antibodies and helminths in field populations of house mice (*Mus domesticus*) in southeastern Australia *Epidemiology and Infection* **110** 399-417

Singleton GR, Smith AL and Krebs CJ (2000) The prevalence of viral antibodies during a large population fluctuation of house mice in Australia *Epidemiology and Infection* **125** 719-727

Singleton GR, Farroway LN, Chambers LK, Lawson MA, Smith AL and Hinds LA (2002) Ecological basis for fertility control in the house mouse (*Mus domesticus*) using immunocontraceptive vaccines *Reproduction Supplement* **60** 31-39

Singleton GR, Brown PR, Pech RP, Jacob J, Mutze GJ and Krebs CJ (2005) One hundred years of eruptions of house mice in Australia - a natural biological curio *Biological Journal Of The Linnean Society* **84** 617-627

Smith LM, Lloyd ML, Harvey NL, Redwood AJ, Lawson MA and Shellam GR (2005) Species-specificity of a murine immunocontraceptive utilising murine cytomegalovirus as a gene delivery vector *Vaccine* **23** 2959-2969

Strive T, Hardy CM, French N, Wright JD, Nagaraja N and Reubel GH (2006) Development of canine herpesvirus based antifertility vaccines for foxes using bacterial artificial chromosomes *Vaccine* **24** 980-988

Subramanian S, Rao J, Jyothi P and Adiga PR (2000) Strain-dependent variability in immune response to chicken riboflavin carrier protein in mice with different haplotypes *Immunological Investigations* **29** 397-409

Subramanian S, Andal S, Karande AA and Adiga PR (2003) Epitope mapping and evaluation of specificity of T-helper sites in four major antigenic peptides of chicken riboflavin carrier protein in outbred rats *Biochemical and Biophysical Research Communications* **311** 11-16

Suri A (2004) Sperm specific proteins-potential candidate molecules for fertility control *Reproductive Biology and Endocrinology* **2** 10

Sutherland DR, Spencer PB, Singleton GR and Taylor AC (2005) Kin interactions and changing social structure during a population outbreak of feral house mice *Molecular Ecology* **14** 2803-2814

Turner JW Jr, Liu IK, Flanagan DR, Bynum KS and Rutberg AT (2002) Porcine zona pellucida (PZP) immunocontraception of wild horses (*Equus caballus*) in Nevada: a 10 year study *Reproduction Supplement* **60** 177-186

Tyndale-Biscoe CH (1991) Fertility control in wildlife *Reproduction, Fertility and Development* **3** 339-343

Tyndale-Biscoe CH (1994) Virus-vectored immunocontraception of feral mammals *Reproduction, Fertility and Development* **6** 281-287

Voigt V, Forbes CA, Tonkin JN, Degli-Esposti MA, Smith HR, Yokoyama WM and Scalzo AA (2003) Murine cytomegalovirus m157 mutation and variation leads to immune evasion of natural killer cells *Proceedings of the National Academy of Sciences USA* **100** 13483-13488

**Williams CK** (1997) Development and use of virus-vectored immunocontraception *Reproduction, Fertility and Development* **9** 169-178

**Zeng W, Ghosh S, Lau YF, Brown LE and Jackson DC** (2002) Highly immunogenic and totally synthetic lipopeptides as self-adjuvanting immunocontraceptive vaccines *Journal of Immunology* **169** 4905-4912

# Functional analyses of the sperm centrosome in human reproduction: implications for assisted reproductive technique

Yukihiro Terada

*Department of Obstetrics and Gynecology, Tohoku University School of Medicine, 1-1 Seiryo-machi, Aoba-ku, Sendai, Miyagi 980-8574, Japan*

Although intracytoplasmic sperm injection (ICSI) is an innovative treatment for male infertility, a significant number of clinical cases of fertilization failure remain. ICSI overcomes the difficulty during fertilization of spermatozoon entry into the egg cytoplasm. The goal of fertilization, however, is the union of the male and female genomes; spermatozoon incorporation into the oocyte is only the initiation of fertilization. During fertilization in most mammalian species, including humans, the spermatozoon introduces the centrosome, which acts as a microtubule organizing center (MTOC). By promoting pronuclear apposition and mitotic spindle formation, the spermatozoon plays the leading part in the induction of "motility" post-ICSI in fertilization. The present review introduces the remaining challenges in functional assessment of the human sperm centrosome and discusses the biparental (for example, rabbit) and maternal (for example, parthenogenesis) centrosomal contributions to microtubule organization during development.

## Introduction

Assisted Reproductive Technique (ART) is an innovative tool for treatment of human sterility. Although intracytoplasmic sperm injection (ICSI) is an innovative treatment for male infertility (Palermo *et al.*, 1992), a significant number of clinical cases of fertilization failure remain (Rawe *et al.*, 2000). Little information exists, however, about the molecular and cellular events occurring in ART in humans (Schatten *et al.*, 1998). From the gamete to the neonate, human reproduction involves a series of cell motility events, including both cell movement and morphological changes. The cytoskeleton plays a critical role in cell motility.

During human fertilization, the sperm introduces a centrosome, a microtubule organizing center (MTOC), into the oocyte. The radial array of microtubules emanating from the sperm centrosome, called the "sperm aster", is essential for pronuclear movement and formation of the first mitotic spindle (Schatten, 1994).

In this article, recent discoveries in the mechanisms governing cytoskeletal dynamics during fertilization are reviewed. Investigations into human sperm centrosomal function are also discussed in detail. With regard to the MTOC in oocyte cytoplasm without spermatozoon centrosome, we examined cytoskeletal dynamics in both rabbit fertilization, which exhibits biparental inheritance of centrosomes, and the parthenogenesis of bovine eggs.

E-mail: terada@mail.tains.tohoku.ac.jp

## Sperm centrosomal function post-ICSI during human fertilization and a functional assay for human sperm centrosomes by heterologous ICSI

While numerous studies have assessed the fertility of human sperm, most only measure the sperm's ability to enter the egg cytoplasm. Fertilization, however, requires union of the male and female genomes at metaphase of the first mitosis, not only spermatozoon entry. During human fertilization, the spermatozoon introduces a centrosome, which serves as the microtubule organizing center (MTOC), allowing microtubules to be nucleated within the inseminated oocyte by the sperm centrosome. Organization of microtubules from the sperm centrosome is essential for both the movement and fusion of male and female pronuclei (Simerly et al., 1995; Terada et al., 2003). The paternal centrosome replicates during the first cell cycle to form the two poles of the mitotic spindle that are required for cleavage.

Abnormal microtubule organization in human zygotes that have been clinically diagnosed as "unfertilized" suggests that centrosomal dysfunction contributes significantly to fertilization failure after proper spermatozoon entry (Asch et al., 1995). Rawe et al. (2000) reported that of 150 human oocytes that failed to fertilize after ICSI, approximately half displayed activation failure; approximately 30% exhibited defects in pronuclear formation/migration. Function of the zygotic centrosome varies among bulls during in vitro fertilization (IVF); this variation affects male fertility (Navara et al., 1996). These results suggest that sperm centrosomal function impacts on fertility in humans. As proper function of the human sperm centrosome is essential for human fertility, an appropriate assay examining centrosomal inheritance and function would likely benefit ART immensely. Direct assessment of human sperm centrosomal function, however, remains difficult. Recently, a novel method examining sperm centrosomal function using the heterologus ICSI system was reported, in which human sperm are microinjected into either rabbit (Terada et al., 2000; 2004) or bovine (Nakamura et al., 2001; 2002; Rawe et al., 2002) eggs. After incorporation of the human spermatozoa into the eggs, we observed that the sperm aster was organized from the sperm centrosome; the sperm aster enlarged as the sperm nuclei underwent pronuclear formation (Terada et al., 2002). The sperm aster formation rate at 6 h post-ICSI was 60.0% in bovine eggs and 36.1% in rabbit eggs (Terada et al., 2000; Nakamura et al., 2001). In these systems, microtubule organization derives from the paternal centrosome during fertilization, which is similar to the events of microtubule organization functioning in human fertilization. In rabbit eggs, human sperm aster formation rate correlated with the rate of cleavage, but did not reflect the rate of pronuclear formation in clinical IVF (Terada et al., 2004). These studies reflect the intimate relationship between infertility and sperm centrosomal dysfunction.

Globozoospermia is a special feature of teratospermia, in which sperm are characterized by a round head, the lack of an acrosome and acrosomal enzymes, and a disorganized mid-piece. Round-headed sperm cannot penetrate the zona pellucida of an egg, so they cannot achieve fertilization, resulting in infertility (Weissenberg et al., 1983). We assessed sperm centrosomal function in round-headed spermatozoa by heterologous ICSI, using bovine eggs. The rate of sperm aster formation in eggs injected with round-headed spermatozoa was 15.8%, which is significantly lower than the rates observed for eggs injected with fertile donor spermatozoa (Nakamura et al., 2002). Ethanol activation after ICSI improved male pronuclear formation in eggs injected with round-headed spermatozoa to 84.9%. In contrast, ethanol activation did not improve the rate of sperm aster formation following ICSI with round-headed spermatozoa (32.3%), suggesting that sperm centrosomal function is independent of the spermotozoa's ability to activate the egg (Nakamura et al., 2002).

Dysplasia of the fibrous sheath (DFS), a rare form of teratospermia, results in infertility. DFS sperm are immotile due to structural deformities from the midpiece to the tail (Chemes et al.,

1988; 1998). These sperm also exhibit centrosomal dysfunction; both abnormalities are potential causes of infertility (Rawe *et al.*, 2002). Even after ICSI, failure of either fertilization or embryonic development continued in several patients (Chemes and Rawe, 2003). The rate of human sperm aster formation in bovine eggs injected with DFS spermatozoa was less than 10% (Rawe *et al.*, 2002).

These results indicate that sperm from men with congenital teratospermia exhibit centrosomal dysfunction. The cause of centrosomal dysfunction in these forms of teratozoospermia is uncertain, although morphological abnormalities in the midpiece of the sperm are reported (Pedersen and Rebbe, 1974; Chemes *et al.*, 1988).

Is it possible to restore defective human sperm centrosomal function? Nakamura *et al.* (2005) reported an attempt to restore defective human sperm centrosomal function using a heterologus ICSI system. Prior to ICSI, sperm were treated with dithiothreitol (DTT), which reduces the disulphide bonds within the head and pericentriolar regions of the sperm, thus unraveling the sperm centrosome. After ICSI, the bovine oocytes were treated with the cytoskeletal stabilizer paclitaxel. The combination of DTT and paclitaxel treatment promoted microtubule organization in heterologous ICSI using dead spermatozoa from a fertile donor, which could not induce microtubule organization without treatment. This treatment, however, was ineffective for DFS sperm. The safety of this method should be discussed carefully. While these initial experiments were unsuccessful at treating the sperm defects, it may be possible in the future to reverse the failure of sperm centrosomal function post-ICSI.

Current methods of ART available do not address sperm centrosomal dysfunction. Further cellular and molecular biological advances are needed to overcome post-ICSI fertilization failure caused by sperm centrosomal dysfunction.

### Is sperm centrosomal function essential for the completion of fertilization?

In human fertilization, centrosomal inheritance is paternal; sperm centrosomal function is important for events in development immediately post-ICSI. The pattern of centrosome inheritance during fertilization differs between species. Appearance of the microtubule organizing center (MTOC) shows its centrosomal inheritance. In contrast to most other mammals, rodents exhibit maternal inheritance of centrosomes during fertilization (Schatten *et al.*, 1985; 1986; 1991; Hewitson *et al.*, 1997). In mice, the paternal centrosome degenerates during spermiogenesis (Manandhar *et al.*, 1998), sperm asters are not formed at the base of the incorporated sperm head (Schatten, 1994). Microtubules are instead organized from multiple centrosomal foci that pre-exist in the cytoplasm of unfertilized eggs (Maro *et al.*, 1985; Schatten *et al.*, 1986). Thus, centrosomal inheritance in rodent fertilization is maternal; the sperm centrosome is dispensable for the completion of fertilization. The number of reports of successful full term births in cloned mammals (Wakayama *et al.*, 1998), even of parthenogenesis (Kono *et al.*, 2004), have been increasing. These techniques make it possible to produce mammalian offspring without any contribution from a sperm centrosome. Thus, we are forced to raise the question, "is sperm centrosomal function crucial for the completion of fertilization?"

### Commonalities and differences in microtubule organization during rabbit ICSI in the presence or absence of a sperm centrosome

Rabbits are lagomorphs, unique animals in which centrosomal inheritance during fertilization is a blend of paternal and maternal. Previously, centrosomes were considered to follow a paternal pattern of inheritance, due to the presence of a monoastral sperm aster during fertilization (Longo, 1976; Yellera-Fernandez *et al.*, 1992). An isolated sperm head lacking a midpiece failed to nucleate sperm asters in rabbit eggs (Pinto-Correia *et al.*, 1994). These reports support the concept that a functioning paternal centrosome is necessary for rabbit fertilization. During *in vivo* rabbit fertilization, microtubules are organized into a radial aster from the sperm head; later, cytoplasmic microtubules are organized around the male and female pronuclei without a distinct nucleation site (Terada *et al.*, 2000). Microtubule distribution at the late pronuclear stage, however, is more similar to that described for the mouse (Schatten *et al.*, 1991) than that described for humans. In parthenogenetically-activated rabbit eggs, microtubule arrays were organized around the single female pronucleus (Terada *et al.*, 2000); *de novo* formation of centrioles was observed at morula or early blastocyst stages, but were not observed in the first cell cycle of fertilization (Szollosi and Ozil, 1991). These reports indicate the possibility of a maternal contribution to fertilization.

These observations suggest the possibility of biparental centrosomal contribution during rabbit fertilization, which contrasts previous reports of a strictly paternal inheritance pattern (Longo, 1976; Yellera-Fernandez *et al.*, 1992). The specific roles of the paternal and maternal centrosomes, however, remain unclear in species with biparental inheritance.

Morita *et al.* (2005) tested a model of paternal centrosome dysfunction to identify the role of the sperm centrosome in rabbit fertilization. The sperm centrosome was removed from the sperm nucleus by sonication. In this model of paternal centrosomal dysfunction, ICSI with an isolated sperm head was performed using a Piezo-driven pipette (Piezo-ICSI, Primtech, Tukuba, Japan) (Kimura and Yanagimachi, 1995; Yanagimachi, 1998). As a control, rabbit Piezo–ICSI using an intact rabbit spermatozoon was also evaluated. To assess the relative contributions of paternal and maternal centrosomes during rabbit fertilization, microtubule organization and early embryonal development in rabbit zygotes following Piezo-ICSI with and without sperm centrosomes were compared. In Piezo-ICSI using intact spermatozoon, the observed microtubule organization and chromatin configuration were similar to that observed in *in vivo* fertilization (Terada *et al.*, 2000). No aster formation could be observed in oocytes following injection with an isolated sperm head. Microtubule organization between male and female pronuclei was observed without a distinct nucleation site. At the late pronuclear stage, ICSI with an isolated sperm head produced a similar microtubule organization as that seen in late pronuclear stage eggs after intact sperm injection. The first mitotic spindle was organized in eggs following ICSI with either an isolated sperm head or an intact sperm head. These results suggest that the maternal centrosome fulfilled the paternal sperm centrosomal function in its absence, as microtubule organization without a clear nucleation site was not observed in oocytes following ICSI with an intact spermatozoon. In rabbit oocytes, the function of the paternally-derived MTOC can be replaced by the maternal cytoplasmic centrosome, suggesting that normal fertilization in rabbits can progress in the absence of a sperm-derived centrosome.

### Microtubule organization during mammalian parthenogenesis: cleavage in the absence of the sperm centrosome

Parthenogenesis is an extraordinary process, in which the activated oocyte initiates full development in the absence of a genetic contribution from a male that can result in a sexually mature adult. Parthenogenesis is observed in many insects, crustaceans, rotifers and reptiles, but is not naturally seen in mammals. Artificial parthenogenesis, however, can be induced in mammals by artificial activation of the egg. In mammals, almost all parthenogenetic embryos die at the early stages of development. Kono *et al.* (2004), however, reported that parthenogenetic mice generated from reconstructed oocytes containing two haploid copies of the maternal genome could develop to adulthood.

Almost mammalian egg loses its centrosome during oogenesis; the spermatozoon reintroduces a centrosome at fertilization to function as the MTOC. The sperm aster plays functions in pronuclear migration and positioning, a process that requires cytoplasmic dynein, a microtubule-based motor protein and its cofactor dynactin (Schatten, 1994; Reinsch and Gönczy, 1998; Payne *et al.*, 2003). Multiple studies have focused on the mechanisms governing microtubule organization, centrosome behavior and pronuclear migration during fertilization. In parthenogenesis, however, the mechanisms of microtubule organization and pronuclear positioning in the absence of a sperm centrosome have not been elucidated.

Fertilization in rodents, unlike that in other mammals, relies on maternal centrosomes. After the spermatozoon enters the rodent oocyte, the cytoplasm fills with microtubules in a disarrayed pattern, which ultimately move the pronuclei into close apposition at the cell center (Schatten *et al.*, 1986). The microtubule and chromatin dynamics of rodent parthenogenesis are very similar to those observed during normal fertilization, with the exception that there is no male contribution. Hewitson *et al.* (1997) reported that cytoplasmic microtubules are first detected at telophase-II stage in hamster parthenogenesis; these structures enlarge throughout the cytoplasm, causing the female pronucleus, which is surrounded by disordered microtubules, to move towards a more central position.

While the paternal inheritance of functional centrosomes has been suggested during fertilization of non-rodent mammals, in some mammals, such as cows (Navara *et al.*, 1994; Shin and Kim, 2003), rabbits (Pinto-Correia *et al.*, 1994), pigs (Kim *et al.*, 1996) and marsupials (Breed *et al.*, 1994), parthenotes exhibit disarrayed microtubules in the cytoplasm shortly after artificial activation. These results suggest that mammalian oocytes may be competent to form functional centrosomes in the absence of any contribution from spermatozoa.

Morito *et al.* (2005) reported the function of maternal centrosomes in bovine parthenotes, which promoted cleavage without a sperm centrosome, by imaging microtubule organization, pronuclear position, and the distribution of γ-tubulin by immunocytochemistry and conventional epifluorescence microscopy. In bovine parthenotes treated with paclitaxel, cytoplasmic microtubule asters became organized shortly after chemical activation, with the microtubules radiating dynamically toward the female pronucleus. The patterns of microtubule localization correlated well with pronuclear movement to the cell center. Microtubules aggregated at regions of high γ-tubulin concentrations, although γ-tubulin did not localize into spots until the first interphase of bovine parthenogenesis. These findings indicate that γ-tubulin serves as the maternal centrosome, promoting cytoplasmic microtubule organization to move the female pronucleus to the cell center. Thus, the maternal centrosome may serve as a functional centrosome in the absence of a sperm contribution, although this structure is less competent for microtubule organization in comparison to centrosomes including sperm centrosomal components.

Mammalian parthenotes can develop to late pre-implantation stages, and even in rare cases, to adulthood, although normal genomic imprinting requires a biparental nuclear contribution to reach full term and birth. Although the mechanisms governing microtubule organization and pronuclear positioning during mammalian parthenogenesis have not been well examined, the observation of cytoplasmic microtubules during parthenogenesis in the absence of a sperm centrosome suggests that maternal centrosomes can organize microtubules. These centrosomes are likely responsible for pronuclear movement in parthenogenesis.

## Acknowledgements

I am grateful to Drs. Takashi Murakami, Soichi Nakamura, Junko Morita, Masahito Tachibana, Yuki Shima-Morito, Tomoko Kakoi-Yoshimoto and Shin-ichi Hayasaka in Department of Obstetrics and Gynecology, Tohoku University School of Medicine, for their collaboration.

# References

Asch R, Simerly C, Ord T, Ord VA and Schatten G (1995) The stages at which human fertilization arrests: microtubule and chromosome configurations in inseminated oocytes which failed to complete fertilization and development in humans *Molecular Human Reproduction* **10** 1897-1906

Breed WG, Simerly C, Navara CS, VandeBerg JL and Schatten G (1994) Microtubule configurations in oocytes, zygotes, and early embryos of a marsupial, *Monodelphis domestica Developmental Biology* **164** 230-240

Chemes HE and Rawe VY (2003) Sperm pathology: a step beyond descriptive morphology. Origin, characterization and fertility potential of abnormal sperm phenotypes in infertile men *Human Reproduction Update* **9** 405-428

Chemes HE, Brugo S, Zanchetti F, Carrere C and Lavieri JC (1988) Dysplasia of the fibrous sheath: an ultrastructural defect of human spermatozoa associated with sperm immotility and primary sterility *Fertility and Sterility* **48** 664-669

Chemes HE, Olmedo SB and Carrere C (1998) Ultrastructural pathology of the sperm flagellum: association between flagellar pathology and fertility prognosis in severely asthenozoospermic men *Human Reproduction* **13** 2521-2526

Hewitson L, Haavisto A, Simerly C and Schatten G (1997) Microtubule organization and chromatin configurations in hamster oocytes during fertilization and parthenogenetic activation, and after insemination with human sperm *Biology of Reproduction* **57** 967-975

Kim N-H, Simerly C, Funahashi H, Schatten G and Day BN (1996) Microtubule organization in porcine oocytes during fertilization and parthenogenesis *Biology of Reproduction* **54** 1397–1404

Kimura Y and Yanagimachi R (1995) Intracytoplasmic sperm injection in the mouse *Biology of Reproduction* **52** 709-720

Kono T, Obata Y and Wu Q (2004) Birth of parthenogenetic mice that can develop to adulthood *Nature* **428** 860-864

Longo FJ (1976) Sperm aster in rabbit zygotes: its structure and function *Journal of Cell Biology* **69** 539-547

Manandhar G, Sutovsky P, Joshi HC, Stearns T and Schatten G (1998) Centrosome reduction during mouse spermiogenesis *Developmental Biology* **203** 424-434

Maro B, Howlett SK and Webb M (1985) Non spindle microtubule organizing centers in metaphase-2 arrested mouse oocytes *Journal of Cell Biology* **101** 1665-1672

Morita J, Terada Y, Hosoi Y, Fujinami N, Sugimoto M, Nakamura S, Murakami T, Yaegashi N and Okamura K (2005) Microtube organization during fertilization by intracytoplasmic sperm injection with and without sperm centrosome *Reproductive Medicine and Biology* **4** 169-177

Morito Y, Terada Y, Nakamura S, Morita J, Yoshimoto T, Murakami T, Yaegashi N and Okamura K (2005) Dynamics of microtubules and positioning of females pronucleus during bovine parthenogenesis *Biology of Reproduction* **73** 935-942

Nakamura S, Terada Y and Horiuchi T (2001) Human sperm aster formation and pronuclear decondensation in bovine eggs following intracytoplasmic sperm injection using a Piezo-Driven Pipette: a novel assay for human sperm centrosomal function *Biology of Reproduction* **65** 1359-1363

Nakamura S, Terada Y and Horiuchi T (2002) Analysis of the human sperm centrosomal function and the oocyte activation ability in a case of globozoospermia, by ICSI into bovine oocytes *Human Reproduction* **17** 2930-2934

Nakamura S, Terada Y, Rawe Y, Uehara S, Morito Y, Yoshimoto T, Tachibana M, Murakami T, Yaegashi N and Okamura K (2005) A trial to restore defective human sperm centrosomal function *Human Reproduction* **20** 1933-1937

Navara CS, First NL and Schatten G (1994) Microtubule organization in the cow during fertilization, polyspermy, parthenogenesis, and nuclear transfer: The role of the sperm aster *Developmental Biology* **162** 29–40

Navara CS, First NL and Schatten G (1996) Phenotypic variations among paternal centrosomes expressed within the zygote as disparate microtubule lengths and sperm aster organization: correlations between centrosome activity and developmental success *Proceedings of the National Academy of Sciences USA* **93** 5384-5388

Palermo G, Joris H, Devroey P and Van Steirteghem AC (1992) Pregnancies after intracytoplasmic injection of single spermatozoon into an oocyte *Lancet* **340** 17-18

Payne C, Rawe V, Ramalho-Santos J, Simerly C and Schatten G (2003) Preferentially localized dynein and perinuclear dynactin associate with nuclear pore complex proteins to mediate genomic union during mammalian fertilization *Journal of Cell Science* **116** 4727-4738

Pedersen H and Rebbe H (1974) Fine structure of round-headed human spermatozoa *Journal of Reproduction and Fertility* **37** 51-54

Pinto-Correia C, Poccia DL, Chang T and Robl JM (1994) Dephosphorylation of sperm midpiece antigens initiates aster formation in rabbit oocytes *Proceedings of the National Academy of Sciences USA* **91** 7894-7898

Rawe VY, Olmedo SB, Nodar FN, Doncel GD, Acosta AA and Vitullo AD (2000) Cytoskeletal organization defects and abortive activation in human oocytes after IVF and ICSI failure *Molecular Human Reproduction* **6** 510-516

Rawe VY, Terada Y, Nakamura S, Chillik CF, Brugo OS and Chemes HE (2002) A pathology of the sperm centriole responsible for defective sperm aster formation, syngamy and cleavage *Human Reproduction* **17** 2344-2349

Reinsch S and Gönczy P (1998) Mechanisms of nuclear

positioning *Journal of Cell Science* **111** 2283-2295

Schatten G (1994) The centrosome and its mode of inheritance: the reduction of the centrosome during gametogenesis and its restoration during fertilization *Developemntal Biology* **165** 299-335

Schatten G, Simerly C and Schatten H (1985) Microtubile configurations during fertilization, mitosis, and early development in the mouse and the requirement for egg microtubule-mediated motility during mammalian fertilization *Proceedings of the National Academy of Sciences USA* **82** 4152-4156

Schatten H, Schatten G, Mazia D, Balczon R and Simerly C (1986) Behavior of centrosome during fertilization and cell division in mouse oocytes and in sea urchin eggs *Proceedings of the National Academy of Sciences USA* **83** 105-109

Schatten G, Simerly C and Schatten H (1991) Maternal inheritence of centrosomes in mammals? Study on parthenogenesis and polyspermy in mice *Proceedings of the National Academy of Sciences USA* **88** 6785-6789

Schatten G, Hewitson L, Simerly C, Sutovsky P and Huszar G (1998) Cell and molecular challenges of ICSI: ART before science? *Journal of Law, Medicine and Ethics* **26** 29-37

Shin M-R and Kim N-H (2003) Maternal gamma (γ)-tubulin is involved in microtubule reorganization during bovine fertilization and parthenogenesis *Molecular Reproduction and Development* **64** 438–445

Simerly C, Wu GJ and Zoran S (1995) The paternal inheritance of the centrosome, the cell's microtubule-organizing center, in humans, and the implications for infertility *Nature Medicine* **1** 47-52

Szollosi D and Ozil JP (1991) *De novo* formation of centrioles in parthenogenetically activated, diploidized rabbit embryos *Biology of the Cell* **72** 61-66

Terada Y, Simerly CR, Hewitson L and Schatten G (2000) Sperm aster formation and pronuclear decondensation during rabbit fertilization and development of a functional assay for human sperm *Biology of Reproduction* **62** 557-563

Terada Y, Nakamura S, Hewitson L, Sinerly CR, Horiuchi T, Murakami T, Okamura K and Schatten G (2002) Human sperm aster formation after intracytoplasmic sperm injection with rabbit and bovine eggs *Fertility and Sterility* **77** 1283-1284

Terada Y, Nakamura S and Morita J (2003) Intracytoplasmic sperm injection (ICSI): Stilleto Conception or a Stab in ths Dark *Archives of Andrology* **49** 169-177

Terada Y, Nakamura S, Simerly C, Hewitson L, Murakami T, Yaegashi N, Okamura K and Schatten G (2004) Centrosomal function assessment in human sperm using heterologous ICSI with rabbit eggs: a new male factor infertility assay *Molecular Reproduction and Development* **67** 360-365

Wakayama T, Perry AC, Zuccotti M, Johnson KR and Yanagimachi R (1998) Full-term development of mice from enucleated oocytes injected with cumulus cell nuclei *Nature* **394** 369-374

Weissenberg R, Eshkol A and Rudak E (1983) Inability of round acrosome-less human spermatozoa to penetrate zona-free hamster ova *Archives of Andrology* **11** 167-169

Yanagimachi R (1998) Intracytoplasmic sperm injection experiments using the mouse as a model *Human Reproduction* **13** 87-98

Yellera-Fernandez MDM, Crozet N and Ahmed-Ali M (1992) Microtubule distribution during fertilization in the rabbit *Molecular Reproduction and Development* **32** 271-276

# Non-genomic membrane progesterone receptors on human spermatozoa

DN Modi, C Shah and CP Puri

*National Institute for Research in Reproductive Health, Indian Council of Medical Research, J. M. Street, Parel, Mumbai 400 012, India*

Progesterone regulates vital sperm functions such as capacitation and motility; it is also considered as one of the physiological initiators of the acrosome reaction. Progesterone binding and progesterone mediated biological effects are crucial for sperm functions; these are reportedly dysfunctional in a subset of infertile males. Acting through a mechanism independent of transcriptional regulation, the sperm membrane progesterone receptor (PR) demonstrates high structural specificity for the steroid and is unable to interact with progesterone analogs and antiprogestins. At present, the identity of the receptor is unknown; the hormone-receptor interactions are facilitated by albumin and disulphide bonds. Antibodies to the nuclear PR recognize a protein of 55 kDa in sperm lysates that localizes on the acrosomal membrane suggesting the immunological identity of the membrane and the nuclear PR. Decoding the identity of the membrane steroid receptor and understanding the basic cascades of non-genomic mechanisms of progesterone action would be useful in drug designing, targeted towards modifying sperm functions for contraceptive use and for the management of male infertility.

## Introduction

Male factor infertility is observed in about 50% of couples experiencing involuntary childlessness. Amongst these, one third have abnormalities at the chromosomal or at the gene level (micro deletions/mutations). However, the underlying cause of infertility in the remaining cases is essentially unknown. Clinical results from human *in vitro* fertilization (IVF) programs have identified a class of men whose infertility is attributed to dysfunctional acrosome reaction. It has been observed that spermatozoa from a sub-group of normozoospermic infertile men are unable to undergo the acrosome reaction or fertilize an ovum (Benoff *et al.*, 1996). It is suggested that the presence of non-deleterious oligogenic or polygenic modifiers in the testis/spermatozoa may be responsible for defective sperm production or sperm functions hence leading to infertility. The identification of the molecular lesion in the spermatozoa of these men will aid in enhancing our understanding of the process of acrosome reaction and developing rational screening and treatment strategies for male infertility. Such information can also be used in reverse for devising newer male contraceptives.

Corresponding author: Dr Deepak N Modi
E-mail: modidn@icmr.org.in, deepaknmodi@yahoo.com

The acrosome reaction is an irreversible exocytotic process consisting of the fusion and fenestration of the outer acrosomal membrane with the plasma membrane resulting in release of acrosomal enzymes that aid the spermatozoon to penetrate various layers of the oocyte. Several physiological and pharmacological agents are known to initiate acrosome reaction. The pharmacological agents include ionophore A 23187 (Tesarik, 1985), platelet activating factor (Krausz et al., 1994), gamma amino butyric acid (Calogero et al., 1999) and cAMP analogues, dibutyryl cAMP and forskolin (de Jonge et al., 1991). The physiological inducers of the acrosome reaction include different egg investments like the zona pellucida (Cross et al., 1988) and follicular fluid (Tesarik, 1985). Zona pellucida-3 (ZP3) a glycoprotein in the zona pellucida is an in vivo initiator of the acrosome reaction (Wasserman et al., 2005). Along with ZP3, human follicular fluid is also known to bring about capacitation and the acrosome reaction. The acrosome inducing capability of follicular fluid is owing to its steroid content, specifically progesterone (reviewed in Baldi et al., 1995).  The capability of progesterone in inducing the acrosome reaction has been demonstrated in a variety of mammalian species including humans (reviewed in Baldi et al., 1995; Calogero et al., 1999; Sirivaidyapong et al., 2001). Interestingly, progesterone is proposed as a primer for ZP3 interactions and ZP mediated acrosome reaction (Roldan et al., 1994; Baldi et al., 1995). The responsiveness of human spermatozoa to progesterone correlates with the fertilization rate in IVF (Forti et al., 1999; Giojalas et al., 2004) and removal of cumulus cells, a source of progesterone, from oocytes significantly reduces the success rate of IVF in most mammals (Tanghe et al., 2002). Along with acrosome reaction, progesterone regulates sperm capacitation, preserves sperm viability and induces hyperactive motility (Calogero et al., 2000).

That, progesterone is a physiological initiator of the acrosome reaction and a modulator of sperm functions is biologically momentous. Progesterone is secreted by the ovary and present in the follicular fluid with highest concentrations around the cumulous oophorus. Spermatozoa, in the female reproductive tract would be exposed to progesterone, initially leading to hyperactive motility; upon reaching the cumulus (the site for initiation of the acrosome reaction) the concentrations of progesterone increase to those required for the acrosome reaction. The exposure of capacitated spermatozoa to high concentrations of progesterone along with the presence of other facilitators around the cumulus (for example, ZP3) would ultimately culminate in acrosomal exocytosis. Indeed, exposing human spermatozoa to a progesterone gradient simulating the in vivo conditions (encountered as spermatozoa approach the oocyte) results in a differential calcium response. Initially, a slow rise in intracellular calcium occurs in response to progesterone, in a pattern that does not permit the acrosome reaction, but instead show an alternating pattern correlating with flagellar activity (Harper and Publicover, 2005). Eventually, another sustained peak of calcium influx follows at high concentrations of progesterone (encountered around the cumulus) that results in the acrosome reaction (Meizel et al., 1997).

## Progesterone and sperm pathophysiology

Considering the facilitatory role of progesterone in sperm functions, specifically the acrosome reaction, it is conceivable that defective progesterone binding or function in spermatozoa might account for some cases of male infertility; particularly, those associated with acrosome reaction defects. Indeed, an association between defective progesterone activity and reduced fertility has been demonstrated. Subnormal responses to progesterone of calcium influx and the acrosome reaction have been demonstrated in spermatozoa from oligozoospermic men and in spermatozoa from male partners of unexplained infertility (Tesarik and Mendoza, 1992; Oehninger et al., 1994; Baldi et al., 1998). Along with defective responsiveness, the binding of progesterone on sperm membranes is also altered in men with abnormal spermiograms. As compared to

normozoospermic men, a significant decrease in the number of progesterone binding spermatozoa was observed in men suffering from oligozoospermia, asthenozoospermia,
oligoasthenozoospermia and teratozoospermia (Gadkar et al., 2002). A weak positive correlation was observed between the percentage of progesterone binding spermatozoa and the traditional semen parameters like sperm motility and morphology. Interestingly, a strong positive
correlation was observed between the number of progesterone binding spermatozoa and the
percentage of hypo-osmotic swelling test positive and acrosome reacting spermatozoa from
normozoospermic males and also in men with abnormal spermiograms (Gadkar et al., 2002).
These observations suggest that assessment of progesterone binding and progesterone-mediated calcium influx in spermatozoa can serve as markers to diagnose men with dysfunctional
acrosome reaction. The priming effects of progesterone on ZP-mediated acrosome reaction
(Roldan et al., 1994) as well as oocyte penetration in infertile patients (Oehninger et al., 1994)
suggests a role of progesterone in fertilization.

## Modes of progesterone action

According to the generally accepted theory of steroid actions, progesterone acts via its intracel-
lular/nuclear receptors that are transcription factors regulating gene transcription to modulate
biological functions. This mode of steroid hormone actions is described as the 'genomic mode'
as it involves action on the DNA. Along this pathway, the effects of steroids are observed only
hours after activation and the effects are generally long lasting.

However, spermatozoa are transcriptionally and translationally inactive cells and the effects
of progesterone are too rapid (ranging from a few seconds to minutes). Furthermore, the effects
can be elicited even by progesterone coupled with albumin that does not pass across the
plasma membrane and enter the cell. It suggests that the effects of progesterone on spermatozoa are not via the "genomic mode" of action. In some tissues/cells, like ovary, testis, brain,
hepatocytes and platelets et cetera, progesterone was found effective in eliciting a biological
response even in the presence of transcription and translation inhibitors (Revelli et al.,1998;
Losel and Wehling, 2003) suggesting an existence of a mechanism, independent of transcription regulation. The mechanism by which progesterone elicits biological effects independent
of the genomic mode is termed as the 'non-genomic mode' of progesterone actions (Thomas et
al., 2002; Losel and Wehling, 2003). The non-genomic mode of progesterone action has the
following characteristics:

- Effects are too rapid (from few seconds to minutes) to be compatible with the involvement of
changes in mRNA and protein synthesis.

- Effects can be observed in highly specialized cells that do not accomplish mRNA and protein
synthesis (such as spermatozoa and platelets) or in cell clones lacking the nuclear receptors.

- Effects can be elicited even by steroid coupled with high molecular weight substances and therefore
do not pass across the plasma membrane and do not enter the cell.

- Inhibitors of mRNA and protein synthesis do not block the effects.

- Effects are generally not blocked by the antagonist of the nuclear receptor.

- Effects are highly specific, as steroids with very similar but not identical, chemical structures may
show various degree of potency in exerting them.

The effects of progesterone on spermatozoa are known to be rapid, insensitive to antiprogestins,
demonstrate high steroid specificity and act via a membrane receptor system (reviewed in
Revelli et al., 1998). Furthermore, sperm are transcriptionally quiescent cells and the absence

of nuclear PR protein in these cells (Shah et al., 2005a; b) is strong evidence supporting the existence of the non-genomic progesterone action in spermatozoa.

## Progesterone receptors in spermatozoa

We and others have previously reported that membrane progesterone receptors (PR) exist on the sperm plasma membrane particularly the acrosome (Blackmore and Lattanzio, 1991; Tesarik and Mendoza, 1993; Ambhaikar and Puri, 1998; Gadkar et al., 2002). Studies have shown that sperm membrane PR are heat labile and masked molecules (Ambhaikar and Puri 1998). Progesterone covalently conjugated to serum albumin (a molecule which does not allow the steroid to cross the sperm membrane) is able to increase the intracellular concentrations of calcium and induce the acrosome reaction (Blackmore and Lattanzio, 1991; Meizel and Turner, 1991). The same ligand conjugated to fluorescein isothiocynate shows hormone binding at the whole or equatorial acrosomal region (Meizel et al., 1997; Gadkar et al., 2002). However, only about 10-30% of spermatozoa demonstrate progesterone binding on their surface probably indicating the existence of this system only in a small sub-population of cells (Blackmore and Lattanzio, 1991; Tesarik and Mendoza, 1993; Revelli et al., 1998). Based on the single cell assessment of effects of progesterone on calcium influx, some authors hypothesize that a greater proportion of spermatozoa in an ejaculate possess membrane PR. Indeed, about 70- 90% of spermatozoa respond to progesterone with an abrupt increase in intracellular calcium (Plant et al., 1995; Kirkman-Brown et al., 2000; Harper and Publicover, 2005). Thus, the size of the sperm population having membrane PR seems to be greater than originally believed. Supporting this observation, our studies have demonstrated that the sperm membrane PR is a masked protein that is exposed when treated with mild detergent like digitonin (Ambhaikar and Puri, 1998). Flow cytometric quantitation revealed binding of progesterone to less than 20% of the cells in the ejaculate, which increased to more than 90% after the treatment with the detergent (Gadkar et al., 2002). An increase in progesterone binding post digitonin treatment has also been recently demonstrated in porcine spermatozoa (Losel et al., 2004).

## Biophysical characteristics of the sperm membrane PR

Extensive studies have reported that the steroid binding ability, steroid specificity and modes of progesterone actions with respect to membrane PR differ in spermatozoa as compared to the nuclear PR. Herein, we present a detailed account of the structural, biochemical and molecular nature of the putative sperm receptor that mediates the rapid non-genomic actions of progesterone.

### Sperm membrane PR shows high structural specificity to progesterone

The sperm PR has specificity for steroids different from that of the nuclear PR. Using a range of progestins, binding kinetic studies have demonstrated the presence of two classes of PR in spermatozoa, one that has an elevated affinity constant (in the nanomolar range) and is specific for progesterone. The other class of PR has an affinity constant in the micromolar range and binds equally well with other hydroxylated progesterone derivatives (Luconi et al., 1998). The progesterone specific, but not the agonist binding receptor, seems to be unmasked by digitonin treatment (Ambhaikar and Puri, 1998). Whether the agonist-binding site in spermatozoa is biologically active needs be evaluated. However, calcium influx and the acrosome reaction is induced only by progesterone but not other progestins, indicating that the high affinity receptor is responsible for the biological effects (Baldi et al., 1991; Luconi et al., 1998; Harper et al., 2003).

Unlike the nuclear PR, sperm membrane PR do not bind to mifepristone, onapristone and other antiprogestins *in vitro* (Ambhaikar and Puri, 1998). It has been proposed that the failure of antiprogestins to bind the sperm membrane PR could be owing to the differences in the structure of the compounds as compared to progesterone (Yang et al., 1994). However, in most studies, the compounds tested have substitutions in the alpha phenyl ring of the progesterone molecule (Fig. 1A). In this context, we argued that compounds demonstrating modifications at sites other than the alpha phenyl ring might permit the binding of the analogs to sperm membrane PR. One such compound is the antiprogestin, J867 (asoprisnil), which has a substitution in the beta phenyl ring (Fig. 1A). This compound, at specified concentrations, has demonstrated antiprogestational activity in conventional bioassays and has high affinity to the nuclear PR (DeManno et al., 2003). However, like other antiprogestins, in a typical radio-receptor assay, a 10 to 1000 fold higher concentration of J867 failed to displace binding of labeled progesterone to digitonin treated spermatozoa (Fig. 1B). These results entice us to conclude that sperm membrane PR has a high structural specificity only for progesterone which can not be mimicked by similar compounds that behave as agonists or antagonist to the nuclear receptors.

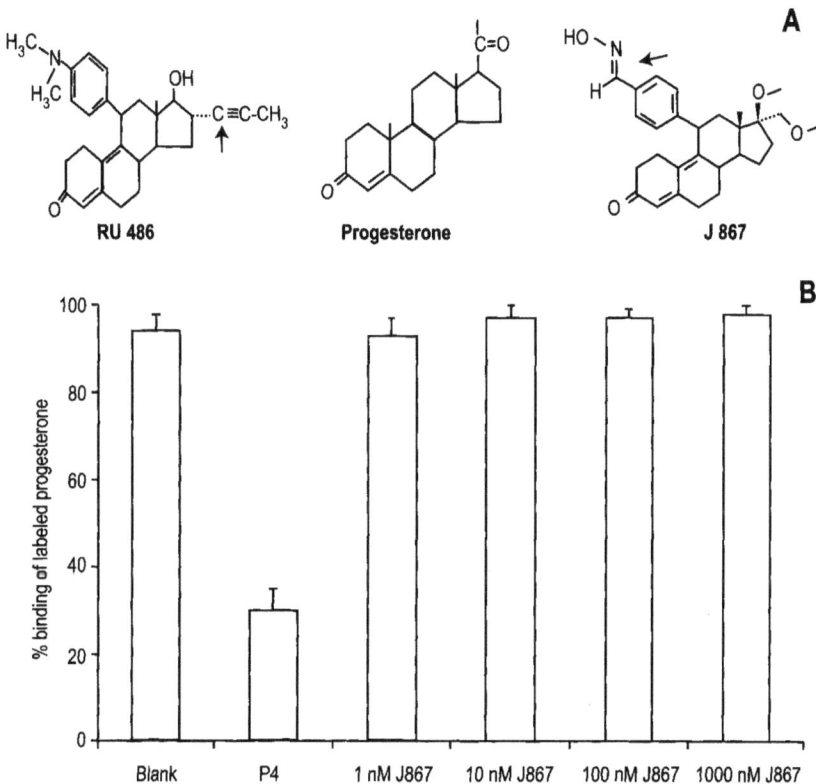

**Fig. 1** Antiprogestin J867 does not interact with sperm progesterone receptor. (A) Comparison of the chemical structures of antiprogestin RU486 and J867 with progesterone. Note the modifications in the beta phenyl ring (arrow) in the J867 compound as opposed to that in the alpha phenyl ring (arrow) in RU486. (B) Radio-receptor assay to study the effect of J867 on progesterone binding. While cold progesterone (P4) significantly displaced the binding of radio labeled progesterone, J867 in a range of concentrations failed to displace progesterone binding to digitonin treated human spermatozoa. The results are expressed as mean ± SD of % binding of labeled progesterone (blank; considered as 100%) observed in three independent experiments.

*Progesterone binds on spermatozoa to a protein containing disulphide bonds*

Considering the requirement of high structural specificity of the ligand to bind the sperm membrane PR, we next investigated if structural modifications in the membrane receptor would affect progesterone binding. Since, disulphide bonds are the basic backbones of proteins for maintaining the tertiary structures, we tested the effects of common disulphide reducing agents, dithioerithrol (DTT) and ß-mercaptoethanol, on progesterone binding to digitonin treated spermatozoa (unpublished results). The binding of tritiated progesterone to sperm membranes was significantly reduced in spermatozoa exposed to varying concentrations of DTT or ß-mercaptoethanol (Fig. 2). These experiments suggest that disulfide bonds must exist in the sperm membrane PR protein. It will be of interest to determine if such modifications in the binding protein could alter the steroid specificity and promote interactions with other progestins and antiprogestins.

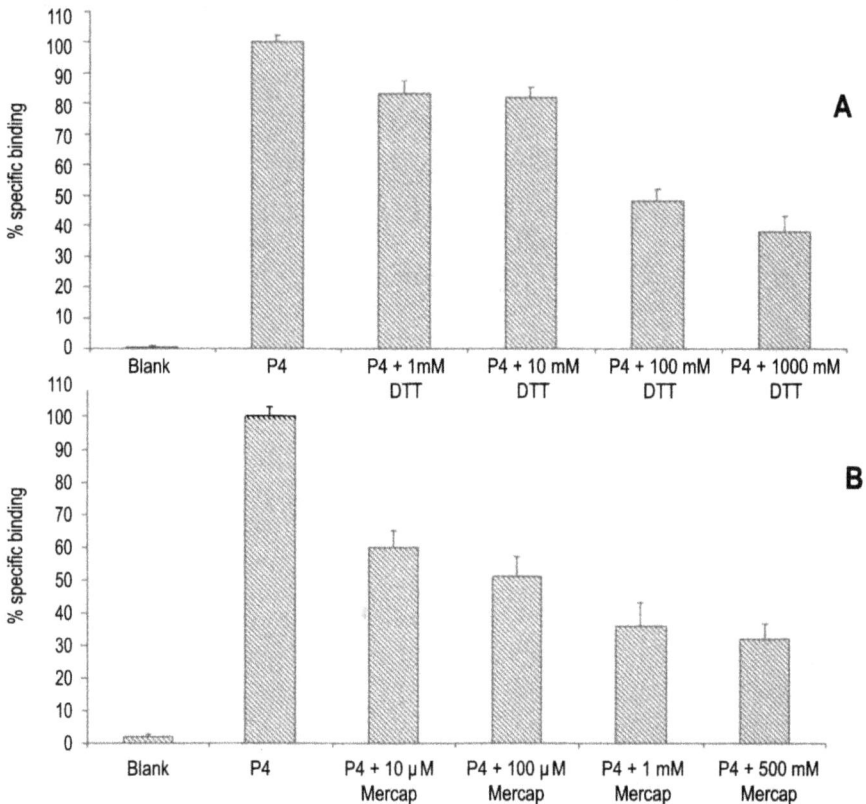

Fig. 2 Effects of disulfide bond modifying agents on binding of progesterone (P4) to human spermatozoa. Digitonin treated spermatozoa were incubated with increasing concentrations of dithioerithrol (DTT, A) or ß-mercaptoethanol (B) and tested for P4 binding in a radio-receptor assay. Both the compounds at different concentrations reduced specific binding of P4. The results are expressed as mean ± SD of % specific binding observed in three independent experiments.

*Binding of progesterone to sperm membrane PR is facilitated by albumin*

In case of the nuclear PR, the diffusion of the steroid in the cell and its binding is facilitated by a number of progesterone binding proteins (Adams, 2005; Hammes et al., 2005). In search of such a facilitator, for the membrane PR, we have earlier shown that follicular fluid enhances binding of progesterone to spermatozoa; this activity is unrelated to its progesterone content (Ambhaikar and Puri, 1998). This indicates existence of some factors in the follicular fluid that promote progesterone binding. To determine the identity of the factor, a strategy outlined in Fig. 3 was employed. Briefly, proteins in steroid-stripped human follicular fluid were fractionated based on size, three fractions namely, F1, F2 and F3 containing proteins in the range of molecular weight <25 kDa, 25-75 kDa and >75 kDa respectively were tested in the radioreceptor assays to check for their ability to facilitate progesterone binding on digitonin treated spermatozoa. Of these, F2 demonstrated "activity" as evident by enhanced binding of progesterone as compared to untreated control (Fig. 4A). Fraction F2 was further separated by eluting proteins from G75 column using increasing concentrations of NaCl (0.1 to 0.5 M). Three fractions F2.1, 2.2 and 2.3 were tested in radioreceptor assays of which F2.1 demonstrated "activity". SDS-PAGE analysis of F2.1 demonstrated presence of a dominant protein with molecular weight of 66 kDa (Fig. 4B). Since this size corresponds to that of albumin, the effect of increasing concentrations of purified human serum albumin were checked in radio-receptor assays. Indeed, human serum albumin enhanced progesterone binding to spermatozoa in a concentration dependent manner (Fig. 4C).

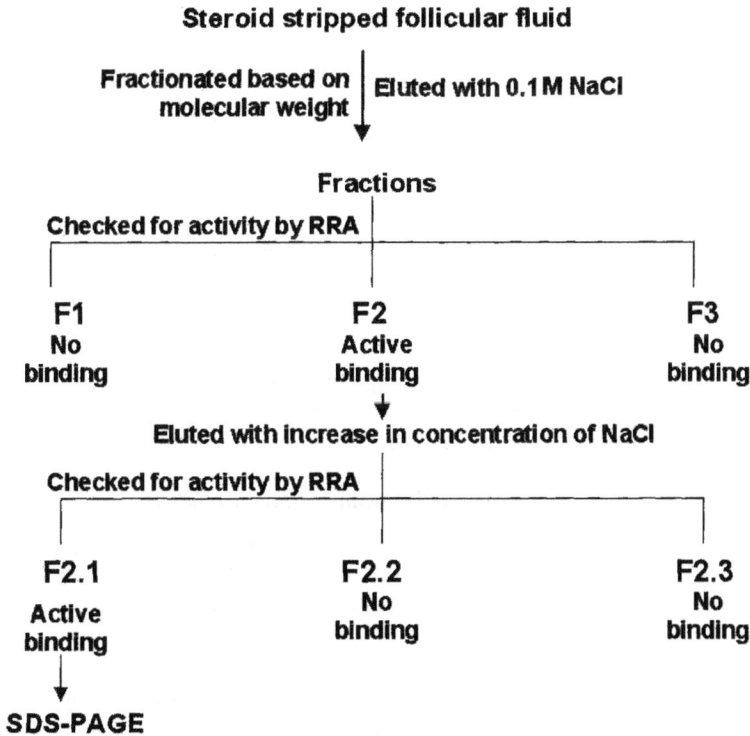

**Steroid stripped follicular fluid**

**Fractionated based on** | **Eluted with 0.1 M NaCl**
**molecular weight**

**Fractions**

**Checked for activity by RRA**

| **F1** | **F2** | **F3** |
| No | Active | No |
| binding | binding | binding |

**Eluted with increase in concentration of NaCl**

**Checked for activity by RRA**

| **F2.1** | **F2.2** | **F2.3** |
| Active | No | No |
| binding | binding | binding |

**SDS-PAGE**

**Fig. 3** Strategy used to identify the protein in human follicular fluid that facilitates progesterone binding on human spermatozoa.

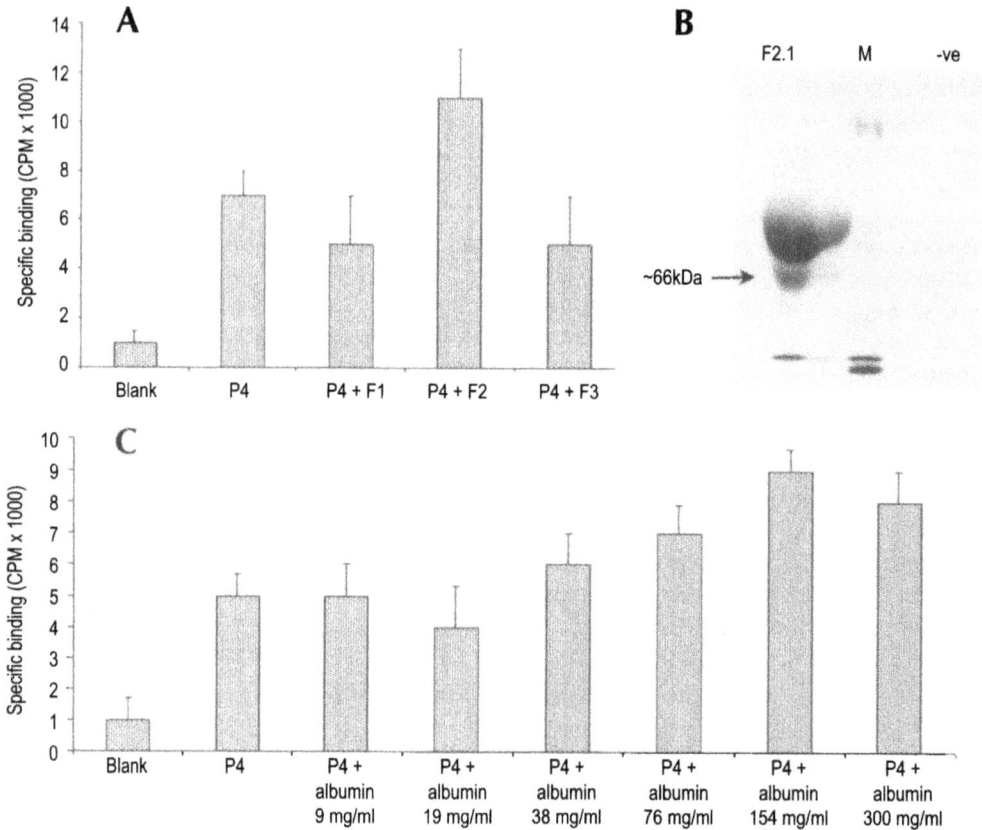

**Fig. 4** Identification of the factor from follicular fluid that facilitates progesterone (P4) binding to human spermatozoa. (A) Results of radio-receptor assay for P4 in presence of fractions F1, F2 and F3 obtained as outlined in Fig 3. Increase in specific binding of progesterone was evident in F2. (B) Commassie blue stained SDS-PAGE of fraction F2.1 (Lane F2.1), lane M is molecular mass ladder, lane –ve is water control. (C) Radio-receptor assay for P4 in presence human serum albumin at varying concentrations. The results of radio-receptor assay are expressed as mean ± SD of specific binding (CPM x 1000) observed in three independent experiments.

*Sperm membrane PR may exist as multiple biochemically distinct isoforms*

Based on the downstream events of progesterone action in spermatozoa, at least three different types of the receptor have been postulated (Revelli et al., 1998). These include: i) plasma membrane $Ca^{2+}$ channel (PR1); ii) membrane-associated protein tyrosine kinase (PR2); and iii) plasma membrane chloride channel (PR3). The tyrosine kinase-associated PR (PR2) seems to be the one visualized by the hormone-binding assay because those spermatozoa that bind to the albumin conjugated progesterone also increase their phosphotyrosine content and undergo the acrosome reaction (Tesarik et al., 1993). PR2 is probably responsible for both the effect of progesterone on the acrosome reaction and on hyperactivated motility (Parinaud and Milhet, 1996). PR1 is probably responsible for the rapid opening of the $Ca^{2+}$ channel and seems to be active in a higher percentage (more than 90%) of spermatozoa (Plant et al., 1995; Kirkman-

Brown *et al.*, 2000; Harper and Publicover, 2005), but has different ligand-binding properties as compared to PR2. The third PR (PR3) is likely to be a γ-aminobutyric acid (GABA) A-receptor/ chloride channel complex, and probably mediates the $Cl_2$- fluxes occurring during acrosomal exocytosis (Wistrom and Meizel, 1993). However, studies have demonstrated that the GABA receptor is not involved in the progesterone induced $Ca^{2+}$ influx in spermatozoa (Baldi *et al.*, 1996; Luconi *et al.*, 1998). At present it is unknown whether the non-genomic effects of progesterone on spermatozoa are mediated by a multi-receptor system or if these PRs are comprised of a single receptor type.

## Molecular identity of membrane receptors for progesterone

For understanding the basic mechanisms of the rapid action of progesterone and a rational use of this system in diagnostics and therapy, it is essential to establish the molecular identity of the membrane PR. Using a variety of experimental systems, from amphibians to primates, a number of investigators are aspiring to solve the enigma of the membrane PR. While most of the studies are directed towards the basic biochemical characterization of the membrane PR; attempts have been made by several investigators to clone and characterize this system (Falkenstein *et al.*, 1996; Buddhikot *et al.*, 1999; Luconi *et al.*, 2002; Saner *et al.*, 2003; Zhu *et al.*, 2003a; b; Losel *et al.*, 2005; Peluso *et al.*, 2005; Shah *et al.*, 2005b). The results emerging from these studies have suggested that

1. The membrane PR is unrelated to nuclear PR and there exists a novel class of protein(s) to mediate the non-genomic action.

2. The nuclear receptor itself, by an unknown mechanism, translocates to the membrane to modulate the non-genomic action.

3. The membrane PR is a truncated/modified version of the nuclear PR to modulate the non-genomic action.

Discussed below is the evidence in support to each of these hypotheses.

*Membrane PR as unrelated to nuclear PR*

This is the most widely accepted theory for membrane PR. Using different strategies and different model systems, proteins that can be designated as the membrane PR have been identified (Falkenstein *et al.*, 1996; Zhu *et al.*, 2003a; b; Peluso *et al.*, 2005). In search of specific steroid membrane-binding sites, two membrane progesterone- binding sites (mPR) from porcine liver microsomes with apparent *K*d values of 11 and 286 nM, have been cloned (Falkenstein *et al.*, 1996; accession no. NM 213911). Interestingly, human spermatozoa incubated with the specific antibody exhibited a significantly reduced progesterone induced calcium increase and an inhibition of the progesterone-mediated acrosome reaction by 62.1% (Buddhikot *et al.*, 1999). However, the site and size of this protein in spermatozoa is different than that speculated for the sperm membrane PR. While the sperm membrane PR is localized exclusively on the acrosome (Sabeur *et al.*, 1996) the mPR antibody recognizes a protein post acrosomally and also in the midpiece (Buddhikot *et al.*, 1999). In Western blots, this antibody identifies a protein of approximately 40 kDa as against 50-60 kDa estimated for sperm PR. These results imply that the progesterone membrane-binding protein in spermatozoa may be closely similar but not identical to the porcine liver mPR.

Another study that has caused a lot of excitement in the area of non-genomic actions of steroids is the identification of a novel class of transmembrane G protein coupled receptor (GPCR, accession no. AF 313615-AF 313620) that bind specifically to progesterone but not to antiprogestins (Zhu *et al.*, 2003a; b). These proteins, originally identified from teleost fish oocytes, have multiple human homologues that are predicted to encode a protein with seven transmembrane domains and one of the isoforms of GPCR activates progesterone mediated signal in cells not responsive to progesterone (Zhu *et al.*, 2003a; b). However, the sperm membrane PR has not been demonstrated to be a GPCR, instead G protein inhibitors failed to block the progesterone mediated acrosome reaction. Thus, the relevance of this class of protein in the context to sperm membrane PR that mediates the acrosome reaction needs to be assessed.

Another candidate membrane PR has been identified in rat granulosa cell membrane extracts. Referred to as either RDA288 or PAI-1 mRNA binding protein 1 (PAIRBP1; accession number XM 216160), it has been implicated as a membrane PR. This protein is localized on the cell membrane and overexpression of PAIRBP1 in granulosa cells increases progesterone binding and its responsiveness (Peluso *et al.*, 2005). In addition, an antibody against PAIRBP1 ablates progesterone mediated biological actions in granulosa cells. However, nothing is known about this protein in humans and its relevance to the sperm membrane PR has not explored.

Although, the above studies demonstrate the existence of a membrane PR which is distinct from the nuclear PR, in all these cases the definition of the role of this molecule as the modulator of progesterone actions awaits experimental investigations using transgenic and knockout approaches. The interpretations of the data derived from the above studies are tantalizing but the obvious caveats including the differences in the experimental systems used, precludes us from making strong conclusions on the identity of the membrane PR that is responsible for the rapid actions of progesterone in spermatozoa. Some support to this misapprehension is further supported by the fact that the putative membrane PR molecules identified by the above independent groups have not been the same and rather belong to a diverse class of proteins.

*Nuclear PR as membrane PR*

An emerging school of thought is that the nuclear PR itself is capable of mediating rapid progesterone induced activation of signal transduction pathways in the absence of gene transcription. The B isoform of the nuclear PR has a unique SH-3 interacting site (AF-3 domain) and is reported to be the motif responsible for the rapid effects of progesterone (Leonhardt *et al.*, 2003; Boonyaratnakornkit *et al.*, 2001). We have recently demonstrated that PR-B mRNA is expressed in the human testis and spermatozoa (Shah *et al.*, 2005a; b). Using RT-PCR, transcript for the B isoform that have 100% sequence homology with the nuclear PR was readily detectable in sperm RNA (accession no. AY382152: Shah *et al.*, 2005b). The existence of sequences encoding for the protein and the hormone binding domains of the nuclear PR have also been demonstrated (accession no. AY382151: Sachdeva *et al.*, 2000; 2005; Luconi *et al.*, 2002). These transcripts are not because of contamination from immature germ cells as *in situ* hybridization demonstrated its localization in the mid piece region of almost all cells (Fig. 5A) as reported for other transcripts (Modi *et al.*, 2005). Thus, nuclear PR mRNA specific to the B isoform is expressed in spermatozoa; whether they translate a functional protein or are just stored, for use post-fertilization, needs to be investigated.

Nevertheless, an antibody against the PR-B isoform blocked progesterone binding to spermatozoa and activation of progesterone mediated protein phosphorylation (Shah *et al.*, 2005b). This protein on immunostaining localizes on the acrosomal region of digitonin treated spermatozoa; ultrastructural studies demonstrated an integral localization on the acrosomal membrane

(Shah et al., 2005b). All these evidences suggest that the nuclear PR, probably the B isoform mediate the rapid non-genomic actions of progesterone in spermatozoa.

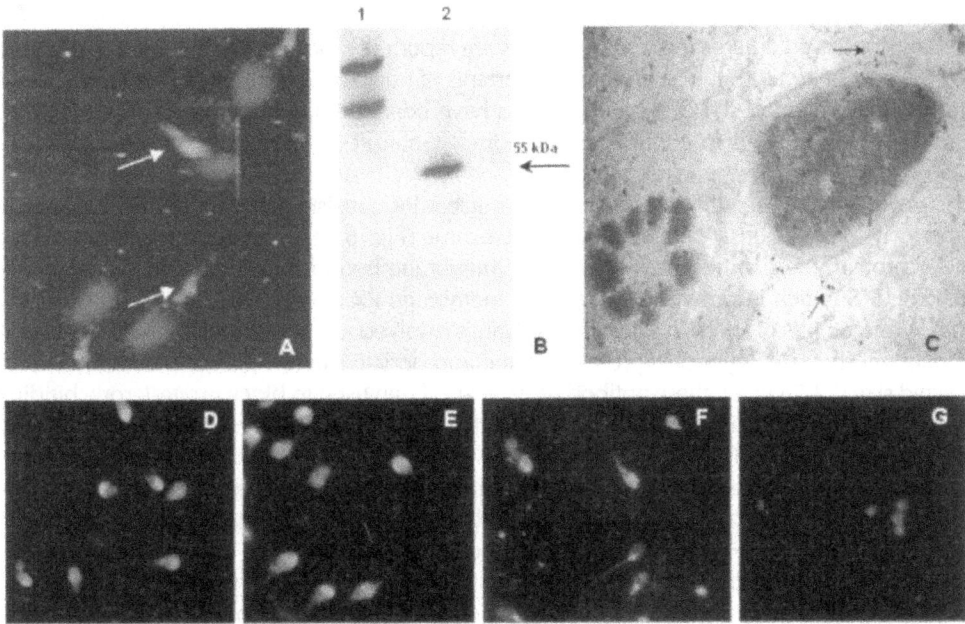

**Fig. 5** Nuclear progesterone receptors and human spermatozoa. (A) By *in situ* hybridization, nuclear PR mRNA was localized in the mid piece region  (arrow, green staining). (B) Western blot demonstrating the immuno-reactivity of the nuclear PR antibody to a 55 kDA protein in spermatozoa (Lane 2). Lane 1 are uterine proteins (+ve control). (C) Ultrastructure immunogold localization of PR using nuclear PR antibody. Along with the nuclear signals, gold particles (arrow) were visible in the membrane of the rat testicular sperm head. The tail showed no signals demonstrating specificity of localization. (D-G) Antibodies against the nuclear PR recognize an acrosomal protein in human spermatozoa. Green fluorescence denote positive staining, red is counterstaining using propidium iodide. (D) Monoclonal antibody against nuclear PR that recognize both the isoform (PR-A and PR-B), (E) Monoclonal antibody specific to the PR-B isoform, (F) Polyclonal antibody against nuclear PR that recognize both the isoform and (G) Buffer control.

However, a major set back to this speculation is derived from our immunoblotting experiments where the PR-B specific monoclonal antibody recognizes a protein of only 55 kDa rather than the expected 120 kDa as in case of uterine lysates (Shah et al., 2005b). Additionally, another antibody that recognizes both A and B isoform of nuclear PR, also recognizes a single 55 kDa protein in sperm lysates (Fig. 5B). A similar observation was also made earlier where antibodies recognizing different epitopes of the nuclear PR identify proteins of 52-58 kDa in sperm lysates (Sabeur et al., 1996; Luconi et al., 1998). This has lead to a speculation that the sperm membrane PR is probably a truncated variant of the nuclear PR.

*Membrane PR as a truncated/modified PR*

Splicing/deletions/insertions in the nuclear PR transcripts may lead to synthesis of a smaller/

truncated version of the protein (with conserved epitope) that may act as the membrane PR. Interestingly, several smaller forms of PR have been identified in multiple tissues that are postulated to result from in-frame initiation of translation (Saner et al., 2003). Many of these spliced variants generally have a deletion of the DNA binding domain, the ligand-binding domain is generally conserved. Interestingly such variants are reportedly localized on the plasma membrane of cells (Saner et al., 2003). In spermatozoa, variants of nuclear PR transcripts which are predicted to translate a protein of lower molecular mass have been identified (Hirata et al., 2000; 2003); although we have failed to confirm these findings (Shah et al., 2005b, Sachdeva et al., 2000; 2005).

However, a panel of antibodies against the nuclear PR consistently identifies a 55 kDa protein in sperm lysates, which localize on sperm acrosome (Fig. 5 D-F). Interestingly, along with the nuclear protein, these antibodies recognize a membrane bound form in ultrathin sections of rat testis (Fig. 5C) which is known to express the membrane PR (Shah et al., 2005a). In two dimensional Western blots of sperm lysates, this protein is resolved as a triplet with a predicted pl of 4.8-5.2 (data not shown). Beyond its presence at the appropriate site (acrosomal membrane) and the expected size (52-58 kDa), these antibodies have also been used to block progesterone binding to sperm membrane, and ablate progesterone mediated calcium mobilization and protein phosphorylation in spermatozoa (Modi et al., unpublished data). Isolation, sequencing and annotation of the cognate protein recognized by these antibodies is currently underway in our laboratory.

Why is the protein recognized by various antibodies against the conventional PR in spermatozoa of a size smaller (~ 55 kDa) than the expected (90 or 120 kDa) ? There could be several explanations for this enigma. As discussed above, it is likely that post transcriptional splicing of the nuclear PR may result in a truncated isoform which has conserved epitopes for antibody binding. Alternate splicing/deletions/insertions in the nuclear PR transcripts may lead to synthesis of a smaller protein (with conserved epitopes) that may bind to nuclear antibodies. Alternately, it is possible that the protein(s) recognized by the antibodies against the nuclear PR may be some other cross-reacting proteins that may only share structural motifs or immunological identity. In this context, it is interesting to note that a monoclonal antibody against the nuclear PR (c262) identified PAIRBP-1 that does not share homology with the nuclear PR (Peluso et al., 2005). Experiments are currently underway to determine which of these possibilities hold relevance for the 55 kDa protein in spermatozoa.

## Summary and future directions

Adequate experimental data exist demonstrating the presence of a membrane bound protein that mediates rapid non-genomic mode of progesterone actions in spermatozoa to regulate the acrosome reaction; studies are needed to decipher the identity of this protein. The discovery of the membrane PR would enable detailed studies to be conducted at the molecular level on the membrane receptor system, its physiological relevance and its interactions with the nuclear receptors. It is envisaged that the non-genomic signalling mechanism would represent the system through which steroids would rapidly activate cellular functionality needed to accommodate dynamic changes in the surrounding milieu.

However, the non-genomic steroid hormone receptor signalling is still a conundrum. Although, the work in recent years has highlighted the molecular basis of some of these actions, a large number of rapid effects exerted in different tissues and cell types still have to be fully characterized. It needs to be determined whether conditions exist where modifications of the local steroid concentrations could trigger the nongenomic signalling *in vivo*. It is presumable that in some ways tissues could be sensitive to either general or local hormonal variations, which would lead to the non-genomic effects. In this context, a possibility of activating or repressing some steroid hormone receptor actions would open newer avenues for designing drugs which could differentially recruit transcriptional and non-transcriptional effects on selected tissues. Thus understanding the molecular mechanism through which the nongenomic action takes place represents an important frontier in engineering newer pharmacological tools for the prevention and treatment of a wide range of diseases as well as infertility.

## Acknowledgements

We express our gratitude to Dr W. Elger (Schering, Germany) for kindly providing asoprisnil. The help provided by Dr S. D'Souza (Electron Microscopy Department, NIRRH) is highly appreciated. The work presented in this manuscript (NIRRH/MS/24/2005) was financially supported by Indian Council of Medical Research, New Delhi, India. CS is also thankful to Council for Scientific and Industrial Research, New Delhi, India for Senior Research Fellowship.

## References

Adams JS (2005) Bound to work: the free hormone hypothesis revisited *Cell* **122** 647-649

Ambhaikar MB and Puri CP (1998) Cell surface binding sites for progesterone on human spermatozoa *Molecular Human Reproduction* **4** 413-421

Baldi E, Casano R, Falsetti C, Krausz C, Maggi M and Forti G (1991) Intracellular calcium accumulation and responsiveness to progesterone in capacitating human spermatozoa *Journal of Andrology* **12** 323–330

Baldi E, Krausz C, Luconi M, Bonaccorsi L, Maggi M and Forti G (1995) Actions of progesterone on human sperm: a model of non-genomic effects of steroids *Journal of Steroid Biochemistry and Molecular Biology* **53** 199–203

Baldi E, Luconi M, Bonaccorsi L, Krausz C and Forti G (1996) Human sperm activation during capacitation and acrosome reaction: role of calcium, protein phosphorylation and lipid remodeling pathways *Frontiers in Bioscience* **1** 189-205

Baldi E, Luconi M, Bonaccorsi L and Forti G (1998) Nongenomic effects of progesterone on spermatozoa: mechanisms of signal transduction and clinical implications *Frontiers in Bioscience* **3** 1051-1059

Benoff S, Barcia M, Hurley IR, Cooper GW, Mandel FS, Heyner BR and Hershlag A (1996) Classification of male factor infertility relevant to *in-vitro* fertilization insemination strategies using mannose ligands, acrosome status and anti-cytoskeletal antibodies *Human Reproduction* **11** 1905-1918

Blackmore PF and Lattanzio F (1991) Cell surface localization of a novel non-genomic progesterone receptor on the head of human sperm *Biochemical and Biophysical Research Communication* **181** 331–336

Boonyaratnakornkit V, Scott MP, Ribbion V, Sherman L, Anderson SM and Miller WT (2001) Progesterone receptor contains proline rich motif that directly interacts with SH3 domains and activates c-Src family tyrosine kinase *Molecular Cell* **8** 269-280

Buddhikot M, Falkenstein E, Wehling M and Meizel S (1999) Recognition of a human surface protein involved in the progesterone initiated acrosome reaction by antisera against an endomembrane progesterone binding protein from porcine liver *Molecular and Cellular Endocrinology* **158** 187-193

Calogero AE, Burrello N, Barone N, Palermo I, Grasso U and D'Agata R (1999) γ–Aminobutyric acid (GABA) A and B receptors mediate the stimulatory effects of GABA on the human sperm acrosome reaction: interaction with progesterone *Fertility and Sterility* **71** 930-936

Calogero AE, Burrello N, Barone N, Palermo I, Grasso U and D'Agata R (2000) Effects of progesterone on sperm function: mechanism of action *Human Reproduction* **15** 28-45

Cross NL, Morales P, Overstreet JW and Hanson FW (1988) Induction of acrosome reaction by the human zona pellucida *Biology of Reproduction* **38** 235-244

de Jonge CJ, Han HL, Mack SR and Zaneveld LJD (1991) Effect of phorbol diesters, synthetic diacylglycerols, and a protein kinase C inhibitor on the human sperm acrosome reaction *Journal of Andrology* **12** 62-70

DeManno D, Elger W, Garg R, Lee R, Schneider B, Hess-Stumpp H, Schubert G and Chwalisz K (2003) Asoprisnil (J867): a selective progesterone receptor modulator for gynecological therapy *Steroids* **68** 1019-1032

Falkenstein E, Meyer C, Eisen C, Scriba PC and Wehling M (1996) Full length cDNA sequence of a progesterone membrane-binding protein from porcine vascular smooth muscle cells *Biochemical Biophysical Research Communication* **229** 86-89

Forti G, Baldi E, Krausz C, Luconi M, Bonaccorsi L, Maggi M, Bassi F and Scarselli G (1999) Effect of progesterone on human spermatozoa: clinical implications *Annals of Endocrinology (Paris)* **60** 107-110

Gadkar S, Shah CA, Sachdeva G, Samant U and Puri CP (2002) Progesterone receptor as an indicator of sperm function *Biology of Reproduction* **67** 1327-1336

Giojalas LC, Iribaren P, Molina R, Ravasio RA and Estifan D (2004) Determination of human sperm calcium uptake mediated by progesterone may be useful for evaluating unexplained sterility *Fertility and Sterility* **82** 738-740

Hammes A, Andersen TK, Spoelgen R, Raila J, Hubner N,

Schulz H, Metzer J, Schweigert FJ, Luppa PB, Nykjaer A and Willnow TE (2005) Role of endocytosis in cellular uptake of sex steroids *Cell* **122** 751-762

Harper CV and Publicover SJ (2005) Reassessing the role of progesterone in fertilization-compartmentalized calcium signaling in human spermatozoa? *Human Reproduction* **20** 2675-2680

Harper CV, Kirkman–Brown JC, Barratt CL and Publicover SJ (2003) Encoding of progesterone stimulus intensity by intracellular (Ca²⁺) ([Ca²⁺]ᵢ) in human spermatozoa *Biochemical Journal* **372** 407-417

Hirata S, Shoda T, Kato J and Hoshi K (2000) The novel isoform of the progesterone receptor cDNA in the human testis and detection of its mRNA in the human uterine endometrium *Oncology* **39** 39-44

Hirata S, Shoda T, Kato J and Hoshi K (2003) Novel isoform of the mRNA for human female sex steroid hormone receptors *Journal of Steroid Biochemistry and Molecular Biology* **83** 25-30

Kirkman-Brown JC, Bray C, Stewart PP, Barratt CLR and Publicover SJ (2000) Biphasic elevation of [Ca²⁺]ᵢ in individual human spermatozoa exposed to progesterone *Developmental Biology* **222** 326-335

Krausz C, Gervasi G, Forti G and Baldi E (1994) Effect of platelet activating factor on motility and acrosome reaction of human spermatozoa *Human Reproduction* **9** 471-476

Leonhardt SA, Boonyaratnakornkit V and Edwards D (2003) Progesterone receptor transcription and non transcription signaling mechanisms *Steroids* **68** 761-770

Losel R and Wehling M (2003) Nongenomic actions of steroid hormones *Nature Molecular and Cell Biology* **4** 46-56

Losel R, Dom-Beineke A, Falkenstein F and Wehling M (2004) Porcine spermatozoa contain more than one membrane progesterone receptor *International Journal of Biochemistry and Cell Biology* **38** 1532-1541

Losel R, Breiter S, Seyfert M, Wehling M and Falkenstein E (2005) Classic and non-classic progesterone receptors are both expressed in human spermatozoa *Hormone and Metabolic Research* **37** 10-14

Luconi M, Bonnaccorsi L, Maggi M, Pecchioli P, Krausz C, Forti G and Baldi E (1998) Identification and characterization of functional nongenomic progesterone receptors on human sperm membrane *Journal of Clinical Endocrinology and Metabolism* **83** 877-885

Luconi M, Bonnaccorsi L, Bini L, Liberatori S, Pallini V, Forti G and Baldi E (2002) Characterization of membrane nongenomic receptor for progesterone in human spermatozoa *Steroids* **67** 505-509

Meizel S and Turner KO (1991) Progesterone acts at the plasma membrane of human sperm *Molecular and Cellular Endocrinology* **77** R1–R5

Meizel S, Turner KO and Nuccitelli R (1997) Progesterone triggers a wave of increased free calcium during the human sperm acrosome reaction *Developmental Biology* **182** 67-75

Modi D, Shah CA, Sachdeva G, Gadkar S, Bhartiya D and Puri C (2005) Ontogeny and cellular localization of SRY transcripts in the human testes and its detec-tion in spermatozoa *Reproduction* **130** 603-613

Oehninger S, Blackmore PF, Morshedi MI, Sueldo C, Acosta A and Alexander NJ (1994) Defective calcium influx and acrosome reaction in spermatozoa of infertile men with severe teratozoospermia *Fertility and Sterility* **61** 349-354

Parinaud J and Milhet P (1996) Progesterone induces Ca11-dependent 39, 59-cyclic adenosine monophosphate increase in human sperm *Journal of Clinical Endocrinology and Metabolism* **81** 1357–1360

Peluso JJ, Pappalardo A, Losel R and Wehling M (2005) Expression and function of PAIRBP1 within gonadotropin primed immature rat ovaries: PAIRBP1 regulation of granulosa cells and luteal cell viability *Biology of Reproduction* **73** 261-270

Plant A, McLaughlin EA and Ford WCL (1995) Intracellular calcium measurements in individual human sperm demonstrate that the majority can respond to progesterone *Fertility and Sterility* **64** 1213–1215

Revelli A, Massobrio M and Tesarik J (1998) Nongenomic actions of steroid hormones in reproductive tissues *Endocrine Reviews* **19** 3-17

Roldan ER, Murase T and Shi QX (1994) Exocytosis in spermatozoa in response to progesterone and zona pellucida *Science* **266** 1578–1581

Sabeur K, Edwards DP and Meizel S (1996) Human sperm plasma membrane progesterone receptor(s) and the acrosome reaction *Biology of Reproduction* **54** 993-1001

Sachdeva G, Shah CA, Kholkute SD and Puri CP (2000) Detection of progesterone receptor transcript in human spermatozoa *Biology of Reproduction* **62** 1610-1614

Sachdeva G, Gadkar S, Shah CA, Kholkute SD and Puri CP (2005) Characterization of a critical region in the hormone binding domain of sperm progesterone receptor *International Journal of Andrology* **28** 120-124

Saner KJ, Welter BH, Zhang F, Hansen E, Dupont B, Wei Y and Price TM (2003) Cloning and expression of a novel, truncated, progesterone receptor *Molecular and Cellular Endocrinology* **200** 155-163

Shah CA, Modi D, Sachdeva G, Gadkar S and Puri CP (2005a) Co-existence of nuclear and membrane bound progesterone receptor in human testis *Journal of Clinical Endocrinology and Metabolism* **90** 474-483

Shah CA, Modi D, Sachdeva G, Gadkar-Sable S, D'Souza S and Puri CP (2005b) N-terminal region of progesterone receptor B isoform in human spermatozoa *International Journal of Andrology* **28** 360-371

Sirivaidyapong S, Bevers MM, Gadella BM and Colenbrander B (2001) Induction of the acrosome reaction in dog sperm cells is dependent on epididymal maturation: the generation of a functional progesterone receptor is involved *Molecular Reproduction and Development* **58** 451-459

Tanghe S, Van Soom A, Sterckx V, Maes D and de Kruif A (2002) Assessment of different sperm quality parameters to predict *in vitro* fertility of bulls *Reproduction of Domestic Animals* **37** 127-132

Tesarik J (1985) Comparison of acrosome reaction-inducing activities of human cumulus oophorus, follicular fluid and ionophore A 23187 in human sperm populations of proven fertilizing ability *in vitro Journal of Reproduction and Fertility* **74** 383-388

Tesarik J and Mendoza C (1992) Defective function of a nongenomic progesterone receptor as a sole sperm anomaly in infertile patients *Fertility and Sterility* **58** 793-797

Tesarik J and Mendoza C (1993) Insights into the functions of a sperm surface progesterone: Evidence of ligand–induced receptor aggregation and the proteolysis *Experimental Cell Research* **205** 111-117

Tesarik J, Moos J and Mendoza C (1993) Stimulation of protein tyrosine phosphorylation by a progesterone receptor on the cell surface of human sperm *Endocrinology* **133** 328-335

Thomas P, Zhu Y and Pace M (2002) Progestin membrane receptors involved in the meiotic maturation of teleost oocytes: a review with some new findings *Steroids* **67** 511-517

Wassarman PM, Jovine L, Qi H, Williams Z, Darie C and Litscher ES (2005) Recent aspects of mammalian fertilization research *Molecular and Cellular Endocrinology* **234** 95-103

Wistrom AC and Meizel S (1993) Evidence suggesting involvement of a unique human sperm steroid receptor/Cl$_2$ channel complex in the progesterone-initiated acrosome reaction *Developmental Biology* **159** 679 - 690

Yang J, Serres C, Philbert D, Robel P, Baulieu EE and Jouanner P (1994) Progesterone and RU486: Opposing effects on human sperm *Proceedings of National Academy of Sciences USA* **91** 529-533

Zhu Y, Bond J and Thomas P (2003a) Identification, classification and partial characterization of genes in humans and other vertebrates homologues to a fish membrane progestin receptor *Proceedings of National Academy of Sciences USA* **100** 2237-2242

Zhu Y, Rice CD, Pang Y, Pace M and Thomas P (2003b) Cloning, expression and characterization of a membrane progestin receptor and evidence it is an intermediary in meiotic maturation in fish oocytes *Proceedings of National Academy of Sciences USA* **100** 2331-2363

# Oocyte growth and acquisition of meiotic competence

Takashi Miyano[1] and Noboru Manabe[2]

[1]Faculty of Agriculture, Kobe University, Kobe 657-8501, Japan; [2]Research Unit for Animal Life Sciences, Animal Resource Science Center, The University of Tokyo, Ibaraki-Kasama 319-0206, Japan

Ovaries contain a huge number of non-growing and growing oocytes. Once non-growing oocytes (pig and cow: 30 μm in diameter) in primordial follicles enter the growth phase, they grow toward their final size (120-125 μm) taking a long period of time. The small oocytes have no ability to mature because of the inability to activate Cdc2 kinase and MAP kinase that are required for maturation. During the final growth phase, oocytes acquire the ability to activate these kinases. The population of the oocytes that grow to the final size is quite small in the ovary. Artificially growing-up of small oocytes could provide a new source of mature eggs for livestock production and assisted reproduction in humans. Baby mice have been produced by *in vitro* grown oocytes from primordial follicles. In large domestic species, only two baby calves have been produced from cultured oocytes that were at the mid-growth phase (90-99 μm) from early antral follicles. A culture system for the oocytes in secondary or smaller follicles has not been established. Xenotransplantation of oocytes to immunodeficient mice is a substitute for the culture. Bovine secondary follicles (oocyte: 55 μm) developed to the antral stage with oocytes reaching their final size in xenografts after 2 months. The grown oocytes matured and were penetrated by spermatozoa. Bovine and porcine primordial follicles (oocyte: 30 μm) developed to the antral stage after 6 months. *In vitro* growth and xenotransplantation systems will provide a new understanding of the mechanisms regulating oogenesis and folliculogenesis in the ovary.

## Introduction

Mammalian oocytes enter the meiotic cell cycle and arrest at prophase I in the fetal ovary. They are small (15-20 μm in diameter in rodents, and 30 μm in pigs, cows, and human beings) and have a large nucleus called the "germinal vesicle". They are surrounded by a layer of flat-shaped granulosa cells (pre-granulosa cells) to make primordial follicles. Oocytes in the primordial follicles are "non-growing immature oocytes" (Fig. 1).

Oocytes must undergo many changes themselves before ovulation and fertilization. The first change is the "growth" which occurs during their arrested period at prophase I. Once oocytes enter the growth phase, they undergo a tremendous increase in volume to 60-100 times their original size (from 15-20 to 75 μm in diameter in rodents, and from 30 to 120-125 μm in large animals).

E-mail: miyano@kobe-u.ac.jp

The first sign showing that the oocytes are entering the growth phase is the morphological change of the surrounding granulosa cells, which change from flat to cuboidal in shape. After this change, the follicle develops by proliferation of granulosa cells. Through a series of mitotic divisions of the granulosa cells, unilaminar primary follicles develop into multilaminar secondary follicles, and the follicles increase in size. Follicles further develop to the antral follicle stage, where a single, large, fluid-filled antral cavity is formed. In pigs and cows, the antrum is formed when oocytes reach around 90 μm in diameter. Oocytes further grow within the antral follicles to become fully-grown oocytes.

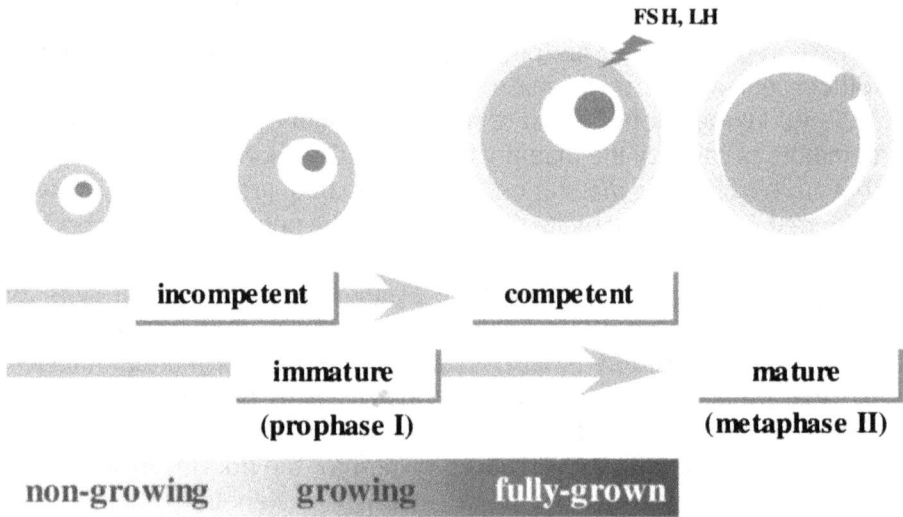

FSH, LH

incompetent    competent

immature    mature
(prophase I)    (metaphase II)

non-growing    growing    fully-grown

**Fig. 1** Oocyte growth and maturation

After animals reach puberty, fully-grown oocytes are subjected to the next change which is "maturation". They resume meiosis in the follicles by the surge of gonadotropic hormones, FSH and LH. Fully-grown immature oocytes enter the maturational phase, undergo germinal vesicle breakdown (GVBD), condense their chromosomes, form metaphase I spindle and release a half set of chromosomes into the first polar body. The oocytes in this maturational process are maturing oocytes. Finally, oocytes reach metaphase II to be mature oocytes. During this maturational change, cumulus granulosa cells surrounding the oocytes synthesize and secrete hyaluronic acid, and they expand. Then mature oocytes and surrounding cumulus granulosa cells are ovulated into oviducts.

### Acquisition of meiotic competence

During the growth phase, oocytes acquire the competence to resume meiosis; for example, porcine oocytes (30 μm in diameter) in primordial follicles have no competence to resume meiosis. Coincident with the initiation of antrum formation, oocytes in the follicles start to acquire the competence to resume meiosis. Beyond this point, when oocytes are released from follicles and are cultured under appropriate conditions, they resume meiosis spontaneously.

Acquisition of meiotic competence proceeds in a stepwise manner. First, oocytes acquire the capacity for resuming meiosis and for progression to metaphase I, and second, they gain the ability to reach metaphase II. Although porcine oocytes from 0.2-0.4 mm in diameter preantral follicles are already about three times their non-growing diameter, they are nevertheless still totally incapable of resuming meiosis. A small number of oocytes of around 100 μm in diameter from 0.5-1.5 mm in diameter early antral follicles undergo GVBD and reach diakinesis or metaphase I, although no significant number of oocytes reach metaphase II. Over 90% of oocytes from middle (2-3 mm in diameter) and large (4-6 mm in diameter) antral follicles undergo GVBD. However, less than half of the oocytes from the middle antral follicles reach the metaphase II, while most of the oocytes which have grown to full size from the large antral follicles reach the metaphase II. This stepwise acquisition of meiotic competence of growing oocytes has been identified in the mouse (Sorensen and Wassarman, 1976; Hirao *et al.*, 1993), rabbit (Jelínková *et al.*, 1994), pig (Motlík *et al.*, 1984a; Hirao *et al.*, 1995; Kanayama *et al.*, 2002) and cow (Fair *et al.*, 1995) using various maturation conditions. Mouse oocytes acquire competence to resume the first meiotic division at a diameter of 65 μm (Sorensen and Wassarman, 1976; Hirao *et al.*, 1993). Bovine and porcine oocytes achieve complete nuclear maturation to metaphase II, at a diameter of 110 μm (Fair *et al.*, 1995) and 110-115 μm (Motlík *et al.*, 1984a; Hirao *et al.*, 1995), respectively.

It has been proposed that during the acquisition of meiotic competence, growing oocytes store the necessary information and materials to carry them through meiosis, fertilization and early development. Oocyte maturation is conducted by two key cell cycle molecules, Cdc2 kinase and MAP kinase. Current evidence suggest that oocyte growth correlates with changes in the accumulation and activation of these cell cycle molecules. Growing porcine oocytes in preantral follicles start to accumulate the catalytic subunit of Cdc2 kinase, $p34^{cdc2}$ molecules (Hirao *et al.*, 1995). Oocytes in early antral follicles (0.5-0.7 mm in diameter) have enough amounts of $p34^{cdc2}$ molecules and synthesize the regulatory subunit of Cdc2 kinase, cyclin B1 in the maturation culture. However, they cannot resume meiosis because the Cdc2 kinase is negatively phosphorylated and thus inactivated (Kanayama *et al.*, 2002). In 1.0-1.5 mm in diameter follicles, growing oocytes that resume meiosis but are arrested at metaphase I, are capable of activating Cdc2 kinase, but they have not established a MAP kinase-activating pathway or the ability to activate MEK (MAP kinase kinase). Thus, the growing oocytes first develop the ability to activate Cdc2 kinase, then the ability to activate MAP kinase during the growth phase. Since oocyte maturation includes complicated serial events, changes in many other molecules are thought to be involved in the process of the acquisition of meiotic competence.

It is noteworthy that the change in nuclear morphology correlates with the acquisition of meiotic competence in the growing oocytes. It has been reported that the morphology of the oocyte chromatin changes from a diffuse to a perinucleolar condensed state in the mouse (Wickramasinghe *et al.*, 1991) and pig (Motlík *et al.*, 1984a; Hirao *et al.*, 1995). Fibrillo-granular and vacuolated nucleoli are compacted in the oocytes in the pig (Crozet *et al.*, 1981) and in cattle (Fair *et al.*, 1977; Crozet *et al.*, 1986). These changes reflect a significant decrease in rRNA synthesis (Crozet *et al.*, 1981; Motlík *et al.*, 1984b). In the chromatin, DNA is tightly bound to histones. The N-terminal tails of histones are covalently modified; acetylated, phosphorylated and methylated. Recently these modifications have been suggested to perform crucial functions in regulating chromatin structure in somatic cells (Jenuwein and Allis, 2001). During oocyte maturation, histone H3 and H4 molecules are deacetylated and phosphorylated (Akiyama *et al.*, 2004; Bui *et al.*, 2004; Endo *et al.*, 2005). During the growth of porcine oocytes, histone H3 molecules are acetylated and become methylated as the oocytes acquire the meiotic competence.

## Artificial oocyte growth

The ovary contains a huge number of non-growing and growing oocytes. Approximately 4,000 primordial follicles are contained in a pair of ovaries in the mouse (Peters, 1969), 100,000 primordial follicles are contained in the sheep and cow (Erickson, 1966; Gosden and Telfer,

1987), and the number is estimated at 420,000 in pig (Black and Erickson, 1968; Oxender et al., 1979). As described above, oocytes in the primordial follicles have no ability to resume meiosis, and furthermore, almost all the oocytes before reaching their full size lack the ability to mature. Oocytes acquire the meiotic competence depending on their growth. Even the porcine and bovine oocytes that had been grown either *in vitro* (Hirao et al., 1994; Harada et al., 1997) or as xenografts (Senbon et al., 2005), showed maturational competence depending on their diameter. If we can manage the oocyte growth, small oocytes in the ovary will provide a new source of mature oocytes for livestock production and for human *in vitro* fertilization (IVF) programs.

The first successful production of live young derived from *in vitro* grown, matured and fertilized mouse oocytes was reported in 1989 (Eppig and Schroeder, 1989) (Table 1). The great advancement in *in vitro* growth (IVG) culture was achieved by the production of a baby mouse by Eppig and O'Brien who cultured oocytes in primordial follicles in newborn mice to grow to their final size (Eppig and O'Brien, 1996). Application of culture systems for mouse oocytes to large domestic species is quite challenging, because of the relatively longer growth phase (2-3 months in pigs and 6 months in cows) and large size (120-125 μm in diameter). Autotransplantation or xenotransplantation of non-growing/growing oocytes can be a substitute for an effective long-term culture system. Lee et al. (2004) reported the production of a female baby monkey derived from ectopically autotransplanted ovarian tissues followed by fertilization and transfer to a surrogate mother. Snow et al. (2002) produced mature mouse oocytes by xenotransplantation of 3-week-old mouse ovaries into immunodeficient nude rats. The mature oocytes could subsequently be fertilized and develop into fertile adult mice. These reports suggest the possibility of utilizing ovarian non-growing/growing oocytes as a potential source of mature oocytes, if they grow in culture or in grafted tissues.

**Table 1.** *In vitro* growth culture systems for mammalian oocytes

| Species | Follicle stage | Oocyte diameter | Culture days | IVM/IVF* | Baby | Reference |
|---------|---------------|-----------------|--------------|----------|------|-----------|
| Mouse | secondary | <60 μm | 10 | Yes | Yes | Eppig and Schroeder, 1989 |
| Pig | preantral | 70-90 μm | 16 | Yes | No | Hirao et al., 1994 |
| Mouse | primordial | 15-20 μm | 22 | Yes | Yes | Eppig and O'Brien, 1996 |
| Cow | early antral | 90-99 μm | 14 | Yes | Yes | Yamamoto et al., 1999 |
| Cow | early antral | 90-99 μm | 14 | Yes | Yes | Hirao et al., 2004 |

* IVM/IVF: *in vitro* maturation/*in vitro* fertilization.

### IVG culture

In the domestic species, no culture system supporting the entire developmental course from the non-growing stage has yet been developed. Hirao et al. (1994) cultured porcine preantral follicles (not including theca cells) for 2 weeks and reported that the follicles made antrum-like structures and the oocytes grew from 70-90 μm to the fully-grown size in the structures (Table 1). Some of the fully-grown oocytes matured to metaphase II and were penetrated by spermatozoa. Neither successful production of blastocysts nor production of piglets derived from ovarian growing oocytes has been reported.

Among farm animals, there are two reports describing the successful production of baby calves derived from growing oocytes 90-99 μm in diameter in early antral follicles (Yamamoto *et al.*, 1999; Hirao *et al.*, 2004). In our laboratory, early antral follicles 0.5-0.7 mm in diameter were dissected from the cortex of bovine ovaries. Using fine forceps and a needle, the oocyte-cumulus complexes containing pieces of parietal granulosa (OCCGs) were collected from the follicles. OCCGs were embedded in collagen gels and cultured for 2 weeks in TCM199 containing 10% fetal calf serum and 4 mM hypoxanthine. The oocytes were further cultured for maturation and subsequently inseminated with spermatozoa. Of the 135 oocytes, 6 developed to the blastocyst stage. Three blastocysts were transferred to 3 recipient cows, and one became pregnant and delivered a live calf (Yamamoto *et al.*, 1999). Hirao *et al.* (2004) cultured OCCGs using insert membranes fitting in culture plates or 96-well culture plates, and sticky culture medium containing a high concentration, 4% (w/v), of polyvinylpyrrolidone (molecular weight of 360 kDa). They produced a calf from one of 4 embryos derived from oocytes grown *in vitro* for 2 weeks, matured, fertilized *in vitro* and then transferred to a recipient cow.

IVG culture for growing oocytes from large animals is not easy. The main problem is disconnection of oocytes from surrounding granulosa cells during the long-term culture. It is essential for the growth of oocytes to maintain the viability of oocytes and granulosa cells, and the metabolic coupling between them. Nevertheless, successful production of calves demonstrate that *in vitro* grown bovine oocytes acquire full developmental competence to produce live young. To date, no culture system supporting the entire developmental course from the primordial to the fully-grown stage has been established for the domestic species. To be more precise, we have not yet obtained successful results in IVG of oocytes from early secondary follicles. It has been reported that sheep preantral follicles grow to the antral stage in serum-free conditions after 1 month, and a small number of the *in vitro* grown oocytes mature to metaphase II (Cecconi *et al.*, 1999). The application of serum-free medium should be taken into consideration for the long-term culture of a month or more for growing much smaller oocytes from domestic species.

## Xenotransplantation

Xenotransplantation of non-growing and growing oocytes to nude mice or SCID (severe combined immune deficiency) mice can be a substitute for an effective long-term culture system. Mice homozygous for the SCID mutation lack both humoral and cell-mediated immunity due to the absence of mature T and B lymphocytes (Bosma *et al.*, 1983). Gosden and his colleagues have developed xenotransplantation of mammalian follicles into SCID mice as a model for investigating early stages of follicular development and for verifying follicle viability after cryopreservation (Gosden *et al.*, 1994). This method has been applied to human oocytes to rescued from ovarian failure and infertility caused by chemotherapy, radiotherapy and radical surgery (Kim *et al.*, 2002). However, production of babies from xenotransplanted oocytes has been reported only in rodents (Snow *et al.*, 2002). The idea is shifting from xeno- to autotransplantation especially for women (Oktay *et al.*, 2004).

We xenotransplanted bovine and porcine follicles at different stages into SCID mice. When bovine secondary follicles (150-200 μm) were transplanted under the kidney capsules of SCID mice, they developed to the antral stage after 4-6 weeks. Some oocytes had grown to be 120 μm or more, matured to metaphase II after maturation culture (Senbon *et al.*, 2003), and fertilized *in vitro* (Senbon *et al.*, 2005). Porcine growing oocytes in secondary follicles (less than 300 μm) also grew in SCID mice and acquired maturational competence (Kagawa *et al.*, 2005). Although we have not assessed the ability of the oocytes to develop into live babies yet, the results so far show that the method is effective for developing the bovine and porcine growing oocytes in the secondary follicles to their final size with maturational competence.

Xenotransplanted bovine and porcine primordial follicles (40 μm in diameter, oocyte: 30

μm) from adult ovaries did not develop even after 2 months (Senbon et al., 2003; Moniruzzaman et al., 2004). We dissected out the primordial follicle-rich regions from ovarian cortex and xenotransplanted the tissues for 8 weeks. The oocytes and follicles survived, although there were no changes in the mean diameter of oocytes before and after transplantation. In contrast, primordial follicles from fetal or newborn animals develop in xenografts. Ovaries from 10-20 days old piglets contain mainly primordial follicles. Several months after transplantation, primordial follicles developed to the antral stage in the grafts (Kaneko et al., 2003; Moniruzzaman et al., 2004). These findings suggest that secondary follicles and their growing oocytes in the adult ovary, and probably some of the primordial follicles and/or their oocytes in the neonatal ovaries have been oriented towards their final growth stage, and the oocytes are able to accomplish the task, when they are cultivated in appropriate conditions. On the other hand, it is likely that the oocytes in primordial follicles in adults have some modifications inhibiting their entry into the growth phase.

## Perspectives

Both IVG culture and xenotransplantation for porcine and bovine non-growing and growing oocytes are still at the experimental stage. However, successful production of calves suggests that *in vitro* grown bovine oocytes acquire maturational competence and developmental competence to produce live young. A combination of cryopreservation and IVG and/or xenotransplantation systems, if successful, would provide the desired number of mature oocytes from a preserved ovarian tissue by using existing assisted reproduction technologies. Non-growing oocytes in primordial follicles are quite useful for these techniques because they are present in huge numbers and are comparatively cryo-resistant due to their small volume.

Our findings suggest that secondary or more developing follicles and their mid-growing oocytes from the adult ovary, and probably some of the primordial follicles and their non-growing oocytes from the neonatal ovary, have been oriented towards their final stage. Such oocytes are able to accomplish their growth phase when they are cultivated in appropriate conditions. However, oocytes in primordial follicles from adult animals hardly entered the growth phase in the artificial condition. Recently, we observed that oocytes in primordial follicles from adult pigs and cows took much longer time before starting their growth. It suggests that oocytes in primordial follicles in adults have some inhibitory modifications for their entrance into the growth phase. Tilly and his colleagues have claimed that adult mouse ovaries possess mitotically active germ cells (Johnson et al., 2004), and that the cells are provided by the bone marrow (Johnson et al., 2005). Although the idea is still controversial (Telfer et al., 2005), mature oocytes would be produced from small volume of ovarian tissue containing such germ cells or small oocytes in combination with IVG and xenotransplantation methods. Finally, it should be emphasized that the methods could also provide the answer to the basic questions in oocyte/follicle physiology including classical questions about "mechanism initiating oocyte/follicle development", "follicular selection" and "acquisition of meiotic competence" in the ovary.

## Acknowledgments

This work was supported in part by a Grant-in-Aid for Creative Scientific Research (13GS0008) to NM and TM and the 21st Century COE Program to TM from the Ministry of Education, Culture, Sports, Science and Technology of Japan.

# References

Akiyama T, Kim JM, Nagata M and Aoki F (2004) Regulation of histone acetylation during meiotic maturation in mouse oocytes *Molecular Reproduction and Development* **69** 222-227

Black JL and Erickson BH (1968) Oogenesis and ovarian development in the prenatal pig *The Anatomical Record* **161** 45-56

Bosma GC, Custer RP and Bosma MJ (1983) A severe combined immunodeficiency mutation in the mouse *Nature* **301** 527-530

Bui HT, Yamaoka E and Miyano T (2004) Involvement of histone H3 (Ser10) phosphorylation in chromosome condensation without Cdc2 kinase and mitogen-activated protein kinase activation in pig oocytes *Biology of Reproduction* **70** 1843-1851

Cecconi S, Barboni B, Coccia M and Mattioli M (1999) *In vitro* development of sheep preantral follicles *Biology of Reproduction* **60** 594-601

Crozet N, Motlík J and Szöllösi D (1981) Nucleolar fine structure and RNA synthesis in porcine oocytes during the early stages of antrum formation *Biology of the Cell* **41** 35-42

Crozet N, Kanka J, Motlík J and Fulka J (1986) Nucleolar fine structure and RNA synthesis in bovine oocytes from antral follicles *Gamete Research* **14** 65-73

Endo T, Naito K, Aoki F, Kume S and Tojo H (2005) Changes in histone modifications during *in vitro* maturation of porcine oocytes *Molecular Reproduction and Development* **71** 123-128

Eppig JJ and O'Brien MJ (1996) Development *in vitro* of mouse oocytes from primordial follicles *Biology of Reproduction* **54** 197-207

Eppig JJ and Schroeder AC (1989) Capacity of mouse oocytes from preantral follicles to undergo embryogenesis and development to live young after growth, maturation, and fertilization *in vitro Biology of Reproduction* **41** 268-276

Erickson BH (1966) Development and senescence of the postnatal bovine ovary *Journal of Animal Science* **25** 800-805

Fair T, Hulshof SCJ, Hyttel P, Greve T and Boland M (1997) Nucleus ultrastructure and transcriptional activity of bovine oocytes in preantral and early antral follicles *Molecular Reproduction and Development* **46** 208-215

Fair T, Hyttel P and Greve T (1995) Bovine oocyte diameter in relation to maturational competence and transcriptional activity *Molecular Reproduction and Development* **42** 437-442

Gosden RG, Boulton MI, Grant K and Webb R (1994) Follicular development from ovarian xenografts in SCID mice *Journal of Reproduction and Fertility* **101** 619-623

Gosden RG and Telfer E (1987) Numbers of follicles and oocytes in mammalian ovaries and their allometric relationships *Journal of Zoology London* **211** 169-175

Harada M, Miyano T, Matsumura K, Osaki S, Miyake M and Kato S (1997) Bovine oocytes from early antral follicles grow to meiotic competence *in vitro*: Effect of FSH and hypoxanthine *Theriogenology* **48** 743-755

Hirao Y, Miyano T and Kato S (1993) Acquisition of maturational competence in *in vitro* grown mouse oocytes *Journal of Experimental Zoology* **267** 543-547

Hirao Y, Nagai T, Kubo M, Miyano T, Miyake M and Kato S (1994) *In vitro* growth and maturation of pig oocytes *Journal of Reproduction and Fertility* **100** 333-339

Hirao Y, Tsuji Y, Miyano T, Okano A, Miyake M, Kato S and Moor RM (1995) Association between p34$^{cdc2}$ levels and meiotic arrest in pig oocytes during early growth *Zygote* **3** 325-332

Hirao Y, Itoh T, Shimizu M, Iga K, Aoyagi K, Kobayashi M, Kacchi M, Hoshi H and Takenouchi N (2004) *In vitro* growth and development of bovine oocyte-granulosa cell complexes on the flat substratum: effects of high polyvinylpyrrolidone concentration in culture medium *Biology of Reproduction* **70** 83-91

Jelínková L, Kubelka M, Motlík J and Guerrier P (1994) Chromatin condensation and histone H1 kinase activity during growth and maturation of rabbit oocytes *Molecular Reproduction and Development* **37** 210-215

Jenuwein T and Allis CD (2001) Translating the histone code *Science* **293** 1074-1080

Johnson J, Canning J, Kaneko T, Pru JK and Tilly JL (2004) Germline stem cells and follicular renewal in the postnatal mammalian ovary *Nature* **428** 145-150

Johnson J, Bagley J, Skaznik-Wikiel M, Lee HJ, Adams GB, Niikura Y, Tschudy KS, Tilly JC, Cortes ML, Forkert R, Spitzer T, Iacomini J, Scadden DT and Tilly JL (2005) Oocyte generation in adult mammalian ovaries by putative germ cells in bone marrow and peripheral blood *Cell* **122** 303-315

Kagawa N, Kuwayama M, Miyano T and Manabe N (2005) Growth and maturation of follicles and oocytes following xenotransplantation of porcine ovarian tissues and *in vitro* maturation *The Journal of Reproduction and Development* **51** 741-748

Kanayama N, Miyano T and Lee J (2002) Acquisition of meiotic competence in growing pig oocytes correlates with their ability to activate Cdc2 kinase and MAP kinase *Zygote* **10** 261-270

Kaneko H, Kikuchi K, Noguchi J, Hosoe M and Akita T (2003) Maturation and fertilization of porcine oocytes from primordial follicles by a combination of xenografting and *in vitro* culture *Biology of Reproduction* **69** 1488-1493

Kim SS, Soules MR and Battaglia DE (2002) Follicular development, ovulation, and corpus luteum formation in cryopreserved human ovarian tissue after xenotransplantation *Fertility and Sterility* **78** 77-82

Lee DM, Yeoman RR, Battaglia DE, Stouffer RL, Zelinski-Wooten MB, Fanton JW and Wolf DP (2004) Live birth after ovarian tissue transplant *Nature* **428** 137-138

Moniruzzaman M, Senbon S and Miyano T (2004) Growth of oocytes in pig primordial follicles xenotransplanted into SCID mice *Reproduction, Fertility and Development* **16** 233-234 (abstract)

Motlík J, Crozet N and Fulka J (1984a) Meiotic competence *in vitro* of pig oocytes isolated from early antral follicles *Journal of Reproduction and Fertility* **72** 323-328

Motlík J, Kopecny V, Travnik P and Pivko J (1984b) RNA synthesis in pig follicular oocytes. Autoradiographic and cytochemical study *Biology of the Cell* **50** 229-236

Oktay K, Buyuk E, Veeck L, Zaninovic N, Xu K, Takeuchi T, Opsahl M and Rosenwaks Z (2004) Embryo development after heterotopic transplantation of cryopreserved ovarian tissue *The Lancet* **363** 837-840

Oxender WD, Colenbrander B, van deWiel DFM and Wensing CJG (1979) Ovarian development in fetal and prepubertal pigs *Biology of Reproduction* **21** 715-721

Peters H (1969) The development of the mouse ovary from birth to maturity *Acta Endocrinologica* **62** 98-116

Senbon S, Ota A, Tachibana M and Miyano T (2003) Bovine oocytes in secondary follicles grow and acquire meiotic competence in severe combined immunodeficient mice *Zygote* **11** 139-149

Senbon S, Ishii K, Fukumi Y and Miyano T (2005) Fertilization and development of bovine oocytes grown in female SCID mice *Zygote* **13** 309-315

Snow M, Cox SL, Jenkin G, Trounson A and Shaw J (2002) Generation of live young from xenografted mouse ovaries *Science* **297** 2227

Sorensen RA and Wassarman PM (1976) Relationship between growth and meiotic maturation of the mouse oocyte *Developmental Biology* **50** 531-536

Telfer EE, Gosden RG, Byskov AG, Spears N, Albertini D, Andersen CY, Anderson R, Braw-Tal R, Clarke H, Gougeon A, McLaughlin E, McLaren A, McNatty K, Schatten G, Silber S and Tsafriri A (2005) On regenerating the ovary and generating controversy *Cell* **122** 821-822

Wickramasinghe D, Ebert KM and Albertini DF (1991) Meiotic competence acquisition is associated with the appearance of M-phase characteristics in growing mouse oocytes *Developmental Biology* **143** 162-172

Yamamoto K, Otoi T, Koyama N, Horikita N, Tachikawa S and Miyano T (1999) Development to live young from bovine small oocytes after growth, maturation and fertilization *in vitro* *Theriogenology* **52** 81-89

# Hyaluronan binding protein-1: a modulator of sperm-oocyte interaction

Ilora Ghosh[1], Ratna Chattopadhaya[2], Vinod Kumar[1], BN Chakravarty[2] and Kasturi Datta[1]

[1]School of Environmental Sciences, Jawaharlal Nehru University, New Delhi 110 067, [2]Institute of Reproductive Medicine, HB 36/A/3, Salt lake City, Kolkata 110 091, India

Hyaluronan (HA), a complex glycosaminoglycan, is an important component in reproductive fluids and regulates several reproductive processes. It is thought that the multifaceted biological function of HA is mediated through hyaladherin family protein that binds with HA. We have reported a novel glycoprotein from human that has specific affinity towards hyaluronan, referred to as Hyaluronan Binding Protein-1 (HABP1, Ac. No. NP-001203) and is localized on human chromosome 17 p13.3. Sequence analysis of this gene has further revealed that HABP1 is synthesized as a precursor protein of 282 amino acids which undergoes a post-translational modification to give rise to the mature form of 209 amino acids by proteolytic cleavage of 73 amino acids at the N-terminal. The localization of mature HABP1 in several organs including the sperm surface and its involvement in fertilization have already been demonstrated. Enhanced phosphorylation of HABP1 in motile spermatozoa suggests its involvement in cellular signalling. Though only the mature form of HABP1 is detected in somatic tissues, the precursor form of HABP1 was detected in testicular tubules in a stage-specific manner in pachytene and round spermatids. To study the role of HABP1 in the fertilization process, we have shown the absence of mature HABP1 in cryptorchidic rats testes and an accumulation of the precursor form of HABP1 in giant cells, generated in infertile cryptorchidic rats. The loss of HABP1 from the sperm surface of a patient with very low sperm motility and the absence of the proprotein form of HABP1 in pachytene and round spermatids from testicular biopsy material with spermatogenic arrest, suggests that male infertility may be associated with the level of HABP1 on spermatozoa. In order to examine the role of HABP1 in sperm-oocyte interaction, we found that the number of spermatozoa bound to an oocyte was reduced significantly in the presence of D-mannosylated albumin, the universal blocker of sperm-oocyte interaction, and that this effect could be reversed by the addition of purified recombinant HABP1. In continuation, we have used spermatozoa of a patient, who had failed in IVF: the spermatozoa were incubated with HABP1 containing IVF medium

Corresponding author: Kasturi Datta
E-mail: datta_k@hotmail.com

for 2 hrs, then washed and allowed them to interact with the oocyte. The fertilized egg thus devloped up to the 16 cell stage, suggesting that HABP1 can modulate the sperm-oocyte interaction, even when sub-fertile spermatozoa are used.

## Introduction

Spermatozoa attain fertilizing ability by interacting with various components of the extracellular matrix in the epididymis (Miranda and Tezon, 1992) as well as components of both the seminal (Geipel *et al.*, 1992) and follicular fluids (Ericksen *et al.*, 1994) during their pre-fertilization journey. Such interactions affect their fertilizing capacity by influencing sperm maturation, capacitation, acrosomal reaction and the gain in forward motility. Hyaluronic acid (HA), or hyaluronan, a polysaccharide and a major component of the extracellular matrix, has been shown to promote sperm motility *in vitro*. This has led to the development of a medium containing HA, known as Sperm Select (Pharmacia AB, Uppsala, Sweden) (Karlstrom *et al.*, 1991; Psalti *et al.*, 1993). This medium is being widely used in human *in vitro* fertilization as it increases the pregnancy rate by improving the recovery of motile spermatozoa from human ejaculates and helps in the retention of sperm motility and velocity (Huszar *et al.*, 1990). HA has been shown to have a viscosity and density similar to cervical mucus, and as a major component of cervical mucus, affects the process of fertilization.

The event of fertilization is multiphasic and involves the interaction of several components on both the gametes in a complex sequential manner. It is already well established that a number of molecules on the sperm surface interact with the extracellular matrix of the oocyte, which is known as the zona pellucida (ZP). Carbohydrate chains of ZP glycoproteins have been identified as the molecules, which bind to receptors on sperm and initiate sperm-oocyte interaction (Wassarman, 1987). More precisely, N- and O-linked oligosaccharides of ZP glycoproteins are required for sperm recognition and binding, since chemical and enzymatic removal of oligosaccharides results in complete inactivation of the glycoproteins in different mammalian species (Tulsiani *et al.*, 1997; Wassarman, 1999). Particularly in sperm-oocyte adhesion, experiments with neoglycoproteins suggest the presence of mannose binding sites on the sperm surface, which can interact with the mannose residues of ZP in a defined spatial arrangement (Saling, 1989). Keeping this in view, several workers in this field (Youssef *et al.*, 1996) consider D-mannosylated albumin (DMA) as a substitute of ZP, which can inhibit sperm-oocyte interaction. A qualitative analysis of mannose binding sites on the sperm surface was done using DMA as a probe in human (Chen *et al.*, 1995; Mori *et al.*, 1997) and the mannose binding sites on the surface of human spermatozoa may be correlated to its ability to bind to human zona in IVF (Tesarik *et al.*, 1991). Recently, the N-linked highly mannosylated carbohydrate chain, the major neutral chain of bovine egg zona glycoproteins, has been shown to possess the ability to bind bovine sperm (Amari *et al.*, 2001). Though the presence of mannose binding sites on spermatozoa is evident, the specific sperm protein, which serves as the receptor of clustered mannose, is yet to be isolated and identified. So, one of the regulations of fertilization could be the availability of mature motile spermatozoa with the ability to interact with ZP properly.

We have already reported on a multiligand multifunctional glycoprotein having the specific affinity towards HA with a Kd of $10^{-9}$ (Kumar *et al.*, 2001). The gene encoding Hyaluronan Binding Protein-1 (HABP1) has been identified from the human fibroblast. HABP1 is an animal lectin, which shows affinity towards the two complex carbohydrates, HA and DMA, both of which are involved with sperm function (Deb and Datta, 1996). Our laboratory is working for the presence of HABP1 on the sperm surface and its critical role in sperm functions including its interaction with ZP through the clustered mannose residues.

## Differential expression of HABPI on sperm during spermatogenesis and abnormal pathological conditions

The gene encoding HABP1 has been identified from a human fibroblast c-DNA library (Deb and Datta, 1996) and its localization confirmed on human chromosome 17p12-p13 (Majumdar and Datta, 1998). Sequence analysis further revealed its multifunctional nature (Das et al., 1997) since it is homologous to a protein named P32, which co-purifies with pre-mRNA alternate splicing factor SF2 (Krainer et al., 1991) and gC1qR, the receptor for globular head of C1q (Ghebrehiwet et al., 1994). The unique feature of the gene encoding HABP1 is that HABP1 is synthesized as a precursor of 282 amino acids, which undergoes a post-translational modification to give rise to the mature protein of 209 amino acids by proteolytic cleavage of the first 73 amino acid residues at the N-terminal (Honore et al., 1993). We reported a long time back that the mature form of HABP1 is present on the sperm surface and that the anti HABP1 antibody inhibits sperm-oocye interaction suggesting its role in fertilization (Ranganathan et al., 1994).

As already reported, HA helps in cellular migration and induces sperm motility so our interest was to examine whether HA induces phosphorylation of HABP1 in ejaculated sperm. We confirmed the enhanced cellular phosphorylation in hyaluronan stimulated cauda spermatozoa and specifically enhanced phosphorylation of HABP1, suggesting its role in cellular signalling (Ranganathan et al., 1995).

HABP1 is synthesized as a proprotein, but its precursor form could not be detected on cauda spermatozoa or any somatic tissue. In testis, spermatogenesis is the most complex differentiation process in higher eukaryotes that comprises of the generation of haploid spermatozoa from testicular stem cells. The entire process depends upon the precise and ordered expression of many genes that are unique to spermatogenesis. We demonstrated high mRNA levels of HABP1 in the testis only. Detailed immunohistochemical studies of seminiferous tubules revealed the expression of the HABP1 proprotein in specific stages of germ cells, like pachytene spermatocytes and round spermatids but not in elongated ones (Bharadwaj et al., 2002) suggesting its role in spermatogenic differentiation. High concentrations of HABP1 proprotein in specific stages of spermatogenesis could be possible by delayed post-translational cleavage due to changes in proteolytic activities.

The presence of a mature form of HABP1 on the sperm surface and the stage specific location of precursor form of HABP1 during spermatogenesis encouraged us to study the expression of HABP1 during pathological conditions such as asthenozoospermic, oligospermic as well as in arrest of spermatogenesis (cryptorchidic condition).

We reported a significant reduction in the concentrations of HABP1 from asthenozoospermic and oligospermic patients compared to normozoospermic ones (Ghosh et al., 2002). Furthermore the absence of HABP1 in spermatozoa with motility < 20% is documented: motility is a determining factor for fertilization. Our data suggests that decreased HABP1 concentrations may be associated with a low motility of spermatozoa which in turn might cause infertility in the patient. HABP1 concentrations might therefore be used as a diagnostic marker.

Cryptorchidism is one reason for spermatogenic dysfunction and is a developmental congenital abnormality. All the main cell types in the testis are probably affected because of exposure of testis to higher temperatures. By surgical induction of cryptorchidism in Wistar rats, we confirmed the loss of spermatogenesis along with apoptosis induction and the accumulation of the proprotein form of HABP1 during the defective spermatogenesis (unpublished observations).

## Probable interaction of sperm surface HABP1 with ZP and its use in IVF

One interesting feature of HABP1 was reported that along with the reported ligand HA, it can bind with DMA as confirmed by overlay assay and purification using DMA affinity column. DMA inhibits the binding of HA to HABP1 in a concentration dependant manner (Kumar et al.,

2001). It is further supported by our studies where we reported that in solution, HABP1 exhibited the structure flexibility which is influenced by ionic changes, under *in vitro* condition near physiological pH (Jha *et al.*, 2003). At low ionic strength, HABP1 exists in a highly expanded and loosely held trimeric structure, similar to that of the molten globule-like state, whereas the presence of salt stabilizes the trimeric structure in a more compact fashion, reflecting in its differential affinity towards the different ligands. Whereas the binding of HABP1 towards HA is enhanced on increasing ionic strength upto 150 mM, no significant effect was observed with DMA. This novel receptor of HABP1 prompted us to study the role of HABP1 to ZP through its mannose residues.

We reported that only N-linked mannosylated zona glycoproteins bind to sperm surface HABP1(Ghosh and Datta, 2003). Labeled HABP1 interacts with ZP of intact oocyte of *Bubalus bubalis*, which can compete with unlabeled HABP1 or excess DMA. This data suggests the specific interaction of HABP1 with ZP, through clustered mannose residues. In order to examine the physiological significance of such an interaction, the capacity of sperm binding to oocytes under IVF conditions was examined either in presence of DMA alone or in combination with HABP1. The number of sperm bound to oocytes was observed to reduce significantly in the presence of DMA, which could be reversed by the addition of purified recombinant HABP1 (rHABP1) in the same plate. This suggests that sperm surface HABP1 may act as mannose binding sites for zona recognition. As already reported in asthenozoospermic patients the level of HABP1 in spermatozoa is lowered significantly. Thus, the use of rHABP1 in intrauterine insemination of asthenozoospermic patients is highly promising.

## References

Amari S, Yonezawa N, Mitsui S, Katsumata T, Hamano S, Kuwayama M, Hashimoto Y, Suzuki A, Takeda Y and Nakano M (2001) Essential role of the non-reducing terminal alpha-mannosyl residues of the N-linked carbohydrate chain of bovine zona pellucida glycoproteins in sperm-egg binding *Molecular Reproduction and Development* 59 221-226

Bharadwaj A, Ghosh I, Sengupta A, Cooper TG, Weinbauer GF, Brinkworth MH, Nieschlag E and Datta K (2002) Stage-specific expression of proprotein form of hyaluronan binding protein 1 (HABP1) during spermatogenesis in rat *Molecular Reproduction and Development* 62 223-232

Chen JS, Doncel GF, Alvarez C and Acosta AA (1995) Expression of mannose-binding sites on human spermatozoa and their role in sperm-zona pellucida binding *Journal of Andrology* 16 55-63

Das S, Deb TB, Kumar R and Datta K (1997) Multifunctional activities of human fibroblast 34-kDa hyaluronic acid-binding protein *Gene* 190 223-225

Deb TB and Datta K (1996) Molecular cloning of human fibroblast hyaluronic acid binding protein confirms its identify with P-32, a protein co-purified with splicing factor SF2 *Journal of Biological Chemistry* 269 2206-2212

Ericksen GV, Malmstrom A, Vldbjerg N and Huszar GA (1994) Follicular fluid chondroitin sulfate proteoglycan improves the retention of motility and velocity of human spermatozoa *Fertility and Sterility* 62 618-623

Geipel V, Kropf K, Krause W and Gressner AM (1992) The concentration pattern of laminin, hyaluranan and aminoterminal propeptide of type III procollagen in seminal fluid *Andrologia* 24 205-211

Ghebrehiwet B, Lim BL, Peerschke EI, Willis AC and Reid KB (1994) Isolation, cDNA cloning, and over-expression of a 33-kDa cell surface glycoprotein that binds to the globular "heads" of C1q *Journal of Experimental Medicine* 179 1809-1821

Ghosh I and Datta K (2003) Sperm surface hyaluronan binding protein (HABP1) interacts with zona pellucida of water buffalo (*Bubalus bubalis*) through its clustered mannose residues *Molecular Reproduction and Development* 64 235-244

Ghosh I, Bharadwaj A and Datta K (2002) Reduction in the level of hyaluronan binding protein 1 (HABP1) is associated with loss of sperm motility *Journal of Reproductive Immunology* 53 45-54

Honore B, Madsen P, Rasmussen HH, Vandekerckhove J and Delis JE (1993) Cloning and expression of a cDNA covering the complete coding region of the P32 subunit of human pre-mRNA splicing factor SF2 *Gene* 134 283-287

Huszar G, Willets M and Corrales M (1990) Hyaluronic acid (sperm select) improves retention of sperm motility and velocity in normospermic and oligospermic specimens *Fertility and Sterility* 54 1127-1134

Jha BK, Salunke DM and Datta K (2003) Structural flexibility of multifunctional HABP1 may be important for its binding to different ligand *Journal of Biological Chemistry* 278 27464-27472

**Karlstrom PO, Bakos O, Bergh T and Lundkvist O** (1991) Intrauterine insemination and comparison of two methods of sperm preparation *Human Reproduction* **6** 390-395

**Krainer AR, Mayeda A, Kozak D and Binns G** (1991) Functional expression of cloned human splicing factor SF2: homology to RNA-binding proteins, U1 70K, and *Drosophila* splicing regulators *Cell* **66** 383-394

**Kumar R, Choudhury NR, Salunke DM and Datta K** (2001) Evidence for clustered mannose as a new ligand for hyaluronan-binding protein (HABP1) from human fibroblast *Journal of Biosciences* **26** 325-332

**Majumdar M and Datta K** (1998) Assignment of cDNA encoding hyaluronic acid-binding protein 1 to human chromosome 17 p12-p13 *Genomics* **51** 476-477

**Miranda PV and Tezon JG** (1992) Characterization of fibronectin as a marker for human epididymal sperm maturation *Molecular Reproduction and Development* **33** 443-450

**Mori E, Mori T and Takasaki S** (1997) Binding of mouse sperm to beta-galactose residues on egg zona pellucida and asialofetuin coupled beads *Biochemical and Biophysical Research Communications* **238** 95-99

**Psalti I, Thomas K and De Cooman S** (1993) Effects of hyaluronate, strontium and prolonged incubation on different sperm parameters *Gynecologic and Obstetrics Investigation* **36** 47-51

**Ranganathan S, Ganguly AK and Datta K** (1994) Evidence for presence of hyaluronan binding protein on spermatozoa and its possible involvement in sperm function *Molecular Reproduction and Development* **38** 69-76

**Ranganathan S, Bharadwaj A and Datta K** (1995) Hyaluronan mediates sperm motility by enhancing phosphorylation of proteins including hyaluronan binding protein *Cellular and Molecular Biology Research* **41** 467-476

**Saling P** (1989) Mammalian sperm interaction with extracellular matrices of the egg *Oxford Reviews of Reproductive Biology* **11** 339-388

**Tesarik J, Mendoza C and Carreras A** (1991) Expression of D-mannose binding sites on human spermatozoa: comparison of fertile donor and infertile patients *Fertility and Sterility* **56** 113-118

**Tulsiani DRP, Yoshida-Komiya H and Araki Y** (1997) Mammalian fertilization: A carbohydrate-mediated event *Biology of Reproduction* **57** 487-494

**Wassarman PM** (1987) Early events in mammalian fertilization *Annual Review of Cell Biology* **3** 109-142

**Wassarman PM** (1999) Mammalian fertilization: molecular aspects of gamete adhesion, exocytosis, and fusion *Cell* **96** 175-183

**Youssef HM, Doncel GF, Bassiouni BA and Acosta AA** (1996) Mannose binding site on human spermatozoa and sperm morphology *Fertility and Sterility* **66** 640-645

# INDEX